Date Palm

A date palm tree (*Phoenix dactylifera* L.) and segments of the fruit bordered by six scenes illustrating traditional use by man (c.1840, lithograph print with watercolor). (From Wellcome Trust, London. Available at: https://wellcomecollection.org/works/r27bsc6w (accessed 12 November 2022). Frontispiece kindly suggested by Jane C. MacKnight of the Cincinnati Museum Center, Cincinnati, Ohio, USA.)

Date Palm

Edited by

Jameel M. Al-Khayri
Department of Agricultural Biotechnology, King Faisal University, Al-Ahsa, Saudi Arabia

S. Mohan Jain
Department of Agricultural Sciences, University of Helsinki, Helsinki, Finland

Dennis V. Johnson
Cincinnati, Ohio, USA

Robert R. Krueger
USDA–ARS National Clonal Germplasm Repository for Citrus & Dates, Riverside, California, USA

CABI is a trading name of CAB International

CABI	CABI
Nosworthy Way	200 Portland Street
Wallingford	Boston
Oxfordshire OX10 8DE	MA 02114
UK	USA
Tel: +44 (0)1491 832111	T: +1 (617)682-9015
E-mail: info@cabi.org	E-mail: cabi-nao@cabi.org
Website: www.cabi.org	

A catalogue record for this book is available from the British Library, London, UK.

ISBN-13: 9781800620186 (paperback)
          9781800620193 (ePDF)
          9781800620209 (ePub)

DOI: 10.1079/9781800620209.0000

Commissioning Editor: Rebecca Stubbs
Editorial Assistant: Lauren Davies
Production Editor: Shankari Wilford

Typeset by Exeter Premedia Services Pvt Ltd, Chennai, India
Printed and bound in the USA by Integrated Books International, Dulles, Virginia

CROP PRODUCTION SCIENCE IN HORTICULTURE SERIES

This series examines economically important horticultural crops selected from the major production systems in temperate, subtropical and tropical climatic areas. Systems represented range from open field and plantation sites to protected plastic and glass houses, growing rooms and laboratories. Emphasis is placed on the scientific principles underlying crop production practices rather than on providing empirical recipes for uncritical acceptance. Scientific understanding provides the key to both reasoned choice of practice and the solution of future problems.

Students and staff at universities and colleges throughout the world involved in courses in horticulture, as well as in agriculture, plant science, food science and applied biology at degree, diploma or certificate level will welcome this series as a succinct and readable source of information. The books will also be invaluable to progressive growers, advisers and end-product users requiring an authoritative, but brief, scientific introduction to particular crops or systems. Keen gardeners wishing to understand the scientific basis of recommended practices will also find the series very useful.

The authors are all internationally renowned experts with extensive experience of their subjects. Each volume follows a common format covering all aspects of production, from background physiology and breeding, to propagation and planting, through husbandry and crop protection, to harvesting, handling and storage. Selective references are included to direct the reader to further information on specific topics.

Titles Available:

1. **Ornamental Bulbs, Corms and Tubers** A.R. Rees
2. **Citrus** F.S. Davies and L.G. Albrigo
3. **Onions and Other Vegetable Alliums** J.L. Brewster
4. **Ornamental Bedding Plants** A.M. Armitage
5. **Bananas and Plantains** J.C. Robinson
6. **Cucurbits** R.W. Robinson and D.S. Decker-Walters
7. **Tropical Fruits** H.Y. Nakasone and R.E. Paull
8. **Coffee, Cocoa and Tea** K.C. Willson
9. **Lettuce, Endive and Chicory** E.J. Ryder
10. **Carrots and Related Vegetable *Umbelliferae*** V.E. Rubatzky, C.F. Quiros and P.W. Simon
11. **Strawberries** J.F. Hancock
12. **Peppers: Vegetable and Spice Capsicums** P.W. Bosland and E.J. Votava
13. **Tomatoes** E. Heuvelink
14. **Vegetable Brassicas and Related Crucifers** G. Dixon
15. **Onions and Other Vegetable Alliums, 2ⁿᵈ Edition** J.L. Brewster
16. **Grapes** G.L. Creasy and L.L. Creasy
17. **Tropical Root and Tuber Crops: Cassava, Sweet Potato, Yams and Aroids** V. Lebot

Contents

Contributors

Larbi Abahmane, Plant Biotechnology Laboratory, Regional Centre of Agricultural Research of Marrakech, National Institute for Agricultural Research, Post Box 533, Marrakech Principale 40000, Morocco
Email: abahmane1@yahoo.fr; larbi.abahmane@inra.ma
ORCID: https://orcid.org/0000-0001-8948-9268

Mohamed Ait-El-Mokhtar, Agro-Food, Biotechnology and Valorization of Plant Bioresources Laboratory, Faculty of Science, Semlalia, Department of Biology, Cadi Ayyad University, PO Box 2390, Marrakech 40000, Morocco; and Laboratory of Biochemistry, Environment & Agri-Food (LBEA) URAC 36, Department of Biology, Faculty of Sciences and Techniques Mohammedia, Hassan II University of Casablanca, Casablanca 28806, Morocco
Email: mohamed.aitelmokhtar@gmail.com
ORCID: https://orcid.org/0000-0001-8096-9927

Fatima-Zahra Akensous, Center of Agrobiotechnology and Bioengineering, Research Unit labelled CNRST (Centre AgroBiotech-URL-CNRST-05), Physiology of Abiotic Stress Team, Cadi Ayyad University, Marrakech 40000, Morocco; and Agro-Food, Biotechnology and Valorization of Plant Bioresources Laboratory, Faculty of Science, Semlalia, Department of Biology, Cadi Ayyad University, PO Box 2390, Marrakech 40000, Morocco
Email: fatimazahra.akensous@gmail.com
ORCID: https://orcid.org/0000-0001-9518-8146

M. Tahir Akram, Department of Horticulture, PMAS-Arid Agriculture University, Rawalpindi 46300, Pakistan
Email: tahiruaf786@gmail.com

Francisco Alcaraz, Universidad de Murcia, Facultad de Biología, Departamento Biología Vegetal, 30100 Murcia, Spain
Email: falcaraz@um.es
ORCID: https://orcid.org/0000-0003-3254-2691

Salah Mohammed Aleid, Department of Food and Nutritional Sciences, College of Agricultural and Food Sciences, King Faisal University, PO Box 400, Al-Ahsa 31982, Saudi Arabia
Email: seid@kfu.edu.sa
ORCID: https://orcid.org/0000-0003-0558-2065

Melkamu Alemayehu, Department of Horticulture, College of Agriculture and Environmental Sciences, Bahir Dar University, PO Box 5501, 6000 Bahir Dar, Ethiopia
Email: melkalem65@gmail.com

Khalid Al-Hashmi, Department of Plant Sciences, College of Agricultural and Marine Sciences, Sultan Qaboos University, PO Box 34, Al-Khoud 123, Oman
Email: kalhashmi@squ.edu.om

Abda Abdalla Emam, Department of Agribusiness and Consumer Sciences, College of Agriculture and Food Sciences, King Faisal University, PO Box 400, Al-Ahsa 31982, Saudi Arabia
Email: aaeali@kfu.edu.sa
ORCID: https://orcid.org/0000-0003-1685-3692

Latifa Al-Kharusi, Department of Plant Sciences, College of Agricultural and Marine Sciences, Sultan Qaboos University, PO Box 34, Al-Khoud 123, Oman
Email: l.kahrusi@squ.edu.om
ORCID: https://orcid.org/0000-0003-2317-214X

Jameel M. Al-Khayri, Department of Agricultural Biotechnology, King Faisal University, Al-Hassa 31982, Saudi Arabia
Email: jkhayri@kfu.edu.sa
ORCID: http://orcid.org/0000-0001-9507-0201

Mohammed Al-Mahish, Department of Agribusiness and Consumer Sciences, College of Agriculture and Food Sciences, King Faisal University, PO Box 400, Al-Ahsa 31982, Saudi Arabia
Email: malmahish@kfu.edu.sa
ORCID: https://orcid.org/0000-0002-8143-1114

Abdulrasoul Mosa Al-Omran, Soil Science Department, King Saud University, Riyadh, 11451 Saudi Arabia
Email: rasoul@ksu.edu.sa
ORCID: htps://orcid.org/0000-0001-8806-3871

Lyutha Al-Subhi, Department of Food Science and Nutrition, College of Agricultural and Marine Sciences, Sultan Qaboos University, Al-Khoud 123, Muscat, Oman
Email: lyutha@squ.edu.om

Rashid Al-Yahyai, Department of Plant Sciences, College of Agricultural and Marine Sciences, Sultan Qaboos University, PO Box 34, Al-Khoud 123, Oman
Email: alyahyai@squ.edu.om; alyahyai@gmail.com
ORCID: https://orcid.org/0000-0001-7628-6218

Asunción Amorós, Universidad Miguel Hernández de Elche, Departamento de Biología Aplicada, 03202 Elche, Alicante, Spain
Email: aamoros@umh.es
ORCID: https://orcid.org/0000-0002-5817-2898

Mohamed Anli, Center of Agrobiotechnology and Bioengineering, Research Unit labelled CNRST (Centre AgroBiotech-URL-CNRST-05), Physiology of Abiotic Stress Team, Cadi Ayyad University, Marrakech 40000, Morocco; and Agro-Food, Biotechnology and Valorization of Plant Bioresources Laboratory, Faculty of Science, Semlalia, Department of Biology, Cadi Ayyad University, PO Box 2390, Marrakech 40000, Morocco
Email: moh1992anli@gmail.com
ORCID: https://orcid.org/0000-0002-7028-709X

Mohamed Marouf Aribi, Higher National School of Biotechnology – Taoufik Khaznadar, University town Ali Mendjeli BP E66 251000, Constantine, Algeria
Email: mar-bio-tp@live.fr

Raja Ben-Laouane, Center of Agrobiotechnology and Bioengineering, Research Unit labelled CNRST (Centre AgroBiotech-URL-CNRST-05), Physiology of Abiotic Stress Team, Cadi Ayyad University, Marrakech 40000, Morocco; and Agro-Food, Biotechnology and Valorization of Plant Bioresources Laboratory, Faculty of Science, Semlalia, Department of Biology, Cadi Ayyad University, PO Box 2390, Marrakech 40000, Morocco
Email: benlaouaneraja@gmail.com
ORCID: https://orcid.org/0000-0002-1656-0462

Houda Besser, Research Unit of Geo-systems, Geo-resources and Geo-environments (UR3G), Department of Earth Sciences, Faculty of Sciences of Gabes, City Campus Erriadh-Zirig, 6072 Gabes, Tunisia
Email: Besserhouda@gmail.com

Neeru Bhatt, Global Science Heritage, Toronto, Ontario, Canada, L3P 0A1.
Email: neerubhattdp@gmail.com

Abderrahim Boutasknit, Center of Agrobiotechnology and Bioengineering, Research Unit labelled CNRST (Centre AgroBiotech-URL-CNRST-05), Physiology of Abiotic Stress Team, Cadi Ayyad University, Marrakech 40000, Morocco; and Agro-Food, Biotechnology and Valorization of

Plant Bioresources Laboratory, Faculty of Science, Semlalia, Department of Biology, Cadi Ayyad University, PO Box 2390, Marrakech 40000, Morocco
Email: abderrahim.boutasknit@gmail.com
ORCID: https://orcid.org/0000-0002-0311-2441

Roshini Brizmohun, Faculty of Agriculture, University of Mauritius, Reduit, Mauritius 80837
Email: r.brizmohun@uom.ac.mu
ORCID: https://orcid.org/0000-0003-4992-3896

Latifa Dhaouadi, Regional Center for Research in Oasis Agriculture, Deguache, Laboratory of Eremology and Combating Desertification, Institut des Régions Arides de Médenine, Route du Djorf Km 22.5, 4119 Médenine, Tunisia
Email: latifa_hydro@yahoo.fr

Gisela Díaz, Universidad Miguel Hernández de Elche, Departamento de Biología Aplicada, 03202 Elche, Alicante, Spain
Email: gdiaz@umh.es
ORCID: https://orcid.org/0000-0003-0250-2517

Raga Elzaki, Department of Agribusiness and Consumer Sciences, College of Agriculture and Food Sciences, King Faisal University, PO Box 400, Al-Ahsa 31982, Saudi Arabia
Email: rmali@kfu.edu.sa
ORCID: https://orcid.org/0000-0002-4993-8505

Abdessamad Fakhech, Agro-Food, Biotechnology and Valorization of Plant Bioresources Laboratory, Faculty of Science, Semlalia, Department of Biology, Cadi Ayyad University, PO Box 2390, Marrakech 40000, Morocco
Email: fakhch-ads@hotmail.com
ORCID: https://orcid.org/0000-0002-7145-5489

Mohamed Abusaa Fennir, Department of Agricultural Engineering, Faculty of Agriculture, University of Tripoli, Sidi Almasri, Furnj Road, PO Box 13438, Tripoli, Libya
E-mail: m.fennir@uot.edu.ly

Ibrahim E. Greiby, Department of Food Science and Technology, Faculty of Agriculture, University of Tripoli, Sidi Almasri, Furnj Road, PO Box 13438, Tripoli, Libya
Email: i.greiby@uot.edu.ly

Ayah R. Hilles, International Institute for Halal Research and Training, International Islamic University Malaysia, 53100 Kuala Lumpur, Malaysia
Email: ayah.hilles90@gmail.com

S. Mohan Jain, Department of Agricultural Sciences, University of Helsinki, PL-27, Helsinki 00014, Finland
Email: mohan.jain@helsinki.fi
ORCID: https://orcid.org/0000-0002-8289-2174

Rhonda Janke, Department of Plant Sciences, College of Agricultural and Marine Sciences, Sultan Qaboos University, PO Box 34, Al-Khoud 123, Oman
Email: rhonda@squ.edu.om
ORCID: https://orcid.org/0000-0002-4530-0782

Muhammad Jafar Jaskani, Institute of Horticultural Sciences, University of Agriculture, Agriculture University Road, Faisalabad 38040, Pakistan
Email: jjaskani@uaf.edu.pk
ORCID: https://orcid.org/0000-0001-5226-1307

Dennis V. Johnson, Consultant, 3726 Middlebrook Ave, Cincinnati, OH 45208, USA
Email: djohn37@aol.com
ORCID: https://orcid.org/0000-0002-7284-1074

Imran Ul Haq, Department of Pathology, University of Agriculture, Faisalabad 38040, Punjab, Pakistan
Email: imran_1614@yahoo.com
ORCID: https://orcid.org/0000-0001-7899-2052

Iqrar Ahmad Khan, Institute of Horticultural Sciences, University of Agriculture, Agriculture University Road, Faisalabad 38040, Pakistan
Email: iqrarahmadkhan2008@gmail.com

M. Mumtaz Khan, Department of Plant Sciences, College of Agricultural and Marine Sciences, Sultan Qaboos University, PO Box 34, Al-Khoud 123, Oman
Email: mumtaz@squ.edu.om
ORCID: https://orcid.org/0000-0002-1423-9667

Rashad Rasool Khan, Department of Entomology, University of Agriculture, Faisalabad 38040, Punjab, Pakistan
Email: rashadkhan@uaf.edu.pk
ORCID: https://orcid.org/0000-0002-5554-2953

Robert R. Krueger, USDA–ARS National Clonal Germplasm Repository for Citrus & Dates, 1060 Martin Luther King Blvd, Riverside, CA 92507, USA
Email: robert.krueger@usda.gov
ORCID: https://orcid.org/0000-0003-2570-7881

Mithlesh Kumar, AICRN on Potential Crops, Agricultural Research Station (ARS), Agriculture University, Jodhpur Rajasthan – 342 304, India

Email: mithleshgenetix@gmail.com
ORCID: https://orcid.org/0000-0003-1862-8683

Abdelilah Meddich, Center of Agrobiotechnology and Bioengineering, Research Unit labelled CNRST (Centre AgroBiotech-URL-CNRST-05), Physiology of Abiotic Stress Team, Cadi Ayyad University, Marrakech 40000, Morocco; and Agro-Food, Biotechnology and Valorization of Plant Bioresources Laboratory, Faculty of Science, Semlalia, Department of Biology, Cadi Ayyad University, PO Box 2390, Marrakech 40000, Morocco
Email: a.meddich@uca.ma
ORCID: https://orcid.org/0000-0001-9590-4405

C.M. Muralidharan, Date Palm Research Station, Sardarkrushinagar Dantiwada Agricultural University, Mundra-Kachchh, Gujarat – 370 421, India
Email: muralidharancm@yahoo.com
ORCID: https://orcid.org/0000-0002-1140-9341

Summar Abbas Naqvi, Institute of Horticultural Sciences, University of Agriculture, Agriculture University Road, Faisalabad 38040, Pakistan
Email: summar.naqvi@uaf.edu.pk
ORCID: https://orcid.org/0000-0001-8186-9055

Taseer Abbas Naqvi, Department of Plant Breeding and Genetics, University of Agriculture, Faisalabad 38040, Pakistan
Email: taseernaqvi110@gmail.com

Concepción Obón, Universidad Miguel Hernández de Elche, Departamento de Biología Aplicada, 30312 Orihuela, Alicante, Spain
Email: cobon@umh.es
ORCID: https://orcid.org/0000-0002-0244-601X

Redouane Ouhaddou, Center of Agrobiotechnology and Bioengineering, Research Unit labelled CNRST (Centre AgroBiotech-URL-CNRST-05), Physiology of Abiotic Stress Team, Cadi Ayyad University, Marrakech 40000, Morocco; and Agro-Food, Biotechnology and Valorization of Plant Bioresources Laboratory, Faculty of Science, Semlalia, Department of Biology, Cadi Ayyad University, PO Box 2390, Marrakech 40000, Morocco
Email: ouhadou.redouan@gmail.com
ORCID: https://orcid.org/0000-0002-4208-6960

Ozcan Ozturk, College of Public Policy, Hamad Bin Khalifa University, PO Box 34110, Education City, Doha, Qatar
Email: OOzturk@hbku.edu.qa
ORCID: https://orcid.org/0000-0002-2187-4327

Ouissame Raho, Center of Agrobiotechnology and Bioengineering, Research Unit labelled CNRST (Centre AgroBiotech-URL-CNRST-05), Physiology of Abiotic Stress Team, Cadi Ayyad University, Marrakech 40000, Morocco; and Agro-Food, Biotechnology and Valorization of Plant Bioresources Laboratory, Faculty of Science, Semlalia, Department of Biology, Cadi Ayyad University, PO Box 2390, Marrakech 40000, Morocco
Email: ouissame.raho@gmail.com
ORCID: https://orcid.org/0000-0002-0306-6019

Diego Rivera, Universidad de Murcia, Facultad de Biología, Departamento Biología Vegetal, 30100 Murcia, Spain
Email: drivera@um.es
ORCID: https://orcid.org/0000-0001-6889-714X

Ricardo Salomón-Torres, Unidad Académica San Luis Rio Colorado, Universidad Estatal de Sonora, Sonora 83500, Mexico
Email: ricardo.salomon@ues.mx
ORCID: https://orcid.org/0000-0002-6486-2131

Kapil Mohan Sharma, Date Palm Research Station, Sardarkrushinagar Dantiwada Agricultural University, Mundra-Kachchh, Gujarat – 370 421, India
Email: k.m.sharma456@sdau.edu.in
ORCID: https://orcid.org/0000-0003-4714-7854

Mostafa I. Waly, Department of Food Science and Nutrition, College of Agricultural and Marine Sciences, Sultan Qaboos University, Al-Khoud 123, Muscat, Oman
Email: mostafa@squ.edu.om
ORCID: https://orcid.org/0000-0002-8197-3378

Muhammad Waseem, Institute of Horticultural Sciences, University of Agriculture, Agriculture University Road, Faisalabad 38040, Pakistan
Email: wasimm45@gmail.com

Glenn C. Wright, Yuma Agriculture Center, University of Arizona, 6425 W. 8th Street, Yuma, AZ 85364, USA
Email: gwright@arizona.edu
ORCID: https://orcid.org/0000-0002-8516-9600

Acknowledgements

The editors extend their appreciation to the Deanship of Scientific Research, Vice Presidency for Graduate Studies and Scientific Research, King Faisal University, Saudi Arabia for supporting this work through Project No. GRANT3109.

Preface

Over the past century, the date palm has become a major commercial fruit crop and a key component of agricultural production in the world's subtropical arid and semiarid regions at elevations below about 1500 m. A crop suited both to the low-input small farmer and the modern high-input commercial plantation, the date palm provides livelihood to millions of people living in marginal land areas where farming options are restricted.

Accelerating scientific research into date palm improvement for fruit production in recent decades has broadened horizons for improved elite cultivars, stress and pathogen resistance, and enhanced postharvest technologies. These developments set the stage for a fresh set of recommendations to date palm producers with regard to the best practices they should follow to achieve optimal fruit production levels of high-quality fruit, as well as the opportunity to promote novel fruit products and uses and for a more complete utilization of the multitude of products the date palm can provide at both the subsistence and commercial level of production.

Several recent books on date palm have addressed various aspects of this tree crop's development but none represents a cultivation manual comparable to that published by the Food and Agricultural Organization of the United Nations in 2002, *Date Palm Cultivation*. The present book, *Date Palm*, is part of the book series: Crop Production Science in Horticulture, which is intended to deliver a new comprehensive standard reference work for date palm producers the world over. Chapters are directed towards formulating research results and providing an applicable set of recommendations and technologies to guide date palm producers. Given the enormous changes that have taken place in date palm science over the past 20 years, the need exists to provide producers with a comprehensive single source of information on the proven modern practices that can carry this important crop forward to a higher level of productivity in the 21st century.

The book is envisioned as a comprehensive book that will provide basic guidelines to date palm growers, processors, and marketers; a summary of the application of the state of science for date palm for researchers and advanced students; and a current information source on the crop for political leaders, planners, and policy makers. It consists of 18 chapters, illustrated with 161 color figures and 53 tables. The book starts with an introductory chapter laying out the discussed topic within the context of the book. Discussed topics include date palm botany and physiology; genetic diversity and conservation; genetic improvement and growth requirements; propagation, plantation establishment, pollination management, irrigation, and salinity management; biofertilizers, pest and disease management; organic fruit production, agroecological practices, harvest and postharvest procedures; food and nonfood products, health benefits, and nutraceutical properties; and economics and marketing.

The chapters are contributed by 54 internationally reputable scientists, from 17 countries, selected based on their extensive experience in date palm research. We are greatly appreciative of their assiduous efforts and diligence. We are also grateful to the editor of the Crop Production Science in Horticulture Series for the invitation and supervision to compile this book and to CABI for the opportunity to publish this book.

Jameel M. Al-Khayri
Al-Ahsa, Saudi Arabia

S. Mohan Jain
Helsinki, Finland

Dennis V. Johnson
Cincinnati, Ohio, USA

Robert R. Krueger
Riverside, California, USA

Introduction: The Date Palm Legacy

1

Robert R. Krueger[1]* ⓘ, Jameel M. Al-Khayri[2] ⓘ, S. Mohan Jain[3] ⓘ and Dennis V. Johnson[4] ⓘ

[1]USDA–ARS National Clonal Germplasm Repository for Citrus & Dates, Riverside, California, USA; [2]King Faisal University, Al-Hassa, Saudi Arabia; [3]University of Helsinki, Helsinki, Finland; [4]Cincinnati, Ohio, USA

Abstract

The date palm (*Phoenix dactylifera* L.) is one of the oldest cultivated fruit crops. It may have originated as a species from a primitive *Phoenix* ancestor and shows genes shared with or derived from other *Phoenix* species. Most probably, the date palm originated in the arid subtropical deserts of the Middle East near current Iraq and Iran. It was probably first domesticated and cultivated in that area approximately 5000 years BP, its cultivation later expanding into Egypt around 3500 years BP (before present era), and into India, western North Africa, and Southern Europe by 2000 years BP. The date palm may be considered "intermediately" domesticated as it maintains some characteristics of undomesticated crops but also has some characteristics of intermediate and fully domesticated crops. In addition to archeological evidence of its cultivation, various artistic representations of date cultivation show both cultivation and simple forms of most cultural practices used in modern date production, particularly artificial pollination. The latter led to an association of the date palm with ancient fertility goddesses, and it was later integrated into the three great monotheistic religions. Date cultivation later expanded outside the traditional area of date cultivation and is today cultivated in over 40 countries around the world, although the ancient center of date culture remains the most import region producing dates.

Keywords: Archeology, Domestication, Fertilization, Introgression, Oasis, *Phoenix*, Pollination

*Corresponding author: robert.krueger@usda.gov

1.1   Introduction

The date palm (*Phoenix dactylifera* L.) is the tree producing the date, a delicious and nutritious fruit that is widely consumed around the world. The date palm most probably originated in the area near the present-day countries of Iraq and Iran. It has been cultivated there since ancient times. Date cultivation spread to many areas with suitable climates in South Asia and North Africa millennia before the present era, and into other areas more recently. This chapter summarizes the early history of the date palm: its domestication, early spread, ancient cultivation, and current status. It will serve to set the stage for the remainder of the chapters, which provide up-to-date information on current practices in date cultivation and other useful information.

1.2   Date Palm and Related Genera

The date palm (*Phoenix dactylifera* L.) is an arborescent monocot and the type species of the genus *Phoenix* of the family *Arecaceae* (Barrow, 1998). *Phoenix* is currently considered to include 13 or 14 species mostly native to the tropical and subtropical areas of Asia (Table 1.1). *P. dactylifera* is native to the arid, subtropical deserts of Western Asia (Zohary and Spiegel-Roy, 1975; Sauer, 1993; Zohary *et al.*, 2012). The date palm is adapted to the climatic and edaphic conditions found in that region, specifically riverbeds and banks and near naturally occurring springs. It tolerates high heat, and indeed requires it to produce high-quality fruit; although able to withstand arid conditions, it requires abundant groundwater to thrive; and it tolerates the saline soil conditions often found in these regions better than any other fruit tree (Krueger, 2021). These attributes led to the famous characterization of the date palm as having its head in the fire and its feet in the water. A somewhat contrasting view of *Phoenix* origins is presented in Rivera *et al.* (2020), who trace the origins of the genus *Phoenix* further back into the past when, due to different climatic conditions, the genus was much more geographically dispersed.

1.3   Domestication and Early Cultivation of Date Palm

The date palm is generally considered to be one of the earliest domesticated fruit trees (Zohary and Spiegel-Roy, 1975; Sauer, 1993; Zohary *et al.*, 2012), although it was domesticated several millennia after grains and pulses were domesticated in the same area (Larson *et al.*, 2014). 'Domestication' is defined by Harlan thus:

> To domesticate means to bring into the household. A domestic is one (servant) who lives in the same house. In the case of domesticated plants and animals, we mean that they have been altered genetically from their wild state

Table 1.1. The genus *Phoenix*: a summary. (Johnson, 1996; From Barrow, 1998; USDA–ARS National Plant Germplasm System, 2022.)

Species	Common name	Distribution	Notes
P. acaulis Roxb.	–	Northern India, Burma	Stemless; fruit edible; sometimes confused with *P. loureiri*; conservation status uncertain; local populations possibly threatened by development
P. andamanensis Sander & C.F. Sander ex R.H. Pearson	–	Bay of Bengal	Single trunk; semidwarf; species status somewhat questionable; rare, may be considered threatened
P. atlantica A. Chev.	–	Cape Verde Islands	Clustering; conservation status unknown
P. caespitosa Chiov.	–	Somalia, Arabian Peninsula	Stemless; fruit edible; habitat: wadis; species status somewhat questionable; restricted area, may be considered threatened
P. canariensis H. Wildpret	Canary (Island) date palm	Canary Islands	Single trunk; fruit edible; widely cultivated as ornamental; wide range of habitats within distribution; genetic erosion from hybridization threatens genetic integrity
P. dactylifera L.	Date palm	Middle East to western India, northern Africa	Habitat: wadis, oases; widely cultivated in suitable climates for fruit; many other plant parts utilized
P. loureirin Kunth	–	India, China, Indochina, Philippines	Dwarf; fruit edible; other plant parts utilized; taxonomy somewhat confused: two varieties (*loureiri*, *humilis*); development threatens local populations but overall, not threatened

Continued

Table 1.1. Continued

Species	Common name	Distribution	Notes
P. paludosa Roxb.	–	Bay of Bengal, Indochina, Malaysia	Semidwarf; habitat: mangrove swamps and estuaries; not considered threatened as a species but specific populations might be threatened
P. pusilla Gaertn.	–	Southern India, Sri Lanka	Fruit edible; other plant parts utilized; conservation status unclear
P. reclinata Jacq.	Senegal date palm	Tropical and subtropical Africa, Madagascar, Comoro Islands	Habitat and morphology variable; fruit edible; other plant parts utilized; widely cultivated as ornamental; not considered threatened
P. roebelenii O'Brien	Pygmy date palm	Laos, Vietnam, southern China	Rheophytic; dwarf; widely cultivated as ornamental; conservation status unclear; use as ornamental may result in removal of native populations
P. rupicola T. Anderson	Cliff date palm	Northern India	Single trunk; semidwarf; fruits eaten by animals but not humans; conservation status unclear
P. sylvestris (L.) Roxb.	Indian date palm	India and Pakistan	Wide range of habitats; utilized for sugar, fruit; not threatened
P. theophrasti Greuter	Cretan date palm	Crete, Turkey	Habitat: coastal areas; species status questionable; restricted growing area, threatened by population pressure

and have come to be at home with man. Since domestication is an evolutionary process, there will be found all degree of plant and animal association with man and a range of morphological differentiations from forms identical to wild races to fully domesticated races. A fully domesticated plant or animal is completely dependent upon man for survival. Therefore, domestication implies a change in ecological adaptation, and this is usually associated with morphological differentiation. There are inevitably many intermediate states.

(Harlan, 1992)

This definition, while useful, is somewhat anthropocentric. Domestication can also be thought of as:

a coevolutionary process that arises from a specialized mutualism, in which one species controls the fitness of another in order to gain resources and/ or services. This inclusive definition encompasses both human-associated domestication of crop plants and livestock as well as other non-human domesticators, such as insects.

(Purugganan, 2022)

With respect to plants, domestication thus is a continuum from 'food procurement' (gathering from the wild), through 'incidental domestication' (the result of human dispersal and protection of plants in a natural environment), to 'specialized domestication' (mediated through the environmental impact of humans, especially at the local level), to 'agricultural domestication' (the culmination of the other two processes, with plants continuing to evolve in response to agroenvironmental conditions), resulting in 'food production' (as contrasted to 'food procurement') (Harris, 1989). These are not distinct stages and are not mutually exclusive (Harris, 1989), and the selection pressures resulting in domestication are sometimes intentional and sometimes unintentional (Larson *et al.*, 2014). Plant and animal domestication related to food production is thought to have begun globally in different areas of the world about 12,000–11,000 years BP, when the climate was transitioning from the most recent Ice Age to the present interglacial period (Larson *et al.*, 2014).

The date palm evolved from a wild phenotype prior to being exploited by humans through stages to the date palm of current times. This would have been a long and gradual process, with many 'dead ends'. Although it is not possible at this time to know precisely when domestication of the date palm began, Zohary and Spiegel-Roy (1975), Terral *et al.* (2012), Tengberg (2012), Gros-Balthazard and Flowers (2021), among other sources, cite and summarize archeological and archeobotanical evidence that suggests that date palms were cultivated since at least approximately 5000 years BP. (Note: dates shown in this chapter as 'BP' indicate years before present, "present" considered to be approximately AD 2000.) Actual cultivation would have been preceded perhaps 6000 to 7000 years BP by a period of exploitation prior to domestication (Larson *et al.*, 2014). The evidence of date palm cultivation is various types of date palm plant fragments associated with archeological sites; however, the presence of plant fragments itself is not definitive proof of exploitation (Terral *et al.*, 2012).

P. dactylifera, as well as the other *Phoenix* species, evolved into their present forms over an extremely long time, and the exact sequence and timing of the reticulation events are now, and may remain forever, unknown. There are various speculations or models as to how the present-day date palm evolved. It has been suggested that *P. dactylifera* originated from a genetically and geographically close species, particularly *P. sylvestris*, *P. canariensis*, *P. atlantica*, or *P. reclinata* (Munier, 1973; Barrow, 1998). More recently, molecular and genomic studies have supported the concept that the current date palm developed from populations of wild date palms. Pintaud *et al.* (2010) surveyed 308 *Phoenix* accessions representing 12 *Phoenix* species using simple sequence repeat (SSR) markers. All members of the same species clustered together and the apparently closely related *P. dactylifera*, *P. sylvestris*, *P. theophrasti*, *P. atlantica*, and *P. canariensis* had few shared alleles. *Phoenix* species are interfertile and hybridize readily (Gros-Balthazard, 2013). Therefore, it is also possible that *P. dactylifera* resulted from admixture between different *Phoenix* species. Mathew *et al.* (2015) conducted genotyping-by-sequencing on 70 accessions collected from across the traditional date-producing region from north-western Africa to eastern Pakistan, along with two samples each from *Phoenix hanceana* (*loureiroi*) and *P. sylvestris*. They identified two major gene pools of *P. dactylifera*, an eastern (West Asia) and a western (North Africa), with some admixture near Egypt and a few other areas. *P. hanceana* is native to South-east Asia and is thus remote geographically and genetically from the main *P. dactylifera* gene pools, yet it shared alleles mostly with the western gene pool, whereas the more genetically and geographically close *P. sylvestris* showed higher overall allele sharing. The authors suggest that the fact that the more distant *P. hanceana* shared more genes with the western *P. dactylifera* gene pool indicates that these are ancestral alleles. Additional evidence of western and eastern gene pools for *P. dactylifera* was reported by Hazzouri *et al.* (2015) from sequencing of 62 genotypes of *P. dactylifera* from across the traditional area of date production. In this case, admixture between the two gene pools was observed in the interface between the two centers in Egypt and Sudan. Higher levels of genetic diversity were observed in the western gene pool. Two possible explanations for the existence of the two gene pools were posited: first, that there were two separate domestication events, one in Western Asia with a second, later event in North Africa; alternatively, domesticated Western Asian date palms spread into North Africa and were later introgressed with wild or semiwild date palms in North Africa. The latter theory is consistent with the higher diversity levels observed in the North African gene pool. Flowers *et al.* (2019) resequenced 71 cultivated *P. dactylifera* cultivars and 2–18 accessions of *P. sylvestris*, *P. theophrasti*, *P. atlantica*, *P. canariensis*, and *P. reclinata*. As with Hazzouri *et al.*

(2015), Flowers *et al.* (2019) found that the North African *P. dactylifera* was admixed with the Western Asian *P. dactylifera*, but also with *P. theophrasti*. These two species, along with *P. sylvestris* and *P. atlantica*, constituted a closely related 'date palm group', while *P. canariensis* and *P. reclinata* were more distant relatives. *P. atlantica* and *P. theophrasti* were sometimes thought to be relict *P. dactylifera* populations isolated on the Cape Verde Islands and Crete due to their phenotypic characteristics (Barrow, 1998), but more current thought posits them as probably being distinct species.

The occurrence of shared alleles between cultivated *P. dactylifera* and other *Phoenix* species does not necessarily indicate hybridization between species but may also indicate that the shared genes were derived from some now nonexistent wild, ancestral, or proto-*Phoenix* species. However, until recently, in contrast to, for instance, olive and grape, wild date palm populations were not known to exist (Zohary and Spiegel-Roy, 1975; Zohary *et al.*, 2012). Differentiation between 'wild' and 'feral' populations or individuals is sometimes difficult, and it is not entirely clear what phenotype a 'wild' date palm, without selection pressure from humans, would be. On the face of it, early human selection would logically involve selection based mostly on the fruit characteristics, with perhaps some directed towards clonal reproductive capacity (number of offshoots) (Clement, 1992). This would be coupled with natural selection pressures such as disease resistance, adaptation to environmental conditions, etc.. Vegetative propagation has the advantage of perpetuating desirable phenotypes and decreasing time to bearing but has the disadvantage of slowing adaptation that might occur via sexual reproduction (Clement, 1992). This is an example of humans possibly retarding diversification and also exerting 'conscious' selection pressure (Zohary and Spiegel-Roy, 1975).

Terral *et al.* (2012) quantified patterns of shape differentiation in *Phoenix* species and concluded that *P. dactylifera* was clearly differentiated from other *Phoenix* species. Morphometric analysis of seeds collected from archeological sites suggested that Western Asian dates were already present in Egypt approximately 16,000 BP and that both wild and cultivated dates were utilized at some sites. Gros-Balthazard *et al.* (2016) studied seed size and shape among cultivated and wild-collected *P. dactylifera* along with other uncultivated *Phoenix* species. Seed size alone was only weakly discriminate, whereas seed size combined with shape provided a good rate of discrimination. Seeds of the uncultivated *Phoenix* species were smaller and more rounded, with a strong correlation between length and width, while seeds of cultivated *P. dactylifera* and seedlings thereof were longer, more elongated, and lacked a correlation between length and width. The wild-collected samples of *P. dactylifera* may have represented either 'wild' (unselected) or 'feral' (cultivated types that escaped) populations. Based upon the seed characteristics, wild-collected *P. dactylifera* were considered to represent 'wild' date palms. Fuller (2018) measured trends in *P. dactylifera* seed length from seeds excavated at various sites in the Middle East and Egypt. The data suggested that a major trend in date seed elongation occurred between 5000 and 3000 BP, this being interpreted as evidence of domestication. However, seeds with lengths indicative of cultivated date palms are also found prior to that in a few cases. This is said to suggest either separate domestication events in Western Asia or delayed increases in seed length due to admixture with a large number of wild populations in that area. The earliest apparent cultivated dates (based on these criteria) were from Western Asia, with appearances in Egypt not occurring until approximately 3500 BP.

The presence of date palms and, in some cases, their cultivation in the ancient Middle East and Egypt is attested to by artistic representations and writing in addition to the archeological evidence (Danthine, 1937) (Fig. 1.1). There is also archeological and artistic evidence of date cultivation as far east as modern south-eastern Iran and south-western Pakistan as early as 4700 BP (summarized in Tengberg, 2012; Rivera *et al.*, 2020; Gros-Balthazard and Flowers, 2021). Although there is less evidence of the spread of date cultivation westward from Egypt and into Iberia, this apparently occurred between about 1500 and 2000 BP (Munier, 1973; Rivera *et al.*, 2020).

A 'completely' domesticated crop is dependent upon humans for its survival (Harlan, 1992). The date palm is not considered completely domesticated by Zohary and Spiegel-Roy (1975) since, if not vegetatively propagated, a single generation of open pollination would perpetuate the species and produce segregating populations. Conversely, Clement (1992) does consider the date palm to be completely domesticated since 'a given cultivar is dependent upon man ... as it can only survive by vegetative propagation.' It would seem that survival of a 'given cultivar' would not define the domestication status of the crop as a whole. Domestication can be considered as something other than a binary status choice, however. Goldschmidt (2013) contrasted the domestication of perennial crops with that of annual crops. He divided this process into three distinct stages: 'wild' (native, seed production, dioecious, extended juvenility, environmentally constrained, irregular fruiting); 'intermediate' (near human settlements, seed and/ or vegetative propagation, mixed-type flowering, reduced juvenility, less environmental constraints, alternate bearing); and 'domesticated' (highly managed habitats, vegetative propagation, hermaphroditic, short or absent juvenility, management of stresses, regular fruiting). Following this schema, current date palms and date production are a mix of the three categories, having 'wild' (dioecious), 'intermediate' (vicinity to human settlements, reduced juvenility, alternate bearing) and 'domesticated' (highly managed habitats, vegetative propagation, management of stresses) characteristics. Along with similar data from other perennial crops, Fuller (2018) postulates a 'domestication syndrome for fleshy fruits', wherein selection for fleshier fruits is accompanied by lengthening of the accompanying seeds and sharpening of their tips. Fuller (2018) regards this as corresponding

(a)

(b)

(c)

Fig. 1.1. (a) Bas-relief of date cultivation recovered from Tell Halaf site near present-day Aleppo, Syria, approximately 4000 BP (US Department of Agriculture archival image). (b) Picture of bas-relief of Ninevah, approximately 2800 BP (from Danthine, 1937). (c) Steatite column from Sumer, approximately 4500 BP (US Department of Agriculture archival image).

to the intermediate state of fruit cultivation of Goldschmidt (2013), with routine vegetative propagation beginning near or after the end of the domestication process or about 3000 BP in the case of *P. dactylifera*. The fact that date palms can be clonally reproduced by removing and transplanting the offshoots, rather than by more complex means, such as grafting, made them 'preadapted' for domestication (Zohary and Spiegel-Roy, 1975).

The date palm, being an 'intermediately' domesticated plant, interestingly has many current cultural practices that were practiced in antiquity. Pruessner (1920), quoting an earlier work by Scheil (1913) and a translation of the Code of Hammurabi (written *c.*3750 BP), notes that: (i) large-scale date plantations existed in the Tigris–Euphrates Delta as early as 4400 BP; (ii) their size was noted as number of trees rather than cultivated area; (iii) female trees were fertilized with pollen from male trees that were physically separated; (iv) production (in volume) was recorded from uniformly sized trees; (v) maximum yields were as high as 105 kg/tree; and (vi) precise records were kept concerning production. Some of the ancient depictions of date palms show cultural practices being carried out, including artificial pollination (Sarton, 1934; Gandz, 1935).

1.4   Cultural Significance of Date Palm

Popenoe (1973) quotes Beccari (1890) as noting four ways in which the date palm had an 'intimate relation' to ancient date growers: (i) it produced excellent, easily transportable fruit; (ii) its various other parts could be used for construction and other uses; (iii) it could produce a fermented, alcoholic beverage; and (iv) the dioecious nature of the date palm led to it 'symboliz[ing] the creative force of nature.' Popenoe seems quite excited at the discovery of fermentation 'earlier than the juice of the grape' but even more so at the idea that the date palm became the 'object as well as the symbol of worship.' Popenoe (1973) states that the date palm 'became identified … with the primitive semitic goddess, later personified as Ishtar or Astarte, who particularly embodied nature's creative forces.' This 'mother-goddess' (who enjoys many variations on the names used by Popenoe, 1973) is supposed to have been a goddess of sex, fertility, maternity, and war (Paton, 1910). She is believed to have been the 'principal, although not the exclusive deity of the ancient Canaanites, as of their primitive Semitic forefathers' (Paton, 1910). It should be noted that some question the interpretation of ancient beliefs by modern society and may '[reflect] modern Western ideas that may not have been valid in the prehistoric world' (Talalay, 2012).

The date palm originated in the same area as the three great mono-theistic religions (Judaism, Christianity, Islam). Domestication of the date palm and cultivation of dates preceded establishment of even the oldest of these (Judaism). Paton (1910) and Popenoe (1973) state that the cult of the 'palm god' (or goddess) spread throughout the ancient Middle East until it was disrupted, initially in just the small area occupied by the Jews, later by the more broadly dispersed Christianity, and eventually by the regional dominance of Islam. Although the three religions superseded the 'palm goddess' cult, along with the other gods scattered throughout the region, the alimentary and cultural significance of the date palm was well established in the region and became incorporated into the three religions to varying degrees.

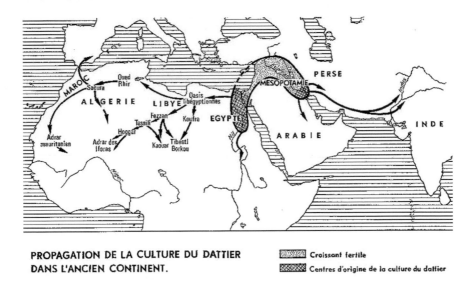

PROPAGATION DE LA CULTURE DU DATTIER
DANS L'ANCIEN CONTINENT.

▒▒▒ Croissant fertile

▒▒▒ Centres d'origine de la culture du dattier

Fig. 1.2. Early spread of the date palm. (From Munier, 1973.)

According to Popenoe (1973), the plain of the Tigris and Euphrates rivers was known as Edin, where grew an 'oracle tree' known as the 'Tree of Life'. This is said to have transformed at some point into the Biblical tale of the Garden of Eden and the Tree of Knowledge of Good and Evil. There are various references to date palms in the Jewish and Christian Bibles. In Jewish religious/cultural use, the tips of emerging date palm leaves, referred to as 'lulav', are used in the Sukkot (Feast of Booths) ritual, and in Christian religious/cultural use, the leaves are used in the Palm Sunday ritual. Islam originated in the area that was and is the center of date production. The date palm is mentioned over 20 times in the Quran, including being present at the birth of Jesus, as is also mentioned in Christian Apocrypha. Date consumption was already deeply engrained in Middle Eastern/Arabic culture by the time that Islam began, and it spread with the spread of Islam (Fig. 1.2). Initially, Islam began and spread into North Africa and Southern Europe on the west, and into India on the east. These areas have climates where date production is possible, although not always conducive to the high quality that dates achieve near their center of origin.

An interesting statement attributed to 'Qazwini' (not further identified but possibly Abu Yahya Zakariya ibn Muhammad al-Qazwini, Arab cosmologist, 1203–1283) by Popenoe, showing the cultural importance of the date palm, has strong links back to pre-monotheistic culture:

The blessed date-palm is found only in countries where Islam is the prevailing religion. The Prophet, in speaking of it, said, "Honor the palm, which is your paternal aunt"; and he gave it this name because it was made from the remains of the earth out of which Adam was created. The date-palm bears a striking resemblance to Man, in the beauty of its erect and lofty stature, its division into two distinct sexes, male and female, and the property which is peculiar to it of being fecundated by a sort of copulation. If its head is cut off, it dies. Its [male or pollen-bearing] flowers have an extraordinary spermatic odor and are enclosed in a case similar to the sac in which the fetus is contained, among animals. If an accident happens to the marrow-like substance at its summit [i.e., the terminal bud] the palm dies just as we see a man dies when his skull is severely injured. Like the members of a man, the leaves which are cut off never grow again; and the mass of fiber in which the palm is surrounded offers an analogy to the hairs which cover the human body.

(Popenoe, 1973)

1.5   Date Palm Oases

Until recently, date production in this traditional area of cultivation occurred in oasis systems (Fig. 1.3). Oasis agriculture became established in ancient times in areas with abundant groundwater, either in the form of springs or fluvial environments. In many oasis agroecosystems, the date palm is the keystone species (Tengberg, 2012). In addition to providing food (fruit) and other useful plant parts, mature date palms serve as an upper-story tree, with lower-statured trees (such as citrus, figs, etc..) and low-statured food and fodder crops being cultivated beneath them. Oasis agriculture became established first around existing naturally occurring date palm populations, which were exploited communally or in very small individual plots. Human exploitation of the water resource allowed expansion of smallholdings outward from the original core area. In more modern times, technological advances (pumps) allowed further expansion and, in some cases, the establishment of larger plantations (Stevens, 1970).

The spread of date palms from their center of origin into other areas of the Middle East, northern Africa and south-eastern Asia resulted in the establishment of new date palm oases in these areas. These human-created oases probably resulted from the introduction of a relatively small range of genotypes propagated from seed. Subsequent clonal propagation by the planting of offshoots of desirable types resulted in distinctive fruit types being associated with the various oases. Even in the center of origin, selection pressure resulting from clonal propagation of desired types over a long period of time would result in a certain amount of genetic erosion and the same association of characteristic varieties with specific oases. The relative isolation of the oases before long-distance transportation was easily accomplished further intensified the association of specific varieties with specific geographical areas. Several oases in close proximity to each other would more likely exchange desirable varieties between themselves as offshoots than they would with more distant oases. Dispersal of offshoots most likely only occurred after the domestication of the camel, or perhaps the ass,

Fig. 1.3. Khattwa Oasis, Oman, 1998. (Photo by R.R. Krueger.)

both of which were domesticated after the date palm (Larson *et al.*, 2014). This does not mean that seedling date palms were not present in different oases. Even when seedlings were present, they may have originated from a relatively narrow genetic base since the initial establishment of the palms may have been from only a few seeds or offshoots.

As an example of this, consider the case of Egyptian date culture in the early years of the 20th century as reported by Mason (1927): "The date varieties are similarly localized, the Samany and Zagloul being largely centered around Edku and Rasheid, the Amri along the eastern Delta border, the Amhat and Saidy (Sewi) in upper Giza Province. The Hayany alone of the commercial varieties enjoying a rather wide range.'

Mason (1927) further described other areas of date production in Egypt further up the Nile Valley as being centered around seedling dates. In some instances, the 'seedlings of a variety ... so nearly resemble the parent variety that they are grown and marketed with it.' Similarly, Brown and Bahgat note that:

> most of the well-known varieties confine themselves to certain localities ... at Rosetta we find the Zaghloul, the Samani, the Hayâni and Bint Eisha. Damietta is exclusively a Hayâni district. At Sharqiya, the Amri, the Aglani and Hayâni are most prominent. At Giza and Fayiûm the Siwi and Amhat predominate; while Marg is mainly a Hayâni district. Siwa and Baharia Oases are occupied mostly with Siwi. Aswân is devoted to dry varieties, namely, Bartamuda, Barakawi, Gondeila, Gargouda and Dagana.
>
> (Brown and Bahgat, 1938)

In addition to modern industrial production, traditional oasis date culture continues in Egypt, apparently with few introductions of new varieties. Nabhan (2007) studied the agricultural crop inventories in Siwa during the period 2004–2006 and compared them with information from 1890 and the early 20th century. Although it was unclear as to whether or not a few varieties had been lost in this period, it is notable that there have apparently been no introductions of new or elite varieties. Similarly, El-Assar *et al.* (2005) listed the traditional varieties Amhat, Feryhy, Shakngobil, Siwi, and Taktakt as still existing in Siwa Oasis. The Siwi accession from Siwa Oasis, along with other Siwi accessions studied, showed a much different amplified fragment length polymorphism (AFLP) profile than the other 44 samples studied. Most of the 47 accessions studied fell into one cluster. This suggests that these were similar to the ancestral types from the Middle East (some of which were also included in the molecular study) and that Siwi represents a type that is distinctive genetically as well as morphologically.

1.6   Later Spread of Date Cultivation

Date consumption later spread with Islam into now Islamic countries not having a climate conducive to date production, such as Indonesia. Date cultivation also spread outside the umma (worldwide community of Muslims) into areas having a climate suitable for its cultivation. These included America (referring to the entire American continent from the Arctic to the Antarctic) and Australia. The climate of central and south Australia is suitable for date production. Seedling dates were planted in the late

19th century but apparently did not establish. The commercial date industry in Australia started in the latter half of the 20th century with propagative material from California and from tissue culture (Johnson, 2010).

The date palm was first brought to America by the Spanish as seed from Spain or the Barbary Coast, whether for religious or food purposes, or both not being certain. Date palms were being grown on various Caribbean islands by 1526, in various places in Mexico by 1550, and in Peru and Chile by the mid-17th century (Rivera *et al.*, 2013). However, climates in these areas are not favourable for date production and, except for a few places in Peru, dates are not currently produced in these areas. The regions of Mexico that are favourable for date production are in the north-west: Sonora and the Baja California peninsula (Ortiz-Uribe *et al.*, 2019). On the Baja California peninsula, seedling date palm populations integrated into native *Washingtonia* spp. oases. Missions were established at oases because these were the sites with available water in this arid region. The date and source of the introduction of date palms into Baja California are not certain. The date was most likely the 18th century (Rivera *et al.*, 2013). There are no established records of date palms in Spanish Mexico and early Republican Mexico, although scattered seedlings may have existed (Johnson, 2010). Although the traditional *Phoenix–Washingtonia* oases still exist in Baja California Sur, the current date industry in Mexico is centered in Baja California and Sonora, based upon cultivars obtained from California starting in the 1960s (Ortiz-Uribe *et al.*, 2019). The date palm was first brought to (Alta) California in the late 18th century by the Spanish, being planted at the coastal missions (Rivera *et al.*, 2013). However, this region is not climatically suited for date production and, although a few date trees may be descended from these original trees, date production did not become established in this area. The 'low desert' of south-eastern California and south-western Arizona is the only region in the USA that has a climate suitable for date production. The commercial US date industry started in the early 20th century with offshoots imported from the Middle East and North Africa (Krueger, 2015).

1.7 Current Status of Date Production

From its origin in a very limited area in the current Middle East, the date palm has spread widely and is now cultivated in 41 countries; this is a limited distribution compared to, for example, citrus, which is produced in 140 countries (FAO, 2022). The limitation on the distribution of date production is the requirement for high heat accumulation for successful production. Global date production has been trending to increase over the last 20 years (Fig. 1.4). Cultivated area has increased about 17% during this period, whereas total production has increased about 42%, from approximately 6300 kg/ha to approximately 7700 kg/ha. This is due to increased

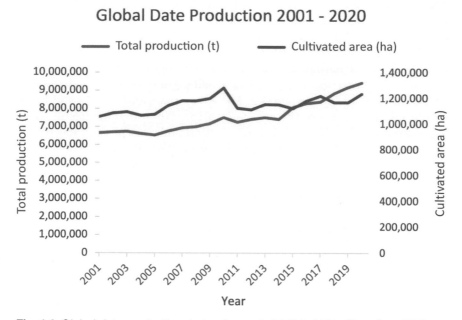

Fig. 1.4. Global date production during the period 2001–2020. (Data from FAO, 2022.)

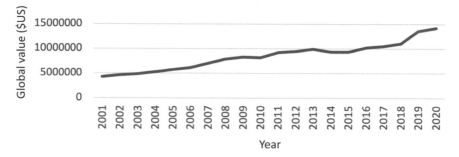

Fig. 1.5. Value of global date production in constant 2020 US dollars. (Data from FAO, 2022.)

production efficiency as unit production has increased 22% during this time period (FAO, 2022). At the same time, the value of global date production has increased 330% in constant 2020 US dollars (Fig. 1.5).

The top ten date producers in terms of cultivated area (average annual), total production (cumulative), and value of production (cumulative) from 2001 to 2020 are shown in Table 1.2. In terms of cultivated area and total production (annual), the top-ranking countries tend to remain the same but change places in the ranks. Cumulative value of production includes some smaller producers, such as Israel and the USA, that either

Table 1.2. Top ten date-producing countries, 2001–2020, in terms of cultivated area (annual average), total production (cumulative), and value of production (cumulative). (Data from FAO, 2022.)

Rank	Country	Cultivated area (ha)	Country	Total production (t)	Country	Value of production (10^3 US$)
1	Iran	189,309	Egypt	27,485,301	Saudi Arabia	67,244,428
2	Iraq	169,300	Iran	21,175,608	Algeria	44,874,247
3	Algeria	163,331	Saudi Arabia	21,088,407	Iran	17,064,946
4	Saudi Arabia	140,676	Algeria	14,882,173	Iraq	10,129,420
5	United Arab Emirates	93,674	Iraq	12,378,017	Egypt	8,000,659
6	Pakistan	93,111	Pakistan	10,454,883	Pakistan	6,401,721
7	Morocco	51,987	Oman	5,901,289	Tunisia	5,292,023
8	Tunisia	51,276	Tunisia	3,728,000	Israel	2,367,362
9	Egypt	42,691	Libya	3,329,690	USA	1,334,180
10	Libya	30,545	Morocco	1,745,624	Yemen	1,022,494

have higher unit production or produce higher-valued fruit, or both. The unit production is not included in Table 1.2 since the figures from the Food and Agriculture Organization of the United Nations (FAO) are somewhat suspect. FAO also does not include some countries known to produce dates (Australia, India) and does include some that are not known to produce dates (mainland China and Turkey).

1.8 Conclusions and Prospects

This, then, is the legacy of the date palm. From its perhaps never-to-be-known evolution, lost in the shadows of ancient time, to its initial exploitation and domestication ten millennia ago, to its current status as one of the world's premier fruit crops, the date has remained a constant companion of humans and served them well as a source of delicious, nutritious fruit and in many other ways. As humankind marches forward, it is certain that the date will continue to be a valued asset.

To that end, this book provides up-to-date information on date production. Chapter 2 provides basic information on date palm biology, while Chapter 3 expands upon the information presented here on domestication. Chapter 4 presents information on how modern genetic techniques can potentially be used to adapt date palms to current production problems. Besides seeds and offshoots, date palms can today be propagated by tissue culture and other techniques as detailed in Chapter 5. The ancient practices of planting, pollinating, irrigating, and fertilizing are updated in Chapters 6, 7, 8 and 9, while current pest problems are discussed in Chapter 10. Chapters 11 and 12 deal with farming practices that can be deployed to harmonize production with the environment. Modern harvesting techniques and postharvest handling, as outlined in Chapters 13 and 14, respectively, have greatly improved the availability and distribution of this tasty delight, while the various food and nonfood uses are detailed in Chapters 15 and 17, respectively. Besides being tasty and nutritious, dates offer serious nutritional value as presented in Chapter 16. Finally, Chapter 18 provides information on making dates available to the public.

The editors have greatly enjoyed developing this book on this ancient and enduring crop. We hope the readers enjoy reading it and are able to utilize the information presented by our knowledgeable and accomplished authors.

References

Barrow, S. (1998) A monograph of *Phoenix* L. (*Palmae: Coryphoidede*). *Kew Bulletin* 53, 513–575. Available at: https://www.jstor.org/stable/pdf/4110478.pdf (accessed 14 July 2022).

Beccari, O. (1890) Rivista monografica delle specie del genere *Phoenix* L. *Malesia* 3, 345–416.

Brown, T.W. and Bahgat, M. (1938) *Date-Palm in Egypt.* Booklet No. 24. Ministry of Agriculture, Egypt, Horticultural Section, Government Press, Cairo.

Clement, C.R. (1992) Domesticated palms. Principes 36(2), 70–78.

Danthine, H. (1937) *Le Palmier-Dattier et Les Arbres Sacrés: Dans l'iconographie de l'Asie Occidentale Ancienne.* Haut-Commissariat de la Republique Framcaise en Syrie et au Liban, Service des Antiquites, Bibliotheque Archeogique et Historique, Tome XXV. Libraire Orientaliste de Paul Geuthner, Paris.

El-Assar, A.M., Krueger, R.R., Devanand, P.S. and Chao, C.C.T. (2005) Genetic analysis of Egyptian date (*Phoenix dactylifera* L.) accessions using AFLP markers. *Genetic Resources and Crop Evolution* 52(5), 601–607. DOI: 10.1007/s10722-004-0583-z.

FAO (2022) FAOSTAT. Food and agriculture data. Food and Agriculture Organization of the United Nations, Rome. Available at: https://www.fao.org/faostat/en/#home (accessed 15 July 2022).

Flowers, J.M., Hazzouri, K.M., Gros-Balthazard, M., Mo, Z., Koutroumpa, K. *et al.* (2019) Cross-species hybridization and the origin of North African date palms. *Proceedings of the National Academy of Sciences of the United States of America* 116(5), 1651–1658. DOI: 10.1073/pnas.1817453116.

Fuller, D.Q. (2018) Long and attenuated: comparative trends in the domestication of tree fruits. *Vegetation History and Archaeobotany* 27(1), 165–176. DOI: 10.1007/s00334-017-0659-2.

Gandz, S. (1935) Artificial fertilization of date-palms in Palestine and Arabia. *Isis* 23(1), 245–250. DOI: 10.1086/346941.

Goldschmidt, E.E. (2013) The evolution of fruit tree productivity: a review. *Economic Botany* 67(1), 51–62. DOI: 10.1007/s12231-012-9219-y.

Gros-Balthazard, M. and Flowers, J.M. (2021) A brief history of the origin of domesticated date palms. In: Al-Khayri, J.M., Jain, S.M. and Johnson, D.V. (eds) *The Date Palm Genome*, Vol. 1. Compendium of Plant Genomes, Springer, Cham, Switzerland, pp. 55–74. DOI: 10.1007/978-3-030-73746-7_3.

Gros-Balthazard, M. (2013) Hybridization in the genus *Phoenix*: a review. *Emirates Journal of Food and Agriculture* 25(11), 831–842. DOI: 10.9755/ejfa.v25i11.16660.

Gros-Balthazard, M., Newton, C., Ivorra, S., Pierre, M.H., Pintaud, J.C, *et al.* (2016) The domestication syndrome in *Phoenix dactylifera* seeds: toward the identification of wild date palm populations. *PloS ONE* 11(3), e0152394. DOI: 10.1371/journal.pone.0152394.

Harlan, J.R. (1992) *Crops and Man*, 2nd edn. American Society of Agronomy, Madison, Wisconsin. DOI: 10.2135/1992.cropsandman.

Harris, D.R. (1989) An evolutionary continuum of people-plant interaction. In: Harris, D.R. and Hillman, G.G. (eds) *Foraging and Farming: The Evolution of Plant Exploitation.* Routledge, London, pp. 11–26.

Hazzouri, K.M., Flowers, J.M., Visser, H.J., Khierallah, H.S.M., Rosas, U. *et al.* (2015) Whole genome re-sequencing of date palms yields insights into diversification of a fruit tree crop. *Nature Communications* 6(1), 8824. DOI: 10.1038/ncomms9824.

Johnson, D. (ed.) (1996) *Palms: Their Conservation Status and Sustained Utilization. Status Survey and Conservation Action Plan.* IUCN, Gland, Switzerland and Cambridge.

Johnson, D.V. (2010) Worldwide dispersal of the date palm from its homeland. *Acta Horticulturae* 882, 369–375. DOI: 10.17660/ActaHortic.2010.882.42.

Krueger, R.R. (2015) Date palm status and perspective in the United States. In: Al-Khayri, J., Johnson, D. and Jain, S.M. (eds) *Date Palm Genetic Resources and Utilization.* Springer, Dordrecht, The Netherlands, pp. 447–485. https://doi.or g/10.1007/978-94-017-9694-1_14

Krueger, R.R. (2021) Date palm (*Phoenix dactylifera* L.) biology and utilization. In: Al-Khayri, J.M., Johnson, D.V. and Jain, S.M. (eds) *The Date Palm Genome*, Vol. 1. Compendium of Plant Genomes, Springer, Cham, Switzerland, pp. 3–28. DOI: 10.1007/978-3-030-73746-7_1.

Larson, G., Piperno, D.R., Allaby, R.G., Purugganan, M.D., Andersson, L. *et al.* (2014) Current perspectives and the future of domestication studies. *Proceedings of the National Academy of Sciences of the United States of America* 111(17), 6139–6146. DOI: 10.1073/pnas.1323964111.

Mason, S.C. (1927) *Date Culture in Egypt and the Sudan.* USDA Bulletin No. 1457. US Department of Agriculture, Washington, D.C. DOI: 10.5962/bhl.title.109002.

Mathew, L.S., Seidel, M.A., George, B., Mathew, S., Spannagl, M, *et al.* (2015) A genome-wide survey of date palm cultivars supports two major subpopulations in *Phoenix dactylifera. G3: Genes, Genomes, Genetics* 5(7), 1429–1438. DOI: 10.1534/g3.115.018341.

Munier, P. (1973) *Le Palmier-Dattier.* Techniques Agricoles et Productions Tropicales 24. G.-P. Maisonneueve & LaRose, Paris.

Nabhan, G.P. (2007) Agrobiodiversity change in a Saharan desert oasis, 1919–2006: historic shifts in Tasiwit (Berber) and Bedouin crop inventories of Siwa, Egypt. *Economic Botany* 61(1), 31–43. DOI: 10.1663/0013-0001(2007)61[31:ACIASD]2.0.CO;2.

Ortiz-Uribe, N., Salomón-Torres, R. and Krueger, R. (2019) Date palm status and perspective in Mexico. *Agriculture* 9(3), 46. DOI: 10.3390/agriculture9030046.

Paton, L.B. (1910) The cult of the mother-goddess in ancient Palestine. *The Biblical World* 36(1), 26–38. DOI: 10.1086/474349.

Pintaud, J.C., Zehdi, S., Couvreur, T., Barrow, S., Henderson, S. *et al.* (2010) Species delimitation in the genus Phoenix (Arecaceae) based on SSR markers, with emphasis on the identity of the date palm (*Phoenix dactylifera* L). In: Seberg, O., Petersen, G., Barfod, A.S. and Davis, J.I. (eds) *Diversity, Phylogeny, and Evolution in the Monocotyledons.* Aarhus University Press, Aarhus, Denmark, pp. 267–286. Available at: http://www.couvreurlab.org/uploads/6/4/8/8/64882181/pinta ud_et_al._2010_monocots_iv_phoenix.pdf (accessed 14 July 2022).

Popenoe, P. Henry, F. (ed.). (1973) *The Date Palm.* Field Research Projects, Coconut Grove, Florida.

Pruessner, A.H. (1920) Date culture in ancient Babylonia. *The American Journal of Semitic Languages and Literatures* 36(3), 213–232. DOI: 10.1086/369904.

Purugganan, M.D. (2022) What is domestication? *Trends in Ecology & Evolution* 37(8), 663–671. DOI: 10.1016/j.tree.2022.04.006.

Rivera, D., Johnson, D., Delgadillo, J., Carrillo, M.H., Obón, C. *et al.* (2013) Historical evidence of the Spanish introduction of date palm (*Phoenix dactylifera* L., Arecaceae) into the Americas. *Genetic Resources and Crop Evolution* 60(4), 1433–1452. DOI: 10.1007/s10722-012-9932-5.

Rivera, D., Abellán, J., Palazón, J.A., Obón, C., Alcaraz, F. *et al.* (2020) Modelling ancient areas for date palms (*Phoenix* species: Arecaceae): Bayesian analysis of biological and cultural evidence. *Botanical Journal of the Linnean Society* 193(2), 228–262. DOI: 10.1093/botlinnean/boaa011.

Sarton, G. (1934) The artifical fertilization of date-palms in the time of Ashur-Nasir-Pal BC 885-860. *Isis* 21(1), 8–13. DOI: 10.1086/346827.

Sauer, J.D. (1993) *Historical Geography of Crop Plants: A Select Roster.* CRC Press, Boca Raton, Florida.

Scheil, V. (1913) De l'exploitation des dattiers dans l'ancienne babylonie. *Revue d'Assyriologie et d'Archéologie Orientale* 10(1/2), 1–9. Available at: https://www.jst or.org/stable/pdf/23284334.pdf (accessed 14 July 2022).

Stevens, J.H. (1970) Changing agricultural practice in an Arabian oasis. *The Geographical Journal* 136(3), 410. DOI: 10.2307/1795193.

Talalay, L. (2012) The mother goddess in prehistory: debates and perspectives. In: James, S.L. and Dillon, S. (eds) *A Companion to Women in the Ancient World.* Wiley-Blackwell, Malden, Massachusetts, pp. 7–10. Available at: https://online library.wiley.com/doi/book/10.1002/9781444355024 (accessed 14 July 2022).

Tengberg, M. (2012) Beginnings and early history of date palm garden cultivation in the Middle East. *Journal of Arid Environments* 86, 139–147. DOI: 10.1016/j. jaridenv.2011.11.022.

Terral, J.-F., Newton, C., Ivorra, S., Gros-Balthazard, M., de Morais, C.T. *et al.* (2012) Insights into the historical biogeography of the date palm (*Phoenix dactylifera* L.) using geometric morphometry of modern and ancient seeds. *Journal of Biogeography* 39(5), 929–941. DOI: 10.1111/j.1365-2699.2011.02649.x.

USDA–ARS National Plant Germplasm System (2022) Germplasm Resources Information Network (GRIN Taxonomy). US Department of Agriculture, Agricultural Research Service, National Germplasm Resources Laboratory, Beltsville, Maryland. Available at: http://npgsweb.ars-grin.gov/gringlobal/ta xon/taxonomygenus?id=9268 (accessed 6 July 2022).

Zohary, D. and Spiegel-Roy, P. (1975) Beginnings of fruit growing in the old world: olive, grape, date, and fig emerge as important bronze age additions to grain agriculture in the near east. *Science* 187(4174), 319–327. Available at: https: //www.science.org/doi/pdf/10.1126/science.187.4174.319 (accessed 14 July 2022).

Zohary, D., Hopf, M. and Weiss, E. (2012) *Domestication of Plants in the Old World: The Origin and Spread of Domesticated Plants in Southwest Asia, Europe, and the Mediterranean Basin.* Oxford University Press, New York. DOI: 10.1093/acprof :osobl/9780199549061.001.0001.

Botany and Physiology of Date Palm

2

Concepción Obón[1] ⓘ, Diego Rivera[2]* ⓘ, Asunción Amorós[1] ⓘ, Gisela Díaz[1] ⓘ, Francisco Alcaraz[2] ⓘ and Dennis V. Johnson[3] ⓘ

[1]Universidad Miguel Hernández de Elche, Orihuela, Spain; [2]Universidad de Murcia, Murcia, Spain; [3]Cincinnati, Ohio, USA

Abstract

The genus *Phoenix* has agricultural and economic importance mainly due to the use of the fruits of *Phoenix dactylifera* as food and to the cultivation as ornamental plants of *P. canariensis, P. roebelenii,* and other species of the genus. *Phoenix* is made up of between 15 and 20 species of dioecious palm trees with pinnate leaves, which extend from the Cape Verde islands in the Atlantic to Hong Kong, presenting two clearly differentiated groups. The subtropical, in which *P. dactylifera, P. canariensis,* and *P. theophrasti* stand out, and whose populations are found between latitudes 20° and 40° in the northern hemisphere, is the one that presents the greatest global economic relevance. The tropical group, whose populations are mainly found between the tropics of Cancer and Capricorn in Africa and Asia, and in which *P. loureiroi, P. reclinata,* and *P. paludosa* stand out, also presents local relevance. This chapter reviews the outstanding aspects of the botany and physiology of the species of the genus *Phoenix* and especially of *P. dactylifera.* The morphological and genetic heterogeneity of currently accepted *Phoenix* taxa is quite uneven, notably found in the various geographic groups of *P. dactylifera* and *P. loureiroi.* From the botanical point of view, detailed morphological studies are still necessary in the field or germplasm banks of some *Phoenix* that have been described at the species level, but which are possibly hybrids or varieties of other species, notably in the Horn of Africa, the Arabian Peninsula, and the Indian Ocean islands. The physiology of *Phoenix* has been widely studied in the aspects that fundamentally affect the development of the plant under stress conditions (hydric, thermal, saline) and the production, quantity, and quality of the fruit. However, there are still many aspects remaining to be clarified, as is the case with metaxenia, which is the influence of the specific pollen that fertilizes the female flower on the morphology and quality

*Corresponding author: drivera@um.es

© CAB International 2023. *Date Palm* (eds J.M. Al-Khayri *et al.*)
DOI: 10.1079/9781800620209.0002

of the fruit. This is despite the fact that in its entirety, except for the embryo in the seed, the fruit is made up of maternal tissues. Studies of symbionts at the level of the roots of the palm tree have provided information of considerable interest, notably in the case of mycorrhizae, which point to the beneficial effect of these symbioses for the plant in terms of improving nutrient availability and resistance to stress conditions.

Keywords: Botany, Date palm, Metaxenia, Physiology, Taxonomy

2.1   Introduction

The genus *Phoenix* (*Arecaceae*) stands out among tree crops due to the importance as a food source of the fruits (dates) of *P. dactylifera*. Date production covered an area of 1,396,727 ha and world production reached 9,248,033 t in 2019 (FAO, 2021). Knowledge of the botany and physiology of the genus *Phoenix* is not only relevant from a scientific point of view but is also especially interesting for various applied fields of agronomy and economics.

The morphological diversity of the genus *Phoenix* presents, for the most part, a gradual pattern of variation, with notable exceptions that are differentiated by exclusive characters, particularly in the seed and stipe morphology (*P. paludosa, P. roebelenii*) (Rivera *et al.*, 2014, 2022). Extensive hybridization with consequent introgressions has determined much of this diversity, with numerous intermediate forms making it very difficult to distinguish well-defined taxonomic units. Therefore, the integrated study of natural populations, accessions, and herbarium material is the only option that, by combining phenotypic and genotyping studies, may allow us in the future to achieve a clearer understanding of the wild and cultivated diversity of the *Phoenix* genus.

The *Phoenix* area spans Africa, Asia, and Europe in a range of about 15° south and north of the Tropic of Cancer (Fig. 2.1). This implies that palm trees have to develop, in most of this area, under conditions of high temperatures, at least during the summer, and irregularity in the availability of rainwater, depending on the contributions of the monsoon, or the existence of oases associated with upwelling of water. This determines various types of water or thermal stress that have been studied as well as their influence on successful pollination, fruit development, and yield.

Outside this area *P. canariensis* is the most widely cultivated as an ornamental palm, although other *Phoenix* spp. (*P. roebelenii, P. reclinata,* and *P. dactylifera* itself) are also widely cultivated (Martínez-Rico, 2017). *P. dactylifera* achieves high yields and fruit quality in the USA and Mexico, where this species was introduced centuries ago, among other countries outside its native range (Rivera *et al.*, 2013).

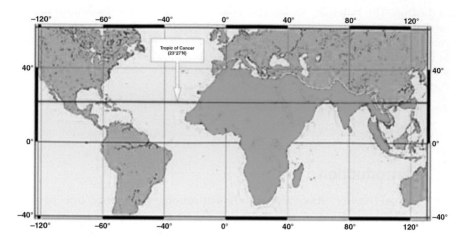

Fig. 2.1. Distribution map of genus *Phoenix*. Green shading indicates the primary distribution area of genus *Phoenix*, excluding recent introductions outside this area. (Image by F. Alcaraz.)

2.2 Botany of *Phoenix*

2.2.1 Date palm morphology

2.2.1.1 *Vegetative morphology*

The date palm and its congeners belong to the genus *Phoenix* (*Arecaceae*). These are dioecious, usually solitary single-stemmed palms (i.e. *P. canariensis*, *P. rupicola*, *P. sylvestris*, and *P. roebelenii*), as well as multistemmed (clustering) palms (i.e. *P. dactylifera*, *P. caespitosa*, and *P. reclinata*), although occasionally branched individuals are reported, especially in *P. dactylifera* (Fig. 2.2) or short-stemmed (*P. acaulis*) types. The *Phoenix* chromosome numbers are 2 *n*=36, including the *P. dactylifera* group, and rarely 32 (Al-Ani *et al.*, 2010).

The following paragraphs summarize the authors' data from the living collections of the *Phoenix* National Collection of Spain at Orihuela (Phoenix-Spain, 2021) and those published by Tisserat and DeMason (1982), Dransfield *et al.* (2008), and Carreño (2017). To facilitate the comparison of the typical date palm (*P. dactylifera*) with other *Phoenix* species in the descriptions that follow, the values for *P. dactylifera* are specified in square brackets and bold font. The morphological description of the root system is fundamentally based on the work of Dreyer *et al.* (2010).

The vegetative morphology of the genus *Phoenix* can be summarized as follows.

ROOTS. Homorhizal root system, formed by the growth of equivalent first-order roots from the base of the stem, from which other roots develop, up

Fig. 2.2. Branched individual of western *Phoenix dactylifera* at Huerto del Cura in Elche (Spain). (Photo by D. Rivera.)

to three root orders. With the development of each root order there is an abrupt change in diameter and length. No root hairs are observed.

The third-order roots are morphologically classified into four root types: long-thin roots, short-thin roots, short-thick roots ('root tubers'), and pneumatorhizas or pneumatophores. The short-thick roots are modified lateral roots, strongly swollen, and bottle-shaped. In addition, mycorrhizal thick roots – a fifth type of swollen, intense yellow, third-order, swollen roots – can be observed in the root systems of mycorrhizal *Phoenix* plants (cf. Dreyer *et al.*, 2010, Fig. 2.1).

Throughout the roots of all orders, numerous aerating root structures, the pneumatodes (pneumatozones or pneumatorings), can be found. In addition, the root system also has numerous aerating roots, e.g. pneumatorhizas and pneumatophores.

Pneumatorhizas are extremely short, modified lateral roots in which the detachment of the rhizodermis and the outer cortex form a cap at the apex, while pneumatorings are present at the base. Pneumatophores are second- or third-order aerial roots that develop with negative geotropic growth, usually with more than one pneumatoring on their surface.

Although palm roots are composed solely of primary tissues, their composition varies, first with root order and, secondly, in the degree of dependence on root diameter within each root order. Thus, multiple transitions between root orders can be found. In general, roots are composed of the following tissues: rhizodermis consisting of cells with thickened outer cell walls, exodermis, outer cortex, inner cortex, endodermis, and vascular cylinder.

STEMS AND OFFSHOOTS.   Stem 0.1–30 [**10–20**] m tall. Sheathed with spirally arranged leaf bases that detach over time, leaving characteristic scars. Stipe, with leaf-base remains, less than 15–150 [**25–35**] cm diameter. Stipe, without leaf-base remains, less than 15–120 [**18–25**] cm diameter. Underground stems: [**not known**] or present. Roots above ground level: nonexistent, rare or [**frequent**].

Offshoots below ground level (suckering offshoots): from nonexistent to 1–40 [**10–25**]. Offshoots on the trunk above ground level: none, [**rarely present**], or numerous. Branched specimens: [**not known or very rare (eastern *P. dactylifera* and most *Phoenix* species), or some (western *P. dactylifera*)**]. Leaf-base remains and scars with a central dome of remnant vascular tissue: annular, oblique, [**diamond-shaped**], or ill defined.

LEAVES.   Leaves pinnate, usually persisting for several years; sheath forming a fibrous network; pseudopetiole very short to well-developed, more or less densely covered with acanthophylls (proximal leaflets modified as spines). Rachis elongated, tapering, usually terminating in a leaflet. Crown with fewer than 20–200 [**60–90**] leaves in adult individuals; leaf 50–700 [**310–400**] cm long; leaf rachis less than 3–35 [**20–25**] cm wide at base. Leaf base color: yellowish-green, or with a glaucous cast, or with a faint yellowish-red tinge, or [**green, bright green, dull green, or old leaves with reddish-maroon discoloration regularly on edges**]. Leaf color: deep green, light green, discolored, or [**subglaucous to glaucous**]; haut color whitish or [**brown**]; pseudopetiole 10–160 [**40–100**] cm long. P/L (pseudopetiole length/leaf length index): 0.01–0.5 [**0.2–0.3**].

SPINES OR ACANTHOPHYLLS. Basal pulvinula of spines prominently swollen or [**not noticeably swollen**]. Divergence between basal pulvinula of spines from 0 to 120° [**90–120°**]; basal neck in the upper spines of adult leaves 0–130 [**0–5**] mm long; spines [**solitary, or paired**], or in groups of three or four; 1–35 [**9–16**] spines on each side of the rachis; spines 1–25 [**5–15**] cm long; spines 0.1–1.5 [**0.4–0.8**] cm wide; spines [**green**] or yellow; distal part of spines [**concolorous**] or black to purple tinged.

LEAFLETS. Leaflets single-fold, induplicate, acute, regularly arranged or variously grouped, parallel-veined, in some species bearing scales (ramenta), emergent leaves frequently with brown, or whitish, or orange, floccose indumentum and/or waxy. Basal pulvinula of leaflets [**not noticeably swollen**] or swollen; divergence between basal pulvinula of leaflets 0–130° [**90–120°**]; leaflets solitary, [**in pairs, or in groups of three**], or in groups of four or five; 5–200 [**50–130**] leaflets on each side of the rachis; leaflets at the middle of fully developed leaves 10–100 [**20–30**] cm long; leaflets 0.5–6 [**3–4**] cm wide; leaflets arranged along the rachis regularly at 180° or [**irregularly in clusters and spreading in different planes (quadrifarious)**]; leaflet consistence grassy, [**leathery, or stiff**]. Leaflet apex unarmed or [**very sharp needle-like**]; apical leaflet [**similar in shape and dimensions to the subapical leaflets**] or clearly larger because of merging with adjacent subapical leaflets.

LEAFLET BLADE ANATOMY. Prominent marginal veins in leaflets present or [**nonexistent**]; tannin-filled sclerotic cells in the margin of leaflets present or [**nonexistent**]. Abaxial ramenta present or [**nonexistent**] in young and adult leaves.

PHYTOLITHS. In *P. dactylifera* palms, silicon deposits are present as aggregates of silica, called phytoliths, in small, specialized cells, called stegmata. Phytolith-bearing stegmata are present on leaves in two anatomically distinct locations. One location is around isolated bundles of sclerenchyma fibers immediately sub-epidermal or deep in the mesophyll, covered with axially arranged rows of stegmata (Fig. 2.3). The second location is around the sheath of the sclerenchyma fibers that surround the veins, with a collateral arrangement of vascular tissue. In roots, stegmata cells containing silica aggregates are located on the outer surface of sclerenchyma bundles, arranged in rows of cells with an average distance between individual phytoliths of approximately 10–12 µm and an average phytolith size of 6–8 µm. Silicon is also present at the apex of the shoot (Bokor *et al.*, 2019).

2.2.1.2 Reproductive morphology

PHOENIX INFLORESCENCE. Pistillate and staminate inflorescences occurring in separate individuals (dioecious). Inflorescences interfoliar, branching to first order (with unbranched rachillae). The staminate and pistillate inflorescences are similar in shape but differ in size, the pistillate being larger. Peduncle flattened; rachis angulate usually shorter than the peduncle.

HV | curr | det | mode | use case | WD | mag ʁ HFW | 5 µm
5.00 kV | 0.20 nA | ETD | SE | Standard | 9.1 mm | 8 000 x 15.9 µm | Murcia University Apreo

Fig. 2.3. Scanning electron microscopy image of stegmata cells containing phytoliths of *Phoenix theophrasti*. (Photo courtesy of M.T. Coronado.)

Rachillae each bearing from a few to more than 20, spirally arranged, low triangular bracts, vestigial, early deciduous, or persistent, each subtending a solitary flower. Inflorescence emerging from a 2-keeled, glabrous, or floccose-hairy prophyll that dries at the beginning or throughout flowering and can persist in this state for months on the plant before falling.

MALE INFLORESCENCE AND FLOWERS. Staminate inflorescence; 1–45 [**15–20**] male inflorescences per palm tree. Male prophyll 10–100 [**35–55**] cm long, less than 5–25 [**10–12**] cm wide; inflorescence peduncle 5–100 [**35–50**] cm long; inflorescence rachillae 5–50 [**20–30**] cm long.

Male flower; subtending bracts [**nonexistent**] or present. With three sepals **1–3** [**2**] mm long, connate in a low cupule; three strongly nerved petals 1–10 [**6–10**] mm long, 1–5 [**3–5**] mm wide; petal margin [**entire**] or jagged; petal apex [**obtuse**] or acute to acuminate; usually six stamens (sometimes fewer or more), anthers linear 1–5 [**4–5**] mm long (Fig. 2.4). Pistillode [**nonexistent**], or rarely three abortive carpels, or seldom a minute trifid vestige.

Pollen ellipsoidal; 17–30 [**19–26**] µm long, 9–16 [**10–16**] µm wide; with aperture a distal sulcus; ectexine tectate, coarsely perforate, finely reticulate, foveolate, or perforate-rugulate; perforations with irregular and semicircular lumina; aperture margin slightly finer, psilate, or scabrate; infratectum columellate.

Fig. 2.4. Male inflorescence (a) and flower (b) of *Phoenix dactylifera*. (Photos by D. Rivera.)

Sometimes 'hermaphrodite' individuals are found with morphologically male flowers, as far as the perianth is concerned, but with incompletely developed stamens and with the three carpels developing in parallel, but without producing any seeds.

FEMALE INFLORESCENCE AND FLOWERS.   Pistillate inflorescence; 1–30 [**10–15**] female inflorescences per palm tree. Female prophyll 10–100 [**40–70**] cm long, 5–20 [**10–15**] cm wide. Peduncle frequently elongating after fertilization, 15–250 [**30–90**] cm long, 0.5–8 [**4.5–5**] cm wide; [**greenish-yellow, yellow, or orange yellow**], or orange-reddish tinged. Peduncle after fertilization and fruit ripening suberect, [**oblique, or pendulous**]. Rachillae less than 5–90 [**20–40**] cm long; bracts subtending female flowers [**nonexistent**], papery, or fleshy.

Pistillate flowers globose; sepals three, connate in a 3-lobed cupule, 1–3 [**2–3**] mm long, 1–5 [**2–4**] mm wide; three petals, imbricate, 2–7 [**3–5**] mm long, 2–6 [**2–4**] mm wide. Staminodes usually six, scale-like or connate in a low cupule; ovary with three carpels (Fig. 2.5), follicular, more or less ovoid, tapering into a short, recurved, exerted stigma, of which only one develops after pollination.

FRUITS.   Fruit morphology: fruits with basal remains of perianth and with apical stigmatic remains. Basal remains of perianth (in ripe fruits) 1–5 [**2.5–4.5**] mm high, 3–10 [**7.5–9**] mm wide and (in fresh fruits) [**yellow**], greenish, orange, brown, or whitish.

For details on the fruit ripening stages, refer to Section 2.3.1.3 of this chapter.

At khalal and bser stages, the fruit shape in lateral view is [**cylindrical, oblong-elliptical, ovoid, subglobose, pear-shaped obovate, or curved**], or globose; and in transversal section is [**circular**] or dorsoventrally flattened. Fruit apex [**truncate**] or obtuse, acute, ovate-oblique, retuse-emarginate.

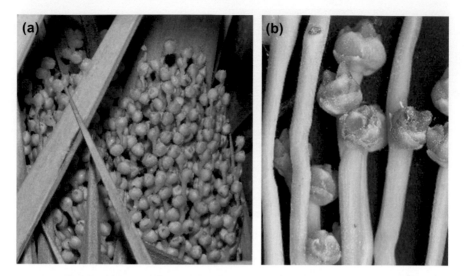

Fig. 2.5. Female inflorescence of *Phoenix dactylifera* (a) and flowers of *Phoenix iberica* (b). (Photos by D. Rivera.)

Apical stigmatic remains [**inconspicuous or very short**] or prominently pointed. Fruit base [**truncate or retuse-emarginate**] or oblique or rounded.

Fruit, at khalal and bser stages, 10–100 [**25–100**] mm long, 5–35 [**18–35**] mm wide. B/L (breadth/length index) in fruits: 0.2–1 [**0.3–0.7**].

Fruit color (fresh fruit, bser stage) green, purple [**red, yellow, or orange**]; and (soft ripe fruit, advanced rutab stage) [**amber brown**], inky black, or bluish-black. Reaching a darker hue in the tamr stage.

Fruit flesh 0.5–10 [**4–10**] mm thick. Epicarp [**smooth, rough, or wrinkled**], mesocarp [**fleshy**], endocarp [**membranous**]. Mesocarp consistency (in advanced rutab stage) [**soft (moisture over 50%), semidry (moisture 20–50%), or dry (moisture 15–20%)**]. Epicarp (peel) [**adherent or nonadherent**] to the flesh. Fruit organoleptic quality: not edible, low to medium (astringent, dry), [**good (sweet, palatable), or very good (sweet, highly palatable)**].

SEEDS. *Phoenix* seed morphology has been extensively investigated by Rivera *et al.* (2014). Seed elongate, terete, or plano-convex, and more or less deeply grooved with intruded seedcoat below the elongate ventral raphe. Seed 5–40 [**15–40**] mm long, 3–12 [**7–9.5**] mm wide, 2–12 [**6–8.5**] mm thick. B/L (breadth/length index) in seed: 0.25–1.2 [**0.46–0.52**]. Seed weight 0.03–1.9 [**0.5–1.4**] g.

Seed shape [**cylindrical-narrow, elliptical-oblong**], elliptical, ovate to triangular, or rounded. Seed apex [**obtuse, acute, or truncate**]; [**mucronate or not**]. Hilum basal; seed base [**truncate, obtuse, acute, or oblique**]; [**mucronate or not**]. Seed surface smooth, [**rough, or irregular**]; longitudinally grooved or [**transversely grooved**]. Seed ventral furrow [**deep U-shaped, deep V-shaped**], or shallow. Dorsal furrow [**nonexistent**] or present. Winged or toothed [**nonexistent or present**]; seed testa color [**greyish-cream, brown**], or blackish; embryo micropyle [**subapical (distal) or equatorial**] or sub-basal. Endosperm homogeneous, [**greyish**] or white; [**non-ruminate**]. Reports of ruminate seed seem erroneous or misapplied: Gaertner (1788) described the invaginations of the seedcoat along the ventral furrow as 'ruminated' and Barrow (1998) interpreted the black spots in the endosperm of some seeds in herbarium sheets of *Phoenix andamanensis* at Kew and Florence as rumination when these are more likely due to fungal infection.

Germination remote-tubular; eophyll undivided, narrowly lanceolate.

2.2.2 Date palm diversity and taxonomy

2.2.2.1 *Genus* Phoenix *diversity*

To estimate morphological diversity, the authors have worked with 595 individuals (353 female and 242 male) belonging to 50 operative taxonomic units that represent the known *Phoenix* species, their varieties, and hybrids (Rivera *et al.*, 2022). The descriptions have been made following a descriptor pattern based on the date palm descriptors (IPGRI *et al.*, 2005) and in the case of the seed in Rivera *et al.* (2014). For the analysis, 42 vegetative characters have been used, 24 from the inflorescence and the flower (11 in male individuals and 13 in female), and 46 from the fruit and the seed. This has allowed the authors to estimate the diversity for each of the characters and the complexity of the different species. Among *Phoenix* spp., inflorescence and floral characteristics (especially in male individuals) represent the least amount of morphological diversity, followed by vegetative characteristics, and finally by fruit and seed characteristics representing the largest amount of morphological diversity. This section presents a summary of *Phoenix* intraspecific heterogeneity (Table 2.1) that can be consulted in more detail in Rivera *et al.* (2022).

Table 2.1. Morphological heterogeneity in *Phoenix* infrageneric taxa.

Morphological heterogeneity	*Phoenix* taxa
Very low	*P. arabica*, *P. caespitosa*, *P. canariensis*, *P. paludosa*, *P. roebelenii*, *P. rupicola*
Intermediate	*P. acaulis*, *P. andamanensis*, *P. atlantica*, *P. iberica*, *P. reclinata*, *P. sylvestris*, *P. theophrasti*
High	*P. loureiroi*, *P. pusilla* – *P. zeylanica*
Very high	*P. dactylifera* western group, *P. dactylifera* eastern group

The maximum morphological heterogeneity is found within the genus *Phoenix* in the western group of *P. dactylifera*, which exhibits maximum diversity in the western oases of the Sahara in Morocco. This is closely followed by the eastern group of *P. dactylifera* (Table 2.1), especially in the populations of the Island of Socotra and those of the Euphrates Valley and neighboring areas of the Persian Gulf region. The Spanish populations and their descendants in Baja California (Rivera *et al.*, 2014; Carreño, 2017), belonging to the western group of the date palm, show high values of diversity although lower than the previously cited ones. The date palm varieties of the Nile River Valley show slightly lower levels of morphological heterogeneity.

It is very likely that a good part of the high morphological heterogeneity detected in *P. dactylifera* is due to hybridizations, not necessarily recent, and to the process of selecting plants, propagated by seed, under highly heterogeneous criteria. Hybrids of *P. dactylifera* × *P. sylvestris* also show a very high heterogeneity as a whole in the areas where they coincide (parts of the Indus Valley), due to the complexity of this hybrid group.

In the rest of the *Phoenix* species, the morphological heterogeneity within each one is much lower than that detected for *P. dactylifera*, except in the case of *P. loureiroi*, which in the aggregate presents comparable values. It is remarkable to note that this heterogeneity of *P. loureiroi* can also be observed in the herbarium sheets, especially in the extensive collection available at the Museum National d'Histoire Naturelle of Paris. *P. reclinata* would present similar values of morphological heterogeneity if the populations called *P. abyssinica*, *P. arabica*, and part of those of *P. caespitosa* were included within the species, but this is something that requires further investigation (Rivera *et al.*, 2022).

The populations of *P. roebelenii*, *P. canariensis* (excluding the variety *porphyrococca*), *P. paludosa*, and *P. theophrasti* are particularly homogeneous.

Carreño (2017) highlighted the low heterozygosity observed for almost all the *Phoenix* species analysed based on DNA from field samples, the Kew DNA bank, and the specimens conserved in the Spanish National Collection of *Phoenix*. The high number of privative alleles detected by Carreño (2017) in *P. loureiroi* stands out; it is a species with a very wide distribution area, extending from the Indus Valley to Taiwan, with a remarkable morphological diversity that could be correlated with the existence of geographically and genetically differentiated taxa.

2.2.2.2   Phoenix dactylifera *diversity and relatives*

Wild populations of *P. dactylifera* peripheral to the main cultivation areas exist, notably in Cape Verde, Oman, and Spain. These populations differ morphologically from the local cultivars but share alleles with them in the categories of microsatellite markers and chlorotypes. This posed the question: to what extent are these populations wild relatives, ancestors, sub-spontaneous, or even a mixture of some of the above? This led to the publication as novel species, such as *P. atlantica* or *P. iberica*, of date palms that appear to be wild. These putative species clearly fall within the range of the western group of *P. dactylifera* but their status merits further investigation (Rivera *et al.*, 2015, 2020; Carreño, 2017).

Chaluvadi *et al.* (2019) presented an intrageneric phylogeny of *Phoenix* derived from 52 shared chloroplast genes and whole plastomes of 29 accessions representing 14 *Phoenix* species. The phylogenetic trees based on the shared plastid gene sequences and the whole-plastome sequences agree in that the cultivated date palm accessions are most closely related to the male and female accessions of *P. sylvestris*, *P. caespitosa*, *P. atlantica*, male accessions of *P. acaulis*, and *Phoenix pusilla*.

The plastome-based phylogenetic trees obtained by Chaluvadi *et al.* (2019) showed marked cytonuclear discordance when compared to the nuclear gene-based (Cyp703) phylogenetic tree reported in Torres *et al.* (2018). However, both these studies agree that *P. sylvestris* may not be the direct progenitor of the cultivated date palm. This is in contrast to the previous finding that *P. sylvestris* is the most likely progenitor to the cultivated date palm (Gros-Balthazard *et al.*, 2017).

The simple sequence repeat (SSR) genotyping data in Chaluvadi *et al.* (2019) and several earlier studies (cf. Carreño, 2017) have shown that the observed heterozygosity was lower than the expected heterozygosity in date palms. Expected heterozygosity increases with an increase in the number of alleles and with an even distribution of alleles. The fact that a high level of heterozygosity remains in most of the analysed accessions indicates that farmer selection or natural selection can still be acting on heterozygote vs homozygote fitness.

Analysing microsatellite markers of date palms from Morocco and Sudan, Elshibli (2009) concluded that although a large amount of diversity exists among date palm germplasm, the role of the biological nature of the tree, isolation by distance, and environmental effects on structuring the date palm genome was highly influenced by human impacts. The identity of date palm cultivars as developed and manipulated by date palm growers may continue to mainly depend on tree morphology and fruit characters.

Elshibli and Korpelainen (2008) in their seminal work observed weak clustering of the cultivars investigated in Sudan using microsatellite markers and suggested that these cultivars are not a result of pure cloning processes. The propagation cycles performed by farmers have been interfered with by mixture with plants that have been derived from seed. Thus, new cultivars are a result of a continuous selection process carried out by farmers in their fields following sexual reproduction. Similar processes were detected for 'Candits' of Elche by Carreño *et al.* (2021).

The analyses by Carreño (2017) revealed three groups. Group one includes most *P. dactylifera* samples from Spain, the samples labelled as *P. iberica*, and those of *P. atlantica*, together with samples originating from

North Africa and the Eastern Mediterranean; group two includes some Spanish samples, a sample from Algeria, and another from Mali, as well as two samples from *P. atlantica*; and group three includes most of the samples of *P. dactylifera* of Asian origin and the rest of the taxa of the genus close to *P. dactylifera*, including those labelled as *P. arabica*, from Djebel Bura, Yemen. In all cases, analysis of the data set grouped the samples of *P. atlantica* as part of the *P. dactylifera* complex.

The SSR genotypes analysed by Chaluvadi *et al.* (2019) classified all date palm (*P. dactylifera*) germplasm into two major groups, with one group predominately enriched with African accessions and the other group enriched with Asian accessions. This agrees with earlier studies, and also suggests that date palm may have been independently domesticated in Asia and North Africa (Hazzouri *et al.*, 2015; Mathew *et al.*, 2015). Furthermore, Chaluvadi *et al.* (2019) found that the Indian accessions were genetically narrow, well differentiated from all other accessions (except Pakistani accessions), and most similar to Asian accessions. However, with the greater differentiation, Indian and Pakistani accessions showed more alleles that are present predominately in African accessions than in Asian accessions. A greater similarity of African and Pakistani accessions was also reported in a previous study (Mathew *et al.*, 2015). These results suggest that India and Pakistan received their date palm germplasm primarily from Africa, and not from the nearer germplasm sources in the Middle East. These results suggest that investigation of date palms in Somalia and Ethiopia may be particularly informative. This is particularly true when considering the likely existence in this area of another hybrid swarm, in this case between *P. reclinata* and *P. dactylifera*.

2.2.2.3 Date palm taxonomy

Taxonomic approaches to the genus *Phoenix* have evolved over time since the 18th century when Linnaeus (1753) published descriptions of *Phoenix dactylifera* and *Elate sylvestris*. This is due to the increase in botanical exploration, the wider diversity of *Phoenix* palms introduced into cultivation in private gardens, botanical gardens, and germplasm collections, and the availability of novel tools and methods. The number of recognized taxa ranged from two (Linnaeus, 1753) to 25 (ten species and 15 varieties: Beccari, 1890) (Table 2.2).

The taxonomic approach for *Phoenix*, currently accepted in the most widely consulted databases (GRIN, 2021; Palmweb, 2021), depends fundamentally on the proposal of Barrow (1998), which is primarily based on herbarium material and with a lack of data regarding the morphology of fruits and/or seeds of some species. For this reason, a review that incorporates the results of the numerous molecular and phylogenetic studies with data from living collections is now highly necessary.

Table 2.2. Historical evolution of the taxonomical treatment of genus *Phoenix*.

Groups	Linnaeus (1753)	Steudel (1841)	Martius (1836–1850)	Beccari (1890)	Miller et al. (1919)	Barrow (1998)	Palmweb (2021)	GRIN (2021)
Mediterranean – West Asian								
Species	1	1	1	2	3	3	4	4
Infraspecific	0	0	7	9	2	0	0	0
East Asian								
Species	1	5	5	7	9	8	8	7
Infraspecific	0	0	0	6	1	2	?	2
Tropical African								
Species	0	2	2	1	1	2	2	2
Infraspecific	0	0	0	0	0	0	?	0
Total	2	8	15	25	16	15	14	15

Concerning the eastern and western groups of *P. dactylifera*, it is evident that the type of *Phoenix dactylifera* L. falls within the area dominated by the eastern group while the earlier available name for the western group at the rank of species is *Phoenix excelsior* Cav. (Rivera *et al.*, 2019b); however, given the wide hybrid zone and the introgressions it seems unwise to separate these groups in terms of species or even subspecies.

2.2.3   Date palm symbionts

Date palm can associate with a large diversity of soil microorganisms in the rhizosphere that influence plant performance and survival. Among them, the mycorrhizal fungi are a key component of the soil microbiota that establish mutualistic symbiosis with plant roots, providing a plant with water and nutrients and receiving carbohydrates in turn, thus promoting plant growth, nutrition and tolerance to stress. The collectively so-called plant growth-promoting rhizobacteria (PGPR) are bacteria associated with roots in a symbiotic or nonsymbiotic way, that are known to be able to benefit plant health and alleviate stresses. Date palm roots can also harbour other endophytes (Mahmoud *et al.*, 2017) that colonize roots without causing disease or conferring benefit.

2.2.3.1   *Mycorrhizal symbioses*

Early observations of the mycorrhizal status of date palm roots (Khudairi, 1969; Khaliel and Abou-Heilah, 1985) revealed that it is a mycotrophic species that forms arbuscular mycorrhiza (i.e. endotrophic, with intracellular arbuscules, vesicles, and extraradical spores, fungi belonging to the phylum *Glomeromycota*). As regards anatomy, the mycorrhiza is intermediate between the morphological types of Arum (or linear) and Paris (or coiling), characterized by intercalary coils and intra- and intercellular hyphal growth, the fungal colonization being restricted to the inner cortex of the third-order roots (Dreyer *et al.*, 2010).

Mycorrhizae occurrences are common in date palm roots under natural conditions (Fig. 2.6). Surveys of mycorrhizae in palm groves from Morocco (Bouamri *et al.*, 2006, Bouamri *et al.*, 2014; Sghir *et al.*, 2014), Tunisia (Zougari-Elwedi *et al.*, 2016), and other countries (Sghir *et al.*, 2016) evidenced that mycorrhizae were prevalent in most samples, with extensive root mycorrhizal colonization of about 50–70% of the root mass, seasonal variation in number of spores, and almost 90 morphospecies reported across the world (mainly *Glomus*, *Scutellospora*, and *Acaulospora* species).

By using molecular approaches, Al-Yahya'ei *et al.* (2011) found unique communities of arbuscular mycorrhizal fungi (AMF) in date plantations in Southern Arabia, characterized by the absence of common ubiquitous species like *Glomus mosseae* or *Glomus intraradices*, but the presence of undescribed species, and with higher species richness and spore abundance

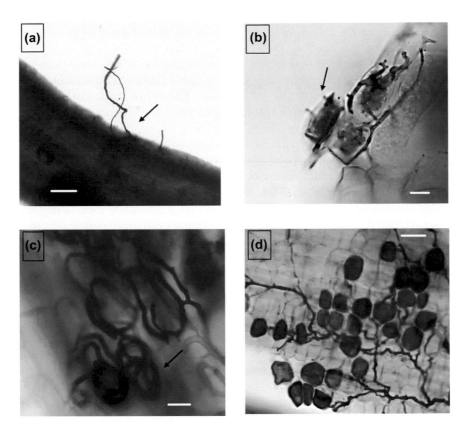

Fig. 2.6. Arbuscular mycorrhiza structures in *Phoenix dactylifera* roots collected from a palm grove at Elche, Alicante, Spain. (a) External hypha at an entry point (scale bar = 100 μm); (b) intracellular arbuscules connected to endophytic longitudinal hyphae (scale bar = 60 μm); (c) hyphal coils (scale bar = 50 μm); (d) vesicles and intercellular hyphae (scale bar = 80 μm). The mycorrhizal colonization appeared in the cortex of second and mainly third-order fine roots. (Photos by G. Díaz.)

than those of native adjacent plants. Salinity, as a common soil feature at sites where date palms grow, could reduce fungal colonization inside the roots; however, a positive correlation between soil salinity along a gradient and spore density has been found (Chebaane *et al.*, 2020). These authors identified species such as *Funneliformis coronarius*, *Rhizoglomus irregulare*, *Dominikia disticha*, or *Albahypha drummondii*, presumably adapted to saline conditions. However, despite the reports of mycorrhiza in date palm, only five references for *P. dactylifera* are currently recorded in the FungalRoot database, a global online database of plant mycorrhizal associations (Soudzilovskaia *et al.*, 2020), while no data exist in the MycoFlor database for Central Europe (Hempel *et al.*, 2013) or the global study by Wang and Qiu (2006).

It is acknowledged that arbuscular mycorrhiza confers many benefits to the host plant by improving plant growth, mineral nutrition, and alleviation of biotic and abiotic stresses. Therefore, inoculation with mycorrhizal fungi has been explored as an option to increase date palm productivity.

In general, the date palm is very responsive to mycorrhizae, and a number of studies report palm growth promotion and positive effects in establishment and survival (Khaliel and Abou-Heilah, 1985; Al-Whaibi and Khaliel, 1994; Khaled *et al.*, 2008; Baslam *et al.*, 2014; Meddich *et al.*, 2015, 2018; Ghadbane *et al.*, 2021). These effects are particularly interesting for tissue-cultured seedlings, which face challenging environmental conditions during the acclimation or transplantation stage, as shown for cvs. Khenizi (Shabbir *et al.*, 2010), Mekfazy (Al-Karaki, 2013) and Feggous (El Kinany *et al.*, 2019).

Moreover, inoculation with mycorrhizal fungi could mitigate the detrimental effects produced by saline soils where date palms are usually cultivated. The mechanisms by which mycorrhizae contribute to salt tolerance are complex and include enhancement of relative water content, photosynthetic efficiency, and pigment content; increase of stomatal conductance (which could improve CO_2 fixation); and improvement of the antioxidant defense system (superoxide dismutase (SOD), catalase, peroxidase, and ascorbate peroxidase) thus preserving cell components from oxidative damage (Meddich *et al.*, 2018; Ait-El-Mokhtar *et al.*, 2019, 2020). Overall, inoculation leads to improved dry matter production and nutrient uptake under saline conditions.

Despite the date palm being rather drought tolerant, water deficit may negatively affect its productivity. Mycorrhizal inoculation is a possible way to alleviate this problem. Mycorrhizal symbiosis improves plant tolerance to drought by modifying biochemical, physiological, and morphological responses such as an increase of cell-wall elasticity, thus maintaining higher water content and water potential, lower values of stomatal resistance, and higher levels of phenols, peroxidase, and polyphenol oxidase activities (Baslam *et al.*, 2014; Meddich *et al.*, 2015, 2018). As a result, mycorrhizal plants are less sensitive and produce higher yields than non-mycorrhizal ones under water deficit.

Mycorrhizal inoculation is also effective in increasing resistance to fungal pathogens, not only in terms of growth but also in terms of final mortality. One of the most aggressive diseases of the date palm is bayoud, caused by *Fusarium oxysporum* f. sp. *albedinis* (Foa) (Abohatem *et al.*, 2011). Mycorrhizae can stimulate defense mechanisms, i.e. by increasing total phenols, phenolic compounds like hydroxycinnamic acid known to be implicit in resistance to Foa (Jaiti *et al.*, 2007, 2008), and peroxidase activities (Khaled *et al.*, 2008; Meddich *et al.*, 2018). In this case, the prior inoculation with mycorrhizal fungi induces activation of the defense system of plants against pathogens known as mycorrhiza-induced resistance (MIR).

However, the impact of mycorrhizal inoculation can vary depending on the particular combination of fungal species/date palm cultivar (Baslam *et al.*, 2014; Diatta *et al.*, 2014; Meddich *et al.*, 2015). The use of indigenous isolates or native consortia that are presumably better adapted to harsh conditions where date palms grow is thus advisable to increase date palm productivity and to overcome drought, salinity, and pathogenetic diseases.

2.2.3.2   Bacterial symbioses

As suggested by several studies, *P. dactylifera* provides a suitable niche for PGPR. Although macroecological factors influence bacterial community structure, the date palm might shape bacterial community services by consistently selecting PGPR able to promote plant growth under harsh conditions, even irrespective of edaphic environments. That is, in desert/oasis agroecosystems, date palms exert a selection pressure and recruit a specific microbiota (in terms of PGPR taxa or in terms of plant growth-promoting functional traits) that provides important benefits for the plant (Ferjani *et al.*, 2015; Yaish *et al.*, 2016; Mosqueira *et al.*, 2019; Thennarasu *et al.*, 2019).

Predominant phyla with promoting traits are Gammaproteobacteria, Alphaproteobacteria, Betaproteobacteria, as well as Actinobacteria and Firmicutes (Ferjani *et al.*, 2015; Yaish *et al.*, 2016; Mosqueira *et al.*, 2019).

In vitro assessments for PGPR activities have revealed a great variety of mechanisms in bacteria isolated from the date palm rhizosphere. For example, the prevalent *Pseudomonas* showed ability for P solubilization, siderophore release, N fixation, or phytohormone production (Cherif *et al.*, 2015). Over 80% of the bacteria isolated from the rhizosphere, among which *Pseudomonas, Pantoea, Microbacterium, Bacillus, Arthrobacter,* and *Enterobacter* are abundant, showed plant growth-promoting activities (auxin production, ammonium synthesis, P solubilization, siderophore production) (Ferjani *et al.*, 2015). *Enterobacter, Bacillus,* and *Klebsiella* showed phosphate and K solubilization and ammonium production activities (El-Maati *et al.*, 2020). *Bacillus, Enterobacter,* and *Paenibacillus* showed ability for K, P, or Zn solubilization, ammonium production, and production of aminocyclopropane carboxylase deaminase implied in the regulation of ethylene and indole-3-acetic acid (IAA) levels that might help plants to grow under saline conditions (Yaish *et al.*, 2015).

These findings suggest a coevolution between *P. dactylifera*, which has been cultivated for a very long time, and associated soil bacteria likely adapted to withstand local harsh conditions (Mosqueira *et al.*, 2019). Therefore, these bacteria can be a good source of native inoculants. Some of the PGPR strains isolated from date palm have been tested experimentally as biofertilizers or biocontrol agents, also showing positive effects under saline and drought conditions.

It is widely recognized that PGPR improve plant nutrition through increasing nutrient solubilization. In this sense, a mixture of *Azotobacter* and *Azospirillum, Bacillus megaterium,* and *Bacillus circulans* as sources of N, P, and K, respectively, resulted in palm growth promotion and resistance to saline stress conditions (El-Sharabasy *et al.*, 2018). Beneficial effects of PGPR have also been shown for promoting rooting and survival of offshoots (El-Taweel *et al.*, 2015) and for improving fruit yield and increasing concentration of valuable phytochemicals, making the fruit more valuable (Abdel-Gawad *et al.*, 2019).

Inoculation with PGPR (*Azotobacter, Azospirillum, Bacillus polymyxa*) may be crucial for the survival and growth of *in vitro*-produced date palm plantlets during the acclimation stage, increasing plant protection against the usual constraints encountered at this stage (Abdel-Galeil *et al.*, 2018).

A step forward is the combined inoculation with PGPR and AMF (Zougari-Elwedi *et al.*, 2018). This strategy can be an effective way to boost *in vitro* plant biomass under water deficit conditions and to improve physiological parameters (stomatal conductance, photosynthetic pigments, photosynthetic efficiency, sugar and protein contents) and antioxidant enzymes (polyphenol oxidase, peroxidase) (Anli *et al.*, 2020). Moreover, mixed inoculations based on native PGPR and AMF may result in mitigation of salt stress expressed as higher growth rates and fruit yield, higher photosynthetic capacities, enhanced oxidative enzymatic system (polyphenol oxidase, peroxidase), phytohormones, physiological traits like leaf water potential, and organic compounds, both in adult plants (Naser *et al.*, 2016) and transplanted *in vitro* plants (Toubali *et al.*, 2020). The observed effects of combined inoculations suggest that beneficial microbial associations operate at multiple levels (nutrition, photosynthesis, antioxidant system, gene regulation, osmolyte biosynthesis).

The use of selected native microbial strains (bacteria, mycorrhizal fungi) provides interesting strategies for the sustainable culture of date palm.

2.3 Physiology of *Phoenix*

2.3.1 Pollination, fruit setting, and fruit stages

2.3.1.1 Pollination, xenia, and metaxenia

NATURAL AND ARTIFICIAL POLLINATION. The fruit set percentage of the racemes is one factor that determines the yield of date palms, together with fruit size, number of bunches, etc. To obtain an abundant harvest good pollination is necessary. Pollen quality, female–male affinity, temperature, irrigation, fertilization, and soil characteristics are factors that affect fruit set (Salomón-Torres *et al.*, 2021).

Under natural sexual reproduction conditions, the percentage of male and female date palm individuals in a population is approximately 50%; good pollination is achieved spontaneously, provided the flowering periods for males and females overlap. However, optimized use of the area under cultivation led to a drastic reduction of the proportion of male individuals via a more or less intentional selection for good pollen donors. Under these circumstances, artificial pollination is essential. Pollination methods are natural pollination; traditional (strand placement); dusting by hand; dusting with a manual, mechanical, or electric pollinator; and liquid pollination (Salomón-Torres *et al.*, 2021).

Natural pollination by wind, bees, and other insects is found to yield a fair fruit set in various areas of date-growing countries (Elche, Spain; Marrakech, Morocco; Ica, Perú; San Ignacio, Baja California, Mexico). All these regions are characterized by their nearly 100% seedling composition with about 50% males (Zaid and de Wet, 2002).

Artificial pollination may have played and still plays a role in the genesis of date palm diversity. Although commercial groves are often exclusively female, purchased pollen can exhibit genotype-dependent variability in its effects on fruit size, quality, and maturity, otherwise known as 'metaxenia' (Swingle, 1928; Crawford, 1936; Nixon, 1936). Thus, chance propagation of the resultant seed and clonal propagation thereafter appear to have shaped the evolutionary dynamics of date palm even after domestication (Chaluvadi *et al.*, 2019).

XENIA AND METAXENIA. The pollen of *P. dactylifera*, and of other *Phoenix* species, exerts a direct influence on the color, shape, and size of the seed, and on the rate of fruit development, the size of the fruit, as well as the fruit ripening time. The particular male used to fecundate the female flowers influences the development of the date fruit. Each male exerts approximately the same effect on fruit of all varieties and the same effect in different years (Swingle, 1928).

The term 'xenia' was coined by Focke (1881) to name variations in the normal appearance or coloration of some parts of a plant through the influence of foreign pollen. After coining this term, Focke went on to describe two different forms of xenia: 'xenoplasms', which are changes in fruit form and which can be assumed to include effects on size as well as on shape; and 'xenochrome' that are changes in fruit color (Denney, 1992).

Later, xenia was used by some authors to exclusively name the effect of the pollen on the endosperm and embryo, while metaxenia was the name given to the direct effect of the pollen on the features of parts of the fruit and seed lying outside the embryo and endosperm (Denney, 1992).

Swingle (1928) hypothesized that the simplest and most likely theory to explain metaxenia is that either the embryo or the endosperm, or both, secrete hormones or soluble substances analogous to them that diffuse into the tissues of the mother plant that constitute the seed and the fruit and exert a specific effect on these tissues that varies according to the particular male parent used to fertilize the embryo and endosperm. Denney reformulated the hypothesis, only for date fruit size, in the following terms:

> In date, differential xenic size effects can be ascribed to different concentrations in one or several of the three hormones most closely associated with fruit growth: auxin, cytokinins, and/or gibberellic acid. Smaller fruits (seed + pericarp) will have lower levels of the hormone(s); larger fruits will have higher levels of the hormone(s). Addition of exogenous hormone(s) will increase size in both large and small fruits.
>
> (Denney, 1992)

The above hypothesis was tested by Abbas *et al.* (2012) using two pollen donors, cvs. Khikri Adi and Ghannami Akhdar, and the female cv. Hillawi. Pollen from cv. Khikri Adi caused an increase in fruit size, fresh weight of the whole fruit, pulp, and seed in comparison with fruit produced by the pollen of cv. Ghannami Akhdar. The pollen parent had a significant effect on the levels of free gibberellins, with fruits produced by pollen from cv. Khikri Adi having the highest level but varying over time and growth phases. The levels of endogenous gibberellins were low at fruit set (5–7 weeks from pollination) but rose to a peak value during maximum fruit growth rate (9–11 weeks from pollination) and then declined as the fruit advanced toward the stage of physiological maturity. However, the differences recorded are not statistically significant.

El-Hamady *et al.* (2010) detected increases in IAA, gibberellic acid (GA), and abscisic acid (ABA) in the first growing phase of cv. Hayany fruits that were fertilized by either of the two studied male palm genotypes. Male one was found to induce a higher hormonal content than Male two. This was related to fruit growth.

The metaxenic effect of using different pollen donors and female cultivars was studied in the USA, Israel, Mexico, Pakistan, Iran, and Morocco (Rahemi, 1998; Zaid and de Wet, 2002; Rezazadeh *et al.*, 2013; Salomon-Torres *et al.*, 2017) as well as other countries (Al-Khayri *et al.*, 2015) with confirming results in terms of obtaining a higher-quality date, at least in size, when using optimal pollen donors. However, this was highly dependent on the paired pollen donor/female cultivar.

However, in some cases no observable effects were reported from pollinations with pollen from different males tested. This has been attributed to the fact that metaxenic effects are less pronounced when climatic conditions are favourable, or the possibility that the specific males tested did not produce metaxenic effects (Chao and Krueger, 2007).

2.3.1.2 Fruit set

Fruit set of dates depends on pollen, as seen in the previous section, but also on temperature. Good fruit is usually obtained when the daily temperature is in the range of 23.9–26.2°C (Lobo *et al.*, 2014). If temperatures are very high, there is a decrease in the fruit set (Shabana and Al Sunbol, 2007). Lower temperatures also reduce fruit set, although, in this case, fruit set can be improved by covering the inflorescences with paper bags (Lobo *et al.*, 2014). In addition, it seems that if pollination takes place around 12:00 pm (midday), there is an increase in fruit set, production, and quality (Iqbal *et al.*, 2014).

Another important factor for good fruit set is the leaves/bunch ratio, which Omar *et al.* (2013) established at 10 leaves/bunch. Fertilization is also important for a good fruit set. Hesami *et al.* (2017) found that fertilization with N and Zn in combination (345 g N in the form of urea/L and 2.30 g Zn in the form of $ZnSO_4.7H_2O/L$) injected into the trunk significantly increased the fruit set of dates. If the fruit set percentage is low, it can be increased with various treatments. Fruit set can be increased by using pollen grains diluted in water between 0.1 and 2.0 g/L compared to traditional pollination (Awad, 2010; Al-Qurashi, 2011). In addition, if 2 ml treacle + 2 g ascorbic acid + 1.0 g boric acid are added to the pollen grains dissolved in water, there is a considerable increase in fruit set and in retention percentage of the fruit (Al-Wasfy, 2014). Treatment with 50 ppm GA_3 + 1000 ppm salicylic acid also significantly increased fruit set and retention of dates (Merwad *et al.*, 2015). Occasionally the dates produced are very small because there has been a very high percentage of fruit set. In these cases, thinning should be done. Fruit thinning is very expensive due to the labour involved, so it is more profitable to thin the flowers. According to Mawloud (2010), with a ¼ thinning of female flower strands from the inflorescence, there is less fruit drop and higher production. Thinning can also be done with auxins, such as naphthalene acetic acid (NAA) or 2,4-dichlorophenoxyacetic acid(2,4-D) or ethephon (ethylene releasing compound), but with variable results so it is not usually used (Lobo *et al.*, 2014).

2.3.1.3 Fruit development and maturation stages

The development and maturation of the fruits (dates) occurs over a period of several months (usually from five to eight), involving changes in their size, morphology, coloration, and composition (Fig. 2.7). The dates, once set, grow until they reach maturity. Dates are given Arabic names at different stages of their growth: hababouk, kimri, khalal, bser, rutab, and tamr, beginning when the fruit is globose and tiny (hababouk).

The hababouk stage, which begins with the fruit set, only lasts 1 to 5 weeks and is characterized by the light green color (Lobo *et al.*, 2014;

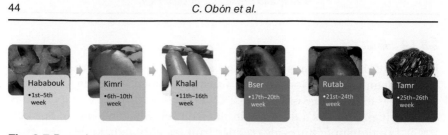

Fig. 2.7. Date ripening stages. (Image by D. Rivera.)

Al-Hajjaj and Ayad, 2018). It is followed by an increase in size and elongation (the various kimri phases) that can reach up to the 16th week after pollination (Al-Hajjaj and Ayad, 2018). In the kimri stage, fruit is approaching maximum size but is still green in color and firm in texture (Amorós *et al.*, 2009). Then the fruits reach a size close to the final one but with a light green color and firm texture (khalal) that gives way to the bser stage (besser or bisir) characterized by the complete replacement of the greenish color by a yellowish, orange, or reddish color and the texture decreasing. In the khalal and bser stages, dates are excessively astringent to eat except in some varieties. Different authors combine the khalal and bser stages under the name khalal that reaches to the 20th week in the total development of the fruit. The decrease in firmness of dates has been correlated with an increase in enzyme activities (Serrano *et al.*, 2001). In this stage, dates have a high amount of sucrose (20 g/100 g), fructose (15 g/100 g), and glucose (15 g/100 g). Also in this stage, the total phenol content and total antioxidant activity are very high, with values between 200 and 400 gallic acid equivalents/100 g and between 400 and 1000 mg Trolox equivalents/100 g, respectively, according to variety (Amorós *et al.*, 2009).

Some dates are consumed at khalal stage, but most date cultivars begin to be ripe and edible in the rutab stage, which extends from the 21st to the 24th week, where there is a change in the texture of the mesocarp and in fruit color. In the rutab stage, the date goes from being soft only in the equatorial region to reaching a uniform amber brown or dark red color, and finally the entire fruit becomes soft. This complete softening of the fruit is also due to the activities of the polygalacturanose and β-galactosidase enzymes, which present maximum activity in the rutab stage (Serrano *et al.*, 2001). In this stage, dates show a decrease in sucrose, while fructose and glucose increase to concentrations of about 25 g/100 g for each sugar. In this stage, dates also show a decrease in total phenol values with values of 50 gallic acid equivalent/100 g and a decrease in total antioxidant activity with values of 30 mg Trolox equivalent/100 g (Amorós *et al.*, 2009).

Full ripening is reached between the 25th and 26th, or eventually 27th weeks in the tamr phase (tamar, tamer), where the fruit darkens and wrinkles as a consequence of the loss of water. In the tamr stage, dates are black in color and lose most of their moisture, making them softer than dates at the rutab stage. Due to this loss of moisture, these dates can be

Fig. 2.8. Unevenly ripening dates along the same rachilla. (Photo by D. Rivera.)

easily preserved. All dates do not reach the tamr stage naturally on the tree. Thus, for example, dates from Spain, where temperatures are not as high as in Middle Eastern countries, remain in the rutab stage until they rot without ever reaching the tamr stage naturally. In this stage, sugar content is maximum, due to the loss of moisture from the fruit.

However, this ripening process does not occur simultaneously in all dates of the same infructescence or even rachilla (Fig. 2.8). Ripening is uneven within the same cluster due to different factors that include nutrient availability (Serrano *et al.*, 2001). The uneven ripening of dates forces date palm growers to repeatedly harvest the same palm tree.

2.3.2 Physiological responses to temperature

2.3.2.1 *Temperature requirements and photosynthetic efficiency and rate*

OPTIMUM TEMPERATURE.   Precisely defining the temperature range of *Phoenix* (tribe *Phoeniceae*) under cultivation is very difficult, as limiting factors such as frosts of considerable intensity can occur with a very low frequency, allowing apparent problem-free development of cultivated *Phoenix* populations for a number of years in a given locality or region until the limiting event occurs. This is the case for *P. dactylifera* in the city of Granada in Spain. The widespread cultivation of *P. dactylifera*, but also *P. canariensis*, *P. reclinata*, *P. roebelenii*, *P. rupicola*, and *P. sylvestris* (Reichgelt *et al.*, 2018), produces this false picture of a wide temperature range that is accentuated when looking at the distribution maps of the different *Phoenix* species in the Global Biodiversity Information Facility (GBIF, 2022). When considering the native range of *Phoenix*, the lowest monthly mean temperature threshold is at 7.4°C. *Phoenix* has similar frost tolerance to *Washingtonia* palms (Reichgelt *et al.*, 2018).

Shabani *et al.* (2012) modelled future optimal areas for date palm (*P. dactylifera*) cultivation, defining the limits as follows: cultivation-limiting low temperature 14°C, lower optimal temperature 20°C, upper optimal temperature 39°C, cultivation-limiting high temperature 46°C.

The cold stress temperature threshold mechanism was used by Shabani *et al.* (2012) to describe the species' response to frost. Generally, the minimum winter temperature that can be tolerated by *P. dactylifera* is 10°C. However, date palms have been recorded in locations with temperatures as low as 4°C. Therefore, intolerance to frost was incorporated by accumulating stress when the average monthly minimum temperature fell below 4°C. A key factor governing the amount of frost damage is the length of time the temperature is at its low point. The heat stress parameter was set at 46°C because it was reported that *P. dactylifera* is able to persist up to this temperature in eastern Pakistan.

Rivera *et al.* (2019a) in their study on date palm (*Phoenix, Arecaceae*) iconography on coins from the Mediterranean and West Asia (485 BC–1189 AD) have shown a correlation between the most favourable climate episodes, due to an increase in temperature, and sometimes humidity, and greater diversity in the minting of coins representing date palm trees in this area and period.

OPTIMUM TEMPERATURE, HEAT PULSES, AND PHOTOSYNTHESIS. As poikilothermic organisms, plants have to cope with potentially large variations in leaf temperature, which strongly influence rates of biochemical reactions – including those that drive photosynthesis (Kruse *et al.*, 2019).

The temperature response of net photosynthesis can be described by a bell-shaped optimum curve, where the peak rate of photosynthesis (optimal rate) is achieved at a distinct optimum temperature (T_{opt}). Instantaneous, temperature-dependent rates of thylakoid electron transport are symmetrical near the optimum temperature. For many species, the optimum temperature is quite variable and differs with changing growth temperature (Kruse *et al.*, 2016). Regarding physiological acclimation in date palm, Kruse *et al.* (2019) hypothesized that optimal leaf temperature for photosynthesis (T_{opt}) would track changes in ambient growth temperature (T_{growth}). In the experiments carried out by Arab *et al.* (2016) acclimation to contrasting growth temperature had little effect on T_{opt}. This optimum was already high in 20°C-grown plants (*c.*29.5°C) and increased by only 2°C in 35°C-grown plants.

The temperature-dependent increase in isoprene biosynthesis might contribute to the stabilization of thylakoid membranes under conditions of short-term heat pulses (Arab *et al.*, 2016).

Calvin cycle activity is controlled in numerous ways, most notably encompassing the thioredoxin system and Rubisco activase activity, itself dependent on ATP/ADP and NADPH/NADP. Kruse *et al.* (2016) argued that constant temperature dependency of 'overall' activation energy is an emergent property of metabolic networks such as the Calvin cycle. Monotonic change of overall activation energy across measurement temperatures, even extending beyond T_{opt}, suggests close coordination between the different processes (Kruse *et al.*, 2019). It has also been shown that rates of CO_2 assimilation correlate with those of thylakoid electron transport and that declining rates above T_{opt} are generally reversible provided environmental temperatures have not exceeded *c.*40–45°C and produced irreversible damage. Thus, date palm exhibits a remarkable capability to coordinate acclimation in leaf-level T_{opt} with whole-plant growth, which can be regarded as 'optimal' under environmental conditions to which this species is adapted.

Climate and soil water deprivation cause clear physiological responses in date palms, such as the reduction of net CO_2 assimilation and isoprene emission. The protein expression plasticity of the date palm contributes to the plant's acclimation to a large fluctuation of environmental conditions. As proteins are the functional macromolecules in cells, adjustments at the protein expression level help plants maintain homeostasis of fundamental metabolic processes such as seen in photosynthesis and are instrumental in achieving cellular stress resistance under environmental changes. Recent data suggest that the one underlying mechanism of date palm's tolerance to heat and drought is the remarkable plasticity of its proteome (Ghirardo *et al.*, 2021).

TEMPERATURES BELOW OPTIMUM AND PHOTOSYNTHESIS. The temperature resistance of 1- and 2-year-old potted seedlings of *P. canariensis* was tested in cold chambers at –6 to –18°C in early, mid-, and late winter by Mekhtiev *et al.* (1990). The threshold temperature after 10 hour of exposure was –12°C. A longer exposure time or lower temperature was injurious. Resistance under similar conditions of temperature and exposure increased toward late winter. The impact of cold on photosynthesis in date palm has not yet been investigated.

PHOTOSYNTHETIC RATE EFFECTS ON GROWTH AND FRUIT DEVELOPMENT. Many studies have reached similar general conclusions regarding the response of plants to water stress. They mainly identify limitation of CO_2 supply to affect plants' metabolic functions and consequently limit photosynthetic capacity (Mohammed *et al.*, 2020). In the experiments of Elshibli *et al.* (2016), reduction of water to 50% of field capacity did not generate significant effects on the photosynthetic ability of the tested plants; this result was also confirmed by slight changes in the biochemical reaction indices. However, at 25 and 10% of field-capacity water availability, changes in photosynthesis rate become more pronounced. The reduction in growth traits was visible mainly at water levels of 25% and/or 10% of field capacity.

Special irrigation systems may help to increase the photosynthetic rate in date palm plants and overcome water stress conditions. Water-use efficiency based on the net photosynthesis and transpiration rate was higher when the date palms were irrigated by a subsurface irrigation system followed by surface drip irrigation and surface bubbler irrigation (Mohammed *et al.*, 2020).

2.3.2.2 Temperature effects on plant growth and fruit development

Baoguo *et al.* (2019) performed leaf metabolome profiling on mature and young leaves of 2-year-old date palm seedlings grown in climate chambers simulating summer and winter conditions in eastern Saudi Arabia. Cultivation under high temperature (summer) resulted in higher H_2O_2 levels in young leaves despite increases in dehydroascorbate reductase activities.

Arab *et al.* (2016) experimentally showed that total N content in leaves decreased due to heat and increased upon water deprivation. In roots, a temperature-related decline of total N was only observed for well-watered plants. These effects of drought and heat on N contents of leaves and roots resulted in changed C/N ratios. In the leaves, the C/N ratios increased with elevated growth temperature irrespective of water availability and decreased with drought stress irrespective of growth temperature.

The levels of raffinose and galactinol, tricarboxylic acid cycle intermediates, and total amino acids were greater under high temperature conditions, particularly in young leaves (Baoguo *et al.*, 2019). The accumulation of unsaturated fatty acids, 9,12-octadecadienoic acid and 9,12,15-octadecatrienoic acid, was lower in mature leaves. In contrast, the amounts of saturated tetradecanoic acid and heptadecanoic acid were increased in young leaves under summer conditions. The accumulation of phenolic compounds was favored under summer conditions, while flavonoids accumulated under lower temperature (winter) conditions. Young leaves displayed values indicating effective scavenging of reactive oxygen species (ROS). These findings demonstrate the substantial metabolic adjustments that facilitate tolerance to high temperatures in the leaves (Baoguo *et al.*, 2019).

Temperature, not only environmental but specifically accumulated heat units at the level of the infructescences, may influence the rate of fruit ripening, independent of the rate of photosynthesis at the leaf level. Awad (2007) showed that bunch bagging with different materials such as black or blue polyethylene bags, white 'agrisafe' (polypropylene fleece), and paper bags during the growing season significantly increased the rate of fruit ripening and increased rutab yield per bunch. In this respect, black and blue polyethylene bags were the most effective. It is conceivable that bunch bagging, especially with black and blue polyethylene bags, accumulated higher heat units than other bags and the controls. Consequently, accumulated heat might induce higher respiration rates and the CO_2

accumulation within bags might lead to more acetaldehyde production and removal of astringency.

2.3.3 Physiological responses to soil salinity

Soil salinity is a global problem that severely affects agricultural production and the total organic matter on our planet. This problem becomes greater in regions with dry climates, sporadic rainfall, and high temperatures, leading to high evapotranspiration rates. Insufficient rainfall causes farmers to overexploit aquifers using brackish or saline groundwater. All these agriculturally devastating factors are common in some arid and semi-arid regions of the Middle East and North Africa, where date palm is a major fruit crop (Yaish *et al.*, 2017). Soil salinity has resulted in desertification of large agricultural areas, particularly in coastal zones where the overharvesting of water has led to the infiltration of seawater into the groundwater aquifer (Yaish and Kumar, 2015). It has been estimated that more than 7% of the arable land is salinized, and this is expected to increase up to 50% during the 21st century (Meddich *et al.*, 2018). Moreover, salinization dramatically limits agricultural yield by more than 20% overall as it negatively affects plant growth and development (Porcel *et al.*, 2012).

Saline environments can severely affect plants primarily by exerting osmotic and ionic effects. Osmotic effects lead to cellular dehydration and reduced accessibility of soil water by the root system, whereas toxic ion effects result from accumulation of Na^+ and/or Cl^- ions in the cell (Yaish *et al.*, 2017). Thus, salinity stress is usually associated with growth inhibition and yield reduction (Tester, 2003) due to the osmotic effect on water uptake, reduced water conductivity of the roots, disrupted ion homeostasis in cells, inhibited metabolism, damaged membranes, and divergence of energy to salt protection (Tester, 2003).

Date palms are better adapted to salinity stress compared to most cultivated trees, with a threshold of 4 dS/m and a reduction of 3.6% yield per unit of electrical conductivity (EC (dS/m)) (Maas and Hoffman, 1990). Electrical conductivity (EC) is the parameter that is most currently used and expressed in decisiemens per meter (dS/m), as an equivalent of the mass of dissolved salts per unit volume of water (TDS). One decisiemens per meter is approximately equal to a total dissolved salts (TDS) of 640 mg per liter of water (1 dS/m = 640 mg/L) (Yaish and Kumar, 2015). However, according to Sperling *et al.* (2014), date palm growth is inhibited by irrigation salinity above 9 dS/m, the yield is reduced by half (EC_{50}) at a salinity of 18 dS/m, and leaf elongation is suppressed at a salinity over 4 dS/m. Thus, date palm is generally considered a relatively salt-tolerant species (Yaish *et al.*, 2017). However, there are many date cultivars and not all behave in the same way against saline stress. Al Kharusi *et al.* (2017) screened ten date palm cultivars and studied their adaptations by mechanisms involved in the exclusion of Na^+ from the leaf and the regulation of oxidative and photosynthetic damage, classifying cultivars into two groups: tolerant and sensitive.

With regards to solute accumulation in date palm tissue, Aljuburi and Al-Masry (2000) reported a correlation between excess solute concentration in leaves, stems, and roots of date palm cv. Lulu seedlings and soil water salinity. More recently, Tripler *et al.* (2007) found elevated Na^+ and Cl^- concentrations in the roots of date palms irrigated with saline water. However, these differences were not evident at the leaf level, where the solute concentrations were remarkably low. Sperling *et al.* (2014) found that treatment with 105 mM NaCl decreased plant growth and increased Na^+ accumulation in the roots and lower stem. However, Na^+ ions were mostly excluded from the sensitive photosynthetic tissues of the leaf. The Cl^- concentration was significantly elevated due to irrigation salinity in the root zone alone. In the canopy, the Cl^- concentrations were high, increasing from leaflets, through stem, to leaf rachis, and did not differ between treatments. Reduction in the CO_2 assimilation rate was primarily attributed to a reduced stomatal conductance. Consistent with this finding, the photosynthetic response to variable intercellular CO_2 concentrations revealed no permanent damage to the photosynthetic apparatus and implicated developed photoprotective mechanisms. Independent of salinity treatment, 80% of the energy absorbed by the leaf was directed to non-photochemical quenching. Functioning at full capacity, the non-photochemical mechanism could not compensate for all the excess irradiance. Thus, of the remaining absorbed energy, a significant portion was directed to photochemical O_2-related processes, rather than CO_2-prevented photoinhibition. The exclusion of toxic ions and O_2-dependent energy dissipation maintained photosynthetic efficiency and supported survival under salt stress (Sperling *et al.*, 2014). These data agree with those of Yaish *et al.* (2017) who found that, under salt stress, all gas exchange parameters decreased but the quantum yield of photosystem II (PSII) was unaffected while non-photochemical quenching was increased.

When the date palm is exposed to saline water, it produces an overaccumulation of the amino acid proline in the roots and leaves (Yaish, 2015). In addition to its well-known role as an osmoprotectant acting as a vital factor in enhancing cellular turgor, proline has various crucial roles in plant growth under abiotic stress conditions. These roles include reducing the possible membranous oxidative damage caused by ROS, enhancing signal transduction pathways, stabilizing DNA and protein complexes (including membrane proteins), and providing an alternative resource for N and C (Yaish, 2015).

Yaish *et al.* (2017) carried out gene ontology analysis in leaves and root tissues of salinized date palms. Gene ontology analysis in leaves revealed an enrichment of transcripts involved in metabolic pathways including photosynthesis, sucrose and starch metabolism, and oxidative phosphorylation, while in roots the genes involved in membrane transport, phenylpropanoid biosynthesis, purine, thiamine, and tryptophan metabolism, and Casparian strip development were enriched. Differentially expressed genes common to both tissues included the auxin responsive gene, GH3, putative potassium transporter 8, and vacuolar membrane proton pump. Therefore, leaf and root tissues respond differentially to salinity stress.

Several studies have been carried out with different treatments to try to reverse the saline effect on date palms. Darwesh (2013) tested different treatments with yeast + amino acids that reversed all the harmful effects caused by NaCl + CaCl$_2$ (2:1 ratio). The effects that were reversed were the decreases in fresh and dry weight, height, and contents of chlorophylls *a* and *b*, iodole, catalase, and peroxidase of date palm seedlings.

El-Khawaga (2013) carried out treatments with various mixtures of compounds to try to reverse the effects of salt on the production and quality of dates. Treatments that obtained the best results were Cal-Mor at 150 ml/palm (contains 9% Ca, 7% glutaric acid, 19% N, 2% citric acid), Uni-Sal at 50 ml/palm (contains 9% polyethylene glycol (PEG), 7.5% Ca, 7% glutaric acid, 19.5% N, 2% citric acid), and citric acid at 500 ppm twice.

Other measures that are very effective in reversing the harmful effects of salt include treatment with AMF. Mycorrhizal fungi also decrease their growth due to the effect of salt, so fungi that are naturally in the roots of palm trees in saline soils should be chosen, since they will be more adapted to these conditions. Mycorrhization has a positive effect on date palm plant growth. Thus, inoculation with AMF increased the number and area of date palm leaves under the salinized conditions. AMF significantly improved the dry matter production compared to non-mycorrhizal plants (Meddich *et al.*, 2018). AMF also promoted the absorption of water and nutrients in saline conditions. Mycorrhizal date palms showed a high stomatal conductance compared to control plants, and this could improve the CO$_2$ fixation in mesophyll, which contributes to an increase in plant photosynthesis rate.

Ait-El-Mokhtar *et al.* (2019) used mycorrhizal fungi to improve the tolerance of date palm seedlings to salt stress. Mycorrhiza mitigated the decrease of K, P, and Ca content induced by salinity. Otherwise, AMF symbiosis improved physiological parameters through elevating stomatal conductance, photosynthetic efficiency, and leaf water potential under salinity stress. Under the same conditions, mycorrhizal inoculation significantly enhanced concentrations of photosynthetic pigments and protein content. Furthermore, salt stress caused high lipid peroxidation and increased H$_2$O$_2$ content; however, the application of AMF reduced these two parameters in salt-affected plants while activities of antioxidant enzymes (SOD, catalase, peroxidase, ascorbate peroxidase) were increased by salt stress and were further enhanced in plants treated with AMF.

2.3.4   Physiological responses to water stress

Global climate change induced by gradually rising atmospheric greenhouse gas concentrations is projected to drive more frequent and prolonged periods of drought and heat (Du *et al.*, 2021). Drought is one of the most acute environmental stresses increased by climate change as a major future risk in many areas of the world, particularly in arid and semi-arid environments, where plants are even more prone to water deficiency (Du *et al.*, 2021).

Drought causes a substantial decline in crop productivity because it influences the transport and availability of soil nutrients, and it affects the morphological, physiological, and nutritional traits of plants, especially water content, leaf water potential, photosynthetic pigmentation, stomatal conductance, and P and N absorption (Anli *et al.*, 2020). However, during evolution, some plants, such as date palm, have survived remarkable seasonal variations of temperature and soil water availability by developing complex adaptation strategies to maintain metabolic homeostasis (Ghirardo *et al.*, 2021).

Date palms can experience temperature extremes and prolonged periods of drought under the arid and semi-arid conditions of their natural environment (Arab *et al.*, 2016). Djibril *et al.* (2005) evaluated date palm (*P. dactylifera* L.) tolerance to osmotic stress induced by PEG or NaCl during the early stages of plant development. Two varieties, Nakhla hamra and Tijib, widely cultivated in Mauritania were tested. These authors concluded that their osmotic stress tolerances differed because the cultivar more tolerant to salinity, Tijib, had Cl⁻ and Na⁺ transporters in the tonoplast and the non-tolerant cultivar did not. The more drought-tolerant cv. Nakhla hamra could synthesize proline for osmotic adjustment and continue to absorb water in drought, a strategy that Tijib, the non-drought-tolerant cultivar, did not possess. The first showed increasing of epicotyl length, primary root length, secondary root number, and proline content when water deficit was induced by PEG. In contrast, on the basis of the same developmental and biochemical characters, the cv. Tijib was more tolerant to salinity stress.

At the cellular level, regarding drought, expansive cell growth is more inhibited by water stress than is cell division. Modification of the cell wall is one of the adaptations that date palm trees undergo under water deficit. During development, cell-wall solubilization and remodeling was characterized by a decrease in the degree of methyl esterification of pectin, an important loss of galactose content, and a reduction of the branching of xylan by arabinose in irrigated conditions (Gribaa *et al.*, 2013).

Water deficit has a profound effect on fruit size, pulp content, cell-wall composition, and cell-wall remodeling. Transcriptomic analyses during fruit development and ripening have shown changes in the expression profile of several cell-wall remodeling proteins (Gribaa *et al.*, 2013). Loss of galactose content was not as important, arabinose content was significantly higher in the pectin-enriched extracts from non-irrigated conditions, and the levels of methyl esterification of pectin and *O*-acetylation of xyloglucan were lower than in irrigated conditions (Gribaa *et al.*, 2013).

The lower levels of hydrophobic groups (methylester and *O*-acetyl) and the less intensive degradation of the hydrophilic galactan, arabinan, and arabinogalactan in the cell wall may be implicated in maintaining the hydration status of the cells under water deficit (Gribaa *et al.*, 2013).

Another strategy to tolerate drought is to have a barrier to cuticular transpiration. In this sense, Bueno *et al.* (2019) compared the date palm with a non-drought-tolerant species. Minimum leaf conductance (g_{min}) at 25°C was six times lower in date palm (1.1×10^{-5} m/s) than in *Citrullus colocynthis* (6.9×10^{-5} m/s). Additionally, g_{min} in the range 25–50°C did not change in the palm leaf but increased by a factor of 3.2 in *C. colocynthis*. Arrhenius formalism applied to the *C. colocynthis* g_{min} led to a biphasic graph with a steep increase at temperatures > 35°C, whereas for date palm the graph was linear over all temperatures. Leaf cuticular wax coverage amounted to $4.2 \pm 0.4 \mu g/cm^2$ for *C. colocynthis* and $29.4 \pm 4.2 \mu g/cm^2$ for *P. dactylifera*. In both species, waxes were mainly composed of very-long-chain aliphatic compounds. Midpoints of the wax melting ranges of *P. dactylifera* and *C. colocynthis* were 80 and 73°C, respectively. Therefore, in date palm a particular wax and cutin chemistry prevents the rise of g_{min} at elevated temperatures (Bueno *et al.*, 2019). In addition, the palm tree also modifies the fatty acids of the membranes in response to drought (independently of the heat) so that they maintain their fluidity and functionality, with a decrease in double bond indices of leaves while in the roots the double bond indices increased (Arab *et al.*, 2016). In the date palm there is a great adaptation of photosynthesis. So, photosynthesis did not decline under drought but even increased slightly compared to well-watered plants (Arab *et al.*, 2016).

Neither heat nor water shortage has any significant effect on stomatal conductance of date palm and stomatal conductance is already low under well-watered conditions compared to temperate woody species. Date palm is a slow-growing species with conservative water usage, which appears to be a major feature of this species' adaptation to xeric environments. There is a drop of intercellular CO_2 concentration (C_i) observed during water shortage caused by enhanced photosynthesis and not by reduced stomatal conductance, and increased leaf-N content values at reduced water availability (Arab *et al.*, 2016). Leaf hydration is increased by drought (Arab *et al.*, 2016). Ghirardo *et al.* (2021) studied proteomic changes produced by the effects of heat and drought on photosynthesis. They found that the date palm, in the summer, induced the expression of proteins related to Rubisco, chlorophyll metabolism, PSII, and the electron transport chain. Hot/drought stress also caused an increase of ATP synthases and downregulation of the H^+-ATPase integrated in plasmatic membranes. ATP synthases are crucial in energy transduction and alleviation of stress. The hot summer weather also increases the expression of *P. dactylifera* isoprene synthase (PdIspS) and the leaf emissions of isoprene. Isoprene can protect the photosynthetic apparatus from abiotic stress as it is effective against cellular oxidative stress occurring during drought (Ghirardo *et al.*, 2021).

ROS formation occurs in cells under normal as well as a wide range of stressful conditions. Surprisingly, in date palm leaves, concentrations of glutathione (GSH) are reduced, and total ascorbate concentrations significantly decreased, due to heat independent of water supply. In the roots, in contrast, drought causes increased levels of GSH independent of growth. Cysteine concentrations in roots increase in response to the combination of drought and heat (Arab *et al.*, 2016). These changes can be considered a mechanism to deal with enhanced ROS production during stress in the roots. In contrast, the reduced levels of ascorbate and GSH in leaves of stressed date palms might suggest an overload of the antioxidative capacity of the plant. Glutathione reductase (GR) activity increases under elevated growth temperature and drought at 35°C. The higher activities of GR indicate that although concentrations of GSH decrease due to stress, the turnover through the antioxidative system was enhanced, thereby ensuring effective scavenging of ROS produced during the stress period (Arab *et al.*, 2016).

Ghirardo *et al.* (2021) also found in their proteomic study a down-regulation of the GR protein and an upregulation of the plastid SOD and L-ascorbate peroxidase (APX) involved in the Foyer–Halliwell–Asada cycle. Du *et al.* (2021) found variations in the responses of date palms to drought in summer and winter. In summer, drought induces significantly decreased leaf hydration and concentrations of ascorbate, most sugars, primary and secondary organic acids, as well as phenolic compounds, while thiol, amino acid, raffinose, and individual fatty acid contents were increased compared to well-watered plants. In winter, drought has no effect on leaf hydration, ascorbate, and fatty acids contents, but results in increased foliar thiol and amino acid levels as observed in summer. Safronov *et al.* (2017) also found that soluble carbohydrates, such as fucose, and glucose derivatives, were increased, suggesting a switch to carbohydrate metabolism and cell-wall biogenesis. In addition, transcriptomics data showed transcriptional activation of genes related to ROS under three conditions (drought, heat, combined heat and drought), suggesting increased activity of enzymatic antioxidant systems in cytosol, chloroplast, and peroxisome. The genes that were differentially expressed in heat and combined heat and drought stresses were significantly enriched for circadian and diurnal rhythm motifs, suggesting new stress avoidance strategies.

Although date palms are considered remarkably tolerant to harsh climatic conditions, their distribution and productivity will be significantly affected by more frequent and prolonged drought events due to climate change (Du *et al.*, 2021). Several authors have tried numerous treatments with AMF to reverse the harmful effects of drought (Baslam *et al.*, 2014;

Meddich *et al.*, 2015, 2018; Anli *et al.*, 2020; Harkousse *et al.*, 2021). These authors concluded that AMF, alone or in combination with PGPR, caused increases in the water status of the plant and the permeability of membranes, along with increases in biomass, photosynthetic efficiency, chlorophylls, and the absorption of minerals, so that the treated palms were able to counteract the damaging effects of drought.

2.4 Conclusions and Prospects

The genus *Phoenix* includes more than 15 species, among which *P. dactylifera* stands out for its importance as a food source. *Phoenix* has a remarkable ability to adapt to life in extreme environments with high temperatures, high salinity, or limitations in the availability of water. However, detailed knowledge of the ability to respond to various types of stress of cultivated populations of *P. dactylifera* is essential to successfully address the challenges posed by global climate change in which we find ourselves immersed. Mycorrhizae are another very relevant aspect as a factor that facilitates the resilience of wild and cultivated populations of *Phoenix*.

Recent molecular studies have allowed to approximate an idea of the diversity within *Phoenix* and especially of the *P. dactylifera* complex. However, exhaustive and detailed morphometric studies are required, understanding the set of diversity described and especially the populations that are not easily accessible due to their geopolitical situation.

References

Abbas, M.F., Abdulwahid, A.H. and Abass, K.I. (2012) Effect of pollen parent on certain aspects of fruit development of Hillawi date palm (*Phoenix dactylifera* L.) in relation to levels of endogenous gibberellins. *Advances in Agriculture & Botanics* 4(2), 42–47.

Abdel-Galeil, L.M., Farrag, H.M.A. and Zayed, E.M.M. (2018) Plant growth promoting rhizobacteria (PGPR) protect date palm (*Phoenix dactylifera* L.) plantlets from fungal attack during acclimatization stage. *American-Eurasian Journal of Agricultural & Environmental Sciences* 18(2), 89–95.

Abdel-Gawad, H., Saleh, A.M., Al Jaouni, S., Selim, S., Hassan, M.O. *et al.* (2019) Utilization of actinobacteria to enhance the production and quality of date palm (*Phoenix dactylifera* L.) fruits in a semi-arid environment. *Science of the Total Environment* 665, 690–697. DOI: 10.1016/j.scitotenv.2019.02.140.

Abohatem, M., Chakrafi, F., Jaiti, F., Dihazi, A. and Baaziz, M. (2011) Arbuscular mycorrhizal fungi limit incidence of *Fusarium oxysporum* f.sp. *albedinis* on date palm seedlings by increasing nutrient contents, total phenols and peroxidase activities. *The Open Horticulture Journal* 4, 10–16. DOI: 10.2174/1874840601104010010.

Ait-El-Mokhtar, M., Ben-Laouane, R., Anli, M., Boutasknit, A., Wahbi, S. *et al.* (2019) Use of mycorrhizal fungi in improving tolerance of the date palm (*Phoenix dactylifera* L.) seedlings to salt stress. *Scientia Horticulturae* 253, 429–438. DOI: 10.1016/j.scienta.2019.04.066.

Ait-El-Mokhtar, M., Baslam, M., Ben-Laouane, R., Anli, M., Boutasknit, A, *et al.* (2020) Alleviation of detrimental effects of salt stress on date palm (*Phoenix dactylifera* L.) by the application of arbuscular mycorrhizal fungi and/or compost. *Frontiers in Sustainable Food Systems* 4, 131. DOI: 10.3389/fsufs.2020.00131.

Al-Ani, B., Zaid, A. and Shabana, H. (2010) On the status of chromosomes of the date palm (*Phoenix dactylifera* L.). *Acta Horticulturae* 882, 253–268. DOI: 10.17660/ActaHortic.2010.882.28.

Al-Hajjaj, H. and Ayad, J. (2018) Effect of foliar boron applications on yield and quality of Medjool date palm. *Journal of Applied Horticulture* 20(3), 181–188. DOI: 10.37855/jah.2018.v20i03.32.

Aljuburi, H.J. and Al-Masry, J.J. (2000) Effects of salinity and indole acetic acid on growth and mineral content of date palm seedlings. *Fruits* 55, 315–323.

Al-Karaki, G.N. (2013) Application of mycorrhizae in sustainable date palm cultivation. *Emirates Journal of Food and Agriculture* 25(11), 854–862. DOI: 10.9755/ejfa.v25i11.16499.

Al Kharusi, L., Assaha, D.V.M., Al-Yahyai, R. and Yaish, M.W. (2017) Screening of date palm (*Phoenix dactylifera* L.) cultivars for salinity tolerance. *Forests* 8(4), 136. DOI: 10.3390/f8040136.

Al-Khayri, J.M., Jain, S.M. and Johnson, D.V. (2015) *Date Palm Genetic Resources and Utilization*, Vol. 2. *Asia and Europe*. Springer, Dordrecht, The Netherlands. DOI: 10.1007/978-94-017-9707-8.

Al-Qurashi, A.D. (2011) Effect of pollen grain-water suspension spray on fruit set, yield and quality of 'Helali' date palm (*Phoenix dactylifera* L.). *Journal of Applied Horticulture* 13(1), 44–47. DOI: 10.37855/jah.2011.v13i01.10.

Al-Wasfy, M. (2014) Yield and fruit quality of Zaghloul date palms in relation to using new technique of pollination. *Stem Cell* 5, 14–17.

Al-Whaibi, M.H. and Khaliel, A.S. (1994) The effect of Mg on Ca, K and P content of date palm seedlings under mycorrhizal and non-mycorrhizai conditions. *Mycoscience* 35(3), 213–217. DOI: 10.1007/BF02268440.

Al-Yahya'ei, M.N., Oehl, F., Vallino, M., Lumini, E., Redecker, D. *et al.* (2011) Unique arbuscular mycorrhizal fungal communities uncovered in date palm plantations and surrounding desert habitats of Southern Arabia. *Mycorrhiza* 21(3), 195–209. DOI: 10.1007/s00572-010-0323-5.

Amorós, A., Pretel, M.T., Almansa, M.S., Botella, M.A., Zapata, P.J. *et al.* (2009) Antioxidant and nutritional properties of date fruit from Elche grove as affected by maturation and phenotypic variability of date palm. *Food Science and Technology International* 15(1), 65–72. DOI: 10.1177/1082013208102758.

Anli, M., Baslam, M., Tahiri, A., Raklami, A., Symanczik, S. *et al.* (2020) Biofertilizers as strategies to improve photosynthetic apparatus, growth, and drought stress tolerance in the date palm. *Frontiers in Plant Science* 11, 516818. DOI: 10.3389/fpls.2020.516818.

Arab, L., Kreuzwieser, J., Kruse, J., Zimmer, I., Ache, P. *et al.* (2016) Acclimation to heat and drought—Lessons to learn from the date palm (*Phoenix dactylifera*). *Environmental and Experimental Botany* 125, 20–30. DOI: 10.1016/j.envexpbot.2016.01.003.

Awad, M.A. (2007) Increasing the rate of ripening of date palm fruit (*Phoenix dactylifera* L.) cv. Helali by preharvest and postharvest treatments. *Postharvest Biology and Technology* 43(1), 121–127. DOI: 10.1016/j. postharvbio.2006.08.006.

Awad, M.A. (2010) Pollination of date palm (*Phoenix dactylifera* L.) cv. Khenazy by pollen grain–water suspension spray. *Journal of Food, Agriculture and Environment* 8, 313–317.

Baoguo, D., Kruse, J., Winkler, J.B., Alfarray, S., Schnitzler, J. *et al.* (2019) Climate and development modulate the metabolome and antioxidative system of date palm leaves. *Journal of Experimental Botany* 70(20), 5959–5969. DOI: 10.1093/jxb/erz361.

Barrow, S. (1998) A revision of *Phoenix*. Kew Bulletin 53, 513–575.

Baslam, M., Qaddoury, A. and Goicoechea, N. (2014) Role of native and exotic mycorrhizal symbiosis to develop morphological, physiological and biochemical responses coping with water drought of date palm, *Phoenix dactylifera*. *Trees* 28, 161–172.

Beccari, O. (1890) Rivista monografica del genere phoenix L. *Malesia* 3, 345–416.

Bokor, B., Soukup, M., Vaculík, M., Vd'ačný, P., Weidinger, M. *et al.* (2019) Silicon uptake and localisation in date palm (*Phoenix dactylifera*) - A unique association with sclerenchyma. *Frontiers in Plant Science* 10, 988. DOI: 10.3389/fpls.2019.00988.

Bouamri, R., Dalpe, Y., Serrhini, M.N. and Bennani, A. (2006) Arbuscular mycorrhizal fungi species associated with rhizosphere of *Phoenix dactylifera* L. in Morocco. *African Journal of Biotechnology* 5(6), 510–516.

Bouamri, R., Dalpe, Y. and Serrhini, M. (2014) Effect of seasonal variation on arbuscular mycorrhizal fungi associated with date palm. *Emirates Journal of Food and Agriculture* 26(11), 977–986. DOI: 10.9755/ejfa.v26i11.18985.

Bueno, A., Alfarhan, A., Arand, K., Burghardt, M., Deininger, A.-C. *et al.* (2019) Effects of temperature on the cuticular transpiration barrier of two desert plants with water-spender and water-saver strategies. *Journal of Experimental Botany* 70(5), 1613–1625. DOI: 10.1093/jxb/erz018.

Carreño, E. (2017) Diversidad genética en especies del género *Phoenix*. PhD thesis, Universidad Miguel Hernández, Orihuela, Spain.

Carreño, E., Rivera, D., Obón, C., Alcaraz, F., Johnson, D. *et al.* (2021) What are candits? Study of a date palm landrace in Spain belonging to the western cluster of *Phoenix dactylifera* L. *Genetic Resources and Crop Evolution* 68(1), 135–149. DOI: 10.1007/s10722-020-00973-w.

Chaluvadi, S.R., Young, P., Thompson, K., Bahri, B.A., Gajera, B. *et al.* (2019) *Phoenix* phylogeny, and analysis of genetic variation in a diverse collection of date palm (*Phoenix dactylifera*) and related species. *Plant Diversity* 41(5), 330–339. DOI: 10.1016/j.pld.2018.11.005.

Chao, C.T. and Krueger, R.R. (2007) The date palm (*Phoenix dactylifera* L.): overview of biology, uses, and cultivation. *HortScience* 42(5), 1077–1082. DOI: 10.21273/HORTSCI.42.5.1077.

Chebaane, A., Symanczik, S., Oehl, F., Azri, R., Gargouri, M. *et al.* (2020) Arbuscular mycorrhizal fungi associated with *Phoenix dactylifera* L. grown in Tunisian Sahara oases of different salinity levels. *Symbiosis* 81(2), 173–186. DOI: 10.1007/s13199-020-00692-x.

Cherif, H., Marasco, R., Rolli, E., Ferjani, R., Fusi, M. *et al.* (2015) Oasis palm endophytes promote drought resistance. *Environmental Microbiology Reports* 7, 668–678.

Crawford, C.L. (1936) Growth rate of Deglet Noor dates in metaxenia. *Proceedings of the American Society for Horticultural Science* 33, 51–54.

Darwesh, R.S.S. (2013) Improving growth of date palm plantlets grown under salt stress with yeast and amino acids applications. *Annals of Agricultural Sciences* 58(2), 247–256. DOI: 10.1016/j.aoas.2013.07.014.

Denney, J.O. (1992) Xenia includes metaxenia. *HortScience* 27(7), 722–728. DOI: 10.21273/HORTSCI.27.7.722.

Diatta, I.L.D., Kane, A., Agbangba, C.E., Sagna, M., Diouf, E. *et al.* (2014) Inoculation with arbuscular mycorrhizal fungi improves seedlings growth of two Sahelian date palm cultivars (*Phoenix dactylifera* L., cv. Nakhla hamra and cv. Tijib) under salinity stresses. *Advances in Bioscience and Biotechnology* 5, 64–72.

Djibril, S., Mohamed, O.K., Diaga, D., Diégane, D., Abaye, B.F. *et al.* (2005) Growth and development of date palm (*Phoenix dactylifera* L.) seedlings under drought and salinity stresses. *African Journal of Biotechnology* 4(9), 968–972.

Dransfield, J., Uhl, N.W., Asmussen, C., Baker, W., Harley, M. *et al.* (2008) *Genera Palmarum. The Evolution and Classification of Palms.* Royal Botanic Gardens Kew, London.

Dreyer, B., Morte, A., López, J.A. and Honrubia, M. (2010) Comparative study of mycorrhizal susceptibility and anatomy of four palm species. *Mycorrhiza* 20(2), 103–115. DOI: 10.1007/s00572-009-0266-x.

Du, B., Kruse, J., Winkler, J.B., Alfarraj, S., Albasher, G. *et al.* (2021) Metabolic responses of date palm (*Phoenix dactylifera* L.) leaves to drought differ in summer and winter climate. *Tree Physiology* 41(9), 1685–1700. DOI: 10.1093/treephys/tpab027.

El-Hamady, M., Hamdia, M., Ayaad, M., Salama, M.E. and Omar, A.K. (2010) Metaxenic effects as related to hormonal changes during date palm (*Phoenix dactylifera* L.) fruit growth and development. *Acta Horticulturae* 882, 155–164. DOI: 10.17660/ActaHortic.2010.882.17.

El-Khawaga, A.S. (2013) Effect of anti-salinity agents on growth and fruiting of different date palm cultivars. *Asian Journal of Crop Science* 5(1), 65–80. DOI: 10.3923/ajcs.2013.65.80.

El Kinany, S., Achbani, E., Faggroud, M., Ouahmane, L., El Hilali, R. *et al.* (2019) Effect of organic fertilizer and commercial arbuscular mycorrhizal fungi on the growth of micropropagated date palm cv. Feggouss. *Journal of the Saudi Society of Agricultural Sciences* 18(4), 411–417. DOI: 10.1016/j.jssas.2018.01.004.

El-Maati, Y., Msanda, F., Eljiati, A., Ouchaou, H., Boubaker, H. *et al.* (2020) Characterization of plant growth promoting rhizobacteria isolated from an arid area soil of date palm in Saudi Arabia. *Journal of Applied Sciences* 20(6), 196–207. DOI: 10.3923/jas.2020.196.207.

El-Sharabasy, S.F., Orf, H.O.M., AboTaleb, H.H., Abdel-Galeil, L.M. and Saber, T.Y. (2018) Effect of plant growth promoting rhizobacteria (PGPR) on growth and leaf chemical composition of date palm plants cv. Bartamuda under salinity stress. *Middle East Journal of Agriculture Research* 7(2), 618–624.

Elshibli, S. (2009) Genetic diversity and adaptation of date palm (*Phoenix dactylifera* L). Dissertation. Helsinki: University of Helsinki. Available at: https://

helda.helsinki.fi/bitstream/handle/10138/20761/geneticd.pdf?sequence=1 (accessed 9 November 2022).

Elshibli, S. and Korpelainen, H. (2008) Microsatellite markers reveal high genetic diversity in date palm (*Phoenix dactylifera* L.) germplasm from Sudan. *Genetica* 134(2), 251–260. DOI: 10.1007/s10709-007-9232-8.

Elshibli, S., Elshibli, E.M. and Korpelainen, H. (2016) Growth and photosynthetic CO2 responses of date palm plants to water availability. *Emirates Journal of Food and Agriculture* 28(1), 58–65. DOI: 10.9755/ejfa.2015.05.189.

El-Taweel, M.M., El-Deep, M.D., Sourour, M.M. and El-Alakmy, H.A. (2015) Effect of plant growth promoting rhizobacteria and some plant extracts on rootability of aerial Hayany date palm offshoots. A – Rooting parameters and survival (%). *Sinai Journal of Applied Sciences* 4(3), 165–176. DOI: 10.21608/sinjas.2015.78583.

FAO (2021) FAOSTAT. Crops and livestock products: dates. Food and Agriculture Organization of the United Nations, Rome. Available at: www.fao.org/faostat/en/#data/QCL (accessed 4 October 2021).

Ferjani, R., Marasco, R., Rolli, E., Cherif, H., Cherif, A. *et al.* (2015) The date palm tree rhizosphere is a niche for plant growth promoting bacteria in the oasis ecosystem. *BioMed Research International* 2015, 153851. DOI: 10.1155/2015/153851.

Focke, W.O. (1881) *Die Pflanzen-Mischlinge: Ein Beitrag Zur Biologie Der Gewächse.* Bomtraeger, Berlin, pp. 510–518. DOI: 10.5962/bhl.title.127428.

Gaertner, J. (1788) *De Fructibus et Seminibus Plantarum: Accedunt Seminum Centuriae Quinque Priores Cum Tabulis Aeneis LXXIX*, Typis Academiae Carolinae, Vol. 1. Stutgardiae, Germany. DOI: 10.5962/bhl.title.102753.

GBIF (2022) *Phoenix* L. Global Biodiversity Information Facility, Copenhagen. Available at: www.gbif.org/species/2732076 (accessed 1 June 2022).

Ghadbane, M., Medjekal, S., Benderradji, L., Belhadj, H. and Daoud, H. (2021) Assessment of arbuscular mycorrhizal fungi status and rhizobium on date palm (*Phoenix dactylifera* L.) cultivated in a Pb contaminated soil. In: Ksibi, M., Ghorbal, A., Chakraborty, S., Chaminé, H.I. and Barbieri, M., *et al* (eds) *Recent Advances in Environmental Science from the Euro-Mediterranean and Surrounding Regions*, 2nd Edition. Springer, Cham, Switzerland, pp. 703–707. DOI: 10.1007/978-3-030-51210-1.

Ghirardo, A., Nosenko, T., Kreuzwieser, J., Winkler, J.B., Kruse, J. *et al.* (2021) Protein expression plasticity contributes to heat and drought tolerance of date palm. *Oecologia* 197(4), 903–919. DOI: 10.1007/s00442-021-04907-w.

Gribaa, A., Dardelle, F., Lehner, A., Rihouey, C., Burel, C. *et al.* (2013) Effect of water deficit on the cell wall of the date palm (*Phoenix dactylifera* 'Deglet nour Arecales) fruit during development. *Plant, Cell & Environment* 36, 1056–1070.

GRIN (2021) *Phoenix*. US Department of Agriculture, Agricultural Research Service, National Germplasm Resources Laboratory, Beltsville, Maryland. Available at: https://npgsweb.ars-grin.gov/gringlobal/taxon/taxonomysearch (accessed 4 October 2021).

Gros-Balthazard, M., Galimberti, M., Kousathanas, A., Newton, C., Ivorra, S. *et al.* (2017) The discovery of wild date palms in Oman reveals a complex domestication history involving centers in the Middle East and Africa. *Current Biology* 27(14), 2211–2218.. DOI: 10.1016/j.cub.2017.06.045.

Harkousse, O., Slimani, A., Jadrane, I., Aitboulahsen, M., Mazri, M.A. *et al.* (2021) Role of local biofertilizer in enhancing the oxidative stress defence systems of date palm seedling (*Phoenix dactylifera*) against abiotic stress. *Applied and Environmental Soil Science* 2021, 6628544. DOI: 10.1155/2021/6628544.

Hazzouri, K.M., Flowers, J.M., Visser, H.J., Khierallah, H.S., Rosas, U. *et al.* (2015) Whole genome re-sequencing of date palms yields insights into diversification of a fruit tree crop. *Nature Communications* 6(1), 8824. DOI: 10.1038/ncomms9824.

Hempel, S., Götzenberger, L., Kühn, I., Michalski, S.G., Rillig, M.C. *et al.* (2013) Mycorrhizas in the Central European flora: relationships with plant life history traits and ecology. *Ecology* 94(6), 1389–1399. DOI: 10.1890/12-1700.1.

Hesami, A., Jafari, N., Shahriari, M.H. and Zolfi, M. (2017) Yield and physico-chemical composition of date palm (*Phoenix dactylifera*) as affected by nitrogen and zinc application. *Communications in Soil Science and Plant Analysis* 48, 1943–1954.

IPGRI, INRAA, INRAM, INRAT, FEM *et al.* (2005) *Descripteurs Du Palmier Dattier (Phoenix Dactylifera L.).* Institut international des ressources phytogénétiques, Institut National de la Recherche Agronomique d'Algérie, Institut National de la Recherche Agronomique du Maroc, Institut National de la Recherche Agronomique de Tunisie, Fonds pour l'Environnement Mondial and Programme des Nations Unies pour le Développement, Rome, Algiers, Rabat, Tunis, Washington, DC and New York.

Iqbal, M., Jatoi, S.A., Niamatullah, M., Munir, M. and Khan, I. (2014) Effect of pollination time on yield and quality of date fruit. *Journal of Animal and Plant Sciences* 24, 760–764.

Jaiti, F., Meddich, A. and El-Hadrami, I. (2007) Effectiveness of arbuscular mycorrhizal fungi in the protection of date palm (*Phoenix dactylifera* L.) against Bayoud disease. *Physiological and Molecular Plant Pathology* 71(4–6), 166–173.

Jaiti, F., Kassami, M., Meddich, A. and El-Hadrami, I. (2008) Effect of arbuscular mycorrhization on the accumulation of hydroxycinnamic acid derivatives in date palm seedlings challenged with *Fusarium oxysporum* f. sp. *albedinis. Journal of Phytopathology* 156, 641–646.

Khaled, L., Pérez-Gilabert, M., Dreyer, B., Oihabi, A., Honrubia, M. *et al.* (2008) Peroxidase changes in *Phoenix dactylifera* palms inoculated with mycorrhizal and biocontrol fungi. *Agronomy for Sustainable Development* 28(3), 411–418. DOI: 10.1051/agro:2008018.

Khaliel, A.S. and Abou-Heilah, A.N. (1985) Formation of vesicular-arbuscular mycorrhiza in *Phoenix dactylifera* L. cultivated in Qassim region, Saudi Arabia. *Pakistan Journal of Botany* 17, 267–270.

Khudairi, A.K. (1969) Mycorrhiza in desert soils. *BioScience* 19(7), 598–599. DOI: 10.2307/1294933.

Kruse, J., Alfarraj, S., Rennenberg, H. and Adams, M. (2016) A novel mechanistic interpretation of instantaneous temperature responses of leaf net photosynthesis. *Photosynthesis Research* 129(1), 43–58. DOI: 10.1007/s11120-016-0262-x.

Kruse, J., Adams, M., Winkler, B., Ghirardo, A., Alfarraj, S. *et al.* (2019) Optimization of photosynthesis and stomatal conductance in the date palm *Phoenix dactylifera* during acclimation to heat and drought. *New Phytologist* 223(4), 1973–1988. DOI: 10.1111/nph.15923.

Linnaeus, C. (1753) *Species Plantarum*, Vol. 2. Laurentii Salvii, Stockholm.

Lobo, M.G., Yahia, E.M. and Kader, A.A. (2014) Biology and postharvest physiology of date fruit. In: Siddiq, M., Aleid, S.M. and Kader, A.A. (eds) *Dates: Postharvest Science, Processing Technology and Health Benefits.* Wiley, Hoboken, New Jersey, pp. 57–80.

Maas, E.V. and Hoffman, G.J. (1990) Crop salt tolerance – current assessment. In: Tanji, K.K. (ed.) *Agricultural Salinity Assessment and Management*. American Society of Civil Engineers, New York, pp. 262–304.

Mahmoud, F.M., Krimi, Z., Maciá-Vicente, J.G., Brahim Errahmani, M. and Lopez-Llorca, L.V. (2017) Endophytic fungi associated with roots of date palm (*Phoenix dactylifera*) in coastal dunes. *Revista Iberoamericana de Micologia* 34(2), 116–120. DOI: 10.1016/j.riam.2016.06.007.

Martínez-Rico, M. (2017) *El género Phoenix en jardinería y paisajismo: el caso de Phoenix canariensis*. Doctoral dissertation, Universidad Miguel Hernández, Orihuela, Spain. Available at: http://dspace.umh.es/handle/11000/4504 (accessed 15 December 2021).

Martius, K. (1836–1850) *Historia Naturalis Palmarum; Volumen Tertium Expositio Systematica*. K. Martius, Munich, Germany.

Mathew, L.S., Seidel, M.A., George, B., Mathew, S., Spannagl, M. *et al.* (2015) A genome-wide survey of date palm cultivars supports two major subpopulations in *Phoenix dactylifera*. *G3: Genes, Genomes, Genetics* 5(7), 1429–1438. DOI: 10.1534/g3.115.018341.

Mawloud, E.A. (2010) The effect of thinning on fruit drop of date palm Khanezi. *Acta Horticulturae* 882, 861–865. DOI: 10.17660/ActaHortic.2010.882.99.

Meddich, A., Jaiti, F., Bourzik, W., Asli, A.E. and Hafidi, M. (2015) Use of mycorrhizal fungi as a strategy for improving the drought tolerance in date palm (*Phoenix dactylifera*). *Scientia Horticulturae* 192, 468–474. DOI: 10.1016/j.scienta.2015.06.024.

Meddich, A., Ait-El-Mokhtar, M., Bourzik, W., Mitsui, T., Baslam, M. and Hafidi, M. (2018) Optimizing growth and tolerance of date palm (*Phoenix dactylifera* L.) to drought, salinity, and vascular Fusarium-induced wilt (*Fusarium oxysporum*) by application of arbuscular mycorrhizal fungi (AMF). In: Giri, B., Prasad, R. and Varma, A. (eds) *Root Biology. Soil Biology*, Vol. 52. Springer, Cham, Switzerland, pp. 239–258.

Mekhtiev, T.A., Maksimov, A.P. and Elmanova, T.S. (1990) The frost resistance of date palm. *Izvestiya Akademii Nauk Azerbaĭdzhanskoĭ SSR, Biologicheskie Nauki* 3, 43–49.

Merwad, M.A., Eisa, R.A. and Mostafa, E.A.M. (2015) Effect of some growth regulators and antioxidants sprays on productivity and some fruit quality of Zaghloul date palm. *International Journal of ChemTech Research* 8, 1430–1437.

Miller, W., Smith, J. and Taylor, N. (1919) Phoenix. In: Bailey, L.H. (ed.) *The Standard Cyclopedia of Horticulture*, Vol. 5. *P–R*, The Macmillan Co, London, pp. 2591–2594.

Mohammed, M.E., Alhajhoj, M., Dinar, H.M. and Munir, M. (2020) Impact of a novel water-saving subsurface irrigation system on water productivity, photosynthetic characteristics, yield, and fruit quality of date palm under arid conditions. *Agronomy* 10(9), 1265. DOI: 10.3390/agronomy10091265.

Mosqueira, M.J., Marasco, R., Fusi, M., Michoud, G., Merlino, G. *et al.* (2019) Consistent bacterial selection by date palm root system across heterogeneous desert oasis agroecosystems. *Scientific Reports* 9(1), 4033. DOI: 10.1038/s41598-019-40551-4.

Naser, H.M., Hanan, E.H., Elsheery, N.I. and Kalaji, H.M. (2016) Effect of biofertilizers and putrescine amine on the physiological features and productivity of date palm (*Phoenix dactylifera*, L.) grown on reclaimed-salinized soil. *Trees* 30(4), 1149–1161. DOI: 10.1007/s00468-016-1353-1.

Nixon, R.W. (1936) Metaxenia and interspecific pollinations in *Phoenix*. *Proceedings of the American Society for Horticultural Science* 33, 21–26.

Omar, A., Soliman, S. and Ahmed, M. (2013) Impact of leaf/bunch ratio and time of application on yield and fruit quality of Barhi date palm trees (*Phoenix dactylifera* L.) under Saudi Arabian conditions. *Journal of Testing and Evaluation* 41(5), 813–817. DOI: 10.1520/JTE20120340.

Palmweb (2021) *Phoenix*. Available at: www.palmweb.org/cdm_dataportal/taxon/f5c5c3ef-0a77-46eb-ac37-9c3867446bc1 (accessed 4 October 2021).

Phoenix-Spain (2021) *Phoenix Spain Colección Nacional*. Available at: www.phoenix-spain.org/ (accessed October 4).

Porcel, R., Aroca, R. and Ruiz-Lozano, J.M. (2012) Salinity stress alleviation using arbuscular mycorrhizal fungi. *Agronomy for Sustainable Development* 32(1), 181–200. DOI: 10.1007/s13593-011-0029-x.

Rahemi, M. (1998) Effects of pollen sources on fruit characteristics of Shahani date. *Iran Agricultural Research* 17(2), 169–174.

Reichgelt, T., West, C.K. and Greenwood, D.R. (2018) The relation between global palm distribution and climate. *Scientific Reports* 8(1), 4721. DOI: 10.1038/s41598-018-23147-2.

Rezazadeh, R., Hassanzadeh, H., Hosseini, Y., Karami, Y. and Williams, R.R. (2013) Influence of pollen source on fruit production of date palm (*Phoenix dactylifera* L.) cv. Barhi in humid coastal regions of southern Iran. *Scientia Horticulturae* 160, 182–188.

Rivera, D., Johnson, D., Delgadillo, J., Carrillo, M.H., Obón, C. *et al.* (2013) Historical evidence of the Spanish introduction of date palm (*Phoenix dactylifera* L., Arecaceae) into the Americas. *Genetic Resources and Crop Evolution* 60(4), 1433–1452. DOI: 10.1007/s10722-012-9932-5.

Rivera, D., Obón, C., García-Arteaga, J., Egea, T., Alcaraz, F. *et al.* (2014) Carpological analysis of *Phoenix* (Arecaceae): contributions to the taxonomy and evolutionary history of the genus. *Botanical Journal of the Linnean Society* 175(1), 74–122. DOI: 10.1111/boj.12164.

Rivera, D., Obón, C., Alcaraz, F., Carreño, E., Laguna, E. *et al.* (2015) Date palm status and perspective in Spain. In: J.M., A.-K., Jain, S.M. and Johnson, D.V. (eds) *Date Palm Genetic Resources and Utilization*, Vol. 2. *Asia and Europe*, Springer, Dordrecht, The Netherlands, pp. 489–526. DOI: 10.1007/978-94-017-9707-8_15.

Rivera, D., Obón, C., Alcaraz, F., Laguna, E. and Johnson, D. (2019a) Date-palm (*Phoenix*, Arecaceae) iconography in coins from the Mediterranean and West Asia (485 BC–1189 AD). *Journal of Cultural Heritage* 37, 199–214. DOI: 10.1016/j.culher.2018.10.010.

Rivera, D., Obón, C., Carreño, E., Laguna, E., Ferrer Gallego, P.P. *et al.* (2019b) La especie *Phoenix* excelsior de Cavanilles y la diversidad del complejo *Phoenix dactylifera* L. (Arecaceae): tipificación de *Phoenix* excelsior Cav. *Flora Montiberica* 73, 81–88.

Rivera, D., Abellán, J., Palazón, J.A., Obón, C., Alcaraz, F. *et al.* (2020) Modelling ancient areas for date palms (*Phoenix* species: Arecaceae): Bayesian analysis of biological and cultural evidence. *Botanical Journal of the Linnean Society* 193(2), 228–262. DOI: 10.1093/botlinnean/boaa011.

Rivera, D., Alcaraz, F., Rivera-Obón, D.J. and Obón, C. (2022) Phenotypic diversity in wild and cultivated date palm (*Phoenix*, Arecaceae): quantitative analysis using information theory. *Horticulturae* 8(4), 287. DOI: 10.3390/horticulturae8040287.

Safronov, O., Kreuzwieser, J., Haberer, G., Alyousif, M.S., Schulze, W. *et al.* (2017) Detecting early signs of heat and drought stress in *Phoenix dactylifera* (date palm). *PloS ONE* 12(6), e0177883. DOI: 10.1371/journal.pone.0177883.

Salomon-Torres, R., Ortiz-Uribe, N., Villa-Angulo, R., Villa-Angulo, C., Norzagaray-Plasencia, S. *et al.* (2017) Effect of pollenizers on production and fruit characteristics of date palm (*Phoenix dactylifera* L.) cultivar Medjool in Mexico. *Turkish of Agriculture and Forestry* 41(5), 338–347. DOI: 10.3906/tar-1704-14.

Salomón-Torres, R., Krueger, R., García-Vázquez, J.P., Villa-Angulo, R., Villa-Angulo, C. *et al.* (2021) Date palm pollen: features, production, extraction and pollination methods. *Agronomy* 11(3), 504. DOI: 10.3390/agronomy11030504.

Serrano, M., Pretel, M.T., Botella, M.A. and Amorós, A. (2001) Physicochemical changes during date ripening related to ethylene production. *Food Science and Technology International* 7(1), 31–36. DOI: 10.1106/Y6MD-JJDH-LT0P-Y9AE.

Sghir, F., Touati, J., Chliyeh, M., Touhami, A.O., Filali-Maltouf, A. *et al.* (2014) Diversity of arbuscular mycorrhizal fungi in the rhizosphere of date palm tree (*Phoenix dactylifera*) in Tafilalt and Zagora regions (Morocco). *International Journal of Pure and Applied Bioscience* 2(6), 1–11.

Sghir, F., Touati, J., Chliyeh, M., Talbi, Z., Selmaoui, K. *et al.* (2016) Bibliographic inventory of endomycorrhizal species associated to the rhizosphere of the date palm (*Phoenix dactylifera*). *International Journal of Innovation and Scientific Research* 26, 503–510.

Shabana, H. and Al Sunbol, A. (2007) Date palm flowering and fruit setting as affected by low temperatures preceding the flowering season. *Acta Horticulturae* 736, 193–198. DOI: 10.17660/ActaHortic.2007.736.16.

Shabani, F., Kumar, L. and Taylor, S. (2012) Climate change impacts on the future distribution of date palms: a modeling exercise using CLIMEX. *PloS ONE* 7(10), e48021. DOI: 10.1371/journal.pone.0048021.

Shabbir, G., Dakheel, A.J. and Al-Naqbi, M. (2010) The effect of arbuscular mycorrhiza (AM) fungi on the establishment of date palm (*Phoenix dactylifera* L.) under saline conditions in the UAE. *Acta Horticulturae* 882, 303–314. DOI: 10.17660/ActaHortic.2010.882.34.

Soudzilovskaia, N.A., Vaessen, S., Barcelo, M., He, J., Rahimlou, S. *et al.* (2020) FungalRoot: global online database of plant mycorrhizal associations. *New Phytologist* 227(3), 955–966. DOI: 10.1111/nph.16569.

Sperling, O., Lazarovitch, N., Schwartz, A. and Shapira, O. (2014) Effects of high salinity irrigation on growth, gas-exchange, and photoprotection in date palms (*Phoenix dactylifera* L., cv. Medjool). *Environmental and Experimental Botany* 99, 100–109. DOI: 10.1016/j.envexpbot.2013.10.014.

Steudel, E. (1841) *Nomenclator Botanicus*. J.G. Cottae, Stuttgart and Tubingen, Germany.

Swingle, W.T. (1928) Metaxenia in the date palm: possibly a hormone action by the embryo or endosperm. *Journal of Heredity* 19(6), 257–268. DOI: 10.1093/oxfordjournals.jhered.a102996.

Tester, M. (2003) Na$^+$ tolerance and Na$^+$ transport in higher plants. *Annals of Botany* 91(5), 503–527. DOI: 10.1093/aob/mcg058.

Thennarasu, S., Natarajan, E. and Muthukumar, B. (2019) Novel plant growth promoting rhizobacteria from rhizospheres of date palm (*Phoenix dactylifera*): 16S rRNA typing and phylogenetic analyses reveal their distinct identity. *Research and Reviews: A Journal of Microbiology and Virology* 9(1), 10–24.

Tisserat, B. and DeMason, D.A. (1982) A scanning electron microscope study of pollen of *Phoenix* (Arecaceae). *Journal of the American Society for Horticultural Science* 107(5), 883–887. DOI: 10.21273/JASHS.107.5.883.

Torres, M.F., Mathew, L.S., Ahmed, I., Al-Azwani, I.K., Krueger, R. *et al.* (2018) Genus-wide sequencing supports a two-locus model for sex-determination in *Phoenix*. *Nature Communications* 9(1), 3969. DOI: 10.1038/s41467-018-06375-y.

Toubali, S., Tahiri, A., Anli, M., Symanczik, S., Boutasknit, A. *et al.* (2020) Physiological and biochemical behaviors of date palm vitroplants treated with microbial consortia and compost in response to salt stress. *Applied Sciences* 10(23), 8665. DOI: 10.3390/app10238665.

Tripler, E., Ben-Gal, A. and Shani, U. (2007) Consequence of salinity and excess boron on growth, evapotranspiration and ion uptake in date palm (*Phoenix dactylifera* L., cv. Medjool). *Plant and Soil* 297(1–2), 147–155. DOI: 10.1007/s11104-007-9328-z.

Wang, B. and Qiu, Y.L. (2006) Phylogenetic distribution and evolution of mycorrhizas in land plants. *Mycorrhiza* 16(5), 299–363. DOI: 10.1007/s00572-005-0033-6.

Yaish, M.W. (2015) Proline accumulation is a general response to abiotic stress in the date palm tree (*Phoenix dactylifera* L.). *Genetics and Molecular Research* 14(3), 9943–9950. DOI: 10.4238/2015.August.19.30.

Yaish, M.W. and Kumar, P.P. (2015) Salt tolerance research in date palm tree (*Phoenix dactylifera* L.), past, present, and future perspectives. *Frontiers in Plant Science* 6, 348. DOI: 10.3389/fpls.2015.00348.

Yaish, M.W., Antony, I. and Glick, B.R. (2015) Isolation and characterization of endophytic plant growth-promoting bacteria from date palm tree (*Phoenix dactylifera* L.) and their potential role in salinity tolerance. *Antonie van Leeuwenhoek* 107(6), 1519–1532. DOI: 10.1007/s10482-015-0445-z.

Yaish, M.W., Al-Harrasi, I., Alansari, A.S., Al-Yahyai, R. and Glick, B.R. (2016) The use of high throughput DNA sequence analysis to assess the endophytic microbiome of date palm roots grown under different levels of salt stress. *International Microbiology* 19(3), 143–155. DOI: 10.2436/20.1501.01.272.

Yaish, M.W., Patankar, H.V., Assaha, D.V.M., Zheng, Y., Al-Yahyai, R. *et al.* (2017) Genome-wide expression profiling in leaves and roots of date palm (*Phoenix dactylifera* L.) exposed to salinity. *BMC Genomics* 18(1), 246. DOI: 10.1186/s12864-017-3633-6.

Zaid, A. and de Wet, P.F. (2002) Date palm propagation. In: Zaid, A. and Arias-Jimenez, E.J. (eds) *Date Palm Cultivation*. FAO Plant Production and Protection Paper no.156 Rev. 1, Food and Agriculture Organization of the United Nations, Rome, pp. 73–105.

Zougari-Elwedi, B., Issami, W., Msetra, A., Sanaa, M., Yolande, D. *et al.* (2016) Monitoring the evolution of the arbuscular mycorrhizal fungi associated with date palm. *Journal of New Sciences, Agriculture and Biotechnology* 31(12), 1822–1831.

Zougari-Elwedi, B., Sameh, S. and Ilhem, O. (2018) Effect of mycorrhiza-associated bacteria on mycorrhization, growth and mineral nutrition of date palm seedlings. In: Zaid, A. and Alhadrami, G. (eds), *Proceedings of the Sixth International Date Palm Conference, Abu Dhabi, UAE*, Khalifa International Award for Date Palm and Agricultural Innovation, March 19–21, 2018, pp. 180–185.

3

Diversity, Conservation, and Utilization of Date Palm Germplasm

Summar Abbas Naqvi* ⓘ, Muhammad Waseem, Taseer Abbas Naqvi, Muhammad Jafar Jaskani ⓘ and Iqrar Ahmad Khan

University of Agriculture, Faisalabad, Pakistan

Abstract

The threat of global climate change is of great concern to scientists as these changes are negatively impacting agriculture and threatening global food security by decreasing germplasm resources of crops. Germplasm of crop species, especially the wild species, contains genetic diversity that is critically important in meeting the challenges of food security and environmental concerns. However, natural genetic diversity is threatened by habitat loss from overpopulation, deforestation, mineral exploitation, wars, and development of dams and other large-scale projects like construction of roads, factories, canals, and housing. For example, new treaties, national legislation, border changes, protection of rights for future uses of germplasm, patent rights for proprietary developed germplasm, and quarantines and trade regulations all make access to and conservation of genetic resources less certain than in the past. Date palm is a naturally cross-pollinated fruit tree with a huge diversity starting from the Atlantic Ocean and ending at the Arabian Sea. Date palm diversity is at risk due to earlier discussed issues. Date palm genetic resources conservation is crucial for sustainable development of date palm production and ensuring food security. This chapter provides information on the status of date palm genetic resources, conservation techniques, and utilization strategies.

Keywords: Cryopreservation, Date palm, Germplasm, *In situ* conservation, Micropropagation, Molecular characterization

*Corresponding author: summar.naqvi@uaf.edu.pk

© CAB International 2023. *Date Palm* (eds J.M. Al-Khayri *et al.*)
DOI: 10.1079/9781800620209.0003

3.1   Introduction

Date is an ancient fruit crop being used as food for centuries. It is a mono-cotyledonous diploid plant having chromosome number 2 *n*=36 (Sallon *et al.*, 2020). According to Dowson (1982), there are about 200 genera and 2500 species in the family *Arecaceae*, which comprises the palms. Among the 200 genera, *Phoenix* is one of the important and consists of 14 species. It is believed that all species of *Phoenix* are native to tropical and subtropical regions of South Asia and Africa (Gros-Balthazard *et al.*, 2021). Due to the long history of date palm cultivation, extensive exchange of germplasm resources, dioecious behavior, and seedling reproduction, there is a certain confusion in the naming of date palm varieties (Hazzouri *et al.*, 2015). There are large number of names associated with single varieties in different countries of the world. Arabic names may have several translations in other languages, and this creates problems in naming cultivars within countries and worldwide (Jaradat and Zaid, 2004).

Genetic resources are living materials having potential value to humans. The genetic resources of plants include crops and their wild relatives having useful characters. The loss of genetic resources is very damaging for species, the habitat of species, and the people who depend on the biodiversity of natural resources directly linked to their livelihood (Ceccarelli, 2014). Date is a very diverse fruit crop, and it grows naturally in oases. For several decades, date palm oases have been declining due to many factors like disease attack, climate change, and increasing population, creating food security threats worldwide (Hazzouri *et al.*, 2019). There is a need to conserve these genetic resources to overcome food security challenges. The effectiveness of genetic resource protection strategies, especially in the long term, depends on how the agricultural environment is managed. Therefore, it is important to understand genetic resource conservation strategies that can sustain farmers' livelihoods and help conserve genetic resources.

The process of genetic erosion and biodiversity loss in date palm is more difficult to document because its center of origin is not confirmed yet (Jaradat, 2011). Several ecological and socio-economic factors are affecting the date palm oases. Factors including floods, drought, rainfall, and fluctuations in weather are environmental factors that may affect the genetic diversity of fruits by creating phenotypic variation (Hassan *et al.*, 2018). Other factors like land degradation, genetic erosion, inappropriate agronomic practices, introduction of exotic cultivars, and frequent droughts may also affect the genetic resources of date palm. Due to crop failures and the loss of varieties, these changes may lead to genetic erosion. As a result of such natural disasters, North Africa lost millions of fruit trees within a decade (Jain, 2010). In addition, biological factors such as diseases and pests attack fruit trees, thereby negatively affecting genetic variation within tree species (Bekheet and Taha, 2013). Moreover, the red palm weevil

(*Rhynchophorus ferrugineus*), Fusarium wilt caused by *Fusarium oxysporum*, salinity, and drought are increasingly serious threats to the expanding date industry in some countries (Jaradat, 2015). Decrease in production and loss of genetic resources of date palm have been seen in the past due to many factors. For example, natural disasters (Baaziz *et al.*, 2000), abiotic stress (Zaid *et al.*, 2002), diseases and pest attacks (Bendiab *et al.*, 1993), human factors (Hussein *et al.*, 1993), and ageing of the trees (El-Juhany, 2010; Jaradat, 2011) have decreased the production as well as export of dates in the Middle East. Unfortunately, the only effective control measure is cutting down infected trees and destroying them at an early stage to prevent spread of diseases. Date palms are among the most salt-tolerant fruit trees. Economic and social factors also affect the diversity of date palm orchards. Socio-economic factors affecting people's livelihoods and oasis agroecosystems include the marginalization of indigenous communities and the rapid loss of local cultures and indigenous knowledge.

Agriculture and climate change are internally interrelated in various ways, with climate change mainly being due to biotic and abiotic pressures, with negative impacts on agricultural areas. Climate change adversely affects agriculture and the global ecosystem in different ways. The variability in average temperature, heat stress, change in annual rainfall, changing behavior of weeds, microorganisms, and insect pests, changes in atmospheric ozone and carbon dioxide levels, and unpredictable variation in sea level are some examples of climate change (Raza *et al.*, 2019). The oases of date palm are also being affected continuously due to changes in climate, and adverse effects in the form of drought have been seen in many areas of the world (Fig. 3.1). If this drought pattern continues, deforestation will

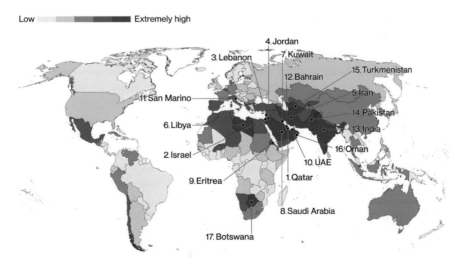

Fig. 3.1. Top countries under water stress. (Used with permission from Ford *et al.*, 2016.)

Fig. 3.2. Old and unattended date palm groves affected by rural migration. (Photo by Summar A. Naqvi.)

increase (Fig. 3.2), and it will lead to loss of these sustainable and productive oases (Jaradat, 2011). Date palm genetic diversity is a significant element of oases and needs to be improved by resilience and by confirming that date palm varieties in oases will not be further reduced. Furthermore, it is important not to allow market forces to dominate or dictate the selection of future planted or preferred varieties. Changes in the environment have lasting impacts on global agriculture and food security. Unfortunately, no attention has been given to adopting measures to resolve this problem. Therefore, there is an urgent need to deal with these climate changes to avoid further losses in agriculture and ensure food security. To adapt crops to changing environmental pressures, follow-up measures are required. There is the need to take several steps like improved cultural practices, phenotypic screening of cultivars, genetic diversity analysis for crop improvement and marker-assisted breeding, identifying genomic variants that show tolerance to several biotic and abiotic stresses based on quantitative trait locus (QTL) mapping and genome-wide association studies, and genome editing strategies like genetic engineering and clustered regularly interspaced short palindromic repeats (CRISPR)/CRISPR-associated protein 9 (Cas 9) for the mitigation of these challenges (Raza *et al.*, 2019).

The collection and conservation of date palm genetic resources helps to ensure the existence of these genetic resources. Morphological estimation determines the relationship between specific varieties and the differences between varieties, as well as helps find the best varieties based on the production status. The physical and chemical analysis of date fruits helps to select the best varieties on the basis of nutritional value and health benefits (Haider *et al.*, 2018). We must attach importance to the protection of biodiversity so that we can appreciate the natural diversity of resources that are the basis for improving quality of life. Activities related to the long-term protection of living organisms require continuity to avoid the loss of valuable materials. Genetic resources contribute to the sustainable development and diversification of agricultural production by maintaining biodiversity and promoting sustainable agricultural production. The collected materials can be used for plant breeding, research, and economic utilization of varieties, breeding lines, or species. The main objective of this chapter is to provide information on the status of date palm genetic resources, exploration of germplasm conservation techniques, and utilization strategies for genetic resources. There are different conservation techniques being used for date palm. Since each conservation method has its own advantages and disadvantages, to ensure optimal sustainable use of date palm genetic diversity, a complementary conservation strategy is required. In addition, date palm genetic resources collected and conserved should be accessible to researchers and other interested parties, and with detailed documentation on characterization, assessment, and utilization.

3.2   Germplasm Collection, Documentation, and Distribution

Breeders believe that the success of any breeding program depends mainly on exploration of the genetic diversity of germplasm resources of crop plants. This exploration is carried out to find the genetic variability among different varieties of any crop present naturally along with that in its relative species (Abul-Soad *et al.*, 2017). Date palm diversity is as high as 5000 cultivars worldwide (Jaradat and Zaid, 2004). Morocco, Tunisia, Iraq, and Iran have approximately 244, 250, 370, and 400 varieties, respectively (Zaid and de Wet, 2002), while Sudan has 400 varieties (Elshibli and Korpelainen, 2009) and Pakistan has 150 varieties (Nadeem *et al.*, 2019); other types of dates in addition to known cultivars are also grown for food purposes. These available genetic resources are an important source for crossbreeding and crop improvement. In the past three decades, people have become more and more aware that for the betterment of our future, we should collect and protect these fast-disappearing, irreplaceable resources. Comprehensive information associated with the preserved germplasm, such as cropping history and its morphological, physiological, biochemical, and genetic characteristics, makes it easier for a breeder to make beneficial decisions. The management of germplasm resources encompasses five important steps, namely: (i) surveying and collection; (ii) identification and evaluation; (iii) preservation; (iv) exchange; and at the end (v) utilization and documentation. Additionally, all these steps depend directly on plant quarantine to prevent the introduction of pests and diseases along with the germplasm resources. It is believed that breeders and/or scientists spend most of their time understanding the data generated from the above steps (Coudy _et al._, 2021). Therefore, documentation and conservation of plant genetic resources are very important for a crop improvement program.

3.3 Conservation of Germplasm

The protection of plant genetic diversity has become an important issue for breeders and agronomists around the world striving to improve food, fiber, and industrial crop plants, and its loss is an issue of concern for current and future agricultural plant scientists. The losses are mainly due to rapid industrialization, fires, deforestation, and environmental pollution. In addition, the lack or absence of ideal genotypes prevents plant breeders from developing new varieties. The most common method to preserve plant genetic resources is seed preservation. It is quite cost-effective, safe, and easy to operate. The second most used method is _in vitro_ cold storage and cryopreservation (Bekheet _et al._, 2007). The success of this method largely depends on maintaining genetic fidelity during plant regeneration so that the genotype is conserved. Currently, international germplasm exchange has become more cumbersome, and it is sometimes difficult to obtain good genotypes. Due to patent and ownership issues, and political treaties such as the Convention on Biological Diversity (CBD), the Convention on International Trade in Endangered Species of Wild Fauna and Flora (CITES), etc. many countries are unwilling to share germplasm resources and have established national germplasm resource protection or gene bank facilities.

3.3.1 _In vitro_ conservation

The protection of date palm germplasm resources has become an important issue relating to the development of date palm production to overcome food security challenges. _In situ_ and _ex situ_ conservation and protection of genetic resources of date palm are important aspects of maintaining genetic diversity of date palm. In addition, some wild gene banks involving the collection of reproductive materials and planting them in diverse areas have been developed for basic and active collections. However, pest and disease attack and any natural adversity are major drawbacks of this type of conservation. Recently, with the rapid advancement in plant biotechnology, traditional methods of protecting date palm genetic resources have been supplemented with new ones. There are different _ex situ_ techniques using _in vitro_ technology, DNA storage, and seed conservation involved in preservation of date palm genetic resources. The _in vitro_ storage technology of date palm plant materials is a slow growth process; germplasm resources are stored as sterile plant tissues or plants on nutrient gels, or at ultra-low temperatures in liquid nitrogen. Slow growth preservation provides short-term storage options, while ultra-low-temperature storage permits long-term storage of plant material. With the development of biotechnology, DNA storage technology is considered one of the emerging technologies for preservation of germplasm resources. A DNA library stores whole genomic DNA for distribution. The application of molecular tools like genome-wide association studies and marker-assisted breeding has further promoted the use of DNA germplasm in date palm breeding and has added new value to existing collections.

In vitro conservation advantages include disease-free cultivation materials, high plant proliferation rate, can be used by growers all year-round, has the potential to produce low-cost cultivation materials, and can maintain the genetic fidelity as verified by molecular markers. High labour costs, destruction of stored genetic material caused by serious events (fires or earthquakes), the potential risk of somaclonal mutation of certain species, contamination caused by bacteria, fungi, viruses, and mites, and power supply interruptions are the main problems associated with *in vitro* conservation (Cruz-Cruz *et al.*, 2013). Therefore, maximum precautions should be taken, using healthy plant material for storage and testing for virus contamination. Unfortunately, there are no published studies on bacterial contamination or somaclonal mutations of dates after thawing. However, slowing down the growth of callus under appropriate conditions can give the callus enough time to mature. Subsequently, the developing somatic embryos will be increased. In addition, the stored callus can be used as explants for rapid propagation as required (Abul-Soad *et al.*, 2017). In addition, *in vitro* cultures can be safely transferred long distances in slow growth media. Since the date palm is cultured *in vitro* for plant regeneration, several groups have been engaged in low-temperature storage of tissues such as date palm apex, nodule culture, callus, and somatic embryo culture (Bekheet *et al.*, 2007). The implementation of the date palm germplasm protection strategy is very important for the livelihood of people living in oases and date-producing areas. Global assessment of the genetic diversity and conservation strategy of current date palm germplasm resources is necessary (Abul-Soad *et al.*, 2017). Therefore, the establishment of a combined system of dynamic *ex situ* and *in situ* conservation is an urgent need for maintaining the diversity of date palms and for responding to climate change and biological and non-biological challenges.

The rapid development of plant biotechnology has supplemented the traditional methods of plant genetic resource protection and led to *in vitro* conservation of plants developed via tissue culture. Explants are kept in a sterile, pathogen-free environment using *in vitro* preservation techniques. It is used to preserve species that produce recalcitrant seeds or preserve materials that maintain specific genotypes through vegetative reproduction (Engelmann, 2011). Since it is a dioecious, heterozygous fruit tree, date palm is often vegetatively propagated through offshoots for commercial purposes, so it is difficult to store and process its germplasm by traditional methods. Tissue culture technology has great capabilities for collection, reproduction, and storage of date palm germplasm resources. The *in vitro* cultures most commonly used include callus culture, cell suspension culture, embryo culture, shoot tip culture, and microspores. *In vitro* conservation is achieved by changing the culture medium or reduction of temperature requirements of germplasm resources. Meristem culture is kept in a temperature range from 0 to 5°C, so it grows slowly. This allows storage for several years, or even more than 10 years, depending on the type of plant. There are two techniques involved for *in vitro* gene banks: (i) *in vitro* basic gene bank; and (ii) *in vitro* active gene bank. The attribute of the *in vitro* basic gene bank is the long-term preservation of germplasm resources, rather than direct normal distribution to users. The *in vitro* active gene bank stores date palms that can be proliferated and distributed immediately. In this regard, previous reports have shown that date palm germplasm cultured *in vitro* from shoot buds (Bekheet, 2011), callus culture (Zaid *et al.*, 2011; El-Dawayati *et al.*, 2012; El-Ashry *et al.*, 2013), pollen (Mortazavi *et al.*, 2010), and somatic embryos (Bekheet *et al.*, 2005) has been preserved in *in vitro* gene banks at low temperature.

3.3.2 *In situ* conservation

The term *in situ* involves protecting plants in their natural habitats, or conservation in the field where they have been cultivated and maintained by farmers in ancient agricultural systems, thus allowing the restoration of genetic resources in their natural environment (Rao and Sthapit, 2012). The protection of date palm varieties in their actual environment and primitive habitat is one of the important components for the conservation of date palm germplasm resources. *In situ* conservation has significant advantages as it protects genetic material and accelerates the processes that produced biodiversity. The continuity of a breeding program can be achieved only through continuous accessibility of genetic diversity achieved via *in situ* conservation at farm level. Date growers and enthusiasts play significant roles in protecting the diversity of conventional date palm orchards. They continue to use ancient practices to maintain the propagation of wild or local date palm varieties and newly developed varieties with unique characteristics. Since this protection of agricultural biodiversity is carried out on the farm, this type of protection is also called farm protection. Under traditional horticultural and cropping systems the genetic diversity of desired plant species is conserved in this way, and it has also been used to maintain wild species (Maxted *et al.*, 1997). It helps the gene pool of species to adapt to variation caused by many environmental hazards such as floods, weather fluctuations, global warming, winds, rainfall, hot and cold temperature, etc. (Heywood and Dulloo, 2005).

As we know, date palm is well adapted in subtropical areas with a long summer season and less rainfall, due to its origin in oases of Middle Eastern deserts in Asia and Africa (Krueger, 2001). Both North African countries and the Arabian Peninsula countries are part of the date palm center of origin and are an important part of the date palm center of diversity; they are the countries with the highest date palm production and consumption in the world (Jaradat, 2015).

Traditional oases play a vital role in sustaining and conserving the biodiversity of date palms. Support for traditional oases must strike a balance between the need to protect their positive aspects and improve some of their less-desirable characteristics. The most important issues in natural growing areas of dates are improvements in agronomic practices, salinization, availability of irrigation water, and disease control. In addition, farmers should be strongly advised to multiply date palm trees from their offshoots to ensure the true-to-type in varieties. In contrast to this, the recording of field cultivars is considered one of the most significant characteristics of *in situ* protection of date palm. However, seed reproduction could result in useful new hybrids to conserve the genotype. Information such as existence/planting patterns of farmer-named varieties, population genetic structure, and identity of varieties selected by farmers is required to record the amount and distribution of genetic diversity at the farm level. In this regard, when it comes to agricultural communities, farm protection documents may be particularly effective (Bekheet and Taha, 2013).

3.3.3 *Ex situ* conservation

Ex situ is the protection of plant germplasm resources externally. It involves storage at a nursery, storage of seeds, DNA storage, conservation at low temperature, and *in vitro* protection (Bekheet and Taha, 2013). *Ex situ* conservation approaches for the improvement and successful breeding of genetic resources of date palm have been practiced previously and scientists are finding new ways for the conservation and utilization of genetic resources of date palm. *Ex situ* conservation comprises advanced strategies for conservation of genetic resources and is far better than *in situ* conservation techniques for plants. It is quite easy to recognize the biodiversity preserved in gene banks or botanical gardens, because these materials are generally fully documented for scientists' and plant breeders' use. In addition, the conservation of genetic diversity achieved by these methods is easily manageable. The chances of material loss are very low so long as the materials are preserved under appropriate conditions and updated regularly. However, losses can and do occur. Another advantage of *ex situ* conservation worth noting is that resources that are maintained *ex situ* are generally more accessible to the user community.

3.3.4 Botanical garden/germplasm repository

Botanical garden conservation involves the collection and then planting of genetic material in fields or orchards, where it is grown. These germ-plasm conservation banks are conventionally used for perennial plants and species that produce few or no seeds; species that are best preserved as clonal materials have a longer life cycle to produce planting materials. It is an ideal choice to store date palms in a field gene bank. In the field gene bank, which is also called a germplasm bank, variety bank, or clone bank, date palm genetic resources maintain the same genetic composition and type through asexual reproduction, which provides long-term preservation of genetic or interspecific variation. When collecting the germplasm for field gene banks, random sampling and nonrandom sampling can be used. Nonrandom samples will select those with vibrant morphological charac-teristics without concern about any physiological or disease resistance char-acters (Hamza *et al.*, 2015). Collection includes gathering samples from the natural habitat or from the populations in farmers' fields, because the crops in the field or nursery need proper cultivation, including proper nutrition, pest control, and irrigation (Yap and Saad, 2001). In this protec-tion strategy, there is always the risk of damage caused by environmental hazards, pests, and pathogens (Bouguedoura *et al.*, 2015). Environmental disasters will cause serious damage to the farmland gene bank. Floods, droughts, strong winds, thunderstorms, and haze are common natural disasters that can destroy healthy plants in gene banks (Aman, 2001). The preservation of germplasm resources in this way has been reported all over the world. In date palm, this type of protection has been reported in different countries like Pakistan (Markhand *et al.*, 2010; Abul-Soad *et al.*, 2016), India (Pareek, 2015), Tunisia (Hamza *et al.*, 2015), Saudi Arabia (Al-Ghamdi, 2001; Aleid *et al.*, 2015), Iraq (Khierallah *et al.*, 2015), and Kuwait (Sudhersan *et al.*, 2015). The establishment of such a collection site allows evaluation and comparison of the fruit quality of exotic varieties at the test station. This will help growers make the right decision about introducing valuable exotic varieties suitable for the local environment.

3.3.5 Seed conservation

There are two ways of seed preservation, namely: (i) community seed bank; and (ii) seed gene bank. In many developing countries, community seed banks are quite common at the village level and are used to preserve local varieties and agricultural production. Farmers rely on an informal seed system based on the practice that local growers retain previously harvested seeds, and store, process, and exchange these seeds within and between communities. In date palm, local germplasm is also stored by preserving seeds. But from seed there will be the disadvantage of varia-tion rather than true-to-type in the progeny, so high quality may be lost.

Among *ex situ* methods of germplasm conservation seed conservation is a valuable and extensive approach as it maintains seed vigour under dry and low-temperature conditions. In this strategy, seeds are dried and then conserved at low temperature in a freezer or using a cold-storage facility. About 90% of the plant genetic resources collected around the world so far are preserved in seed gene banks (Bekheet, 2011). Moreover, the reason that the recalcitrant seeds cannot withstand storage at very low temperatures because low temperature hinders the germination rate of the seeds or survival of the seeds is severely affected and seeds eventually die, limits the date palm seed conservation strategy. However, date palm seeds are orthodox in nature and can be stored for a long period.

3.3.6 Cryopreservation

Cryopreservation techniques store plant tissues at ultra-low temperature ($-196°C$) in liquid nitrogen for long-term preservation and to prevent genetic variation in tissue culture. Storage is carried out in liquid nitrogen at the temperature of $-196°C$ to stop all metabolic activities and cell division (Engelmann, 2004). With continuous subculture, there is always danger of contamination and handling issues (Cruz-Cruz *et al.*, 2013). However, compared to continuous subculture, the cells will not change genetically during cryo-storage. Plant species like date palm in which vegetative propagation is practiced and whose seeds are stubborn are ideal material for cryopreservation technology.

The main advantage of low-temperature storage is that metabolic processes and biological deterioration are greatly slowed down or even stopped. Therefore, the material stored at low temperature remains genetically stable and thus low-temperature storage has advantages over other storage methods. Since date palm is a dioecious plant, ultra-low-temperature preservation is the best solution as it needs little space and a large number of genetic resources can be stored without worrying about natural disasters and disease outbreaks. Several explants used corpus callosum meristem (Fki *et al.*, 2013), fragile callus (Subaih *et al.*, 2007; Al-Bahrany and Al-Khayri, 2012), embryonic mass (Fki *et al.*, 2011), somatic embryo (Bekheet *et al.*, 2005), stem apex (Bagniol *et al.*, 1992), and pollens (Mortazavi *et al.*, 2010). Several researchers have reported various cryopreservation methods such as cell suspension of date palm cv. Khalas (Al-Bahrany and Al-Khayri, 2012), standard vitrification of cv. Khenizi explants (Fki *et al.*, 2013), and cv. Sakkoty embryogenic cultures cryopreservation in a mixture of different salts and polyethylene glycol (Bekheet, 2015).

3.3.7 DNA storage

Modern biotechnology techniques, especially DNA technology, are increasingly being used for the conservation of plant biodiversity and assisted plant protection programs. The amplification of genes from entire genomic DNA has become easy with the development of polymerase chain reaction (PCR) technology, and international DNA storage banks have been established to store genomic DNA (Coudy *et al.*, 2021). DNA storage, in principle, is simple to operate, inexpensive, and has a wide range of applications. The stored DNA can be used for research or can provide references for future research. In the meantime, the genetic diversity of many species that are difficult to preserve or highly threatened in the wild will be protected using DNA storage technology until an effective method is developed.

DNA can be used for variety identification, date palm gene bank information sources, and variety genetic diversity research. With the development of DNA sequencing technology and the reduction of sequencing reagent costs, date palm genome analysis has made significant progress. Researchers have started to sequence total genomic DNA of date palm. In this way, they are identifying the genes controlling the important developmental stages of date fruits (Hazzouri *et al.*, 2019). Studies have revealed that the genome of date palm consists of 18 pairs of chromosomes, almost 350 Mb in size (Bekheet and Taha, 2013). In addition, the genome size of the date palm is 1.2 billion bp according to flow cytometry estimates (Gros-Balthazard *et al.*, 2021). In this regard, Yang *et al.* (2010) have reported the complete sequence of the chloroplast genome of the high-quality variety Khalas from Al-Hassa Oasis, Saudi Arabia. DNA barcode technology is becoming an important means of species identification. The DNA barcode project provides access to high-quality and long-term preservation of DNA. The universal *psbK-psbI* primer was used as a DNA barcode to identify date palm species. There are five haplotypes in 30 date palm varieties, and the haplotype diversity is 0.685 (Jatt *et al.*, 2019). Effective methods for isolating, sequencing, processing, and preserving date palm DNA from collected materials will help as a complementary strategy to preserve date palm germplasm resources.

3.3.8   Plant tissue culture

Traditionally, date palms reproduce sexually through seeds and vegetatively through offshoots produced at the base of the trunk. Offshoots production is limited depending upon the variety, so there is a limit in increasing plant number by offshoot production. So far, there is no available technology to accelerate the increase in number of plants while reducing the development time. Date palm trees have a long life cycle, strong dioeciousness, and are difficult to breed. The US Department of Agriculture's Date and Citrus Station in Indio, California, initiated a breeding program in 1948 to develop male varieties by continuous backcrossing for use in breeding programs (Krueger, 2015). Many male backcrosses that resemble the named female species in morphology were produced. However, due to the suppression, sterility, and low vigour of the inbred lines, no useful offspring were released. These obstacles in breeding date palms are overcome somewhat by improved biotechnological approaches such as plant tissue culture (Singh, 2009; Cruz-Cruz *et al.*, 2013). *In vitro* technology has been successfully applied to the plant propagation of a wide range of crops including date palms. The success of this method largely depends on the genotype (Jain, 2011).

The date palm is propagated *in vitro* by two main methods, the first is by the production of auxiliary buds (organogenesis) and the second is propagation by somatic embryogenesis. A relatively short period *in vitro* and large amount of plant proliferation define the somatic embryogenesis cycle. The downside is the possibility of mutations and abnormalities that occur later on during *in vitro* development (Bekheet *et al.*, 2001). In comparison, organogenesis micropropagation has the greatest advantage that *in vitro* plantlets have strong genetic and vegetative identity with the mother plant. The organogenesis technique comprises four stages: initiation of the vegetative bud; multiplication of the bud; elongation of the shoot; and rooting. Success of this technique depends heavily on success in the first step. Consequently, direct organogenesis provides the benefit of using low concentrations of plant growth regulator and therefore a callus phase is avoided. A number of meristematically induced explants (including shoot tips and lateral flower explants) were used to research *in vitro* date palm propagation of several genotype forms using organogenesis. Similarly, the *in vitro* multiplication of date palm buds was established by an efficient quick method (Bekheet *et al.*, 2001). A promising protocol has recently been developed for the mass propagation by direct organogenesis from the date palm, namely from cvs. Zaghlool, Samany, Hayany, Amhat, Siwy, Selmy, and Malakaby (Bekheet and Taha, 2013).

3.4 Germplasm Assessment

For commercialization of the most promising date palm cultivars, precise description and characterization is essential. Different markers such as morphological, biochemical, and molecular markers have been used for phenotypic and genetic characterization of horticultural crops. Moreover, genetics is one of the most important factors influencing the composition and quality of fruits (Asma and Ozturk, 2005).

3.4.1 Date palm germplasm characterization through morphological characters

The cultivars of date palm are commonly identified and described by different morphological features of the tree (trunk, crown, leaf), fruits, and seeds (Elhoumaizi *et al.*, 2002; Simozrag *et al.*, 2016). It is very hard to identify the cultivar when fruit is not present on the tree. The identification of such

Fig. 3.3. Diversity of date palm based on variation in trunk and leaf. (Photos by Summar A. Naqvi.)

cultivars can be carried out by using different vegetative and agronomic traits (Bedjaoui and Benbouza, 2018). Ahmed *et al.* (2011) studied 30 traits of stem, leaf, peduncle, and fruit in 20 date palm varieties of diverse origin and found variability in the studied parameters when the data were subjected to multivariate analysis. The screening of morphological characters like leaf, spine, and fruit characteristics can be accomplished well in mature trees (Figs 3.3 and 3.4). Morphological parameters are common tools used for identification of cultivars and to evaluate diversity levels in date palm (Djerouni *et al.*, 2015). A morphometric study on date palm trees of 16 varieties from different agroecological zones of Pakistan was carried out to evaluate similarity levels and morphological variability on the basis of 42 different qualitative and quantitative traits. The data were subjected to multivariate analysis and results depicted differences in morphological characters of the trees. However, some similarities were also found within cultivars of the same origin (Haider *et al.*, 2015).

3.4.2 Date palm germplasm characterization by biochemical profiling

The chemical composition of date fruits is different in different cultivars. Many factors like variety, maturation time, growing conditions, geographic

Fig. 3.4. Diversity of date palm based on variation in fruit and seed. (Photos by Summar A. Naqvi.)

origin, and soil type are responsible for variation in antioxidant activity and phytochemical compounds of date fruits (Biglari *et al.*, 2008; Haider *et al.*, 2013). Biochemical studies, proximate analysis, isozyme analysis using biomarkers, antioxidant enzyme (like peroxidase, catalase) activity, and mineral analysis are used to differentiate the date palm cultivars in the world (Awad *et al.*, 2011; Sadiq *et al.*, 2013; Al-Harrasi *et al.*, 2014; Martín-Sánchez *et al.*, 2014; Eoin, 2016).

Date palm is a rich source of nutrients as it contains carbohydrates, polyphenols, vitamins, protein, and minerals. The radical scavenging activity of date fruits is high due to the presence of polyphenols like phenolics and flavonoids and other antioxidant compounds like ascorbic acid, anthocyanins, and carotenoids (Assirey, 2015). Phytochemical analysis of five date palm varieties from Tunisia revealed that total phenolic content, total flavonoid content, and 2,2-diphenyl-1-picrylhydrazyl (DPPH) scavenging activity were in the range of 455.51–602.38 mg/100 g, 204.55–307.59 mg/100 g, and 1.68–3.18 gallic acid equivalents, respectively, at the khalal stage (Amir *et al.*, 2011). In another study, Haider *et al.* (2018) characterized ten date palm varieties for their biochemical composition. The results showed that antioxidant activity, total phenolics, flavonoids, and activity of antioxidant enzymes decreased with increased ripening of fruit, while increases in reducing sugars and pH were observed in all studied varieties. Date fruits are also a rich source of minerals and excellent for inclusion in human diets as minerals are necessary for the structure, maintenance, and proper functioning of living cells (Haider *et al.*, 2014; Mohamed *et al.*, 2014). The results of a study conducted to assess mineral contents in fruits of 21 date varieties from Pakistan revealed significant amounts of potassium, magnesium, manganese, calcium, iron, sodium, copper, and zinc (Nadeem *et al.*, 2019).

3.4.3 Molecular diversity

Genetic diversity of organisms is due to variations in chromosomes, nucleotides, and genes (alleles), or differences in the whole genome of organisms. Molecular markers are the most reliable tools to reveal genetic diversity at the molecular level. They also play a significant role in identifying QTLs associated with specific traits in crops. Molecular markers quickly expose genetic variation without the influence of environment. There are a number of molecular markers such as random amplified polymorphic DNA (RAPD), restriction fragment length polymorphism (RFLP), single sequence repeat (SSR), and single-nucleotide polymorphism (SNP) available for genetic characterization, cultivar identification, genetic diversity assessment, phylogenetic analysis, and gene mapping (Semagn *et al.*, 2006). Variable growth conditions and genetics affect the quality and appearance of dates (Ismail *et al.*, 2008). Different researchers in different countries such as Sudan (Elshibli and Korpelainen, 2009), Qatar (Ahmed and Al-Qaradawi, 2009), Tunisia (Hamza *et al.*, 2012), Pakistan (Mirbahar *et al.*, 2014), Nigeria (Yusuf *et al.*, 2015), Saudi Arabia (Al-Faifi *et al.*, 2016), and Ethiopia (Ahmed *et al.*, 2021) have assessed genetic diversity using different markers. A similar study has also been carried out using 195 date palm accessions from 13 different countries and the results depicted that the differences were due to diverse genetic origins (Srinivasa *et al.*, 2019).

3.5   Utilization of Genetic Resources

3.5.1   Conventional propagation (sexual/asexual)

Plant propagation refers to the propagation of new plants from seeds or vegetative parts (such as leaves, stems, or roots). The traditional method of reproduction of date palms is through offshoots or seeds. At present, there are three main propagation techniques for date palms: (i) seed propagation techniques; (ii) vegetative propagation techniques (traditional methods); and (iii) newly developed tissue culture techniques.

3.5.1.1   Propagation by seed

Seed propagation is the oldest propagation method for date palms. On the other hand, due to heterozygosity, the plants produced from seed are not exactly the same, their quality is lower than that of the mother plant, and about 50% are male (Othmani *et al.*, 2009).

3.5.1.2 Propagation by offshoots

The use of offshoots (suckers) to propagate date palm trees is a common method of asexual reproduction in date-palm-growing areas all over the world, with the goal of propagating the best varieties. Transplanting date palm offshoots to propagate is still the best and most commonly used method (Al-Khateeb, 2008). These offshoots develop slowly and are limited in number, being produced only during a certain period of the mother palm's life from the base of the trunk. To meet the demand for new planting material, new nurseries should be developed to increase the production of offshoots for consistent availability of plant material to farmers for further propagation (Sudhersan and AboEl-Nil, 2004).

3.5.1.3 Micropropagation of date palm

Plant tissue culture techniques have been used to produce a variety of economically important palm trees. Through these techniques, date palms can be micropropagated by somatic embryogenesis, where embryos are produced from embryogenic callus that then germinate to form complete plants (McCubbin and Zaid, 2007), or by organogenesis, where embryos are not at the callus stage and plants are produced from proliferating shoots (Al-Khateeb, 2008). Micropropagation refers to the production of whole plants from small parts such as shoot tips, nodes, meristems, embryos, and even seeds. Shoot tip culture is the most widely used method for date palm population reproduction by organogenesis or embryogenesis mediated by callus stage. Mass reproduction can be achieved through organogenesis or somatic embryogenesis (Mazri and Meziani, 2013). Propagation by organogenesis involves the production of meristematic shoots, which then develop into complete seedlings (Mazri and Meziani, 2015). Somatic embryogenesis has been used for mass dissemination of date palms and has been reported in many cultivars such as Barhee, Zardai, Khalasah, Muzati, Shishi, and Zart (Mazri *et al.*, 2018). In date palm trees, the auxin 2,4-dichlorophenoxyacetic acid (2,4-D) has been widely used to induce somatic embryogenesis (Eshraghi *et al.*, 2015; Al-Khayri and Al-Bahrany, 2012). However, major barriers to scaling up include synchronization of embryogenesis, transformation of mature somatic embryos, low rate of new shoot proliferation, and high plant production costs. In recent years, liquid culture systems have achieved great success in automating many plant micropropagation steps. This achievement is another solution to all the above problems. For example, suspension culture produced 17 times the yield of cotyledonary somatic embryos compared to agar-solidified media (Almusawi *et al.*, 2017). However, it is very costly due to the requirement for expensive chemicals, continuous electricity usage, high labour cost, long acclimation time, and the intensive care needed while using plant tissue culture techniques.

3.6 Conclusions and Prospects

Date palm has played a vital role in the livelihood of growers and provided shelter and food security for nomads in deserts in the Middle East and North Africa since ancient times. Date palm is a highly cross-pollinated and heterozygous tree with almost 5000 reported varieties and numerous varietal strains. Genetic diversity of date palm is at risk of genetic erosion due to biotic and abiotic stresses; replacement by elite/superior cultivars of local types; and duplications of the cultivars with different names from one place/region to another. Earlier, many researchers worked on different aspects of characterizing date palm genetic resources and proposed different strategies for conservation (*in situ* or *ex situ* conservation).

There is an urgent need to understand and conserve date palm genetic resources in two components: (i) seedling-based diversity in natural regions; and (ii) domesticated/cultivated regions where date palm is propagated clonally. With this, male diversity estimation and conservation would be another area of research interest. However, comprehensive and collective strategies for the conservation of date palm genetic resources are the need of the hour. For example, *in situ* and *ex situ* conservation simultaneously, gene banks, DNA and RNA storage, DNA libraries, cryopreservation, etc. should all be adopted concurrently. Similarly, genetic resource conservation can be achieved by adopting strategies to: (i) strengthen national date palm genetic resource conservation programs; (ii) develop collaborative research and development activities with international gene banks and researchers; and (iii) improve and adopt conservation technologies.

References

Abul-Soad, A.A., Emara, K.S., Abdallah, A.S. and Mahdi, S.M. (2016) Conservation of superior Egyptian date palm genotypes through *in vitro* culture of inflorescence explants. *Presented at the 2nd International Conference for Date Palm (ICDP), Qassim, Kingdom of Saudi Arabia*, October 10–12, 2016.

Abul-Soad, A.A., Jain, S.M. and Jatoi, M.A. (2017) Biodiversity and conservation of date palm. In Ahuja, M.R. and Jain, S.M. (eds) *Biodiversity and Conservation of Woody Plants*, Vol. 17. *Sustainable Development and Biodiversity*. Springer, Cham, Switzerland, pp. 313–353.

Ahmed, M., Bouna, Z.E.O., Lemine, F.M.M., Djeh, T.K.O., Mokhtar, T. *et al.* (2011) Use of multivariate analysis to assess phenotypic diversity of date palm (*Phoenix dactylifera* L.) cultivars. *Scientia Horticulturae* 127, 367–371.

Ahmed, T.A. and Al-Qaradawi, A. (2009) Molecular phylogeny of Qatari date palm genotypes using simple sequence repeats markers. *Biotechnology* 8, 126–131.

Ahmed, W., Feyissa, T., Tesfaye, K. and Farrakh, S. (2021) Genetic diversity and population structure of date palms (*Phoenix dactylifera* L.) in Ethiopia using microsatellite markers. *Journal, Genetic Engineering & Biotechnology* 19(1), 64. DOI: 10.1186/s43141-021-00168-5.

Al-Bahrany, A.M. and Al-Khayri, J.M. (2012) Optimizing in vitro cryopreservation of date palm (*Phoenix dactylifera* L.). *Biotechnology* 11(2), 59–66. DOI: 10.3923/ biotech.2012.59.66.

Aleid, S.M., Al-Khayri, J.M. and Al-Bahrany, A.M. (2015) Date palm status and perspective in Saudi Arabia. In: Al-Khayri, J.M., Jain, S.M. and Johnson, D.V. (eds) *Date Palm Genetic Resources and Utilization*, Vol. 2. *Asia and Europe.* Springer, Dordrecht, The Netherlands, pp. 125–168.

Al-Faifi, S.A., Migdadi, H.M., Algamdi, S.S., Khan, M.A., Ammar, M.H. *et al.* (2016) Development, characterization and use of genomic SSR markers for assessment of genetic diversity in some Saudi date palm (*Phoenix dactylifera* L.) cultivars. *Electronic Journal of Biotechnology* 21, 18–25. DOI: 10.1016/j. ejbt.2016.01.006.

Al-Ghamdi, A.S. (2001) Date palm (*Phoenix dactylifera* L.) germplasm bank in King Faisal University, Saudi Arabia. Survival and adaptability of tissue cultured plantlets. *Acta Horticulturae* 450(560), 241–244. DOI: 10.17660/ ActaHortic.2001.560.46.

Al-Harrasi, A., Rehman, N.U., Hussain, J., Khan, A.L., Al-Rawahi, A. *et al.* (2014) Nutritional assessment and antioxidant analysis of 22 date palm (*Phoenix dactylifera*) varieties growing in Sultanate of Oman. *Asian Pacific Journal of Tropical Medicine* 7, 591–598. DOI: 10.1016/S1995-7645(14)60294-7.

Al-Khateeb, A.A. (2008) The problems facing the use of tissue culture technique in date palm (*Phoenix dactylifera* L). *Scientific Journal of King Faisal University* 9, 85–104.

Al-Khayri, J.M. and Al-Bahrany, A.M. (2012) Effect of abscisic acid and polyethylene glycol on the synchronization of somatic embryo development in date palm (*Phoenix dactylifera* L.). *Biotechnology* 11, 318–325.

Almusawi, A.H.A., Sayegh, A.J., Alshanaw, A.M.S. and Griffis, J.L. Jr. (2017) Plantform bioreactor for mass micropropagation of date palm. In: Al-Khayri, J., Jain, S. and Johnson, D. (eds) *Date Palm Biotechnology Protocols*, Vol. I. *Tissue Culture Applications.* Methods in Molecular Biology, Vol. 1637, Humana Press, New York, pp. 251–265.

Aman, R. (2001) Problems and challenges in managing field genebank. In: Saad, M.S. and Rao, V.R. (eds) *Establishment and Management of Field Genebank, A Training Manual.* International Plant Genetic Resources Institute–Regional Office for Asia, the Pacific and Oceania, Serdang, Malaysia, pp. 77–80.

Amir, E.A., Guido, F., Behija, S.E., Manel, I., Nesrine, Z. *et al.* (2011) Chemical and aroma volatile compositions of date palm (*Phoenix dactylifera* L.) fruits at three maturation stages. *Food Chemistry* 127(4), 1744–1754. DOI: 10.1016/j. foodchem.2011.02.051.

Asma, B.M. and Ozturk, K. (2005) Analysis of morphological, pomological and yield characteristics of some apricot germplasm in Turkey. *Genetic Resources and Crop Evolution* 52(3), 305–313. DOI: 10.1007/s10722-003-1384-5.

Assirey, E.A.R. (2015) Nutritional composition of fruit of 10 date palm (*Phoenix dactylifera* L.) cultivars grown in Saudi Arabia. *Journal of Taibah University for Science* 9(1), 75–79. DOI: 10.1016/j.jtusci.2014.07.002.

Awad, M.A., Al-Qurashi, A.D. and Mohamed, S.A. (2011) Biochemical changes in fruit of an early and a late date palm cultivar during development and ripening. *International Journal of Fruit Science* 11(2), 167–183. DOI: 10.1080/15538362.2011.578520.

Baaziz, M., Majourhat, K. and Bendiab, K. (2000) Date palm culture in the Maghreb: constraints and scientific research. In: *Proceedings of the Date Palm International Symposium, Windhoek, Namibia, 22–25 February 2000*, Food and Agriculture Organization of the United Nations, Rome, pp. 306–311.

Bagniol, S., Engelmann, F. and Michaux-Ferriè, N. (1992) Histo-cytological study of apices from *in vitro* plantlets of date palm during a cryopreservation process. *CryoLetters* 13, 405–412.

Bedjaoui, H. and Benbouza, H. (2018) Assessment of phenotypic diversity of local Algerian date palm (*Phoenix dactylifera* L.) cultivars. *Journal of the Saudi Society of Agricultural Sciences* 19, 1012–1016.

Bekheet, S.A. (2011) *In vitro* conservation of date palm germplasm. In: Jain, S.M., Al-Khayri, J.M. and Johnson, D.V. (eds) *Date Palm Biotechnology*. Springer, Dordrecht, The Netherlands, pp. 337–360.

Bekheet, S.A. (2015) Effect of cryopreservation on salt and drought tolerance of date palm cultured *in vitro*. *Scientia Agriculturae* 9, 142–149.

Bekheet, S.A. and Taha, H.S. (2013) Complementary strategy for conservation of date palm germplasm. *Global Journal of Biodiversity Science and Management* 3, 96–107.

Bekheet, S.A., Taha, H.S. and Saker, M.M. (2001) *In vitro* long-term storage of date palm. *Biologia Plantarum* 45, 121–124.

Bekheet, S.A., Taha, H.S., Solliman, M.E. and Hassan, N.A. (2007) Cryopreservation of date palm (*Phoenix dactylifera* L.) cultured *in vitro*. *Acta Horticulturae* 736, 283–291. DOI: 10.17660/ActaHortic.2007.736.26.

Bekheet, S.A., Taha, H.S. and Elbahr, M.K. (2005) Preservation of date palm cultures using encapsulated somatic embryos. *Arabian Journal of Biotechnology* 8, 319–328.

Bendiab, K., Baaziz, M., Brakez, Z. and Sedra, H. (1993) Correlation of isoenzyme polymorphism and Bayoud-disease resistance in date palm cultivars and progeny. *Euphytica* 65, 23–32.

Biglari, F., AlKarkhi, A.F.M. and Easa, A.M. (2008) Antioxidant activity and phenolic content of various date palm (*Phoenix dactylifera*) fruits from Iran. *Food Chemistry* 107(4), 1636–1641. DOI: 10.1016/j.foodchem.2007.10.033.

Bouguedoura, N., Bennaceur, M., Babahani, S. and Benziouche, S.E. (2015) Date palm status and perspective in Algeria. In: Al-Khayri, J.M., Jain, S.M. and Johnson, D.V. (eds) *Date Palm Genetic Resources and Utilization*, Vol. 1. *Africa and the Americas*. Springer, Dordrecht, The Netherlands, pp. 125–168.

Ceccarelli, S. (2014) Drought. In: Jackson, M., Ford-Lloyd, B. and Parry, M. (eds) *Plant Genetic Resources and Climate Change*. CAB International, Wallingford, UK, pp. 221–235.

Coudy, D., Colotte, M., Luis, A., Tuffet, S. and Bonnet, J. (2021) Long term conservation of DNA at ambient temperature. Implications for DNA data storage. *PLoS ONE* 16, e0259868. DOI: 10.1371/journal.pone.0259868.

Cruz-Cruz, C.A., González-Arnao, M.T. and Engelmann, F. (2013) Biotechnology and conservation of plant biodiversity. *Resources* 2(2), 73–95. DOI: 10.3390/resources2020073.

Djerouni, A., Chala, A., Simozrag, A., Benmehaia, R. and Mebarek, B. (2015) Evaluation of male palms used in pollination and the extent of its relationship with cultivars of date-palms (*Phoenix dactylifera* L.) grown in region of Oued Righ, Algeria. *Pakistan Journal of Botany* 47, 2295–2300.

Dowson, V.H.W. (1982) *Date Production and Protection*. FAO Technical Paper no.35. Food and Agriculture Organization of the United Nations, Rome.

El-Ashry, A.A., Shaltout, A.D., El-Bahr, M.K., Abd El-Hamid, A. and Bekheet, S.A. (2013) *In vitro* preservation of embryogenic callus cultures of two Egyptian dry date palm cultivars at darkness and low temperature conditions. *Journal of Horticultural Sciences and Ornamental Plants* 5, 118–126.

El-Dawayati, M.M., Zaid, Z.E. and Elsharabasy, S.F. (2012) The effect of conservation on steroids contents of callus explants of date palm cv. Sakkoti. *Australian Journal of Basic and Applied Sciences* 6, 305–310.

Elhoumaizi, M.A., Saaidi, M., Oihabi, A. and Cilas, C. (2002) Phenotypic diversity of date-palm cultivars (*Phoenix dactylifera* L.) from Morocco. *Genetic Resources and Crop Evolution* 49(5), 483–490. DOI: 10.1023/A:1020968513494.

El-Juhany, L.I. (2010) Degradation of date palm trees and date production in Arab countries: causes and potential rehabilitation. *Australian Journal of Basic and Applied Sciences* 4, 3998–4010.

Elshibli, S. and Korpelainen, H. (2009) Biodiversity of date palms (*Phoenix dactylifera* L.) in Sudan: chemical, morphological and DNA polymorphisms of selected cultivars. *Plant Genetic Resources* 7(02), 194–203. DOI: 10.1017/S1479262108197489.

Engelmann, F. (2004) Plant cryopreservation: progress and prospects. *In Vitro Cellular & Developmental Biology - Plant* 40(5), 427–433. DOI: 10.1079/IVP2004541.

Engelmann, F. (2011) Use of biotechnologies for the conservation of plant biodiversity. *In Vitro Cellular & Developmental Biology - Plant* 47(1), 5–16. DOI: 10.1007/s11627-010-9327-2.

Eoin, L.N. (2016) Systematics: blind dating. *Nature Plants* 2(5), 16069. DOI: 10.1038/nplants.2016.69.

Eshraghi, P., Zaghami, R. and Mirabdulbaghi, M. (2015) Somatic embryogenesis in two Iranian date palm cultivars. *African Journal of Biotechnology* 4, 1309–1312.

Fki, L., Bouaziz, N., Sahnoun, N., Swennen, R., Drira, N. and Panis, B. (2011) Palm cryobanking. *CryoLetters* 32, 451–462.

Fki, L., Bouaziz, N., Chkir, O., Benjemaa-Masmoudi, R., Rival, A. *et al.* (2013) Cold hardening and sucrose treatment improve cryopreservation of date palm meristems. *Biologia Plantarum* 57(2), 375–379. DOI: 10.1007/s10535-012-0284-y.

Ford, J.D., Cameron, L., Rubis, J., Maillet, M., Nakashima, D. *et al.* (2016) Including indigenous knowledge and experience in IPCC assessment reports. *Nature Climate Change* 6(4), 349–353. DOI: 10.1038/nclimate2954.

Gros-Balthazard, M., Flowers, J.M., Hazzouri, K.M., Ferrand, S., Aberlenc, F. *et al.* (2021) The genomes of ancient date palms germinated from 2,000 y old seeds. *Proceedings of the National Academy of Sciences of the United States of America* 118(19), e2025337118. DOI: 10.1073/pnas.2025337118.

Haider, M.S., Khan, I.A., Naqvi, S.A., Jaskani, M.J. and Khan, R.W. (2013) Fruit developmental stages effects on biochemical attributes in date palm. *Pakistan Journal of Agricultural Sciences* 50, 577–583.

Haider, M.S., Khan, I.A., Jaskani, M.J., Naqvi, S.A. and Khan, M.M. (2014) Biochemical attributes of dates at three maturation stages. *Emirates Journal of Food and Agriculture* 26(11), 953–962. DOI: 10.9755/ejfa.v26i11.18980.

Haider, M.S., Khan, I.A., Jaskani, M.J., Naqvi, S.A., Hameed, M. *et al.* (2015) Assessment of morphological attributes of date palm accessions of diverse agro-ecological origin. *Pakistan Journal of Botany* 47, 1143–1151.

Haider, M.S., Khan, I.A., Jaskani, M.J., Naqvi, S.A., Mateen, S. *et al.* (2018) Pomological and biochemical profiling of date fruits (*Phoenix dactylifera* L.) during different fruit maturation phases. *Pakistan Journal of Botany* 50, 1069–1076.

Hamza, H., Benabderrahim, M.A., Elbekkay, M., Ferdaous, G., Triki, T. *et al.* (2012) Investigation of genetic variation in Tunisian date palm (*Phoenix dactylifera* L.) cultivars using ISSR marker systems and their relation with fruit characteristics. *Turkish Journal of Biology* 36, 449–458. DOI: 10.3906/biy-1107-12.

Hamza, H., Jemni, M., Benabderrahim, M.A., Mrabet, A., Touil, S. *et al.* (2015) Date palm status and perspective in Tunisia. In: Al-Khayri, J.M., Jain, S.M. and Johnson, D.V. (eds) *Date Palm Genetic Resources and Utilization*, Vol. 1. *Africa and the Americas*. Springer, Dordrecht, The Netherlands, pp. 193–221.

Hassan, S., Bhat, K.M., Jan, A., Mehraj, S. and Wani, S.A. (2018) Managing genetic resources in temperate fruit crops. *Economic Affairs* 63(4), 987–996. DOI: 10.30954/0424-2513.4.2018.23.

Hazzouri, K.M., Flowers, J.M., Visser, H.J., Khierallah, H.S.M., Rosas, U. *et al.* (2015) Whole genome re-sequencing of date palms yields insights into diversification of a fruit tree crop. *Nature Communications* 6, 8824. DOI: 10.1038/ncomms9824.

Hazzouri, K.M., Gros-Balthazard, M., Flowers, J.M., Copetti, D., Lemansour, A. *et al.* (2019) Genome-wide association mapping of date palm fruit traits. *Nature Communications* 10(1), 4680. DOI: 10.1038/s41467-019-12604-9.

Heywood, V.H. and Dulloo, M.E. (2005) In Situ *Conservation of Wild Plant Species: A Critical Global Review of Good Practices*. IPGRI Technical Bulletin 11. International Plant Genetic Resources Institute, Rome.

Hussein, F., El-Kholy, M.H. and Ahmed, T.A. (1993) Organic-chemical constituents of some Egyptian dry-date cultivars grown at Aswan. *Zagazig Journal of Agriculture Research* 20, 1313–1321.

Ismail, B., Haffar, I., Baalbaki, R. and Henry, J. (2008) Physico-chemical characteristics and sensory quality of two date varieties under commercial and industrial storage conditions. *Lebensmittel-Wissenschaft Technologie* 41(5), 896–904. DOI: 10.1016/j.lwt.2007.06.009.

Jain, S.M. (2010) Date palm genetic diversity conservation for sustainable production. *Acta Horticulturae* 882, 785–791. DOI: 10.17660/ActaHortic.2010.882.89.

Jain, S.M. (2011) Prospects of *in vitro* conservation of date palm genetic diversity for sustainable production. *Emirates Journal of Food and Agriculture* 23(2), 110–119. DOI: 10.9755/ejfa.v23i2.6344.

Jaradat, A.A. (2011) Biodiversity of date palm. Soils, plant growth and crop production. In: UNESCO-EOLSS (ed.) *Encyclopedia of Life Support Systems: Land Use, Land Cover and Soil Sciences*. EOLSS Publishers, Oxford, pp. 1–31.

Jaradat, A.A. (2015) Genetic erosion of *Phoenix dactylifera* L.: perceptible, probable, or possible. In: Ahuja, M.R. and Jain, S.M. (eds) *Genetic Diversity and Erosion in Plants*, Vol. 8. *Sustainable Development and Biodiversity*. Springer, Cham, Switzerland, pp. 131–213.

Jaradat, A.A. and Zaid, A. (2004) Quality traits of date palm fruits in a center of origin and center of diversity. *Journal of Food Agriculture and Environment* 2, 208–217.

Jatt, T., Lee, M.S., Rayburn, A.L., Jatoi, M.A. and Mirani, A.A. (2019) Determination of genome size variations among different date palm cultivars (*Phoenix dactylifera* L.) by flow cytometry. *3 Biotech* 9, 457.

Khierallah, H.S., Bader, S.M., Ibrahim, K.M. and Al-Jboory, I.J. (2015) Date palm status and perspective in Iraq. In: Al-Khayri, J.M., Jain, S.M. and Johnson, D.V. (eds) *Date Palm Genetic Resources and Utilization*, Vol. 2. *Asia and Europe*. Springer, Dordrecht, The Netherlands, pp. 97–152.

Krueger, R.R. (2001) Date palm germplasm: overview and utilization in USA. In: Afifi, M.A.R. and Al-Badawy, A.A. (eds), *Proceedings of the First International Conference on Date Palms, Al-Ain, United Arab Emirates*, UAE University, Al-Ain, March 8–10, 1998, pp. 2–37.

Krueger, R.R. (2015) Date palm status and perspective in the United States. In: Al-Khayri, J.M., Jain, S.M. and Johnson, D.V. (eds) *Date Palm Genetic Resources and Utilization*, Vol. 1. *Africa and the Americas*. Springer, Dordrecht, The Netherlands, pp. 447–485.

Markhand, G.S., Abul-Soad, A.A., Mirbahar, A.A. and Kanahr, N.A. (2010) Fruit characterization of Pakistani dates. *Pakistan Journal of Botany* 42, 3715–3722.

Martín-Sánchez, A.M., Cherif, S., Ben-Abda, J., Barber-Vallés, X., Pérez-Álvarez, J.Á. *et al.* (2014) Phytochemicals in date co-products and their antioxidant activity. *Food Chemistry* 158, 513–520. DOI: 10.1016/j.foodchem.2014.02.172.

Maxted, N., Ford-Lloyd, B.V. and Hawkes, J.G. (eds) (1997) Complementary conservation strategies. In: *Plant Genetic Conservation: The* In Situ *Approach*. Springer, Dordrecht, The Netherlands, pp. 15–39. DOI: 10.1007/978-94-009-1437-7.

Mazri, M.A. and Meziani, R. (2013) An improved method for micropropagation and regeneration of date palm (*Phoenix dactylifera* L.). *Journal of Plant Biochemistry and Biotechnology* 22(2), 176–184. DOI: 10.1007/s13562-012-0147-9.

Mazri, M.A. and Meziani, R. (2015) Micropropagation of date palm: a review. *Cell & Developmental Biology* 4, 160.

Mazri, M.A., Meziani, R., Belkoura, I., Mokhless, B. and Nour, S. (2018) A combined pathway of organogenesis and somatic embryogenesis for an efficient large-scale propagation in date palm (*Phoenix dactylifera* L.) cv. Mejhoul. *3 Biotech* 8(4), 215. DOI: 10.1007/s13205-018-1235-x.

McCubbin, M.J. and Zaid, A. (2007) Would a combination of organogenesis and embryogenesis techniques in date palm micropropagation be the answer? *Acta Horticulturae* 736, 255–260. DOI: 10.17660/ActaHortic.2007.736.23.

Mirbahar, A.A., Markhand, G.S., Khan, S. and Abul-Soad, A.A. (2014) Molecular characterization of some Pakistani date palm (*Phoenix dactylifera* L.) cultivars by RAPD markers. *Pakistan Journal of Botany* 46, 619–625.

Mohamed, R., Fageer, A.S., Eltayeb, M.M. and Mohamed Ahmed, I.A. (2014) Chemical composition, antioxidant capacity, and mineral extractability of Sudanese date palm (*Phoenix dactylifera* L.) fruits. *Food Science & Nutrition* 2(5), 478–489. DOI: 10.1002/fsn3.123.

Mortazavi, S.M.H., Arzani, K. and Moieni, A. (2010) Optimizing storage and *in vitro* germination of date palm (*Phoenix dactylifera*) pollen. *Journal of Agricultural Sciences and Technology* 12, 181–189.

Nadeem, M., Qureshi, T.M., Ugulu, I., Riaz, M.N., Ain, Q.U. *et al.* (2019) Mineral, vitamin and phenolic contents and sugar profiles of some prominent date palm (*Phoenix dactylifera*) varieties of Pakistan. *Pakistan Journal of Botany* 51(1), 171–178. DOI: 10.30848/PJB2019-1(14).

Othmani, A., Bayoudh, C., Drira, N. and Trifi, M. (2009) *In vitro* cloning of date palm *Phoenix dactylifera* L., cv. Deglet Bey by using embryogenic suspension and temporary immersion bioreactor (TIB). *Biotechnology & Biotechnological Equipment* 23(2), 1181–1188. DOI: 10.1080/13102818.2009.10817635.

Pareek, S. (2015) Date palm status and perspective in India. In: Al-Khayri, J.M., Jain, S.M. and Johnson, D.V. (eds) *Date Palm Genetic Resources and Utilization*, Vol. 2. *Asia and Europe*. Springer, Dordrecht, The Netherlands, pp. 441–485.

Rao, V.R. and Sthapit, B.R. (2012) Tropical fruit tree genetic resources: status and effect of climate change. In: Sthapit, B.R., Rao, V.R. and Sthapit, S.R. (eds) *Tropical Fruit Tree Species and Climate Change*. Bioversity International, New Delhi, pp. 97–137.

Raza, A., Razzaq, A., Mehmood, S.S., Zou, X., Zhang, X. *et al.* (2019) Impact of climate change on crops adaptation and strategies to tackle its outcome: a review. *Plants (Basel, Switzerland)* 8(2), 34. DOI: 10.3390/plants8020034.

Sadiq, S., Thompson, I., Shuaibu, M., Dogoyaro, A.I., Garba, A. *et al.* (2013) The nutritional evaluation and medicinal value of date palm (*Phoenix dactylifera*). *International Journal of Modern Chemistry* 4, 147–154.

Sallon, S., Cherif, E., Chabrillange, N., Solowey, E., Gros-Balthazard, M. *et al.* (2020) Origins and insights into the historic Judean date palm based on genetic analysis of germinated ancient seeds and morphometric studies. *Plant Science* 6, 0384.

Semagn, K., Bjornstad, A. and Ndjiondjop, M.N. (2006) An overview of molecular marker methods for plants. *African Journal of Biotechnology* 5, 2540–2568.

Simozrag, A., Adel, C., Djerouni, A. and Mouhamad, B. (2016) Phenotypic diversity of date palm cultivars (*Phoenix dactylifera* L.) from Algeria. *Gayana Botánica* 73(1), 42–53. DOI: 10.4067/S0717-66432016000100006.

Singh, B.P. (2009) Germplasm introduction, exchange, collection/evaluation and conservation of medicinal and aromatic plants – their export potential. In: Trivedi, P.C. (ed.) *Medicinal Plants: Utilization and Conservation*. Aavishkar Publishers, Jaipur, India, pp. 8–11.

Srinivasa, R.C., Younga, P., Thompson, K., Bahribc, B.A., Gajerad, B. *et al.* (2019) Phoenix phylogeny, and analysis of genetic variation in a diverse collection of date palm (*Phoenix dactylifera*) and related species. *Plant Diversity* 41(5), 330–339. DOI: 10.1016/j.pld.2018.11.005.

Subaih, W.S., Shatnawi, M.A. and Shibli, R.A. (2007) Cryopreservation of date palm (*Phoenix dactylifera*) embryogenic callus by encapsulation-dehydration, vitrification and encapsulation-vitrification. *Jordan Journal of Agricultural Sciences* 3, 156–171.

Sudhersan, C. and AboEl-Nil, M. (2004) Axillary shoot production in micropropagated date palm. *Current Science* 86, 771–773.

Sudhersan, C., Sudhersan, J., Ashkanani, J. and Al-Sabah, L. (2015) Date palm status and perspective in Kuwait. In: Al-Khayri, J.M., Jain, S.M. and Johnson, D.V. (eds) *Date Palm Genetic Resources and Utilization*, Vol. 2. *Asia and Europe*. Springer, Dordrecht, The Netherlands, pp. 299–321.

Yang, M., Zhang, X., Liu, G., Yin, Y., Chen, K. *et al.* (2010) The complete chloroplast genome sequence of date palm (*Phoenix dactylifera* L.). *PLoS ONE* 5(9), e12762. DOI: 10.1371/journal.pone.0012762.

Yap, T.C. and Saad, M.S. (2001) Factors in field genebank layout. In: Saad, M.S. and Rao, V.R. (eds) *Establishment and Management of Field Genebank: A Training*

Manual. International Plant Genetic Resources Institute–Regional Office for Asia, the Pacific and Oceania, Serdang, Malaysia, pp. 73–76.

Yusuf, A.O., Culham, A., Aljuhani, W., Ataga, C.D., Hamza, A.M. *et al.* (2015) Genetic diversity of Nigerian date palm (*Phoenix dactylifera*) germplasm based on microsatellite markers. *International Journal of Bio-Science and Bio-Technology* 7(1), 121–132. DOI: 10.14257/ijbsbt.2015.7.1.12.

Zaid, A. and de Wet, P.F. (2002) Origin, geographical distribution and nutritional values of date palm. In: Zaid, A. and Arias Jimenez, E.J. (eds) *Date Palm Cultivation.* Food and Agriculture Organization of the United Nations, Rome, pp. 29–44.

Zaid, A., de Wet, P.F., Djerbi, M. and Oihabi, A. (2002) Diseases and pests of date palm. In: Zaid, A. and Arias Jimenez, E.J. (eds) *Date Palm Cultivation.* FAO Plant Production and Protection Paper no.156 Rev. 1, Food and Agriculture Organization of the United Nations, Rome, pp. 223–278.

Zaid, A., El-Korchi, B. and Visser, H.J. (2011) Commercial date palm tissue culture procedures and facility establishment. In: Jain, S.M., Al-Khayri, J.M. and Johnson, D.V. (eds) *Date Palm Biotechnology.* Springer, Dordrecht, The Netherlands, pp. 137–180.

Genetic Improvement to Produce Value-Added Date Palm Cultivars

4

Larbi Abahmane*

National Institute for Agricultural Research, Marrakech, Morocco

Abstract

Date palm (*Phoenix dactylifera* L.) is a woody monocot tree producing fruit with high nutritional and economic importance. Currently, date production suffers from several biotic and abiotic stresses that reduce its profitability and threaten its genetic diversity. Moreover, climate change due to global warming has negatively affected crop production worldwide including dates. Despite the high resilience of the date palm tree, there is a need to develop new cultivars that are higher yielding and possess highly valued agronomic traits, such as fruit quality, disease and pest resistance, and tolerance of extreme environmental conditions. Conventional breeding, applied in most countries to improve date palm cultivars, is a time-consuming and tedious practice. Therefore, new technologies based on molecular markers, such as marker-assisted selection (MAS), are promising tools to overcome this key limitation, speed up this process, and make it more efficient. More advanced technologies have been recently developed and can be useful in date palm breeding. Genome editing (GE) is among the necessary tools in modern biology and comprises technologies to modulate the DNA of living organisms. In addition, genome-wide association study (GWAS) is a method for the study of associations between a genome-wide set of single-nucleotide polymorphisms (SNPs) and desired phenotypic traits. Moreover, advances in whole-genome sequencing and genome assembly computational tools hold considerable advantage for cultivar identification, identification of genome variants, and genome annotation. In the present chapter, a survey of the main advances in date palm breeding tools to produce value-added cultivars is presented.

Keywords: Agronomic traits, Breeding, Genetic diversity, Genetic improvement, Molecular markers, *Phoenix dactylifera* L

*abahmanel@yahoo.fr; larbi.abahmane@inra.ma

© CAB International 2023. *Date Palm* (eds J.M. Al-Khayri *et al.*)
DOI: 10.1079/9781800620209.0004

4.1   Introduction

The date palm (*Phoenix dactylifera* L.) is grown throughout the arid and semi-arid regions of the world, particularly in West Asia and North Africa. It is well adapted to the desert environment, where a dry and warm climate is important for fruit maturity and ripening. Interlinked ecological and socio-economic factors are affecting the delicate equilibrium of oasis agroecosystems. Ecological factors include land degradation, genetic erosion, inappropriate agronomic practices, frequent droughts, aquifer depletion, and sand encroachment. In addition, date palm orchards in North Africa are ageing. Indeed, almost one-third of productive date palm trees in Algeria are beyond the limits of their productive years, and almost half of productive Tunisian date palms are over 50 years old. Moreover, date palm biodiversity and date production are threatened by a number of biotic and abiotic stresses (Jaradat, 2011). Indeed, two serious biotic threats are currently menacing millions of date palms in the Middle East and North Africa (MENA) region, specifically: bayoud disease, caused by *Fusarium oxysporum* f. sp. *albedinis*; and the red palm weevil (*Rhynchophorus ferrugineus* Olivier). In addition, date palms suffer from a variety of abiotic stresses such as salinity, drought, and extreme temperatures, which cause significant yield losses. Abiotic stresses lead to dehydration or osmotic stress through reduced availability of water for vital cellular functions (Marco *et al.*, 2015). Most importantly, water shortage and groundwater salinity are abiotic stresses that decrease date production. In consequence, adaptation of crop plants to water deficit and salt stress is of high priority in worldwide programs for breeding modern varieties (El Rabey *et al.*, 2015). Plants have evolved with various mechanisms such as changes in cellular and metabolic processes to cope with stress conditions. According to Marco *et al.* (2015), recent developments in molecular genetics have contributed greatly to our understanding of the biochemical and genetic basis of abiotic stress tolerance. This has led to the development of abiotic stress-tolerant plants with sustained yield by modulation of the expression of genes that encode for enzymes involved in the biosynthesis of osmoprotectants. All these challenges are exacerbated in arid and semi-arid regions where climate change and chronic water shortages have reduced arable land area and reduced crop yields (Hazzouri *et al.*, 2020). Unfortunately, date palms inhabit harsh desert environments, but they remain viable even in areas with saline soils and survive long periods with limited water supply (Elshibli *et al.*, 2016; Müller *et al.*, 2017). Despite a high tolerance to abiotic stress, date palms require large volumes of water to produce commercial-grade fruit and suffer from lower productivity and reduced fruit quality when subject to drought and salinity stresses (Hussain *et al.*, 2012). The phenomenon of global climatic change occurring over the years is partly responsible for modifying the crop production environment. Allbed *et al.* (2017) modeled a significant reduction in climatic suitability for date palm cultivation in

Saudi Arabia by 2100. This means that new cultivars of crops need to be bred for new production environments (Acquaah, 2007; Jain, 2019).

Most of the wild date palm germplasm is already lost, and only a few natural populations are believed to exist (Gros-Balthazard *et al.*, 2017; Wales and Blackman, 2017). Genetic diversity existing in the cultivated date palm is also being lost because of shifts to fewer varieties (El-Juhany, 2010). This will increase the vulnerability of date palm to sudden changes in climate, diseases, and insect pests (Chaluvadi *et al.*, 2019). Therefore, the exploitation of biodiversity is essential to select more resilient genotypes employable in more sustainable cropping systems (Mercati and Sunseri, 2020).

4.2   Agronomic Traits in Date Palm Breeding

The specific characters that farmers consider for continuous and sustained improvement in crop plants are known as agronomic traits of the taxa. These agronomic traits include high yield, disease resistance, tolerance to both biotic and abiotic stresses, and better physiological and morphological features. Improvement in agricultural quality traits shares the long history of selection and domestication of wild plants by humans. Seeking better yield coupled with resistance to various stresses has been done as a continuous process resulting in a fine selection of cultivars and landraces. This resulted in the process of elimination, retention, and acquisition of certain traits that proved to be beneficial (Ranjisha *et al.*, 2020).

A plant breeding program is carried out to improve certain aspects of plants to perform new roles or enhance existing ones (Acquaah, 2012, 2015). The practice of plant breeding has advanced from the cynical view of *crossing the best with the best and hoping for the best,* to the now carefully planned and thought-out strategies for developing high-performing cultivars with high predictability (Acquaah, 2012). The methods and tools employed, which keep changing with advances in science and technology, provide the basis for categorizing plant breeding approaches into two basic types, specifically: conventional; and modern or molecular (Acquaah, 2015).

Date palms are grown under adverse climatic conditions and the plants experience extreme heat shock, prolonged periods of drought, and high soil and/or water salinity. The productivity of the date palm is affected, so there is a need to develop new date cultivars with resistance to abiotic stresses. Obtaining resistance to bayoud disease and the red palm weevil through conventional breeding is a time-consuming and tedious process. There is enormous potential in transformation technology to introduce disease resistance traits into the date plant. In addition, date fruits face a storage problem, as the shelf life of dates declines in normal environmental conditions, affecting the date-marketing potential internationally. There is a need to improve date cultivars to give them improved shelf life (Al-Khayri *et al.*, 2018).

According to the context of date palm cultivation, the targeted goals of breeding can be different. Indeed, where some biotic problems (bayoud, red palm weevil, etc.) are widespread, breeders are called upon to develop biotic stress-resistant/tolerant genotypes. In the Moroccan context, the main breeding criteria were:

- resistance/tolerance to bayoud disease, which is essential for the survival of palm trees in infested areas; and
- good fruit quality, a criterion imposed by market constraints as well as consumer preferences.

Generally, date palm breeding had as its main objectives (Ibraheem, 2008):

- *Resistance to biotic stresses:* The main disease threatening date palms worldwide is incontestably the bayoud disease caused by a soilborne fungus, *F. oxysporum* f. sp. *albedinis* (Foa), which destroyed millions of date palm trees in North Africa. Use of chemicals for bayoud control is not suitable as all palm groves have been infested and because of probable negative effects on the environment. Indeed, genetic control based on the selection of resistant cultivars was, until now, the most effective means to control this vascular disease in Morocco. The second threat to date palm cultivation is the red palm weevil (*R. ferrugineus*), which causes serious damages and losses of date palms in the Arabian Gulf countries.
- *Resistance to abiotic stresses:* Date palm is one of major crops growing in regions where abiotic stress conditions are extreme. Abiotic stress affects plant growth, physiology, and biochemical processes. Salinity and drought are major adverse environmental conditions, particularly in the arid and semi-arid regions of the MENA region where the date palm is predominately cultivated (Al-Khayri *et al.*, 2017). The adverse effects of these stresses are intensified under climate change. Other environmental stresses can reduce the area of date palm cultivation such as rain, high humidity, and low temperature for long periods.
- *Early-ripening cultivars:* This character is important in marginal regions known with autumnal rainfalls and low temperatures. Under such environment, some cultivars as Kenta, Ajwa, and Abou Hatem from Tunisia produce mature dates.

4.3   Mass Selection

Plant breeding focuses on improvement of cultivated crops by developing new varieties that are higher yielding and possess highly valued agronomic traits such as fruit quality, disease resistance, and tolerance of extreme environmental conditions. Plant breeding success depends on the collection and selection of germplasm, its evaluation for desired characters, multiplication of germplasm, and release and distribution of new varieties. Date palm is one of the oldest cultivated crops and most countries have followed conventional breeding methods up until now. Date palm breeding objectives mostly depend on regional problems such as abiotic and biotic stresses (Al-Khayri *et al.*, 2018).

Selection is simply discriminating among the available or created variability to identify individuals with the desired combination of genes or expressed traits. The final selection cycle in breeding results in a small number of genotypes that are potential candidates for advancing as cultivars for release to producers (Acquaah, 2015). The pool of genetic variation is the basis for selection as well as for date palm improvement as a functional genetic resource within the oasis agroecosystem. Therefore, conservation of genetic diversity is essential for present and future functioning of these ecosystems and for the survival of human communities therein (Jaradat, 2011). Besides the known date palm cultivars, there is a wide range of elite germplasm of date palm originating from seedlings. Farmers usually rely on the physical traits of the fruits that have a high degree of consumer preference to select superior seedling progeny of date palms. Over the years, the present well-known cultivars of date have evolved from seedlings selected by date farmers for good fruit qualities (Al-Abdoulhadi, 2015).

The mode of propagation has impacted the level of diversity and varietal composition in the oasis ecosystems. In the case of elite selections, propagation by transplanted offshoots resulted in dissemination of genetically identical cultivars to various parts of a country or region. In the case of noncommercial or less desirable cultivars, seed dissemination resulted in the establishment of more localized and adapted cultivars. However, some natural selection pressure in the new oases may have occurred due to resistance or susceptibility to biotic and abiotic stresses. Similarly, natural selection could have been applied on the nonelite cultivars that originated from seeds (Jaradat, 2011).

4.3.1 Methods

The selection process followed by farmers in their palm groves was focused for centuries not only on certain fruit criteria (taste, quality, appearance, color, use of fruit, etc.) but also on the local adaptation of cultivars. These criteria were usually subjective and took into consideration the famers' own preferences and those of tradesmen: mainly large brown dates that can be stored at ambient conditions. In Morocco, as in some other countries of North Africa, sexually propagated palms from natural crosses were important and exhibited variability according to region and the technological level of the farmer. Hence, cultivars of agronomic interest were selected and multiplied by offshoots in surrounding lands (Sedra, 2011).

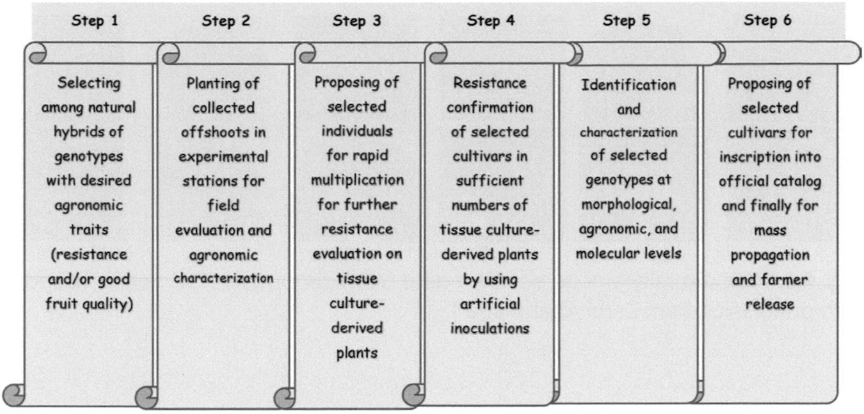

Step 1	Step 2	Step 3	Step 4	Step 5	Step 6
Selecting among natural hybrids of genotypes with desired agronomic traits (resistance and/or good fruit quality)	Planting of collected offshoots in experimental stations for field evaluation and agronomic characterization	Proposing of selected individuals for rapid multiplication for further resistance evaluation on tissue culture-derived plants	Resistance confirmation of selected cultivars in sufficient numbers of tissue culture-derived plants by using artificial inoculations	Identification and characterization of selected genotypes at morphological, agronomic, and molecular levels	Proposing of selected cultivars for inscription into official catalog and finally for mass propagation and farmer release

Fig. 4.1. Selection steps for bayoud-resistant genotypes adopted in the Moroccan context. (Adapted from Sedra, 2011.)

The selection program conducted in Morocco aimed to find bayoud-resistant varieties producing fruits of great commercial value. Taking into account that these two characters were never found together at the highest level in the most widespread cultivars, the first research works were focused on the selection of average-quality dates in varieties with a high level of resistance or partial resistance in varieties with high fruit quality (Saaidi, 1990). The selection first took place within common varieties, then among natural populations of individuals from natural hybridization. Such a program required a rigorous methodology, especially at the stage of the resistance assessment (Fig. 4.1). Indeed, reliable and rapid methods of cultivar screening have been developed in the field using artificial inoculation of palm trees with the pathogen, in the laboratory on young plants (Sedra and Besri, 1994), or by the use of pathogen toxins (El Fakhouri *et al.*, 1997).

Prospecting for resistance to bayoud and/or for good fruit quality genotypes in Moroccan palm groves was carried out in successive stages from 1967 to 1986. Collected plant materials consisting of genotypes producing fruits of good quality and surviving in areas infested by Foa were transferred to experimental stations for further evaluation. Since being planted, these date palms have been monitored for over 25 years in the field for their susceptibility/tolerance to bayoud and productivity (Saaidi, 1990; Sedra, 2011).

4.3.2 Results

Among 32 Moroccan cultivars tested at the Zagora Experiment Station of the National Institute for Agricultural Research (INRA-Morocco), six varieties (Black Bousthami, Tadment, White Bousthami, Iklane, Saïr-Layalate, and Boufeggous ou Moussa) were found to be resistant to bayoud

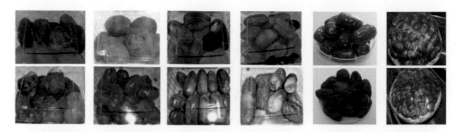

Fig. 4.2. Fruit diversity among selected date palm genotypes in Morocco. (Used with permission from Larbi Abahmane.)

disease (Saaidi, 1992). A seventh bayoud-resistant cultivar (Boukhani) was selected 20 years later (Sedra, 1995). In Algeria, the first identified bayoud-resistant cultivar was Takerbouchte (Bulit *et al.*, 1967). Later, Tirichine (1991) added another resistant cultivar, Akerbouch, selected in the Mzab region of Algeria. Among these resistant cultivars, only Boukhani, Saïr-Layalate, and Takerbouchte were relatively acceptable fruit quality-wise but were not as good as elite varieties like Mejhoul and Deglet Noor. In the same framework, many foreign cultivars, particularly from Iraq (Barhee, Hallawy, Khastawy, Khadrawy, Sair, and Zahdi) and Tunisia (Boufeggous, Besser Lahlou, Gondi, Horra, Kenta, and Kentichi), have all shown high susceptibility to bayoud disease (Djerbi and Sedra, 1982; Sedra, 1995).

In a second step, the search for high-performance palms was set up within the naturally occurring hybrid palms estimated to 2.5 million and representing 55% of the total population of date palms in Morocco. About 2337 genotypes resulted from mass selection of which 1130 showed wide genetic diversity at the level of agronomic and phenological characteristics. Nevertheless, about 60% of these genotypes were susceptible to bayoud disease. Interestingly, several genotypes (females and males) were selected (Fig. 4.2) either for their tolerance to bayoud or for their good fruit quality (Sedra, 2013). Some of these sought-after individuals were multiplied and disseminated to farmers to control bayoud disease and to improve date production in Morocco.

The main quantitative and qualitative agro-morphological characters of some selected cultivars compared to the most famous variety Mejhoul as well as to the most bayoud-resistant variety Black Bousthami are summarized in Table 4.1. All the susceptible cultivars have been mass-propagated and distributed to farmers for evaluation of their fruit quality. Farmers were advised to plant these cultivars, as well as the susceptible cultivars Mejhoul and Boufeggous, in areas free of bayoud disease to better valorize them without risk. After the first selected cultivar Najda, seven other selected cultivars were identified to control bayoud disease: Deraouia, Sedrat, Al-Amal, Al-Fayda, Bourihane, Mabrouk, and INRA-3010.

Table 4.1. Main agro-morphological characteristics of selected female cultivars in Morocco. (Adapted from Sedra, 2005.)

Cultivar	Fruit quality	Resistance to bayoud disease	Heat requirement (°C)	Fruit consistency	Fruit stalk length (cm)
Najda	Good	Resistant	4000–4500	Semisoft	77
Mabrouk	Very good	Resistant	4500–5000	Semisoft	63
Ayour	Very good	Susceptible	4500–5000	Semisoft	97
Hiba	Good	Susceptible	3500–4000	Dry	65
Tanourte	Good	Susceptible	3500–4000	Semisoft	75
Tafoukte	Fair	Susceptible	<3500	Semidry	74
Khair	Good	Susceptible	4500–5000	Semisoft	65
Al-Fayda	Excellent	Resistant	3500–4000	Semisoft	170
Bourihane	Excellent	Resistant	3500–4000	Semisoft	61
Al-Amal	Excellent	Resistant	3500–4000	Dry	65
Deraouia	Excellent	Resistant	4000–4500	Semisoft	121
Sedrat	Excellent	Resistant	<3500	Soft	96
INRA-3010	Excellent	Resistant	4000–4500	Semidry	93
Mejhoul	Excellent	Very susceptible	>5000	Semisoft	143
Black Bousthami	Poor	Resistant	4500–5000	Soft	120

Table 4.2. Main agro-morphological characteristics of selected male cultivars in Morocco. (Adapted from Sedra, 2005.)

Male cultivar	Resistance to bayoud disease	Pollen production	Pollen germination	Leaf length (cm)
Nebch-Bouskri, NP3	Resistant	Moderate	Very high	383
Nebch-Boufeggous, NP4	Resistant	High	Very high	438

Moreover, two male cultivars, Nebch-Bouskri (INRA-NP3) and Nebch-Boufeggous (INRA-NP4), have been selected and characterized as resistant genotypes (Table 4.2).

In other countries, such as Saudi Arabia, several seedling date palm progenies have been identified by farmers and propagated commercially. Recently, a study has been undertaken with the main objectives to identify, evaluate, and safeguard valuable genetic resources (Al-Abdoulhadi, 2015). This survey has permitted selection of 12 potential seedling date palms, from the Al-Zulfi and Riyadh regions of the Kingdom, found to produce dates that meet standards for commercial date farming and were much sought after by the farmers of these regions. Among them, fruits of Adbah had the highest total sugars at 72.9%, followed by Duhiba with a total sugar content of 71.9%. Fruit length was maximum (49.1 mm) in the variety Aliah, while the fruits of the variety Hussan recorded the maximum weight of 19.5 g. The selected cultivars can serve as a genetic resource for future improvement programs for date palm in the Kingdom and elsewhere and have been conserved in the gene bank of the Date Palm Research Centre, Al-Hassa, Saudi Arabia.

In Tunisia, date palm cultivation is oriented toward the monoculture of only the Deglet Noor variety, which could lead to a decline in diversity and the extinction of several cultivars in the long term. With a view to identifying and enhancing the existing genetic diversity richness in Tunisia, a research study to select varieties of agronomic and commercial interest among spontaneous palms in the continental oases was undertaken (Sghairoun and Ferchichi, 2013). The prospecting carried out has allowed the selection of a new variety called Rhaymi, which presents strong and interesting characters, particularly its semisoft consistency, late maturity, dark brown color, seed/fruit ratio (18.12%), Brix rate (64.5), moisture content (30.49%), low sucrose level (13.17%), and better levels of glucose (15.71%) and fructose (23.48%). This genotype could have pride of place in creation of new oases.

In Niger, a mass selection study has been conducted by Zango *et al.* (2016). Out of 200,000 palm population, 19 seed-propagated genotypes of different fruit colors were selected in the Manga region. Research and sustainable development of date palm cultivation in the Sahel, taking account of adaptation to climate change, must rely on date palm improvement in relation to early production and humidity tolerance during fruit maturation. The problem to be solved is in storage, for which the solution would seem to lie in breeding varieties that start bearing early or are more humidity-tolerant during ripening, like the Garouda seed-propagated variety grown in just one of the surveyed villages. This selection, combined with technical capacity building for farmers, would help to improve the self-subsistence of the local populations.

According to Johnson *et al.* (2013), date palm germplasm collections in the USA contain 11 American elite cultivars, including the Empress, Amber Queen, and Honey cultivars. All of them were selected from seedlings obtained from introduced foreign cultivars (Nixon and Furr, 1965; Hodel and Johnson, 2007).

In a large-scale breeding experiment in the United Arab Emirates, several thousands of date seeds have been planted in two separate locations at Al Ain city in Abu Dhabi. The seedlings were subjected to continuous observation from the earliest stages of development. Among the interesting new cultivars selected was one that produces seedless fruits. Of the many other selected cultivars were ones that produced seeded fruit of good quality according to the known standards. Some cultivars resembling Barhi, Khalas, Mejhoul, Sultana, and other elite cultivars were among the selected individuals. These selections will be subjected to further studies including fruit quality characteristics, physicochemical properties, and morphological, anatomical, and biotechnological investigations (Shabana *et al.*, 2018).

4.3.3 Advantages and disadvantages of mass selection

The main advantage of mass selection is that it is rapid, simple, and the least expensive method of plant breeding. Although significant progress has been made in crop improvement through phenotypic selection for agronomically important traits, considerable difficulties are often encountered during this process. A new variety in conventional breeding could take more than 10 years to be developed. Therefore, breeders are very interested in new technologies to speed up this process or make it more efficient. The development of molecular markers was seen as a major breakthrough to overcome this key limitation. Marker-assisted selection (MAS) is likely to become more valuable as a larger number of genes is identified and their functions and interactions are elucidated. However, reduced costs and optimized strategies for integrating MAS with phenotypic selection are needed before the technology can reach its full potential. Overall, MAS has proven to be useful in plant breeding (Lema, 2018). MAS is discussed further in Section 4.9.

4.4   Crossbreeding

Despite the great spontaneous diversity that can be found in landraces, simply applying selection to preexisting diversity is an eroding process that eventually comes to a limit. The true creative power of plant breeding resides in promoting recombination for shuffling favourable alleles. The combination of different alleles in many loci results in a virtually infinite number of genotypes (Breseghello and Coelho, 2013). In crossbreeding programs, starting from a broad and very diverse genetic base, selection criteria other than bayoud resistance and fruit quality can be retained: tree vigour, productivity, capacity to produce offshoots, speed of growth, precocity of fruit setting and maturation of dates, hardiness, and soil and climate adaptation.

4.4.1   Methods

In controlled hybridization, pollen is collected from elite male trees, carefully dusted on unopened mature female inflorescences of the selected date palm cultivars, and immediately covered with paper bags to avoid foreign pollen. In general, cultivars are selected on the basis of possessing one or more outstanding characters that might be desirable in a new cultivar. Beyond biotic and abiotic stress tolerance, primary considerations are fruit characteristics such as large size, attractive color and appearance, good texture and flavor, good shipping and storing quality, time of ripening, high yield, and rain tolerance (Nixon and Furr, 1965). Some other sought-after traits can be adaptation to the mechanization of harvesting and packaging and low tree height (Krueger, 2015).

In Algeria, genetic selection of the date palm started in 1973, especially for creation of new commercial varieties adapted to different climatic areas of the Sahara. After that, programs for selecting interesting genotypes, in palm groves or by hybridization, were launched in Algeria and Tunisia to find high-performance clones that are resistant to bayoud disease (Saaidi, 1990).

In Morocco, Pereau-Leroy (1958) carried out the first breeding work between 1949 and 1956. This work aimed to select varieties resistant to Fusarium wilt (bayoud) in palm groves. Later, the controlled hybridization program aimed at creating palm trees resistant to bayoud disease and producing dates of better quality than those resulting from natural hybridization. The descendants of controlled crosses as well as selected genotypes were assessed for their resistance to bayoud by artificial inoculation and by planting in naturally infested lands. The survivors were maintained until fruit set for examining the quality of dates (Saaidi, 1990; Sedra, 2011).

As much as possible, hybridization programs were started using a wide range of parents, especially palm trees with a high level of resistance and palm trees producing good-quality fruit. Male parents are selected on the basis of their pollen production, timing of flowering, metaxenic effects, and resistance or tolerance to biotic stresses (Saaidi, 1990).

Female palms used in the Moroccan program included:

- *Resistant females:* Eight varieties were used, of which six were resistant to bayoud disease (Black Bousthami, White Bousthami, Saïr-Layalate, Tadment, Iklane, Boufeggous ou Moussa) (Saaidi *et al.*, 1981), one was fairly resistant (Bouzeggar), and one was an Algerian cultivar (Takerbouchte) reported as resistant by Toutain and Louvet (1974).
- *Females of good fruit quality:* Seven Moroccan varieties (Boufeggous, Mejhoul, Bouskri, Jihel, Ahardane, Aguellid, Outoukdim) were used, as well as the best Algerian-Tunisian variety (Deglet Noor) and two Alligs of Tunisia (Ftimi and Oukhouet). During 1986–1987, some high-quality and presumed resistant Saïrs (natural hybrids) were also introduced in this program as well as certain individuals already resulting from controlled crosses (Saaidi, 1990).

Male palms used in the Moroccan program included:

- *Resistant males:* They were chosen among the individuals selected in bayoud-infested lands, based upon their presumed resistance. A total of 40 males from the Draa, Tafilalet, and Bani Moroccan regions were used.
- *Quality males:* These are males obtained after three to five successive backcrosses carried out in the USA from the world's best varieties. Therefore, 19 males were used from five varieties: Deglet Noor, Mejhoul, Halawy, Khadrawy, and Barhi.
- *Susceptible local males:* During each pollination campaign, one or two susceptible males (showing symptoms of bayoud and pollen in the same year) were used. Thus, 24 males from this category were integrated in this program (Saaidi, 1990).

4.4.2   Results

In the USA, crosses made by the US Department of Agriculture (USDA) in 1971 yielded nine seedlings considered worth saving. These included dry, semidry, and soft-fruited selections. Some of the soft or dry selections had the potential to compete with commercial cultivars. However, no semidry selection was found that appeared to equal Deglet Noor in quality, although some may have lent themselves to mechanical harvesting and processing. The nine selections were incorporated into the USDA National Date Palm Germplasm Repository when the USDA Date Station was closed (Carpenter, 1979; Krueger, 2001). One selection, Mejhoul × (Dayri × Deglet Noor BC3), has recently received some interest due to its large size (approximately 50% larger than Mejhoul). It has a mild flavor but probably would not ship well due to its soft skin (Krueger, 2015). Moreover, there are currently 27 lines of backcrossed males representing 11 cultivars in the USDA National Date Palm Germplasm Repository. These represent the lines considered most valuable at the time that the USDA Date Station was closed.

From the crossbreeding and backcrosses done at Indio experimental station (USA), the following main results were obtained (Ferry, 2010):

- Good transmission of the size of fruit trait, with large fruit size for the descendants of the Mejhoul variety and reduced fruit size for descendants of the Khadrawy variety.
- Good resemblance of the fruits for Barhi but not for Deglet Noor; mainly its flavor, which makes it so special, has not been preserved even if this variety has benefited from five backcrosses.
- Among the crosses, most of descendants were of inferior quality, having insufficient tolerance to rain at the khalal stage and poor storage characteristics. Only 1% of new genotypes obtained were retained.
- Interestingly, about 70% of the seedlings had flowered when less than 4 years old from seed. Limited data suggested that cultivars Empress, Khadrawy, and Thoory females and Khadrawy BC3, Tadala BC1, and Thoory BC3 males induced early flowering in a high proportion of the crosses in which they were used (Ream, 1975; Carpenter and Ream, 1976).

In Morocco, bayoud has raged since the end of the 19th century. This disease has exerted strong natural selection pressure for resistance to bayoud, sometimes to the detriment of fruit quality. It generates inevitably, in palm groves, a genetic impoverishment of quality genes. Therefore, several cultivars of good quality have disappeared (Pereau-Leroy, 1958) and others are threatened. The investigations conducted on crossbreeding have achieved the following results (Saaidi, 1990):

- Resistance tests in the nursery and in the field highlight the relationship between the resistance of the genotypes used as parents and the resistance of their progeny. The absence of a dominant resistance in the homozygous state in the resistant parents constitutes a certain guarantee against the possible risks of mutation of the parasite (Foa). The resistance does not appear to be linked to sex as both males and females transmitted resistance to their progeny.
- The quality of the fruits constitutes an inheritable characteristic, but in much lower proportions than resistance. Segregation in this case is very important. Some female individuals producing good-quality dates were found in all cross categories, but at different frequencies.
- The Deglet Noor variety, much appreciated on the international market, seems to constitute a good parent since it can transmit its fruit quality, the vegetative characteristics of the tree, and probably even some minor recessive resistance genes.

- A total of 15 hybrids of good fruit quality and not showing symptoms of bayoud disease have been selected and proposed for the reconstitution of the palm groves as well as for the creation of new palm groves.

Further evaluation of these controlled crosses (Zaher and Sedra, 1998) during the 1990s additionally showed that:

- Only 5% among crossbreeding descendants produced fruits of acceptable quality equivalent to Boufeggous variety fruits.
- The fruit quality of the mother plant did not influence the frequency of progeny producing fruits of good quality.
- Some male palms transmitted good fruit quality in some crossbreeding cases.
- Some crossbreeding (e.g. female Jihel × resistant male) produced 10% resistant progeny with good fruit quality. This means that males transmitted resistance and females transmitted fruit quality.
- Some other crossbreeding (e.g. female Boufeggous × resistant male) gave a high frequency of susceptible progeny. Hence, females do contribute towards resistance transmission.

4.4.3   Advantages and disadvantages of crossbreeding

Although crossbreeding can make certain crosses between genotypes that can never occur naturally, this technique has some limitations. Indeed, it is time-consuming, requiring 30 years to complete three backcrosses due to the palm's long life cycle. Another constraint is gender identification, which is not possible before the onset of fruiting, 5–7 years after planting (Al-Khayri *et al.*, 2018). An additional drawback is the time required to produce enough offshoots for trials (5 years minimum or more if a large number is needed, which would entail several generations of offshoot production). Finally, date palms do not reach full production until they are 10–15 years of age. All these factors make date palm breeding a long-term project (Krueger, 2015).

4.5   Induced Mutagenesis

Natural genetic diversity is gradually eroded and consequently there is a loss of valuable genetic resources, which hinders genetic improvement of crops including date palm. Moreover, the rate of spontaneous mutations is extremely slow and makes the availability of wide genetic diversity more limited for plant breeders (Jain, 2010). Induced mutations hasten the rate of genetic diversity in a short period and are readily available to plant breeders for developing new cultivars. Mutation breeding can be described as the purposeful application of mutations in plant breeding (Carimi *et al.*, 2012). A mutation is a sudden heritable change in the DNA of a living cell that is not caused by genetic segregation or genetic recombination (van Harten, 1998). Mutations are generally induced by physical and chemical mutagenic treatments in seed and vegetatively propagated crops including the date palm. Mechanistically, mutations in plant DNA have the same effect on plant phenotype whether they result from natural or human-directed processes (Jain *et al.*, 2010). In both cases, gene activity can be altered by nucleotide substitution, the deletion or insertion of DNA sequences, or modification of cis regulation. Mutagenesis is a useful tool for plant breeding that can be made more precise when combined with molecular technology and bioinformatics. Alternatively, mutagenesis can be conducted with single cells or small cell clusters that are competent for plant regeneration (Wilde, 2015). The Food and Agriculture Organization of the United Nations/International Atomic Energy Agency (FAO/IAEA) Mutant Varieties Database collects information on plant mutant varieties (cultivars) released officially or commercially worldwide (https://www. iaea.org/resources/databases/mutant-varieties-database   (accessed   14 November 2022)). It contains more than 3200 plant varieties developed from induced mutations (Jain *et al.*, 2010). Thus, induced mutations can provide a useful alternative or complement to natural variation as well as to hybridization (Chakraborty and Paul, 2013).

4.5.1 Methods

Gamma irradiation breaks DNA into small fragments and then the DNA starts a repair mechanism. During this second step, new variations develop, or mutations occur. Mutation induction in date palm is feasible now due to a reliable plant regeneration system via somatic embryogenesis and organogenesis. The optimal mutagen dose is determined empirically to minimize the impact on growth and fertility. Hence, the most important subject of *in vitro* mutagenesis is to select a suitable radiation dose (LD_{50}) to obtain the maximum viability while the regenerative capacities of irradiated plant material are preserved (Jain, 2013).

In date palm, induced mutagenesis has been undertaken in the framework of FAO/IAEA Regional Technical Cooperation projects including Morocco, Algeria, and Tunisia, particularly:

- control of bayoud disease in date palm: RAF/5/035(1995–2001); and
- field evaluation of bayoud disease resistance of date palm mutants: RAF/5/049(2001–2006).

The major components of these projects were: *in vitro* culture, radiation-induced mutations, screening of mutants and their field evaluation, and molecular techniques (random amplified polymorphic DNA (RAPD), restriction fragment length polymorphism (RFLP)).

In the framework of these projects, the following methodology was implemented (Bougerfaoui *et al.*, 2006):

- determination of efficient radiation dose for date palm tissue culture-derived plant material (gamma radiation doses tested: 0, 15, 30, 45, 60 Gy);
- use of a suitable dose for irradiation of sufficient date palm plant material;
- multiplication of irradiated plant material, regeneration of complete plantlets, and acclimation under controlled conditions;
- determination of discriminant dose of Foa toxins between susceptible and resistant cultivars (Foa toxin doses tested: 0, 50, 75, 100, 150 µg/ml);
- screening of produced plant material and evaluation of tolerance/resistance by using Foa toxins in comparison to controls of susceptible and resistant plant material;
- field transfer of putative mutants for further evaluation in hotspots of Foa-infected fields and final valuation of bayoud resistance; and
- evaluation of the resulting plant material for other traits such as dwarfism, early flowering, and fruit quality.

4.5.2 Results

4.5.2.1 Determination of optimal radiation dose

Results from testing of numerous gamma radiation doses on date palm plant material in three Moroccan date palm varieties (Boufeggous, Saïr-Layalate, INRA-16-bis) showed that 15 Gy dose had no negative effects on date palm tissues and their multiplication rate was similar to that of non-irradiated tissues. However, 45 and 60 Gy doses drastically affected the capability of plant material to regenerate new tissues and the incidence of tissue browning was increased. Moreover, plantlet regeneration from irradiated material was very low beyond 30 Gy dose. So, 15 Gy was retained as the optimal dose for irradiation of date palm tissues at the beginning of the project. Further comparison of 20 and 30 Gy doses led to adopting 20 Gy as the recommended dose for irradiation of date palm buds (Bougerfaoui *et al.*, 2006). For the Deglet Noor date palm variety, somatic embryogenic cell suspension cultures were used in Tunisia for induced mutagenesis. The radiation dose of 20 Gy was used for mutation induction (Jain, 2012).

During the duration of the FAO/IAEA project, vegetative buds of the Boufeggous variety were regularly subjected to 20 Gy of gamma radiation at the Tangier irradiation station, Morocco. After four multiplication cycles, the buds were individually separated and transplanted on media that promoted their elongation, rooting, and development into complete plantlets. The resulting plants with two or three leaves and a good root system were transferred to a glasshouse for acclimation under controlled conditions.

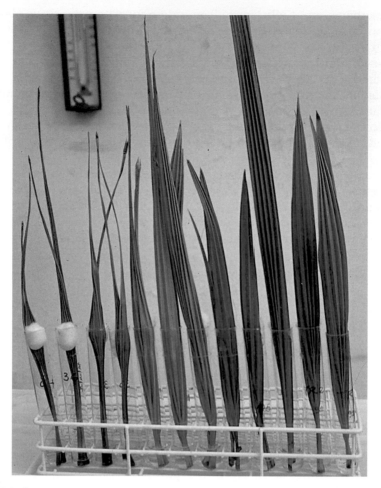

Fig. 4.3. Screening of irradiated-derived plant material against bayoud toxins. (Used with permission from Larbi Abahmane.)

4.5.2.2   Determination of discriminating dose of Foa toxins

To determine the discriminating dose of Foa toxins between susceptible and resistant cultivars, detached leaves (25–35 cm length) excised from susceptible (Boufeggous) and resistant (Saïr-Layalate) date palm varieties were used. After 1 week of incubation, leaf-browning severity was noted to increase with increasing toxin dose (Fig. 4.3). Leaves treated with 50 and 75 µg/ml toxin doses did not show any symptoms. However, using 100 µg/ml dose, symptoms of petiole browning and leaf dryness appeared only in the susceptible cultivar. The 150 µg/ml dose caused browning in both the resistant and susceptible cultivars. Hence, 100 µg/ml toxin dose was adopted as being the discriminating dose against bayoud toxins. This dose

Fig. 4.4. Selected resistant mutants screened against bayoud toxins before field transfer. (From Jain, 2012; published in *Emirates Journal of Food and Agriculture* under the terms of the Creative Commons Attribution 4.0 license.)

was later used for evaluation of plantlets produced from irradiated plant material (Bougerfaoui *et al.*, 2006).

4.5.2.3   Evaluation of irradiated plant material

Collected data indicated that 81% of plantlets produced from irradiated plant material showed similar symptoms to the susceptible control. About 14% showed a much higher level of tolerance to toxins and 5% showed a level of tolerance comparable to that of the resistant control Saïr-Layalate (Bougerfaoui *et al.*, 2006). Overall, ten mutant plants were identified in the greenhouse (Fig. 4.4). The selected mutant plants were later transferred to the 'hotspot' in the field infested with bayoud disease. All of them are maintained to date in the experimental station of Zagora in Morocco. The performance of these mutant plants is under evaluation for other agronomic traits.

4.5.3   Advantages and disadvantages of induced mutagenesis

Mutation induction has become a proven way to create variation in plants and induce desired traits not found in nature or those lost during evolution. Moreover, induced mutagenesis enlarges genetic diversity, which is fundamental to date palm breeding (Al-Khayri *et al.*, 2018). However, there are also limitations of mutation breeding to be considered. One issue is that large mutant populations must be generated due to the random nature of mutation induction. Theoretical calculations of the population size needed to find induced mutations in diploid plants suggest that tens of thousands of M1 lines are necessary (Roychowdhury and Tah, 2013). Producing and maintaining such numbers of line progenies can sometimes be impractical for species that are large or long-lived. Therefore, genetic screening could help to focus efforts on lines with functional mutations. Indeed, early line selection, based on genetic screening for candidate gene mutations, can be used to increase the efficiency of mutation breeding (Wilde, 2015). One more drawback of mutation breeding is the undesirable effect produced by the pleiotropic action of the mutant gene or simultaneous mutation of closely linked genes. In addition, most of the induced mutations can be deleterious, but when an appropriate selection technique is applied, useful mutants can be recovered (Chakraborty and Paul, 2013).

4.6 Interspecies Crossbreeding

Interspecific hybridization is the process whereby two distinct species are crossed, creating hybrid individuals. Gene flow, i.e. the incorporation of a gene from one species to the gene pool of another, is called introgression and is the result of interspecific hybridization (Gros-Balthazard, 2013). *Phoenix* species are interfertile and crossing distinct species leads to fertile hybrid offspring. In cultivation, such crossings may be spontaneous or the result of artificial pollination. Crossing the date palm with other *Phoenix* is of great interest for its cultivation. Indeed, because of metaxenia effects, the selection of pollen from other species to pollinate females could improve yield, fruit size, and even produce seedless fruits. More research and experiments in metaxenia are necessary to assess the effect of different male genotypes and understand the basis of these effects. This could lead to the selection of male cultivars specific to the pollination of given female cultivars. Moreover, other *Phoenix* species appear to be genetic reservoirs and hybrid cultivars could be of great interest, in terms of cultivation and disease resistance, and therefore more research should be carried out in this area (Gros-Balthazard, 2013).

4.6.1 Interspecies crosses in *Phoenix*

The cultivated date palm is closely related to a variable aggregate of wild and feral palms distributed over the southern, warm, and dry Middle East as well as the North-eastern Saharan and North Arabian deserts. These spontaneous palms show close morphological similarities and parallel climatic requirements with the cultivated clones. In addition, they are interfertile with the cultivars and are interconnected with them through occasional hybridization (Jaradat, 2011; Flowers *et al.*, 2019). In Senegal, hybrids between the sub-Saharan *Phoenix reclinata* and the date palm have been reported (Munier, 1973). In the Canary Islands, interspecific hybridization between the endemic *Phoenix canariensis* and the widespread *P. dactylifera* has been confirmed (Gonzalez-Perez *et al.*, 2004; Gros-Balthazard *et al.*, 2018).

Investigation of the evolutionary history of *P. dactylifera* and its wild relatives by sequencing the genomes of date palm varieties and five of its closest relatives indicates that the North African population has mixed ancestry with components from Middle Eastern *P. dactylifera* and *Phoenix theophrasti*, a wild relative endemic to the Eastern Mediterranean. Introgressive hybridization is supported by tests of admixture, reduced subdivision between North African date palms and *P. theophrasti*, sharing of haplotypes in introgressed regions, and a population model that incorporates gene flow between these populations. Analysis of ancestry proportions indicates that as much as 18% of the genome of North African varieties can be traced to *P. theophrasti* and a large percentage of loci in this population are segregating for single-nucleotide polymorphisms (SNPs) that are fixed in *P. theophrasti* and absent from date palms in the Middle East. Results from previous investigations suggest that hybridization with *P. theophrasti* was of central importance in the diversification history of the cultivated date palm (Flowers *et al.*, 2019).

4.6.2 Controlled interspecies crossbreeding

Date palm breeding can be achieved using either conventional breeding or biotechnology techniques. Due to the high genetic diversity in date palm, it is easy to create genetic variability in the new generation. A large number of varieties has been developed through natural crossbreeding. However, the process is very slow and takes more than 30 to 40 years to establish a new improved cultivar for release to the growers. Creation of new variants through genetic engineering is also at its infancy in date palm. Crossbreeding integrated with *in vitro* techniques is more achievable and reduces the time required for crop improvement (Sudhersan *et al.*, 2009). All the initial date palm breeding trials were based on fruit quality and resistance to pests and diseases. However, in date palm cultivation, tree height is one of the major constraints to good-quality date production. Pruning, pollination, fruit thinning, bunch removal, and fruit picking contribute to the cost of date production, particularly when the trees grow taller. Moreover, frequent climbing for pollination and fruit picking is highly dangerous in the case of taller old trees. In order to develop dwarf date palms, a dwarf species, *Phoenix pusilla*, was crossed with selected cultivars of female date palms at the Kuwait Institute for Scientific Research (Sudhersan *et al.*, 2009).

4.6.3 Methods

Female date palm cultivars Barhi, Mejhoul, and Sultana were pollinated with male pollen of the dwarf palm (*P. pusilla*). The unopened female flowers of the selected date palm cultivars were opened with a surgical knife and the dwarf palm pollen was dusted over the female flowers and covered immediately with paper to avoid pollen mixing. Fruits were carefully observed during their different stages of development. Seed development was also observed frequently by opening the fruit at different stages (Sudhersan and Al-Shayji, 2011).

4.6.4   Results

The main results from the interspecific crossing carried out were as follows (Al-Sabah *et al.*, 2018):

- The crossing between date palms and the dwarf palm was successful and fruit set was observed. Interestingly, the pollen from the dwarf palm affected the fruit development and morphology at the khalal and tamar stages. Hence, the size of fruit in Barhi was smaller than the fruit size attained by normal date palm pollen, while in the other two cultivars, Mejhoul and Sultana, fruits were larger in size than normal fruits. Seed development was noted early but was later arrested due to less endosperm development. In the early stages, seeds showed embryo development, but the embryos were aborted at the final stage due to the total arrest in endosperm development.
- The seeds from different stages kimri, khalal, rutab, and tamar placed on MS (Murashige and Skoog, 1962) basal medium showed different responses according to their stage. The seeds collected from kimri stage fruit swelled 100%, while those collected at the other stages failed to swell. After 2 weeks, the embryo came out of the seedcoat. The highest percentage of embryo germination was obtained in the case of the Mejhoul hybrid. All hybrid plantlets rescued from the embryos produced adventitious roots and elongated to about 15 cm height after 30 days in MS medium. All the plantlets were acclimatized and 100% survived the acclimation procedure.
- The hardened hybrid of the cultivar Sultana was planted in the field for observation. The plant showed normal growth and development in the field.
- After 4 years of field growth, the hybrids started producing flowers. Some of them were males and others were females. The female flowers were pollinated using date palm pollen and fruits were developed. The new interspecific hybrid date palm fruits were entirely different from the mother date palm in fruit color, fruit shape, and size. The hybrid palms were taller than the male parent and shorter than the female parent. The field evaluation, yield characteristic features, and fruit quality analysis are not yet completed and are ongoing in Kuwait (Al-Sabah *et al.*, 2018).
- This work was the first successful trial carried out to develop hybrids between normal date palms and the dwarf species *P. pusilla.*

Other attempts for interspecific crosses have been reported in the literature. Crossing of *P. reclinata*, *P. canariensis*, *Phoenix sylvestris*, *Phoenix roebelenii*
and *Phoenix rupicola* with date palm aimed to produce fruits of best quality.
However, only the cross between the date palm and *P. sylvestris* produced
slightly larger fruits than the normal (Nixon, 1935).

4.7   *In Vitro* Culture

Plant tissue culture (PTC) plays a pivotal role in agriculture and plant
breeding as it complements crop production through micropropagation,
somaclonal variation, hybridization, cybridization, synthetic seed production, haploid culture, hairy root culture, preservation of germplasm,
pathogen eradication, etc. (Yeole *et al.*, 2016; Tazeb, 2017). With the growth
of agriculture and plantation crops, the demand for high-yielding, high-
quality, disease-free planting stocks, including date palms, has increased
in the last 20 years. Tissue culture has been successfully used to meet the
high demand for these elite stocks of plants. Somatic embryogenesis can be
used for large-scale propagation (Othmani *et al.*, 2009), producing artificial
seeds (Pintos *et al.*, 2008), gene transfer (Li *et al.*, 2002), *in vitro* selection
for various biotic and abiotic stresses (Ahmed *et al.*, 2021), providing potential models for studying molecular, regulatory, and morphogenetic events
during plant embryogenesis (Kamle *et al.*, 2011; Zein El Din *et al.*, 2021),
and for cryopreservation of plant material (Fki *et al.*, 2017).

Together with molecular techniques, PTC significantly contributed to
altering genetic traits through gene transfer (Brown and Thorpe, 1995).
Despite being at its preliminary stage, currently nanotechnology is being
utilized in tissue culture and heralds a new phase in this field (Gulzar
et al., 2020). Moreover, genetic engineering would not be possible without
the involvement of *in vitro* PTC regeneration procedures (Bridgen *et al.*,
2018). Genetically transformed cells in all cases have been regenerated
into full plants via PTC. In addition, development of distant hybrids and
cybrids, which were not possible easily through the conventional breeding
methods, has successfully been done by using *in vitro* techniques.

4.7.1   Micropropagation

The date palm is the dominant component in the oasis agroecosystem. The
tremendous advantage of the date palm tree is its resilience, its long-term
productivity, and its multi-purpose attributes. However, some of its unique
characteristics (slow growth, dioecy, the slow offshoot-based propagation,
difficulty of predicting adult characteristics of the seedlings) have severely
restricted its improvement. Vegetative propagation by offshoots is the only
conventional method to maintain genetic integrity of date palm cultivars.
Due to its slow growth and limited number of offshoots, PTC remains the

ideal means for propagating and disseminating selected cultivars among growers (Abahmane, 2011; Jaradat, 2011). The application of tissue culture techniques to date palm has many advantages: propagation of healthy selected female cultivars, large-scale multiplication, avoiding seasonal effects, production of genetically uniform plants, fast exchange of plant material, and economic profitability. Looking at these advantages, micropropagation techniques have been developed for rapid mass propagation of date palm (Reddy, 2015; Abahmane, 2017). Furthermore, the use of bioreactors in micropropagation of date palm can hasten multiplication and regeneration of well-formed plantlets from either selected or improved genotypes (Abahmane, 2020b). The use of bioreactors in date palm resulted in an improved multiplication rate and reduced micropropagation time. It also reduces the cost of saleable units and thus improves economic return for commercial micropropagation (Fki *et al.*, 2011; Almusawi *et al.*, 2017).

4.7.2 Hybridization and embryo rescue

The most important and critical thing in any crop improvement program is the act of bringing the vital and beneficial traits together in a crop of interest. This is done either by genetic transformation or by hybridization. Mostly, a single gene is preferred for transfer by genetic engineering. To transfer more than one gene of interest, hybridization is preferred. Tissue culture has been key in aiding hybridization when the embryo aborts and is unable to establish the plant. Embryo rescue has been used successfully to overcome embryo abortion or lack of seed development (Tazeb, 2017). The term 'embryo rescue' is used to describe the *in vitro* techniques aiming to encourage the development of immature embryos into complete plants (Zulkarnain *et al.*, 2015). Embryo rescue is a non-genetic modification (GM) biotechnological *in vitro* technique that has become a tool for breeders and cytogeneticists to achieve interspecific and intergeneric hybridization where embryo abortion, degeneration, self-, and cross-incompatibility are commonly witnessed to occur after hybridization, leading to drastic failure in transferring of desirable genes from wild species to elite cultivars (Pramanik *et al.*, 2021). Indeed, distant crosses are often sought to transfer genetic traits from secondary and tertiary gene pools to the cultivated, primary gene pool of crop plants (Sahijram and Rao, 2015). The embryo rescue technique was successfully used to surmount embryo abortion and hybrid plants (*P. dactylifera* × dwarf palm (*P. pusilla*)) were produced for the first time with the objective of obtaining date palms with reduced height (Al-Sabah *et al.*, 2018).

4.7.3 Genetic transformation and somaclonal variation

Many genetically modified plants have been developed in the last two decades by genetic engineering techniques (Bawa and Anilakumar, 2013).

After the successful transformation, the engineered cells have been recovered and regenerated *in vitro* into complete plants. Tissue culture has been playing a very significant role in regeneration, mass multiplication, and propagation of whole plants from transformed or genetically modified cells or tissues. This is one of the most important steps in tissue culture-based genetic engineering programs (Gulzar *et al.*, 2020).

Genetic variability has a huge role in the success of any plant breeding program. With the advent of new technologies such as PTC and recombinant DNA technology, a lot of progress has been made in increasing food production. Indeed, somaclonal variation is a natural phenomenon occurring during *in vitro* tissue culturing and can produce useful genetic variations in plants (Rajan and Singh, 2021). It includes DNA-related genetic or epigenetic variations, which induce phenotypic changes distinguishable from the original parent (Sattar *et al.*, 2021a). Major causes include, but are not limited to, prolonged *in vitro* culturing, media composition, and the presence of plant growth regulators and certain other mechanical factors during culturing (Ranghoo-Sanmukhiya, 2021). These variations may include DNA methylation, rearrangement of chromosomes, and/or point mutations (Krishna *et al.*, 2016). Somaclonal variations lead to various genetic changes in plants that can result in changes in various plant characteristics such as plant height, yield, and fruit quality, and provide resistance against diseases, pests, and drought (Patnaik *et al.*, 1999). Somaclones can be detected through morphological assessments of the off-type regenerants, biochemical response of explants, fingerprinting with protein- or isozyme-based markers, and cytogenetic assessment (Cevallos-Cevallos *et al.*, 2018). In addition, more advanced DNA or transposon-based molecular markers (Henao-Ramírez *et al.*, 2021) and next-generation sequencing (NGS) screening have also been successfully applied to detect somaclonal variation in fruit tree breeding (Sattar *et al.*, 2021a).

Date palm micropropagation has largely been carried out by using different plant parts as the explant sources (Abahmane, 2013, 2020a). However, during plant regeneration *in vitro* culture, unsolicited somaclonal variation may occur. Despite these somaclonal variations, tissue culturing has been extensively used to enhance date palm propagation (Al-Khateeb *et al.*, 2019). The undesirable reported phenotypes resulting from somaclonal variation include: plant dwarfism (Al-Wasel, 2001), abnormal growth of leaves or fruit strands and inability to form inflorescences (McCubbin *et al.*, 2004), leaf albinism and dryness of apical bud (Al-Khateeb, 2008), changes in fruit quality and delayed flowering time (Al-Khateeb and Ali-Dinar, 2002), production of deformed offshoots and twisted inflorescences (Zaid and Al-Kaabi, 2003), production of abnormal parthenocarpic fruits (Cohen *et al.*, 2004), and excessive vegetative growth and plants with poor establishment (Abutalebi, 2010). The screening of such deformations and their molecular nature using a simple, yet reproducible marker system is important to optimize tissue culture protocols and for the selection of true-to-type clonal materials for field planting (Mirani *et al.*, 2020). Somaclonal variation, even if it is a burden for the propagation of true-to-type clones for commercial plantations, has a specific interest in nonconventional date palm breeding. This variation, if selected for, could lead to the creation of new varieties with traits affecting nutrient uptake and overall agronomic performance, dwarfism, the characteristics and arrangement of leaves on the stem, fertilization proprieties, tolerance/resistance to various stresses, as well as yield, fruit size, shape, and texture (El Hadrami *et al.*, 2011).

4.7.4   *In vitro* selection

In vitro selection represents a useful biotechnology tool in date palm breeding. Depending on the selective agent, *in vitro* selection can be conducted using regenerative and embryogenic calli, cell suspensions, zygotic and somatic rescued embryos, fused protoplasts (Yatta-El Djouzi *et al.*, 2020), and cybrids, but also at later stages during the regeneration of shoot and root meristems. The method of choice often depends on the advanced control of the micropropagation technique as well as the ease of application and the efficiency of the selective agent in inducing high levels of variation. The regeneration method of cells is also important to preserve the inheritance of the desired trait (El Hadrami *et al.*, 2011). *In vitro* selection can considerably shorten the time needed for the selection of desirable traits, suffers minimal influence from exterior environmental conditions, and can precede and complement field selections. *In vitro*-selected putative variants should be tested in the field to confirm the genetic stability of the selected trait (Jain, 2001) and genetically stable somaclones or mutants can then be used directly as elite varieties or introduced into plant breeding programs. Moreover, *in vitro* selection of somaclonal variants and induced mutants is not protected under intellectual property regulations, nor it is subject to public safety concerns that currently hamper transgenic (i.e. genetically modified organism (GMO)) approaches for the development of new crop cultivars. Thus, *in vitro* selection is a promising, non-transgenic approach, which offers an attractive alternative method for producing improved cultivars (Jayasankar *et al.*, 2003; Lebeda and Svabova, 2010). Recently, *in vitro* selection has been reported in date palm improvement. Studying various salt levels on somatic embryogenic calli (cv. Sukary), Al-Khateeb *et al.* (2019) reported that obtained somaclones showed positive morphological and physiological responses to the high salinity levels. Nevertheless, this response seemed unrelated to the genetic changes incurred during salt stress. Besides major changes at the chromosomal level or the duplication and deletion of specific regions, abiotic stress tolerance in the somaclones may also arise due to base changes in the respective genes. Further studies to detect such point mutations in stress-related genes through exome capture sequencing or whole-genome sequencing may provide better insight into the *in vitro* stress response of somaclones in the future. It may also help to devise better strategies against salt stress in date palm (Al-Khateeb *et al.*, 2019).

4.7.5   Conservation of date palm germplasm

In vitro technology offers a potential solution for the conservation of date palm germplasm. Slow growth induced by low temperature allows storage from several months up to a few years. *In vitro* conservation techniques have many pros compared to *in vivo* conservation as they lessen the space and time required and allow conservation of endangered plant species (Engelmann, 2011). In cold storage, 70% of shoot buds remain healthy after storing for 12 months at 5°C, and callus cultures remain fully viable after 12 months of storage (Bekheet, 2017). Furthermore, cryopreservation is suitable for long-term *in vitro* conservation. Indeed, this technique is one of the cost-effective and long-term storage methods for germplasm, in which biological material is stored at a very low temperature (−196°C). At this temperature, all the metabolic activities of cells cease, which helps in storing the material for a longer time without any alterations or modifications (Gulzar *et al.*, 2020). Protocols and suitable conditions for date palm cryopreservation are fully described in the literature (Bekheet, 2017; Fki *et al.*, 2017). Moreover, many of the risks linked to field conservation, like erosion due to climatic, edaphic, and biotic constraints, are circumvented by cryopreservation.

4.7.6   Advantages and disadvantages of *in vitro* culture

Despite numerous applications of *in vitro* techniques in plant breeding and production, there are many limitations that need to be addressed to make the technique more efficient, particularly where the response is poor. Undesirable somaclonal variation coupled with unsuccessful regeneration from androgenesis and gynogenesis are among the main drawbacks of PTC in date palm. However, integration of PTC techniques and other advanced biotechnological tools with classical plant breeding and plant improvement is well within reality (Gulzar *et al.*, 2020).

4.8   Genetic Engineering

With climate change predicted to alter conditions in various parts of the world, researchers have started developing models to forecast adverse effects of various stress factors on crop productivity. Although conventional strategies like breeding for resistant varieties and agrochemicals and biocontrol agents for control of diseases and pests have been in use for a long time, they have met with limited success (Sree and Rajam, 2015).

In addition, increasing desertification and decreasing water resources pose serious threats to agricultural biodiversity. Presently, one of the most effective means to develop new cultivars is to identify inherent genetic resistance traits in cultivars and interbreed these with cultivars with other desired traits (El Modafar, 2010). As date palm trees have long generation times of 6 years or more, it is of paramount importance to identify cultivars with desirable features early on, using molecular and genetic signatures to aid the selection process (Thareja *et al.*, 2018). Molecular tools now permit monitoring the dynamics of genomic recombination, making possible a gene-by-gene breeding approach. The most modern methods will be necessary for going back to what remains of the ancient material to find the genes that will be the building blocks of the cultivars that will solve the problems of the 21st century (Breseghello and Coelho, 2013). Like other plants, the date palm genome is accompanied by *SNP deserts*, which contain important genes for abiotic and biotic stress resistance. Targeting these SNP deserts can be used to regulate tolerance to these stresses in date palm (Al-Mssallem *et al.*, 2013).

Generally, date palm is considered a halophytic species that tolerates electrical conductivity (EC) of 10 mS/cm. Nevertheless, both date quality and yield are affected by soil salinity and water stress. Due to the serious threat of global warming, these problems are expected to be intensified in the date-palm-growing regions (Jain, 2012). There is a dire need to improve the adaptability, productivity, and robustness of date palms cultivated under stressful environments (Sattar *et al.*, 2021a). For these reasons, breeding efforts geared toward the improvement of tolerance to abiotic stress conditions are of paramount importance for sustainable date palm production (Al-Khayri and Ibraheem, 2014).

During the last 10 years, technological advancements in genetic engineering have led to the development of transgenic crop varieties resistant to various biotic stresses (Sree and Rajam, 2015). Genome editing (GE) is one of the most important tools in modern biology and comprises technologies that enable scientists to change/edit an organism's DNA. These technologies rely on nucleases to cut specific genomic target sequences. This process can be used to generate direct modifications in plants and is a potential tool for targeted GE (Tyagi *et al.*, 2020). A number of engineered nucleases such as homing endonucleases, zinc-finger nucleases (ZFNs), transcription activator-like effector nucleases (TALENs), oligonucleotide-directed mutagenesis (ODM), and clustered regularly interspaced short palindromic repeats (CRISPR) nucleases have been identified that make targeted breaks to activate gene conversion (Bisht *et al.*, 2019). The latest revolutionary technology for GE is based on the RNA-guided engineered nucleases called CRISPR/CRISPR-associated protein 9 (CRISPR/Cas9), which holds great promise because of its specificity, simplicity, efficiency, and versatility (Vats *et al.*, 2019; Tyagi *et al.*, 2020). Several important advantages of CRISPR/Cas9-based approaches such as ease in functioning, multiplexing of several genes, cost-effectiveness, and robustness make it a reliable technique in a breeder's arsenal. Most importantly, the use of CRISPR/Cas9-based approaches may bypass the current GMO regulations for transgenic plants due to absence of exogenous genetic elements during its application (Woo *et al.*, 2015; Sattar *et al.*, 2020). The versatile CRISPR/Cas9-based approaches have been successfully engineered to target plant genomes to improve multiple genetic traits (Puchta, 2017) and to configure resistance against plant viruses (Iqbal *et al.*, 2016), fungal diseases including rice blast and powdery mildew (Kanda *et al.*, 2017), and bacterial pathogens such as Citrus canker (Peng *et al.*, 2017). Modern GE tools have advanced from biomedical sciences to modern agriculture. Under these circumstances, CRISPR/Cas9-based GE tools can pave the way toward innovations in date palm breeding. This technique is versatile in single base editing (BE), multiplexing to target gene families, gene knockout or knock-in, gene transfer/replacement, epigenetic modifications, DNA barcoding, genotyping, pathogen profiling, and many other expanding applications beyond GE (Sattar *et al.*, 2021b).

Lessons from genetic engineering of African oil palm (*Elaeis guineensis*) may help with developing strategies for *P. dactylifera*. Both micro-projectile bombardment (Kadir *et al.*, 2015) and *Agrobacterium* (Budiani *et al.*, 2018) have been used to transform oil palm. Moreover, CRISPR/Cas9 technology has been used successfully in oil palm breeding. Indeed, Budiani *et al.* (2018) used *Agrobacterium* to introduce the CRISPR/Cas9 constructs for editing isoflavone reductase and metallothionein-like protein in an effort to introduce resistance to *Ganoderma*. Given the success in other crops, CRISPR/Cas9 will soon provide a means for creating stable, site-directed gene edits in date palm and may provide the best chance at modification of date palm characteristics (Hazzouri *et al.*, 2020).

Recently, Yaish (2015) reported that date palm seedlings accumulate proline not only in response to drought and salinity stress, but also in response to extreme temperatures and abscisic acid treatments. They concluded that proline production is a common response for multiple stressors, which makes it a possible marker in date palm breeding programs that aim to improve drought and salt tolerance. There are presently few -omics studies of drought in date palm. A recent proteomics study identified genes involved in salt and drought tolerance in *P. dactylifera* (Rabey *et al.*, 2016). Safronov *et al.* (2017) used transcriptomic and metabolomic profiling to characterize the response to heat and drought stress in *P. dactylifera*. The two stresses had similar effects including the upregulation of soluble carbohydrates and increased antioxidant activity in the cytosol, chloroplasts, and peroxisomes. Then, researchers described how genes discovered by genetic mapping can be targeted by gene editing (e.g. by Cas9) toward the goal of engineering varieties with improved stress tolerance.

The generation of genetic variability through irradiated mutagenesis has been successfully used to create variations against various abiotic stresses and bayoud disease in date palm. However, the major limitation of random mutagenesis is the overloaded mutation background (Braatz *et al.*, 2017). Furthermore, the use of classical ways of genetic mutation is not always suitable to generate gene knockouts. Under such circumstances, CRISPR/ Cas9-based genetic approaches can help to study the gene expression of individual genes or multiple genes in the complex date palm genome through site-specific mutagenesis and multiplexing, respectively. Moreover, this technique can also be used to study pathogen effector proteins, generation of secondary metabolites during fruit development, and various quantitative and qualitative traits in date palm. Despite the availability of various genetic markers, determination of male and female individual plants exactly in the date palm progeny is very tedious until maturity. The dioecious status of each plant can be determined exactly by employing CRISPR/Cas9 gene-knocking ability to identify sex-linked genetic markers. Moreover, it can be a throughput tool in date palm functional genomics to improve the genetic background by enhancing the traits related to yield, plant architecture, nutrient uptake, disease resistance, and plant adaptation (Sattar *et al.*, 2020).

To decrease the time for analysis of the expression of any gene of interest, transient expression of foreign genes in plant tissues is a valuable tool for plant transformation technology. An efficient transformation protocol for gene delivery in date palm was recently reported (Solliman, 2017). This method, called fruit agro-injection, utilizes *Agrobacterium tumefaciens* strain LBA4404 harboring binary vector pRI201-AN-GUS carrying the β-glucuronidase (GUS) gene. This protocol has proven the efficiency of this reliable technology as a tool for transgene expression in date palm. In addition, *Agrobacterium*-mediated genetic transformation was successfully carried out in date palm, cv. Khalasah, using mature somatic embryos (Aslam *et al.*, 2015). For genetic transformation, morphologically advanced mature somatic embryos developed on MS medium were co-cultured with *A. tumefaciens* strain LBA4404 harboring binary vector pBI121, containing *uidA* (GUS) and *nptII* (neomycin phosphotransferase gene) genes, and incubated for 4 days and later inoculated on germinating and plantlet conversion MS medium. Prolific shoots developed from putatively transformed mature embryos showed 47.5% transformation efficiency. A large number of transgenic plants was obtained and later established in plastic bags. A strong GUS activity was detected in the putatively transformed plant leaves by histochemical assay, and the integration of *uid*A (GUS) and *nptII* genes into transgenic plants was confirmed by polymerase chain reaction (PCR) and Southern hybridization analysis. Moreover, Mousavi *et al.* (2014) described an efficient transformation system for gene delivery in date palm. The effects of different physical and biological parameters were optimized for transient transformation of *uidA* gene in somatic embryos of the Estamaran cultivar. The tissues were bombarded with constructs harboring the *uidA* gene driven by CaMV 35S or rice Act1 promoter. Regenerated plantlets were checked by PCR using gene-specific primers. About 16% of the plantlets were reported to be stably transformed.

Currently, advanced -omics and related biotechnological approaches are effectively assisting traditional plant breeding through the use of NGS, molecular markers, genomics, proteomics, transcriptomics, and GE tools (Sattar *et al.*, 2017). The sustainability of date palm cultivation could possibly be attained through genetic modification of the key genes of the existing elite date palm cultivars having a role against various biotic and abiotic stresses (Al-Khateeb *et al.*, 2019) or by employing genetically engineered resistance against various pests and diseases. Even if date palm has long been an ignored crop for the application of modern approaches, the modern -omic tools coupled with NGS have progressed in date palm genomics during the current decade (Sattar *et al.*, 2021c).

4.9   Molecular Marker-Assisted Selection and Breeding

Although significant progresses have been made in crop improvement through phenotypic selections for agronomically important traits, considerable difficulties are often encountered during this process. A new variety in conventional breeding could take more than 8 to 10 years to be developed. Breeders are very interested in new technologies to speed up this process or make it more efficient. The development of molecular markers was therefore greeted with great enthusiasm as it was seen as a major breakthrough promising to overcome this key limitation. MAS is likely to become more valuable as a larger number of genes is identified and their functions and interactions are elucidated (Lema, 2018). Recently, new strategies based on MAS have been used to reduce time and effort in plant breeding. Genetic markers are suitable entities that are associated with economically important traits and have been used by plant breeders as selection tools. However, to obtain the fingerprints of closely related varieties or to have some characteristic differences, it is essential to obtain a very high level of polymorphism (Kharb and Singh, 2020). Nevertheless, the advent of DNA marker technology has revolutionized crop breeding research as it has enabled the breeding of elite cultivars with targeted selection of desirable gene or gene combinations in breeding programs. DNA markers are considered better over traditional morphological- and protein-based markers because they are abundant, neutral, reliable, convenient to automate, and cost-effective. Over the years, DNA marker technology has matured from restriction-based to PCR-based to sequence-based and eventually to the sequence itself with the emergence of novel genome sequencing technologies (Kadirvel *et al.*, 2015).

In date palm, molecular markers are a good way to study genetic diversity, which plays an important role in plant improvement programs. Many molecular markers, such as RAPD, inter-simple sequence repeat (ISSR), simple sequence repeat (SSR), etc., have been used for several purposes. The most important of these purposes are molecular identification of date palm varieties, study of the genetic convergence between varieties, identification of varieties resistant to bayoud disease, and determination of true-to-type tissue culture-derived plants (Guettouchi, 2018). According to the targeted goals, different molecular marker types (Kharb and Singh, 2020) have been used in date palm as follows:

- *Cultivar identification:* RFLP (Corniquel and Mercier, 1994), amplified fragment length polymorphism (AFLP) (Diaz *et al.*, 2003; El-Khishin *et al.*, 2003; Khierallah *et al.*, 2011), SSR (Khierallah *et al.*, 2011; Ahmed *et al.*, 2021), and ISSR (Ayesh, 2017).
- *Somaclonal variation in tissue culture-derived plants:* AFLP (Saker *et al.*, 2006).
- *Genetic relationships and genetic diversity:* RAPD (Mirbahar *et al.*, 2014; Al-Khalifah and Shanavaskhan, 2017), SSR (Khierallah *et al.*, 2017), ISSR (Karim *et al.*, 2010), and SNP (Al-Dous *et al.*, 2011).
- *Collection and annotation of gene models:* Expressed sequence tag (EST) (Zhang *et al.*, 2012).
- *Sex determination:* AFLP, RAPD, conserved DNA-derived polymorphism (CDDP), intron-targeted amplified polymorphism (ITAP), start codon-targeted polymorphism (SCoT) (Atia *et al.*, 2017), and sequence-characterized amplified region (SCAR) (Dhawan *et al.*, 2013).

The combined use of the AFLP, RAPD, and RFLP techniques is costly in time and manpower, and thus precludes the analysis of large numbers of individuals. A concentration on using more markers and especially the most reliable markers (e.g. SSRs and SNPs) may alleviate these problems (Kharb and Singh, 2020; Ahmed *et al.*, 2021). Genetic studies using SSR markers were first carried out for analysis of the genetic diversity of *P. dactylifera* in Tunisia (Zehdi *et al.*, 2004), Sudan (Elshibli and Korpelainen, 2008), Oman (Al-Ruqaishi *et al.*, 2008), Qatar (Elmeer *et al.*, 2011; Elmeer and Mattat, 2015), Iraq (Khierallah *et al.*, 2011), Iran (Arabnezhad *et al.*, 2012), and the United Arab Emirates (Chaluvadi *et al.*, 2014). However, all these studies were based on a relatively small number of accessions centered on single countries, and consequently are not representative of overall date palm genetic diversity. A wider analysis of date palm diversity is required to unravel the genetic relationships between the geographic groups distributed in the Old World from west to east, and to identify the potential backgrounds of genetic diversity useful for breeding programs and for the selection of adaptability traits to biotic and abiotic stresses (Kharb and Singh, 2020). Recently, genetic diversity levels and population genetic structure were investigated through the genotyping of a collection of 295 date palm accessions ranging from Mauritania to Pakistan using a set of 18 SSR markers and a plastid minisatellite (Zehdi-Azouzi *et al.*, 2015). The results showed that date palm displayed high genetic diversity and that the genetic variation is geographically structured. Accordingly, date palm genotypes can be structured into two different gene pools: an eastern pool, consisting of accessions from Asia and Djibouti; and a western pool, consisting of accessions from Africa. The presence of admixed genotypes was noted, which points to gene flows between the eastern and western pools, mostly from east to west. This information can be used to establish core collections useful in breeding strategies for agronomically interesting traits, such as fruit quality or resistance to biotic and abiotic stresses, through genome-wide association study (GWAS) (Mathew *et al.*, 2015). Recently, by utilizing organellar genome sequencing, Mohamoud *et al.* (2019) identified two additional haplotypes representing subpopulations beyond the currently known North Africa/Arabian Gulf separation. According to the SNP analysis, these authors suggest that there were at least three major centers of date palm cultivation, two in the Arabian Gulf and one in North Africa. A fourth one that derived from one of the Arabian Gulf cultivars and includes the famous North African Deglet Noor and Egyptian Zaghlool cultivars was also postulated.

To provide useful genomic information for the date palm, Mokhtar *et al.* (2016) established the Date Palm Molecular Markers Database (http://dpmmd.easyomics.org/ (accessed 14 November 2022)). This database includes information on more than 3,611,400 DNA markers (SSR, SSR-SNPs, SNP markers), genetic linkage maps, KEGG maps, DNA barcodes, and all previously published date palm articles in PubMed-indexed journals from 1976 to 2017 (Kharb and Singh, 2020). Moreover, a research team from the genomics laboratory of Weill Cornell Medical College, Qatar, sequenced the whole date palm genome using the NGS approach, utilizing the Solexa (Illumina) sequencer based on a shotgun sequencing method. The date palm genome was made available in 2009 (https://qatar-weill.cornell.edu/research/research-highlights/date-palm-research-program/date-palm-draft-sequence (accessed 14 November 2022)). The genome size reported was 500 Mbp. Later, Al-Mssallem *et al.* (2013) reported the successful sequencing of the nuclear genome of date palm cv. Khalas using pyrosequencing. Based on their results, the genomic size is 605.4 Mbp, which consists of about 90% of the genome and 96% of its genes. Then, Mathew *et al.* (2014) reported the first genetic map for date palm and identification of a putative sex chromosome. They developed the genetic map from the date palm cv. Khalas, with the draft genome. From this draft genome, for the first time, 19% of sequence scaffolds were placed on to linkage groups. The comparison of linkage groups in the date palm showed remarkable long-range synteny to oil palm.

Recently, in order to provide a useful basis for assessing genetic diversity of Foa, the causal agent of date palm vascular wilt (bayoud disease), Khayi *et al.* (2020) reported the genome assembly of the Foa 133 strain, which consists of 3325 contigs with a total length of 56,228,901 bp and 3684 predicted genes. This first draft genome sequence will uncover the molecular mechanisms underlying pathogenicity in the Foa–date palm interaction. Furthermore, this sequence will allow comparative genomics studies with other *F. oxysporum* formae speciales. Radwan and Al-Naemi (2016) studied the genetic control of date palm to *Thielaviopsis punctulata*, the causal agent of black scorch, through understanding molecular interactions between the plant host and the pathogen. Genomic and bioinformatics tools were used to sequence, assemble, and annotate the whole genome of the pathogen and to decipher the molecular mechanisms of date palm resistance to the pathogen. Results showed that cv. Khalas is less affected by the pathogen compared to cv. Kinzy, reflecting a degree of resistance in Khalas. Data from RNA sequencing (RNA-Seq) are being analysed to identify regulatory genes involved in resistance to black scorch disease.

Hazzouri *et al.* (2019) reported that date palm fruits vary in color from deep red to pale yellow, and this is controlled by the *VIRESCENS* (*VIR*) gene, which encodes an R2R3-MYB transcription factor. Analysis showed that red coloration was associated with a wild-type *VIR*[+] allele, while yellow fruits had either the dominant *VIR*[IM] or were homozygous for the recessive *vir*[saf] allele. To identify the genes controlling high sucrose content, Malek *et al.* (2020) analysed date fruit metabolomics for association with genotype data from 120 date fruits. They found a significant association between dried date sucrose content and a genomic region that contains three tandem copies of the β-fructofuranosidase (invertase) gene in the reference Khalas genome, a low-sucrose fruit. High-sucrose cultivars including the popular Deglet Noor had a homozygous deletion of two of the three copies of the invertase gene. In addition, to assess genetic variability of date palm in Qatar, Thareja *et al.* (2018) conducted genotyping-by-sequencing (GBS) on 179 DNA samples representing arguably the most genetically diverse collection of date samples available to date. Results showed the limited genetic variability and predominance of eastern cultivars across municipalities in Qatar. Meanwhile, the findings can be used in the future for date palm cultivar identification, to aid selecting suitable cultivars for targeted breeding, to improve a country's date palm genetic diversity, and to certify the origin of date fruits and trees.

Gender discrimination before flowering is highly desirable for speeding up breeding and screening for gender at an early developmental stage would save time, cost, and other resources. Dhawan *et al.* (2013) used RAPD markers for gender screening in date palm and found one RAPD primer, OPA-02, that amplified a fragment of ~1.0 kb in all the individual samples of male genotypes, whereas this fragment was absent in all the female genotypes. This male-specific fragment was cloned and sequenced (GenBank accession no. JN123357), and a SCAR primer pair was designed that amplified a 406 bp fragment in both female and male genotypes and a unique fragment of 354 bp in only male genotypes. Later, in order to identify sex-determining genes, Torres *et al.* (2018) sequenced the genomes of 15 female and 13 male trees representing all 14 species of the genus *Phoenix*.

They identified male-specific sequences and extended them using phased single-molecule sequencing or bacterial artificial chromosome (BAC) clones. They observed that only four genes contained sequences conserved in all analysed *Phoenix* males. Most of these sequences showed similarity to a single genomic locus in the closely related monoecious oil palm. CYP703 and GPAT3, two single-copy genes present in males and critical for male flower development in other monocots, were absent in females. A LOG-like gene appears translocated into the Y-linked region and was suggested to play a role in suppressing female flowers (Torres *et al.*, 2018). Recently, Solliman *et al.* (2019) identified in the Y chromosome a gene allowing identification of male plants regardless of the origin, variety, or cultivar of the date palm. The differences between the two sexes were confirmed by the presence of the SRY gene in males. The complete sequence of the DNA has been registered and deposited in GenBank (BankIt1598036 DPSRY1 KC577225 then KJ873056). According to Naqvi *et al.* (2021), development of more reliable molecular markers is needed to differentiate male and female seedlings.

Finally, the cost/benefit ratio of MAS will depend on several factors, such as the inheritance of the trait, the method of phenotypic evaluation, the cost of field and glasshouse trials, and labour costs. It is also worth noting that large initial capital investments are required for the purchase of equipment, and regular expenses will be incurred for maintenance. Intellectual property rights, e.g., licensing costs due to patents, may also affect the cost of MAS. One approach to this problem is to contract the marker work out to larger laboratories that can benefit from economies of scale and high-throughput equipment (Lema, 2018).

4.10 Prospects of Gene Editing, Transgenics, and Genomics

The rapid advances in plant genome sequencing and phenotyping have enhanced trait mapping and gene discovery in crops. Increasing adoption of machine learning algorithms is crucial to derive meaningful inferences from complex multidimensional phenotyping data. Emerging breeding approaches like optimal contribution selection, alone or in combination with genomic selection, will enhance the genetic base of breeding pro-grams while accelerating genetic gain. Integrating speed breeding with new-age genomic breeding technologies holds promise to relieve the long-standing bottleneck of lengthy crop breeding cycles (Varshney *et al.*, 2021). Mapping studies are used to determine loci that control heritable variation in phenotypic traits. These *forward genetic* approaches comprise a powerful set of methods to identify genes that control variation in phenotypic traits and dissect their genetic basis. Genetic mapping can yield candidate genes and mutations that control variation in a trait and suggest strategies for its modification using genetic engineering. In other cases, genetic mapping can lead to the discovery of linked markers that can be used in MAS and crop breeding (Das *et al.*, 2017). Linkage mapping is possible in date palm as hundreds to thousands of seedlings from controlled crosses can be generated to produce full-sib progeny for standard linkage mapping or half-sib progeny for use in pedigree-based mapping designs (Hazzouri *et al.*, 2020).

Previously, biotechnological approaches, such as PTC, marker-assisted breeding, and DNA fingerprinting, have been used in date palm genomics but failed to produce significant improvement. For the sustainability of date palm cultivation, employment of new techniques in date palm breeding programs is needed to develop tolerant varieties and enrich the existing germplasm. This can be achieved by modifying the date palm genome against various biotic and abiotic stresses by overexpressing or downregulating the key genes involved in biochemical pathways, or by engineering resistance against various pests and diseases. Additionally, the dissection of genetic information in date palm would help in understanding the role of various genes involved in sex determination, enzymatic reactions controlling fruit ripening, fruit sweetness, and fruit quality (Sattar *et al.*, 2017).

GE strategies have evolved during the last three decades, and nowadays four types of 'programmable' nucleases are available in this field: mega nucleases, ZFNs, TALENs, and the CRISPR/Cas9 system. Each group has its own characteristics, and it is necessary that researchers select the most suitable gene editing tool for their applications. Genome engineering technology will revolutionize the creation of precisely manipulated genomes of cells or organisms to modify a specific characteristic. Introducing constructs into target cells or organisms is the key step in genome engineering (Khalil, 2020).

The long generation times to first flowering, multiple years to generate offshoots for propagation, and 10–15 years to reach maximum yield all complicate multi-generation experiments in date palm (Hazzouri *et al.*, 2020). This has hindered development of quantitative trait locus (QTL) mapping populations and efforts at breeding date palms for crop improvement. The difficulty of applying traditional genetic approaches has spurred the development of genomic resources to address problems in date palm cultivation and accelerate the discovery of important trait genes. GWAS of tree crops provides an attractive alternative to QTL mapping approaches to identify loci controlling important traits (Teh *et al.*, 2016). GWAS identifies or studies the correlation between the genetic variants/traits/phenotypes in a population of any organism based on SNPs in the sequence data. GWAS explores the complete genome and can identify more variants associated with traits (Challa and Neelapu, 2018). Several groups have employed GWAS on different plant species to identify drought response genes. The case studies include proline accumulation induced by drought response genes under low water potential (Verslues *et al.*, 2013), response to collective stresses (Thoen *et al.*, 2017), and understanding drought resistance genes in *Aegilops tauschii* (Qin *et al.*, 2016). In addition, case studies were reported by various groups on usage of GWAS to identify salt tolerance genes (Challa and Neelapu, 2018).

Mapping abiotic stress-related traits in adult date palms is currently intractable owing to the long juvenile stage and cost of growing and maintaining large mapping populations. An alternative solution would be to map such traits in early-stage seedlings where environmental conditions can be carefully controlled. Another alternative approach would be to conduct GWAS on early-stage plants from diverse varieties propagated in tissue culture. Use of micropropagated varieties would (Hazzouri *et al.*, 2020):

- allow expensive genotyping steps to be performed only once on a clonal lineage followed by phenotyping of many traits;
- facilitate phenotyping of replicates of a clone, thereby reducing the effects of measurement error and plant-to-plant variability in phenotypic measures;
- allow experiments to be replicated in different environments or treatments to ensure stability of QTLs; and
- allow more complex experimental designs to be adopted.

Actually, there is a need for continued improvement to the genome assembly and gene annotation of date palm. There are presently three draft assemblies including two females (Al-Dous *et al.*, 2011; Al-Mssallem *et al.*, 2013) and one male genome (Hazzouri *et al.*, 2019). The two female draft genomes are fragmented assemblies of the Khalas cultivar with low contiguity, while the male assembly is derived from a fourth-generation backcross (BC4) male of a cross with the Barhee cultivar as the recurrent parent. This BC4 male assembly is the only one of the three genomes to include long-read sequencing technology (i.e. Pacific Biosciences) and integrate a genetic map to place contigs on linkage groups (Mathew *et al.*, 2014). The primary sequence and gene models can be accessed at the Date Palm Genome Hub website (https://datepalmgenomehub.abudhabi.nyu. edu (accessed 14 November 2022)). Currently, factors limiting improvement in date palm are the absence of a high-density genetic map and the heterozygosity of date palm cultivars including the BC4 male (Hazzouri *et al.*, 2019).

4.11   Conclusions and Prospects

Production of disease resistance to bayoud and red palm weevil through conventional breeding is a time-consuming and tedious process. There is enormous scope in the transformation technology to introduce disease resistance traits in the date plant. The combined application of conventional breeding and applied biotechnological tools can generate genetic variations and produce fast-propagating new genotypes with superior fruit quality and resistance to biotic and abiotic stresses. Well-established PTC systems can be used for the rapid propagation of elite cultivars, to establish germplasm banks, and for mutagenic studies *in vitro* to isolate useful mutants. Application of protoplast techniques can be utilized in somatic hybridization and introduction of desired genes through genetic transformation of date palm. Protocols for protoplast isolation from embryogenic callus of date palm have been developed.

Generally, date palm is considered a halophytic species. Nevertheless, both date quality and yield are affected by soil salinity and water stress. Due to the serious threat of global warming, these problems are expected to be intensified in date palm cultivation regions. For these reasons, breeding efforts geared toward the improvement of tolerance to abiotic stress conditions are of paramount importance for sustainable date production. The identification and sequencing of the genomic regions related to disease resistance, fruit quality, and productivity are necessary for breeding and improvement programs. The application of molecular breeding, genomics, proteomics, and transcriptomic studies is needed to meet these objectives. Currently, cultivar identification mostly relies on phenotypic traits rather than genotyping. Still, limited information is available on date palm genotyping and genome mapping. Nevertheless, advancements in whole-genome sequencing and genome assembly computational tools hold a substantial advantage to boost the identification of date palm cultivars, genome variants, and gene/genome annotation. The science of bioinformatics is emerging as a tool to understand genome sequences. Computational tools are useful in discovering new molecular markers and analysing gene function, mutations, microRNAs, and structural biology. The genes, promoters, or genomic regions analysed using bioinformatics approaches in date palm include the traits of abiotic stress tolerance, sex determination, invertase enzyme activity, color of fruit peel, phytochelators, and metal-responsive genes. Every year, the percentage of crop genomes sequenced has increased. The incredible rate at which DNA samples become accessible is mainly due to the reduction in cost and enhancement in speed of sequencing techniques. Modern sequencing techniques enable the sequencing of plant genomes at realistic prices. Although many of the published genomes are deemed incomplete, they have nevertheless proven to be useful instruments for understanding significant plant characteristics.

In vitro mutagenesis has been an effective strategy for genetic improvements in several traits of crop plants. Several conventional approaches including physical and chemical mutagens, insertional and somaclonal mutations have been practiced creating the desired traits. However, contemporary site-directed mutation approaches like TALENS, ZNFs, and CRISPR/Cas9 have not yet been put into practice for the date palm. New breeding tools for targeted mutagenesis through CRISPR/Cas9-based GE and its BE versions can be very effective to engineer date palm genomes.

However, with a large and complex genome, heterozygosity and outcrossing, somaclonal variation during *in vitro* regeneration, the presence of SNPs, and ultimately genetic instability caused by these SNPs pose challenges. Such challenges could be addressed effectively by the execution of site-specific CRISPR/Cas9 versions, like BE, coupled with high-throughput screening techniques. Moreover, GWAS can identify more variants associated with traits and correlations between the genetic variants/traits/phenotypes in a population based on SNPs in the sequence data.

References

Abahmane, L. (2011) Date palm micropropagation via organogenesis. In: Jain, S.M., Al-Khayri, J.M. and Johnson, D.V. (eds) *Date Palm Biotechnology*. Springer, Dordrecht, The Netherlands, pp. 69–90. DOI: 10.1007/978-94-007-1318-5_5.

Abahmane, L. (2013) Recent achievements in date palm (*Phoenix dactylifera* L.) micropropagation from inflorescence tissues. *Emirates Journal of Food and Agriculture* 25(11), 863–874. DOI: 10.9755/ejfa.v25i11.16659.

Abahmane, L. (2017) Cultivar-dependent direct organogenesis of date palm from shoot tip explants. In: Al-Khayri, J.M., Jain, S.M. and Johnson, D.V. (eds) *Date Palm Biotechnology Protocols*, Vol. I. *Tissue Culture Applications*. Methods in Molecular Biology, Vol. 1637, Humana Press, New York, pp. 3–15.

Abahmane, L. (2020a) Means of date palm (*Phoenix dactylifera* L.) propagation. In: Bolduc, B. (ed.) *Date Palm: Composition, Cultivation and Uses*. Nova Science Publishers, New York, pp. 31–90.

Abahmane, L. (2020b) A comparative study between temporary immersion system and semi-solid cultures on shoot multiplication and plantlets production of two Moroccan date palm (*Phoenix dactylifera* L.) varieties *in vitro*. *Notulae Scientia Biologicae* 12(2), 277–288. DOI: 10.15835/nsb12210610.

Abutalebi, A. (2010) Report of the survey studied on somaclonal variations in *in vitro* propagated date palm plants. *Acta Horticulturae* 882, 803–810. DOI: 10.17660/ActaHortic.2010.882.91.

Acquaah, G. (2007) *Principles of Plant Genetics and Breeding*, 1st edn. Blackwell, Oxford.

Acquaah, G. (2012) *Principles of Plant Genetics and Breeding*, 2nd edn. Wiley-Blackwell, Oxford. DOI: 10.1002/9781118313718.

Acquaah, G. (2015) Conventional plant breeding principles and techniques. In: Al-Khayri, J.M., Jain, S.M. and Johnson, D.V. (eds) *Advances in Plant Breeding Strategies: Breeding, Biotechnology and Molecular Tools*. Springer, Cham, Switzerland, pp. 115–158. DOI: 10.1007/978-3-319-22521-0_5.

Ahmed, W., Feyissa, T., Tesfaye, K. and Farrakh, S. (2021) Genetic diversity and population structure of date palms (*Phoenix dactylifera* L.) in Ethiopia using microsatellite markers. *Journal, Genetic Engineering & Biotechnology* 19(1), 64. DOI: 10.1186/s43141-021-00168-5.

Al-Abdoulhadi, I.A. (2015) Evaluation and selection of date palm varieties from seedling progeny in Saudi Arabia. *Journal of Agricultural Science and Technology* 5(1), 1–7. DOI: 10.17265/2161-6256/2015.01A.001.

Al-Dous, E.K., George, B., Al-Mahmoud, M.E., Al-Jaber, M.Y., Wang, H. *et al.* (2011) *De novo* genome sequencing and comparative genomics of date palm (*Phoenix dactylifera*). *Nature Biotechnology* 29(6), 521–527. DOI: 10.1038/nbt.1860.

Allbed, A., Kumar, L. and Shabani, F. (2017) Climate change impacts on date palm cultivation in Saudi Arabia. *Journal of Agricultural Science* 155(8), 1203–1218. DOI: 10.1017/S0021859617000260.

Al-Khalifah, N.S. and Shanavaskhan, A.E. (2017) Molecular identification of date palm cultivars using random amplified polymorphic DNA (RAPD) markers. In: Al-Khayri, J.M., Jain, S.M. and Johnson, D.V. (eds) *Date Palm Biotechnology Protocols*, Vol. II. *Germplasm Conservation and Molecular Breeding*. Methods in Molecular Biology, Vol. 1638, Humana Press, New York, pp. 185–196. DOI: 10.1007/978-1-4939-7159-6_16.

Al-Khateeb, A.A. (2008) A review of the problems facing the use of tissue culture technique in date palm (*Phoenix dactylifera* L.). *Scientific Journal of King Faisal University Basic and Applied Sciences* 9, 85–104.

Al-Khateeb, A.A. and Ali-Dinar, H. (2002) Date palm in Kingdom of Saudi Arabia: cultivation production and processing. *Translation Authorship and Publishing Center, King Faisal University* 188, 281–292.

Al-Khateeb, S.A., Al-Khateeb, A.A., Sattar, M.N., Mohmand, A.S. and El-Beltagi, H.S. (2019) Assessment of somaclonal variation in salt-adapted and non-adapted regenerated date palm (*Phoenix dactylifera* L.). *Fresenius Environmental Bulletin* 28(5), 3686–3695.

Al-Khayri, J.M. and Ibraheem, Y. (2014) *In vitro* selection of abiotic stress tolerant date palm (*Phoenix dactylifera* L.): a review. *Emirates Journal of Food and Agriculture* 26(11), 921–933. DOI: 10.9755/ejfa.v26i11.18975.

Al-Khayri, J.M., Naik, P.M. and Alwael, H.A. (2017) *In vitro* assessment of abiotic stress in date palm: salinity and drought. In: Al-Khayri, J.M., Jain, S.M. and Johnson, D.V. (eds) *Date Palm Biotechnology Protocols*, Vol. I. *Tissue Culture Applications*. Methods in Molecular Biology, Vol. 1637, Humana Press, New York, pp. 333–346. DOI: 10.1007/978-1-4939-7156-5_27.

Al-Khayri, J.M., Naik, P.M., Jain, S.M. and Johnson, D.V. (2018) Advances in date palm (*Phoenix dactylifera* L.) breeding. In: Al-Khayri, J.M., Naik, P.M., Jain, S.M. and Johnson, D.V. (eds) *Advances in Plant Breeding Strategies: Fruits*. Springer, Cham, Switzerland, pp. 727–771. DOI: 10.1007/978-3-319-91944-7_18.

Al-Mssallem, I.S., Hu, S., Zhang, X., Lin, Q., Liu, W. *et al.* (2013) Genome sequence of the date palm *Phoenix dactylifera* L. *Nature Communications* 4, 2274. DOI: 10.1038/ncomms3274.

Almusawi, A.H.A., Sayegh, A.J., Alshanaw, A.M.S. and Griffis, J.L. (2017) Plantform bioreactor for mass micropropagation of date palm. In: Al-Khayri, J.M., Jain, S.M. and Johnson, D.V. (eds) *Date Palm Biotechnology Protocols*, Vol. I. *Tissue Culture Applications*. Methods in Molecular Biology, Vol. 1637, Humana Press, New York, pp. 251–265.

Al-Ruqaishi, I.A., Davey, M., Alderson, P. and Mayes, S. (2008) Genetic relationships and genotype tracing in date palms (*Phoenix dactylifera* L.) in Oman, based on microsatellite markers . *Plant Genetic Resources* 6(1), 70–72. DOI: 10.1017/S1479262108923820.

Al-Sabah, L., Sudhersan, C., Jibi, S. and Al-Melhem, S. (2018) A new interspecific date palm hybrid. In: Zaid, A. and Alhadrami, G. (eds) *Proceedings of the 6th*

International Date Palm Conference, Khalifa International Award for Date Palm and Agricultural Innovation, Abu Dhabi, pp. 154–157.

Al-Wasel, A. (2001) Field performance of soma-clonal variants of tissue culture-derived date palm (*Phoenix dactylifera* L.). *Plant Tissue Culture* 11, 97–105.

Arabnezhad, H., Bahar, M., Mohammadi, H.R. and Latifian, M. (2012) Development, characterization and use of microsatellite markers for germplasm analysis in date palm (*Phoenix dactylifera* L.). *Scientia Horticulturae* 134, 150–156. DOI: 10.1016/j.scienta.2011.11.032.

Aslam, J., Khan, S.A. and Azad, M.A.K. (2015) *Agrobacterium*-mediated genetic transformation of date palm (*Phoenix dactylifera* L.) cultivar "Khalasah" via somatic embryogenesis. *Plant Science Today* 2(3), 93–101. DOI: 10.14719/pst.2015.2.3.119.

Atia, M.A.M., Sakr, M.M., Mokhtar, M.M. and Adawy, S.S. (2017) Development of sex-specific PCR-based markers in date palm. In: Al-Khayri, J.M., Jain, S.M. and Johnson, D.V. (eds) *Date Palm Biotechnology Protocols*, Vol. II. *Germplasm Conservation and Molecular Breeding*. Methods in Molecular Biology, Vol. 1638, Humana Press, New York, pp. 227–244. Available at: doi.org/10.1007/978-1-4 939-7159-6_19

Ayesh, B.M. (2017) Genotyping and molecular identification of date palm cultivars using inter-simple sequence repeat (ISSR) markers. In: Al-Khayri, J.M., Jain, S.M. and Johnson, D.V. (eds) *Date Palm Biotechnology Protocols*, Vol. II. *Germplasm Conservation and Molecular Breeding*. Methods in Molecular Biology, Vol. 1638, Humana Press, New York, pp. 173–183. Available at: doi.org/10.1007/978-1-4 939-7159-6_15

Bawa, A.S. and Anilakumar, K.R. (2013) Genetically modified foods: safety, risks and public concerns-a review. *Journal of Food Science and Technology* 50(6), 1035–1046. DOI: 10.1007/s13197-012-0899-1.

Bekheet, S.A. (2017) *In vitro* conservation of date palm tissue cultures. In: Al-Khayri, J.M., Jain, S.M. and Johnson, D.V. (eds) *Date Palm Biotechnology Protocols*, Vol. II. *Germplasm Conservation and Molecular Breeding*. Methods in Molecular Biology, Vol. 1638, Humana Press, New York, pp. 15–24. Available at: doi.org/10.1007 /978-1-4939-7159-6_2

Bisht, D.S., Bhatia, V. and Bhattacharya, R. (2019) Improving plant-resistance to insect-pests and pathogens: The new opportunities through targeted genome editing. *Seminars in Cell & Developmental Biology* 96, 65–76. DOI: 10.1016/j.semcdb.2019.04.008.

Bougerfaoui, M., Abahmane, L., Anjarne, M. and Sedra, M.H. (2006) Utilisation de la mutagenèse induite pour l'amélioration de la résistance aux toxines du Bayoud chez le palmier dattier (*Phoenix dactylifera* L.). In: Abed, F. (ed.) *Actes de La Conférence Régionale Sur La Mutagenèse Induite et Biotechnologies Dappui Pour La Protection Du Palmier Dattier Contre Le Bayoud*. AIEA, RAF/5/049, Algiers, pp. 13–15.

Braatz, J., Harloff, H.J., Mascher, M., Stein, N., Himmelbach, A. *et al.* (2017) CRISPR-Cas9 targeted mutagenesis leads to simultaneous modification of different homoeologous gene copies in polyploid oilseed rape (*Brassica napus*). *Plant Physiology* 174(2), 935–942. DOI: 10.1104/pp.17.00426.

Breseghello, F. and Coelho, A.S.G. (2013) Traditional and modern plant breeding methods with examples in rice (*Oryza sativa* L.). *Journal of Agricultural and Food Chemistry* 61(35), 8277–8286. DOI: 10.1021/jf305531j.

Bridgen, M.P., Van Houtven, W. and Eeckhaut, T. (2018) Plant tissue culture techniques for breeding. In: Huylenbroeck, J.V. (ed.) *Ornamental Crops*, Vol. 11. Handbook of Plant Breeding. Springer, Cham, Switzerland, pp. 127–144. DOI: 10.1007/978-3-319-90698-0_6.

Brown, D.C. and Thorpe, T.A. (1995) Crop improvement through tissue culture. *World Journal of Microbiology & Biotechnology* 11(4), 409–415. DOI: 10.1007/BF00364616.

Budiani, A., Putranto, R.A., Riyadi, I., Sumaryono, H. and Faizah, R. (2018) Transformation of oil palm calli using CRISPR/Cas9 system: toward genome editing of oil palm. *IOP Conference Series: Earth Environmental Science* 183, 012003. DOI: 10.1088/1755-1315/183/1/012003.

Bulit, J., Louvet, J., Bouhot, D. and Toutain, G. (1967) Recherches sur les fusarioses. I. Travaux sur le Bayoud, fusariose vasculaire du palmier dattier en Afrique du Nord. *Annales Des Épiphyties* 18, 231–239.

Carimi, F., Pathirana, R. and Carra, A. (2012) Biotechnologies for grapevine germplasm management and improvement. In: Szabo, P.V. and Shojania, J. (eds) *Grapevines: Varieties, Cultivation*. Nova Science Publishers, New York, pp. 199–249.

Carpenter, J.B. (1979) The national date palm germplasm repository. *Date Growers' Institute Report* 54, 29–32.

Carpenter, J.B. and Ream, C.L. (1976) Date palm breeding: a review. *Date Growers' Institute Report* 53, 25–33.

Cevallos-Cevallos, J.M., Jines, C., Maridueña-Zavala, M.G., Molina-Miranda, M.J., Ochoa, D.E. *et al.* (2018) GC-MS metabolite profiling for specific detection of dwarf somaclonal variation in banana plants. *Applications in Plant Sciences* 6(11), e01194. DOI: 10.1002/aps3.1194.

Chakraborty, N.R. and Paul, A. (2013) Role of induced mutations for enhancing nutrition quality and production of food. *International Journal of Bio-Resource and Stress Management* 4(1), 91–96.

Challa, S. and Neelapu, N.R.R. (2018) Genome-wide association studies (GWAS) for abiotic stress tolerance in plants. In: Wani, S.H. (ed.) *Biochemical, Physiological and Molecular Avenues for Combating Abiotic Stress in Plants*. Academic Press, London, pp. 135–150. DOI: 10.1016/B978-0-12-813066-7.00009-7.

Chaluvadi, S.R., Khanam, S., Aly, M.A.M. and Bennetzen, J.L. (2014) Genetic diversity and population structure of native and introduced date palm (*Phoenix dactylifera*) germplasm in the United Arab Emirates. *Tropical Plant Biology* 7(1), 30–41. DOI: 10.1007/s12042-014-9135-7.

Chaluvadi, S.R., Young, P., Thompson, K., Bahri, B.A., Gajera, B. *et al.* (2019) *Phoenix* phylogeny, and analysis of genetic variation in a diverse collection of date palm (*Phoenix dactylifera*) and related species. *Plant Diversity* 41(5), 330–339. DOI: 10.1016/j.pld.2018.11.005.

Cohen, Y., Korchinsky, R. and Tripler, E. (2004) Flower abnormalities cause abnormal fruit setting in tissue culture-propagated date palm (*Phoenix dactylifera* L.) *The Journal of Horticultural Science and Biotechnology* 79(6), 1007–1013. DOI: 10.1080/14620316.2004.11511853.

Corniquel, B. and Mercier, L. (1994) Date palm (*Phoenix dactylifera* L.) cultivar identification by RFLP and RAPD. *Plant Science* 101(2), 163–172. DOI: 10.1016/0168-9452(94)90252-6.

Das, G., Patra, J.K. and Baek, K.H. (2017) Insight into MAS: a molecular tool for development of stress resistant and quality of rice through gene stacking. *Frontiers in Plant Science* 8, 985. DOI: 10.3389/fpls.2017.00985.

Dhawan, C., Kharb, P., Sharma, R., Uppal, S. and Aggarwal, R.K. (2013) Development of male-specific SCAR marker in date palm (*Phoenix dactylifera* L.). *Tree Genetics & Genomes* 9(5), 1143–1150. DOI: 10.1007/s11295-013-0617-9.

Diaz, S., Pire, C., Ferrer, J. and Bonete, M.J. (2003) Identification of *Phoenix dactylifera* L. varieties based on amplified fragment length polymorphism (AFLP) markers. *Cellular & Molecular Biology Letters* 8, 891–899.

Djerbi, M. and Sedra, M.H. (1982) Screening commercial Iraqi date varieties to bayoud. In: *Nenadates News* 2. United Nations Development Programme/Food and Agricultural Organization of the United Nations, Rome.

El Fakhouri, R., Lotfi, F., Sedra, M.H. and Lazrek, H.B. (1997) Production et caractérisation chimique des toxines sécrétées par *Fusarium oxysporum* f. sp. *albedinis*, agent causal du Bayoud. *Al Awamia* 93, 83–93.

El Hadrami, F., Daayf, S., Elshibli, M.S., Jain, S.M. and El Hadrami, I. (2011) Soma-clonal variation in date palm. In: Jain, S.M., Al-Khayri, J.M. and Johnson, D.V. (eds) *Date Palm Biotechnology*. Springer, Dordrecht, The Netherlands, pp. 183–203. DOI: 10.1007/978-94-007-1318-5_9.

El-Juhany, L.I. (2010) Degradation of date palm trees and date production in Arab countries: causes and potential rehabilitation. *Australian Journal of Basic and Applied Sciences* 4, 3998–4010.

El-Khishin, D.A., Adawy, S.S., Hussein, E.H.A. and El-Itriby, H.A. (2003) AFLP fingerprinting of some Egyptian date palm (*Phoenix dactylifera* L.) cultivars. *Arab Journal of Biotechnology* 6, 223–234.

Elmeer, K. and Mattat, I. (2015) Genetic diversity of Qatari date palm using SSR markers. *Genetics and Molecular Research* 14(1), 1624–1635. DOI: 10.4238/2015. March.6.9.

Elmeer, K., Sarwath, H., Malek, J., Baum, M. and Hamwieh, A. (2011) New microsatellite markers for assessment of genetic diversity in date palm (*Phoenix dactylifera* L.). *3 Biotech* 1(2), 91–97. DOI: 10.1007/s13205-011-0010-z.

El Modafar, C. (2010) Mechanisms of date palm resistance to Bayoud disease: current state of knowledge and research prospects. *Physiological and Molecular Plant Pathology* 74(5–6), 287–294. DOI: 10.1016/j.pmpp.2010.06.008.

El Rabey, H.A., Al-Malki, A.L., Abulnaja, K.O. and Rohde, W. (2015) Proteome analysis for understanding abiotic stress (salinity and drought) tolerance in date palm (*Phoenix dactylifera* L.). *International Journal of Genomics* 2015, 407165. DOI: 10.1155/2015/407165.

Elshibli, S. and Korpelainen, H. (2008) Microsatellite markers reveal high genetic diversity in date palm (*Phoenix dactylifera* L.) germplasm from Sudan. *Genetica* 134(2), 251–260. DOI: 10.1007/s10709-007-9232-8.

Elshibli, S., Elshibli, E. and Korpelainen, H. (2016) Growth and photosynthetic CO_2 responses of date palm plants to water availability. *Emirates Journal of Food and Agriculture* 28(1), 58–65. DOI: 10.9755/ejfa.2015.05.189.

Engelmann, F. (2011) Use of biotechnologies for the conservation of plant biodiversity. *In Vitro Cellular & Developmental Biology - Plant* 47(1), 5–16. DOI: 10.1007/s11627-010-9327-2.

Ferry, M. (2010) Etat des lieux sur l'amélioration génétique du palmier dattier. In: Aberlenc-Bertossi, F. (ed.) *Biotechnologies Du Palmier Dattier*. IRD Editions, Montpellier, France, pp. 217–226.

Fki, L., Masmoudi, R., Kriaa, W. and Mahjoub, A. (2011) Date palm micropropagation via somatic embryogenesis. In: Jain, S.M., Al-Khayri, J.M. and Johnson, D.V. (eds) *Date Palm Biotechnology*. Springer, Dordrecht, The Netherlands, pp. 47–68.

Fki, L., Chakir, O., Kriaa, W., Nasri, A., Baklouti, E. *et al.* (2017) *In vitro* cryopreservation of date palm caulogenic meristems. In: Al-Khayri, J.M., Jain, S.M. and Johnson, D.V. (eds) *Date Palm Biotechnology Protocols*, Vol. II. *Germplasm Conservation and Molecular Breeding*. Methods in Molecular Biology, Vol. 1638, Humana Press, New York, pp. 39–48. Available at: https://doi.org/10.1007/9 78-1-4939-7159-6_4

Flowers, J.M., Hazzouri, K.M., Gros-Balthazard, M., Mo, Z., Koutroumpa, K. *et al.* (2019) Cross-species hybridization and the origin of North African date palms. *Proceedings of the National Academy of Sciences of the United States of America* 116(5), 1651–1658. DOI: 10.1073/pnas.1817453116.

Gonzalez-Perez, M.A., Caujape-Castells, J. and Sosa, P. (2004) Molecular evidence of hybridisation between the endemic *Phoenix canariensis* and the widespread *P. dactylifera* with random amplified polymorphic DNA (RAPD) markers. *Plant Systematics and Evolution* 247(3–4), 165–175. DOI: 10.1007/ s00606-004-0166-7.

Gros-Balthazard, M. (2013) Hybridization in the genus *Phoenix*: a review. *Emirates Journal of Food and Agriculture* 25(11), 831–842. DOI: 10.9755/ejfa.v25i11.16660.

Gros-Balthazard, M., Galimberti, M., Kousathanas, A., Newton, C., Ivorra, S. *et al.* (2017) The discovery of wild date palms in Oman reveals a complex domestication history involving centers in the Middle East and Africa. *Current Biology* 27, 2211–2218. DOI: 10.1016/j.cub.2017.06.045.

Gros-Balthazard, M., Hazzouri, K. and Flowers, J. (2018) Genomic insights into date palm origins. *Genes* 9(10), 502. DOI: 10.3390/genes9100502.

Guettouchi, A. (2018) Date palm: application of molecular markers. In: Zaid, A. and Alhadrami, G. (eds) *Proceedings of the 6th International Date Palm Conference*, Khalifa International Award for Date Palm and Agricultural Innovation, Abu Dhabi, pp. 158–161.

Gulzar, B., Mujib, A., Malik, M.Q., Mamgain, J., Syeed, R. *et al.* (2020) Plant tissue culture: agriculture and industrial applications. In: Kiran, U., Zainul Abdin, M. and Kamaluddin, K. (eds) *Transgenic Technology Based Value Addition in Plant Biotechnology*. Academic Press, New Delhi, pp. 25–49. DOI: 10.1016/ B978-0-12-818632-9.00002-2.

Hazzouri, K.M., Gros-Balthazard, M., Flowers, J.M., Copetti, D., Lemansour, A.V. *et al.* (2019) Genome-wide association mapping of date palm fruit traits. *Nature Communications* 10, 4680. DOI: 10.1038/s41467-019-12604-9.

Hazzouri, K.M., Flowers, J.M., Nelson, D., Lemansour, A., Masmoudi, K. *et al.* (2020) Prospects for the study and improvement of abiotic stress tolerance in date palms in the post-genomics era. *Frontiers in Plant Science* 11, 293. DOI: 10.3389/fpls.2020.00293.

Henao-Ramírez, A.M., Salazar Duque, H.J., Calle Tobón, A.F. and Urrea Trujillo, A.I. (2021) Determination of genetic stability in cacao plants (*Theobroma cacao* L.) derived from somatic embryogenesis using microsatellite molecular markers (SSR). *International Journal of Fruit Science* 21(1), 284–298. DOI: 10.1080/15538362.2021.1873219.

Hodel, D.R. and Johnson, D.V. (2007) *Imported and American Varieties of Dates in the United States*. University of California Agriculture and Natural Resources Publication No. 3498. University of California, Oakland, California.

Hussain, N., Al-Rasbi, S., Al-Wahaibi, N.S., Al-Ghanum, G. and El-Sharief, A.O.A. (2012) Salinity problems and their management in date palm production. In: Manickavasagan, A., Essa, M.M. and Sukumar, E. (eds) *Dates: Production, Processing, Food, and Medicinal Value*. CRC Press, Boca Raton, Florida, pp. 87–112.

Ibraheem, A.O. (2008) The date palm (*Phoenix dactylifera* L.). In: *Tree of Life*. Arab Center for Studies on Arid Zones and Dry Lands (ACSAD), Damascus.

Iqbal, Z., Sattar, M.N. and Shafiq, M. (2016) CRISPR/Cas9: a tool to circumscribe cotton leaf curl disease. *Frontiers in Plant Science* 7, 475. DOI: 10.3389/fpls.2016.00475.

Jain, S.M. (2001) Tissue culture-derived variation in crops. *Euphytica* 118(2), 153–166. DOI: 10.1023/A:1004124519479.

Jain, S.M. (2010) Date palm genetic diversity conservation for sustainable production. *Acta Horticulturae* 882, 785–791. DOI: 10.17660/ActaHortic.2010.882.89.

Jain, S.M. (2012) *In vitro* mutagenesis for improving date palm (*Phoenix dactylifera* L.). *Emirates Journal of Food and Agriculture* 24(5), 400–407.

Jain, S.M. (2013) Date palm improvement with innovative technologies. In: Yahia, E. (ed.) *Regional Workshop on the Improvement of the Dates Value Chain in the Near East and North Africa Region*, December 9–12, 2013. Available at: http://hdl.ha ndle.net/10138/42984

Jain, S.M. (2019) Date palm (*Phoenix dactylifera* L.) genetic diversity and conservation under climate change. *The Blessed Tree* 11(01), 6–21.

Jain, S.M., Ochatt, S.J., Kulkarni, V.M. and Predieri, S. (2010) *In vitro* culture for mutant development. *Acta Horticulturae* 865, 59–68. DOI: 10.17660/ActaHortic.2010.865.6.

Jaradat, A.A. (2011) Biodiversity of date palm. Soils, plant growth and crop production. In: UNESCO-EOLSS (ed.) *Encyclopedia of Life Support Systems: Land Use, Land Cover and Soil Sciences*. EOLSS Publishers, Oxford, pp. 1–31.

Jayasankar, S., Li, Z. and Gray, D.J. (2003) Constitutive expression of *Vitis vinifera* thaumatin-like protein after *in vitro* selection and its role in anthracnose resistance. *Functional Plant Biology* 30(11), 1105–1115. DOI: 10.1071/FP03066.

Johnson, D.V., Al-Khayri, J.M. and Jain, S.M. (2013) Seedling date palms (*Phoenix dactylifera* L.) as genetic resources. *Emirates Journal of Food and Agriculture* 25(11), 809–830. DOI: 10.9755/ejfa.v25i11.16497.

Kadir, A.P.G., Bahariah, B., Ayub, N.H., Yunus, A.M.M., Rasid, O. *et al.* (2015) Production of polyhydroxybutyrate in oil palm (*Elaeis guineensis* Jacq.) mediated by microprojectile bombardment of PHB biosynthesis genes into embryogenic calli. *Frontiers in Plant Science* 6, 598. DOI: 10.3389/fpls.2015.00598.

Kadirvel, P., Senthilvel, S., Geethanjali, S., Sujatha, M. and Varaprasad, K.S. (2015) Genetic markers, trait mapping and marker-assisted selection in plant breeding. In: Bahadur, B., Sahijram, M.V.R.L. and Krishnamurthyet, K.V. (eds) *Plant Biology and Biotechnology*, Vol. II. *Plant Genomics and Biotechnology*. Springer, New Delhi, pp. 65–89. DOI: 10.1007/978-81-322-2283-5_4.

Kamle, M., Bajpai, A., Chandra, R., Kalim, S. and Kumar, R. (2011) Somatic embryogenesis for crop improvement. *GERF Bulletin of Biosciences* 2(1), 54–59.

Kanda, N., Ichikawa, M., Ono, A., Toyoda, A., Fujiyama, A. *et al.* (2017) CRISPR/ Cas9-based knockouts reveal that CpRLP1 is a negative regulator of the sex pheromone PR-IP in the *Closterium peracerosum-strigosum-littorale* complex. *Scientific Reports* 7(1), 17873. DOI: 10.1038/s41598-017-18251-8.

Karim, K., Chokri, B., Amel, S., Wafa, H., Richid, H. *et al.* (2010) Genetic diversity of Tunisian date palm germplasm using ISSR markers. *International Journal of Botany* 6(2), 182–186. DOI: 10.3923/ijb.2010.182.186.

Khalil, A.M. (2020) The genome editing revolution: review. *Journal, Genetic Engineering & Biotechnology* 18(1), 68. DOI: 10.1186/s43141-020-00078-y.

Kharb, P. and Singh, R. (2020) Molecular markers as tools to improve date palms. In: Tuteja, N., Tuteja, R., Passricha, N. and Saifi, S. (eds) *Advancement in Crop Improvement Techniques.* Elsevier, Amsterdam, pp. 319–327. DOI: 10.1016/ B978-0-12-818581-0.00019-X.

Khayi, S., Khoulassa, S., Gaboun, F., Abdelwahd, R., Diria, G. *et al.* (2020) Draft genome sequence of *Fusarium oxysporum* f. sp. *albedinis* strain Foa 133, the causal agent of bayoud disease on date palm. *Microbiology Resource Announcements* 9(29), e00462-20. DOI: 10.1128/MRA.00462-20.

Khierallah, H., Bader, S., Baum, M., Hamwieh, A. *et al.* (2011) Assessment of genetic diversity for some Iraqi date palms (*Phoenix dactylifera* L.) using amplified fragment length polymorphisms (AFLP) markers. *African Journal of Biotechnology* 10(47), 9570–9576. DOI: 10.5897/AJB11.055.

Khierallah, H.S.M., Bader, S.M., Hamwieh, A. and Baum, M. (2017) Date palm genetic diversity analysis using microsatellite polymorphism. In: Al-Khayri, J.M., Jain, S.M. and Johnson, D.V. (eds) *Date Palm Biotechnology Protocols*, Vol. II. *Germplasm Conservation and Molecular Breeding.* Methods in Molecular Biology, Vol. 1638, Humana Press, New York, pp. 113–124. Available at: https://doi.org /10.1007/978-1-4939-7159-6_11

Krishna, H., Alizadeh, M., Singh, D., Singh, U., Chauhan, N. *et al.* (2016) Somaclonal variations and their applications in horticultural crops improvement. *3 Biotech* 6(1), 54. DOI: 10.1007/s13205-016-0389-7.

Krueger, R.R. (2001) Date palm germplasm: overview and utilization in USA. In: Kaakeh, W.A. (ed.) *Proceedings of the First International Conference on Date Palm, Al-Ain University, Abu Dhabi*, UAE University Printing Press, Abu Dhabi, pp. 2–37.

Krueger, R.R. (2015) Date palm status and perspective in the United States. In: Al-Khayri, J.M., Jain, S.M. and Johnson, D.V. (eds) *Date Palm Genetic Resources and Utilization*, Vol. 1. *Africa and the Americas*, Springer, Dordrecht, The Netherlands, pp. 447–485. DOI: 10.1007/978-94-017-9694-1_14.

Lebeda, A. and Svabova, L. (2010) *In vitro* screening methods for assessing plant disease resistance. In: IAEA (ed.) *Mass Screening Techniques for Selecting Crops Resistant to Diseases. Joint FAO/IAEA Program of Nuclear Techniques in Food and Agriculture.* International Atomic Energy Agency, Vienna, pp. 5–47. Available at: www-pub.iaea.org/MTCD/publications/PDF/TDL-001_web.pdf (accessed 14 November 2022).

Lema, M. (2018) Marker assisted selection in comparison to conventional plant breeding: review article. *Agricultural Research & Technology: Open Access Journal* 14(2), 2. DOI: 10.19080/ARTOAJ.2018.14.555914.

Li, S., Dean, S., Li, Z., Horecka, J., Deschenes, R.J. *et al.* (2002) The eukaryotic two-component histidine kinase Sln1p regulates OCH1 via the transcription factor, Skn7p. *Molecular Biology of the Cell* 13(2), 412–424. DOI: 10.1091/mbc.01-09-0434.

Malek, J.A., Mathew, S., Mathew, L.S., Younuskunju, S., Mohamoud, Y.A. *et al.* (2020) Deletion of beta-fructofuranosidase (invertase) genes is associated with sucrose content in date palm fruit. *Plant Direct* 4, e00214. DOI: 10.1002/pld3.214.

Marco, F., Bitrian, M., Carrasco, P., Rajam, M.V., Alcázar, R. *et al.* (2015) Genetic engineering strategies for abiotic stress tolerance in plants. In: Bahadur, B., Sahijram, M.V.R. and Krishnamurthyet, K.V. (eds) *Plant Biology and Biotechnology*. Vol. II. *Plant Genomics and Biotechnology*. Springer, New Delhi, pp. 579–610. DOI: 10.1007/978-81-322-2283-5_29.

Mathew, L.S., Spannagl, M., Al-Malki, A., George, B., Torres, M.F. *et al.* (2014) A first genetic map of date palm (*Phoenix dactylifera*) reveals long-range genome structure conservation in the palms. *BMC Genomics* 15, 285. DOI: 10.1186/1471-2164-15-285.

Mathew, L.S., Seidel, M.A., George, B., Mathew, S., Spannagl, M. *et al.* (2015) A genome-wide survey of date palm cultivars supports two major subpopulations in *Phoenix dactylifera*. *G3: Genes, Genomes, Genetics* 5, 1429–1438. DOI: 10.1534/g3.115.018341.

McCubbin, M., Zaid, A. and Stade, J. (2004) A southern african survey conducted for off-types on date palms produced using somatic embryogenesis. *Emirates Journal of Food and Agriculture* 16(1), 8. DOI: 10.9755/ejfa.v12i1.5213.

Mercati, F. and Sunseri, F. (2020) Genetic diversity assessment and marker-assisted selection in crops. *Genes* 11(12), 1481. DOI: 10.3390/genes11121481.

Mirani, A.A., Teo, C.H., Markhand, G.S., Abul-Soad, A.A. and Harikrishna, J.A. (2020) Detection of somaclonal variations in tissue cultured date palm (*Phoenix dactylifera* L.) using transposable element-based markers. *Plant Cell, Tissue and Organ Culture (PCTOC)* 141(1), 119–130. DOI: 10.1007/s11240-020-01772-y.

Mirbahar, A.A., Markhand, G.S., Khan, S. and Abul-Soad, A. (2014) Molecular characterization of some Pakistani date palm (*Phoenix dactylifera* L.) cultivars by RAPD markers. *Pakistan Journal of Botany* 46(2), 619–625.

Mohamoud, Y.A., Mathew, L.S., Torres, M.F., Younuskunju, S., Krueger, R. *et al.* (2019) Novel subpopulations in date palm (*Phoenix dactylifera*) identified by population-wide organellar genome sequencing. *BMC Genomics* 20(1), 498. DOI: 10.1186/s12864-019-5834-7.

Mokhtar, M.M., Adawy, S.S., El-Assal, S.E. and Hussein, E.H.A. (2016) Genic and intergenic SSR database generation, SNPs determination and pathway annotations, in date palm (*Phoenix dactylifera* L.). *PloS ONE* 11(7), e0159268. DOI: 10.1371/journal.pone.0159268.

Mousavi, M., Mousavi, A., Habashi, A.A. and Dehsara, B. (2014) Genetic transformation of date palm (*Phoenix dactylifera* L.) cv. Estamaran via particle bombardment. *Molecular Biology Reports* 41(12), 8185–8194. DOI: 10.1007/s11033-014-3720-6.

Müller, H.M., Schäfer, N., Bauer, H., Geiger, D., Lautner, S. *et al.* (2017) The desert plant *Phoenix dactylifera* L. closes stomata via nitrate-regulated SLAC1 anion channel. *New Phytologist* 216(1), 150–162. DOI: 10.1111/nph.14672.

Munier, P. (1973) *Le Palmier Dattier*. Techniques Agricoles et Productions Tropicales 24. G.-P. Maisonneuve & LaRose, Paris.

Murashige, T. and Skoog, F. (1962) A revised medium for rapid growth and bio assays with tobacco tissue cultures. *Physiologia Plantarum* 15(3), 473–497. DOI: 10.1111/j.1399-3054.1962.tb08052.x.

Naqvi, S.A., Shafqat, W., Haider, M.S., Awan, F.S., Khan, I.A. *et al.* (2021) Gender determination of date palm. In: Al-Khayri, J.M., Jain, S.M. and Johnson, D.V. (eds) *The Date Palm Genome*, Vol. 1. *Phylogeny, Biodiversity and Mapping.* Compendium of Plant Genomes, Springer, Cham, Switzerland, pp. 161–177. DOI: 10.1007/978-3-030-73746-7_7.

Nixon, R.W. (1935) Meta-xenia in dates. *Proceedings of the American Society for Horticultural Science* 32, 221–226.

Nixon, R.W. and Furr, J.R. (1965) Problems and progress in date breeding. *Date Growers' Institute Report* 42, 2–5.

Othmani, A., Bayoudh, C., Drira, N. and Trifi, M. (2009) *In vitro* cloning of date palm *Phoenix dactylifera* L., cv. Deglet Bey by using embryogenic suspension and temporary immersion bioreactor (TIB). *Biotechnology & Biotechnological Equipment* 23(2), 1181–1188. DOI: 10.1080/13102818.2009.10817635.

Patnaik, J., Sahoo, S. and Debata, B.K. (1999) Somaclonal variation in cell suspension culture-derived regenerants of *Cymbopogon martinii* (Roxb.) Wats var. motia. *Plant Breeding* 118(4), 351–354. DOI: 10.1046/j.1439-0523.1999.00383.x.

Peng, A., Chen, S., Lei, T., Xu, L., He, Y. *et al.* (2017) Engineering canker-resistant plants through CRISPR/Cas9-targeted editing of the susceptibility gene CsLOB1 promoter in citrus. *Plant Biotechnology Journal* 15(12), 1509–1519. DOI: 10.1111/pbi.12733.

Pereau-Leroy, P. (1958) *Le Palmier Dattier Au Maroc*. Ministère de l'Agriculture-Institut Français de Recherche, Rabat.

Pintos, B., Bueno, M.A., Cuenca, B. and Manzanera, J.A. (2008) Synthetic seed production from encapsulated somatic embryos of cork oak (*Quercus suber* L.) and automated growth monitoring. *Plant Cell, Tissue and Organ Culture* 95(2), 217–225. DOI: 10.1007/s11240-008-9435-4.

Pramanik, K., Sahoo, J.P., Mohapatra, P.P., Acharya, L.K. and Jena, C. (2021) Insights into the embryo rescue – a modern *in vitro* crop improvement approach in horticulture. *Plant Cell Biotechnology and Molecular Biology* 22(15–16), 20–33. Available at: www.ikprress.org/index.php/PCBMB/article/view/6025 (accessed 15 November 2022).

Puchta, H. (2017) Applying CRISPR/Cas for genome engineering in plants: the best is yet to come. *Current Opinion in Plant Biology* 36, 1–8. DOI: 10.1016/j.pbi.2016.11.011.

Qin, P., Lin, Y., Hu, Y., Liu, K., Mao, S. *et al.* (2016) Genome-wide association study of drought-related resistance traits in *Aegilops tauschii. Genetics and Molecular Biology* 39(3), 398–407. DOI: 10.1590/1678-4685-GMB-2015-0232.

Rabey, H.E., Al-Malki, A. and Abulnaja, K. (2016) Proteome analysis of date palm (*Phoenix dactylifera* L.) under severe drought and salt stress. *International Journal of Genomics* 2016, 7840759. DOI: 10.1155/2016/7840759.

Radwan, O. and Al-Naemi, F.A. (2016) Implementation of genomics tools for genetic improvement of date palm to black scorch disease. *Qatar Foundation Annual Research Conference Proceedings* 2016(1), EEPP1164. DOI: 10.5339/qfarc.2016.EEPP1164.

Rajan, R.P. and Singh, G.A. (2021) Review on application of soma-clonal variation in important horticulture crops. *Plant Cell Biotechnology and Molecular Biology* 22, 161–175.

Ranghoo-Sanmukhiya, V.M. (2021) Somaclonal variation and methods used for its detection. In: Siddique, I. (ed.) *Propagation and Genetic Manipulation of Plants.* Springer, Singapore, pp. 1–18.

Ranjisha, K.R., Surendran, K., Nair, R.A. and Pillai, P.P. (2020) Bio-prospecting of biodiversity for improvement of agronomic traits in plants. In: Kiran, U., Zainul Abdin, M. and Kamaluddin, K. (eds) *Transgenic Technology Based Value Addition in Plant Biotechnology.* Elsevier, New Delhi, pp. 1–24. DOI: 10.1016/B978-0-12-818632-9.00001-0.

Ream, C.L. (1975) Date palm breeding: a progress report. *Annual Report Date Growers' Institute* 52, 8–9.

Reddy, M.P. (2015) Desert plant biotechnology: jojoba, date palm, and acacia species. In: Bahadur, B., Sahijram, M.V.R. and Krishnamurthyet, K.V. (eds) *Plant Biology and Biotechnology.* Vol. II. *Plant Genomics and Biotechnology.* Springer, New Delhi, pp. 725–741. DOI: 10.1007/978-81-322-2283-5_36.

Roychowdhury, R. and Tah, J. (2013) Mutagenesis: a potential approach for crop improvement. In: Hakeen, K.R., Ahmad, P. and Öztürk, O. (eds) *Crop Improvement.* Springer, New York, pp. 149–187.

Saaidi, M. (1990) Amélioration génétique du palmier dattier. Critères de sélection, techniques et résultats. In: Dollé, V. and Toutain, G. (eds) *Les Systèmes Agricoles Oasiens.* CIHEAM, Montpellier, France, pp. 133–154.

Saaidi, M. (1992) Comportement au champ de 32 cultivars de palmier dattier vis-à-vis du bayoud : 25 années d'observations. *Agronomie* 12(5), 359–370. DOI: 10.1051/agro:19920502.

Saaidi, M., Toutain, G., Bennerot, H. and Louvet, J. (1981) La sélection du palmier dattier *Phoenix dactylifera* L. pour la résistance au Bayoud. *Fruits* 36(4), 241–249.

Safronov, O., Kreuzwieser, J., Haberer, G., Alyousif, M., Schulze, W. *et al.* (2017) Detecting early signs of heat and drought stress in *Phoenix dactylifera* L. (date palm). *PLoS ONE* 12, e0177883. DOI: 10.1371/journal.pone.0177883.

Sahijram, L. and Rao, B.M. (2015) Hybrid embryo rescue in crop improvement. In: Bahadur, B., Sahijram, M.V.R. and Krishnamurthyet, K.V. (eds) *Plant Biology and Biotechnology,* Vol. II. *Plant Genomics and Biotechnology.* Springer, New Delhi, pp. 363–384. DOI: 10.1007/978-81-322-2283-5_18.

Saker, M.M., Adawy, S.S., Mohamed, A.A. and El-Itriby, H.A. (2006) Monitoring of cultivar identity in tissue culture-derived date palms using RAPD and AFLP analysis. *Biologia Plantarum* 50(2), 198–204. DOI: 10.1007/s10535-006-0007-3.

Sattar, M.N., Iqbal, Z., Tahir, M.N., Shahid, M.S., Khurshid, M. *et al.* (2017) CRISPR/Cas9: a practical approach in date palm genome editing. *Frontiers in Plant Science* 8, 1469. DOI: 10.3389/fpls.2017.01469.

Sattar, M.N., Iqbal, Z., Sarwar, S., Hassan, J., Al-Khateeb, S.A. *et al.* (2020) Date palm cultivation in the post-genomic era. In: Bolduc, B. (ed.) *Date Palm: Composition, Cultivation and Uses.* Nova Science Publishers, New York, pp. 161–184.

Sattar, M.N., Iqbal, Z., Al-Khayri, J.M. and Jain, S.M. (2021a) Induced genetic variations in fruit trees using new breeding tools: food security and climate resilience. *Plants* 10(7), 1347. DOI: 10.3390/plants10071347.

Sattar, M.N., Iqbal, Z. and Al-Khayri, J.M. (2021b) CRISPR-Cas based precision breeding in date palm: future applications. In: Al-Khayri, J.M., Jain, S.M. and

L. Abahmane

Johnson, D.V. (eds) *The Date Palm Genome*, Vol. 2. *Omics and Molecular Breeding*. Compendium of Plant Genomes, Springer, Cham, Switzerland, pp. 169–199. DOI: 10.1007/978-3-030-73750-4_9.

Sattar, M.N., Iqbal, Z., Naqqash, M.N., Jain, S.M. and Al-Khayri, J.M. (2021c) Induced mutagenesis in date palm (*Phoenix dactylifera* L.) breeding. In: Al-Khayri, J.M., Jain, S.M. and Johnson, D.V. (eds) *The Date Palm Genome*, Vol. 2. *Omics and Molecular Breeding*. Compendium of Plant Genomes, Springer, Cham, Switzerland, pp. 121–154. DOI: 10.1007/978-3-030-73750-4_7.

Sedra, M.H. (1995) Triage d'une collection de génotypes de palmier dattier pour la résistance au bayoud causé par *Fusarium oxysporum* f. sp. *albedinis*. *Al Awamia* 90, 9–18.

Sedra, M.H. (2005) Characterization of selected date palm clones and promotors for bayoud disease control. In: Boulanouar, B. and Kradi, C. (eds) *Proceedings of the International Symposium on the Sustainable Development of Oasis Systems*. INRA Editions, Rabat, pp. 72–79.

Sedra, M.H. (2011) Development of new Moroccan selected date palm varieties resistant to bayoud and of good fruit quality. In: Jain, S.M., Al-Khayri, J.M. and Johnson, D.V (eds) *Date Palm Biotechnology*. Springer, Dordrecht, The Netherlands, pp. 513–531. https://doi.org/10.1007/978-94-007-1318-5_24

Sedra, M.H. (2013) Analyse de la diversité génétique des cultivars marocains du palmier dattier utilisant certains traits agronomiques, phénologiques et marqueurs moléculaires. In: Bensalah, M. (ed.) *Premier Meeting International Sur Le Palmier Dattier: Ressources Phytogénétiques Du Palmier Dattier: Etat, Caractérisation et Défis de Gestion*. Institute of Arid Regions, Medenine, Tunisia, p. 20.

Sedra, M.H. and Besri, M. (1994) Evaluation de la résistance au Bayoud du palmier dattier causé par *Fusarium oxysporum* f. sp. *albedinis*: recherche d'une méthode de discrimination des vitroplants acclimatés en serre. *Agronomie* 14, 467–472.

Sghairoun, M. and Ferchichi, A. (2013) Sélection d'une nouvelle variété de palmier dattier dans l'oasis de Nefzaoua. In: Bensalah, M. (ed.) *Premier Meeting International Sur Le Palmier Dattier: Ressources Phytogénétiques Du Palmier Dattier: Etat, Caractérisation et Défis de Gestion*. Institute of Arid Regions, Medenine, Tunisia, p. 47.

Shabana, H., Al-Ani, B., Zaid, A. and Irshad, M. (2018) Investigation of new cultivars of date palm (*Phoenix dactylifera* L.) raised from seed (pit) germination. In: Zaid, A. and Alhadrami, G. (eds), *Proceedings of the Sixth International Date Palm Conference*, Khalifa International Award for Date Palm and Agricultural Innovation, Abu Dhabi, p. 556.

Solliman, M.E.D.M. (2017) Towards production of transgenic date palm using agroinjection or transient expression of plant and agrobacterium-mediated transformation. *Plant Cell Biotechnology and Molecular Biology* 18(7–8), 423–431. Available at: https://www.ikprress.org/index.php/PCBMB/article/view/1736 (accessed 15 November 2022).

Solliman, M.E.D.M., Mohasseb, H.A.A., Al-Khateeb, A.A., Al-Khateeb, S.A., Chowdhury, K. *et al.* (2019) Identification and sequencing of Date-SRY gene: a novel tool for sex determination of date palm (*Phoenix dactylifera* L.). *Saudi Journal of Biological Sciences* 26(3), 514–523. DOI: 10.1016/j.sjbs.2017.08.002.

Sree, K.S. and Rajam, M.V. (2015) Genetic engineering strategies for biotic stress tolerance in plants. In: Bahadur, B., Sahijram, M.V.R. and

Krishnamurthyet, K.V. (eds) *Plant Biology and Biotechnology*, Vol. II. *Plant Genomics and Biotechnology*. Springer, New Delhi, pp. 611–622. DOI: 10.1007/978-81-322-2283-5_30.

Sudhersan, C. and Al-Shayji, Y. (2011) Interspecific hybridization and embryo rescue in date palm. In: Jain, S.M., Al-Khayri, J.M. and Johnson, DV. (eds) *Date Palm Biotechnology*. Springer, Dordrecht, The Netherlands, pp. 567–584. DOI: 10.1007/978-94-007-1318-5.

Sudhersan, C., Al-Shayji, Y. and Jibi Manuel, S. (2009) Date palm crop improvement via T × D hybridization integrated with *in vitro* culture technique. *Acta Horticulturae* 829, 219–224. DOI: 10.17660/ActaHortic.2009.829.31.

Tazeb, A. (2017) Plant tissue culture technique as a novel tool in plant breeding: a review. *American-Eurasian Journal of Agricultural & Environmental Sciences* 17(2), 111–118. DOI: 10.5829/idosi.aejaes.2017.111.118.

Teh, C.-K., Ong, A.-L., Kwong, Q.-B., Apparow, S., Chew, F.-T. *et al.* (2016) Genome-wide association study identifies three key loci for high mesocarp oil content in perennial crop oil palm. *Scientific Reports* 6(1), 19075. DOI: 10.1038/srep19075.

Thareja, G., Mathew, S., Mathew, L.S., Mohamoud, Y.A., Suhre, K. *et al.* (2018) Genotyping-by-sequencing identifies date palm clone preference in agronomics of the State of Qatar. *PloS ONE* 13(12), e0207299. DOI: 10.1371/journal.pone.0207299.

Thoen, M.P.M., Davila Olivas, N.H., Kloth, K.J., Coolen, S., Huang, P.-P. *et al.* (2017) Genetic architecture of plant stress resistance: multi-trait genome-wide association mapping. *New Phytologist* 213(3), 1346–1362. DOI: 10.1111/nph.14220.

Tirichine, M. (1991) Caractérisatiques des palmeraies du Mzab et de Metlili. Ressources génétiques du palmier dattier: comportement vis-à-vis du Bayoud. *Bulletin du Réseau Maghrébin de Recherche sur la Phoeniciculture et la Protection du palmier dattier* 1(3), 7–10.

Torres, M.F., Mathew, L.S., Ahmed, I., Al-Azwani, I.K., Krueger, R. *et al.* (2018) Genus-wide sequencing supports a two-locus model for sex-determination in *Phoenix*. *Nature Communications* 9(1), 3969. DOI: 10.1038/s41467-018-06375-y.

Toutain, G. and Louvet, J. (1974) Lutte contre le Bayoud: orientations de la lutte au Maroc. *Al Awamia* 53, 114–162.

Tyagi, S., Kesiraju, K., Saakre, M., Rathinam, M., Raman, V. *et al.* (2020) Genome editing for resistance to insect pests: an emerging tool for crop improvement. *ACS Omega* 5(33), 20674–20683. DOI: 10.1021/acsomega.0c01435.

van Harten, A.M. (1998) *Mutation Breeding: Theory and Practical Applications*. Cambridge University Press, Cambridge.

Varshney, R.K., Bohra, A., Roorkiwal, M., Barmukh, R., Cowling, W.A. *et al.* (2021) Fast-forward breeding for a food-secure world. *Trends in Genetics* 37, 12. DOI: 10.1016/j.tig.2021.08.002.

Vats, S., Kumawat, S., Kumar, V., Patil, G.B., Joshi, T. *et al.* (2019) Genome editing in plants: exploration of technological advancements and challenges. *Cells* 8(11), 1386. DOI: 10.3390/cells8111386.

Verslues, P.E., Lasky, J.R., Juenger, T.E., Liu, T.W. and Kumar, M.N. (2013) Genome-wide association mapping combined with reverse genetics identifies new effectors of low water potential-induced proline accumulation. *Plant Physiology* 164(1), 144–159. DOI: 10.1104/pp.113.224014.

Wales, N. and Blackman, B.K. (2017) Plant domestication: wild date palms illuminate a crop's sticky origins. *Current Biology* 27(14), 702–704. DOI: 10.1016/j. cub.2017.05.070.

Wilde, H.D. (2015) Induced mutations in plant breeding. In: Al-Khayri, J.M., Jain, S.M. and Johnson, D.V. (eds) *Advances in Plant Breeding Strategies: Breeding, Biotechnology and Molecular Tools.* Springer, Cham, Switzerland, pp. 329–344. DOI: 10.1007/978-3-319-22521-0_11.

Woo, J.W., Kim, J., Kwon, S.I., Corvalán, C., Cho, S.W. *et al.* (2015) DNA-free genome editing in plants with preassembled CRISPR-Cas9 ribonucleoproteins. *Nature Biotechnology* 33(11), 1162–1164. DOI: 10.1038/nbt.3389.

Yaish, M.W. (2015) Proline accumulation is a general response to abiotic stress in the date palm tree (*Phoenix dactylifera* L.). *Genetics and Molecular Research* 14(3), 9943–9950. DOI: 10.4238/2015.August.19.30.

Yatta-El Djouzi, D., Al-Khayri, J.M. and Bouguedoura, N. (2020) Plant regeneration from cell suspension-derived protoplasts of three date palm cultivars. *Propagation of Ornamental Plants* 20(1), 28–38.

Yeole, M.P., Gholse, Y.N., Gurunani, S.G. and Dhole, S.M. (2016) Plant tissue culture techniques: a review for future view. *Critical Review in Pharmaceutical Sciences* 5, 16–24.

Zaher, H. and Sedra, M.H. (1998) Evaluation of some date palm hybrids for fruit quality and bayoud disease resistance. In: *Proceedings of Date Palm Research Symposium,* Arab Center for Studies on Arid Zones and Dry Lands (ACSAD), Damascus, pp. 181–192.

Zaid, A. and Al-Kaabi, H. (2003) Plant-off types in tissue culture-derived date palm (*Phoenix dactylifera* L.). *Emirates Journal of Food and Agriculture* 15, 17–35.

Zango, O., Rey, H., Bakasso, Y., Lecoustre, R., Aberlenc, F. *et al.* (2016) Local practices and knowledge associated with date palm cultivation in southeastern Niger. *Agricultural Sciences* 07(09), 586–603. DOI: 10.4236/as.2016.79056.

Zehdi-Azouzi, S., Cherif, E., Moussouni, S., Gros-Balthazard, M., Abbas Naqvi, S. *et al.* (2015) Genetic structure of the date palm (*Phoenix dactylifera* L.) in the Old World reveals a strong differentiation between eastern and western populations. *Annals of Botany* 116(1), 101–112. DOI: 10.1093/aob/mcv068.

Zehdi, S., Sakka, H., Rhouma, A., Ould Mohamed Salem, A., Marrakchi, M. *et al.* (2004) Analysis of Tunisian date palm germplasm using simple sequence repeat primers. *African Journal of Biotechnology* 3(4), 215–219. DOI: 10.5897/ AJB2004.000-2040.

Zein El Din, A.F.M., Abd Elbar, O.H., Al Turki, S.M., Ramadan, K.M.A., El-Beltagi, H.S. *et al.* (2021) Morpho-anatomical and biochemical characterization of embryogenic and degenerative embryogenic calli of *Phoenix dactylifera* L. *Horticulturae* 7(10), 393. DOI: 10.3390/horticulturae7100393.

Zhang, G., Pan, L., Yin, Y., Liu, W., Huang, D. *et al.* (2012) Large-scale collection and annotation of gene models for date palm (*Phoenix dactylifera*, L.). *Plant Molecular Biology* 79(6), 521–536. DOI: 10.1007/s11103-012-9924-z.

Zulkarnain, Z., Tapingkae, T. and Taji, A. (2015) Applications of *in vitro* techniques in plant breeding. In: Al-Khayri, J.M., Jain, S.M. and Johnson, D.V. (eds) *Advances in Plant Breeding Strategies: Breeding, Biotechnology and Molecular Tools.* Springer, Cham, Switzerland, pp. 293–328. DOI: 10.1007/978-3-319-22521-0_10.

Growth Requirements and Propagation of Date Palm

5

Melkamu Alemayehu*

Bahir Dar University, Bahir Dar, Ethiopia

Abstract

The date palm (*Phoenix dactylifera* L.) produces dates, an important fruit crop in arid and semiarid areas that contributes to food security and economic development in developing countries. The present chapter aims at providing updated information on climatic requirements and propagation to contribute to the development of date production. Research has shown that temperatures ranging from 18 to 22°C during flowering and from 25 to 29°C during fruiting are best suited for successful production of date fruits. Areas with sufficient irrigation water or a shallow water table are needed for date palm production. Rainfall at flowering is detrimental, reduces pollination efficiency, and causes rotting and physiological damages at late khalal, rutab, and tamar stages of dates. Date palms also need abundant sunshine and well-drained, deep, sandy loam soils with a pH range of 8.0 to 10.0 and free from calcium carbonate. The production site should be protected from strong winds that may reduce fruit quality through blown dust and mechanical damage. Although seed propagation is possible, it is not recommended for propagation of commercial elite date palm cultivars because of its heterozygous nature and production of off-type progeny. Similarly, offshoot propagation is not appropriate for large-scale multiplication of commercial elite cultivars due to limited number of offshoots per tree. Micropropagation, however, permits large-scale multiplication of elite commercial cultivars. Organogenesis and somatic embryogenesis are the main micropropagation techniques implemented. However, in somatic embryogenesis technique, problems associated with callus induction, tissue browning, genetic variability, bacterial contamination, and abnormal somatic embryos have been observed and require further research, while in organogenesis, excessive time is needed for production of *in vitro* plantlets.

Keywords: Arid climate, Offshoots, Organogenesis, Rutab stage, Somatic embryogenesis

*melkalem65@gmail.com

© CAB International 2023. *Date Palm* (eds J.M. Al-Khayri *et al.*)
DOI: 10.1079/9781800620209.0005

5.1 Introduction

Date, the product of the date palm (*Phoenix dactylifera* L.), is one of the oldest known fruit crops. Dates have been cultivated in North Africa and the Middle East and contributed to the food security and livelihoods of the population of the region for over 5000 years (Zohary and Hopf, 2000). The crop has been linked to ancient civilizations and had a very important influence on the history of the Middle East, where dates have great spiritual and cultural significance to the people. The date palm also has great religious significance. In Islam, the date palm is cited 21 times in the Holy Quran and 300 times in the Hadith of the Prophet Mohammed. Dates are usually used to break the long fasting days in the month of Ramadan (Al-Farsi and Lee, 2008). Similarly, the date palm is praised in the Christian and Jewish faiths and has been linked to numerous religious ceremonies (Musselman, 2007).

The date palm tree is well adapted to desert environments characterized by extreme high temperatures and water shortage. It is grown as a food source or as an ornamental plant in different continents of the world. The majority of the date-palm-growing areas are located in developing countries, where dates serve as a primary food crop. Palm trees and their different parts are also used for different purposes including construction of houses, bed frames, and bridges. Palm leaves are used as a raw material for weaving baskets, fans, ropes, sacks, and other materials (Al-Yahyai and Manickavasagan, 2012; Lemlem *et al.*, 2019).

5.2 Global Distribution

Date palms have been cultivated since ancient times. The earliest record from Iraq (Mesopotamia) shows that date palm cultivation began probably as early as 3000 BC. Due to a long history of cultivation and exchange of germplasm, it is difficult to identify the exact center of origin of the date palm, although evidence suggests the Mesopotamia area (current Iraq) as the center of origin. From the Mesopotamia area, date palm cultivation spread to the Arabian Peninsula, North Africa, and the Middle East. The spread of date cultivation later accompanied the expansion of Islam and reached southern Spain and Pakistan (Chao and Krueger, 2007). Spanish missionaries in the 18th and early 19th centuries facilitated dissemination of the date palm and it is currently cultivated in many regions and countries around the globe (Fig. 5.1).

Date consumption is at its peak level during the Holy Month of Ramadan since Muslims all over the world break their fast with dates. During Ramadan, 250,000 t of dates are consumed in Saudi Arabia, which is equivalent to one-quarter of its annual production (FAO, 2021). Another peak of consumption is during the annual Pilgrimage Holy Days to Saudi Arabia, observed by millions of Muslims from all over the world.

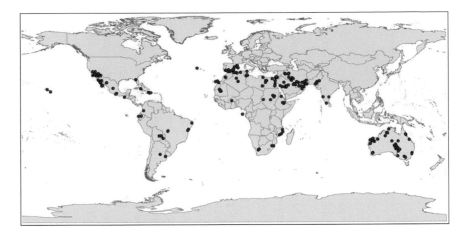

Fig. 5.1. Global distribution of date palm. Red dots indicate areas of production. (From Shabani *et al.*, 2012.)

Globally date palm production covers an area of 1,381,434 ha with a total production of 9,075,446 t. Asia is the leading continent, with 5,167,425 t produced on 930,467 ha of land, followed by Africa (3,825,538 t). Asia and Africa account for 56.9 and 42.2% of the total world harvest, respectively (FAO, 2019). The share of the Americas and Europe in date production was 68,448 and 14,035 t, respectively. The top ten date-producing countries in the world in 2020 were Egypt, Saudi Arabia, Iran, Algeria, Iraq, Pakistan, Sudan, Oman, Libya, and China (Table 5.1).

5.3   Climate Change Impacts on Date Palm Production

Evidence shows that climate is changing globally. Land surface temperatures are expected to increase by 4°C by 2100. Seasonal rainfall patterns are also changing worldwide (Jeffrey and Harold, 1999). Consequently, changes in climate will have serious impacts on the production and productivity of crops, including date palm, mainly through influencing climatic suitability and the occurrence of diseases and insect pests. Some areas that are now climatically suitable for date palm cultivation may become unsuitable, while some areas that are currently unsuitable may become suitable due to climate change (Shabani *et al.*, 2012). According to Shabani *et al.* (2012), many areas in North Africa with currently suitable climate are projected to become climatically unsuitable by 2100, while south-eastern Bolivia and northern Venezuela will become more suitable. Saudi Arabia, Iraq, and western Iran are projected to have a reduction in suitability by 2070. Similarly, Farooq *et al.* (2021) reported that most of the climatically suitable areas for date production in Saudi Arabia in the early 2030s will

Table 5.1. Top ten date producing countries in the 2020 production season. (Data from FAO, 2020.)

Rank	Country	Area (ha)	Production (t)	Climatic conditions
1	Egypt	50,834	1,690,959	Dry and hot, dominated by desert
2	Saudi Arabia	152,705	1,541,769	Desert climate, except south-western part (semi-arid climate)
3	Iran	154,145	1,283,499	Hot and dry climate characterized by long, hot, dry summers and short, cool winters
4	Algeria	170,500	1,151,909	Most part is desert, but coastal areas have a mild Mediterranean climate with hot summers and cool and rainy winters
5	Iraq	245,033	735,353	Hot and dry climate characterized by long, hot, dry summers and short, cool winters
6	Pakistan	106,488	543,269	Climatic conditions vary with the country's topography: dry and hot near the coast and along the lowland plains of the Indus River, becoming cooler in the northern uplands and Himalayas
7	Sudan	37,000	465,323	Tropical climate with high summer temperature, negligible rainfall, and frequent dust storms in desert zones
8	Oman	25,630	368,577	Subtropical and dry climate with summer monsoons and hot, dusty winds
9	Libya	32,868	177,629	Dominated by hot and arid climate, but Mediterranean climate along the coast characterized by cool rainy winter season and hot dry summer
10	China	12,400	158,671	Varied climate with freezing cold winter in the north and mild in the south; hot summer except in highlands; scarce and irregular rainfall in western desert

become unsuitable by 2100. The authors also project the northward shift of date production in Saudi Arabia due to climate changes in the future. The increase in temperature will also have impacts on irrigation water demand in that country by 2100. Crop irrigation water demands are projected to rise by about 602 million and 3122 million m³ at 1 and 5°C increase, respectively, and about 5 to >25% of the expected yield of fruit trees and crops are expected to be lost (Zatari, 2011).

Climate change may also impact the economy directly by affecting agricultural outputs. Reports estimate that maize production in Africa and South America could decline by 10% by 2055, causing a loss of $US2 billion per year due to climate change (Jones and Thornton, 2003). The total annual income from date palms in Middle Eastern countries declined from 1990 to 2000 due to plant diseases and water shortages resulting from climate change (Botes and Zaid, 2002). Generally, climate change is expected to have negative impacts on food security, productivity, and crop distribution (Wheeler and Von Braun, 2013). Therefore, preparing management strategies that address climate change impacts is critical to achieve long-term sustainable cash crop production for date palms.

5.4 Date Palm Climatic Requirements

The date palm is adapted to arid and semi-arid regions where high temperatures and low precipitation and humidity prevail. The fact that the date palm's natural habitat is in oases indicates that its water requirements are not necessarily low. However, the crop can be grown in areas with scarcity of rainfall if irrigation water is available. The date palm tolerates a temperature of 56°C for several days if irrigated, and temperatures below 0°C are also endured by the crop (Krueger, 2021).

5.4.1 Temperature requirement

Like any other crop, the growth and production of date palm are influenced by temperature. Dates can be successfully cultivated in areas having a long hot summer and mild temperatures during flowering (18–22°C) and relatively high temperatures (25–29°C) during fruit ripening, depending on the variety and production region. At temperatures between 7 and 32°C growth increases and remains constant at temperatures between 32 and 40°C. At temperatures above 40°C the growth of date palm starts to decrease (Krueger, 2021).

The date palm fruiting period, which ranges from 120 to 200 days, is also influenced by temperature and variety (Zaid and de Wet, 2002a; Krueger, 2021). The length of the fruiting period required for maturity of a specific cultivar is influenced by accumulation of heat units. Approximately 3000–4000 heat units are required for successful production of dates based on the sum of the daily temperature maximums starting from 18°C (Zaid and de Wet, 2002a; Krueger, 2021).

Although a thermophilic species, date palm withstands large temperature fluctuations. At a temperature of 7°C generally growth is inactive. Temperatures below 0°C cause metabolic disorders and damage leaf tissue. At –6°C pinnae margins become yellow and desiccated. At temperatures between –9 and –15°C leaves in the middle and outside canopy will be damaged and dry out. If these low temperatures persist for a long period (12 hour to 5 days), all leaves will show damage and fruit quality is reduced. Inflorescences of date palm can also be damaged by low temperatures (Zaid and de Wet, 2002a; Krueger, 2021).

5.4.2 Rainfall and water table requirements

The date palm needs sufficient water of acceptable quality to reach its potential yield. If well established, date palm tolerates drought. Most date-producing areas are found in arid and semi-arid climates with little rainfall, the rainfall occurring during the winter season. Water requirements of date palms can be supplied from groundwater or through irrigation. Although scarce, rainfall can have a negative effect on pollination and fruit set depending on the time of occurrence and the physiology of the date tree. Rainfall that occurs during flowering reduces the effectiveness of pollination during fertilization by washing off pollen. Any pollination operation immediately followed by rain (4–6 hour) must be therefore repeated. Rainfall also reduces temperature, which in turn reduces fruit set in date palm. Rainfall also increases relative humidity, creating favourable conditions for different diseases that reduce the yield and quality of dates. Moreover, high relative humidity is also associated with disintegration of pollen grains and reduced pollen supply (Zaid and de Wet, 2002a).

Rainfall at the early khalal stage of date fruit development is beneficial as it removes dust and sand particles from succulent fruits. However, rainfall during the late khalal, rutab, and tamar stages may cause rotting of fruits and physiological damage. The major damage caused by rainfall occurs when either the rain is early and occurs at the flowering stage, or when the dates are late in ripening due to low temperature. Generally, the amount of any particular rainfall event is of less importance than the conditions under which it occurs (Nixon and Carpenter, 1978). A light shower accompanied by prolonged periods of cloudy weather and high relative humidity may cause more damage than heavy rain followed by clear weather and dry winds. Rainfall accompanied with cool weather also delays ripening of date fruits (Zaid and de Wet, 2002a).

Zaid and de Wet (2002a) classified the growing season of a given area into four categories based on the chance of damaging rainfall during fruit maturity (August, September, and October for the northern hemisphere; January, February, and March for the southern hemisphere): (i) good to average, if less than 50 mm of rain fell in each of the three months; (ii) average to poor, if more than 50 mm of rain fell in one of the three months; (iii) poor, if more than 50 mm of rain fell in two of the three months; and (iv) very poor, if more than 50 mm of rain fell in each of the three months.

Rainfall is also beneficial in the reduction of salinity, which is common in arid and semi-arid regions, by leaching surface salt to the deeper soil layers. Selection of appropriate growing sites and varieties and protection of fruit bunches using physical barriers like craft paper help to minimize damage caused by rainfall and associated high relative humidity (Bashir *et al.*, 2015; Harhash *et al.*, 2020).

Date palms need a relatively shallow groundwater table of 0.8–1.5 m depth for successful production. This is because 85% of the roots are found within 2 m of the trunk and in the upper 2 m of the soil profile. With underground water close to the surface and high temperature requirements, date palm is described as having its feet in running water and its head in the fire of the sky. Shallow groundwater is also a valuable resource for irrigation water (Carr, 2012; Al-Muaini *et al.*, 2019).

5.4.3   Irrigation requirement

The date palm is believed to be a relatively high water-use crop where substantial water resources are needed for date production in highly productive areas (Montazar *et al.*, 2020). Therefore, irrigation is essential for successful production of dates since they are produced in arid and semi-arid climates where rainfall is scarce. Flooding is the dominant irrigation method used to supply water in most date-palm-growing countries, except for Israel, which uses drip irrigation. Ground- and surface waters are the sources of irrigation water (Elfeky and Elfaki, 2019). Water requirements for date production depend on the season of application, growing area, and application method. For instance, the summer water requirement (July, August, and September) of a date palm in Tunisia is estimated to be about 7154 m^3/ha, while only 4372 m^3/ha is needed for the winter period (December, January, and February). Summer requirements are almost double the winter ones and constitute one-third of the total annual consumption (Liebenberg and Zaid, 2002). In research conducted at eight date-palm-growing regions in Saudi Arabia based on the Penman–Monteith method, the average irrigation water requirement was 8342.41 m^3/ha per year (1 ha = 100 trees) (Al-Omran *et al.*, 2019). According to the results of experiments conducted in Riyadh, the average amount of water consumption per date palm tree per year was 2396 mm (Al-Amoud *et al.*, 2000).

Irrigation should be done at the time when the crop is in need of water using an appropriate method. Late and excessive irrigation is wasteful, although date palms do not suffer from excess water. Application of irrigation water for date palm trees should be based on the distribution of the root system where 40% of all roots are found in depth of 50 cm, 30% in

100 cm, 20% in 150 cm, and the remaining 10% in 150–200 cm. Thus, for mature date palm trees, 40% of all water is extracted from the first 50 cm, 70% from the first 100 cm, 90% from the top 150 cm, and only 10% from below 150 cm (Liebenberg and Zaid, 2002). Irrigation water for young date palm plants must be applied to a 10 to 30 cm radius around the trunk and at 25 to 50 cm depth (Liebenberg and Zaid, 2002). Due to the shallow root depth of young plants, frequent irrigation is also necessary to avoid water deficiency (for more details, refer to Al-Yahyai *et al.*, 2023 Chapter 6, this volume; Al-Omran *et al.*, 2023 Chapter 8, this volume).

5.4.4 Light requirement

Like for any other crop, light is necessary for the growth and development of the date palm. Light is used as the energy source for photosynthesis. However, strong sunlight increases air temperature that in turn creates heat stress to the growing plants. Heat stress, especially coupled with soil water deficit, causes physiological disorders including scorching and sunburn of leaves, twigs, branches, and stems; leaf senescence and abscission; shoot and root growth inhibition; and fruit discoloration and damage.

The date palm requires abundant sunshine and grows best in full sunlight, although light shade is tolerated. Both the intensity and the quality of light influence the growth of date palms. Ultraviolet-B (UV-B) radiation influences the growth and development of crops including soybean (Liu *et al.*, 2013), black gram (Shaukat *et al.*, 2013; Rajendiran *et al.*, 2015), strawberry (Valkama *et al.*, 2003), date palm (Rekab *et al.*, 2013; Niazwali *et al.*, 2020), and rice (Britto *et al.*, 2011). The effects generally encompass reduced photosynthesis, decreased vegetative growth, decreased developmental times, reduced fruit numbers and retention, and reduced biomass and yield (Kakani *et al.*, 2003). In the study conducted on five commercial cultivars of date palm by Niazwali *et al.* (2020), UV-B radiation reduced plant height and leaf number, shoot and root fresh and dry weights, and total chlorophyll and carotenoid concentrations while increasing the proline and phenolic compound concentrations. These changes were more pronounced under 8 hour exposure to UV-B compared to 4 hour exposure. Similarly, Rekab *et al.* (2013) found significantly reduced root dry weight in date palm varieties treated with UV-B for 4 and 8 hour. On the other hand, exposing shoots to red laser radiation for a short duration recorded the highest length and number of shoots and roots. The extent of light intensity can be reduced by clouds and shade, which may have a negative effect of growth and development of date palm. As the sky in the major date-growing countries of the northern hemisphere (July to October) as well as the southern hemisphere (February to May) is mostly cloudless during the ripening period (Zaid and de Wet, 2002a), scarcity of sunlight should not be a problem for the growth and production of date palms in the major date-palm-growing areas.

5.4.5 Wind and soil erosion

Date palm is not damaged by windy conditions compared to other plant species. It can withstand strong, hot, and dusty summer wind. However, dusts and sand carried by wind may adhere to the date fruits in their soft stage and reduce fruit quality. At early stages, date fruits can be also damaged by wind that beats fruits against the hard leaves. Strong winds have also a negative effect on the pollination of date palm. Severe hot and dry wind dries out the stigmas of the female flowers and shortens the time for the pollen to reach the ovule, while cold winds disturb pollen germination (Abul-Soad, 2011). While light winds are beneficial and favor pollination, high-velocity winds will blow away pollen and break the inflorescence's fruit stalk (rachis), damage the vascular system, and finally cause the death of bunches.

Date-palm-growing areas of arid and semi-arid regions are prone to wind, where it causes removal of soil fine particles in the form of wind erosion. Soil drifting is a fertility-depleting process that can lead to poor crop growth and yield reductions in areas where wind erosion is a recurrent problem. Continual drifting of an area gradually causes a textural change in the soil. Loss of fine sand, silt, clay, and organic particles from sandy soils lowers the moisture-holding capacity of the soil (Adams, 2021).

Covering fruit bunches with physical barriers like craft paper, cloth bags, and others protects fruits from damage caused by windblown fine sand, rainfall, sunburn, diseases, and insect pests. Moreover, such covering practices improve yield and quality parameters of date fruits (Mostafa *et al.*, 2014; Abul-Soad *et al.*, 2015; Bashir *et al.*, 2015; Harhash *et al.*, 2020). Planting appropriate species of windbreak trees against the wind direction also helps to reduce damage to crops and soil erosion caused by strong winds (Osorio *et al.*, 2018; Smith *et al.*, 2021).

5.4.6 Soil quality

Date palms can be grown in a wide range of soils. However, for best growth date palm requires well-drained, deep, sandy loam soils with pH of 8.0 to 10.0 (Dates Farming, 2022). Although date palm is claimed to be highly tolerant to salinity, its growth is negatively affected starting at electrical conductivity (EC) of 4–6 dS/m (Hussain *et al.*, 2012). The soils should have the ability to hold water and should not have a hard pan at above 2.5 m for better root development. Date palm soil should be free from calcium carbonate.

5.5 Propagation of Date Palm

Date palm can be propagated either by seed or vegetatively using offshoot and micropropagation technologies. Among the micropropagation

Fig. 5.2. Date palm seeds (a) and seedlings (b). (Photo (a) from https://www.
feedipedia.org/content/date-palm-seeds (accessed 15 November 2022),
available for use under Attribution 3.0 Unported (CC BY 3.0); photo (b) from
https://commons.wikimedia.org/wiki/File:Date_palm_seedlings.jpg (accessed
15 November 2022), available for use under CC0 1.0 Universal (CC0 1.0) Public
Domain Dedication.)

methods, organogenesis and somatic embryogenesis have been widely
researched by scientific communities.

5.5.1   Seed propagation

Seed propagation, also called sexual propagation, is the oldest means of
date palm propagation. It is by far the easiest and quickest method and
enlarges date palm genetic diversity. Hence, the technique is useful in
breeding programs for selection of progeny that can lead to the development of elite date cultivars with interesting traits. Some date-palm-growing
countries like the United Arab Emirates, Kuwait, Pakistan, Yemen, Ethiopia,
etc. are traditionally using seeds for date palm propagation (Abahmane,
2011; Lemlem *et al.*, 2018).

For seed propagation, seeds are removed from date fruits of desirable
quality and cleaned with fresh water to remove any excess flesh. Seeds
are then soaked in water for 24–48 hour to initiate water imbibition. Date
palm seeds can be sown in containers filled with soil or seedbeds having
desirable substrate characteristics (Fig. 5.2). Seeds are sown at appropriate
depth (1.8–2.9 cm) with the rough side down. Light watering is necessary
to ensure moisture at the seed's depth.

Germination of date palm seeds is influenced by temperature, humidity, light, types of soil mixes, and variety. Seeds germinate best at temperatures ranging from 21 to 24°C and relative humidity between 60 and 70%.
It is necessary to protect the germinating seeds from direct hot sunlight.
Moreover, the germination soil/medium should have relatively high
water-holding capacity and be aerated and free of pathogens. According
to Mohammed (2016), date palm seeds sown in a 2:1 sand–clay soil mix

Fig. 5.3. Ground offshoot of date palm (indicated by arrow). (Photo by M. Alemayehu.)

germinated quickly and recorded the highest germination percentage. Under ideal conditions viable seeds germinate within 3–8 weeks after sowing.

Seed propagation is, however, not suitable for the propagation of elite commercial cultivars because they produce off-type progeny associated with the highly heterozygous nature of the date palm. In seed propagation, desirable characteristics of the parent palm may be lost in the progeny (Abahmane, 2011). As date palm is a dioecious species, half of the progeny will be composed of male trees, which cannot be distinguished before flowering. Moreover, female plants derived from seeds usually produce late-maturing fruits of variable and generally inferior quality (Salomón-Torres *et al.*, 2021). In date palm plantations established using seedlings, about 10% of the palm trees produce fruit of satisfactory quality.

5.5.2 Vegetative propagation using offshoots

Offshoots are developed from axillary buds on the trunk of the mother plant. The use of offshoots is the most conventional vegetative technique used for propagation of date palm in different countries (Fig. 5.3). This method results in true-to-type progeny that produce fruits of the same quality as the mother palm and ensure uniformity in the produce. Offshoot plants will bear fruits about 2–3 years earlier than those plants sourced from seeds (Zaid and de Wet, 2002b).

The use of offshoots for date palm propagation is restricted due to that fact that the number of offshoots produced per palm tree per lifetime is very low (20–30 at most) and is restricted to the early life of the palm (10–15 years from the date of its planting); these effects vary with variety and management practices (Nixon and Carpenter, 1978). Although 20 to 30 offshoots are produced by a palm, only three or four offshoots are suitable for transplanting in the field within 1 year while the remaining must go into the nursery for 1 to 2 years before field planting. In addition, offshoots are difficult to root and their success in being established is relatively low (Eke *et al.*, 2005; Asemota *et al.*, 2007). Offshoot propagation can also spread dangerous diseases and insect pests of date palms.

The success of offshoot propagation varies with the type, source, and size of the offshoot; the time of removal from the mother plant; the rooting capacity of the cultivar; and the rooting medium used. Offshoots must be free from disease and insect pests and at least 3 to 5 years old with a base diameter of 20–35 cm, weighing 10–25 kg, showing signs of maturity (Nixon and Carpenter, 1978), and well connected to the mother tree (Abahmane, 2011). The signs of offshoot maturity are expressed in the buildup of roots, start of first fruiting, and production of a second generation of off-shoots (Nixon and Carpenter, 1978). Ground-level offshoots of large size (10–35 cm in diameter) are usually used for date palm propagation. They have the highest survival rates, this being associated with the availability of more stored carbohydrates for root growth and increased levels of naturally occurring root-promoting substances (Hodel and Pittenger, 2003). Small offshoots weighing 5 kg or less can be also used but they must be kept in a nursery for at least 2 years in a mist bed in a greenhouse or under a shade net structure. As fungi are usually a serious problem in a mist bed, offshoots must be treated twice monthly with a wide-spectrum fungicide (Zaid and de Wet, 2002b).

The time of offshoot removal and transplanting in the nursery is criti-cal as it directly influences the rooting capacity. Late spring (September/October in the southern hemisphere) or early summer (March/April in the northern hemisphere) is the best time for removal of offshoots, while February/March or September/October is the most suitable period for field planting in the southern and northern hemisphere, respectively (Zaid and de Wet, 2002b). Offshoot rooting in the nursery is also influenced by the offshoot's size, position of origin on the mother plant, the method of its removal and preparation for planting, and its treatment after plant-ing (Nixon and Carpenter, 1978). Lower and older offshoots are physi-ologically more active than the upper and younger offshoots and they grow faster as the number of leaves produced increases with age. The upper and younger offshoots have less carbohydrate than lower offshoots resulting in lower root production and consequently lower survival rate (Zaid and de Wet, 2002b).

Rooting medium in the nursery also affects the extent of root formation of offshoots and consequently the survival rate. In a study conducted by Al-Mana *et al.* (1996), the highest rooting percentages were obtained using a perlite–peat moss (3:1) medium followed by wood shavings–peat moss (1:1) and perlite–peat moss (1:1) media. The combination of peat moss and tuff also promoted growth and development of roots and shoots of date palm offshoots (Bitar *et al.*, 2019). Similarly, different researchers reported the positive effects of plant hormones on improvement of offshoot rooting capacity. Application of naphthalene acetic acid (NAA) at 500 ppm to aerial offshoots gave the highest rooting percentage (24.5%), total number of roots per offshoot (82.33), and total weight of roots per offshoot (607.4 g). Similarly, a high number of smaller roots (diameter <0.5 cm) was also achieved in offshoots treated with 500 ppm NAA (Tayyab *et al.*, 2021). Mansour and Khalil (2019) also reported that application of indole-3-butyric acid (IBA) at 8000 mg/L to ground offshoots recorded the highest success in establishment (100%) and the highest number of roots (60.0 roots per offshoot), new palm fronds (4.0 leaves per offshoot), and leaf dry matter (62.82%).

5.5.3 Micropropagation of date palm

Micropropagation is the process of asexually producing plants through tissue or cell culture techniques. Micropropagation is a promising alternative to satisfy the ever-increasing demand for date palm planting materials. The technology is used to clone a wide range of economically important palms. Somatic embryogenesis and organogenesis are the two major techniques used to produce micropropagated date palms. In the somatic embryogenesis method, embryos are produced from embryogenic callus and germinated to form complete plantlets (McCubbin *et al.*, 2000), while in organogenesis, plantlets are produced from multiplied buds without passing through the callus stage (Al-Khateeb, 2006). These date palm propagation methods are widely accepted throughout the world, and date palm commercial tissue culture laboratories are established in a number of countries (Table 5.2).

Date palm micropropagation techniques have tremendous benefits compared to conventional propagation methods such as large-scale multiplication of commercial cultivars; propagation of elite cultivars with desirable characters; production of homogeneous, vigorous, and disease-free plants; and exchange of plant material without any risk of spreading diseases or pests. Successful plant regeneration in micropropagation of date palm depends on the nature of the explant used (source, age, size); intensity and quality of light; temperature; pH of the medium; plant hormones; culture medium; and age of culture (Mazri, 2015; Mazri *et al.*,

Table 5.2. Date palm commercial tissue culture laboratories. (Compiled using information from the respective websites.)

Country	Laboratory name	Address	Website
Egypt	EFC Plants	PO Box 28, 44971 El Obour City, Cairo, Egypt	https://efcplants.com/ (accessed 15 November 2022)
France	Marionnet GFA	21 route de Courmemin, Soings-en-Sologne, France	https://www.zipmec.eu/en/marionnet-gfa_25966.html (accessed 15 November 2022)
India	ACE Date Palm Tissue Culture Laboratory	Plot No. 12, Angels Colony, Kompally Village, Secunderabad-500014, India	http://adptl.com/ (accessed 15 November 2022)
	Atul Rajasthan Date Palms Ltd	Atul 396020, Gujarat, India	https://www.ardp.co.in/ (accessed 15 November 2022)
Israel	Rahan Meristem Ltd	Kibbutz Rosh HaNikra, Western Galilee 22825, Israel	http://www.rahan.co.il/?page_id=6&lang=en (accessed 15 November 2022)
	Zemach Tissue Culture Laboratory	Zemach, Jordan Valley 15132, Israel	http://www.israelexporter.com/zemach-tissue-culture (accessed 15 November 2022)
Morocco	MDPP Marrakesh Date Palm Project	Résidence Marrakesh Plaza, Immeuble B5, Apt. No. B, 40,000 Morocco	https://mdpp.ma/ (accessed 15 November 2022)
	Les Domaine Agricole	Km 5, Azemmour Road, Dar Bouazza, Casablanca 21000, Morocco	https://pepiniereslesdomaines.com/ (accessed 15 November 2022)

Continued

Table 5.2. Continued

Country	Laboratory name	Address	Website
Oman	Date Palm Research Center	Ministry of Agriculture, PO Box 467, Muscat 113, Sultanate of Oman	https://omanportal.gov.om/wps/wcm/connect/en/site/home/gov/gov1/gov5governmentorganizations/maf/moa (accessed 15 November 2022)
Saudi Arabia	Al Rajhi Tissue Culture Laboratory, Clone Biotech	PO Box 55155, Riyadh 11534, Kingdom of Saudi Arabia	https://www.clonebiotech.com/ (accessed 15 November 2022)
	Sapad Tissue Culture Date Palm Co.	PO Box 1806, Dammam 31441, Kingdom of Saudi Arabia	https://www.sapad.com.sa (accessed 15 November 2022)
United Arab Emirates	Date Palm Tissue Culture Laboratory	United Arab Emirates University, PO Box 15551, Al Ain, Abu Dhabi, United Arab Emirates	https://www.uaeu.ac.ae/en/dvcrgs/research/centers/dpdrud/dptcl.shtml (accessed 15 November 2022)
	Al Wathba Marionnet LLC, Tissue Cultured Date Palms	PO Box 41522, Abu Dhabi, United Arab Emirates	https://www.awmpalms.com/ (accessed 15 November 2022)
UK	D.P.D. Ltd (Date Palm Developments)	Ham Street, Baltonsborough, Glastonbury BA6 8QG, UK	http://www.date-palm.co.uk/ (accessed 15 November 2022)
USA	Phoenix Agrotech LLC	2880A South Fairview Street, Santa Ana, CA 92704, USA	https://www.phoenixagrotech.com/ (accessed 15 November 2022)

2016). The challenges of date palm micropropagation include a long *in vitro* cycle, latent contamination, browning, somaclonal variation, *ex vitro* acclimation, and transplanting (Al-Khayri and Naik, 2017).

5.5.3.1 Date palm propagation by organogenesis

Organogenesis techniques are widely adopted for propagation of elite cultivars of date palm. Date palm micropropagation favors the direct organogenesis of an explant without callus formation stage where low concentrations of plant hormones are used (Khierallah and Bader, 2007). The organogenesis technique consists of five steps: (i) selection and preparation of offshoots; (ii) initiation of vegetative buds; (iii) bud multiplication; (iv) shoot elongation; and (v) rooting. The success of this technique is highly dependent on the success of vegetative bud initiation (Abahmane, 2011).

SELECTION AND PREPARATION OF OFFSHOOTS. Selection of suitable offshoots and their removal and preparation influence the success of the following steps of organogenesis. Offshoots with a weight of 2.5–6 kg, 3–5 years old, 60–80 cm in height or bearing eight to 12 leaves (Badawy *et al.*, 2005), and free from diseases and insect pests are suitable for organogenesis. Skilled manpower is required to cut and remove an offshoot properly without damage to its base. Once removed, the offshoot should be cleaned of soil and roots and leaves severely cut back. The offshoot is prepared by removing, gradually (one by one), the outer leaves and fibrous tissues from the base using a sharp knife until exposure of the shoot tip zone. The shoot tip about 3–4 cm in width and 6–8 cm in length is then excised from the offshoot by a circle cut around the base. The shoot tip is then transferred to an antioxidant solution containing 100 mg ascorbic acid and 150 mg citric acid to avoid tissue browning due to the phenolic compounds (Abahmane, 2011). The shoot tips are then disinfected using the following steps: (i) clean the shoot tips with distilled water to remove any organic debris; (ii) soak in a fungicide solution (benomyl, mancozeb) for 10–15 min; (iii) rinse three times with sterile distilled water; (iv) soak again, for 20 min, in a bleach solution (sodium hypochlorite) supplemented with potassium permanganate at 0.3 g/L; and (v) rinse three times with sterile distilled water under aseptic conditions (Abahmane *et al.*, 1999).

After sterilization, the root tip is dissected and the young leaves surrounding the apical dome are gradually removed. The entire shoot tip is cut into four to six pieces and transferred immediately to a culture medium to avoid desiccation. In general, an average of 15–25 explants can be extracted from each offshoot shoot tip (Abahmane, 2011).

INITIATION OF VEGETATIVE BUD. Bud initiation occurs after 3–6 months' incubation in dark conditions, which prevents oxidation of phenolic compounds. Each month, explants should be transferred to fresh basal MS (Murashige and Skoog, 1962) medium at full strength or diluted to half strength depending on the date palm cultivar (Abahmane, 2011). After bud initiation, shoots are transferred to a growth chamber with a photoperiod of 16 hour. Air temperature in the growth chamber is maintained at 27 ± 1°C during the illuminated period and 22 ± 1°C during the dark period (Abahmane *et al.*, 1999). Bud initiation generally requires 6–12 months depending on the genotype, culture medium, and the time of offshoot collection.

Vegetative bud initiation is also influenced by plant hormones where low concentrations improve the formation of vegetative buds (Al-Khateeb, 2006). The hormone application rate depends on the date palm cultivar. According to Khierallah and Bader (2007), cv. Maktoom initiated buds in the MS medium supplemented with 6-benzylamino purine (BAP) at 1 mg/L, 2-isopentyl adenine (2iP) at 2 mg/L, NAA at 1 mg/L, and 2-naphthoxyacetic acid (NOA) at 1 mg/L. Bekheet (2013) reported that MS medium fortified with 2iP at 2 mg/L and NAA at 1 mg/L promotes buds in cv. Zaghlool, while Al-Mayahi (2014) induced buds in cv. Hillawi in the MS medium supplemented with BAP at 1 mg/L and thidiazuron (TDZ) at 0.5 mg/L.

SHOOT MULTIPLICATION. Shoot multiplication is influenced by various factors including medium composition, genotype, and plant hormones. Similar to bud initiation, MS medium at full or half-strength is usually used for shoot multiplication. Cultures are transferred to fresh media at 4 to 6 week intervals (Abahmane, 2011). Aslam and Khan (2009) reported highest frequencies of shoot regeneration in explants of date palm cv. Khalas using 6-benzyladenine (BA) and kinetin (KIN), where increasing the concentration above 7.84 and 9.28 μM, respectively, decreased shoot regeneration. Research conducted by Khierallah and Bader (2007) showed that higher shoot-bud multiplication of date palm cv. Maktoom occurred in MS medium supplemented with a hormone combination of NAA at 1 mg/L, NOA at 1 mg/L, 2iP at 4 mg/L, and BAP at 2 mg/L. Similarly, Al-Mayahi (2014) reported the production of an average of 18.2 buds per culture of cv. Hillawi in MS medium containing BAP at 1 mg/L and TDZ at 0.5 mg/L. On the other hand, Mazri and Meziani (2013) used half-strength MS medium supplemented with NOA at 0.5 mg/L and KIN at 0.5 mg/L and produced 23.5 shoot buds per explant after 3 months of multiplication of cv. Najda. In general, the most common plant growth regulators (PGRs) used at the multiplication stage are auxins NAA, IAA, and NOA, and cytokinins BA, 2-iP, and KIN (Abahmane, 2011).

Carbon sources and their concentration in the culture medium also influence shoot multiplication of date palm. According to Al-Khateeb (2008), sugar concentration of 30 and 60 g/L was optimal for shoot growth of cv. Khanezi while 90 and 120 g/L resulted in abnormal growth. Al-Khateeb (2008) also reported that maltose, fructose, and glucose were almost equally effective as a carbon source for date palm tissue culture as sucrose.

SHOOT ELONGATION AND ROOTING. Shoot elongation and rooting of explants vary with the genotype, medium, and hormones used, although the importance of plant hormones is controversial. Earlier reports showed that the transfer of shoots from the multiplication medium to another medium with a high auxin/cytokinin ratio was required for shoot elongation (Al-Khateeb and Al-Khateeb, 2016). On the other hand, Al-Khayri and Naik (2017) reported that the medium with or without hormone promoted shoot elongation and rooting in date palm. According to Mazri and Meziani (2013), shoot elongation and frequency of root formation in cv. Najda were faster in the medium supplemented with hormones than in a hormone-free medium. However, plantlets produced in the hormone-free medium were wider and had green leaves with optimum survival rates. Meziani *et al.* (2015) also reported cv. Mejhoul shoots with an average length of 13.4 cm and 4.6 roots per shoot with wide and green leaves at 3 months on hormone-free half-strength MS medium.

The combinations of NAA or IBA, KIN, and 2-iPA or BA are the most effective PGRs at shoot elongation and rooting stage (Abahmane, 2011). Successful shoot growth and elongation have been achieved in different date cultivars using different combinations of plant hormones (Aslam and Khan, 2009; Abahmane, 2011). MS salt at three-quarters strength was found best for root formation as compared to one-quarter, one-half and full strength (Al-Kaabi *et al.*, 2001; Taha *et al.*, 2001). In the same studies, light intensity of 8000 lux and sucrose at 40 g/L produced the best results on root number and length.

PLANT ACCLIMATION. Acclimation of plantlets is the final step of micropropagation. This step is essential for tissue-cultured plantlets as they are easily impaired by sudden changes in the field environmental conditions. During the acclimation process, optimization of soil mixtures, relative humidity, temperature, and other greenhouse conditions as well as the quality of plantlets should be given due attention. For maximum survival rate in a greenhouse, plantlets must be at least 12–15 cm long with a well-formed and closed crown, have two or three fully opened leaves, and have more than three roots (Abahmane, 2011). Researchers have reported that application of chemicals during *in vitro* propagation like polyethylene glycol (PEG) to the culture medium (Sidky *et al.*, 2009), γ-aminobutyric acid (Awad *et al.*, 2006), or the growth retardant paclobutrazol to the rooting medium (Seelye *et al.*, 2003) helped to enhance survival rates of plantlets in the greenhouse.

Soil mixture is also the other factor that influences the survival rate of plantlets during the acclimation stage in the greenhouse. The soil mix should have optimal water-holding capacity. Water content that is too high favors fungal rot diseases and too low causes plantlet desiccation. According to Tisserat (1982), date palm plantlets having 10–12 cm height transferred to a peat moss–vermiculite mixture in 1:1 ratio (v/v) and covered with transparent plastic achieved the best survival rate. Using plastic micro-tunnels

Fig. 5.4. Micropropagation of date palm. (a) Shoot tip culture; (b) embryogenic callus induction; (c) embryogenic callus proliferation; (d) embryo formation; (e) rooting; and (f) transplanted plant. (From Al-Khayri and Naik, 2017, used under Attribution 4.0 International (CC BY 4.0) license.)

for newly transferred plantlets in the greenhouse during the first 3 weeks of acclimation maintains high relative humidity around plantlets and thus enhances survival rate. In addition, plantlets should be protected against fungi that cause crown and leaf rot by application of wide-spectrum fungicides twice weekly. Water supply during the first month of acclimation should be also carefully monitored (Abahmane, 2011). According to El Kinany *et al.* (2019), incorporation of compost into the substrate and its inoculation with the arbuscular mycorrhizal fungus, *Glomus iranicum*, increased biomass production (root and shoot biomass) and chlorophyll and mineral nutrient contents of date palm plantlets compared to plantlets grown in substrates amended with compost only or without compost.

5.5.3.2 *Date palm propagation by somatic embryogenesis*

Somatic embryogenesis is the process by which somatic cells develop into somatic embryos after a series of biochemical and morphological changes where the formed embryos are morphologically similar to zygotic embryos. It is widely used for mass propagation of date palm (Fki *et al.*, 2011). The process of somatic embryogenesis encompasses a series of steps including embryogenic callus induction, somatic embryo formation, somatic embryo maturation, and conversion into plantlet (Fig. 5.4). The success of somatic embryogenesis generally depends on genotype, explant type, and PGRs (Mazri and Meziani, 2015).

EMBRYONIC CALLUS INDUCTION AND SOMATIC EMBRYO FORMATION. Embryonic callus can be induced from explants sourced from different date palm parts including zygotic embryos (Fki *et al.*, 2011), shoot tips (Zayed, 2017), leaves (Fki *et al.*, 2017), lateral buds, or inflorescences (Zayed *et al.*, 2015; Al-Ali *et al.*, 2017; Hassan *et al.*, 2021). MS (Murashige and Skoog, 1962) culture medium is commonly used for date palm tissue culture. Solid media have been extensively used to produce somatic embryos from embryogenic calli. However, it is not suitable for large-scale mass propagation of date palms (Fki *et al.*, 2003). The success of callus induction is influenced by PGRs, genotype, explant type, and induction period (Fki *et al.*, 2011).

Among the PGRs, 2,4-dichlorophenoxyacetic acid (2,4-D) is the most effective auxin used for embryogenic callus induction in date palm, the concentration differing with genotype. Research findings suggest that 2,4-D at 100 mg/L induces embryogenic calli on different date palm culti-vars (Eshraghi *et al.*, 2005; Al-Khayri, 2011; Mazri and Meziani, 2015). On the other hand, Fki *et al.* (2011) reported that high doses of 2,4-D may induce somaclonal variation. The use of relatively low concentrations of 2,4-D for callus induction of different date palm cultivars is also reported by researchers. For example, El Hadrami *et al.* (1995) used a concentration of 5 mg/L in cvs. Iklane and Jihel, while Othmani *et al.* (2009a) suggested 2,4-D at 10 mg/L for cv. Boufeggous. Aslam *et al.* (2011) also used a 2,4-D concentration of 1.5 mg/L for somatic callus induction in date palm culti-vars. Concentration of 2,4-D < 1 mg/L was also efficient for inducing callus (Fki *et al.*, 2003). According to Fki *et al.* (2011), cytokinins such as 2iP may not be necessary for induction of callus.

The process of embryogenic callus induction in date palm is gener-ally very slow, varying from a few to several months depending on the genotype. In this regard, Othmani *et al.* (2009a) recorded somatic embryo formation after 6–7 months in cv. Boufeggous. Similarly, mature date palm inflorescence explants (cv. Shaishee) incubated at $25 \pm 2°C$ in dark condi-tion induced callus in 6–8 months (Al-Ali *et al.*, 2017). Hassan and Taha (2012) observed somatic embryos in different cultivars after 9 months of induction while Eshraghi *et al.* (2005) recorded an induction period of 12 months for cvs. Khanizi and Mordarsing.

In addition, the use of biotin and thiamin (Al-Khayri, 2001) and silver nitrate (Al-Khayri and Al-Bahrany, 2003) may promote the quantity and quality of embryogenic calli. Date palm syrup extract at the concentration of 6% can be also used as a sucrose substitute for somatic embryogenesis induction (Al-Khateeb, 2008) and meristematic tissue extracts of dates enhanced somatic embryogenesis (El-Assar *et al.*, 2004).

MULTIPLICATION, DEVELOPMENT, AND MATURATION OF SOMATIC EMBRYO. Many factors reportedly affect date palm somatic embryo multiplication, devel-opment, and maturation. Although solid media can be used to produce somatic embryos, embryogenic suspension cultures using liquid media are preferable for ease of large-scale commercial propagation of date palm (Fki *et al.*, 2011). Moreover, the number of matured embryos is considerably increased in liquid medium (Fki *et al.*, 2003). The authors reported the production of about 200 embryos from 100 mg fresh weight of callus after culturing for 1 month while only ten embryos were recovered from the same amount of callus on a solid medium.

Chopping of the embryogenic callus into small pieces has also a positive effect on the differentiation of date palm somatic embryos (Fki *et al.*, 2003). Half-strength liquid MS medium (Murashige and Skoog, 1962) as well as addition of 2,4-D at 1 mg/L and activated charcoal at 300 mg/L promoted somatic embryo differentiation. PEG at a concentration of 10% stimulated production of embryos while the addition of as low as 1 µM abscisic acid (ABA) to the suspension inhibited growth (Al-Khayri and Al-Bahrany, 2012). Similarly, addition of thiamin and biotin increased the number of embryos and promoted their elongation in cv. Khalas (Al-Khayri, 2001). Other factors such as yeast extract and casein hydrolysate (Al-Khayri, 2011) and paclobutrazol (Khierallah *et al.*, 2015) reportedly influenced somatic embryogenesis in date palm.

GERMINATION OF SOMATIC EMBRYOS. Germination of somatic embryos can be achieved by application of different supplements and techniques. For instance, a significantly improved germination rate of somatic embryos was recorded by reducing the water content of matured somatic embryos from 90 to 75% (Fki *et al.*, 2003). Similarly, cutting back the cotyledon leaf to about half of its length stimulated embryo germination to an 80% germination rate compared to 25% with an intact cotyledon leaf.

Supplementing the medium with different plant hormones also improves the germination of somatic embryos. The combined use of BAP, IBA, and NAA promoted embryo germination in date palm cvs. Jihel and Bousthami Noir (Zouine and El Hadrami, 2007). Supplying NAA at 1 mg/L has shown the conversion of 81% of somatic embryos into plantlets (Othmani *et al.*, 2009a). Moreover, other factors like the strength and texture of the culture medium (Boufis *et al.*, 2014), genotype used, and size of the somatic embryo (small, medium, large, very large) influence somatic embryo germination (Al-Khayri and Al-Bahrany, 2012).

ACCLIMATION OF SOMATIC EMBRYO-DERIVED PLANTLET. Acclimation is the last stage of micropropagation. If it is not properly done, it leads to losses of plantlets. In order to have a relatively high survival rate, plants transplanted into the greenhouse should acquire certain quality parameters such as resistance to higher light intensity, lower relative humidity, fluctuation of temperature, and biotic stresses. Optimal acclimation conditions should ensure 70% survival rate of 12-month-old plantlets at the temperature range of 25–30°C, under moderate light intensity, and in well-drained substrates (Fki *et al.*, 2011).

Survival rates in the greenhouse can be improved by additives such as 5-aminolevulinic acid-based fertilizer (Awad, 2008) and ammonium nitrate and gibberellic acid (Darwesh *et al.*, 2011), as well as by inoculation with rhizosphere bacteria (Farrag *et al.*, 2011).

5.6   Constraints of Date Palm Micropropagation

Although there are tremendous benefits, date palm micropropagation has faced various major problems such as tissue browning, genetic variability, bacterial contamination, and abnormal somatic embryo differentiation (Mazri and Meziani, 2015; Al-Khayri and Naik, 2017). Browning of *in vitro*-cultured explants is associated with high levels of caffeoylshikimic acids found in date palm tissues (Loutfi and El-Hadrami, 2005), which can be reduced by application of activated charcoal into the culture medium (Boufis *et al.*, 2014). The use of ascorbic acid and/or citric acid in disinfectant solution also prevents browning (Hassan and Taha, 2012; Khierallah *et al.*, 2015). Transfer of a culture to a fresh medium every 7 days may also help to reduce tissue browning (Othmani *et al.*, 2009a).

Morphological and genetic variation of micropropagated plants from their originating plant is another constraint observed by researchers. The use of morphological parameters and molecular markers for confirmation of the genetic stability of tissue-cultured date plants is a common tool in breeding programs. Accordingly, Sedra (2005) found similarities between commercial date cvs. Mejhool, Sairlaylate, and Najda (INRA-3014) and their tissue-cultured progeny based on their distinctive phenological descriptors. On the other hand, Al-Mazroui *et al.* (2007) reported a high incidence of dwarfism (18%) in cv. Sukkari propagated by indirect embryogenesis. Scarcity in flowering, fertilization failure, and off-type offshoots were also observed in five important date cultivars, which did not exceed 5%. Excessive vegetative growth, uneven ripening, delay in flowering, and abnormal morphology and structure occurred with a very low incidence of less than 1%. Dwarfism and abnormal flower development were also observed in tissue-cultured cvs. Barhee and Khalas (Zaid and Al-Kaabi, 2003; Al-Kaabi *et al.*, 2007).

Molecular characterization provides an assurance of genetic conformity of micropropagated plantlets (Al-Khayri and Naik, 2017; Mirani *et al.*, 2022), where different molecular markers including amplified fragment length polymorphism (AFLP), random amplified polymorphic DNA (RAPD), restriction fragment length polymorphisms (RFLP), inter-simple sequence repeat (ISSR), and simple sequence repeat (SSR) have been used by different researchers (Saker *et al.*, 2006; Kumar *et al.*, 2010; Othmani *et al.*, 2010; Khanam *et al.*, 2012). According to Moghaieb *et al.* (2011), a high level of variability (37.8%) in micropropagated date cv. Ferhi was observed using RAPD primers. A significantly high level of genetic variation was also detected among cv. Medjool plants generated from tissue culture (Gurevich *et al.*, 2005). On the other hand, Kumar *et al.* (2010) evaluated micropropagated date plants with RAPD and ISSR primers and suggested that the micropropagated plants were true-to-type. Similarly, Aslam *et al.* (2015) confirmed the genetic stability of micropropagated plants of six cultivars (Barhee, Zardai, Khalasah, Muzati, Shishi, and Zart) and their mother plants using RAPD markers. Mirani *et al.* (2022) evaluated the genetic stability of micropropagated date cv. Dedhi using inter-retrotransposon amplified polymorphism (IRAP) markers and found reproducible bands that were all monomorphic and showed no variation among the tissue culture-derived plants. Moreover, there was no significant variation in the fruit characteristics of the tissue culture-derived plants when compared to the mother plants, confirming the effectiveness of the current micropropagation protocol for producing true-to-type plants.

As indicated above, there is controversial information about the genetic variability of micropropagated date palm cultivars. Genetic stability analysis of tissue culture-derived plants using genetic markers is therefore necessary to verify that the plantlets are true-to-type. Researchers have also suggested that the use of juvenile explants, low concentrations of auxins (2,4-D), and minimum subculturing can reduce the risk of genetic variability of date palm clones (El Hadrami *et al.*, 2011; Fki *et al.*, 2011).

5.7 Bioreactor for Date Palm Propagation

Although several protocols of *in vitro* propagation using agar-solidified media have been optimized successfully, the protocols are difficult to utilize for mass clonal propagation of date palm and have high production costs. The low shoot multiplication rate induced in agar-solidified media is also another problem that reduces the success of commercial tissue culture. Therefore, the use of a bioreactor with a liquid medium is necessary for commercial date palm propagation (Othmani *et al.*, 2011).

A bioreactor is a self-contained and sterile environment with liquid/air inflow and outflow systems. The system ensures the control of pH, oxygen, ethylene, carbon dioxide concentration, aeration rate, and temperature. The system is also compatible with the automation of micropropagation procedures and reduces labour costs.

Bioreactors were first used to culture microorganisms. Later, they were used for accumulation of cell biomass from cell suspension for secondary metabolite production (Georgiev *et al.*, 2014). Currently, bioreactors play an important role in commercialization of somatic embryogenesis and multiplication of clusters of meristem- and bud-based plant micropropagation (Othmani *et al.*, 2011; Jain, 2012). According to Etienne *et al.* (2006), bioreactors can be classified into the four categories shown in Table 5.3.

Table 5.3. Classification of bioreactors for micropropagation.

Bioreactor category	Type of bioreactor
Mechanically agitated bioreactors	Aeration-agitation bioreactors Rotating-drum bioreactors Spin-filter bioreactors
Pneumatically agitated bioreactors	Air-lift bioreactors Bubble column bioreactors Simple aeration bioreactors
Non-agitated bioreactors	Gaseous phase (mist) bioreactors Oxygen-permeable membrane bioreactors Overlay aeration bioreactors Perfusion bioreactors
Temporary-immersion bioreactors	Systems with temporary complete immersion by pneumatic-driven transfer of liquid medium

Research findings have indicated that wide ranges of plant species can be regenerated using bioreactors including banana (Kosky *et al.*, 2002; Uma *et al.*, 2021), coffee (Etienne *et al.*, 2006), guava (Akhtar, 2012), and date palm (Othmani *et al.*, 2009b, 2017; Almusawi *et al.*, 2017). As a source of explants, different parts of plants including micro-cuttings, somatic embryos, apical shoots, flowers, roots, leaves, seeds, rootstocks, or shoot clusters can be used in bioreactor regeneration.

Temporary immersion bioreactors consist of two compartments where the upper part is the growth chamber and holds the plant material on a polyurethane filter and the lower part contains the growth medium/ nutrient solution (Othmani *et al.*, 2011; Uma *et al.*, 2021). The temporary immersion system provides the most satisfactory conditions for date palm regeneration via shoot organogenesis and allows a significant increase of multiplication rate (5.5-fold) in comparison with regeneration on agar-solidified medium (Othmani *et al.*, 2017). Almusawi *et al.* (2017) also reported that the Plantform bioreactor improved multiplication rate, reduced micropropagation time, improved weaning success, and reduced unit cost, thus improving economic return for the commercial propagation of date palm. Its use also addressed bottlenecks that hampered the scale-up of date palm micropropagation including asynchrony of somatic embryos, limited maturation of somatic embryos, and highly variable germination frequencies of embryos.

5.8   Synthetic Seed Technology

Synthetic seeds are defined as artificially encapsulated somatic embryos, shoot buds, cell aggregates, or any other tissue that can be used as a seed. They possess the ability to produce plants under *in vitro* or *ex vitro*

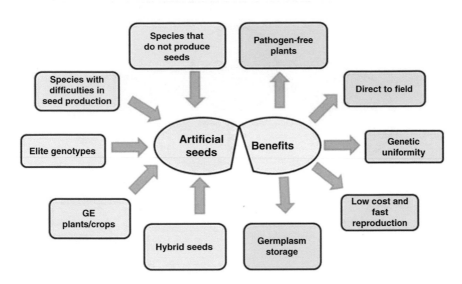

Fig. 5.5. Potential uses and benefits of synthetic seeds. (GE, genetically engineered.) (Rihan *et al.*, 2017, used under Attribution 4.0 International (CC BY 4.0) license.)

conditions as well as after storage (Magray *et al.*, 2017). The production of synthetic seed is one of the applications of somatic embryogenesis that is most appropriate especially for dioecious and vegetatively propagated plant species such as the date palm (Fki *et al.*, 2011). Synthetic seed encapsulation is an effective technique for rapid and large-scale *in vitro* multiplication of elite cultivars of date palm (Magray *et al.*, 2017; Poddar and Poddar, 2021). The synthetic seed technology is developed for different economically important plant species such as vegetable and fruit crops, industrially important crops, cereals, spices, plantation crops, ornamental plants, medicinal plants, etc. (Reddy *et al.*, 2012).

Potential use of synthetic seeds in plant production is tremendous. They can be used for propagation of seedless plant species or species with difficulties in seed production and production of true-to-type elite genotypes. Artificial seeds can be also used to produce male or female sterile plants for hybrid seed production (Saiprasad, 2001) as well as transgenic plants (Daud *et al.*, 2008) and for germplasm conservation (Standardi and Micheli, 2012). Synthetic seeds with protective coating may also increase the success rate of micropropagules in the field (Ara *et al.*, 2000). As synthetic seeds are produced in aseptic conditions, they are free of pathogens and avoid the spread of plant diseases (Nyende *et al.*, 2005; Daud *et al.*, 2008). Moreover, the technique economizes on the space, medium, and time required by the traditional tissue culture methods (Mathur *et al.*, 1989). Furthermore, artificial seeds are more durable, and easy to handle,

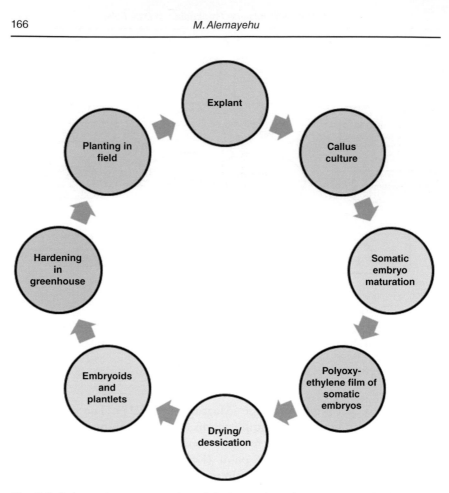

Fig. 5.6. Schematic representation of desiccated synthetic seed production. (Adapted from Poddar and Poddar, 2021.)

transport, and store. Generally, potential uses of synthetic seeds and their benefits are summarized in Fig. 5.5.

Depending on the production procedures, two types of synthetic/ artificial seeds (encapsulated somatic embryos) are commonly produced: desiccated and hydrated (Rihan *et al.*, 2017; Poddar and Poddar, 2021). Desiccated artificial seeds are produced from somatic embryos either packed or encapsulated in polyoxyethylene glycol followed by their desiccation. Desiccation can be applied either rapidly by leaving artificial seeds in unsealed Petri dishes on the bench overnight to dry, or slowly over a more controlled period of reducing relative humidity (Ara *et al.*, 2000). These types of artificial seeds can only be produced from somatic embryos that are desiccation tolerant (Sharma *et al.*, 2013). The schematic representation of desiccated synthetic seed production from somatic embryos is presented in Fig. 5.6.

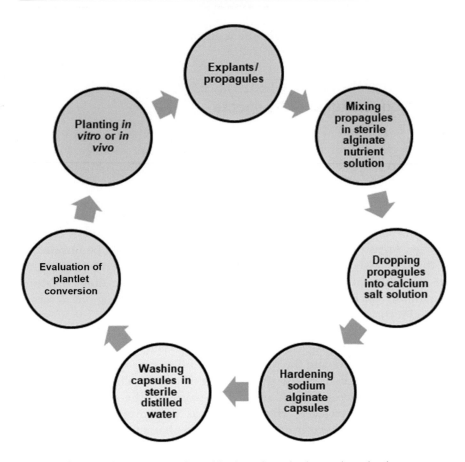

Fig. 5.7. Schematic representation of hydrated synthetic seed production. (Adapted from Poddar and Poddar, 2021.)

Hydrated artificial seeds can be produced by encapsulating somatic embryos in hydrogel capsules. They are produced in plant species with recalcitrant somatic embryos that are vulnerable to desiccation (Ara *et al.*, 2000; Poddar and Poddar, 2021). The schematic representation of hydrated synthetic seed production is presented in Fig. 5.7.

Although having various benefits, synthetic seed technology has also its limitations. Production of an inefficient amount of synthetic seed may increase the unit cost of micropropagules. Limitations in storage caused by lack of dormancy, synchronic deficiency in somatic embryo development, improper maturation, and low levels of conversion into plantlets are the other limitations of synthetic seed technology that need to be solved. Furthermore, as gelling agents dry out easily, they may hinder respiration of embryos (Reddy *et al.*, 2012; Rihan *et al.*, 2017; Poddar and Poddar, 2021). Therefore, synthetic seeds need to be stored under humid storage conditions or coated with a hydrophilic membrane, which help in embryo respiration (Poddar and Poddar, 2021).

5.9   Conclusion and Prospects

Date is one of the major fruit crops grown in arid and semi-arid areas characterized by extreme high temperatures and water shortage. Like any other crop, the growth and production of the date palm is influenced by climatic and edaphic conditions, such as temperature, humidity, sunlight, and wind. Suitability of a given area for date production is therefore mainly determined by its climatic and edaphic characteristics. Reports project that suitability of a given area for date production could be affected by climate change. Areas currently suitable for date production could be unsuitable in the future and vice versa. Due attention should be therefore given to the climatic requirements of the date palm when selecting sites for future expansion, which is necessary to support life in the desert regions in terms of food security and economic development.

Date palms can be propagated either by seed or vegetatively using offshoot and micropropagation technologies. Due to the date palm's heterozygous nature and the production of off-type progeny, seed propagation is not suitable for the propagation of commercial elite cultivars. Offshoot and micropropagation of date palm, on the other hand, are desirable for propagating elite cultivars as they produce true-to-type genotypes and produce fruits with the same characteristics and quality as the mother palm. However, due to limitations like limited number of offshoots, high risk of disease spread, limited rooting capacity, etc. offshoot propagation is not well suited for large-scale multiplication of commercial cultivars.

Micropropagation, on the other hand, permits large-scale multiplication of commercial date palm cultivars, which has a tremendous positive impact on the development of the fruit sector in arid and semi-arid regions. Organogenesis and embryogenesis are the two major micropropagation techniques used for date palm micropropagation. Since plantlets produced by the organogenesis technique are obtained directly from mother tissue, they are true-to-type and identical to the mother tree. In the embryogenesis technique, somatic embryos are regenerated from embryogenic callus. This technology can be used for large-scale propagation of commercial varieties of date palm, thereby opening the way for production of synthetic seeds using bioreactors. Generally, micropropagation is an appropriate technology for date palm propagation. However, further research is required to overcome problems associated with callus induction, tissue browning, genetic variability, bacterial contamination, and abnormal somatic embryos.

References

Abahmane, L. (2011) Date palm micropropagation via organogenesis. In: Jain, S.M., Al-Khayri, J.M. and Johnson, D.V. (eds) *Date Palm Biotechnology*. Springer, Dordrecht, The Netherlands, pp. 69–90.

Abahmane, L., Bougerfaoui, M. and Anjarne, M. (1999) Use of tissue culture techniques for date palm propagation and rehabilitation of palm groves devastated by bayoud disease. In: *Proceedings of the International Symposium on Date Palm*, Assiut University, Assiut, Egypt, November 9–11, 1999, pp. 385–388.

Abul-Soad, A.A. (2011) *Date Palm in Pakistan, Current Status and Prospective*. USAID Firms Project. Available at: www.researchgate.net/profile/Anoop_Srivastava7/post/Date-palm-cultivation/attachment/59d64ef079197b80779a8301/AS%3A495243864416256%401495086746794/download/pnaea333.pdf (accessed April 2022).

Abul-Soad, A.A., Mahdi, S.M. and Markhand, G.S. (2015) Date palm status and perspective in Pakistan. In: Al-Khayri, J.M., Jain, S.M. and Johnson, D.M. (eds) *Date Palm Genetic Resources and Utilization*. Springer, Dordrecht, The Netherlands, pp. 153–205.

Adams, T. (2021) Wind erosion and land degradation: an overview. Available at: https://globalroadtechnology.com/wind-erosion-types-factors-causes/ (accessed 5 March 2022).

Akhtar, N. (2012) Somatic embryogenesis for efficient micropropagation of guava (*Psidium guajava* L.). In: Lambardi, M., Ozudogru, E. and Jain, S. (eds) *Protocols for Micropropagation of Selected Economically-Important Horticultural Plants*, Vol. 994. Methods in Molecular Biology, Humana Press, Totowa, New Jersey, pp. 161–177.

Al-Ali, A.M., Ko, C.Y., Al-Sulaiman, S.A., Al-Otaibi, S.O., Al-Khamees, A.U.H. *et al.* (2017) Indirect somatic embryogenesis from mature inflorescence explants of date palm. In: Al-Khayri, J.M., Jain, S.M. and Johnson, D.V. (eds) *Date Palm Biotechnology Protocols*, Vol. 1. *Tissue Culture Applications*. Methods in Molecular Biology, Vol. 1637, Humana Press, New York, pp. 89–97.

Al-Amoud, A.I., Bacha, M.A. and Al-Darby, A.M. (2000) Seasonal water use of date palms in central region of Saudi Arabia. *Agricultural Engineering Journal* 9, 51–62.

Al-Farsi, M.A. and Lee, C.Y. (2008) Nutritional and functional properties of dates: a review. *Critical Reviews in Food Science and Nutrition* 48(10), 877–887. DOI: 10.1080/10408390701724264.

Al-Kaabi, H.H., Rhiss, A. and Hassan, M.A. (2001) Effect of auxins and cytokinins on the in vitro production of date palm bud generative tissues and on the number of differentiated buds. In: Al-Badawy, A.A. (ed.) *Proceedings of the Second International Conference on Date Palm*, UAE University, Al Ain, United Arab Emirates, March 25–27, 2001, pp. 47–86.

Al-Kaabi, H.H., Zaid, A. and Ainsworth, C. (2007) Plant-off-types in tissue culture-derived date palm (*Phoenix dactylifera* L.) plants. *Acta Horticulturae* 736, 267–274. DOI: 10.17660/ActaHortic.2007.736.25.

Al-Khateeb, A.A. (2006) Role of cytokinin and auxin on the multiplication stage of date palm (*Phoenix dactylifera* L.) cv. Sukry. *Biotechnology* 5, 349–352.

Al-Khateeb, A.A. (2008) Regulation of in vitro bud formation of date palm (*Phoenix dactylifera* L.) cv. Khanezi by different carbon sources. *Bioresource Technology* 99(14), 6550–6555. DOI: 10.1016/j.biortech.2007.11.070.

Al-Khateeb, A.A. and Al-Khateeb, S.A. (2016) *In vitro* role of hormones at multiplication stage of date palm (*Phoenix dactylifera* L.) cvs Khalas and Sukary. *Research Journal of Biotechnology* 11, 58–63.

Al-Khayri, J.M. (2001) Optimization of biotin and thiamine requirements for somatic embryogenesis of date palm (*Phoenix dactylifera* L.). *In Vitro Cellular & Developmental Biology - Plant* 37(4), 453–456. DOI: 10.1007/s11627-001-0079-x.

Al-Khayri, J.M. (2011) Influence of yeast extract and casein hydrolysate on callus multiplication and somatic embryogenesis of date palm (*Phoenix dactylifera* L.). *Scientia Horticulturae* 130(3), 531–535. DOI: 10.1016/j.scienta.2011.07.024.

Al-Khayri, J.M. and Al-Bahrany, A.M (2003) Genotype-dependent in vitro response of date palm (*Phoenix dactylifera* L.) cultivars to silver nitrate. *Scientia Horticulturae* 99, 153–162.

Al-Khayri, J.M. and Al-Bahrany, A.M (2012) Effect of abscisic acid and polyethylene glycol on the synchronization of somatic embryo development in date palm (*Phoenix dactylifera* L.). *Biotechnology* 11, 318–325.

Al-Khayri, J.M. and Naik, P.M. (2017) Date palm micropropagation: advances and applications. *Ciência e Agrotecnologia* 41(4), 347–358. DOI: 10.1590/1413-70542017414000217.

Al-Mana, F.A., El-Hamady, M.A., Bacba, M.A. and Abdelrahman, A.A. (1996) Improving root development on ground and aerial date palm offshoots. *Agricultural Research Center King Saud University, Research Bulletin* 60, 5–19.

Al-Mayahi, A.M.W. (2014) Thidiazuron-induced *in vitro* bud organogenesis of the date palm (*Phoenix dactylifera* L.) cv. Hillawi. *African Journal of Biotechnology* 13(35), 3581–3590. DOI: 10.5897/AJB2014.13762.

Al-Mazroui, H.S., Zaid, A. and Bouhouche, N. (2007) Morphological abnormalities in tissue culture-derived date palm (*Phoenix dactylifera* L.). *Acta Horticulturae* 736, 329–335. DOI: 10.17660/ActaHortic.2007.736.31.

Al-Muaini, A., Green, S., Dakheel, A., Abdullah, A.-H., Abou Dahr, W.A. *et al.* (2019) Irrigation management with saline groundwater of a date palm cultivar in the hyper-arid United Arab Emirates. *Agricultural Water Management* 211, 123–131. DOI: 10.1016/j.agwat.2018.09.042.

Almusawi, A.H.A., Sayegh, A.J., Alshanaw, A.M.S. and Griffis, J.L. (2017) Plantform bioreactor for mass micropropagation of date palm. In: Al-Khayri, J.M., Jain, S.M. and Johnson, D.V. (eds) *Date Palm Biotechnology Protocols*, I. *Tissue Culture Applications*. Methods in Molecular Biology, Vol. 1637, Humana Press, New York, pp. 251–265.

Al-Omran, A., Eid, S. and Alshammari, F. (2019) Crop water requirements of date palm based on actual applied water and Penman–Monteith calculations in Saudi Arabia. *Applied Water Science* 9(4), 69. DOI: 10.1007/s13201-019-0936-6.

Al-Omran, A.M., Dhaouadi, L. and Besser, H. (2023) Irrigation and salinity management of date palm in arid regions. In: Al-Khayri, J.M., Jain, S.M., Johnson, D.V. and Krueger, R.R. (eds) *Date Palm*. CAB International, Wallingford, UK, pp. 241–265.

Al-Yahyai, R. and Manickavasagan, A. (2012) An overview of date palm production. In: Manickavasagan, A., Mohamed Essa, M. and Sukumar, E. (eds) *Dates:*

Production, Processing, Food, and Medicinal Value. CRC Press, Boca Raton, Florida, pp. 3–11.

Al-Yahyai, R., Khan, M.M., Al-Kharusi, L., Naqvi, S.A. and Akram, M.T. (2023) Date palm plantation establishment and maintenance. In: Al-Khayri, J.M., Jain, S.M., Johnson, D.V. and Krueger, R.R. (eds) *Date Palm.* CAB International, Wallingford, UK, pp. 179–208.

Ara, H., Jaiswal, U. and Jaiswal, V. (2000) Synthetic seed: prospects and limitation. *Current Science* 78, 1438–1444.

Asemota, O., Eke, C.R. and Odewale, J.O. (2007) Date palm (*Phoenix dactylifera* L.) *in vitro* morphogenesis in response to growth regulators, sucrose and nitrogen. *African Journal of Biotechnology* 6(20), 2353–2357. DOI: 10.5897/AJB2007.000-2369.

Aslam, J. and Khan, S.A. (2009) *In vitro* micropropagation of Khalas date palm (*Phoenix dactylifera* L.), an important fruit plant. *Journal of Fruit and Ornamental Plant Research* 17, 15–27.

Aslam, J., Khan, S.A., Cheruth, A.J., Mujib, A., Sharma, M.P. *et al.* (2011) Somatic embryogenesis, scanning electron microscopy, histology and biochemical analysis at different developing stages of embryogenesis in six date palm (*Phoenix dactylifera* L.) cultivars. *Saudi Journal of Biological Sciences* 18(4), 369–380. DOI: 10.1016/j.sjbs.2011.06.002.

Aslam, J., Khan, S.A. and Naqvi, S.H. (2015) Evaluation of genetic stability in somatic embryo derived plantlets of six date palm (*Phoenix dactylifera* L.) cultivars through RAPD based molecular marker. *Science, Technology and Development* 34(1), 1–8. DOI: 10.3923/std.2015.1.8.

Awad, M.A. (2008) Promotive effects of a 5-aminolevulinic acid-based fertilizer on growth of tissue culture-derived date palm plants (*Phoenix dactylifera* L.) during acclimatization. *Scientia Horticulturae* 118(1), 48–52. DOI: 10.1016/j.scienta.2008.05.034.

Awad, M.A., Soaud, A.A. and ElKonaissi, S.M. (2006) Effect of exogenous application of anti-stress substances and elemental sulphur on growth and stress tolerance of tissue culture derived plantlets of date palm (*Phoenix dactylifera* L.) cv. 'Khala' during acclimatization. *Journal of Applied Horticulture* 08(2), 129–134. DOI: 10.37855/jah.2006.v08i02.30.

Badawy, E.M., Habib, A.M.A., El Bana, A. and Yosry, G.M. (2005) Propagation of date palm (*Phoenix dactylifera* L.) plants by using tissue culture technique. *Arabian Journal of Biotechnology* 8, 343–354.

Bashir, M.A., Ahmad, M. and Shabir, K. (2015) Effect of different bunch covering materials on Shamran date for enhancement of economical yield. *Journal of Animal and Plant Sciences*, 25(2), 417–421.

Bekheet, S.A. (2013) Direct organogenesis of date palm (*Phoenix dactylifera* L.) for propagation of true-to-type plants. *Scientia Agriculturae* 4, 85–92.

Bitar, A.D., Abu-Qaoud, H.A. and Isaid, H.M. (2019) Studies on date palm propagation by offshoots. *Palestinian Journal of Technology and Applied Sciences* 2, 61–68.

Botes, A. and Zaid, A. (2002) The economic importance of date production and international trade. In: Zaid, A. and Arias-Jimenez, E.J. (eds) *Date Palm Cultivation.* FAO Plant Production and Protection Paper No.156 Rev. 1, Food and Agriculture Organization of the United Nations, Rome. Available at: www.fao.org/3/Y4360E/y4360e00.htm#Contents (accessed 15 October 2021).

Boufis, N., Khelifi-Slaoui, M., Djillali, Z., Zaoui, D., Morsli, A. *et al.* (2014) Effects of growth regulators and types of culture media on somatic embryogenesis in date palm (*Phoenix dactylifera* L. cv. Degla Beida). *Scientia Horticulturae* 172, 135–142. DOI: 10.1016/j.scienta.2014.04.001.

Britto, A.J.D., Sujin, R.M. and Sebastian, S.R. (2011) Morphological and molecular variation of five rice varieties to ultra violet-B radiation stress. *Journal of Stress Physiology & Biochemistry* 7, 80–86.

Carr, M.K.V. (2012) The water relations and irrigation requirements of the date palm (*Phoenix dactylifera* L.): a review. *Experimental Agriculture* 49, 91–113.

Chao, C.T. and Krueger, R.R. (2007) The date palm (*Phoenix dactylifera* L.): overview of biology, uses, and cultivation. *HortScience* 42(5), 1077–1082. DOI: 10.21273/HORTSCI.42.5.1077.

Darwesh, R.S.S., Zaid, Z.E. and Sidky, R.A. (2011) Effect of ammonium nitrate and GA3 on growth and development of date palm plantlets in vitro and acclimatization stage. *Research Journal of Agriculture and Biological Sciences* 7, 17–22.

Dates Farming (2022) Dates farming information guide. Available at: https://www.agrifarming.in/dates-farming (accessed 29 April 2022).

Daud, N., Taha, R.M. and Hasbullah, N.A. (2008) Artificial seed production from encapsulated micro shoots of *Saintpaulia ionantha* Wendl. (African Violet). *Journal of Applied Sciences* 8(24), 4662–4667. DOI: 10.3923/jas.2008.4662.4667.

Eke, C.R., Akomeah, P. and Asemota, O. (2005) Somatic embryogenesis in date palm (*Phoenix dactylifera* L.) from apical meristem tissues from 'Zbia' and 'Loko' landraces. *African Journal of Biotechnology* 4, 244–246.

El-Assar, A.M., El-Messeih, W.M. and El-Shenawi, M.R. (2004) Applying of some natural extracts and growth regulators to culture media their effects on 'Sewi' cv. date palm tissues grown *in vitro*. Assuit Journal of Agricultural Sciences 35, 155–168.

Elfeky, A. and Elfaki, J. (2019) A review: date palm irrigation methods and water resources in the Kingdom of Saudi Arabia. *Journal of Engineering Research and Reports* 9(2), 1–11. DOI: 10.9734/jerr/2019/v9i217012.

El Hadrami, I., Cheikh, R. and Baaziz, M. (1995) Somatic embryogenesis and plant regeneration from shoot-tip explants in *Phoenix dactylifera* L. *Biologia Plantarum* 37, 205–211.

El Hadrami, A., Daayf, F., Elshibli, S., Jain, S.M. and El Hadrami, I. (2011) Somaclonal variation in date palm. In: Jain, S.M., Al-Khayri, J.M. and Johnson, D.V. (eds) *Date Palm Biotechnology.* Springer, Dordrecht, The Netherlands, pp. 183–203.

El Kinany, S., Achbani, E., Faggroud, M., Ouahmane, L., El Hilali, R. *et al.* (2019) Effect of organic fertilizer and commercial arbuscular mycorrhizal fungi on the growth of micropropagated date palm cv. Feggouss. *Journal of the Saudi Society of Agricultural Sciences* 18(4), 411–417. DOI: 10.1016/j.jssas.2018.01.004.

Eshraghi, P., Zaghami, R. and Mirabdulbaghi, M. (2005) Somatic embryogenesis in two Iranian date palm cultivars. *African Journal of Biotechnology* 4, 1309–1312.

Etienne, H., Dechamp, E., Barry-Etienne, D. and Bertrand, B. (2006) Bioreactors in coffee micropropagation. *Brazilian Journal of Plant Physiology* 18(1), 45–54. DOI: 10.1590/S1677-04202006000100005.

FAO (2019) FAOSTAT Data. Food and Agriculture Organization of the United Nations, Rome. Available at: www.fao.org/faostat/en/#data (accessed 12 October 2021).

FAO (2020) FAOSTAT Data. Food and Agriculture Organization of the United Nations, Rome. Available at: www.fao.org/faostat/en/#data (accessed 25 February 2022).

FAO (2021) COAG/2020/21 – Proposal for an international year of date palm. Food and Agriculture Organization of the United Nations, Rome. Available at: www.fao.org/3/nd415en/ND415EN.pdf (accessed 10 November 2021).

Farooq, S., Maqbool, M.M., Bashir, M.A., Ullah, M.I., Shah, R.U. *et al.* (2021) Production suitability of date palm under changing climate in a semi-arid region predicted by CLIMEX model. *Journal of King Saud University – Science* 33, 101394. DOI: 10.1016/j.jksus.2021.101394.

Farrag, H.M.A., Abeer, H.E. and Darwesh, R.S.S. (2011) Growth promotion of date palm plantlets *ex vitro* by inoculation of rizosphere bacteria. *Journal of Horticultural Science and Ornamental Plants* 3, 130–136.

Fki, L., Masmoudi, R., Drira, N. and Rival, A. (2003) An optimized protocol for plant regeneration from embryogenic suspension cultures of date palm (*Phoenix dactylifera* L.) cv. Deglet Nour. *Plant Cell Reports* 21, 517–524.

Fki, L., Masmoudi, R., Kriaâ, W., Mahjoub, A., Sghaier, B. *et al.* (2011) Date palm micropropagation via somatic embryogenesis. In: Al-Khayri, J.M. and Johnson, D.V. (eds) *Date Palm Biotechnology*. Springer, Dordrecht, The Netherlands, pp. 47–68.

Fki, L., Kriaa, W., Nasri, A., Baklouti, E., Chkir, O. *et al.* (2017) Indirect somatic embryogenesis of date palm using juvenile leaf explants and low 2,4-D concentration. In: Jain, S.M. and Johnson, D.V. (eds) *Date Palm Biotechnology Protocols*, Vol. I. *Tissue Culture Applications*. Methods in Molecular Biology, Vol. 1637, Humana Press, New York, pp. 99–106.

Georgiev, V., Schumann, A., Pavlov, A. and Bley, T. (2014) Temporary immersion systems in plant biotechnology. *Engineering in Life Sciences* 14, 607–621. DOI: 10.1002/elsc.201300166.

Gurevich, V., Lavi, U. and Cohen, Y. (2005) Genetic variation in date palms propagated from offshoots and tissue culture. *Journal of the American Society for Horticultural Science* 130(1), 46–73. DOI: 10.21273/JASHS.130.1.46.

Harhash, M.M., Mosa, W.F.A., El-Nawam, S.M., Hassan, R.H. and Gattas, H.R.H. (2020) Effect of bunch covering on yield and fruit quality of 'Barhee' date palm cultivar. *Middle East Journal of Agriculture Research* 9(1), 46–51.

Hassan, M.H. and Taha, R.A. (2012) Callogenesis, somatic embryogenesis and regeneration of date palm *Phoenix dactylifera* L. cultivars affected by carbohydrate sources. *International Journal of Agricultural Research* 7, 231–242.

Hassan, M.M., Allam, M.A., El Din, M.S., Malhat, M.H. and Taha, R.A. (2021) High-frequency direct somatic embryogenesis and plantlet regeneration from date palm immature inflorescences using picloram. *Journal of Genetic Engineering and Biotechnology* 19, 33.

Hodel, D.R. and Pittenger, D.R. (2003) Studies on the establishment of date palm (*Phoenix dactylifera*) 'Deglet Noor' offshoots. Part II. Size of Offshoots. *Palms* 47, 201–205.

Hussain, N., Al-Rasbi, S., Al-Wahaibi, N.S., Al-Ghanum, G. and Abdalla, O.A.E. (2012) Salinity problems and their management in date palm production. In: Mohamed Essa, M. and Sukumar, E. (eds) *Dates: Production, Processing, Food, and Medicinal Values*. CRC Press, Boca Raton, Florida, pp. 88–114.

Jain, S.M. (2012) Date palm biotechnology: current status and prospective – an overview. *Emirates Journal of Food and Agriculture* 24, 386–399.

Jeffrey, S. and Harold, A. (1999) Does global change increased the success of biological invaders? *Trends in Ecology and Evolution* 14, 135–139.

Jones, P.G. and Thornton, P.K. (2003) The potential impacts of climate change on maize production in Africa and Latin America in 2055. *Global Environmental Change* 13, 51–59. DOI: 10.1016/S0959-3780(02)00090-0.

Kakani, V.G., Reddy, K.R., Zhao, D. and Sailaja, K. (2003) Field crop responses to ultraviolet-B radiation: a review. *Agricultural and Forest Meteorology* 120, 191–218. DOI: 10.1016/j.agrformet.2003.08.015.

Khanam, S., Sham, A., Bennetzen, J.L. and Aly, M.A.M. (2012) Analysis of molecular marker-based characterization and genetic variation in date palm (*Phoenix dactylifera* L.). *Australian Journal of Crop Science* 6(8), 1236–1244.

Khierallah, H.S. and Bader, M.S.M. (2007) Micropropagation of date palm (*Phoenix dactylifera* L.) var. Maktoom through organogenesis. *Acta Horticulturae* 736, 213–223.

Khierallah, H.S.M., Muna, H.S., Al-Hamdany, M.H.S., Abdulkareem, A.A. and Saleh, F.F. (2015) Influence of sucrose and paclobutrazol on callus growth and somatic embryogenesis in date palm cv. Bream. *International Journal of Current Research and Academic Review* 1, 270–276.

Kosky, R.G., Silva, M. d. F., Pérez, L.P., Gilliard, T., Martínez, F.B. *et al.* (2002) Somatic embryogenesis of the banana hybrid cultivar FHIA-18 (AAAB) in liquid medium and scaled-up in a bioreactor. *Plant Cell, Tissue and Organ Culture* 68, 21–26.

Krueger, R.R. (2021) Date palm (*Phoenix dactylifera* L.) biology and utilization. In: Al-Khayri, J.M., Jain, S.M. and Johnson, D.V. (eds) *The Date Palm Genome: Phylogeny, Biodiversity and Mapping.* Springer, Cham, Switzerland, pp. 3–28.

Kumar, N., Modi, A.R., Singh, A.S., Gajera, B.B., Patel, A.R. *et al.* (2010) Assessment of genetic fidelity of micropropagated date palm (*Phoenix dactylifera* L.) plants by RAPD and ISSR markers assay. *Physiology and Molecular Biology of Plants* 16(2), 207–213.

Lemlem, A., Alemayehu, M. and Endris, M. (2018) Date palm production practices and constraints in the value chain in Afar Regional State, Ethiopia. *Advances in Agriculture* 2016, 6469104. Available at: www.hindawi.com/journals/aag/2018/6469104/

Lemlem, A., Alemayehu, M. and Endris, M. (2019) Diversification of livelihoods through date palm production in agro-pastoral areas of Afar region, Ethiopia. *Interdisciplinary Description of Complex Systems* 17, 162–176. DOI: 10.7906/indecs.17.1.16.

Liebenberg, P.J. and Zaid, A. (2002) Date palm irrigation. In: Zaid, A. and Arias-Jimenez, E.J. (eds) *Date Palm Cultivation.* FAO Plant Production and Protection Paper No. 156 Rev. 1, Food and Agriculture Organization of the United Nations, Rome. Available at: www.fao.org/3/Y4360E/y4360e00.htm#Contents (accessed 20 October 2021).

Liu, B., Liu, X., Li, Y. and Herbert, S.J. (2013) Effects of enhanced UV-B radiation on seed growth characteristics and yield components in soybean. *Field Crop Research* 154, 158–163.

Loutfi, K. and El-Hadrami, I. (2005) *Phoenix dactylifera* date palm. In: Litz, R.E. (ed.) *Biotechnology of Fruit and Nut Crops.* CAB International, Wallingford, UK, pp. 144–156.

Magray, M.M., Wani, K.P., Chatto, M.A. and Ummyiah, H.M. (2017) Synthetic seed technology. *International Journal of Microbiology and Applied Sciences* 6, 662–674.

Mansour, H.A. and Khalil, N.H. (2019) Effect of wounding and IBA on rooting of aerial and ground offshoots of date palm (*Phoenix dactylifera* L.) Medjool cultivar. *Plant Archives* 19(Suppl. 2), 685–689.

Mathur, J., Ahuja, P.S., Lal, N. and Mathur, A.K. (1989) Propagation of *Valeriana wallichii* DC. using encapsulated apical and axial shoot buds. *Plant Science* 60, 111–116.

Mazri, M.A. (2015) Role of cytokinins and physical state of the culture medium to improve *in vitro* shoot multiplication, rooting and acclimatization of date palm (*Phoenix dactylifera* L.) cv. Boufeggous. *Journal of Plant Biochemistry and Biotechnology* 24, 268–275.

Mazri, M.A. and Meziani, R. (2013) An improved method for micropropagation and regeneration of date palm (*Phœnix dactylifera* L). *Journal of Plant Biochemistry and Biotechnology* 22, 176–184.

Mazri, M.A. and Meziani, R. (2015) Micropropagation of date palm: a review. *Cell & Developmental Biology* 4, 3.

Mazri, M.A., Meziani, R., El Fadile, J. and Ezzinbi, A. (2016) Optimization of medium composition for *in vitro* shoot proliferation and growth of date palm cv. Mejhoul. *3 Biotech* 6, 111.

McCubbin, M.J., Van Staden, J. and Zaid, A. (2000) A southern African survey conducted for off-types on date palm production using somatic embryogenesis. In: *Proceedings of the Date Palm International Symposium, 22–25 February 2000*, Ministry of Agriculture, Water and Rural Development, Windhoek, Namibia and Food and Agriculture Organization of the United Nations, Rome, pp. 68–72.

Meziani, R., Jaiti, F., Mazri, M.A., Anjarne, M., Chitt, M.A. *et al.* (2015) Effects of plant growth regulators and light intensity on the micropropagation of date palm (*Phoenix dactylifera* L.) cv. Mejhoul. *Journal of Crop Science and Biotechnology* 18, 325–331. DOI: 10.1007/s12892-015-0062-4.

Mirani, A.A., Jatoi, M.A., Bux, L., Teo, C.H., Kabiita, A.I. *et al.* (2022) Genetic stability analysis of tissue culture derived date palm cv. Dedhi plants using IRAP markers. *Acta Ecologica Sinica* 42, 76–81. DOI: 10.1016/j.chnaes.2021.02.011.

Moghaieb, R.E.A., Abdel-Hadi, A.A. and Ahmed, M.R.A. (2011) Genetic stability among date palm plantlets regenerated from petiole explants. *African Journal of Biotechnology* 10, 14311–14318.

Mohammed, N.M.I. (2016) Acceleration of date palm *(Phoenix dactylifera* L) seeds germination. MSc. thesis, Sudan University of Science and Technology, Khartoum.

Montazar, A., Krueger, R., Corwin, D., Pourreza, A., Little, C. *et al.* (2020) Determination of actual evapotranspiration and crop coefficients of California date palms using the residual of energy balance approach. *Water* 12(8), 2253. DOI: 10.3390/w12082253.

Mostafa, R.A.A., El-Salhy, A.M., El-Banna, A.A. and Diab, Y.M. (2014) Effect of bunch bagging on yield and fruit quality of Seewy date palm under New Valley Conditions (Egypt). *Middle East Journal of Agriculture Research* 3(3), 517–521.

Murashige, T. and Skoog, F. (1962) A revised medium for rapid growth and bioassays with tissue cultures. *Physiologia Plantarum* 15, 473–497.

Musselman, L.J. (2007) *Figs, Dates, Laurel and Myrrh: Plants of the Bible and the Quran.* Timber Press, Portland, Oregon.

Niazwali, S.A., Senthilkumar, A., Karthishwaran, K. and Salem, M.A. (2020) The growth and tissue mineral concentrations of date palm (*Phoenix dactylifera* L.) cultivars in response to the ultraviolet-B radiation. *Australian Journal of Crop Science* 14(2), 354–361.

Nixon, R.W. and Carpenter, J.B. (1978) *Growing Dates in the United States.* US Department of Agriculture, Washington, DC.

Nyende, A.B., Schittenhelm, S., Mix-Wagner, G. and Greef, J.M. (2005) Yield and canopy development of field grown potato plants derived from synthetic seeds. *European Journal of Agronomy* 22, 175–184. DOI: 10.1016/j.eja.2004.02.003.

Osorio, R.J., Barden, C.J. and Ciampitti, I.A. (2018) GIS approach to estimate windbreak crop yield effects in Kansas–Nebraska. *Agroforestry Systems* 93, 1567–1576.

Othmani, A., Bayoudh, C., Drira, N., Marrakchi, M. and Trifi, M. (2009a) Somatic embryogenesis and plant regeneration in date palm *Phoenix dactylifera* L., cv. Boufeggous is significantly improved by fine chopping and partial desiccation of embryogenic callus. *Plant Cell, Tissue and Organ Culture* 97, 71–79.

Othmani, A., Bayoudh, C., Drira, N., Marrakchi, M. and Trifi, M. (2009b) Regeneration and molecular analysis of date palm (*Phoenix dactylifera* L.) plantlets using RAPD markers. *African Journal of Biotechnology* 8, 813–820.

Othmani, A., Rhouma, S., Bayoudh, C., Mzid, R., Drira, N. *et al.* (2010) Regeneration and analysis of genetic stability of plantlets as revealed by RAPD and AFLP markers in date palm (*Phoenix dactylifera* L) cv. Deglet Nour. *International Research Journal of Plant Science* 1, 048–055.

Othmani, A., Mzid, R., Bayoudh, C., Trifi, M. and Drira, N. (2011) Bioreactors and automation in date palm micropropagation. In: Al-Khayri, J.M. and Johnson, D.V. (eds) *Date Palm Biotechnology.* Springer, Dordrecht, The Netherlands, pp. 119–136.

Othmani, A., Bayoudh, C., Sellemi, A. and Drira, N. (2017) Temporary immersion system for date palm micropropagation. In: Jain, S.M. and Johnson, D.V. (eds) *Date Palm Biotechnology Protocols,* Vol. 1. *Tissue Culture Applications.* Methods in Molecular Biology, Vol. 1637, Humana Press, New York, pp. 239–249.

Poddar, S. and Poddar, S. (2021) Synthetic seed technology: an overview. *Agriculture & Food: E-Newsletter* 3, 57–60.

Rajendiran, K., Vidya, S., Gowsalya, L. and Thiruvarasan, K. (2015) Impact of supplementary UV-B radiation on the morphology, growth and yield of *Vigna mungo* (L.) Hepper var. ADT-3. *International Journal of Agricultural and Veterinary Sciences* 5, 104–112.

Reddy, M.R., Murthy, K.S.R. and Pullaiah, T. (2012) Synthetic seeds: a review in agriculture and forestry. *African Journal of Biotechnology* 11, 14254–14275.

Rekab, Z., Khater, M.S. and Farrag, H.M.A. (2013) Effect of red laser on growth *in vitro*, chemical composition and genome of date palm. *Research Journal of Agriculture and Biological Sciences* 9, 170–175.

Rihan, H.Z., Kareem, F., El-Mahrouk, M.E. and Fuller, M.P. (2017) Artificial seeds (principle, aspects and applications). *Agronomy* 7, 71. DOI: 10.3390/agronomy7040071.

Saiprasad, G. (2001) Artificial seeds and their applications. *Resonance* 6, 39–47. DOI: 10.1007/BF02839082.

Saker, M.M., Adawy, S.S., Mohamed, A.A. and El-Itriby, H.A. (2006) Monitoring of cultivar identity in tissue culture-derived date palms using RAPD and AFLP analysis. *Biologia Plantarum* 50, 198–204. DOI: 10.1007/s10535-006-0007-3.

Salomón-Torres, R., Krueger, R., García-Vázquez, J.P., Villa-Angulo, R., Villa-Angulo, C. *et al.* (2021) Date palm pollen: features, production, extraction and pollination methods. *Agronomy* 11, 504. DOI: 10.3390/agronomy11030504.

Sedra, M.H. (2005) Phenological descriptors and molecular markers for the determination of true-to-type of tissue culture-derived plants using organogenesis of some Moroccan date palm (*Phoenix dactilyfera* L.) varieties. *Al Awamia* 113(1), 85–101. Available at: https://www.inra.org.ma/sites/default/files/11307.pdf (accessed 4 October 2022).

Seelye, J.F., Burge, G.K. and Morgan, E.R. (2003) Acclimatizing tissue culture plants: reducing the shock. *Combined Proceedings International Plant Propagators' Society* 53, 85–90.

Shabani, F., Kumar, L. and Taylor, S. (2012) Climate change impacts on the future distribution of date palms: a modeling exercise using CLIMEX. *PLoS ONE* 7, e48021. DOI: 10.1371/journal.pone.0048021.

Sharma, S., Shahzad, A. and da Silva, J.A.T. (2013) Synseed technology: a complete synthesis. *Biotechnology Advances* 31, 186–207.

Shaukat, S.S., Farooq, M.A., Siddiqui, M.F. and Zaidi, S. (2013) Effect of enhanced UV-B radiation on germination, seedling growth and biochemical responses of *Vigna mungo* (L.) Hepper. *Pakistan Journal of Botany* 45, 779–785.

Sidky, R.A., Zaid, Z.E. and El-Bana, A. (2009) Optimized protocol for *in vitro* rooting of date palm (*Phoenix dactylifera* L.). *Egyptian Journal of Agricultural Research* 87, 277–288.

Smith, M.M., Bentrup, G., Kellerman, T., MacFarland, K., Straight, R. *et al.* (2021) Windbreaks in the United States: a systematic review of producer-reported benefits, challenges, management activities and drivers of adoption. *Agricultural Systems* 187, 103032. DOI: 10.1016/j.agsy.2020.103032.

Standardi, A. and Micheli, M. (2012) Encapsulation of in vitro-derived explants: an innovative tool for nurseries. In: Ozudogru, E. and Jain, S. (eds) *Protocols for Micropropagation of Selected Economically-Important Horticultural Plants.* Methods in Molecular Biology, Vol. 994, Humana Press, Totowa, New Jersey, pp. 397–418.

Taha, H.S., Bekheet, S.A. and Saker, M.M. (2001) Factors affecting *in vitro* multiplication of date palm. *Biologia Plantarum* 44, 431–433.

Tayyab, M., Ahmad, S., Abbas, M.M., Khan, M.I., Aslam, K. *et al.* (2021) Root initiation in aerial offshoots of date palm (*Phoenix dactylifera* L.) by exogenous application of auxins. *Journal of Applied Research in Plant Science* 2, 159–168.

Tisserat, B. (1982) Factors involved in the production of plantlets from date palm callus cultures. *Euphytica* 31, 201–214. DOI: 10.1007/BF00028323.

Uma, S., Karthic, R., Kalpana, S., Backiyarani, S. and Saraswathi, M.S. (2021) A novel temporary immersion bioreactor system for large scale multiplication of banana (Rasthali AAB–Silk). *Scientific Reports* 11, 20371. DOI: 10.1038/s41598-021-99923-4.

Valkama, E., Kivimäenpää, M., Hartikainen, H. and Wulff, A. (2003) The combined effects of enhanced UV-B radiation and selenium on growth, chlorophyll fluorescence and ultrastructure in strawberry (*Fragaria ananassa*) and barley (*Hordeum vulgare*) treated in the field. *Agricultural and Forest Meteorology* 120, 267–278.

Wheeler, T. and Von Braun, J. (2013) Climate change impacts on global food security. *Science* 341, 508–513. DOI: 10.1126/science.1239402.

Zaid, A. and Al-Kaabi, H. (2003) Plant-off types in tissue culture-derived date palm (*Phoenix dactylifera* L.). *Emirates Journal of Food and Agriculture* 15, 17–35.

Zaid, A. and de Wet, P.F. (2002a) Climatic requirements of date palm. In: Zaid, A. and Arias-Jimenez, E.J. (eds) *Date Palm Cultivation*. FAO Plant Production and Protection Paper No. 156 Rev. 1, Food and Agriculture Organization of the United Nations, Rome. Available at: www.fao.org/3/Y4360E/y4360e00.htm# Contents (accessed 15 October 2021).

Zaid, A. and de Wet, P.F. (2002b) Date palm propagation. In: Zaid, A. and Arias-Jimenez, E.J. (eds) *Date Palm Cultivation*. FAO Plant Production and Protection Paper No. 156 Rev. 1, Food and Agriculture Organization of the United Nations, Rome. Available at: www.fao.org/3/Y4360E/y4360e00.htm#Contents (accessed 19 October 2021).

Zatari, T.M. (2011) Second National Communication: Kingdom of Saudi Arabia. A Report Prepared by the Presidency of Meteorology and Environment (PME), Riyadh, Saudi Arabia. Available at: unfccc.int/resource/docs/natc/saunc2. pdf (accessed 16 October 2021).

Zayed, M.M.E., Ola, H. and Elbar, A. (2015) Morphogenesis of immature female inflorescences of date palm *in vitro*. *Annals of Agricultural Sciences* 60, 113–120.

Zayed, Z.E. (2017) Enhanced indirect somatic embryogenesis from shoot-tip explants of date palm by gradual reductions of 2,4-D concentration. In: Jain, S.M. and Johnson, D.V. (eds) *Date Palm Biotechnology Protocols*, Vol. I. *Tissue Culture Applications*. Methods in Molecular Biology, Vol. 1637, Humana Press, New York, pp. 77–88.

Zohary, D. and Hopf, M. (2000) *Domestication of Plants in the Old World: The Origin and Spread of Cultivated Plants in West Asia, Europe, and the Nile Valley*. Oxford University Press, Oxford.

Zouine, J. and El Hadrami, I. (2007) Effect of 2,4-D, glutamine and BAP on embryogenic suspension culture of date palm (*Phoenix dactylifera* L.). *Scientia Horticulturae* 112, 221–226.

Date Palm Plantation Establishment and Maintenance

6

Rashid Al-Yahyai[1]*, M. Mumtaz Khan[1], Latifa Al-Kharusi[1], Summar Abbas Naqvi[2] and M. Tahir Akram[3]

[1]Sultan Qaboos University, Al-Khoud, Oman; [2]University of Agriculture, Faisalabad, Pakistan; [3]PMAS-Arid Agriculture University, Rawalpindi, Pakistan

Abstract

Date palm (*Phoenix dactylifera* L.) cultivation is expanding worldwide, particularly in regions where this crop has been recently introduced. Therefore, it is essential that proper practices are employed to ensure successful date production. Establishment of a date palm plantation necessitates knowledge of the climate, particularly temperature and rainfall, and field conditions, such as soil texture and structure and water quality and quantity. Suitable female date palm cultivars and male palms as source of pollen must be chosen, as the date palm is one of the few metaxenic crops. Furthermore, the cost associated with date palm plantation projects can be very high when cultural practices, including planting design, irrigation systems, fertilizer delivery units, pollination methods, bunch and fruit care, and harvest and postharvest operations, are not well planned for at the establishment phase of the project. This chapter covers the operations that are essential for successful date palm plantation establishment and maintenance. Recent advances in date palm cultivation and the use of technology and modern techniques in date palm cultivation are discussed. Prospects for date palm cultivation and the industry as it continues to expand globally, and the challenges associated with cultivating dates in new regions, are highlighted within the scope of climate change, food security, and socioeconomic development.

Keywords: Cultivars, Date palm, Plantation project, Planting design, Pollination, Site selection

*Corresponding author: alyahyai@squ.edu.om; alyahyai@gmail.com

© CAB International 2023. *Date Palm* (eds J.M. Al-Khayri *et al.*)
DOI: 10.1079/9781800620209.0006

179

6.1   Introduction

The date palm (*Phoenix dactylifera* L.) is the source of date fruits, a pivotal staple fruit that has absolute supremacy in the hearts of the people of the Middle East in terms of numbers, prevalence, and integrated ecological and agricultural systems. In many Arab countries, it is considered as national wealth, and its economic, water, soil, traditional, and environmental returns are very important because it represents the basis of sustainable agriculture in multiple aspects. It is the oldest cultivated fruit tree in the Arabian Peninsula and plays a major role in the life of its people (Erskine *et al.*, 2004).

Several publications and reviews report the history of the date palm, its agrobiodiversity, and its relationship to other cultivars of the genus *Phoenix* (Al-Yahyai, 2007; Al-Yahyai and Al-Khanjari, 2008; El Hadrami and Al-Khayri, 2012; Jain, 2012; Al-Khayri *et al.*, 2015, 2018; Gros-Balthazard and Flowers, 2021). The date palm is considered one of the oldest cultivated crops and has been grown in the Middle East since 6000 BC; however, the origin of the tree is not known and is debatable (Al-Khayri *et al.*, 2018; Krueger, 2021). Zaid and de Wet (2002a) reported that the date palm may have descended from both the African date palm (*Phoenix reclinata* Jacq.) and the Indian date palm (*Phoenix sylvestris* (L.) Roxb.), although the cultivars of *P. dactylifera* are distinguished from the African and Indian species in terms of production of offshoots, stature, cylindrical and thick trunk, and dark green leaves (Al-Yahyai and Al-Khanjari, 2008; Pintaud *et al.*, 2013). Al-Khayri *et al.* (2018) reported that it is most likely related to the Eastern Mediterranean Cretan date palm (*Phoenix theophrasti* Greu.) and to the Canary Islands date palm (*Phoenix canariensis* Chab.) and the South Asian species (*P. sylvestris*).

Zohary and Spiegel-Roy (1975) reported that the date palm may have originated in the southern Near East region. However, it is more likely that the date palm originated from the Arabian Gulf region, mainly determined by its climatic requirements and its cultivation in the dry and semidry lands of that region (Munier, 1973; Chao and Krueger, 2007; Al-Yahyai and Al-Kharusi, 2012a; Krueger, 2021; Mohamed *et al.*, 2021). Archaeological evidence shows that the date palm was cultivated in eastern Arabia in 4000 BC in lower Mesopotamia (currently Iraq) and in the Nile Valley (Erskine *et al.*, 2004). Recent work has also indicated that wild date palms may still exist in certain parts of Oman (Gros-Balthazard *et al.*, 2017). Currently, the date palm is cultivated between latitudes of 10 and 35°N and beyond the Middle East and North Africa in countries to the east such as Thailand (Hinkaew *et al.*, 2021) and in countries to the west such as Chile and Peru (Escobar and Valdivia, 2015) (Fig. 6.1).

The global expansion of date palm cultivation and the growing interest in dates as healthy fruits, as well as the economic and nutritional opportunities associated with dates, warrant a fresh look at the current practices in

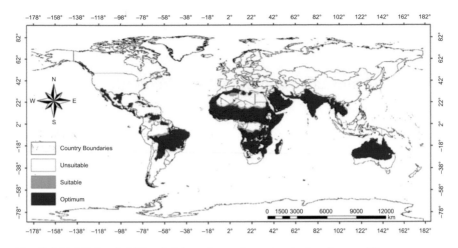

Fig. 6.1. Habitat suitability of date palm under current climatic conditions. (Used with permission from Farooq *et al.*, 2021.)

date palm cultivation. This is especially important in the current technological age, where technologies such as drones, robotics, and the Internet of Things (IoT) are playing an increasing role in crop cultivation (Canlas *et al.*, 2022). The establishment of a date palm plantation requires a set of criteria including climate, cultivars, temperature, rainfall, and field conditions such as soil texture and structure and water quality and quantity. Additionally, cultivation practices of date palms, including planting design, irrigation systems, fertilizer delivery units, pollination methods, bunch and fruit care, and harvest and postharvest operations, are important aspects of the establishment of date palm plantations. This chapter covers these subjects that are related to the establishment of date palm plantations.

6.2   Date Palm Fruit Production

Date production is widespread in arid and semi-arid areas located between latitudes 10 and 35°N (Zohary and Spiegel-Roy, 1975) numbering about 100 million palms. World date fruit production reached over 9.4 million t in 2020 (FAO, 2022). The top three date-producing countries are Egypt, Saudi Arabia, and Iran. About 100 million date palms are present in the world's estimated area of 1 million ha, out of these 60% are present in North Africa and the Middle East (Al-Khayri *et al.*, 2018). There are more than 2000 cultivars spread worldwide; more than 250 cultivars are in Oman alone (Al-Yahyai, 2007). Considerable high-quality dates are produced in Arizona and California in the USA (Nixon, 1951). The date palm is also found in Spain but does not produce good fruits as in the major production areas. In Asia, the date palm is grown up to latitude 39°N (Turkmenistan) and in Africa, it is found at latitude 20°S and disappears with the presence of oil palm at latitude 10°N.

6.3 Date Palm Environmental Requirements

6.3.1 Temperature

Date palm is grown between 10 and 35°N and does not extend beyond latitude 44°24′N. Date cultivars can tolerate a maximum temperature of up to 60°C and a short period of freezing temperatures, with an optimum of 32°C (Safronov *et al.*, 2017). However, cell division and growth may become impaired at low temperatures (8.8°C). Diurnal variation in the growing point temperature does not exceed 9.4–10°C due to the shading of the palm fronds. The temperature at the growing point varies greatly from the ambient temperature. This difference is highest at sunrise and lowest at 2 to 4 p.m. These differences in temperature between the growing point and ambient temperature are mainly because of the following:

- the apical meristem and lateral buds are far below the visible parts of white young leaves and are well shaded;
- the trunk is enclosed by the bases of the old leaves and a fibrous sheath which provide insulation against heat exchange with ambient air; and
- rapid transpiration rate tends to keep the temperature low at the growing point, similar to that at the root system.

The date palm is well adapted to hot and arid conditions due to the above factors and because:

- the leaves have sunken stomata and are coated by a layer of white waxy powder, which reflects direct sunlight; and
- leaf and leaflet widths, angles, and direction help to minimize water loss by shading the palm canopy.

Temperature also affects the growth and development of date fruits. A long hot summer is necessary for fruit ripening; however, the average daily temperature should not exceed 35°C for an extended period during pollination and fruit development stages as it may result in fruit shriveling. Kruse *et al.* (2019) studied the response to leaf temperature of photosynthesis (*A*) and stomatal conductance and found that optimum leaf temperature varied between 20 and 33°C in winter and between 28 and 45°C in summer.

6.3.2 Moisture

Despite the best date production areas being confined to arid and semi-arid areas, the date palm transpires copious amounts of water from its leaves (Allen *et al.*, 2020). Under such conditions, the water loss is estimated as

15,232 m³/ha of mature date plantation per year. About 20% of the applied water is lost through evaporation from the soil surface before being absorbed by the tree.

To obtain high yield and quality fruits at the rutab and tamar stages there should be low rainfall during the periods of pollination and fruit development. When high rainfall and high humidity occur during the kimri and khalal stages, checking and blacknose damage occur. Checking is characterized by the development of minute cracks in the apical third of the fruit. The checked area will turn a solid dark color, a condition known as blacknose. In addition to checking and blacknose, rainfall may cause fruit splitting and rotting during the rutab and tamar stages. High humidity may also cause premature fruit drop, and thus may necessitate early harvesting at the khalal stage.

6.3.3 Other climatic and edaphic factors

The date palm is grown in different varieties of soil media. Dates grow in very hot and dry environmental conditions and require an intensive long hot summer with low humidity during the period from pollination to harvest (Chao and Krueger, 2007). Date palms can grow from 12.7°C up to 50°C and can withstand short periods with temperature as low as –5°C.

Globally, despite the date palm being a very tolerant crop to multiple climate threats, the current global climate change being witnessed now may compromise ecosystems and date palm cultivation. Thus, many researchers have addressed and reported on climate change that may cause or has already caused changes in biotic and abiotic tolerance mechanisms in crop plants. Impact of the climate and environment includes impairment of date palm growth, reduction of photosynthesis and physiological processes, and change in chemical and molecular responses as well as growth and development of date palm (Munns and Tester, 2008). The date palm can grow in different types of soils; however, higher productivity is obtained in light deep soils. The date palm can tolerate saline and alkaline soils to a greater extent than other fruit trees (Alhammadi and Kurup, 2012; Al-Muaini *et al.*, 2019). It is relatively resistant to salinity, but this varies among different cultivars (Furr and Armstrong, 1975; Youssef and Awad, 2008; Ramawat, 2009). Ramoliya and Pandey (2003) observed that some cultivars may tolerate high concentrations of soil salinity (up to 12.8 dS/m) with no significant deficiency symptoms, but photosynthesis capability and growth generally decline at soil salinity levels above 12.8 dS/m (1 dS/m = 640 mg/L) (Yaish and Kumar, 2015). Some date cultivars can grow at soil salinity levels of 22,000 ppm of dissolved salts, but their fruit yield is affected (Erskine *et al.*, 2004). However, date palms die at soil salinity levels of 48,000 ppm.

6.4 Land Preparation

Numerous activities must be considered when starting a new date plantation to ensure long-term success. Initial land preparation is one of these processes, and it should be accomplished before the date palm offshoots are transplanted. The basic objective of land preparation is to establish a favourable growing environment for the effective development of new plants. It enables the grower to plan and organize the process ahead of time, ensuring the date palm's successful establishment. Planning is a vital component of the initial land preparation process since it helps to avoid unnecessary delays during the implementation phase.

The following are important deliberations to be taken care of during the planning process of date palm plantation.

6.4.1 Site selection

The location chosen for the date plantation might have a significant impact on the cost of land preparation. Therefore, a detailed survey of the area must be conducted to get detailed data about the location/lay of the land such as longitude, latitude, and elevation, vegetation cover (if any), medium of communications, and other utilities. The meteorological information from the nearest weather station (proximity of 30 to 50 km) for the last 10–15 years will augment understanding of climatic patterns and their impacts on the plantation (Botes and Zaid, 2002). The basic objective is to highlight the most important factors to consider when selecting land for a new date plantation.

6.4.2 Water availability

For long-term growth and a successful plantation, date palm requires a reliable supply of good-quality water. Saline water is a limiting factor for healthier date trees; therefore, salinity levels should not exceed 21 dS/m, although fully grown mature trees can tolerate somewhat higher levels but these will compromise yield and fruit quality (Botes and Zaid, 2002). The following are a few important facts regarding irrigation that are essential to consider throughout:

- the long-term availability of the water source;
- the total amount of water that can be readily available for irrigation purposes;
- the quality of the water available for irrigation; and
- a well-planned irrigation system design.

6.4.3 Soils and soil testing

Date palms can grow in diverse types of soils. The best soils for successful cultivation must have basic plant nutrients and good water-holding capacity with good drainage properties. Sandy soils are good for date palms, but they lack nutrients and water retention; therefore, such soils need frequent irrigation and fertilization. Water and nutrient depletion through leaching is a very common problem in such soils. None the less, sandy soils with a layer of finer-textured, more absorbent soil in the upper 2m are valued. Plant growth and date fruit quality are restricted significantly under extremely saline soil environments. It is recognized that soil is a principal component in making successful date palm plantations and soil testing is critical for several factors as it helps to optimize crop production by determining the need for future fertilization and irrigation (Aleid *et al.*, 2015). In agriculture, soil testing usually refers to the analysis of soil samples to determine soil pH, salinity/acidity, and nutrient composition. Since soil nutrients change with depth and soil composition changes with time, the depth and time of sampling will also affect the results (Kirkby *et al.*, 2016). Moreover, soil testing helps in diagnosing soil problems, is helpful in improving the nutritional balance, and saves the amount of fertilizer required, ultimately saving money (Cuevas *et al.*, 2019). Pre-plant soil analysis, as a result, identifies soil issues such as nutritional shortfalls, pH imbalances, and soluble salt excesses, and is an important approach for managing crop nutrition and soluble salt levels (Shrivastava and Kumar, 2015).

In general, the following three methods are used for soil testing at large:

- *Saturated media extract (SME):* SME is a universal method used in commercial scientific laboratories. This approach involves making a paste of soil and water, then separating the liquid portion (the extract) from the solid portion for pH, soluble salts, and nutrient analyses (Warnke, 2011). However, this method requires special laboratory techniques and skills.
- *1:2 Dilution method:* In this test, an air-dried sample of soil is combined with water at a 1:2 ratio of soil to water, then the liquid is extracted from the solids by using either laboratory-grade filter paper or an ordinary coffee filter. It is the simplest and ideal method to assess pH and soluble salts (Fisher *et al.*, 2008).
- *Leachate PourThru Method:* One of the major benefits of leachate pour-through is that no media sampling or preparation is required, unlike the SME and 1:2 dilution procedures, as the extract is the leachate collected from a container during regular irrigation (Cavins *et al.*, 2008). pH and electrical conductivity (EC) meters can be used to examine the leachate on site, or the leachate can be transferred to a commercial laboratory for a complete nutritional analysis.

6.4.4 Soil amendments

Date palms are hardy plants and can thrive on clayey, compacted soils with little or no amendments. However, sandy to clay loam soils with good drainage and aeration of the soil are important as the palm can withstand and grow well in areas where the groundwater level is relatively close to the surface (Chao and Krueger, 2007). Mostly date palm is grown in oases in arid and semi-arid regions; however, degradation of these regions is occurring due to soil salinization and use of groundwater for irrigation (Karbout et al., 2021). Moreover, reduction of soil organic matter and soil fertility is observed in these agroecosystems due to mismanagement of cultural practices (Mlih et al., 2016; Plaza et al., 2018). Traditionally, animal waste and plant debris are used as organic matter sources to enhance date palm production (Karbout et al., 2019). Similarly, manures are applied to the soil near the palm as they function as slow-release fertilizers to enhance productivity, provide nutrients to plants, and help in maintaining soil moisture levels (Mlih et al., 2019). Manures are enriched with macro- and microelements that are required for plant growth and help improve soil physical properties such as aeration, porosity, water infiltration, and bulk density (Lehmann and Kleber, 2015).

Some of the other advantages of organic matter in the soil are summarized as:

- helps in increasing the soil's water-holding capacity;
- improves the soil's water infiltration rate;
- recovers soil structure by lowering soil compaction;
- reduces soil crust formation; and
- minimizes side effects of soil alkalinity/salinity and improves salt leaching.

6.5 Selection of Cultivars

Date palm is one of the very few fruit crops that has evolved to have hundreds of cultivars distributed throughout the date-producing countries. However, the majority of cultivars are not of high economic value and thus are not being actively propagated and cultivated. For example, in Oman, out of over 200 cultivars, ten cultivars constitute 70% of the total date palm cultivation area (Al-Yahyai and Al-Khanjari, 2008). In addition to their color, dates can be classified based on their sugar and moisture content (MC) into soft (>30% MC, high reducing sugars), semisoft (20–30% MC, reducing sugars and sucrose), and dry (<20% MC, high sucrose content) (Ashraf and Hamidi-Esfahani, 2011). Additional considerations for the selection of cultivars include harvesting time and stage of harvest. Although most dates are harvested at the rutab or tamar stage, few cultivars can be harvested at the khalal stage, either fresh (such as cv. Barhi) or to cook in a process called *tabseel* (such as the cv. Mabsli in Oman) (Al-Yahyai and Khan, 2015). Of the over 800 cultivars grown across the date-producing countries, the international date palm market recognizes about four major cultivars. These are Majdool, Deglet Noor, Barhi, and Khalas. Table 6.1 provides a list of the most popular cultivars and their characteristics, while Table 6.2 details the most common cultivars grown in the top ten date-producing countries (Świąder *et al.*, 2020).

In addition to the female cultivars listed above (Tables 6.1 and 6.2), male palms also play a significant role in the fruit quality of date palms (metaxenia). The general requirement for pollination ratio is 1 male palm to 50 female palms. Although they have not gained much attention to date, male cultivars have been identified in several date-growing countries. For example, Al-Yahyai and Khan (2015) mentioned that there are around 48 male date cultivars that have been characterized in Oman, with the most common ones used for pollination called Khori and Bahlani. No studies have been conducted about specific cultivation requirements for male cultivars that are different from those being practiced for female date cultivars.

6.6 Planting Operations

6.6.1 Plant spacing and designs

This is the most critical stage in establishing a new date plantation. The purpose is to assist the producer in conducting the planting operation in such a way that the newly established plantation has a high survival rate. Planting distance and the selection of planting designs are essential for plant growth and development, which affect plant yields (Yudianto *et al.*, 2015). Although it is challenging to specify the exact plant spacing, many factors affect plant spacing, including height of the date palm, root area, space available for cultural practices, and availability of enough sunlight for plant functions.

Plant spacing in a commercial date palm plantation was previously assumed to be 10 m × 10 m (100 palms/ha). Modern plantations, on the other hand, use a plant spacing of 9 m × 9 m (121 palms/ha; in Israel) or 10 m × 8 m (125 palms/ha; in Namibia). Different planting designs for date palm plantings are presented in Fig. 6.2.

Planting density is also affected by environmental conditions, particularly temperature, humidity, and the selection of varieties (Raza *et al.*, 2019). In general, commercial date palms are planted at the distance of 10 m × 10 m, 9 m × 9 m, or 10 m × 8 m (Fig. 6.3), apart from a few varieties that are planted at a higher density. In date-growing areas that have a hot and dry climate, plants are planted at a closer distance. In areas having higher humidity levels, especially at ripening, the 10 m × 10 m spacing is preferred as the additional space counteracts the effects of the humidity by providing

Table 6.1. Characteristics of the most commonly grown date palm cultivars.

Cultivar	Color	Type	Characteristics
Amari	Dark brown	Soft	Sweet, medium-size fruit, eaten as dried
Barhi	Amber to red brown	Soft	Shape broadly ovate to rounded, skin medium-thick, smooth and translucent, sweet, very delicious and luscious
Deglet Nour	Dark brown	Semidry	Shape oblong-ovate, skin medium-thick, firm, soft with a unique taste
Fard	Dark brown	Semidry	Fruit sweet and pungent, shape thick cylindrical, skin medium-thick
Hadrawi	Dark brown	Dry	Sweet and fleshy
Hallawy	Golden brown	Soft	Sweet, caramel-like, translucent, shape oblong with rounded apex, skin thin
Hayani	Black and shiny	Soft	Not too sweet taste, fruit with an oblong shape
Kabkab	Dark brown to black	Soft	Elongated shape, can be cooked and dried when unripe
Khadrawi	Red to brown	Soft	Melting and caramel-like, elliptical to ovate fruit shape, skin medium-thick and tender
Khalas	Amber to red brown	Soft	Delicious, oblong-oval shape, thin skin, fruit tender, melting, translucent
Khasab	Red to brown-black	Soft	Shape of a rounded oval, tough separating skin, fruit thick
Lulu	Dark amber	Soft	Sweet, shape of oblong-oval, fruit thick, less fibrous flesh
Mazafati	Dark brown to black	Soft	Fleshy, cylinder shape, desirable taste
Medjool	Light brown to dark brown	Soft	Large and sweet with an attractive appearance, can be consumed soft or dried
Piarom	Dark brown to black	Semidry	Fleshy, long and thin shape, one of the most expensive and desirable in the world
Rabbi	Red to dark brown	Semidry	Fleshy, long and thin shape
Zahidi	Yellow to brown	Semidry	Oblong-ovate shape, skin thick, fruit firm, not very sweet with smooth consistency

Table 6.2. The most common cultivars grown in the top ten countries in date production. (Modified from Świąder *et al.*, 2020.)

Country	Most common cultivars
Algeria	Deglet Nour, Iteema, Thoory
Egypt	Amhat, Hayany, Samany, Siwi, Zoghloul
Iran	Allmehtari, Barhi, Dayri, Estamaran, Gantar, Halawi, Kabkab, Khassui, Khazravi, Mazafati, Mordarsang, Piarom, Pyarom, Rabbi, Sayer, Shahani, Shakkar, Sowaidani, Zahedi
Iraq	Amir Hajj, Barhi, Dayri, Halawy, Khadrawi, Khastawi Maktoom, Sayer, Zahidi
Oman	Bomaan, Bu Narenja, Fard, Khalas, Khasab, Khunaizi, Naghal, Um Sella
Pakistan	Basra, Dhakki, Gulistan, Hsaini, Kajur, Khadrawi, Mobini, Mozafati, Obaidullah, Sabzo, Shakri, Zaidi
Saudi Arabia	Ajwa, Al-Barakah, Al-Qaseem, Berhi, Gur, Helwet El-Goof, Hiladi, Hulwa, Khalasah, Khasab, Majnaz, Mishriq, Miskani, Nabbut Ghrain, Nabtat Seyf, Rothanat, Ruzeiz, Sag'ai, Sebakat Al-Riazh, Sahal, Sellaj, Shashi, Sokkary, Tanjeeb, Tayyar, Thamani, Umelkhashab, Um Rahim, Zamil, Zaghloul
Sudan	Abid Rahim, Barakawi, Bentamoda, Birier, Gondaila, Jawa, Khatieb, Kulma Suda, Medina, Mishriq, Mishriq Wad, Mishriq Wad Lagi, Zughloul
Tunisia	Ammari, Angou, Arichti, Bejjou, Bisr Helou, Bouhattam, Brance de dates, Deglet Nour, Eguiwa, Ftimi, Garn Ghazel, Gounda, Gousbi, Hamraya, Hissa, Kenta, Kentichi, Ksebba, Korkobbi, Lagou, Lemsi, Mattata, Mermella, Rouchdi, Touzerzayet
United Arab Emirates	Berhi, Bomaan, Khalas, Khasab, Lolo, Fard

proper sunlight and air circulation (Latifian *et al.*, 2012). Similarly, wider spacing is also used in date-palm-growing areas that are affected by rain during the ripening stage (Johnson, 2011).

6.6.2 Intercropping

Date palms provide sufficient space for intercropping of different fruits (e.g. citrus, guava, banana, olives), vegetables (maize), and field crops (lucerne (alfalfa), pulses, sesbania) (Fig. 6.4a and b), as date palms are tall, occupy less space, and have lower canopy density (Akyurt *et al.*, 2002; Mansour *et al.*, 2012; Rahnama and Latifian, 2013). Intercropping is also considered a sustainable agriculture technique in enhancing land biodiversity and increasing net profit without affecting the yield of main crops (Abouziena *et al.*, 2010).

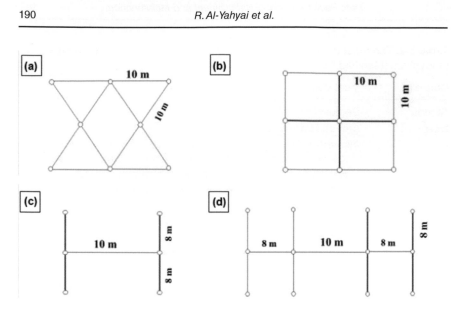

Fig. 6.2. Date palm planting designs: (a) triangular; (b) square; (c) single hedge; (d) double hedge.

6.7   Planting Methods

6.7.1   Physical field preparation

After selecting a suitable location for a plantation, field preparation is started at the most appropriate time based on the specific environmental conditions. The first step in soil preparation involves plowing the land two or three times to bring the soil to a fine tilth (Fig. 6.5a and b). After plowing, stones, rocks, pebbles, and any plant debris should be removed, followed by soil leveling. After field preparation, fertilizers are applied and organic matter is added to the soil for its improvement depending upon soil conditions (Benabderrahim *et al.*, 2018).

6.7.2   Propagation

Date palms are propagated by seed and offshoots (suckers that grow around the base and occasionally on the stem of the palm tree). Seedlings are simple to raise but are not preferred due to the nonidentification of males and females owing to their dioecious nature and not producing true-to-type offspring (Gurevich *et al.*, 2005). The propagation of date palms through suckers produces clonal plants that are identical to their parents and have similar attributes. Sometimes the grower's expertise in offshoot selection determines the success or failure of offshoots (Krueger, 2021). Recently, tissue culture-derived plants are becoming popular because of

Fig. 6.3. Different planting densities in date palm: (a) 10 m × 10 m; (b) 10 m × 8 m; (C) 10 m × 10 m; (d) 10 m × 8 m. (Photos courtesy of Dr Rashid Waseem Khan Qadri, Institute of Horticultural Sciences, University of Agriculture, Faisalabad, Pakistan and Sami Al-Riyami, The Million Date Palm Plantation Project, Oman.)

Fig. 6.4. Date palm intercropping systems: (a) intercropping with pulses; (b) intercropping with sesbania.

their true-to-type nature and excellent plant growth and development. Date palm micropropagation methods have been developed and utilized for commercial propagation (Hoop, 2000) and are employed as a successful method of date palm propagation globally. State-of-the-art tissue culture facilities exist in top date-palm-growing countries; for example, the United Arab Emirates has developed a robust date palm micropropagation facility (Rajmohan, 2011).

Fig. 6.5. Date palm orchard land preparation: (a) plowing of land; (b) leveling operations for preparation of the field. (Photos courtesy of Dr Rashid Waseem Khan Qadri, Institute of Horticultural Sciences, University of Agriculture, Faisalabad, Pakistan.)

Fig. 6.6. Preparation of the planting hole: (a) layout; (b) hole digging. (Photos Courtesy of Dr Rashid Waseem Khan Qadri, Institute of Horticultural Sciences, University of Agriculture, Faisalabad, Pakistan.)

6.7.3 Preparation of the planting hole

After land preparation, the next step is to lay out and dig holes for planting palms as shown in Fig. 6.6a. Normally, a hole with dimension of $1\,m^3$ is appropriate for planting offshoots (Fig. 6.6b). To improve soil structure and fertility, manure is mixed with soil, returned to the hole, and the plant position is marked. Well-rotted manure can also be used in planting holes and irrigated just before planting, but the manure must be placed very carefully and deep enough so that there is a layer of soil 15–20 cm thick between the manure and the roots of the date palm (Klein and Zaid, 2002). The date plant root system develops faster in soils where planting holes are prepared 1–2 months before planting.

Fig. 6.7. (a) Transplanting offshoot into the hole; (b) after transplant (Photos courtesy of Dr Rashid Waseem Khan Qadri, Institute of Horticultural Sciences, University of Agriculture, Faisalabad, Pakistan.)

6.7.4 Transplanting

When offshoots are ready for transplantation, they are pruned 3–5 days before being removed from the frame (Fig. 6.7). The new leaf growth is cut down to half its previous height, leaving three to five leaf stubs to support the increased canopy of leaves. The rooted offshoot should be covered in a soil medium that should be large enough to shield the small fibrous roots from the sun and dry winds (Hodel *et al.*, 2009). The typical planting depth is about 40 cm; however, it might vary depending on the size of the offshoot.

The key component is to transplant the new offshoots at the appropriate time to achieve maximal survival and proper establishment, well before the 'hard' season starts. Spring and autumn are the best times to plant offshoots. Autumn offers the new offshoot more time to develop itself before the onset of summer heat, while spring has the benefit of warm weather that fosters quick growth. Planting should usually begin early in the morning, to give transplants enough time to adjust and to avoid stress. Bags should be removed with care, giving as little disruption to the surrounding substrate as possible. The planting depth is important since the plant's 'heart' should never be submerged in water. When a plant's growth tip is submerged in water, it rots and the plant dies, whereas the roots of an offshoot that is planted too shallowly may dry out (Ata *et al.*, 2012).

The following factors may contribute to offshoot failure:

- placement of the nursery bed in an inappropriate area;
- failure to build the incubation facility nearly airtight enough to maintain the required humidity and high temperature;
- improper cutting and pruning techniques, as well as a lack of acclimation before putting the plants in the nursery bed; and
- failure to apply timely irrigation.

All the points mentioned above are necessary for the success and development of date palms. Ignoring or overlooking these factors may cause losses; 90 to 95% success is obtained in offshoots that are planted with proper care. The young palms should be wrapped properly to avoid freezing in winter. For this purpose, newspapers, burlap, canvas, and palm leaves may be used.

6.7.5 Aftercare

A basin is prepared around the palm as soon as it is transplanted to avoid any runoff. A basin with a diameter of roughly 3m and a depth of 20–30 cm is recommended when employing a micro-irrigation system. Mulching is accomplished by placing a layer of organic substance (e.g. hay, dried plant leaves, etc.) at the base of the palm to minimize surface water loss and crust formation (Iqbal *et al.*, 2020). Moreover, new offshoots should be protected from severe environmental conditions (summer heat and wind, cold in winter) as well as some animals and rodents (rabbits, rats, etc.). Therefore, it is recommended that a hessian wrap, shade net cover, or date-leaf tent be used to wrap palms against hazardous conditions; however, the top portion should be left open for the plant's growth and development (Robinson *et al.*, 2012).

6.8 Nutritional Requirements and Fertilization

6.8.1 Fertilization

Date palm fertilization produces variable results according to location, type of fertilizer, cultivar, and time of application. Organic manuring can play a valuable role as organic materials supply all plant nutrients. However, their nutrient content may be low if the organic material was derived from a deficient soil. Furthermore, organic supplies are often limited, which may be exacerbated as farm size diminishes, and organic manuring is often laborious and costly. In the future, most of the nutrients required for high-yielding agriculture will have to come from mineral fertilizers that provide a range of nutrients, not merely one or two. This supposes availability, understanding by farmers and extension workers, availability of reliable information, etc. Therefore, alternative sources of nutrients to commercial fertilizers will continue to be a subject of interest. Many researchers report that organic manure is not sufficient to provide the palm its requirements to produce good yield (Marzouk and Kassem, 2011).

The application of nitrogen and potassium in date improves its yield and quality (Hussein, 2008). It has been found that the addition of both nitrogen and potassium to the fertilizer regime increases total yield and bunch weight, whereas application of slight amounts of nitrogen and copious amounts of potassium increases fruit weight, length, and diameter

(Idris *et al.*, 2012). Other studies revealed the date palm requirements for major macronutrients (NPK) to produce higher yield and better-quality fruit (Elsadig *et al.*, 2017).

Furr and Armstrong (1957) compared the effect of synthetic and organic-based nitrogen on date palm and found no significant impact on date yield. They suggested that heavy fertilization improved the growth and yield of date palm trees; however, crop thinning is required to improve date quality. Application of nitrogen fertilizers greatly over the need of the trees is largely wasted because of the relatively rapid decomposition of organic matter and conversion of nitrogen into other forms (Furr and Armstrong, 1957).

The purpose of a fertilization program before and after transplantation is to meet the nutrient requirements of the new offshoots to support plant growth. At earlier stages, an underdeveloped plant is not able to reach its full output potential. Date palms require adequate soil fertilization to produce commercial yields and nutrient application is critical for high-quality date production (Benabderrahim *et al.*, 2018). Normally, organic manures are added to the soil at the start of the winter season, while other fertilizers such as nitrogen and potassium are applied in March/April (Cohen *et al.*, 2016).

In the winter, nitrogen-fixing cover crops such as lentils, gram, and peas may be added to enhance soil fertility, while in the summer, mash bean, green gram (mung), and black gram are used (Peoples *et al.*, 2009). With sufficient manuring, suitable vegetable crops and fruit trees such as pomegranate, phalsa, and papaya may be grown as intercrops. In addition, wheat straw, manures, and other loose materials may be used to improve soil composition. Generally, fertilizers are applied once a year and on sandy soil, the concentrations of fertilizers and organic matter may vary according to climatic conditions (Mahal *et al.*, 2019).

6.8.2 Micronutrients

In date palms, deficiency of microelements causes reduction of female spathes, flower and fruit abortion that ultimately reduces crop yield (Holanda *et al.*, 2007). Boron deficiency may be responsible for the mortality of date plants, as it affects the root system and the terminal buds, affects the activity of several enzymes, increases cell membrane permeability, improves carbohydrate transfer, and has a role in lignin synthesis (Hamza *et al.*, 2015). Boron regulates the ratio of potassium that is essential for protein synthesis and cell division (Krueger, 2021).

According to Djerbi (1994), a deficiency of manganese was also discovered in numerous Tunisian date plantations, resulting in palm death in a 5 to 7 year period as manganese acts as a catalyst for enzymatic and physiological processes. Similarly, iron deficiency in date palms causes yellowing of the older leaves.

Table 6.3. Fertilization program per tree at different growth stages. (Modified from Abul-Soad, 2010.)

Growth stage	N (g)	P (g)	K (g)	Organic manure (g)
After transplanting	300–400	200–300	500–800	5,000–10,000
After spathe emergence	500–1000	300–400	800–1000	10,000–15,000
Productive tree	1000–1500	400–500	1500–2000	15,000–25,000

6.8.3 Annual fertilization program

6.8.3.1 Time of application

The application of macro- and microelements is essential to improve yield and production of dates, however they must be applied at the proper stages (Shaaban and Mahmoud, 2012). There are two growth phases throughout the date season: vegetative and reproductive. The latter is further divided into two stages: bloom formation and fruit development. Fertilizer application should be timed to coincide with these phases to ensure an increase in the number of blooms and a potential increase in yield (Elamin *et al.*, 2017). The major and minor elements are generally applied in split doses with varying concentrations at different planting stages to avoid root burn. A fertilizer application program at different stages of plant growth is presented in Table 6.3.

6.8.3.2 Methods of application

MANUAL. In this method, fertilizer is measured in small amounts and applied to an individual palm tree by hand. The essential thing in this strategy is to make sure that the fertilizers are evenly distributed around the palm. However, there may be disadvantages to manual application as uneven distribution may increase root burn or the optimum dose may not be applied according to the plant requirements.

FERTIGATION. Fertigation is applying fertilizers via drip irrigation systems. In this system, soluble fertilizers are applied with optimum doses. The main advantage of this method is that the correct amount of fertilizer with even distribution is applied to plants.

6.9 Post-Planting Care and Maintenance

6.9.1 Pruning

Pruning of date palms refers to the regular removal of yellow or nearly dead leaves. This includes leaves that have died, withered, or been broken.

An adult date palm has a height of some 5m and may consist of 100 to 120 leaves, depending upon the cultivar (Carr, 2013). Date palm leaves are hard and require careful pruning as they do not fall off on their own. Normally, leaves are removed when the pinnae become yellow from the tip to the base. Pruning is recommended to improve the quality of the date fruit as it provides proper sunlight and ventilation and makes harvesting and other production practices involving the bunches easier.

The pruning practices for date palms vary with climatic conditions and location. In Pakistan, adult date palms trees are normally pruned after harvest in August or before spathes appear in January (Abul-Soad *et al.*, 2015). A third pruning is occasionally performed in March and April during pollination to enhance fertilization. Unwanted offshoots should be removed and transplanted during the pruning process to encourage the growth of rooted offshoots for future propagation. This also makes cultural practices easier and improves the parent palm's growth and bearing (Abul-Soad, 2010).

In winter, if there is a danger of frost, no trimming is recommended, and the leaves are left to protect the young vulnerable leaves from the cold. Likewise, no pruning is recommended in the initial stages up to 2 years and only large leaves that touch the ground should be trimmed (Zaid and de Wet, 2002a). Wounded areas resulting from the removal of leaves or bunches are treated with fungicides to reduce scent emissions that attract insects, particularly the red palm weevil (Abdulla *et al.*, 1983).

6.9.2 Pollination

In nature, pollination in date palms occurs by the wind or by insects but is not very effective. In commercial production, a few male trees are planted, and pollen is extracted for artificial pollination. This practice is critical for successful production and has been adopted for several years in many countries (Salomón-Torres *et al.*, 2021). It is argued that 60–80% pollination is sufficient for an adequate fruit set. However, fruit set varies with pollen compatibility or partial compatibility between female and male cultivars, sometimes resulting in low yields; this mechanism is still poorly understood (Nixon and Carpenter, 1978; Zaid and de Wet, 2002b). Drones equipped with artificial intelligence and other robotic devices are recent innovations that are being integrated into date palm pollination with an efficient, high-precision transfer of pollen.

6.9.3 Fruit thinning

Sometimes, thinning is also required to manage fruiting in date palms. It is performed to minimize alternate bearing, increase fruit size, improve fruit quality, speed up ripening, and simplify bunch management. Thinning is performed in three different ways: (i) removing whole bunches; (ii) reducing the number of strands per bunch; and (iii) reducing the number of fruits on each strand. The cultivar, environment, and cultural practices all influence the level of fruit thinning. After fruit set, bagging is used to protect fruit bunches from high humidity, rain, sunburn, and birds (Nixon and Carpenter, 1978; Zaid and de Wet, 2002b). In general, normal brown craft paper, white paper, or cotton or nylon mesh bags are used to wrap the dates (Alshariqi, 2010).

6.9.4   Pest and disease management

Date palms are susceptible to a variety of diseases and pests, the severity varying with cultivar, production area, weather, and cultural approaches (Carpenter and Elmer, 1978). Mostly, date palms are affected by fungal diseases. However, phytoplasma-related disorders have been reported in some recent cases (Abhary and Al-Baity, 2018).

Some major diseases of dates include bayoud disease, black scorch disease, brown leaf spot, Diplodia disease, Graphiola leaf spot, khamedj disease, Omphalia root rot, and fruit rot (Hussain and Ismailli, 2020). Date palms have suffered major losses as a result of these ailments. White scale, red scale, carob moth, and red palm weevil are among the insects that damage date palms, varying with the geographic region (Dalbon *et al.*, 2021). Physical, chemical, biocontrol, pheromone trapping, quarantine, and sanitation measures are commonly used to control date palm insect pests (Howard *et al.*, 2001). In recent decades, the red palm weevil has spread from India and Pakistan to North Africa, the Middle East, and Southern Europe, becoming the most serious date palm pest. Herbicides are commonly used to control weeds; however, hand weeding is also used in less developed areas. Moreover, a blend of the mass-trapping system and aggregation pheromones help capture palm weevils (Dalbon *et al.*, 2021).

6.10   Irrigation

Irrigation of date palms is influenced by the climate, soil, cultivation practices, and growth stage of the date palm. Several methods have been used to deliver irrigation water for date palms, from traditional flood or basin irrigation to modern drip and sprinkler systems (Elfeky and Elfaki, 2019). The source of water may be surface water, river water, or groundwater that is delivered to date palm plantations through duct systems (also called water canal, qanat, or falaj) or through main irrigation plastic or metal pipes. The traditional irrigation system of falaj in Oman is of three types depending on the source of water: (i) ghaili falaj, where water is diverted to canals from a flowing wadi; (ii) aini falaj, spring water; and (iii) daudi falaj, where water is sourced from an underground aquifer (Al-Ghafri, 2006). Besides fresh above- or underground waters, alternative waters such as treated sewage

Table 6.4. The irrigation quantities of water for date palm in various countries. (Modified from Liebenberg and Zaid, 2002.)

Country	Quantity (m³/ha)
Algeria	15,000–35,000
California, USA	27,000–36,000
Egypt	22,300
India	22,000–25,000
Iraq	15,000–20,000
Jordan Valley, Israel	25,000–32,000
Morocco	13,000–20,000
South Africa	25,000
Tunisia	23,600

water have also been used for irrigation of date palm (El Mardi *et al.*, 1995, 1998). The frequency and quantity of irrigation water greatly influence fruit yield and quality (Al-Yahyai and Al-Kharusi, 2012a); deficit irrigation was recommended as it enhanced chemical attributes of the fruit such as total and reducing sugar contents, total soluble solids (TSS), pectin, and dry matter. A similar conclusion was reached by Mohammed *et al.* (2021), who found no significant difference in date yield and fruit quality due to deficit irrigation, especially using subsurface irrigation. Date palm irrigation requirements vary greatly among locations; Table 6.4 presents the average quantities of water requirements of date palm in various countries.

Several methods of calculating irrigation requirements of date palm are available, including utilization of the evapotranspiration (ET)/Class A pan method or one of the following equations: Penman's equation, Blaney–Criddle equation, and Solomon and Kodama's equation (Liebenberg and Zaid, 2002).

6.11 Fruit Handling

6.11.1 Flower and fruit handling

The fruit can be harvested at various stages of maturity including unripe khalal (e.g. cv. Barhi), partially ripe rutab, and fully ripe tamar stages (Al-Yahyai and Khan, 2015). Mechanized harvesters are commonly used and consist of vehicles equipped with several long booms, where laborers stand in a basket to pick the fruits (Mazloumzadeh *et al.*, 2008).

Date harvesting is a critical process that eventually affects fruit quality and ultimately market price as it greatly influences other downstream processes including sorting, processing, packing, and marketing.

6.11.2   Postharvest operations

The harvesting season of dates in many parts of the world ranges from May until November. Selected cultivars such as Barhi, Mabsli, Abounarengah, and Madloky are harvested at the khalal stage. These cultivars may further undergo *tabseel*, a hot-water treatment before drying (Al-Yahyai and Khan, 2015). Dates are normally boiled in large quantities in a large boiling pan for 45 min to 1 hour followed by sun-drying.

Before final packing, dates are graded, the spoiled and deformed dates are removed, and the best are kept for packaging and marketing. Other postharvest operations include checks for the presence of any pests or diseases and uniformity; growers receive payment according to their product quality.

Dates can be stored for an extended period of time reaching 1 year in traditional leaf-made baskets, clay containers, and plastic or metal containers. The date industry usually stores tamar stage dates at −3°C for up to 1 year (Al-Yahyai and Al-Kharusi, 2012b). Packaged dates are expected to have a shelf life of up to 2 years at room temperature (25°C) (Ismail *et al.*, 2008). Various storage methods have been implemented to maintain high date fruit quality and extend their shelf life, including low/high-temperature storage and use of controlled-atmosphere environments, refrigeration, freezing, drying, and climate-controlled storage techniques and preservation methods (Falade and Abbo, 2007; Dehghan-shoar *et al.*, 2009; Al-Yahyai and Al-Kharusi, 2012b). Various postharvest approaches have been applied to increase date fruit quality (Falade and Abbo, 2007; Dehghan-shoar *et al.*, 2009). The main considerations for postharvest handling of dates (Sarraf *et al.*, 2021) are as follows:

- *Handling:* Sorting; cleaning; drying; grading; and packing.
- *Physiological parameters:* Respiration and ethylene production; nutrient composition and biochemical reactions; and protein composition and degradation.
- *Disorders:* Freezing and chilling injuries; antioxidant activities; fermentation; animal and insect damage; and fungal and bacterial infection.

6.12   Conclusions and Prospects

Date is an expanding crop both in cultivation and consumer demand. It has recently been utilized in a variety of nonfood products, including household, medicinal, and cosmetics. The rapid expansion of date cultivation requires an adequate understanding of the palm climatic, environmental, and management practices to ensure the highest yield and fruit quality. The prospect of future challenges, including loss of cultivar biodiversity, extreme climatic anomalies, global warming, erratic precipitation, and spread of exotic pests and diseases, requires global efforts towards ensuring adequate supplies of dates. The increase in demand is also driven by studies that show dates as a good source of biofuel, animal feed, and health products such as natural sugars. Further advancement in technology driven by the Fourth Industrial Revolution is expected to increase reliance on machines for various date palm management operations. There have been many initiatives to employ drones for pollination of date palms, computer vision technology for sorting dates, and robotics in harvesting. Further development will facilitate and reduce the operational costs associated with date palm cultivation.

Date production could benefit greatly from research and development in areas of the Fourth Industrial Revolution. Research on the applications of the IoT, drones, and robotics will facilitate the still largely laborious, human-conducted date production practices such as pollination, pruning, and harvesting. Postharvest losses are still considerable and there is ample room for research in the areas of extending the shelf life, packaging, storage, and transport of fresh dates. Traceability is still lacking in many date-producing countries. None the less, the international interest in dates continues to grow and more countries are adopting date cultivation as a key fruit crop that will ensure food security, employment, and diversified local and national economies, especially in regions with extreme environmental conditions.

References

Abdulla, K.M., Meligi, M.A. and Risk, S.Y. (1983) Influence of crop load and frond/bunch ratio on yield and fruit properties of Hayany dates. In: Makki, Y.M. (ed.) *Proceedings of the First Symposium on Date Palm in Saudi Arabia, Al-Hassa, Saudi Arabia*, King Faisal University, Al-Hassa, Saudi Arabia, March 23–25, 1982, pp. 222–232.

Abhary, M.K. and Al-Baity, O.A. (2018) Occurrence and distribution of date palm phytoplasma disease in Al-Madinah region, KSA. *Journal of Taibah University for Science* 12(3), 266–272. DOI: 10.1080/16583655.2018.1465274.

Abouziena, H.F., Abd El-Motty Elham, H.Z., Youssef, R.A. and Salah, A.F. (2010) Efficacy of intercropping mango, mandarin and Egyptian clover plants with date palm on soil properties, rhizosphere microflora and quality and quantity of date fruits. *The Journal of American Science* 6(12), 230–238.

Abul-Soad, A.A. (2010) *Date Palm in Pakistan, Current Status and Prospective*. USAID Firms Project, pp. 9–11. Available at: www.researchgate.net/profile/Anoop_Srivastava7/post/Date-palm-cultivation/attachment/59d64ef079197b80779a8301/AS%3A495243864416256%401495086746794/download/pnaea333.pdf (accessed 17 September 2021).

Abul-Soad, A.A., Mahdi, S.M. and Markhand, G.S. (2015) Date palm status and perspective in Pakistan. In: Al-Khayri, J.M., Jain, S.M. and Johnson, D.V. (eds) *Date Palm Genetic Resources and Utilization*, Vol. 2. *Asia and Europe*. Springer, Dordrecht, The Netherlands, pp. 153–205.

Akyurt, M., Rehbini, E., Bogis, H. and Aljinaidi, A.A. (2002) A survey of mecha-nization efforts on date palm crown operations. In: *Proceedings of the 6th Saudi Engineering Conference*, KFUPM, Dhahran, December 2002, (Vol. 5). pp. 475–489.

Aleid, S.M., Al-Khayri, J.M. and Al-Bahrany, A.M. (2015) Date palm status and per-spective in Saudi Arabia. In: Al-Khayri, J.M., Jain, S.M. and Johnson, D.V. (eds) *Date Palm Genetic Resources and Utilization*, Vol. 2. *Asia and Europe*. Springer, Dordrecht, The Netherlands, pp. 49–95.

Al-Ghafri, A. (2006) Aflaj irrigation D/S ratio. In: *Proceedings of the International Conference on Economics Incentives and Water Demand Management*, Muscat, Oman, March 8–22, 2006. Available at: www.researchgate.net/publication/259931409_Equitability_of_Water_Distribution_in_Aflaj_Irrigation_Systems_of_Oman (accessed 20 August 2021).

Alhammadi, M.S. and Kurup, S.S. (2012) Impact of salinity stress on date palm (*Phoenix dactylifera* L.) – a review. *Crop Production Technologies* 9, 169–173.

Al-Khayri, J.M., Jain, S.M. and Johnson, D.V. (2015) In: *Date Palm Genetic Resources and Utilization*. Vol. 1. *Africa and the Americas*. Springer, Dordrecht, The Netherlands.

Al-Khayri, J.M., Naik, P.M., Jain, S.M. and Johnson, D.V. (2018) Advances in date palm (*Phoenix dactylifera* L.) breeding. In: Al-Khayri, J.M., Jain, S.M. and Johnson, D.V. (eds) *Advances in Plant Breeding Strategies: Fruits*. Springer, Cham, Switzerland, pp. 727–771.

Allen, R.D., Yaish, M., Rolli, E., Hazzouri, M.K., Flowers, J.M. *et al.* (2020) Prospects for the study and improvement of abiotic stress tolerance in date palms in the post-genomics era. *Frontiers in Plant Science* 11, 293. Available at: doi.org/10.3389/fpls.2020.00293.

Al-Muaini, A., Green, S., Dakheel, A., Abdullah, A.-H., Sallam, O. *et al.* (2019) Water requirements for irrigation with saline groundwater of three date-palm cultivars with different salt-tolerances in the hyper-arid United Arab Emirates. *Agricultural Water Management* 222, 213–220. DOI: 10.1016/j.agwat.2019.05.022.

Alshariqi, A.K. (2010) *Study Covering the Bunches with Punched Paper Bags on the Fertilization and the Specifications of Dates in the Early and Late Ripening Stages*. Agricultural Research Station, Al Hemrania, Northern Agricultural Region, United Arab Emirates.

Al-Yahyai, R. (2007) Improvement of date palm production in the Sultanate of Oman. *Acta Horticulturae* 736, 337–343. DOI: 10.17660/ActaHortic.2007.736.32.

Al-Yahyai, R. and Al-Khanjari, S. (2008) Biodiversity of date palm in the Sultanate of Oman. *African Journal of Agricultural Research* 3, 389–395.

Al-Yahyai, R. and Al-Kharusi, L. (2012a) Sub-optimal irrigation affects chemi-cal quality attributes of dates during fruit development. *African Journal of Agricultural Research* 7(10), 1498–1503. DOI: 10.5897/AJAR11.1553.

Al-Yahyai, R. and Al-Kharusi, L. (2012b) Physical and chemical quality attributes of freeze-stored dates. *International Journal of Agriculture and Biology* 14, 97–100.

Al-Yahyai, R. and Khan, M.M. (2015) Date palm status and perspective in Oman. In: Al-Khayri, J.M., Jain, S.M. and Johnson, D.V. (eds) *Date Palm Genetic Resources and Utilization*, Vol. 2. *Asia and Europe*. Springer, Dordrecht, The Netherlands, pp. 207–240.

Ashraf, Z. and Hamidi-Esfahani, Z. (2011) Date and date processing: a review. *Food Reviews International* 27(2), 101–133. DOI: 10.1080/87559129.2010.535231.

Ata, S., Shahbaz, B. and Ahmad, M. (2012) Factors hampering date palm production in the Punjab: a case study of DG Khan district. *Pakistan Journal of Agricultural Sciences* 49, 217–220.

Benabderrahim, M.A., Elfalleh, W., Belayadi, H. and Haddad, M. (2018) Effect of date palm waste compost on forage alfalfa growth, yield, seed yield and minerals uptake. *International Journal of Recycling of Organic Waste in Agriculture* 7(1), 1–9. DOI: 10.1007/s40093-017-0182-6.

Botes, A. and Zaid, A. (2002) The economic importance of date production and international trade. In: Zaid, A. and Arias-Jimenez, E.J. (eds) *Date Palm Cultivation*. FAO Plant Production and Protection Paper no.156 Rev. 1, Food and Agriculture Organization of the United Nations, Rome. Available at: www.fao.org/3/y4360e/y4360e07.htm#bm07 (accessed 17 November 2022).

Canlas, F.Q., Falahi, M.A. and Nair, S. (2022) IoT based date palm water management system using case-based reasoning and linear regression for trend analysis. *International Journal of Advanced Computer Science and Applications* 13(2), 549–556. DOI: 10.14569/IJACSA.2022.0130264.

Carpenter, J.B. and Elmer, H. (1978) *Pests and Diseases of the Date Palm*. US Department of Agriculture, Washington, DC.

Carr, M.K.V. (2013) The water relations and irrigation requirements of the date palm (*Phoenix dactylifera* L.): a review. *Experimental Agriculture* 49(1), 91–113. DOI: 10.1017/S0014479712000993.

Cavins, T., Whipker, B., Fonteno, W., Harden, B., McCall, I. *et al.* (2008) *Monitoring and Managing PH and EC Using the PourThru Extraction Method*. North Carolina State University, Raleigh, North Carolina.

Chao, C.T. and Krueger, R.R. (2007) The date palm (*Phoenix dactylifera* L.): overview of biology, uses, and cultivation. *HortScience* 42(5), 1077–1082. DOI: 10.21273/HORTSCI.42.5.1077.

Cohen, Y., Slavkovic, F., Birger, D., Greenberg, A., Sadowsky, A. *et al.* (2016) Fertilization and fruit setting in date palm: biological and technological challenges. *Acta Horticulturae* 1130, 351–358. DOI: 10.17660/ActaHortic.2016.1130.53.

Cuevas, J., Daliakopoulos, I.N., del Moral, F., Hueso, J.J. and Tsanis, I.K. (2019) A review of soil-improving cropping systems for soil salinization. *Agronomy* 9(6), 295. DOI: 10.3390/agronomy9060295.

Dalbon, V.A., Acevedo, J.P.M., Ribeiro Junior, K.A.L., Ribeiro, T.F.L., da Silva, J.M. *et al.* (2021) Perspectives for synergic blends of attractive sources in South American palm weevil mass trapping: waiting for the red palm weevil Brazil invasion. *Insects* 12(9), 828. DOI: 10.3390/insects12090828.

Dehghan-shoar, Z., Hamidi-esfahani, Z. and Abbasi, S. (2009) Effect of temperature and modified atmosphere on quality preservation of Sayer date fruits (*Phoenix dactylifera*). *Journal of Food Processing and Preservation* 34(2), 323–334. DOI: 10.1111/j.1745-4549.2008.00349.x.

Djerbi, M. (1994) *Précis de Phoeniciculture*. Food and Agriculture Organization of the United Nations, Rome, pp. 23–191.

Elamin, A., Elsadig, E., Aljubouri, H. and Gafar, M. (2017) Improving fruit quality and yield of Khenazi date palm (*Phoenix dactilifera* L.) grown in sandy soil by application of nitrogen, phosphorus, potassium and organic manure. *International Journal of Development and Sustainability* 6, 862–875.

Elfeky, A. and Elfaki, J. (2019) A review: date palm irrigation methods and water resources in the Kingdom of Saudi Arabia. *Journal of Engineering Research and Reports* 9, 1–11. DOI: 10.9734/jerr/2019/v9i217012.

El Hadrami, A. and Al-Khayri, J.M. (2012) Socioeconomic and traditional importance of date palm. *Emirates Journal of Food and Agriculture* 24, 371–385.

El Mardi, M.O., Salama, S.B., Consolacion, E. and Al-Shabibi, M.S. (1995) Effect of treated sewage water on vegetative and reproductive growth of date palm. *Communications in Soil Science and Plant Analysis* 26(11–12), 1895–1904. DOI: 10.1080/00103629509369416.

El Mardi, M.O., Salama, S.B., Consolacion, E.C. and Al-Solomi, M. (1998) Effect of treated sewage water on the concentration of certain nutrient elements in date palm leaves and fruits. *Communications in Soil Science and Plant Analysis* 29(5–6), 763–776. DOI: 10.1080/00103629809369983.

Elsadig, E.H., Aljuburi, H.J., Elamin, A.H.B. and Gafar, M.O. (2017) Impact of organic manure and combination of N P K S, on yield, fruit quality and fruit mineral content of Khenazi date palm (*Phoenix dactylifera* L.) cultivar. *Journal of Scientific Agriculture* 1, 335–346. DOI: 10.25081/jsa.2017.v1.848.

Erskine, W., Moustafa, A.T., Osman, A.E., Lashine, Z., Nejatian, A. *et al.* (2004) Date palm in the GCC countries of the Arabian Peninsula. In: *Proceedings of the Regional Workshop on Date Palm Development in the Arabian Peninsula*, Abu Dhabi, United Arab Emirates, pp. 29–31. Available at: www.researchgate.net/publication/267714856_Date_Palm_in_the_GCC_countries_of_the_Arabian_Peninsula (accessed 29 July 2021).

Escobar, H.A. and Valdivia, R.G. (2015) Date palm status and perspective in South American countries: Chile and Peru. In: Al-Khayri, J.M., Jain, S.M. and Johnson, D.V. (eds) *Date Palm Genetic Resources and Utilization*. Vol. 1. *Africa and the Americas*. Springer, Dordrecht, The Netherlands, pp. 487–506.

Falade, K.O. and Abbo, E.S. (2007) Air-drying and rehydration characteristics of date palm (*Phoenix dactylifera* L.) fruits. *Journal of Food Engineering* 79(2), 724–730. DOI: 10.1016/j.jfoodeng.2006.01.081.

FAO (2022) FAOSTAT. Food and agriculture data. Food and Agricultural Organization of the United Nations, Rome. Available at: www.fao.org/faostat/en/#home (accessed 20 November 2022).

Farooq, S., Maqbool, M.M., Bashir, M.A., Ullah, M.I., Shah, R.U. *et al.* (2021) Production suitability of date palm under changing climate in a semi-arid region predicted by CLIMEX model. *Journal of King Saud University - Science* 33(3), 101394. DOI: 10.1016/j.jksus.2021.101394.

Fisher, P., Douglas, A. and Argo, W. (2008) Use the 1:2 testing method for media-pH and EC. Available at: www.greenhousemag.com/article/use-the-1-2-testing-method-for-media-ph-and-ec/ (accessed 17 July 2021).

Furr, J.R. and Armstrong, W.W. (1957) Nitrogen fertilization of dates – a review and progress report. *Date Growers' Institute Annual Report* 34, 6–9.

Furr, J. and Armstrong, W. (1975) Water and salinity problems of Abadan island date gardens. *Date Growers' Institute Annual Report* 52, 14–17.

Gros-Balthazard, M. and Flowers, J.M. (2021) A brief history of the origin of domesticated date palms. In: Al-Khayri, J.M., Jain, S.M. and Johnson, D.V. (eds) *The Date Palm Genome*, Vol. 1. *Phylogeny, Biodiversity and Mapping*. Compendium of Plant Genomes, Springer, Cham, Switzerland, pp. 55–74.

Gros-Balthazard, M., Galimberti, M., Kousathanas, A., Newton, C., Ivorra, S. *et al.* (2017) The discovery of wild date palms in Oman reveals a complex domestication history involving centers in the Middle East and Africa. *Current Biology* 27(14), 2211–2218. DOI: 10.1016/j.cub.2017.06.045.

Gurevich, V., Lavi, U. and Cohen, Y. (2005) Genetic variation in date palms propagated from offshoots and tissue culture. *Journal of the American Society for Horticultural Science* 130(1), 46–53. DOI: 10.21273/JASHS.130.1.46.

Hamza, H., Jemni, M., Benabderrahim, M.A., Mrabet, A., Touil, S. *et al.* (2015) Date palm status and perspective in Tunisia. In: Al-Khayri, J.M., Jain, S.M. and Johnson, D.V. (eds) *Date Palm Genetic Resources and Utilization*, Vol. 1. *Africa and the Americas*. Springer, Dordrecht, The Netherlands, pp. 193–211.

Hinkaew, J., Aursalung, A., Sahasakul, Y., Tangsuphoom, N. and Suttisansanee, U. (2021) A comparison of the nutritional and biochemical quality of date palm fruits obtained using different planting techniques. *Molecules* 26(8), 2245. DOI: 10.3390/molecules26082245.

Hodel, D.R., Downer, A.J. and Pittenger, D.R. (2009) Transplanting palms. *HortTechnology* 19(4), 686–689. DOI: 10.21273/HORTSCI.19.4.686.

Holanda, J.S., Oliveira, M.T., Sobrinho, E.E. and Dantas, T.B. (2007) *Tecnologias Para Produção Intensiva de Coco Anão*, Boletim de Pesquisa No 34. Empresa de Pesquisa Agropecuária do Rio Grande do Norte (EMPARN), Natal, Brazil.

Hoop, B.M.D. (2000) Date palm micropropagation in Saudi Arabia: policies and technology transfer. *International Journal of Biotechnology* 2(4), 333–341. DOI: 10.1504/IJBT.2000.000143.

Howard, F.W., Moore, D., Giblin-Davis, R. and Abad, R. (2001) *Insects on Palms*. CAB International, Wallingford, UK. DOI: 10.1079/9780851993263.0000.

Hussein, A.H.A. (2008) Impact of nitrogen and potassium fertilization on Khalas date palm cultivar yield, fruit characteristics, leaves and fruit nutrient content in Al-Hassa oasis, KSA. *Journal of Environmental Sciences* 35, 33–48.

Hussain, M. and Ismailli, N.J. (2020) Isolation and characterization of *Fusarium oxysporum* causing bayoud disease of date palm tree from Gambat, district Khairpur, Sindh, Pakistan. *Plant Protection* 4(1), 43–47. DOI: 10.33804/pp.004.01.3234.

Idris, T.I.M., Khidir, A.A. and Haddad, A.M. (2012) Growth and yield responses of a dry date palm (*Phoenix dactylifera* L.) cultivar to soil and foliar fertilizers. *International Research Journal of Agricultural Science and Soils* 2(9), 390–394.

Iqbal, R., Raza, M.A.S., Valipour, M., Saleem, M.F., Zaheer, M.S. *et al.* (2020) Potential agricultural and environmental benefits of mulches—a review. *Bulletin of the National Research Centre* 44(1), 75. DOI: 10.1186/s42269-020-00290-3.

Ismail, B., Haffar, I., Baalbaki, R. and Henry, J. (2008) Physico-chemical characteristics and sensory quality of two date varieties under commercial and industrial storage conditions. *LWT - Food Science and Technology* 41(5), 896–904. DOI: 10.1016/j.lwt.2007.06.009.

Jain, S.M. (2012) Date palm biotechnology: current status and prospective – an overview. *Emirates Journal of Food and Agriculture* 24, 386–399.

Johnson, D.V. (2011) Introduction: date palm biotechnology from theory to practice. In: Jain, S.M., Al-Khayri, J.M. and Johnson, D.V. (eds) *Date Palm Biotechnology*. Springer, Dordrecht, The Netherlands, pp. 1–11.

Karbout, N., Dhaouidi, L., Boughdiri, A., Jaoued, M., Moussa, M. *et al.* (2019) Qualitative analysis of the indicators of degradation of the Nefzaoui oases and quantitative study of their impacts on the socioeconomic level of the region farmers. *Journal of New Sciences* 65, 4114–4124.

Karbout, N., Mlih, R., Latifa, D., Bol, R., Moussa, M. *et al.* (2021) Farm manure and bentonite clay amendments enhance the date palm morphology and yield. *Arabian Journal of Geosciences* 14(9), 818. DOI: 10.1007/s12517-021-07160-w.

Kirkby, C.A., Richardson, A.E., Wade, L.J., Conyers, M. and Kirkegaard, J.A. (2016) Inorganic nutrients increase humification efficiency and C-sequestration in an annually cropped soil. *PloS ONE* 11(5), e0153698. DOI: 10.1371/journal. pone.0153698.

Klein, P. and Zaid, A. (2002) Land preparation, planting operation and fertilisation requirements. In: Zaid, A. and Arias-Jimenez, E.J. (eds) *Date Palm Cultivation*. FAO Plant Production and Protection Paper no.156 Rev. 1, Food and Agriculture Organization of the United Nations, Rome. Available at: www.fao. org/3/Y4360E/y4360e0a.htm (accessed 10 August 2021).

Krueger, R.R. (2021) Date palm (*Phoenix dactylifera* L.) biology and utilization. In: Al-Khayri, J.M., Jain, S.M. and Johnson, D.V. (eds) *The Date Palm Genome*, Vol. 1. *Phylogeny, Biodiversity and Mapping*. Compendium of Plant Genomes, Springer, Cham, Switzerland, pp. 3–28.

Kruse, J., Adams, M., Winkler, B., Ghirardo, A., Alfarraj, S. *et al.* (2019) Optimization of photosynthesis and stomatal conductance in the date palm *Phoenix dactylifera* during acclimation to heat and drought. *New Phytologist* 223(4), 1973–1988. DOI: 10.1111/nph.15923.

Latifian, M., Rahnama, A.A. and Sharifnezhad, H. (2012) Effects of planting pattern on major date palm pests and diseases injury severity. *International Journal of Agriculture and Crop Sciences* 4, 1443–1451.

Lehmann, J. and Kleber, M. (2015) The contentious nature of soil organic matter. *Nature* 528(7580), 60–68. DOI: 10.1038/nature16069.

Liebenberg, P. and Zaid, A. (2002) Date palm irrigation. In: Zaid, A. and Arias-Jimenez, E.J. (eds) *Date Palm Cultivation*. FAO Plant Production and Protection Paper no.156 Rev. 1, Food and Agriculture Organization of the United Nations, Rome. Available at: www.fao.org/3/Y4360E/y4360e0b.htm# (accessed 11 August 2021).

Mahal, N.K., Osterholz, W.R., Miguez, F.E., Poffenbarger, H.J., Sawyer, J.E. *et al.* (2019) Nitrogen fertilizer suppresses mineralization of soil organic matter in maize agroecosystems. *Frontiers in Ecology and Evolution* 7, 59. DOI: 10.3389/fevo.2019.00059.

Mansour, A., Mohamed, A., Ahmed, F. and Eissa, R. (2012) Benefits of intercropping Samany date palms with some fruit crops. *Journal of Applied Sciences Research* 8, 2045–2049.

Marzouk, H.A. and Kassem, H.A. (2011) Improving fruit quality, nutritional value and yield of Zaghloul dates by the application of organic and/or mineral fertilizers. *Scientia Horticulturae* 127(3), 249–254. DOI: 10.1016/j.scienta.2010.10.005.

Mazloumzadeh, S.M., Shamsi, M. and Nezamabadi-pour, H. (2008) Evaluation of general-purpose lifters for the date harvest industry based on a fuzzy inference system. *Computers and Electronics in Agriculture* 60(1), 60–66. DOI: 10.1016/j. compag.2007.06.005.

Mlih, R., Bol, R., Amelung, W. and Brahim, N. (2016) Soil organic matter amendments in date palm groves of the Middle Eastern and North African region: a mini-review. *Journal of Arid Land* 8(1), 77–92. DOI: 10.1007/s40333-015-0054-8.

Mlih, R.K., Gocke, M.I., Bol, R., Berns, A.E., Fuhrmann, I. *et al.* (2019) Soil organic matter composition in coastal and continental date palm systems: insights from Tunisian oases. *Pedosphere* 29(4), 444–456. DOI: 10.1016/S1002-0160(19)60814-3.

Mohamed, H.I., El-Beltagi, H.S., Jain, S.M. and Al-Khayri, J.M. (2021) Date palm (*Phoenix dactylifera* L.) secondary metabolites: bioactivity and pharmaceutical potential. In: Bhat, R.A., Hakeem, K.R. and Dervash, M.A. (eds) *Phytomedicine: A Treasure of Pharmacologically Active Products from Plants*. Academic Press, London, pp. 483–531.

Mohammed, M., Sallam, A., Munir, M. and Ali-Dinar, H. (2021) Effects of deficit irrigation scheduling on water use, gas exchange, yield, and fruit quality of date palm. *Agronomy* 11(11), 2256. DOI: 10.3390/agronomy11112256.

Munier, P. (1973) *The Date Palm*. Techniques Agricoles et Productions Tropicales 24. G.-P. Maisonneuve & LaRose, Paris.

Munns, R. and Tester, M. (2008) Mechanisms of salinity tolerance. *Annual Review of Plant Biology* 59, 651–681. DOI: 10.1146/annurev.arplant.59.032607.092911.

Nixon, R.W. (1951) The date palm—"Tree of Life" in the subtropical deserts. *Economic Botany* 5(3), 274–301. DOI: 10.1007/BF02985151.

Nixon, R.W. and Carpenter, J.B. (1978) *Growing Dates in the United States*, USDA Bulletin No. 207. US Department of Agriculture, Washington, DC.

Peoples, M.B., Brockwell, J., Herridge, D.F., Rochester, I.J., Alves, B.J.R. *et al.* (2009) The contributions of nitrogen-fixing crop legumes to the productivity of agricultural systems. *Symbiosis* 48(1–3), 1–17. DOI: 10.1007/BF03179980.

Pintaud, J.C., Ludeña, B., Aberlenc-Bertossi, F., Zehdi, S., Gros-Balthazard, M. *et al.* (2013) Biogeography of the date palm (*Phoenix dactylifera* L., Arecaceae): insights on the origin and on the structure of modern diversity. *Acta Horticulturae* 994, 19–38. DOI: 10.17660/ActaHortic.2013.994.1.

Plaza, C., Zaccone, C., Sawicka, K., Méndez, A.M., Tarquis, A. *et al.* (2018) Soil resources and element stocks in drylands to face global issues. *Scientific Reports* 8(1), 13788. DOI: 10.1038/s41598-018-32229-0.

Rahnama, A. and Latifian, M. (2013) Intercropping relative efficiency and its effects on date palm pests and disease control. *International Journal of Agriculture: Research and Review* 3, 617–623.

Rajmohan, K. (2011) Date palm tissue culture: a pathway to rural development. In: Jain, S.M., Al-Khayri, J.M. and Johnson, D.V. (eds) *Date Palm Biotechnology*. Springer, Dordrecht, The Netherlands, pp. 29–45.

Ramawat, K.G. (ed.) (2009) *Desert Plants: Biology and Biotechnology*. Springer, Berlin/Heidelberg, Germany.

Ramoliya, P.J. and Pandey, A.N. (2003) Soil salinity and water status affect growth of *Phoenix dactylifera* seedlings . *New Zealand Journal of Crop and Horticultural Science* 31(4), 345–353. DOI: 10.1080/01140671.2003.9514270.

Raza, A., Razzaq, A., Mehmood, S.S., Zou, X., Zhang, X. *et al.* (2019) Impact of climate change on crops adaptation and strategies to tackle its outcome: a review. *Plants* 8(2), 34. DOI: 10.3390/plants8020034.

Robinson, M., Brown, B. and Williams, C. (2012) *The Date Palm in Southern Nevada*. Extension Publication no.SP-02-12. University of Nevada, Reno, Nevada.

Safronov, O., Kreuzwieser, J., Haberer, G., Alyousif, M.S., Schulze, W. *et al.* (2017) Detecting early signs of heat and drought stress in *Phoenix dactylifera* (date palm). *PloS ONE* 12(6), e0177883. DOI: 10.1371/journal.pone.0177883.

Salomón-Torres, R., Krueger, R., García-Vázquez, J.P., Villa-Angulo, R., Villa-Angulo, C. *et al.* (2021) Date palm pollen: features, production, extraction and pollination methods. *Agronomy* 11(3), 504. DOI: 10.3390/agronomy11030504.

Sarraf, M., Jemni, M., Kahramanoğlu, I., Artés, F., Shahkoomahally, S. *et al.* (2021) Commercial techniques for preserving date palm (*Phoenix dactylifera*) fruit quality and safety: a review. *Saudi Journal of Biological Sciences* 28(8), 4408–4420. DOI: 10.1016/j.sjbs.2021.04.035.

Shaaban, S. and Mahmoud, M.M. (2012) Nutritional evaluation of some date palm (*Phoenix dactylifera* L.) cultivars grown under Egyptian conditions. *The Journal of American Science* 8, 135–139.

Shrivastava, P. and Kumar, R. (2015) Soil salinity: a serious environmental issue and plant growth promoting bacteria as one of the tools for its alleviation. *Saudi Journal of Biological Sciences* 22(2), 123–131. DOI: 10.1016/j.sjbs.2014.12.001.

Świąder, K., Białek, K. and Hosoglu, I. (2020) Varieties of date palm fruits (*Phoenix dactylifera* L.), their characteristics and cultivation. *Postępy Techniki Przetwórstwa Spożywczego (Technological Progress in Food Processing)* 173–179.

Warnke, D. (2011) Recommended test procedures for greenhouse growth media. In: *Recommended Soil Testing Procedures for the Northeastern United States*, 3rd edn. Northeastern Regional Publication No. 493, Agricultural Experiment Stations of Connecticut, Delaware, Maine, Maryland, Massachusetts, New Hampshire, New Jersey, New York, Pennsylvania, Rhode Island, Vermont, and West Virginia. Available at: www.udel.edu/content/dam/udelImages/canr/pdfs/extension/factsheets/soiltest-recs/CHAP13.pdf (accessed 15 November 2021).

Yaish, M.W. and Kumar, P.P. (2015) Salt tolerance research in date palm tree (*Phoenix dactylifera* L.), past, present, and future perspectives. *Frontiers in Plant Science* 6, 348. DOI: 10.3389/fpls.2015.00348.

Youssef, T. and Awad, M.A. (2008) Mechanisms of enhancing photosynthetic gas exchange in date palm seedlings (*Phoenix dactylifera* L.) under salinity stress by a 5-aminolevulinic acid-based fertilizer. *Journal of Plant Growth Regulation* 27(1), 1–9. DOI: 10.1007/s00344-007-9025-4.

Yudianto, A.A., Fajriani, S. and Aini, N. (2015) Pengaruh jarak tanam dan pembumbunan terhadap pertumbuhan dan hasil tanaman garut (*Marantha arundinaceae* L.). *Journal of Produkti Tanaman* 3(3), 172–181.

Zaid, A. and de Wet, P. (2002a) Origin, geographical distribution and nutritional values of date palm. In: Zaid, A. and Arias-Jimenez, E.J. (eds) *Date Palm Cultivation*. FAO Plant Production and Protection Paper no.156 Rev. 1, Food and Agriculture Organization of the United Nations, Rome. Available at: www.fao.org/3/Y4360E/y4360e06.htm#bm06 (accessed 2 September 2021).

Zaid, A. and de Wet, P. (2002b) Pollination and bunch management. In: Zaid, A. and Arias-Jimenez, E.J. (eds) *Date Palm Cultivation*. FAO Plant Production and Protection Paper no.156 Rev. 1, Food and Agriculture Organization of the United Nations, Rome. Available at: www.fao.org/3/Y4360E/y4360e0c.htm#bm12 (accessed 2 September 2021).

Zohary, D. and Spiegel-Roy, P. (1975) Beginnings of fruit growing in the old world. *Science* 187(4174), 319–327. DOI: 10.1126/science.187.4174.319.

Date Palm Pollination Management

7

Kapil Mohan Sharma[1] , Mithlesh Kumar[2*] , C.M. Muralidharan[1] and Ricardo Salomón-Torres[3]

[1]*Sardarkrushinagar Dantiwada Agricultural University, Mundra, India;* [2]*Agriculture University, Jodhpur, India;* [3]*Universidad Estatal de Sonora, Sonora, Mexico*

Abstract

Pollination is one of the most important and laborious agronomical practices in date cultivation, without which it is difficult to get the desired yield. Date palm pollen (DPP) has a major influence on fruit quality, development, and yield, which are widely influenced by pollen structure, viability, and germination capability. Under commercial cultivation, growers need to manage pollen by storing it at a cool temperature and mixing it with various adjuvants to dilute the quantity of the pollen used for pollination. The success of the pollination is further influenced by the female flowers and their receptivity, and thus proper vigilance is required to note the time of anthesis of the female inflorescence. Over the years, methods of pollination have improved and with the usage of mechanized tools the laborious pollination process has been made easier. Apart from pollination, pollen also has secondary uses for various medical ailments such as male sterility. DPP is also a potential cause of various respiratory ailments such as asthma and thus requires careful handling. This chapter discusses the importance of pollen, its structure, its method of utilization, xenia and metaxenia effects, alternative uses, and associated probable problems.

Keywords: Metaxenia, Nutritional property, *Phoenix dactylifera*, Pollen allergen, Pollen extraction, Pollen storage, Pollination, Xenia

*Corresponding author: mithleshgenetix@gmail.com

© CAB International 2023. *Date Palm* (eds J.M. Al-Khayri *et al.*)
DOI: 10.1079/9781800620209.0007

7.1 Introduction

Date fruit, borne on the date palm (*Phoenix dactylifera* L.), is one of the oldest cultivated fruit crops in the world. Being a dioecious crop, the flowering of both male and female inflorescences occurs on separate plants. Date palms are pollinated naturally by wind, but for commercial cultivation, they are mostly artificially pollinated. Owing to the palm's structure, long life, asynchronized flowering, continuous increase in height, and thorny leaves, pollination is one of the most important but laborious agronomical practices in date cultivation. Fruit development is dependent on the pollination and various fruit developmental stages are considered starting from the day of pollination. Without proper pollination, high percentages of abortion occur or parthenocarpic fruits are developed that do not possess commercial value due to inferior quality (Sharma *et al.*, 2021). Over the years, pollination methods and pollen management have been widely optimized, from manual application to mechanized use, which is discussed in this chapter.

The importance of pollen and its management has been understood since the date palm's early domestication, which is estimated to have occurred around 4000 BCE (Zohary and Hopf, 2000). The earliest domestication was probably in the Arabian Gulf region near its center of origin, Mesopotamia (present Iraq) (Zaid and de Wet, 2002a), where the crop was considered sacred. The importance of pollination for the successful production of date palm fruits can be understood from a stone sculpture extracted from the Assyrian Empire, near Mesopotamia (Fig. 7.1), where a human-faced genie and an eagle-faced genie are shown pollinating the Sacred Tree, which is believed to be a date palm. Similar concepts have also been supported by Paszke (2019) and Ziffer (2019), suggesting date cultivation and knowledge of artificial pollination in 3300 BC.

In modern-day cultivation, pollination is considered one of the most important agronomical practices. Its importance is critical as it directly influences the fruit set percentage and yield. If pollination is unsuccessful, fruits develop into parthenocarpic fruit that are small sized and do not possess any desired commercial characters (Cohen *et al.*, 2016). In the wild groves that are present in oases or by riversides, date palms mostly grow in clusters and natural pollination is carried out by the wind. However, for successful pollination, the population of both males and females should ideally be 1:1 and closely spaced, as occurs in oasis ecosystems, but this is impractical for commercial cultivation. Internationally, in most commercial orchards of date palms, the male plants are either planted along the roadside or the farm boundaries, or a separate dedicated area is devoted to the male date palm plants. It is estimated that on average one male plant is sufficient to pollinate around 50 female plants, but this is dependent on the number of flowers and the quantity of pollen available in each spathe (Zaid and de Wet, 2002b).

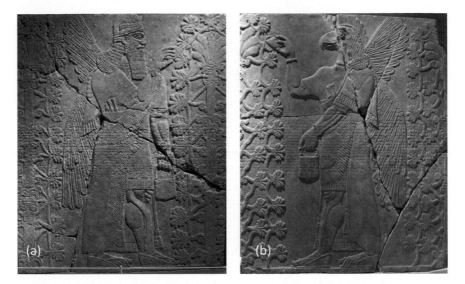

Fig. 7.1. (a) Assyrian Apkallu-figure fertilizing the Sacred Tree, *c*.883–859 BCE. (b) Assyrian Apkallu-figure between two Sacred Trees, *c*.883–859 BCE. (Images from Brooklyn Museum, used under Creative Commons BY license.)

The source and quality of the pollen play a vital role in the successful pollination and fruit set. Essentially, any male date pollen can be used for successful pollination of the female flowers. However, it has been noted that the quality of the pollen influences both the fruit and seed characteristics of the date palm.

Under natural conditions, the synchronized emergence of both male and female inflorescences allows sufficient movement of pollen from the male to the female inflorescences. This synchronization is also beneficial if hand pollination is done by using two or three male strands and placing them upside down in the female inflorescence after initial dusting of male strands on the female inflorescence. This is the traditional method of pollination and is still followed in most parts of the world (Baidiyavadra *et al.*, 2019; Sharma *et al.*, 2021). However, there are two other instances where pollen needs to be managed: (i) if the male inflorescences emerge before the female inflorescences; and (ii) if the female inflorescences emerge before the male inflorescences. In the first case, pollen is extracted manually from the male flowers by drying them in the shade or using a mechanical extraction procedure. In the second case, pollen scarcity can be compensated for by using stored, extracted, dried pollen from the previous season. After extraction, the pollen is stored or used fresh for pollination. Close observation is needed at the time of female spathe cracking and of the temperature during the time of the pollination. With new advances, the method of pollination has been mechanized and with newer innovations like drones, it can be upgraded further. Date palm pollen (DPP) contains nutritional properties and also has secondary use as a medical treatment to treat problems like human male sterility. It has been observed that with regular exposure to pollen, date palm workers often experience respiratory problems like asthma.

7.2 Flower Structures

The date palm is a dioecious plant and flowers are borne in an inflorescence, called a spadix, which is covered by a thick greenish-brown hard and fibrous covering called a spathe, which protects the inflorescence during its early stage of development. The spadix consists of a central axis or stem called a rachis with several strands or spikelets (50–150 in number) arranged in a rosette. A few days before flowering, the inflorescence emerges from the axis of the leaves enclosed in the spathe, emerging from the fibrous sheaths that form leaf bases. On the day of anthesis, the spathe splits, allowing the inflorescence to emerge and eventually to release pollen. In the northern hemisphere flowering mostly occurs between February and April, while in the southern hemisphere flowering occurs in July to September. However, a few variations are observed based on environmental factors (Krueger, 2021).

In general, the male inflorescence is comparatively shorter and broader, having a greater number of strands per inflorescence, compared to the female inflorescence. Each female inflorescence consists of 8000 to 10,000 flowers; and in the male there are even more. Male flowers of the date palm are waxy-white in color and consist of three sepals, three petals, six stamens, and three carpels. Each stamen contains two small pollen sacs that open within an hour or two after splitting of the spathe. The structure of the male flowers resembles rice panicles and the inflorescence possesses a broom-like structure with sweet-scented flowers (Zaid and de Wet, 2002b) (Fig. 7.2).

The female flowers have rudimentary stamens, a tri-carpel structure pressed together, and a superior ovary (hypogynous). The flower contains three sepals and three petals fused together. Female flowers appear more yellowish than male ones and only the tips diverge. The diameter of the flower ranges from 3 to 4 mm and only one ovule per flower is fertilized, leading to development of only one carpel out of three, while the other two are aborted during fruit development. If the flowers remain unpollinated, all three carpels develop into small, undesirable parthenocarpic fruits (Fig. 7.3). On average, one palm produces 15–25 female inflorescences and more than 20 are produced in male plants; however, this is dependent upon orchard management practices, age, vigor, and varietal character (Salomón-Torres et al., 2021).

The appearance of bisexual flowers is also well known and is observed in male plants. Although the same male plant may or may not bear bisexual flowers every year, the possibilities of such occurrences in the same plant are higher. This is mostly due to hormonal imbalance and higher gibberellic

Fig. 7.2. (a, c) Male inflorescence of date palm. (b, d) Female inflorescence of date palm. (e) Male strand of date palm. (f) Female strand of date palm. (g) Male date palm flowers; frontal (top) and profile (bottom) views. (h) Female date palm flowers; frontal (top) and profile (bottom) views. S, sepal; P, petal; St, stamen; C, carpel. Scale bar = 1600 μm. (Photos (a, b, d) by Kapil Mohan Sharma; photos (c, e, f) by C.M. Muralidharan; photos (g, h) from Masmoudi-Allouche *et al.*, 2009 with permission.)

acid (GA)-like activity that promotes female induction in male trees. During the initial stage of flower differentiation, both male and female flowers have low GA activity but at later stages higher GA production in flowers on male trees may result in sex conversion (Leshem and Ophir, 1977; Reuvani, 1985). In such cases the flowers are spirally arranged and have more male than female flowers (Kgazal *et al.*, 1990). The characteristics of male and female flowers are compared in Table 7.1.

7.3 Pollen Structure, Characteristics, and Diversity

7.3.1 Pollen structure

Pollen is the fine dust-like grain material produced by the male flowers and is the source of male gametes. Pollen grains are ellipsoidal, bisymmetric, sometimes asymmetric or rhomboid or spherical in shape, with length

Fig. 7.3. Pollinated and unpollinated fruits in date palm. (Photos by Kapil Mohan Sharma.)

Table 7.1. Comparison of male and female flowers of date palm.

Male flowers	Female flowers
Creamy/waxy-white color	Yellowish-white color
Spathes are short and wide	Spathes are narrow and long
Inflorescences are densely packed with strands and have a greater number of strands (broom-like)	Inflorescences are comparatively less densely packed with strands
Flowers resemble rice panicles	Flowers are globose and resemble sorghum seeds
Flowers are composed of three sepals, three petals, six stamens, and three carpels	Flowers are composed of three sepals and three petals (fused together)
Flowers have a distinct sweet-scented aroma that attracts insects	Flowers have a mild aroma but it does not widely attract insects

ranging from 17.20 to 21.40 µm and width from 6.97 to 10.30 µm (Fig. 7.4). In other studies, sizes were noted as 20.38–21.94 µm × 16.32–16.96 µm and 18.56–18.64 µm × 18.51–18.55 µm (Al-Khalifah, 2006; Soliman and Al-Obeed, 2013).

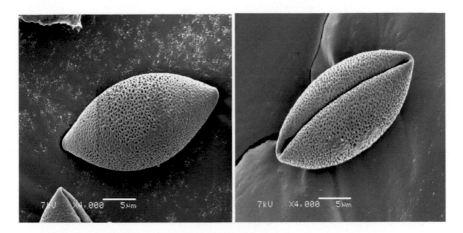

Fig. 7.4. Scanning electron microscopic image of date palm pollen. (From Almehdi *et al.*, 2005 reproduced with permission.)

Pollen is very lightweight and is easily carried/transported by the wind, which is the major natural method of its dispersal. Successful wind pollination is possible if the population is closely planted and possesses around 50% male and 50% female plants. The date palm is not successfully pollinated by bees and insects, because the sweet smell and nectar are produced more by male flowers, therefore these pollinating agents visit the female flowers less, resulting in an un-uniform or unsuccessful pollination. Visually, success of pollination can only be confirmed after around 25–30 days when one of the carpels starts developing into fruits while the other two fall. This is considered the first phase of fruit drop or abscission (Reuvani, 1986).

7.3.2 Pollen viability evaluation

In the date palm, success of pollination and fruit set is more dependent upon pollen viability than in other fruit crops like peach, nectarine, apple, etc. (Soliman and Al-Obeed, 2013). The application of high amounts of pollen determines the success of the pollination and resultant high fruit set percentages, provided that the pollen is viable. For commercial cultivation it is important to ascertain pollen viability, which can be done through various tests such as fluorescein diacetate (FDA), Alexander's staining, absolute pollen viability, *in vitro* germination, acetocarmine staining, and others (Maryam *et al.*, 2017). Among them, acetocarmine is very common and the most widely used. A detailed method of pollen viability and germination test has been elucidated by Maryam *et al.* (2017). Fresh pollen is highly effective and has its highest germination rate, which reduces as the pollen ages (Fig. 7.5). The quality of the pollen and its germination ability also vary with the genotype and time of inflorescence emergence. Karim

Fig. 7.5. Fresh date palm pollen from current season (left) and 1-year-old pollen stored at room temperature (right). (Photo by Kapil Mohan Sharma.)

et al. (2021) found that out of six different male genotypes, an early male flowering genotype ABD1 had the highest pollen quality both at its fresh stage and after storage. They also reported that pollen collected from the inflorescence during the middle of the flowering stage has highest viability and germinability compared to those earlier or later in the flowering period. It is important to store pollen properly to keep its viability intact.

To test pollen viability, 1% (w/v) acetocarmine is used as a stain. Use one drop of 1% (w/v) acetocarmine solution on a glass slide followed by a drop of pollen suspension on it. Cover them with the coverslip, remove the extra stain using tissue paper or filter paper. Observe the pollen under 200× magnification. If the pollen is stained red, it is viable. The higher the percentage of viable pollen, the higher the chances of fruit set. It can be further confirmed using the pollen germination test. For pollen germination, pollen is dusted using a camel-hair brush on the germinating medium composed of 1% (w/v) agar, 10% (w/v) sucrose, and 500 ppm boron with a pH of 6.0 (pH can be modified based on types of cultivars) in a Petri dish. The pollen in the media is incubated at 25–27°C for 24 h and then pollen tube germination is observed under the microscope (Fig. 7.6).

7.3.3   Xenia and metaxenia effects

Metaxenia is the term for the direct influence of pollen on portions of the fruit, while xenia is the term for its influence on the seed. Both the embryo and the endosperm in metaxenia are made up of cells whose nuclei contain chromosomes from the male parent. As a result, the male

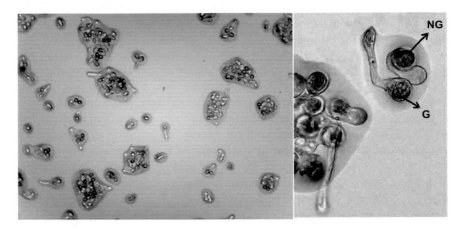

Fig. 7.6. *In vitro* germination of date palm pollen grains. G, germinated; NG, non-germinated. (Reproduced from El-Kadri and Mimoun, 2020, with permission.)

parent effect should be expected to be seen in these two fertilized parts of the seed. The seed of a young developing date fruit is in close contact with endosperm tissues, and it is highly likely that hormones or soluble substances secreted by the embryo or endosperm, or both, could and would diffuse freely from the seed into the nearby tissues, influencing the development of the adjacent ovarian tissue of the mother plant in a definite manner, and exerting a specific effect on this tissue (Swingle, 1928). The pollen of the date palm and other *Phoenix* species has been found to have a direct effect on the size, shape, and color of the seed, as well as on the size of the fruit, its rate of development, and the time of ripening of the fruit, which is made up of the ovarian tissue of the date palm's mother plant (Swingle, 1928). Denney (1992) suggested that different amounts of one or more of the three hormones most closely connected with fruit growth (auxins, cytokinin, and/or GA) could explain the varied metaxenic size effects in date fruits and suggested smaller fruits (seed + pericarp) will have lower hormone levels, whereas larger fruits will have higher hormone levels. El-Hamady *et al.* (2010) reported high indole-3-acetic acid (IAA; a type of auxin) content was correlated with the rapid development of date fruits after pollination from two male plants, M_1 and M_2. The M_1 male maintained a higher IAA level as compared to the M_2, resulting in higher fruit weight of fruit resulting from pollination with M_1 pollen. Further, Abbas *et al.* (2014) elucidated those differences in free IAA levels in date fruit produced by pollinating female flowers with different pollen sources are likely to be responsible for various manifestations of metaxenia.

Farmers choose pollen sources that have a beneficial impact on yield and the physical/chemical quality of the fruit. Various studies have been conducted in this area in order to determine the best male pollinators for date palms. Helail and El-Kholey (2000) observed variation in yield and fruit weight using different pollen sources, while Rezazadeh *et al.* (2013) reported that 12 different pollen sources had different influences on the maturity period of the fruit and fruit size. The use of five DPP sources on the maternal tissues of fruits of cvs. Barhy and Nabtet-Saif showed that the phases of development were considerably impacted by all pollen strains employed and pollen from the cv. Heet promoted early fruit maturity in both cultivars. Conversely, pollen from the cv. Dilim delayed maturity in cv. Barhy and pollen from cv. Muzahmiya delayed ripening in cv. Nabtet-Saif (Al-Khalifah, 2006). Hence, the genetic differences in pollen can be used to select males to improve yields, fruit quality, and change the ripening time of the fruit based on market demand. According to a study conducted by Iqbal *et al.* (2012), the male pollinizer utilized had a substantial impact on fruit set percentage. A multilocational experiment showed that fruit set, bunch weight, fruit weight, flesh weight, fruit volume, and most physical and chemical characters were significantly improved with pollen from cv. Khalas date palm (Omar *et al.*, 2014). Similarly, Outghouliast *et al.* (2020) proposed that the best pollen source (male palm) for cv. Mejhoul dates be chosen in order to obtain the most desired features in the form of fruit set, yield, and fruit quality. Effects of pollen on the tissue culture-derived female cv. Barhee was investigated by assessing the changes in the levels of malic acid, citric acid, fructose, glucose, sucrose, L-proline, and *myo*-inositol during the fruit development. The concentrations of these compounds were significantly influenced by the pollen source (metaxenic effect). In comparison to other pollen sources, pollen from the genotype Tanunda improved fruit size, fresh weight, and seed size considerably (Al-Najm *et al.*, 2021). The above findings point toward identifying the best male pollinators across the zone or specific location that are compatible with or produce the best outcomes with the commercial female cultivar being produced.

7.4   Pollen Extraction and Storage

During the establishment of the orchard, pre-planning of the male block or section helps to make sure there is sufficient pollen available for effective pollination. In general, one male is sufficient to pollinate 50 female plants (Zaid and de Wet, 2002b). However, additional male palms help to overcome pollen scarcity arising from unsynchronized flowering of the female inflorescence earlier than the male, which reduces the supply of fresh pollen. To overcome both cases, it is advisable to extract the pollen whenever it is available and store it or use it as needed.

7.4.1   Pollen extraction methods

The male inflorescence is harvested when the spathe cracks open. This is the anthesis period and the male inflorescence has reached maturity.

Fig. 7.7. (a) Harvested male inflorescence of date palm under shade drying.
(b) Presence of bees around the male inflorescence. (c) Shade drying in covered
boxes. (Photos by Kapil Mohan Sharma.)

The harvested inflorescence is dried in the shade for 3–7 days (Fig. 7.7a),
avoiding direct sunlight or high temperature as they have negative effects
on the pollen and the pollen may lose viability. However, the freshly
harvested inflorescence may be visited by honeybees making the operation
difficult. It is advised to harvest the inflorescence early in the morning to
prevent loss from bees, wind, or heat. Delayed harvest sometimes results in
the high presence of honeybees on the inflorescence, which makes it very
difficult to harvest (Fig. 7.7b). Sometimes, the inflorescence is harvested
a day or two before the spathe cracks open, but this needs experience to
decide. The method is to press the middle part of the spathe using the
thumb and the fingers. Soon after the harvest of the spathe, it is best to
remove the spathe cover over the inflorescence as sometimes it contains
moisture that may spoil the flowers as well as the pollen. To overcome this
problem, a simple box for pollen drying can be used (Fig. 7.7c). The shade
drying of the inflorescence allows opening of all the flowers and eases the
extraction of the pollen. It is necessary to reduce the moisture in the pollen
as it may be spoiled or destroyed by molds.

    To facilitate the pollen extraction and drying, three major methods
are followed.

1. *Manual extraction using a sieve:* This is the most common and traditional
   method. Strands from the inflorescence are separated, spread, and
   shade-dried over a sheet of paper. Then strands are further strained
   manually, which separates the pollen from flowers (Fig. 7.8a).
2. *Manual extraction by beating:* In this method the inflorescences are tied
   upside down on a wire, below which paper or a tray is placed to collect
   the flowers and pollen. Beating of the tied inflorescence is performed
   with the help of stick to get the pollen in sufficient quantities. This is
   suitable to handle a large quantity of pollen (Fig. 7.8b).
3. *Mechanical extraction:* In this case a mechanical pollen extraction
   machine is used (Fig. 7.9). It consists of a cylindrical barrel that rotates
   horizontally and has a rotating screen disk, a cyclone separator, and a
   suction pump that allows removal of pollen from the flowers, as well as

Fig. 7.8. Methods of pollen extraction. (a) Manual extraction using a sieve.
(b) Manual extraction by beating. (Photo (a) by Kapil Mohan Sharma; photo
(b) adapted from https://www.facebook.com/groups/1431837897104498
(accessed 23 November 2022).)

Fig. 7.9. Date palm pollen extraction process. (a) Handcrafted pollen extraction
machine. (b, c) Industrial and automated pollen extraction equipment. (From
Salomón-Torres *et al.*, 2021. Photo (a) from Anonymous, 2020; photos (b,
c) courtesy of Agrom Agricultural Machinery Ltd, www.agrommachine.com
(accessed 22 July 2021).)

a container to collect the pollen. They are more efficient and able to
extract 40% more pollen than the traditional methods. Further, they
also reduce direct exposure of the grower to the pollen thus reducing
the chances of allergies associated with it.

During the period of shade drying (3–7 days), the color of the flowers turns
dark brown; however, this does not have any impact on pollen quality and
quantity (Zaid and de Wet, 2002b).

7.4.2   Pistillate receptivity and pollen storage methods

Asynchronized flowering of male and female creates a problem of pollen
scarcity. Close observation is needed regarding the day of female spathe
cracking as they remain receptive for only 3–12 days after spathe cracking
depending on the cultivar (Zaid and de Wet, 2002b). Delay in pollinating

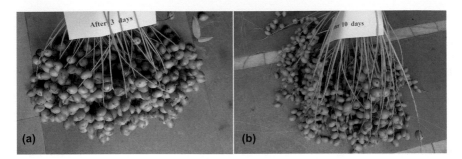

Fig. 7.10. Pollen receptivity in date palm: (a) day 3; (b) day 10. (Photos by C.M. Muralidharan.)

Table 7.2. Pollen receptivity in date palm.

Cultivar	Days after anthesis	Reference
Barhee	2–3	Muralidharan *et al.* (2020)
Gulistan	1	Iqbal *et al.* (2018)
Rothana	1	Ahmed *et al.* (2013)
Sewy	3	El-Salhy *et al.* (2011)
Najda	7–10	Zirari (2010)
Dhakki	4	Iqbal *et al.* (2004)
Khalas	2–4	Shabana *et al.* (2001)

may result in poor fruit set and parthenocarpic fruits (Fig. 7.10). Various workers have observed different periods of pollen receptivity as presented in Table 7.2.

In general, if the male emerges early, the pollen is extracted and can be preserved at room temperature (24°C) for around 1 month or more (Denney, 1992). Commonly, DPP is viable for 2–3 months, after which it begins to lose viability very rapidly. When the female inflorescence emerges after the male inflorescence, it can be pollinated with the freshly harvested pollen of the current season (Sharma *et al.*, 2021). The female inflorescences that emerge before the male inflorescence can be pollinated using stored pollen from the previous season. Several studies have been made on effectively storing the pollen at different temperatures: room temperature, 4°C, –20°C, and –196°C; this last has the best results in storing pollen. However, for practical reasons and lack of availability of cryopreservation units, storing at –20°C is generally a better option for commercial use. Pollen is suitable for pollination on the same day as being taken out of cold storage to reduce quality loss due to reduced viability (Karim *et al.*, 2021). A few research results on pollen germinability are presented in Table 7.3.

Table 7.3. Date palm pollen germination under different temperature/storage time regimes.

Temperature (°C)	Length of storage (months)	Pollen germination (%)	Reference
24	Fresh	94.37	Anushma *et al.*
4	2	61.64	(2018)
4	12	0.00	
−20	12	73.61	
−196	12	90.29	
28	Fresh	96.30	Kadri *et al.* (2021)
28	12	14.70	
4	12	42.10	
−30	12	52.20	
4	12	24.39	Maryam *et al.* (2015)
−20	12	27.40	
−80	12	24.64	

7.5 Pollination Methods

One of the most important activities in the cultivation of the date palm is the pollination process, which will be successful according to the technique adopted and the viability of the pollen. Generally, the male inflorescence emerges earlier in most parts of the world, but due to climatic influences, the flowering period may vary. Further, all the inflorescences on a plant do not emerge or open on the same day, and it may take around three or four sets of pollination activity to pollinate all bunches of a palm. If the recipient female palm has a greater number of inflorescences to pollinate, the consumption of pollen increases, which causes an increase in the production cost for the farmer. On average a single male date palm can produce 20 to 30 spathes per season. The production of pollen per male spathe may vary from 5 to 50 g and may total more than 1000 g per palm in a good-quality male. On many commercial farms, pollination is done using males that have been pre-identified for better production. Pollen germination takes place at 25–28°C on the stigma (Bernstein, 2004; Slavkovia *et al.*, 2016). During February–March, when in most countries date palm flowering has started, usual morning temperatures are low and efficient pollination takes place when there is sufficient warmth. Sharma *et al.* (2019) reported 8 a.m. to 12 noon in India, Iqbal *et al.* (2014) reported 10 a.m. to 2 p.m. in Pakistan, and Hajian (2005) reported 10 a.m. to 3 p.m. in Iran as effective time periods for pollination for higher fruit set.

Fig. 7.11. (a) Closely spaced date palm in natural groves suitable for natural pollination. (b) A well-managed and wide-spaced orchard. (Photos by Kapil Mohan Sharma.)

Broadly, pollination in date palms is by three major methods: (i) natural pollination; (ii) artificial pollination through manual labor; and (iii) mechanized pollination.

1. *Natural pollination:* Natural pollination is very common in wild groves or oases, where it can take place by wind. Close planting and almost equal male and female populations, which are commonly available in natural groves, are optimal for successful pollination (Fig. 7.11a) but are not ideal for a commercial orchard. Unlike other dioecious crops like papaya, where the male plants are planted in between the female plants in a ratio of 1:9, the same is not applicable in date palm due to inefficient wind pollination (Sharma *et al.*, 2021). Moreover, in widely spaced plants ranging from 6 m × 6 m to 10 m × 10 m, replacing a female plant with an unproductive male plant may greatly impact the overall yield per unit area. Further, inefficient wind pollination may lead to parthenocarpic fruits (Mostaan *et al.*, 2010; Siddiq and Greiby, 2014) (Fig. 7.11b).

2. *Artificial pollination through manual labor:* Since natural pollination is not a viable option for commercial cultivation, artificial pollination is the alternative. Artificial pollination in date palm is not new and has been used since time immemorial. A few of the methods are discussed below.

   a. *Using male strands:* One of the most common methods is fixing inverted male strands in the female inflorescences after dusting the pollen on them (Fig. 7.12a). This method is easy to follow and is effective if fresh male strands are used. However, it is laborious, requires well-trained personnel, and is costly.

   b. *Using a cotton swab:* In this method, the pre-collected pollen is swabbed on the female inflorescence using a small cotton ball. This acts as an alternative to the usage of male strands and can also be used with pollen stored from the last season. Similar to the use of male strands,

Fig. 7.12. Methods of pollination in date palm. (a) Fixing inverted male strands on the female inflorescence. (b) Pasting pollen on the female inflorescence using a cotton swab. (c) Spraying a pollen mixture solution. (Photos by Kapil Mohan Sharma.)

it is also laborious and costly (Fig. 7.12b). As an alternative to cotton, sponge strips are also used.

c. *Using a liquid suspension:* In this method, the pollen is mixed with water or a water-based suspension and is sprayed manually on the inflorescence (Fig. 7.12c). It is effective if the pollen is freshly mixed and sprayed. The pollen can also be mixed with GA_3 at 20 mg/L + 10% (w/v) sucrose for better fruit set (Iqbal *et al.*, 2012). A high pressurized liquid suspension is also used for mechanical pollination.

d. *Using a cotton cloth bag:* In this method, the dried pollen is concentrated in the central part of the cloth and is tied. The bag is dusted on the inflorescence and, due to its small size, the pollen emerges out of the bag to achieve pollination. For taller trees, the cotton bag is tied on to a long stick.

3. *Mechanized pollination:* Mechanical pollination has been adopted in most commercial orchards for efficient pollination and improved productivity, overcoming labor shortages, and reducing working costs (Shapiro *et al.*, 2008; Mostaan, 2012); however, expensive methods may not be suitable everywhere (El-Mardi *et al.*, 1995).

a. *Mechanical pollen duster/sprayer:* One of the mechanical types of pollen duster/sprayer is tractor mounted with a hydraulic air cannon for blowing the pollen (Fig. 7.13a). They are effective up to 22 m in height and can pollinate 350 palms per hour. They are popular in highly industrialized countries like Israel and the USA (AGROM, 2020).

b. *Manual pollen duster:* A manual pollinator, in which pumping/blowing air into the tube from the ground level into a bottle containing pollen to eject pollen dust on the inflorescence, is a cheap but effective method of pollination; however, cannot be used for very tall trees (Fig. 7.13b and c).

c. *Motorized pollen duster:* These are mechanical devices with pressurized sprayers. They are more effective than manual pollen dusters and can reach a height of up to 10 m, reducing the effort to climb the

Fig. 7.13. Mechanized methods of pollination in date palm. (a) Pollination machine with a hydraulic mechanism. (b, c) Manual pollen duster. (d) Motorized duster. (Photo (a) courtesy of Agrom Agricultural Machinery Ltd; photos (b, d) from Salomón-Torres *et al.*, 2021, used under CC-BY 4.0 license; photo (c) by Kapil Mohan Sharma.)

tree. They efficiently pollinate 150 to 200 palms per day (Fig. 7.13d) (Al-Wusaibai *et al.*, 2012).

  d. *Electrical pollen duster:* The design is similar to a manual pollen duster with a telescopic pole, a dispenser, and a pollen tank that can be operated using a remote control. It is effective in reducing the human effort to climb the tree (Mostaan *et al.*, 2010).

  e. *Aircraft:* Use of aircraft to pollinate date palms is one of the methods used in the early 1960s and 1970s in the USA and is still followed in some parts of Israel, but it is not efficient and may result in poor fruit set (Cohen and Glasner, 2015).

f. *Drones and artificial intelligence (AI):* These devices are designed to reduce the human effort in the pollination and reduce the quantity of pollen used during pollination. Under the One Million Date Palm Trees Project, drones enabled with a camera and a Global Positioning System (GPS) tracker were developed that have a range of operation of approximately 8 km and are efficient enough to pollinate a large number of date palms with a minimum quantity of pollen and manpower requirements (Oman Daily Observer, 2018). Wakan Tech, an Omani startup, developed drones that are also AI-enabled and 30 times faster than the conventional methods (WIPO, 2020). Researchers of the University of Arizona have developed an AI-enabled Drone Date Pollination System that can identify date palm inflorescences and dispense pollen to them.

7.5.1 Pollen mixture

For efficient pollination, the required amount of pollen needs to reach the pistil and develop the pollen tube in it. However, in mechanical pollination, the quantity of pollen required rises as there is limited control on the amount of pollen used per spathe. In other cases, when the pollen is already scarce, it is advised to blend the dried pollen with a diluting material to reduce the pollen quantity. Table 7.4 shows multiple studies that have been carried out to evaluate the efficiency of the pollination method used.

7.5.2 Bunch care after pollination

After pollination, success can be confirmed after an average of 20–30 days. It is necessary to understand the climate of the cultivation area to properly manage the bunch post-pollination. If the location has a history of heavy mist, rain, or chance of very low temperatures during the flowering or pollination period, it is advisable to cover the recently pollinated inflorescence with a paper bag (Zaid and de Wet, 2002b). This helps to maintain a suitable temperature as well as reduce the chances of pollen washing off due to rain (Fig. 7.14a). Generally, as the bunch develops, the paper bag deteriorates and falls off by itself. Covering of the fruits for a longer period may cause etiolation of fruits (degreening), partial fruit drop, and, if the bag is carrying any fungus, may damage the fruit inside the bag (Fig. 7.14b). However, the covered fruits have less bruising (mostly from leaves), which increases their market value.

A month after pollination, when the fruits attain pea size, they are thinned to reduce the fruit load and attain greater size of the ripe fruit, which commands a higher market value. Thinning can be done by: (i) removal of a third of stands from the inner circle, this helps for better aeration of fruits, especially during the time of ripening (Fig. 7.15a and d);

Table 7.4. Efficiency comparison of the different pollination methods on various date palm cultivars. (From Salomón-Torres *et al.*, 2021, used under CC-BY 4.0 license.)

Pollination method	Treatments	Fruit set (%)	Cultivars	Country	Reference
Hand spraying of pollen suspension	Liquid pollen (1.5 g/L) + 0.2 or 2 g boric or ascorbic acid + 10% (w/w) Egyptian treacle or 10% (w/w) vinasse	48.19	Zaghloul	Egypt	Abdalla *et al.* (2011)
	Liquid pollen (3 g/L) + 0.2 or 2 g boric or ascorbic acids + 10% (w/w) Egyptian treacle or 10% (w/w) vinasse	54.44			
Strands placement	10 strands/bunch	59.72			
Strands placement	5 strands/bunch	26.71	Hayany	Egypt	El-Dengawy (2017)
Hand pollen duster	Dry pollen + wheat flour (1:20 ratio)	26.99			
	Dry pollen + wheat flour (1:10 ratio)	28.95			
	Dry pollen + wheat flour (1:5 ratio)	32.30			
Dusting by hand	2 g pollen + 3 g filler material	54.70	Zaghloul, Samany	Egypt	Hafez *et al.* (2013)
	1 g Milagro + 4 g filler material	59.20			
	1 g pollen + 1 g Milagro + 3 g filler material	65.10			
Strands placement	2–3 strands/bunch	69.10	Khalas	Saudi Arabia	Ben Abdallah *et al.* (2014)
Strips of sponge	Dry pollen + wheat flour (1:4 ratio)	67.03			

Continued

Table 7.4. Continued

Pollination method	Treatments	Fruit set (%)	Cultivars	Country	Reference
Strands placement	5 strands/bunch	44.75	Khalas,	Saudi Arabia	Al-Wusaibai
Portable pollen duster	Dry pollen + wheat flour (1:5 ratio)	37.5	Sheshi		*et al.* (2012)
	Dry pollen + wheat flour (1:10 ratio)	32.25			
	Dry pollen + wheat flour (1:15 ratio)	33.55			
Natural pollination	By wind	26.03	Khalas	Saudi Arabia	Munir (2019)
Strands placement by hand	5 strands/bunch	68.67			
Dusting hand	Dry pollen + wheat flour (1:9 ratio)	82.07			
Hand spraying of pollen suspension	Liquid pollen (3 g/L)	85.71			
Natural pollination	By wind	32.50	Hillawi	Pakistan	Ullah *et al.*
Strands placement	5 strands bunch	51.93			(2018)
Dusting by hand	Dry pollen	45.94			
Dusting with pollinator	Dry pollen	60.76			
	Dry pollen + talc	40.93			
Hand spraying of pollen suspension	Liquid pollen (1 g/L)	74.67	Khadrawy,	Pakistan	Munir (2020)
	Liquid pollen (2 g/L)	79.67	Zahidi		
	Liquid pollen (3 g/L)	84.50			
	Liquid pollen (4 g/L)	86.00			

Continued

Table 7.4. Continued

Pollination method	Treatments	Fruit set (%)	Cultivars	Country	Reference
Natural pollination	By wind	12.65	Zahidi	Iraq	Hamood and Mawlood (1986)
Strands placement	2–3 strands/bunch	62.43			
Dusting mechanical	Dry pollen (8%, w/w)	53.59			
	Dry pollen (16%, w/w)	57.41			
Dusting by hand	Dry pollen	100.00	Medjool, Barhee	Jordan	Abu-Zahra and Shatnawi (2019)
Hand spraying of pollen suspension	Liquid pollen (1 g/L)	72.50			
	Liquid pollen (2 g/L)	81.00			
	Liquid pollen (3 g/L)	81.00			
	Liquid pollen (4 g/L)	91.50			
Dusting hand with brush	Dry pollen + wheat flour (1:1 ratio)	31.74	Medjool	Mexico	Salomon-Torres *et al.* (2017)
Dusting by hand	Dry pollen	52.59	Hallawy	India	Sharma *et al.* (2021)
	Dry pollen + talc (1:9)	49.69			
	Dry pollen + talc (1:19)	47.72			

Fig. 7.14. (a) Covering with paper bags after pollination. (b) Etiolation of fruits due to longer days of covering. (Photos by Kapil Mohan Sharma.)

Fig. 7.15. Fruit thinning in date palm. (a) Thinning of one-third of inner strands. (b) Thinning of one-third of strands by length. (c) Alternate thinning of fruits. (d) Fruit bunch with thinned fruits in the middle, providing better aeration. (Photos by Kapil Mohan Sharma.)

(ii) removal of a third of strands by length, this helps to reduce the fruit load and ultimately helps to increase fruit size (Fig. 7.15b); or (iii) alternate thinning in which, as the name suggests, the alternate fruits are removed

Table 7.5. Composition of date palm pollen grains. (From El-Kholy *et al.*, 2019, used under Creative Commons Attribution CC-BY 4.0 license.)

Component	Content
Energy (kcal)	310.88
Nutrients[a]	
Total solids (g/100 g)	91.11 ± 0.43
Ash (g/100 g)	10.23 ± 0.02
Crude fat (g/100 g)	10.80 ± 0.03
Crude protein (g/100 g)	35.28 ± 0.57
Crude fiber (g/100 g)	8.09 ± 0.13
Total sugars (g/100 g)	6.50 ± 0.69
Carbohydrate (g/100 g)	17.14 ± 0.47
Minerals	
Ca (mg/100 g)	560.00
K (mg/100 g)	750.00
Mg (mg/100 g)	318.70
Fe (mg/100 g)	226.50
Zn (mg/100 g)	124.40
Mn (mg/100 g)	70.00

[a]Mean values with their standard deviation.

(Fig. 7.15c). This is done manually and is the most laborious practice. It is popular in high-value cultivars like Medjool, Barhee, etc.

7.6 Nutritional Properties and Secondary Uses of Pollen

DPP also has a role as a highly nutritious food with health benefits (Waly, 2020) (Table 7.5). It can be an excellent source of minerals and amino acids that exceed the recommended daily values (Sebii *et al.*, 2019; Karra *et al.*, 2020). Traditionally, DPP was used as a rejuvenating medicinal agent by early Egyptian and Chinese people (Otify *et al.*, 2019). It is known to have aphrodisiac and fertility-enhancing properties and can be used as a treatment for male infertility (Abdi *et al.*, 2017). It contains estrogenic gonad-stimulating compounds, estrogen, sterols, and other beneficial macro- and microelements, which promote its aphrodisiac property and improve sperm quality (Bahmanpour *et al.*, 2006). The bioactive compounds in DPP that are responsible for such activities are listed in Table 7.6. DPP is also capable of collecting reactive oxygen species (ROS) (highly reactive chemicals formed from molecular oxygen) of the body and neutralizing them due to its antioxidant properties (Fallahi *et al.*, 2015). Further, it can

Table 7.6. The bioactive components of date palm pollen (DPP) and their mechanism. (From Shehzad *et al.*, 2021, used under CC-BY 4.0 license.)

Active component	Mechanism	Reference
Estrogenic compounds: estradiol, estriol, and estrone	Alleviate infertility through their gonadotropic activity in male rats	El-Ridi *et al.* (1960); Abbas and Ateya (2011); El-Neweshy *et al.* (2013)
Estrogen compounds	Estrogen compounds increase the hormone estrogen. These compounds transfer to embryo and offspring via the placenta and lactation, and affect the reproductive system in adult mice	Abedi *et al.* (2013)
Saponins	Saponins encourage the Leydig cells of the testes to increase the testosterone production system	Anger *et al.* (2004); Gakunga *et al.* (2014)
Carbohydrates, saponins, and gallic tannins	DPP has an aphrodisiac potential and may increase the reproductive parameters of male adult rats	Selmani *et al.* (2017)
Estradiol components	Play a role in regulating the renewal of spermatogenic cells and male reproductive tissues that possess estrogen receptors	Al-Kuran *et al.* (2011)
Phytochemicals: alkaloids, saponins, and flavonoids	Phytochemicals have engorgement and androgen-enhancing properties that improve sexual behavior in male rats	Anger *et al.* (2004)
Estrogenic materials	Gonad-stimulating compounds that improve male infertility	Joshi (2000)

partially inhibit bacterial stains development and can be used as a natural preservative (Karra *et al.*, 2020).

7.7 Pollen Allergens

As indicated above, date growers come into direct contact with pollen, which can result in mild to severe allergies in some cases. According to Radwan *et al.* (2006), DPP should be considered an important allergen in countries where it is cultivated. Pollen allergens are also found in *Phoenix sylvestris* in addition to *P. dactylifera* (Saha and Bhattacharya, 2017). The pollen may cause asthma, rhinitis, and conjunctivitis, and an individual may experience skin sensitivity (Serhane *et al.*, 2017). With the rapid mechanization in pollen extraction and pollination methods, there is a reduction in the exposure period to pollen; however, care must be taken when handling pollen to avoid health problems.

7.8 Conclusions and Prospects

Pollen and pollination management has been an important and vital agronomical practice in date palm production. Proper management of DPP includes: (i) identification of desirable males, to take advantage of metaxenic effects and better germination ability; (ii) timely storage, for utilization in the next season (if needed); (iii) timely pollination, for higher pistillate receptivity and greater success of pollination; and (iv) effective pollination methodology, to reduce labor and improve pollination efficiency. Proper and timely planning and effective execution reduce the chances of unpollinated fruit and improve overall yield. Over the years, there have been many advances to improve the effectiveness of pollen extraction, pollen storage, and pollination methods, but there is still room for improvement in the development of low-cost technologies. Additional research should be conducted to investigate the efficiency of the mechanized pollination methods used. The viability of the pollen source and the agroclimatic conditions in each region should be considered. Furthermore, depending on the pollination method used, the demand for pollen for pollination will necessitate one to two male palms for every 50 females. The use of new-age technologies and AI such as drones is necessary to reduce human labor and increase effective pollination because they can identify ready-to-pollinate inflorescences, the amount of pollen to use, and be operated with little manpower. Furthermore, the use of DPP as a functional food, as well as other secondary uses such as medicine or nutritional supplements, needs to be investigated further to create an additional utility for DPP.

References

Abbas, F.A. and Ateya, A.M. (2011) Estradiol, esteriol, estrone and novel flavonoids from date palm pollen. *Australian Journal of Basic and Applied Sciences* 5, 606–614.

Abbas, M.F., Abdul-Wahid, A.H. and Abass, K.I. (2014) Metaxenic effect in date palm (*Phoenix dactylifera* L.) fruit in relation to level of endogenous auxins. *AAB Bioflux* 6(1), 40–44.

Abdalla, M.G.M., El-Salhy, A. and Mostafa, R.A.A. (2011) Effect of some pollination treatments on fruiting of Zaghloul date palm cultivar under Assiut climatic condition. *Assiut Journal of Agricultural Sciences* 42, 350–362.

Abdi, F., Roozbeh, N. and Mortazavian, A.M. (2017) Effects of date palm pollen on fertility: research proposal for a systematic review. *BMC Research Notes* 10(1), 363. DOI: 10.1186/s13104-017-2697-3.

Abedi, A., Parviz, M., Karimian, S.M. and Rodsari, H.R.S. (2013) Aphrodisiac activity of aqueous extract of *Phoenix dactylifera* pollen in male rats. *Advances in Sexual Medicine* 3, 1–7.

Abu-Zahra, T.R. and Shatnawi, M.A. (2019) New pollination technique in date palm (*Phoenix dactylifera* L.) cv. 'Barhee' and 'Medjool' under Jordan Valley conditions. *American-Eurasian Journal of Agriculture & Environmental Sciences* 19, 37–42.

AGROM (2020) Development, design and manufacture of agricultural and industrial machines. Available at: www.agrommachine.com/home-new-en (accessed 22 July 2020).

Ahmed, M.A., El-Saif, A.M., Soliman, S.S. and Omar, A.K.H. (2013) Effect of pollination date on fruit set, yield and fruit quality of 'Rohtana' date palm cultivar under Riyadh conditions. *Journal of Applied Sciences Research* 9(4), 2797–2802.

Al-Khalifah, N.S. (2006) Metaxenia: influence of pollen on the maternal tissue of fruits of two cultivars of date palm (*Phoenix dactylifera* L.). *Bangladesh Journal of Botany* 35, 151–161.

Al-Kuran, O., Al-Mehaisen, L., Bawadi, H., Beitawi, S. and Amarin, Z. (2011) The effect of late pregnancy consumption of date fruit on labour and delivery. *Journal of Obstetrics and Gynaecology* 31(1), 29–31. DOI: 10.3109/01443615.2010.522267.

Almehdi, A.M., Maraqa, M. and Abdulkhalik, S. (2005) Aerobiological studies and low allerginicity of date-palm pollen in the UAE. *International Journal of Environmental Health Research* 15(3), 217–224. DOI: 10.1080/09603120500105745.

Al-Najm, A., Brauer, S., Trethowan, R., Merchant, A. and Ahmad, N. (2021) Optimization of *in vitro* pollen germination and viability testing of some Australian selections of date palm (*Phoenix dactylifera* L.) and their xenic and metaxenic effects on the tissue culture–derived female cultivar "Barhee." *In Vitro Cellular & Developmental Biology - Plant* 57(5), 771–785. DOI: 10.1007/s11627-021-10206-z.

Al-Wusaibai, N.A., Ben Abdallah, A., Al-Husainai, M.S., Al-Salman, H. and Elballaj, M. (2012) A comparative study between mechanical and manual pollination in two premier Saudi Arabian date palm cultivars. *Indian Journal of Science and Technology* 5(4), 2487–2490. DOI: 10.17485/ijst/2012/v5i4.4.

Anger, J.T., Wang, G.J., Boorjian, S.A. and Goldstein, M. (2004) Sperm cryopreservation and in vitro fertilization/intracytoplasmic sperm injection in men with congenital bilateral absence of the vas deferens: a success story. *Fertility and Sterility* 82(5), 1452–1454. DOI: 10.1016/j.fertnstert.2004.05.079.

Anonymous (2020) School of Palm and Date Lovers. Available at: www.facebook.com/groups/1431837897104498 (accessed 20 July 2020).

Anushma, P.L., Vincent, L., Rajasekharan, P.E. and Ganeshan, S. (2018) Pollen storage studies in date palm (*Phoenix dactylifera* L.). *International Journal of Chemical Studies* 6(5), 2640–2642.

Bahmanpour, S., Talawi, T., Vojdani, Z., Panjehshahin, M., Poostpasanad, Z.S. *et al.* (2006) Effect of *Phoenix dactylifera* pollen on sperm parameters and reproductive system of adult male rats. *Iranian Journal of Medical Sciences* 31(4), 208–212.

Baidiyavadra, D.A., Muralidharan, C.M. and Sharma, K.M. (2019) Fresh date production in Kachchh (India): challenges and future prospects. *Medicinal Plants* 11(3), 218–227. DOI: 10.5958/0975-6892.2019.00029.7.

Ben Abdallah, A., Al-Wusaibai, N.A. and Al-Fehaid, Y. (2014) Assessing the efficiency of sponge and traditional methods of pollination in date palm. *Journal of Agricultural Science and Technology* 4, 267–271. DOI: 10.17265/2161-6264/2014.04B.003.

Bernstein, Z. (2004) *The Date Palm.* Israeli Fruit Board, Tel Aviv, Israel (in Hebrew).

Cohen, Y. and Glasner, B. (2015) Date palm status and perspective in Israel. In: Al-Khayri, J.M., Jain, S.M. and Johnson, D.V. (eds) *Date Palm Genetic Resources and Utilization*, Vol. 2. *Asia and Europe*. Springer, New York, pp. 265–298.

Cohen, Y., Slavkovic, F., Birger, D., Greenberg, A., Sadowsky, A, *et al.* (2016) Fertilization and fruit setting in date palm: biological and technological challenges. *Acta Horticulturae* 1130, 351–358. DOI: 10.17660/ActaHortic.2016.1130.53.

Denney, J.O. (1992) Xenia includes metaxenia. *HortScience* 27(7), 722–728. DOI: 10.21273/HORTSCI.27.7.722.

El-Dengawy, E.R. (2017) Improvement of the pollination technique in date palm. *Journal of Plant Production* 8(2), 307–314. DOI: 10.21608/jpp.2017.39627.

El-Hamady, M., Hamdia, M., Ayaad, M., Salama, M.E. and Omar, A.K.H. (2010) Metaxenic effects as related to hormonal changes during date palm (*Phoenix dactylifera* L.) fruit growth and development. *Acta Horticulturae* 882, 155–164. DOI: 10.17660/ActaHortic.2010.882.17.

El-Kadri, N. and Mimoun, M.B. (2020) *In vitro* germination of different date palm (*Phoenix dactylifera* L.) pollen sources from Southern Tunisia under the effect of three storage temperatures. *International Journal of Fruit Science* 20(suppl. 3), S1519–S1529. DOI: 10.1080/15538362.2020.1815116.

El-Kholy, W.M., Soliman, T.N. and Darwish, A.M.G. (2019) Evaluation of date palm pollen (*Phoenix dactylifera* L.) encapsulation, impact on the nutritional and functional properties of fortified yoghurt. *PloS ONE* 14(10), e0222789. DOI: 10.1371/journal.pone.0222789.

El-Mardi, M., Labiad, S., Consolacion, E. and Addelbasit, K. (1995) Effect of pollination methods and pollen dilution on some chemical constituents of Fard dates at different stages of fruit development. *Emirates Journal of Food and Agriculture* 7, 1–19. DOI: 10.9755/ejfa.v7i1.5351.

El-Neweshy, M.S., El-Maddawy, Z.K. and El-Sayed, Y.S. (2013) Therapeutic effects of date palm (*Phoenix dactylifera* L.) pollen extract on cadmium-induced testicular toxicity. *Andrologia* 45(6), 369–378. DOI: 10.1111/and.12025.

El-Ridi, M.S., El Mofty, A., Khalifa, K. and Soliman, L. (1960) Gonadotrophic hormones in pollen grains of the date palm. *Zeitschrift Für Naturforschung B* 15(1), 45–49. DOI: 10.1515/znb-1960-0109.

El-Salhy, A., Abdel-Galil, H.A., El-Akkad, M.M. and Diab, Y.M. (2011) Effect of delaying pollination on yield and fruit quality of Sewy date palm under new valley conditions in Egypt. *Research Journal of Agriculture and Biological Sciences* 7(6), 408–412.

Fallahi, S., Rajaei, M., Malekzadeh, K. and Kalantar, S.M. (2015) Would *Phoenix dactyflera* pollen (palm seed) be considered as a treatment agent against males' infertility? A systematic review. *Electronic Physician* 7(8), 1590–1596. DOI: 10.19082/1590.

Gakunga, N.J., Mugisha, K., Owiny, D. and Waako, P. (2014) Effects of crude aqueous leaf extracts of *Citropsis articulata* and *Mystroxylon aethiopicum* on sex hormone levels in male albino rats. *International Journal of Pharmaceutics* 3, 5–17.

Hafez, O.M., Saleh, M.A., Mostafa, E., El-Shamma, M. and Maksoud, M. (2013) Improving pollination efficiency, yield and fruit quality of two date palm cultivars using growth activator. *International Journal of Agricultural Research* 9, 29–37. DOI: 10.3923/ijar.2014.29.37.

Hajian, S. (2005) Fundamentals of pollination in date palm plantations in Iran. In: Malik, A.U., Pervez, M.A. and Ziad, K. (eds) *Proceedings of the International Conference on Mango and Date Palm: Culture and Export, Lahore, Pakistan,* University of Agriculture, Faisalabad, Pakistan, June 20–23, 2005, pp. 252–259.

Hamood, H.H. and Mawlood, E.A. (1986) The effect of mechanical pollination on fruit set, yield and fruit characteristics of date palm (*Phoenix dactylifera* L.) Zahidi cultivar. *Date Palm Journal* 4, 175–184.

Helail, B.M. and El-Kholey, L.A. (2000) Effect of pollen grain sources on palm fruiting and date quality of Hallawy and Khadrawy date palms. *Annals of Agricultural Science, Moshtohor* 38(1), 479–494.

Iqbal, M., Ghaffor, A. and Rehman, S. (2004) Effect of pollination times on fruit characteristics and yield of date palm cv. Dhakki. *International Journal of Agriculture & Biology* 6(1), 96–99.

Iqbal, M., Niamatullah, M. and Munir, M. (2012) Effect of various dactylifera males pollinizer on pomological traits and economical yield index of cv's Shakri, Zahidi and Dhakki date palm (*Phoenix dactylifera* L.). *The Journal of Animal & Plant Sciences* 22(2), 376–383.

Iqbal, M., Jatoi, S.A., Niamatullah, M., Munir, M. and Khan, I. (2014) Effect of pollination time on yield and quality of date palm. *The Journal of Animal & Plant Sciences* 23(4), 760–764.

Iqbal, M., Usman, K., Munir, M. and Khan, M.S. (2018) Quantitative and qualitative characteristics of date palm cv. Gulistan in response to pollination times. *Sarhad Journal of Agriculture* 34(1), 40–46.

Joshi, S.G. (2000) *Medicinal Plants.* Oxford and IBH Publishing, New Delhi.

Kadri, K., Elsafy, M., Makhlouf, S. and Awad, M.A. (2021) Effect of pollination time, the hour of daytime, pollen storage temperature and duration on pollen viability, germinability, and fruit set of date palm (*Phoenix dactylifera* L.) cv "Deglet Nour." *Saudi Journal of Biological Sciences* 29(2), 1085–1091. DOI: 10.1016/j.sjbs.2021.09.062.

Karim, K., Awad, M.A., Manar, A., Monia, J., Karim, A. *et al.* (2021) Effect of flowering stage and storage conditions on pollen quality of six male date palm genotypes. *Saudi Journal of Biological Sciences* 29(4), 2564–2572. DOI: 10.1016/j.sjbs.2021.12.038.

Karra, S., Sebii, H., Jardak, M., Bouaziz, M.A., Attia, H. *et al.* (2020) Male date palm flowers: valuable nutritional food ingredients and alternative antioxidant source and antimicrobial agent. *South African Journal of Botany* 131, 181–187. DOI: 10.1016/j.sajb.2020.02.010.

Kgazal, M.A., Salbi, M.I., Alsaadawi, I.S., Fattah, F.A. and Al-Jibouri, A.A.M. (1990) Bisexuality in date palm in Iraq. *Journal of the Islamic Academy of Sciences* 3(2), 131–133.

Krueger, R.R. (2021) Date palm (*Phoenix dactylifera* L.) biology and utilization. In: Al-Khayri, J.M., Jain, S.M. and Johnson, D.V. (eds) *The Date Palm Genome*, Vol. 1. *Phylogeny, Biodiversity and Mapping.* Compendium of Plant Genomes, Springer, Cham, Switzerland, pp. 3–38.

Leshem, Y. and Ophir, D. (1977) Differences in endogenous levels of gibberellin activity in male and female partners of two dioecious tree species. *Annals of Botany* 41(2), 375–379. DOI: 10.1093/oxfordjournals.aob.a085300.

Maryam, M., Jaskani, J., Fatima, B., Haidaer, M.S., Naqvi, S.A. *et al.* (2015) Evaluation of pollen viability in date palm cultivars under different storage temperatures. *Pakistan Journal of Botany* 47(1), 377–381.

Maryam, M., Jaskani, J. and Naqvi, S.A. (2017) Storage and viability assessment of date palm pollen. In: Al-Khayri, J.M., Jain, S.M. and Johnson, D.V. (eds) *Date Palm Biotechnology Protocols*, Vol. II. *Germplasm Conservation and Molecular Breeding.* Methods in Molecular Biology, Vol. 1638, Humana Press, New York, pp. 1–13.

Masmoudi-Allouche, F., Châari-Rkhis, A., Kriaâ, W., Gargouri-Bouzid, R., Jain, S.M. *et al.* (2009) *In vitro* hermaphrodism induction in date palm female flower. *Plant Cell Reports* 28(1), 1–10. DOI: 10.1007/s00299-008-0611-0.

Mostaan, A., Marashi, S.S. and Ahmadizadeh, S. (2010) Development of a new date palm pollinator. *Acta Horticulturae* 882, 315–320. DOI: 10.17660/ActaHortic.2010.882.35.

Mostaan, A. (2012) Mechanization in date palm pollination. In: Manickavasagan, A., Essa, M.M. and Sukumar, E. (eds) *Dates: Production, Processing, Food and Medicinal Values*, 1st edn. CRC Press, London, pp. 129–140.

Munir, M. (2019) Influence of liquid pollination technique on fruit yield and physico-chemical characteristics of date palm cultivars Khadrawy and Zahidi. *Journal of Biodiversity and Environmental Sciences* 15, 41–49.

Munir, M. (2020) A comparative study of pollination methods effect on the changes in fruit yield and quality of date palm cultivar Khalas. *Asian Journal of Agriculture and Biology* 8(2), 147–157. DOI: 10.35495/ajab.2019.11.537.

Muralidharan, C.M., Panchal, C.N., Baidiyavadra, D.A., Sharma, K.M. and Verma, P. (2020) Pistillate receptivity of date palm (*Phoenix dactylifera* L.) cv. Barhee. *Sugar Tech* 22(6), 1166–1169. DOI: 10.1007/s12355-020-00859-2.

Oman Daily Observer (2018) Technology to be used for pollinating date palms. *Oman Daily Observer.* Available at: www.omanobserver.om/ampArticle/48930 (accessed 24 November 2021).

Omar, A.K., Al-Obeed, R.S., Soliman, S. and Al Saif, A.M. (2014) Effect of pollen source and area distribution on yield and fruit quality of 'Khalas' date palm (*Phoenix dactylifera* L.) under Saudi Arabia conditions. *Acta Advances in Agricultural Sciences* 2(3), 7–13.

Otify, A.M., El-Sayed, A.M., Michel, C.G. and Farag, M.A. (2019) Metabolites profiling of date palm (*Phoenix dactylifera* L.) commercial by-products (pits and pollen) in relation to its antioxidant effect: a multiplex approach of MS and NMR metabolomics. *Metabolomics* 15(9), 119. DOI: 10.1007/s11306-019-1581-7.

Outghouliast, H., Messaoudi, Z., Touhami, A.O., Douira, A. and Haddou, L.A. (2020) Effect of pollen source on yield and fruits quality of date palm (*Phoenix dactylifera* L.) cv. 'Mejhoul' in Moroccan oases. *Plant Cell Biotechnology and Molecular Biology* 21, 60–69.

Paszke, M.Z. (2019) Date palm and date palm inflorescences in the late Uruk period (*c.*3300 BC): botany and archaic script. *IRAQ* 81, 221–239. DOI: 10.1017/irq.2019.15.

Radwan, R.A., Barakat, M.M., Selim, M.A. and Fouda, E.E. (2006) Date palm pollen: a significant asthma and allergy inducer. *The Journal of Allergy and Clinical Immunology* 117(2), S111. DOI: 10.1016/j.jaci.2005.12.445.

Reuvani, O. (1985) *Phoenix dactylifera*. In: Halevy, A.H. (ed.) *Handbook of Flowering*. CRC Press, Boca Raton, Florida, pp. 343–349.

Reuvani, O. (1986) Date. In: Monseliese, S.P. (ed.) *Handbook of Fruit Set and Development*. CRC Press, Boca Raton, Florida, pp. 119–144.

Rezazadeh, R., Hassanzadeh, H., Hosseini, Y., Karami, Y. and Williams, R.R. (2013) Influence of pollen source on fruit production of date palm (*Phoenix dactylifera* L.) cv. Barhi in humid coastal regions of southern Iran. *Scientia Horticulturae* 160, 182–188.

Saha, B. and Bhattacharya, S.G. (2017) Charting novel allergens from date palm pollen (*Phoenix sylvestris*) using homology driven proteomics. *Journal of Proteomics* 165(8), 1–10. DOI: 10.1016/j.jprot.2017.05.021.

Salomon-Torres, R., Ortiz-Uribe, N., Villa-Angulo, R., Villa-Angulo, C., Norzagaray-Plasencia, S. *et al.* (2017) Effect of pollenizers on production and fruit characteristics of date palm (*Phoenix dactylifera* L.) cultivar Medjool in Mexico. *Turkish Journal of Agriculture and Forestry* 41, 338–347. DOI: 10.3906/tar-1704-14.

Salomón-Torres, R., Krueger, R., García-Vázquez, J.P., Villa-Angulo, R., Villa-Angulo, C. *et al.* (2021) Date palm pollen: features, production, extraction and pollination methods. *Agronomy* 11(3), 504. DOI: 10.3390/agronomy11030504.

Sebii, H., Karra, S., Bchir, B., Ghribi, A.M., Danthine, S.M. *et al.* (2019) Physio-chemical, surface and thermal properties of date palm pollen as a novel nutritive ingredient. *Advances in Food Technology and Nutrition Sciences* 5(3), 84–91. DOI: 10.17140/AFTNSOJ-5-160.

Selmani, C., Chabane, D. and Bouguedoura, N. (2017) Ethnobotanical survey of *Phoenix dactylifera* L. pollen used for the treatment of infertility problems in Algerian oases. *African Journal of Traditional, Complementary, and Alternative Medicines* 14(3), 175–186. DOI: 10.21010/ajtcam.v14i3.19.

Serhane, H., Amro, L., Sajiai, H. and Alaoui Yazidi, A. (2017) Prevalence of skin sensitization to pollen of date palm in Marrakesh, Morocco. *Journal of Allergy* 2017, 6425869. DOI: 10.1155/2017/6425869.

Shabana, H.R., Saeed, A.A. and Hamoodi, A.H. (2001) Determination of the optimal pollination period for Khalas date palm cultivar. In: Al-Badawy, A.A. (ed.) *Proceedings of the Second International Conference on Date Palms, Al-Ain, United Arab Emirates*, UAE University, Al-Ain, United Arab Emirates, March 25–27, 2001, pp. 13–20.

Shapiro, A., Korkidi, E., Rotenberg, A., Furst, G., Namdar, H. *et al.* (2008) A robotic prototype for spraying and pollinating date palm trees. In: *Proceedings of the*

ASME 2008 9th Biennial Conference on Engineering Systems Design and Analysis, Haifa, Israel, ASME International, Haifa, Israel, July 7–9, 2008, pp. 431–436.

Sharma, K.M., Muralidharan, C.M., Baidiyavadra, D.A. and Panchal, C.N. (2019) Effect of different time of pollination in date palm (*Phoenix dactylifera* L.) cv. Barhee. *Environment and Ecology* 37(4B), 1568–1570.

Sharma, K.M., Muralidharan, C.M., Baidiyavadra, D.A., Bardhan, K. and Panchal, C.N. (2021) Evaluation of potentiality of different adjuvants for date palm pollination and fruit set. *Sugar Tech* 23(1), 139–145. DOI: 10.1007/s12355-020-00885-0.

Shehzad, M., Rasheed, H., Naqvi, S.A., Al-Khayri, J.M., Lorenzo, J.M. *et al.* (2021) Therapeutic potential of date palm against human infertility: a review. *Metabolites* 11(6), 408. DOI: 10.3390/metabo11060408.

Siddiq, M. and Greiby, I. (2014) Overview of date fruit production, post-harvest handling, processing and nutrition. In: Siddiq, M., Aleid, S.M. and Kader, A.A. (eds) *Dates: Postharvest Science, Processing Technology and Health Benefits,* 1st edn. Wiley-Blackwell, Oxford, pp. 1–28.

Slavkovia, F., Greenberg, A., Sadowsky, A., Zemach, H., Ish-Shalom, M. *et al.* (2016) Effects of applying variable temperature conditions around inflorescences on fertilization and fruit set in date palms. *Scientia Horticulturae* 202, 83–90. DOI: 10.1016/j.scienta.2016.02.030.

Soliman, S.S. and Al-Obeed, R.S. (2013) Investigations on the pollen morphology of some date palm male (*Phoenix dactylifera* L.) in Saudi Arabia. *Australian Journal of Crop Science* 7(9), 1355–1360.

Swingle, W.T. (1928) Metaxenia in the date palm: possibly a hormone action by the embryo or endosperm. *Journal of Heredity* 19(6), 257–268. DOI: 10.1093/oxfordjournals.jhered.a102996.

Ullah, M., Ahmad, F., Iqbal, J., Imtiaz, M. and Raza, M.K. (2018) Effects of different pollination methods on fruit quality and yield of date palm candidate line Hillawi. *Journal of Environmental and Agricultural Sciences* 17, 55–62.

Waly, M.I. (2020) Health benefits and nutritional aspects of date palm pollen. *Canadian Journal of Clinical Nutrition* 8(1), 1–3.

WIPO (2020) Oman: pollinating date palms with AI and drones. World Intellectual Property Organization, Geneva, Switzerland. Available at: www.wipo.int/wipo_magazine/en/ip-at-work/2021/oman_wakan.html#:~:text=About%20the%20technology,faster%20than%20conventional%20pollination%20methods (accessed 23 November 2022).

Zaid, A. and de Wet, P.F. (2002a) Origin, geographical distribution and nutritional values of date palm. In: Zaid, A. and Arias-Jimenez, E.J. (eds) *Date Palm Cultivation.* FAO Plant Protection Paper no.156 Rev.1, Food and Agriculture Organization of the United Nations, Rome. Available at: www.fao.org/3/Y4360E/y4360e06.htm#bm06 (accessed 21 November 2022).

Zaid, A. and de Wet, P.F. (2002b) Pollination and bunch management. In: Zaid, A. and Arias-Jimenez, E.J. (eds) *Date Palm Cultivation.* FAO Plant Protection Paper no.156 Rev.1, Food and Agriculture Organization of the United Nations, Rome. Available at: www.fao.org/3/Y4360E/y4360e0c.htm#bm12 (accessed 21 November 2022).

Ziffer, I. (2019) Pinecone or date palm male inflorescence – metaphorical pollination in Assyrian art. *Israel Journal of Plant Sciences* 66(1–2), 19–33. DOI: 10.1163/22238980-00001064.

Zirari, A. (2010) Effects of time of pollination and of pollen source on yield and fruit quality of 'Najda' date palm cultivar (*Phoenix dactylifera* L.) under Draa Valley conditions in Morocco. *Acta Horticulturae* 882, 89–94. DOI: 10.17660/ActaHortic.2010.882.9.

Zohary, D. and Hopf, M. (2000) *Domestication of Plants in the Old World: The Origin and Spread of Cultivated Plants in West Asia, Europe and the Nile Valley*, 3rd edn. Oxford University Press, Oxford.

Irrigation and Salinity Management of Date Palm in Arid Regions

8

Abdulrasoul Mosa Al-Omran[1]* ⓘ, Latifa Dhaouadi[2] and Houda Besser[3]

[1]King Saud University, Riyadh, Saudi Arabia; [2]Institut des Régions Arides (IRA), Médenine, Tunisia; [3]Faculty of Sciences of Gabes, Gabes, Tunisia

Abstract

In the agro-based countries characterized by arid to desert climate conditions such as North Africa and Arabian countries, flexible management strategies to optimize and rationalize the water resources are required to secure sustainable agricultural development that supports huge socioeconomic pressures. This hinges principally on finding different alternatives to using irrigation water resources. Among various possibilities for sustainable natural resources that optimize water extraction, within limited manageable issues of water withdrawal and quality deterioration, is dealing with water salinization. For an effective planning of remediation actions in these agro-based countries, evaluation of the current situation of the agricultural sector and the consequences of the adopted management strategies is required to outline the feasible strategies for sustainable oasis development. Consequently, this chapter constitutes a reexamination of the different factors influencing prolific date productivity and highlights, at the same time, the driving forces governing the variable spatial efficiency. Field investigations, literature review, and laboratory analyses prove that natural resources have undergone progressive degradation amplified by the mismanagement of oasis development. The calculated and measured data of several environmental criteria suggest that the water quality is poor to unsuitable to be used for irrigation purposes. Irrigation water quality is, furthermore, an important factor challenging fruitful agriculture development under different climate conditions. This chapter outlines that, besides climate and natural circumstances, social, financial, and engineering constraints are of primary importance to overcome water- and climate-related issues.

*Corresponding author: rasoul@ksu.edu.sa

© CAB International 2023. *Date Palm* (eds J.M. Al-Khayri *et al.*)
DOI: 10.1079/9781800620209.0008

Keywords: Crop water requirements, Irrigation water management, Middle East and North Africa, Sustainability, Water quality

8.1   Introduction

Sustainable agricultural development relies on several natural and human-induced factors, namely climatic conditions, farming practices, and irrigation management. Indeed, the irrigation scheme used and the scheduling timing adopted are of paramount importance for both the conservation of water resources and the amelioration of salinity issues. Previous studies indicate that the efficiency of the delivered quantity of water for irrigation is more productive for smart and local techniques such as smart irrigation, while the old-school methods have unfortunately induced progressive loss of land fertility and productivity regarding the emerging salinization and alkalization issues.

In fact, soil salinity is a dynamic issue threatening more than 40% of agricultural lands all over the world. It depends, in addition to natural conditions, on the quantity of water delivered for irrigation as well as the suitability of these resources for irrigation purposes, on the one hand, and on the techniques adopted to facilitate the absorption of this quantity without significant loss of water due to evaporation or runoff, on the other. Land salinization is, consequently, a fragile balance relying on water resources and irrigation management, two huge challenges for the agro-based countries characterized by water deficiency and inefficient farming practices.

The growth in the demand for water greatly affects scarce and dwindling water supplies in different areas with harsh climatic conditions. Thus, groundwater resources are considered the most important available resource for the agricultural sector, which accounts for more than 70% of the total available resources (MAW, 2006; CRDA, 2020). Different alternatives have been adopted during recent decades, namely reuse of recycled wastewater and drainage water, desalination, and use of brackish water for irrigation. The effectiveness of these efforts seems to be insufficient, thus constraining safe agricultural development for sustainable productivity. In agriculturally based countries, the increasing and unregulated exploitation of water resources (ground and surface) worldwide, especially in arid and semi-arid areas, is causing the deterioration of water quality. There are many specific variables affecting sustainable agricultural development, such as the quality of irrigation water and poor management of soil resources, in addition to the negative effects of climate change, i.e., rising temperatures and fluctuating rainfall (Besser *et al.*, 2017, 2019; Dhaouadi *et al.*, 2020a, b, Dhaouadi *et al.*, 2021a, b).

The critical issue is how to rapidly reconcile growing demand with scarce and diminishing resources. There is an urgent need to adopt water conservation, which requires management and rationalization programs to achieve an acceptable balance between water needs and availability. There is great potential for improving water-use efficiency (WUE) in various sectors. Saudi Arabia, Tunisia, as well as many areas in the Middle East and North Africa (MENA) region and the Persian Gulf must shift their focus from increasing supply to managing demand and rationalizing water use, especially in agriculture, to avoid depleting nonrenewable water resources.

Knowing the water needs of crops is one of the important factors in agricultural production, because the lack of water during the growth period may cause physiological drought. The application of water in quantities exceeding requirements induces soil waterlogging problems and increases quantities lost to evaporation. Climate change also affects water needs, which are determined by the climatic parameters of the area such as solar radiation, temperature, rain, relative humidity, wind, and evaporation.

The efficiency of water management and conservation efforts relies on scientifically based actions coupled with detailed economic, political, and social and institutional supports to help sustain agricultural production and to meet the food needs of future generations. It has been suggested that over the next decades, the world could witness water-related conflicts in different regions. The Middle East region is the major candidate for such conflict owing to political, demographic, and economic reasons (Malekian *et al.*, 2017). In a region haunted by extreme water scarcity, Saudi Arabia and other countries in the region cannot afford to turn a blind eye to the shortage of water and the shortsighted misuse of stored underground water supplies. The limited water resources in the MENA region, coupled with continuous population growth, produces a pressure on the region's water resources, requiring huge efforts to optimize WUE, overcome water-related issues (overexploitation, water degradation, pollution), and adopt a national policy for using brackish water and recycled wastewater (e.g. agricultural drainage, industrial wastewater, and municipal wastewater). Traditionally, agriculture has viewed water management from the demand side in terms of irrigation, or from the surplus perspective with regards to drainage. Increasingly, tactical management, such as the application of brackish and recycled water in agriculture, will be necessary. In addition, greater advantage will be taken of aquifer storage and groundwater banking.

The alternatives of nonconventional water resources usage and the efficient management of resources are highly required in the Arab countries where the quasi-permanent water deficit constrains the sustainable development of the different water-based sectors, namely agricultural purposes related principally to the date palm. Thus, this chapter evaluates the different issues challenging safe date production in two different countries (Saudi Arabia and Tunisia) to outline the main factors controlling the scarcity and degradation of water resources and to determine the main constraints for agricultural development.

8.2 Date Palm: Importance and Productivity

The date palm (*Phoenix dactylifera* L.) is a monocotyledonous angiosperm plant that is well adapted to the harsh climatic conditions found in arid and semi-arid regions (Munier, 1973). It is a tree of great importance in the Arab world, having social, economic, cultural, and religious value. Consequently, the oasis agro-system represents one of the most important cropping systems in terms of production and cultivated area in these countries, namely Saudi Arabia and Tunisia. In fact, the total cultivated area has reached about 116,125 ha and about 40,000 ha in Saudi Arabia and Tunisia, respectively, for a production estimated at 1,539,775 and 190,000 t, respectively (Table 8.1, Fig. 8.1). The productivity of these oases reveals contradictory indices.

Concomitant to the increase of productivity in recent decades, a significant reduction of crop quality has been manifested by the reduction of the available productive varieties, the various fungal diseases of date palms, and the decreasing economic value of the products in the international market (Zaid and Arias-Jiménez, 2002; CRDA, 2019; Besser *et al.*, 2021b; Dhaouadi *et al.*, 2021b). Close causal links have been reported between these productivity issues and the environmental problems related to agricultural management, essentially water resource degradation, soil salinization, and irrigation scheduling.

8.3 Water Resources Overexploitation

The increased expansion of these agro-systems in such hot dry areas of arid and desert climate relies principally on the permanent use of groundwater resources for irrigation, as the largest consumer of freshwater resources of more than 70% (FAO, 2020) leading to several water-related environmental issues. Indeed, owing to the scarce intermittent surface water resources given the rare rainfall potential and the high temperature and evaporation values summarized in Table 8.2, rainfed agriculture is limited to some seasonal vegetable crops and some tolerant trees. Agricultural projects depend essentially on the accessibility of groundwater resources via springs, wells, and atmospheric cooling systems (Besser *et al.*, 2017). Thus, simultaneously to the continuous agro-systems expansion, the growing exploitation of groundwater resources has reached a higher level inducing progressive degradation of these resources in both qualitative and quantitative terms. These emerging issues have been reported by previous studies in different regions of Arab countries, including Saudi Arabia and Tunisia (Ministry of Environment, Water and Agriculture, 2018; CRDA, 2020).

Over recent decades, the intensive exploitation of groundwater resources in these regions has reached excessive levels exceeding the natural replenishment abilities of the aquifers used for irrigation (Table 8.2). The indices of aquifer decompression have progressively

Table 8.1. Total number of date palms and production quantity in some regions of Saudi Arabia, 2019. (Adapted from NCDP, 2020.)

Region	Total number of date palms	Number of bearing date palms	Total amount of production (t)	Quantity of marketed production (t)
Aseer	1,117,738	983,955	55,209.7	37,272.0
Eastern	4,042,524	3,431,533	213,515.9	182,293.0
Hail	1,973,528	1,740,970	97,390.3	66,041.7
Jazan	8,822	3,813	131.1	923.0
Makkah	1,234,909	1,044,284	59,825.5	45,235.0
Medina	4,751,040	3,812,367	213,668.3	189,643.8
Najran	526,333	391,492	27,180.3	19,314.9
Northern	24,918	9,001	401.7	276.8
Qassim	7,542,914	6,187,301	372,827.1	345,057.7
Riyadh	7,924,947	6,290,624	402,411.6	351,073.0
Tabuk	1,012,499	875,468	50,639	42,489.2
Total	30,160,172	24,770,808	1,493,200.5	1,279,620.1

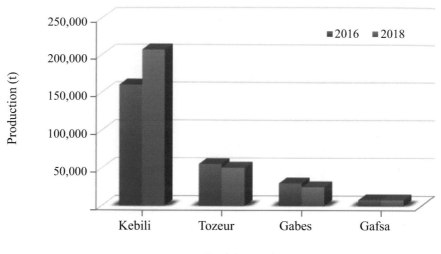

Tunisian oasis

Fig. 8.1. Evolution of date palm production in Tunisia, 2016–2018. (Modified from CRDA, 2016, 2018.)

Table 8.2. Climate conditions of the Tunisian oases, 2018. (Adapted from CRDA, 2018.)

Location	Climate		Water resources (10⁶ m³)		Water exploitation (10⁶ m³)	
	Mean temperature (°C)	Mean rainfall (mm)	Shallow	Deep	Shallow	Deep
Gabes	21.3	136.6	23.2	270.9	27.40	259.50
Tozeur	23.1	60.0	48.1	189.0	41.21	156.08
Kebili	21.7	50.2	5.49	236.7	0.37	418.27
Gafsa	20.4	179.0	33.3	95.1	52.25	121.22

been evaluated by spring extinction, pumping generalization, and drainage (Besser *et al.*, 2017, 2018, 2019, 2021a, b). Despite these commonly observed and reported consequences, the continuously increasing extraction rate amplified by the effects of climate variability and frequent shortages has induced more harmful local effects like land subsidence in the Douz area (Tunisia) as reported by Besser *et al.* (2021a) and increasing drainage and leakage within the hydrogeological system between the water and petroleum resources (Besser *et al.*, 2017, 2019; Besser and Hamed, 2019). These variables impact and highlight the influence of degradation driving forces in the different resources at small-scale level, consequently indicating the necessity of a flexible management plan that may consider these variable features of local regions and agricultural systems. The multi-decade overexploitation of these limitedly renewable resources exhibits high risks of being the main driver of agriculture unsustainability. In these agro-based countries, to deal with the permanent water deficit amplified by the overexploitation of these limited water resources, different supplementary alternatives of nonconventional resources have been used.

In Saudi Arabia, there are currently 35 desalination plants located on the coast of the Arabian Gulf and the Red Sea coast, and the latest data indicate that the total production capacity of desalinated water amounted to 6.28 million m^3/day at the end of 2015 and 7.4 million m^3/day in 2020. The Saline Water Conversion Corporation owns most desalination plants, representing 73% of the total existing production capacity (Ministry of Environment, Water and Agriculture, 2018). The use of treated water in Saudi Arabia for irrigation has become a necessity to preserve the groundwater resources, as they start to deplete. The amount of wastewater in 1994 was estimated at about 1017 million m^3, and about 418 million m^3 of it was treated. It was used with the aim of reaching acceptable levels for the disposal of various types of public sewage water and achieving safe levels for its reuse in the areas of agricultural irrigation, irrigation of public gardens, recreational locations, and cooling. In the last report of the Ministry of Environment, Water and Agriculture (2018), the number of sewage plants was 92, producing an estimated 1.6 billion m^3, and the amount of reused water amounted to 216 million m^3. The utilization of treated sewage water at the national level is estimated at 13.4%. The percentage in the Eastern region is 16% and in the Riyadh region 16.1%, dropping to 9% in the Makkah region.

Regarding the Tunisian experience, however, the desalination of ocean water has been developed since the early 2000s, principally for domestic water consumption during periods of water shortage, whereas the availability of these resources for irrigation is a lower priority. In addition, the use of treated wastewaters has been an alternative in Tunisia since 1965, to (i) reduce the amount of pollution into the environment and (ii) define additional alternatives to be used during periods of water shortage. However, the use of these resources for irrigation is still limited in Tunisia and mostly absent in the agro-systems in the south-west of the country. In fact, these resources are used locally for agricultural activities in the Gafsa region (southern Tunisia) for the irrigation of about 145 ha of oasis agro-systems. For 2 years, treated wastewaters have been used for fruit trees after the deployment of a third treatment station (Dhaouadi *et al.*, 2020a, b).

Consequently, in Tunisia, the principal alternative representing a non-conventional water resource is drainage runoff. Indeed, the huge volume delivered to cultivated crops is above the recommended value for safe production and is coupled with the nonlocalized system of excess of irrigation water collected in a drainage network. Unfortunately, the total volume of the runoff released to the environment from the irrigation system of the oasis agro-systems has not been evaluated with accurate data. There is a dearth of published works about the collected volume amount, and these were only generally evaluated for some localities. These waters often contain high levels of trace elements resulting from the excess use of pesticides or fertilizers. Despite their important volume, which may constitute a real resource that may be used during periods of water shortage, these resources are released directly into the environment without treatment and without reuse process measures (Dhaouadi *et al.*, 2015, 2017; Besser *et al.*, 2017, 2018, 2019).

The evaluation of these resources as an important support of water during these times of unpredictable climate variability is crucial and during the last years, different works have discussed the feasibility of these remediation actions based on critical scientific assessment of the quality of these waters and their potential impacts on soil fertility and crop production (Dhaouadi *et al.*, 2015; Karbout *et al.*, 2019). However, the implementation of different stations and networks to treat and redistribute these waters is expensive. Thus, for the Tunisian oases, the lack of national support from governmental institutions and organizations means that local residents cannot access these resources. There have been several internationally funded projects on runoff reutilization, which are generally local and temporary (Hamdane, 2016). Currently, different projects are showing relevant results that can be applied and adopted, although requiring high levels of engineering. The first steps of these efforts have been observed in the Hezoua region (Tozeur area).

The efficiency of these solutions is still limited to local and regional scales. Besides overexploitation and aquifer replenishment, irrigation water quality suitability for irrigation purposes represents another important factor influencing safe fruitful date palm production.

8.4 Water Quality Degradation

The groundwater resources used for domestic purposes, agriculture, and industry face continuous quality degradation, constraining their use for specific purposes. Numerous investigations have been conducted to highlight the maximum acceptable limits of the different physicochemical water parameters to be used in irrigation without environmental risks. Based on the thresholds of the monitored indices summarized in Table 8.3, waters analysed from the different regions in Saudi Arabia and Tunisia indicate an almost poor to unsuitable quality for irrigation given the high risks of alkalization, salinization, and loss of permeability of agricultural lands. Besides the spatial distribution, these values show important evolution in years relative to exploitation rates and extreme weather events (Table 8.2). The use of these waters for the long and short term should be associated

Table 8.3. Irrigation water quality in Saudi Arabia and Tunisia, 2018–2020.

	Alkalization indices	Salinization indices	Permeability indices	Water quality indices
Saudi Arabia	$2 < SAR < 15$ $20 < \%Na < 40$	$2 < EC < 6$	–	$25 < WQI < 120$
Tunisia	$1.8 < SAR < 32$ $42 < \%Na < 86$	$2 < EC < 9$ $1.5 < TDS < 20$	$62 < PI < 12$ $0.8 < KR < 1$	$8 < IWQI < 58$ $31 < CWQI < 91$

EC, electrical conductivity; CWQI, Canadian water quality index; IWQI, irrigation water quality index; KR, Kelly ratio; %Na, sodium percentage; PI, permeability index; SAR, sodium adsorption ratio; TDS, total dissolved salts; WQI, water quality index.

with high to severe restrictions related to the type of cultivated crop, farming practices, irrigation schemes, and drainage systems.

8.5   Land Degradation and Soil Salinization

Soil salinization is a major aspect of land degradation, leading to reduction of crop yield and the loss of land from production. In addition to naturally salt-affected soils, human-induced salinization is often cited as a major contributor to desertification processes in the world's dry lands (Al-Omran *et al.*, 2004, 2006, 2008, 2012, Al-Omran *et al.*, 2018). This problem is particularly associated with irrigation schemes in the world's dry lands, reducing productivity or even eliminating crops totally from an area. These effects have been observed and measured for some varieties of date palm in Saudi Arabia, along with other fruits such as pomegranates and figs.

Soil salinity occurs in agricultural areas when saline water is used for irrigation in the absence of good drainage systems. Saline soils cannot be reclaimed by any chemicals, conditioners, or fertilizers. The reclamation of these soils consists of simply applying enough high-quality water to thoroughly leach the salts. The use of saline water for leaching is the main problem in soil salinity removal. The water applied for this purpose should be low in salinity (1500–2000 ppm total salt). Application of excessively saline water can exacerbate management problems due to the threat of saline high water tables, increased expense of irrigation water, and difficulty in maintaining adequate levels of soil nitrates for crop growth. Soil salinity is a very common issue in today's irrigated agriculture because even good-quality irrigation water adds salts to the soil. For instance, fresh irrigation water with electric conductivity (EC) of 0.5 dS/m (corresponding to a salt concentration of $0.3\,g/L$ or $0.3\,kg/m^3$) applied in irrigations of $1000\,m^3/$ha per year would add $3000\,kg$ salt per year to the soil, mainly in the form of sodium chloride.

Table 8.4. Classification of salt-affected soils based on analysis of saturation extracts. (Modified from James *et al.*, 1982.)

Criterion	Normal	Saline	Sodic	Saline/Sodic
EC (dS/m)	< 4	> 4	< 4	> 4
SAR	< 13	< 13	> 13	> 13

EC, electrical conductivity; SAR, sodium adsorption ratio.

The classification of salt-affected soils is based on both EC and sodium adsorption ratio (SAR) (James *et al.*, 1982) (Table 8.4). Cultivated soils in the largest oasis in Saudi Arabia have an average value of organic carbon, EC, and clay content of 0.71%, 4.4 dS/m, and 15.4%, respectively, whereas these values for coastal areas are 10%, 55 dS/m, and 4%, respectively (Al-Barrak, 1997).

Al-Hassoun (2010) concluded that due to agricultural development in Saudi Arabia, many soil problems have been recorded in different areas over the country. The main problem the Saudi soil suffers from is soil salinity. Using remote sensing he found that the areas around Skaka City, in northern Saudi Arabia, are considered an example of affected areas that are developing a serious salinity crisis. Moreover, it has been proven that most of the cultivated lands of the date palm agro-systems were affected by salinity at different levels. Temporal analysis revealed that a rapid increase in salinity level and extent had been detected during the short 6 year study period.

In Tunisia, however, despite the large number of studies carried out to highlight the importance of soil salinization issues in the oasis agro-systems, only a limited number have been evaluated accurately and quantitatively via EC, SAR, permeability index (PI), and Kelly ratio (KR) indices in these oases. The results indicate that soil salinization of oasis agricultural lands has reached unrecoverable levels.

8.5.1 Irrigation management

For mature date palms, the depth of the root system is about 5 m, and it extends about 3m out from the trunk. Thus, it is seen for dates that 40% of all water is extracted from the first 50 cm, 70% is from the first 100 cm, and 90% is from the top 150 cm; only 10% is from deeper than 150 cm. For young date plants rooting depth can vary from 25 to 50 cm and the radius from 10 to 30 cm, depending on the size of the plant.

These impacts are related to various factors, namely soil topology, irrigation, water quality, and irrigation management. Indeed, the use of an appropriate irrigation technique that reduces the loss by evaporation to inhibit salt precipitation is required.

Traditional irrigation systems prevail throughout Saudi Arabia except for some farms that have introduced low-volume drip or bubbler systems. The drip and bubbler systems adopted have been modified to prevent clogging. However, this practice results in high discharge rates in addition to likely disequilibrium in pressure and discharge distribution along laterals. Some water savings are acknowledged by farmers in comparison to conventional surface irrigation methods. The common practice by farmers using low-volume irrigation is to fill basins around trees to a depth of 15–20 cm, but the amounts applied are still lower than those reached through the traditional practices. The latter include three types of systems: (i) flooding the entire planted area; (ii) borders with non-irrigated strips of land with the date trees in the middle of a large basin; and (iii) large basins from 3m to over 5m in diameter around individual trees. With these systems, the flooded area represents 70–100% of the total and the depth applied ranges from 8 cm to over 15 cm (CRDA, 2020).

There are many ways to solve the problems and concerns regarding water use. The most important solution is to improve water conservation. The two best water management approaches are decreasing demand and increasing supply. Demand management objectives include reducing water delivery (nonbeneficial evapotranspiration) and reducing water requirements (beneficial evapotranspiration).

Table 8.5 lists the primary technologies available for irrigation water demand management. The areas listed for reducing water use include irrigation scheduling, increased irrigation efficiency, reduced evaporation, and reduced water use by noneconomic crops. Technologies available for reducing crop water use include limiting irrigation and/or cropland hectarages, crop/variety selection, and decision making models and technical assistance (Bucks *et al.*, 1992).

8.5.2   Irrigation systems

Given the importance of agricultural development in these areas, different agricultural management actions have been adopted and principally various irrigation techniques have been used in recent decades in both Saudi Arabia and Tunisia.

In Saudi Arabia, for example, the governmental institutions responsible for agriculture are striving to use modern irrigation systems, such as drip irrigation or subsurface irrigation, to irrigate date palms and avoiding traditional methods, such as flood irrigation. However, despite all these encouragements, most traditional date palm farms are still irrigated by flooding or furrow systems. In modern farms, drip systems are becoming the systems of choice due to high irrigation/fertigation efficiency, ease of operation, suitability for automation, prevention of weed growth, and lower system costs. The drip irrigation systems can apply irrigation water

Table 8.5. Irrigation demand management for water conservation. (Modified from Bucks *et al.*, 1992.)

Objective	Technology
Reduce water delivery	• Irrigation scheduling: Provide information on when, how much, and how to apply irrigation water • Increase irrigation efficiency: Install physical measures such as canal lining, piping, land leveling, control structures, improved and automated irrigation systems • Reduce water evaporation: Use chemical films, conditioners, and floating objects to impede the escape of water molecules • Reduce soil evaporation: Use crop residues, plastic mulches, synthetic or natural soil conditioners, etc.., and install surface or subsurface drip irrigation systems • Reduce water use by noneconomic and phreatophytic vegetation: Clearing and thinning of phreatophytes • Limit irrigation: Apply less water than maximum evapotranspiration demand • Limit irrigated cropland acreage: Convert irrigated cropland in water-short areas to dryland farming
Reduce crop water requirements	• Crop selection and modification: Select drought-resistant strains that can withstand dry periods • Decision making models and systems: Use data on water availability and other factors to recommend optimum management of energy, salinity, fertilizer, insects

by different types of emitters such as orifice emitters, bubblers, valves, and sprayers (Fig. 8.2).

In Tunisia, however, almost all agricultural lands are irrigated by the basin and flood technique, while the application of smart and local techniques is still limited to some project pilot stations. The distribution of the techniques used seems to be one of the major factors constraining an effective management of natural resources, as the various techniques define different requirements and different practices, consequently inducing different environmental impacts (Fig. 8.3).

8.5.2.1 Irrigation scheduling

The purpose of irrigation scheduling is to determine when and how much water to apply in the field, keeping in mind that overirrigation can have negative effects on quality and quantity of yield (Deumier *et al.*, 1995; Al-Omran, 2002; Alnaim *et al.*, 2022). To determine this, knowledge is

Fig. 8.2. Irrigation of date palm trees. (a) By a drip system using the orifice emitter method. (b) By a drip system using the bubbler method. (c) Using the valves method. (d) By a drip system using the sprayer method. (From Al-Omran and Louki, 2020.)

required of the type of soil and how deep it is. This gives an indication of how much water is in the soil and how much is available for the date palm. This information, combined with the daily usage of water by the dates, enables the determination of when the next irrigation cycle is due. The following figures are mean values of available water for the three major soil types:

- light soils: 100 mm/m;
- medium soils: 140 mm/m; and
- heavy soils: 180 mm/m.

The best approach is to determine the water-holding capacity of the specific soil under consideration through laboratory tests and then to establish an effective scheduling program. To ensure that the dates will not be put under water stress, it is the normal practice to allow for only a fraction of the available water to be extracted. The exact amount of water to apply to the field and the exact timing for application depend on the strategy adopted. The amount and the timing of water application are determined by using several methods based on soil water measurement, soil water balance, and

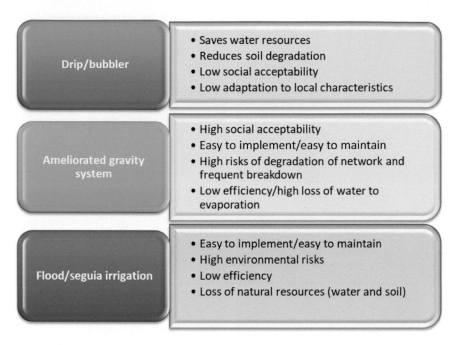

Fig. 8.3. The different techniques of irrigation in Tunisian agro-systems. (From Besser *et al.*, 2021a.)

soil water potential, or even plant stress indicators. Thus, the objectives of irrigation scheduling are to maximize irrigation efficiencies by applying the exact amount of water needed to replenish the soil moisture to the desired level, minimize irrigation cost, and maximize yield.

8.5.2.2   Irrigation water-use efficiency

WUE represents one of the most reliable indices for accurate evaluation of the adopted irrigation scheme. However, the dearth of published and measured data concerning the parameters related to efficiency estimation has gained less attention in Tunisia. In Saudi Arabia, by contrast, the estimated water consumption by the date palm varies depending on irrigation method, region, and age of the tree (Al-Ghobari, 2000, 2002; Hosny *et al.*, 2022). Al-Khatib and Al-Jabor (2006) estimated the consumptive use of water and WUE for date palms in different regions of Saudi Arabia for different irrigation methods (Tables 8.6 and 8.7). Al-Amoud (2006) determined water consumption (range 86.4–102.5 m³/tree per year), WUE (range 0.87–1.59 kg/m³), and yield (range 75.5–163.1 kg/tree) for 17-year-old date palms in the Qassim region for period of 3 years.

Table 8.6. Consumption of water by date palm according to irrigation system in some regions of Saudi Arabia. (Adapted from Al-Khatib and Al-Jabor, 2006.)

Region	Water consumption (m³/ha)		
	Surface	Bubbler	Drip
Abaha	25,107	18,289	15,061
Aseer	25,107	18,289	15,061
Eastern	34,782	26,120	20,865
Hail	35,254	25,680	21,148
Jouf	35,204	25,647	21,121
Madenah	43,305	31,545	25,978
Makkah	34,451	25,095	20,667
Najran	28,868	21,028	17,317
Northern	34,976	25,647	21,121
Qassim	35,204	25,647	21,121
Riyadh	34,343	25,046	20,602
Tabuk	32,157	23,424	19,290

Table 8.7. Water-use efficiency (WUE) of date palm according to irrigation system in some regions of Saudi Arabia. (Adapted from Al-Khatib and Al-Jabor, 2006.)

Region	WUE (kg/m³)		
	Surface	Bubbler	Drip
Aseer	0.383	0.525	0.638
Baha	0.165	0.226	0.275
Eastern	0.294	0.392	0.490
Hail	0.168	0.231	0.280
Jouf	0.252	0.346	0.421
Madenah	0.122	0.168	0.203
Makkah	0.204	0.280	0.340
Najran	0.170	0.234	0.284
Northern	0.173	0.236	0.286
Qassim	0.120	0.165	0.200
Riyadh	0.177	0.242	0.294
Tabuk	0.223	0.306	0.372

8.6 Date Oasis Management

This situation requires the implementation of a participatory approach of integrated and sustainable irrigation water management by landowners and agricultural companies to optimize the utilization of irrigation water resources. This approach is based principally on the application of the actual water requirements of agricultural crops, use of the most efficient irrigation methods, and adoption of the concept of *more crop per less drop.*

8.6.1 Water requirements estimation

Research in Saudi Arabia, where most water is pumped from wells, shows that irrigation scheduling provides average savings of 20–30% in water and energy (Abderrahman, 2001; Abderrahman and Al-Nabulsi, 2001; Al-Amoud, 2006). A field experiment was conducted by Al-Amoud *et al.* (2000) to investigate the response of date palm trees of the Seleg cultivar to different water regimes (50, 100, and 150% of the pan evaporation rate) using three irrigation methods (basin, bubbler, and trickle irrigation systems). Results of their study showed the general trend of yield increasing as irrigation quantity increased. The maximum yield and WUE was with the trickle irrigation system followed by the basin method.

Al-Amoud *et al.* (2012) conducted several field experiments to measure reference evapotranspiration (ET_r) and date palm water use in seven regions of date cultivation in Saudi Arabia. Actual ET_r was obtained through lysimetric measurements, while net date palm water use (ET_c) was measured by the soil water balance approach. The water requirements were also computed by the Penman–Monteith method. The results indicated that there were differences in ET_r and ET_c in various regions due to different geographical locations and the impact of climatic factors. Comparisons between measured and computed ET_r showed higher levels for measured rates in all regions with an average of about 7.04 mm/day and a maximum of 11.7 mm/day. The average daily date palm water use was 184.41/day for all regions. Total net annual date palm water use ranged between 59.4 and 80.0 m³/tree. Date palm crop coefficients ranged between 0.80 and 0.99 for all regions, and the average values of crop coefficients for each month were not equal in the studied regions.

In the study of Al-Shammari (2015) to estimate crop evapotranspiration (ET_c) in several regions in Saudi Arabia (Medina, Tabuk, Makkah, Jouf, Riyadh, Qassim, Hail, Eastern), it was found that the value of ET_c ranged from 2418.75 to 1837.76 mm/year, with a planting density of 100 palm trees/ha. The largest water need was during the month of July, and the total irrigation requirements ranged from 7928.93 to 9495.24 m³/ha. The total annual irrigation requirements for a single palm in these regions ranged from 73 to 95 m³/year, given that the canopy radius of the palms was 3.5 m. It was found that the average total irrigation requirements at the level of the Saudi Arabian regions averaged 8342.41 m³/ha per year with a density of 100 palms/ha. The productivity per hectare in the study ranged between 5406 kg/ha in the Makkah region and 8400 kg/ha in the Al-Ahsa region and the WUE ranged between 0.55 and 0.83 kg/m³. It is obvious that Eastern region, which includes Al-Ahsa Oasis, has the highest

WUE and the oases more suitable for date cultivation compared to another coastal region such as Makkah or Najran.

Al-Amoud *et al.* (2012) studied the estimation of the total water needs of date palms in several regions of Saudi Arabia, between Najran in the south to Qassim in the north, over three seasons (2005–2007). The standard evapotranspiration was obtained from the use of lysimeters, while evapotranspiration of date palms was calculated using the water equilibrium method and compared with the water needs calculated using the Penman–Monteith equation. The researchers found that the highest value of reference evapotranspiration using lucerne (alfalfa) crop (ET_0) was in the Wadi Al-Dawasir region in June, while the lowest value was in January in the Najran region. It was also found that the annual average of ET_0 was between 2253 and 3024 mm/year. The study documented a similarity between the measured values and those calculated using the Penman–Monteith equation and a difference between the regions due to the weather conditions. It was found that the annual ET_c values ranged between 2100 and 2829 mm/year, and from this it was concluded that the water needs of the date palm ranged from 200 to 218 L/day in Saudi Arabia.

In a comparative study to estimate the crop coefficient (K_c) for date palms in the Qassim region, central Saudi Arabia, the Penman–Monteith cumulative evaporation rate was 26.87% higher than the value obtained using the Blaney–Criddle equation and the Penman-Monteith potential evapotranspiration rate was 26.7% higher than the value determined using the Blaney–Criddle equation (Al-Harbi *et al.*, 2015).

In a study conducted by Abdul Salam and Al Mazrooei (2007) on estimating the water needs of date palms in Kuwait, the values of reference evapotranspiration (ET_0), crop evapotranspiration (ET_c), irrigation requirements (IR), and amount of added water (NIR) were 28,830, 26,830, 25,530, and 25,639 m^3/ha per year, respectively.

In Tunisia, however, a few works have been carried out to highlight the real water requirements of date palm and these studies have been conducted on special varieties based on local conditions. In fact, the delivered water quantity for the irrigation of oasis agro-systems is largely above the recommended amounts for the cultivated crops. Taking the example of date palm, the used quantity is 18,000 to 24,000 m^3/ha per year, while according to previously published data, the real requirements of the date do not exceed 15,000 m^3/ha per year (Dhaouadi *et al.*, 2021b). This excess is one of the principal factors reducing soil fertility and increasing loss of land productivity. The issues related to actual water requirements and the applied irrigation volume are largely discussed by different researchers who have confirmed that the adoption of effective water-use techniques is constrained principally by local residents and landowners who refuse to modify the accepted practices and the delivered volume.

8.6.2 Field experiences

Values of ET_c and crop water requirements (CWR), according to the Penman–Monteith equation based on climate data, are attributed to the meteorological conditions of each site. However, the reduction in the estimated CWR in the study by Al-Omran *et al.* (2019) to an average of 8342 m³/ha per year compared to an overall average of 20,000 m³/ha per year as reported by many researchers (Al-Amoud *et al.*, 2012; Ismail *et al.*, 2014; Dewidar *et al.*, 2015; Mihoub *et al.*, 2015) is mainly attributed to the percentage of canopy area or shaded area (Se) of the date palm, as Al-Omran *et al.* (2019) calculated the Se values as 0.33 times the actual area of the tree. Therefore, the practice of 10 m × 10 m planting distance between palms in the date farms of Saudi Arabia is not adequate in all sites. This area of 100 m² for each tree overestimates the CWR and therefore it must be changed to 7 m × 7 m to have a higher canopy cover in date palm farms.

In the same study of Al-Omran *et al.* (2019), the water balance method further showed that water consumption values for the two sites of Jouf and Qassim regions were very low compared to ET_c estimated by the Penman–Monteith equation or compared to the water added to the field. This reduction in the total amount of water consumption is mainly due to the shallow depth of the sensors installed in the two sites (120 cm). It seems that 50% of water added to date palm should be considered as leaching water.

The increased amounts of irrigation used by farmers on adjacent farms were mainly due to poor knowledge of irrigation requirements (Table 8.8). Before installing the system for monitoring irrigation water in the study sites, the farmers used to add triple the amount and it might reach 35,000 m³/ha per year.

Based on Maas and Hoffman (1977), a reduction of yield would be expected when using saline water at all sites of Al-Omran *et al.*'s (2019) study. Results showed that salinity reduced date production by 25.0% at the farm in the Qassim region and by 7.3% at the farm in the Eastern region (Al-Omran *et al.*, 2019).

8.7 Smart Technology

Smart irrigation technology can be installed on existing irrigation systems by exchanging the current controller for a smart one. In some cases, small add-on weather stations or soil moisture sensors can be used with existing controllers and systems. The weather station measures temperature, wind, solar radiation, and relative humidity and then calculates potential evapotranspiration needs. Soil moisture technology uses probes or sensors inserted into the soil to measure actual soil moisture content. The smart watering technology was developed in recent years to apply irrigation mostly to turf and landscape plants, and then extended to agricultural crops. These

Table 8.8. Water requirement of the different methods, the irrigation water applied, and the increase in water usage compared to the Penman–Monteith method in some regions of Saudi Arabia. (From Al-Omran *et al.*, 2019, used under the terms of the Creative Commons CC-BY license.)

Site	Water requirement (m³/ha per year)		Applied irrigation water (m³/ha per year)		Increase in water usage (%) compared to Penman–Monteith method	
	Penman–Monteith method	Water balance method	Study field	Adjacent farm	Study field	Adjacent farm
Medina	9495.24	–	11,305.0	13,717.0	16.0	30.8
Tabuk	7340.18	–	9463.9	12,277.0	22.4	40.2
Makkah	7298.93	–	9692.0	12,220.0	24.7	40.3
Jouf	8913.59	3515.25	11,252.8	13,340.0	20.8	33.2
Riyadh	8614.96	–	10,007.4	12,050.0	13.9	28.5
Qassim	8568.68	3604.31	10,035.0	12,880.0	14.6	33.5
Hail	7996.99	–	10,272.5	12,620.0	21.2	36.6
Eastern	8510.72	–	10,082.8	12,610.0	15.6	32.5

technologies are an evapotranspiration-based irrigation controller, which calculates evapotranspiration values for the local microclimate (Al-Ghobari and Mohammad, 2011; Al-Ghobari et al., 2013). Then, the controller (ET-System) creates a program for loading and communicating automatically with irrigation system controllers and soil moisture sensors.

Previous works (Al-Ghobari et al., 2013, 2015) reported on the use of intelligent irrigation system (IIS) application technology. The IIS for automating irrigation scheduling was implemented and tested with sprinkle and drip irrigation systems to irrigate wheat and tomato crops under open-field conditions. The experiments were conducted for one growing season, 2009/10. The initial results indicated up to 25% water saving by intelligent irrigation (ET-System) compared to the control method (conventional irrigation), while maintaining a similar yield. Results showed that ET_c values for control experiments were higher than those for the ET-System in consistent trend during the whole growth season. The analysis revealed that the ET_c values of the control and ET-System were somewhat close to each other only in the initial plant development stages. Generally, the ET-System, with some modification, was precise in controlling irrigation water under open-field conditions and has been proven to be a good way to determine the water requirements for crops and to schedule irrigation automatically.

8.8 Conclusions and Prospects

The spatial distribution of date palm productivity over the last years within the same country, or between different production zones, highlights the high sensitivity of these plants to local conditions and small-scale farming practices despite their great tolerance to harsh and difficult conditions, namely dryness and salinity. This variable evolution of date palm agro-systems with respect to local, regional, institutional, and social constraints under huge economic pressure threatens the food security of these agro-based countries.

Indeed, sustainable date palm cultivation and increasing land productivity in agro-based countries are facing numerous challenges of which natural resources exploitation and updated science-based practices play pivotal roles. Integrated efforts by irrigation researchers, technicians, managers, and farmers must be associated in a participatory approach to overcome natural conditions and social, economic, and institutional constraints. Given the partial efficiency of the application of individual techniques, different techniques such as irrigation scheduling, advanced irrigation systems, limited irrigation methods, soil moisture management, and wastewater irrigation can be simultaneously used to reduce the problems associated with date palm irrigation. Indeed, improvement in water management is required at all levels of irrigation including planning and design, project implementation, and operations and maintenance. These management improvements require comprehensive changes in institutions and organizations, water policy, and law rehabilitation, or the introduction of new irrigation systems, education and training, and research and development. Increasing date palm cultivation must address improvements in water use and management as well as environmental protection.

References

Abderrahman, W.A. (2001) *Water Demand Management in Saudi Arabia.* IDRC Bulletin No. 12. International Development Research Centre, Ottawa.

Abderrahman, W.A. and Al-Nabulsi, Y. (2001) Irrigation water use of date palm. *Presented at the Fourth Symposium on Date Palm in Saudi Arabia,* Al Hassa, Saudi Arabia, May 5–8, 2007.

Abdul Salam, M. and Al Mazrooei, A. (2007) Crop water and irrigation water requirements of date palm (*Phoenix dactylifera*) in the loamy sands of Kuwait. *Acta Horticulturae* 736, 309–315.

Al-Amoud, A.I. (2006) Date palm response to subsurface drip irrigation. In: *Proceedings of the CSBE/SCGAB 2006 Annual Conference,* Edmonton, Alberta, Canada, July 16–19, 2006. Available at: library.csbe-scgab.ca/docs/meetings/2006/CSBE06204.pdf (accessed 22 November 2022).

Al-Amoud, A.I., Bacha, M.A. and Al-Darby, A.M. (2000) Seasonal water use of date palms in central region of Saudi Arabia. *Agricultural Engineering Journal* 9(2), 51–62.

Al-Amoud, A.I., Mohammed, F.S., Saad, A.A. and Alabdulkader, A.M. (2012) Reference evapotranspiration and date palm water use in the Kingdom of Saudi Arabia. *International Research Journal of Agricultural Science and Soil Science* 2(4), 155–169.

Al-Barrak, S.A. (1997) Characteristics and classification of some coastal soils of Al-Hassa, Saudi Arabia. *Journal of King Saud University, Agricultural Sciences* 9(2), 319–333.

Al-Ghobari, H.M. (2000) Estimation of reference evapotranspiration for southern region of Saudi Arabia. *Irrigation Science* 19(2), 81–86. DOI: 10.1007/s002710050004.

Al-Ghobari, H.M. (2002) Gross irrigation requirements for major crops in Najran region. *Journal of the Saudi Society of Agricultural Sciences* 1(2), 84–106.

Al-Ghobari, H.M. and Mohammad, F.S. (2011) Intelligent irrigation performance: evaluation and quantifying its ability for conserving water in arid region. *Applied Water Science* 1(3–4), 73–83. DOI: 10.1007/s13201-011-0017-y.

Al-Ghobari, H.M., Mohammad, F.S. and El Marazky, M.S.A. (2013) Effect of intelligent irrigation on water use efficiency of wheat crop in arid region. *Journal of Animal and Plant Sciences* 23(6), 1691–1699.

Al-Ghobari, H.M., Mohammad, F.S. and El Marazky, M.S.A. (2015) Assessment of smart irrigation controllers under subsurface and drip-irrigation systems for tomato yield in arid regions. *Crop and Pasture Science* 66(10), 1086. DOI: 10.1071/CP15065.

Al-Harbi, A.R., Al-Omran, A.M., Alenazi, M.M. and Wahb-Allah, M.A. (2015) Salinity and deficit irrigation influence tomato growth, yield and water use efficiency at different developmental stages. *International Journal of Agriculture & Biology* 17(2), 241–250.

Al-Hassoun, S.A. (2010) Remote sensing of soil salinity in arid areas in Saudi Arabia. *International Journal of Civil & Environmental Engineering* 10(2), 11–20.

Al-Khatib, A.A. and Al-Jabor, A.M. (2006) *Date Palm in Saudi Arabia.* Ministry of Agriculture, National Research Center for Date Palm, Al-Hassa, Saudi Arabia (in Arabic).

Alnaim, M.A., Mohamed, M.S., Mohammed, M. and Munir, M. (2022) Effects of automated irrigation systems and water regimes on soil properties, water productivity, yield and fruit quality of date palm. *Agriculture* 12(3), 343. DOI: 10.3390/agriculture12030343.

Al-Omran, A.M. (2002) Irrigation water conservation in Saudi Arabia. *Journal of the Saudi Society of Agricultural Sciences* 1(1), 1–50.

Al-Omran, A.M., Mohammad, F.S., Alghobari, H.M. and Alazba, A.A. (2004) Determination of evapotranspiration of tomato and squash using lysimeters in central Saudi Arabia. *International Agricultural Engineering Journal* 13(1–2), 27–36.

Al-Omran, A.M., Falatah, A.M., Sheta, A.S. and Al-Harbi, A.R. (2006) Use of clay deposits in water management of calcareous sandy soils under-surface and subsurface drip irrigation. *Arab Gulf Journal of Scientific Research* 24(3), 138–143.

Al-Omran, A.M., Abdel-Nasser, G., El-Damry, S., Nadeem, M. and Choudhary, M.I. (2008) Impact of irrigation regime and emitters depth on tomato production and growth. *Journal of King Saud University, Agricultural Sciences* 9(1), 25–38.

Al-Omran, A.M., Al-Harbi, A.A., Wahb-Allah, M.A., Alwabel, M.A., Nadeem, M.E.A. *et al.* (2012) Management of irrigation water salinity in greenhouse tomato production under calcareous sandy soil and drip irrigation. *Journal Agricultural Science and Technology* 14, 939–950.

Al-Omran, A.M., Mousa, M.A., Al-Harbi, M.M. and Nadeem, M.E.A. (2018) Hydrogeochemical characterization and groundwater quality assessment in Al-Hasa, Saudi Arabia. *Arabian Journal of Geosciences* 11(4), 79. DOI: 10.1007/s12517-018-3420-y.

Al-Omran, A., Eid, S. and Alshammari, F. (2019) Crop water requirements of date palm based on actual applied water and Penman–Monteith calculations in Saudi Arabia. *Applied Water Science* 9(4), 69–74. DOI: 10.1007/s13201-019-0936-6.

Al-Omran, A.M. and Louki, I. (2020) *Water Conservation Using Deficit Irrigation and Partial Root Drying System.* Saudi Society for Agricultural Sciences, Bulletin #35, King Saud University, Riyadh.

Al-Shammari, F. (2015) *Determination of date palm water requirements in Saudi Arabia.* Master thesis, King Saud University, Riyadh.

Besser, H. and Hamed, Y. (2019) Causes and risk evaluation of oil and brine contamination in the Lower Cretaceous Continental Intercalaire aquifer in the Kebili region of southern Tunisia using chemical fingerprinting techniques. *Environmental Pollution* 253, 412–423. DOI: 10.1016/j.envpol.2019.07.020.

Besser, H., Mokadem, N., Redhouania, B., Rhimi, N., Khlifi, F. *et al.* (2017) GIS-based evaluation of groundwater quality and estimation of soil salinization and land degradation risks in an arid Mediterranean site (SW Tunisia). *Arabian Journal of Geosciences* 10(16), 350. DOI: 10.1007/s12517-017-3148-0.

Besser, H., Mokadem, N., Redhaounia, B., Hadji, R., Hamad, A. *et al.* (2018) Groundwater mixing and geochemical assessment of low-enthalpy resources in the geothermal field of southwestern Tunisia. *Euro-Mediterranean Journal for Environmental Integration* 3(1), 16. DOI: 10.1007/s41207-018-0055-z.

Besser, H., Redhaounia, B., Bedoui, S., Ayadi, Y., Khelifi, F. *et al.* (2019) Geochemical, isotopic and statistical monitoring of groundwater quality: assessment of the potential environmental impacts of the highly polluted water in Southwestern Tunisia. *Journal of African Earth Sciences* 153, 144–155. DOI: 10.1016/j. jafrearsci.2019.03.001.

Besser, H., Dhaouadi, L., Hadji, R., Hamed, Y. and Jemmali, H. (2021a) Ecologic and economic perspectives for sustainable irrigated agriculture under arid climate conditions: an analysis based on environmental indicators for southern Tunisia. *Journal of African Earth Sciences* 177, 104134. DOI: 10.1016/j. jafrearsci.2021.104134.

Besser, H., Dhaouadi, L. and Hamed, Y. (2021b) *Agriculture Resilience During Challenging Times: Agricultural Sustainability.* LAP Lambert Academic Publishing, Chişinău.

Bucks, D.A., Sammis, T.W. and Dickey, G.L. (1992) Irrigation for arid areas. In: Hoffman, G.J., Howell, T.A. and Solomon, K.H. (eds) *Management of Farm Irrigation Systems.* Monograph no.9, American Society of Agricultural Engineers, St. Joseph, Michigan.

CRDA (2016) *Kebili En Chiffres.* Rapport interne. Commissariat Régional du Développement Agricole, Kebili, Tunisia.

CRDA (2018) Rapport interne. *Kebili En Chiffres.* Commissariat Régional du Développement Agricole, Kebili, Tunisia.

CRDA (2019) *Kebili En Chiffres.* Rapport interne. Commissariat Régional du Développement Agricole, Kebili, Tunisia.

CRDA (2020) *Kebili En Chiffres.* Rapport interne. Commissariat Régional du Développement Agricole, Kebili, Tunisia.

Deumier, J.M., Leroy, P. and Peyremorte, P. (1995) Des outils pour une maîtrise des systèmes irrigués. In: *La gestion de l'eau dans les exploitations de grande culture. Actes des Journées Techniques Nationales de l'AFEID, Toulouse.* Association Francaise pour l'Etude des Irrigations et du Drainage, Toulouse, France, October 12–13.

Dewidar, A.Z., Ben Abdallah, A., Al-Fuhaid, Y. and Essafi, B. (2015) Lysimeter based water requirements and crop coefficient of surface drip-irrigated date palm in Saudi Arabia. *International Research Journal of Agricultural Science and Soil Science* 5(7), 173–183.

Dhaouadi, L., Ben Maachia, S., Mkademi, C., Oussama, M. and Daghari, H. (2015) Etude comparative des techniques d'irrigations sous palmier dattier dans les oasis de Deguache du Sud Tunisien. *Journal of New Sciences: Agriculture & Biotechnology* 18(3), 658–667.

Dhaouadi, L., Boughdiri, A., Daghari, I., Slim, S., Ben Maachia, S. and Mkadmic, C. (2017) Localised irrigation performance in a date palm orchard in the oases of Deguache. *Journal of New Sciences: Agriculture & Biotechnology* 42(1), 2268–2277.

Dhaouadi, L., Besser, H., Wassar, F., Kharbout, N., Ben Brahim, N. *et al.* (2020a) Agriculture sustainability in arid lands of southern Tunisia: ecological impacts of irrigation water quality and human practices. *Irrigation and Drainage Journal* 69(5), 974–996. DOI: 10.1002/ird.2492.

Dhaouadi, L., Besser, H., Wassar, F., Kharbout, N., Ben Brahim, N. *et al.* (2020b) Comprehension of the kinetics of water in the soil from an irrigation test with a bubbler under date palm tree. *Journal of New Sciences: Agriculture & Biotechnology* 77(6), 4533–4542.

Dhaouadi, L., Besser, H., Karbout, N., Wassar, F. and Al-Omrane, A. (2021a) Assessment of natural resources in Tunisian oases: degradation of irrigation water quality and continued overexploitation of groundwater. *Euro-Mediterranean Journal for Environmental Integration* 6(1), 36. DOI: 10.1007/s41207-020-00234-3.

Dhaouadi, L., Besser, H., Karbout, N., El Khaldi, R., Haj-Amor, Z. *et al.* (2021b) Environmental sensitivity and risk assessment in the Saharan Tunisian oasis agro-systems using the deepest water table source for irrigation: water quality and land management impacts. *Environment, Development and Sustainability* 24, 10695–10727. DOI: 10.1007/s10668-021-01878-z.

FAO (2020) *The State of Food and Agriculture 2020: overcoming water challenges in agriculture.* Food and Agriculture Organization of the United Nations, Rome.

Hamdane, A. (2016) Changement climatique et sécurité alimentaire: cas des oasis de Tunisie. Available at: https://www.researchgate.net/publication/3089694 75_Changement_climatique_et_securite_alimentaire_cas_des_oasis_de_la_Tu nisie (accessed 23 November 2022).

Hosny, S.S., El-Kholy, M.F., Khairy, H. and Madboly, E.A. (2022) Productivity and irrigation water use efficiency of Sewi date palm under different irrigation systems. *European Journal of Agriculture & Food Sciences* 4(1), 27–32. DOI: 10.24018/ejfood.2022.4.1.433.

Ismail, S.M., Al-Qurashi, A.D. and Awad, M.A. (2014) Optimization of irrigation water use, yield, and quality of 'Nabbut-Saif' date palm under dry land conditions. *Irrigation and Drainage* 63(1), 29–37. DOI: 10.1002/ird.1823.

James, D.W., Hanks, R.J. and Jurinak, J.H. (1982) *Modern Irrigated Soils.* Wiley, New York.

Karbout, N., Bol, R., Brahimc, N., Moussa, M. and Bousnina, H. (2019) Applying biochar from date palm waste residues to improve the organic matter, nutrient status and water retention in sandy oasis soils. *Journal of Research in Environmental and Earth Sciences* 07, 203–209.

Maas, E.V. and Hoffman, G.J. (1977) Crop salt tolerance: current assessment (Proceedings Paper 12993). *Journal of the Irrigation and Drainage Division, ASCE* 103(IRZ), 115–134.

Malekian, D., Hayati, A. and Noelle, A. (2017) Conceptualizations of water security in the agricultural sector: perceptions, practices and paradigms. *Journal of Hydrology* 544, 224–232. DOI: 10.1016/j.jhydrol.2016.11.026.

MAW (2006) *Agricultural Statistical Year Book*, Vol. 17 and 19. Ministry of Agriculture and Water, Department of Economic Studies and Statistics, Riyadh.

Mihoub, A., Helimi, S., Mokhtari, S., Kharaz, E., Koull, N. *et al.* (2015) Date palm (*Phoenix dactylifera* L.) irrigation water requirements as affected by salinity in Oued Righ conditions, North Eastern Sahara, Algeria. *Asian Journal of Crop Science* 7(3), 174–185.

Ministry of Environment, Water and Agriculture (2018) *National Strategy for Water 2030.* Riyadh.

Munier, P. (1973) *Le Palmier-Dattier.* G.-P. Maisonneuve & LaRose, Paris.

NCDP (2020) *Annual Report.* National Centre of Dates and Palms, Ministry of Environment, Water and Agriculture, Riyadh.

Zaid, A. and Arias-Jiménez, E.J. (eds) (2002) *Date Palm Cultivation*, FAO Plant Production and Protection Paper no.156 Rev. 1. Food and Agriculture Organization of the United Nations, Rome.

Biofertilizers in Date Palm Cultivation

9

Mohamed Anli[1,2]* , Mohamed Ait-El-Mokhtar[3,4] , Fatima-Zahra Akensous[1,3] , Abderrahim Boutasknit[1,3] , Raja Ben-Laouane[1,3] , Abdessamad Fakhech[3] , Redouane Ouhaddou[1,3] , Ouissame Raho[1,3] and Abdelilah Meddich[1,3]

[1]Center of Agrobiotechnology and Bioengineering, Cadi Ayyad University, Marrakech, Morocco; [2]Department of Biology, Patsy University Center, Comoros University, Comoros; [3]Agro-Food, Biotechnology and Valorization of Plant Bioresources Laboratory, Cadi Ayyad University, Marrakech, Morocco; [4]Laboratory of Biochemistry, Hassan II University of Casablanca, Casablanca, Morocco

Abstract

Abiotic and biotic stresses are significant environmental constraints that have detrimental effects on date palm growth and productivity. Climate change and agricultural mismanagement, including the excessive use of chemical fertilizers and pesticides, have exacerbated these constraints, harmed date palm productivity, and so adversely affected oases ecosystems. The improvement of date palm productivity has become a global concern for a growing population in recent years. To meet this challenge, more investigations on sustainable tools to increase date palm cultivation and production are required. Implementing environmentally friendly strategies such as biofertilizers to improve date palm productivity under extreme conditions is promising. To improve date palm growth and productivity while also increasing its tolerance to environmental stresses, biofertilizers such as arbuscular mycorrhizal fungi (AMF), plant growth-promoting rhizobacteria (PGPR), and organic fertilizers like compost and seaweed extracts (SWE) are used. These biofertilizers may help in the upregulation of date palm protective mechanisms by providing essential mineral nutrients while also improving growth and productivity

*Corresponding author: moh1992anli@gmail.com

© CAB International 2023. *Date Palm* (eds J.M. Al-Khayri *et al.*)
DOI: 10.1079/9781800620209.0009

under harsh conditions. By activating the antioxidant defense system and increasing photosynthetic activity, these biofertilizers may also help date palm to limit the harmful effects of environmental stresses.

Keywords: Abiotic stress, Biofertilizers, Climate change, Date palm productivity, Oasis ecosystem

9.1 Introduction

Due to its highly nutritional fruits and its adaptation to harsh environments, the date palm (*Phoenix dactylifera* L.) is considered one of the main fruit crops in regions with arid and semi-arid climate (Chao and Krueger, 2007). Its cultivation is very old, and it still constitutes an essential food supply for local populations in the Middle East and North Africa. Currently, and in addition to the Middle East and North Africa areas, its production is very widespread in other countries with suitable environments, such as Australia, India, Pakistan, Mexico, southern Africa, South America, and the USA (Chao and Krueger, 2007; Arias *et al.*, 2016). The date palm provides protection to the oasis vegetation against the impacts of the desert since it promotes an adequate microclimate to adjacent crops (Arias *et al.*, 2016). Date fruits are rich in minerals and antioxidants including carotenoids, vitamins, and phenolic compounds, which play a vital role in the human diet and may prevent cancer and cardiovascular disease (Benmeddour *et al.*, 2013; Benidir *et al.*, 2020; Hussain *et al.*, 2020). The degradation of date palm cultivated areas is the result of many natural, human, and living factors as well as urban expansion and the lack of appropriate government policies (Kadhim and Kadhim, 2021). Furthermore, date palm growth and productivity are threatened by both biotic and abiotic factors in recent decades (Baslam *et al.*, 2014; Meddich and Boumezzough, 2017; Al-Khateeb *et al.*, 2019a; Sulaiman *et al.*, 2021). In recent decades, date groves have suffered intense degradation from various environmental stresses including drought, salinity, diseases, and pests. These constraints have negatively affected date palm growth and productivity in many parts of the world, particularly in arid and semi-arid regions (Mirzaee *et al.*, 2014; El Rabey *et al.*, 2016; Saeed *et al.*, 2016; Saleh *et al.*, 2017; Patankar *et al.*, 2018).

In this context, the search for sustainable biological techniques like organic matter supplementation and microorganism inoculation could constitute an eco-friendly way to alleviate date palm environmental stresses. Microbial activities and organic matter supplementation in the rhizosphere may potentially promote plant survival under environmental stressors (Eden *et al.*, 2017; Kaushal, 2019). Plant growth-promoting rhizobacteria (PGPR) and arbuscular mycorrhizal fungi (AMF) may stimulate root development and activity, resulting in enhanced water balance and nutrient absorption during stress periods (Kaushal, 2019; Begum *et al.*, 2020). The use of AMF and PGPR is one of the biological processes to alleviate environmental stresses in plants (Begum *et al.*, 2019; Anli *et al.*, 2020a). AMF and PGPR inoculations mitigate environmental stresses by improving nutrient acquisition (Ullah *et al.*, 2016; Symanczik *et al.*, 2018), water uptake (Quiroga *et al.*, 2017), photosynthetic activity and antioxidant system (Begum *et al.*, 2020), and soil rhizosphere characteristics (Kaushal, 2019; Yaseen *et al.*, 2019; Ben-Laouane *et al.*, 2020; Gałązka *et al.*, 2020). However, AMF and PGPR can themselves be negatively affected by natural stressors (Sandhya *et al.*, 2010). For this reason, the selection of appropriate AMF and PGPR to increase plant performance in drylands is important for plant productivity improvement and rehabilitation of fragile and degraded lands (Meddich *et al.*, 2015; Symanczik *et al.*, 2018; Khan *et al.*, 2019). Several studies reported that native AMF and PGPR species are the most effective in alleviating biotic and abiotic stresses compared to exotic ones (Baslam *et al.*, 2014; Symanczik *et al.*, 2018; Anli *et al.*, 2020a; Ait Rahou *et al.*, 2021).

In addition to AMF and PGPR application, compost and seaweed extracts (SWE) supplements as organic matter sources in soils are a promising eco-friendly approach to improve crop productivity and soil fertility (Debode *et al.*, 2020; Vafa *et al.*, 2021). Recent studies have focused on the use of compost in the restoration of drought-affected soils (Hirich and Jacobsen, 2014; Kanwal *et al.*, 2017). The use of compost and SWE can mitigate the negative effects of environmental stresses in plants through improving soil fertility and productivity (Kanwal *et al.*, 2017; Yakhin *et al.*, 2017; Abbas *et al.*, 2020), increasing mineral element availability and plant growth (Duo *et al.*, 2018; Abd El-Mageed *et al.*, 2019; Anli *et al.*, 2020b; Pačuta *et al.*, 2021), and stimulating various functions such as respiration and photosynthesis as well as the antioxidant defense system (Duo *et al.*, 2018; Pačuta *et al.*, 2021; Puglia *et al.*, 2021).

This chapter provides an in-depth understanding of innovative strategies based on biological inputs including the exploitation of the biological potential of AMF rhizosphere microorganisms, PGPR, and organic fertilizers, as well as their effects on the growth stages and productivity of date palm, their applications and benefits, and the potential mechanisms involved to improve date palm tolerance to environmental stresses. The chapter is articulated first on date palm cultivation and constraints and secondly on biofertilizers, and then provides conclusions and perspectives.

9.2   Date Palm Cultivation and Constraints

9.2.1   Date palm cultivation

P. dactylifera L. is a widely distributed species found in diverse geographic, soil, and climatic zones (El Hadrami and Al-Khayri, 2012). Date palms grow in the hot, arid regions of the world (encompassing the dry desert region of the world between 10 and 39°N in the northern hemisphere and between 7 and 33°51′S in the southern hemisphere) and in the nearly rainless regions at 9–39°N latitude, which are represented by the Sahara and the southern edge of the Near East (Arabian Peninsula, southern Iraq, Jordan) (Ibrahim *et al.*, 2013a). Date palms are traded worldwide as a high-value sweet fruit crop. It is considered an important subsistence crop in most desert areas of the world (Mahmoudi *et al.*, 2008). Beyond arid climates, date palm can also be grown in many other countries for food or as an ornamental plant, including the Americas, southern Europe, Asia, Africa, and Australia. The majority of date-palm-growing areas are in developing or underdeveloped countries where date palm is a major food crop, thus playing a major role in the nutritional status of these communities (Siddiq and Greiby, 2013).

The date palm is both a food source and a cash crop for many farmers (Arias *et al.*, 2016). Its industry is a labor-intensive one providing economic benefits to both male and female farming households. It helps to reduce sprawling migration to cities by creating job opportunities in rural areas (Niazi *et al.*, 2017; Ibrahim and Hamzah, 2019). During propagation (using traditional methods or *in vitro* techniques) and post-production phases (including packaging and marketing), women play a critical role (Arias *et al.*, 2016). The date palm is known to create a microclimate suitable for crops to grow beneath its shade in countries where it is traditionally grown. It shields other crops from extreme heat and overexposure to sunlight. Programs to cultivate date palms have increased awareness of their value and contribution to ecological sustainability and desertification control (Ait-El-Mokhtar *et al.*, 2022a). The date palm was used for the Great Green Wall of Africa initiative, which aimed to combat the negative social, economic, and ecologic effects of desertification as well as land degradation in the Sahel and Sahara. To promote date palm cultivation, the Food and Agriculture Organization of the United Nations (FAO) has responded to member countries' requests for technical assistance by providing regional support in the promotion and introduction of date palm cultivation in small-scale farmer fields; the advancement of the date palm industry; the improved performance, rehabilitation, and modernization of commercial farms; and the development of packaging and processing plants (Arias *et al.*, 2016).

9.2.2 Constraints affecting date palm cultivation

Over the last six decades, human population growth has been among the factors resulting in the deterioration of date palm groves due to intensive exploitation. In addition, the cultivation of date palms is subject to other stressors such as reduced yields and marketing problems. The red palm weevil (*Rhynchophorus ferrugineus*) now menaces date palms grown in the Middle East and North Africa (Al-Dosary *et al.*, 2016; Goldshtein *et al.*, 2020). Furthermore, abiotic factors, such as prolonged episodes of drought caused by a lack of rainfall and wells gone dry, leading to water and soil salinization, continue to impede the expansion of date palm cultivation

(El Rabey *et al.*, 2015; Yaish and Kumar, 2015). In addition, many other issues that are seriously threatening date palm cultivation are essentially the bayoud disease, caused by *Fusarium oxysporum* f. sp. *albedinis*, climate change, the problem of maintenance, the plantation age factor, and a lack of a clear marketing strategy (El-Juhany, 2010).

9.2.2.1 Abiotic stresses

Under field conditions, a multitude of unpredictable adverse conditions can occur at once, affecting the plant's overall performance. Such abiotic factors include drought, salinity, heat, and cold. These act synergistically, leading to a disruption of the photosynthetic apparatus and thereby a decrease in the amount of photosynthetic pigments and damage to the chloroplast membranes, among others. To deal with abiotic stress, date palms regulate their growth and development, depending on the period of contact with the abiotic stress and its intensity (Shareef *et al.*, 2020).

DROUGHT.   Among adverse abiotic stresses, water scarcity stands as the major abiotic stress factor, severely affecting date fruit production (El Rabey *et al.*, 2015). The unpredictability of weather events, such as precipitation and temperature, affects soil quality and crop yields. For the plant to grow and be productive, a suitable supply of water is indispensable. The plant functions through photosynthesis, respiration, and nutrition processes, and water is required for all three. Roots are the first organs to absorb water that is transported to the stems, leaves, and flowers. However, when water becomes limiting, plants undergo water stress, which noticeably affects their growth and development (Fig. 9.1). Detrimental plant–water interactions occur, leading to disturbance of the physiology of the plant. Responses such as a decline in cell water potential, cell turgor, and relative water content take place (Hazzouri *et al.*, 2020). In addition, drought stress has a negative impact on the nutritional content of plants, particularly the date palm (Ghirardo *et al.*, 2021).

SALINITY.   In addition to water stress, salt stress can have a negative impact on date palm production. In a broader sense, soil salinity has become a global concern, affecting nearly 20% of the world's arable land (Al-Khateeb *et al.*, 2019b) (Fig. 9.2). Soil salinity is an important abiotic physical stress that affects the Arabian Peninsula, North Africa, and the Saharan zones, and date palm growth and productivity are significantly reduced because of it (Patankar *et al.*, 2018; Al Kharusi *et al.*, 2019). In arid and semi-arid environments, irrigation-induced salinization of agricultural land is of widespread concern. Plant growth is hampered by high salt levels in the soil and in irrigation water, leading to soil degradation. Research has been done to assess the impact of high soil salinity on date palm growth and productivity (Yaish and Kumar, 2015; Hazzouri *et al.*, 2020; Zhen *et al.*, 2020). Plants in dry and semi-arid climates, on the other hand, are subjected to several stressors at the same time, such

Fig. 9.1. Date palm death induced by severe drought in the oases of southern Morocco. (Photo by Mohamed Ait-El-Mokhtar.)

as salt spray, harsh temperatures, low relative humidity, high winds, and increased evaporation rates. All these variables increase the accumulation of salt in the soil (Al-Mulla *et al.*, 2013; Yaish and Kumar, 2015). Salt tolerance in plants is often age dependent; date palm seedlings are more vulnerable to salt stress than mature plants (Al-Abdoulhadi *et al.*, 2012). Salinity can have a direct impact on nutrient absorption; for example, Na$^+$ can reduce plant K$^+$ uptake (Ait-El-Mokhtar *et al.*, 2019; Ben-Laouane *et al.*, 2020) and can trigger a cascade of complicated interactions that influence plant metabolism. The presence of high levels of Na$^+$ and Cl$^-$ in the soil may inhibit nutrient-ion mobility (Cui *et al.*, 2020). Plants suffer

Fig. 9.2. Saline soil of the oases of southern Morocco. (Photo by Mohamed Ait-El-Mokhtar.)

from ion imbalance and hyperosmotic stress because of high salt stress, which affects homeostasis in water potential, stomatal conductance, and ion distribution, resulting in molecular damage, cell elongation, and even death (Al-Abdoulhadi *et al.*, 2012; Al-Khayri *et al.*, 2018; Hazzouri *et al.*, 2020; Ait-El-Mokhtar *et al.*, 2022a).

SOIL QUALITY.   Despite the fact that the date palm is a salt-tolerant plant, excessive salt concentrations in the soil continue to have a significant impact on its development and productivity. Hence, evaluating the degree of soil salinization is of utmost importance, to set up soil recovery programs to reduce this drastic abiotic stress (Allbed *et al.*, 2014). Another problem influencing soil quality is the humic acid fraction and substances (Ben *et al.*, 2019). Soils in dry zones are often characterized by low water retention and depletion, making them unsuitable for agricultural activities (Khalifa and Yousef, 2015). Another problem that hinders development of date palm plantations is soil desertification. This occurs in arid and semi-arid regions as well as subhumid regions, mainly due to climate change and anthropogenic effects. The date palm's natural ecosystem, which is the desert, is continuously under the detrimental effects of desertification and constantly becomes more fragile from land degradation due to geographic location, making the Mediterranean region and its drylands the most affected on the globe (Mihi *et al.*, 2017).

9.3 Biofertilizers

Over the last few years, the interest in biofertilizers as innovative and eco-friendly tools for a sustainable agriculture has increased considerably. Biofertilizers have the potential to meet the agricultural requirements of an ever-growing population without harming the plant and its ecosystem, contrasting with the detrimental effects of excessive chemical fertilizer usage, and ensuring increased yield year-round, extreme periods included. A plethora of biofertilizers, ranging from humic substances, organic material, SWE, chitin, and chitosan derivatives to beneficial microorganisms such as AMF and PGPR, have been shown to benefit plant growth and yield and improve soil quality (Rouphael and Colla, 2020; Fasusi *et al.*, 2021). Some of these studies are shown in Table 9.1. Biofertilizers hold the potential to improve the overall performance of date palms via multiple mechanisms (Salama *et al.*, 2012; Ibrahim *et al.*, 2013b) (Fig. 9.3).

9.3.1 Compost

9.3.1.1 Effects on plants

Compost is a stable product of organic matter degradation that is rich in nutrients essential to plants. Compost can improve overall plant growth, development, yield, and performance, as well as improving soil structure. Compost has repeatedly been proven to ameliorate both soil health and soil microorganism activity (Adugna, 2016).

Many studies have highlighted the effects of organic amendments on date palm. Anli *et al.* (2021) reported that growth parameters, notably root length, leaf area, and shoot and root dry weights of date palm seedlings, were significantly improved with the application of a grass waste-based compost. Furthermore, they recorded an improvement in nutritional status, notably N, P, and K contents, and physiological and root anatomical traits. In addition, date palm seedlings affected by salt stress exhibited a notable growth improvement and increased P, N, K, and Ca^{2+} uptake through the application of a green waste-based compost (Ait-El-Mokhtar *et al.*, 2022b). Compost also positively affected the photosynthetic efficiency, leaf water potential, stomatal conductance, photosynthetic pigments, metabolites such as sugar and proline, and the antioxidant defense machinery (Ait-El-Mokhtar *et al.*, 2020a).

9.3.1.2 Effects on soil

Compost is known to improve soil quality, giving it a granular structure and increasing its water-holding capacity, helping it resist compaction, and facilitating root penetration in the substrate. In a study conducted by Toubali *et al.* (2020) on saline stress, the application of compost on a soil

Table 9.1. Role of biofertilizers in mitigating cultivation constraints.

Abiotic stress	Type of biofertilizer(s)	Impacts	Reference
Drought	PGPR	Improvement of: • Morphological, physiological, and biochemical parameters • Soil physical and chemical characteristics: N, P, and organic matter	Anli *et al.* (2020a)
Drought	PGPR	Increment of date palm biomass and its resilience to cope with drought damage	Cherif *et al.* (2015)
Drought	AMF	Improvement of biomass, relative water content, and leaf water potential	Meddich *et al.* (2015)
Salt stress	AMF	Boosting K, P, and Ca contents, Ca/Na and K/Na ratios, stomatal conductance, photosynthetic efficiency, leaf water potential, photosynthetic pigments concentration, and protein content	Ait-El-Mokhtar *et al.* (2019)
Salt stress	Compost	Improvement of: • Biomass accumulation • Mineral nutrition: K, P, and Ca contents • Physiology: stomatal conductance, leaf water potential, and photosynthetic pigment concentrations • Antioxidant enzymes production	Ait-El-Mokhtar *et al.* (2022b)
Drought	AMF + PGPR	Activation of antioxidant machinery: SOD, CAT, POX, and GST	Harkousse *et al.* (2021)

Continued

Table 9.1. Continued

Abiotic stress	Type of biofertilizer(s)	Impacts	Reference
Salt stress	AMF + compost	Enhancement of plant growth, stomatal conductance, leaf water potential, antioxidant enzyme activities, N, P, K^+, and Ca^{2+} uptake, proline and soluble sugars contents Decrease of stress markers content linked with membrane peroxidation and ROS production	Ait-El-Mokhtar *et al.* (2020a)
Poor substrate in organic matter	AMF + compost	Increment of: • Plant growth: root biomass • Mineral nutrients: N, P, and K contents • Physiological traits: stomatal conductance and chlorophyll fluorescence • Histological parameters: vascular bundles (number of xylem and phloem), number of sclerenchyma fibers, and endoderm lignification	Anli *et al.* (2021)
Poor substrate in organic matter	AMF + SWE	Improvement of: • Plant growth: root length and leaf area • Mineral nutrients: N, P, K, and Ca contents • Physiological traits: stomatal conductance and chlorophyll fluorescence • Histological parameters: vascular bundles (number of xylem and phloem)	Anli *et al.* (2020b)

Continued

Table 9.1. Continued

Abiotic stress	Type of biofertilizer(s)	Impacts	Reference
Drought	AMF + PGPR + compost	Increment of: • Plant growth: total biomass and leaf area • Physiology: leaf water potential, stomatal conductance, chlorophyll and carotenoids contents, photosynthetic efficiency of PSII • Biochemistry: sugar and protein concentrations, antioxidant activities (polyphenol oxidase and POX) • Mineral nutrition: N and P uptake	Anli *et al.* (2020a)
Salt stress	AMF + PGPR + compost	Improvement of: • Shoot dry weight, plant growth, accumulation of osmotic adjustment compounds, and antioxidant enzyme activity • Soil characteristics: organic carbon and P	Toubali *et al.* (2020)

AMF, arbuscular mycorrhizal fungi; CAT, catalase; GST, glutathione S-transferase; PGPR, plant growth-promoting rhizobacteria; POX, peroxidase; PSII, photosystem II; ROS, reactive oxygen species; SOD, superoxide dismutase.

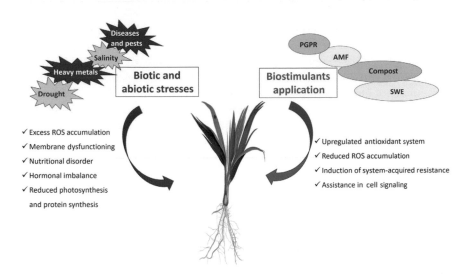

Fig. 9.3. Mitigation strategies for biotic and abiotic stress tolerance in date palm through application of biostimulants. AMF, arbuscular mycorrhizal fungi; PGPR, plant growth-promoting rhizobacteria; ROS, reactive oxygen species; SWE, seaweed extracts. (Prepared by Raja Ben-Laouane.)

characterized by its lack of organic matter and essential nutrients reduced soil pH under saline conditions induced by 120 mM NaCl. Organic matter was increased in soil treated with compost and soil P content was increased under all salinity levels. Moreover, compost application has the capacity to increase soil aggregation, reduce soil pH, and improve soil organic matter and nutrient uptake, especially N, P, K, and Fe in dryland ecosystems (Badawi, 2020; Karbout *et al.*, 2021). In addition, Karbout *et al.* (2021) reported the beneficial effect of compost application as organic fertilizer on oasis soil water retention. They showed also that the improvement in water retention in the soil was linked to an increase in macro and inter-mediate size pores in soils amended with organic fertilizers. Another study found that adding organic fertilizers to the soil improved water retention capacity in a sandy soil and consequently decreased the amount of water needed to irrigate date palms (Badawi, 2020).

9.3.2 Seaweed extracts (SWE)

9.3.2.1 Effects on plants

SWE represent biofertilizers easily absorbed by plants that can improve growth, development, and yield (Khan *et al.*, 2009). Furthermore, such beneficial effects have been proven over time on different economic crops of interest, ranging from cereals to legumes to vegetables (Ali *et al.*, 2021).

In a study conducted on date palm seedlings of Boufgouss variety, it was revealed that a brown seaweed extract of *Fucus spiralis* and an exotic AMF strain (*Rhizophagus irregularis*), applied alone or in combination, significantly improved the growth of date palm seedlings, for instance leaf area and root elongation (Anli *et al.*, 2020b). Furthermore, *F. spiralis* extract increased mineral nutrient content and improved physiological traits and histological parameters, suggesting better growth and development of date palms (Anli *et al.*, 2020b). Another study in date palms reported the positive impact of SWE on plant growth and physiological traits, especially leaf dry biomass and photosynthetic pigment concentration (Taha and Abood, 2018).

9.3.2.2   Effects on soil

Because seaweed extracts contain plant growth-stimulating compounds, they have been used as tools to amend soils for better crop production on a large scale. Seaweed extracts are rich in organic matter and can supplement essential nutrients for better growth and development of the plant, leading to a better production (Khan *et al.*, 2009; Anli *et al.*, 2020b). They play a key role in influencing the rhizospheric soil health status by improving its structure and enhancing the beneficial microbiome for the plants (Ali *et al.*, 2021).

9.3.3   Plant growth-promoting rhizobacteria (PGPR)

9.3.3.1   Effects on plants

PGPR have proven to be effective bioinoculants as an alternative to chemical inputs to ensure good, safe, and sustainable agriculture (Ferjani *et al.*, 2015; Zhang *et al.*, 2019; Abdel Latef *et al.*, 2020; Anli *et al.*, 2020a). Due to their unique and beneficial traits, these microorganisms can significantly boost growth performance of many agricultural crops through direct and/or indirect effects on phytohormone production (El-Sharabasy and Ragab, 2009), aminocyclopropane-1-carboxylic acid (ACC) deaminase (Barnawal *et al.*, 2012; Meena and Meena, 2017), organic matter mineralization and degradation, and improved bioavailability of mineral nutrients such as Fe and P (Valencia-Cantero *et al.*, 2007; Berg, 2009). Bacteria adhering to the roots of date palms have been reported to contribute to the improvement of biomass as well as the protection of the palm from environmental stress (Anli *et al.*, 2020a; Harkousse *et al.*, 2021). Toubali *et al.* (2020) showed that application of PGPR combined with other biofertilizers such as AMF increased growth parameters of date palm tissue culture-derived plants. A recent study showed that inoculation of date palm seeds with PGPR (*Pseudomonas fluorescens* and *Bacillus megaterium*) stored at 4°C allowed the bacteria to settle in the rhizosphere and to carry out their beneficial functions to promote plant growth (Thwaini *et al.*, 2021). El-Sharabasy *et al.* (2018) revealed that four bacteria strains had the ability to promote date palm growth and increase photosynthetic pigments and mineral nutrients uptake. Moreover, Anli *et al.* (2020a) demonstrated that inoculation of date palm plants with native PGPR significantly promoted AMF root colonization compared to uninoculated plants. To improve vegetative growth, PGPR bacteria can improve nutrient uptake in the date palm and other crops and have the power to stimulate enzymes in the plant's antioxidant system to cope with abiotic stresses (Anli *et al.*, 2020a; Ben-Laouane *et al.*, 2020; Toubali *et al.*, 2020).

9.3.3.2 Effects on soil

Soil bacteria (PGPR) have long been widely used in the rhizosphere of plants for their productivity through the improvement of soil quality. Plant–bacteria relationships in the rhizosphere influence plant development and health as well as soil fertility and productivity (Yaseen *et al.*, 2019; Anli *et al.*, 2020a). These microorganisms can, for example, limit the development of many plant pathogens in the soil by restricting access to Fe through siderophore production and consequently reducing Fe-dependent spore germination (Valencia-Cantero *et al.*, 2007; Ribeiro and Cardoso, 2012; Mouloud *et al.*, 2017). They can regulate plant mineral nutrients uptake by solubilizing inaccessible nutrients, namely P and N, in the soil (Nadeem *et al.*, 2014; Yu *et al.*, 2019). Various studies have found that PGPR improve soil organic matter and mineral nutrient levels in the rhizosphere via diverse mechanisms including atmospheric nitrogen (N_2) fixation by symbiosis, siderophore and exopolysaccharides (EPS) production, and phosphate and K solubilization (Bhattacharyya and Jha, 2012; Sharma *et al.*, 2013). Baumert *et al.* (2018) demonstrated that plant root exudates have a positive effect on the prevailing microbial community in the rhizosphere, which facilitates biotic macroaggregation. Recently a study was carried out on date palms in which PGPR treatment improved soil organic matter and total organic carbon, which could have been related to the PGPR's ability to metabolize several compounds secreted by the palm root system including organic acids and carbohydrates (Anli *et al.*, 2020a). After harvesting date palm plants inoculated with a PGPR consortium, Toubali *et al.* (2020) observed an improvement in the soil physicochemical characteristics by increasing the organic matter and P content. Previous works indicated that PGPR inoculation combined with biochar amendment increased nitrate, total K, and water content in the soil (Ren *et al.*, 2020), as well as the richness and diversity of soil bacteria (Ju *et al.*, 2019). According to Ferjani *et al.* (2015), the rhizosphere of date palms constitutes a reservoir of PGPR that influences soil characteristics, thereby maintaining its sustainability. They also act as beneficial agents of heavy metal phytoremediation in polluted ecosystems (Ghadbane *et al.*, 2021).

9.3.4 Arbuscular mycorrhizal fungi (AMF)

9.3.4.1 Effects on plants

AMF provide many advantages including increased plant growth, nutrient absorption, water uptake, and tolerance to such stresses as temperature variability and heavy metal toxicity. They also play a crucial role in the stable soil aggregate formation, building a porous soil structure that allows water absorption and air circulation and prevents soil erosion and degradation. These advantages give AMF the potential capacity to boost agricultural productivity and make them essential for sustainable agricultural ecosystem functioning. Inoculation of date palm seedlings with AMF results in an increase in their growth (Anli *et al.*, 2021). Similarly, Khaliel *et al.* (1994) showed that P and K nutritional status was improved in mycorrhized plants compared to uninoculated seedlings. While desert soils are generally poor in N, P, and K, these ions were increased in the presence of AMF (Al-Karaki, 2006). AMF inoculation of date palm seedlings in combination with other organic practices proved to benefit seedlings even further. Date palm seedlings transplanted into a compost-adjusted substrate treated with AMF had higher root and shoot biomass, photosynthetic activity, and nutrient concentrations compared to seedlings planted with or without compost (Anli *et al.*, 2021). As a result, this inoculum strengthened the compost's stimulating function and was effective in colonizing the root system. Furthermore, date palm AMF co-inoculation with PGPR strains like *Pseudomonas striata* and *Bacillus subtilis* significantly improved growth parameters such as root elongation and dry weight, and leaf elongation and dry weight, in addition to increased plant P uptake (Zougari-Elwedi *et al.*, 2019). Date palms inoculated with AMF combined with SWE exhibited increased dry biomass and leaf area, increased stomatal conductance, and increased N, P, K, and Ca concentrations more than either treatment alone (Anli *et al.*, 2020b). The same study also revealed an increase in chlorophyll fluorescence (Fv/Fm) and number of sclerenchyma fibers and vascular bundles. The AMF protective impact against abiotic stresses on inoculated date palms is probably the most important one, since date palms generally grow in harsh environments and under constraints such as water scarcity and salinity. Under water stress, AMF increased the biomass production of date palms and allowed the maintenance of a high level of leaf water potential (Anli *et al.*, 2020a). Physiological parameters including stomatal conductance and water potential were enhanced by AMF application (Anli *et al.*, 2020a). In a recent study, Harkousse *et al.* (2021) reported that water status parameters were considerably better in plants treated with AMF + PGPR. In response to extreme water stress, the inoculated plants had significantly decreased superoxide dismutase (SOD), catalase (CAT), peroxidase (POX), and glutathione *S*-transferase (GST) activity (Harkousse *et al.*, 2021). In another study, under water stress, co-inoculation with AMF and PGPR increased plant biomass, leaf water potential, stomatal conductance, photosynthetic pigments (chlorophyll and carotenoid concentrations), and photosynthetic efficiency. Co-inoculation also considerably increased sugar and protein concentrations and activities of antioxidant enzymes (polyphenol oxidase and POX). P absorption was increased in date palms that had been treated with AMF and/or PGPR compared to untreated controls, both under well-watered and drought stress conditions (Anli *et al.*, 2020a). On the other hand, AMF showed a positive effect on date palms under salt stress, increasing the aerial dry weight and plant height (Ait-El-Mokhtar *et al.*, 2019). In combination with PGPR and compost, AMF improved shoot dry weight, plant growth and development, osmotic adjustment compounds accumulation, and antioxidant enzymes activity (Toubali *et al.*, 2020). AMF were also found to reduce salinity-induced decreases in K, P, and Ca contents, improve Ca/Na and K/Na ratios, and improve physiological parameters by increasing stomatal conductance, Fv/Fm, relative water content, leaf water potential, and photosynthetic pigment and protein concentrations (Ait-El-Mokhtar *et al.*, 2019). In another study and under saline conditions, the application of AMF alone and combined with compost significantly increased date palm seedling growth and vigour, Fv/Fm, leaf water potential, stomatal conductance, antioxidant enzyme activity, proline and soluble sugar concentrations, and P, K^+, N, and Ca^{2+} uptake (Ait-El-Mokhtar *et al.*, 2020a). The same authors reported, however, decreased Na^+ and Cl^- absorption and oxidative stress marker concentrations.

9.3.4.2 *Effects on soil*

Because of their ability to influence plant growth both indirectly and directly, AMF are crucial in improving soil mineral availability and soil structure (Elias *et al.*, 2017). AMF serve as biofertilizers to improve plant nutrients such as N, P, K, Ca, S, and micronutrients, preventing nutrient loss and improving soil cation exchange capacity under drylands conditions (Chen *et al.*, 2017; Ait-El-Mokhtar *et al.*, 2019; Anli *et al.*, 2021). AMF inoculation can boost soil aggregation and carbon sequestration as well as N, P, K, and Fe availability in arid and semi-arid regions (Ben-Laouane *et al.*, 2020; Harkousse *et al.*, 2021). The purpose of AMF inoculation is to increase water-holding capacity and improve soil structure, nutrient availability, and the living conditions for other soil organisms like PGPR that are necessary for plant development (Nadeem *et al.*, 2014). In addition, AMF are known as a key biological component in improving soil fertility, productivity, water-holding capacity, and microbe populations (Nadeem *et al.*, 2014; Coutinho *et al.*, 2019).

9.3.5   Role of biofertilizers in mitigating cultivation constraints

Many studies have reported that biotic stresses (diseases and pests of date palm), climate-related abiotic stresses (heat, drought, flooding), and soil-related abiotic stresses (salinity, heavy metals, fertility loss) hinder normal plant growth and development (Saleh *et al.*, 2017; Raza *et al.*, 2019). Climatic changes are exacerbating the problem, as are poor farming practices such as overuse of inorganic fertilizers and pesticides (Liu *et al.*, 2017). Agricultural output is reduced, and the ecosystem deteriorates as a result (Begum *et al.*, 2019). Thus, innovative management strategies that are environmentally friendly to the soil and its surrounding ecosystem are needed. Biofertilizers such as PGPR, AMF, compost, and SWE have been effectively utilized to sustain plant growth and development in stressful environments (Abo-Kora and Maie Mohsen, 2016; Van Oosten *et al.*, 2017; Ben-Laouane *et al.*, 2020; Basu *et al.*, 2021; Del Buono, 2021).

Biofertilizers play a crucial role in alleviating environmental stressors in different trees, including date palms, through several physiological, biochemical, and molecular mechanisms (Ait-El-Mokhtar *et al.*, 2019, 2020a, b; Anli *et al.*, 2020a). Biofertilizers enhance plant nutrient availability by increasing mineral nutrient mobility in the rhizosphere (Schütz *et al.*, 2018); releasing organic acids, signaling compounds, and enzymes (Del Buono, 2021); and inducing antioxidant machinery and modulation of plant stress markers and osmolyte synthesis (Van Oosten *et al.*, 2017). They also have an important role in bioremediation technologies. For example, the coexistence of microorganisms like bacteria and fungi in the soil rhizosphere zone can actively sequester metals in contaminated soils and serve as growth-promoting bioinoculants for crops (Mahmud *et al.*, 2021).

When plants are treated with biofertilizers under drought stress, their physiological and biochemical statuses change as an adaptive mechanism. Inoculation with AMF changes plant water status by enhancing hydraulic conductivity via upregulation of root aquaporin (AQP) genes, activating osmolyte synthesis or hormonal signaling including ABA (abscisic acid)-mediated stomatal conductance, jasmonic acid, and strigolactones (Xie *et al.*, 2018). Biofertilizers like PGPR boost the production of phytohormones and the synthesis of compounds including auxins, cytokinins, gibberellins, and ABA (Kaushal, 2019; Ilyas *et al.*, 2020). As a result, plants witnessed an improvement in root elongation and architecture, growth, and lateral and fine root formation for increased plant water uptake (Raza *et al.*, 2019).

A high quantity of salt in the soil of arid and semi-arid regions affects plant agro-physiological processes and is classified as osmotic, ionic, and oxidative stress (Ben-Laouane *et al.*, 2020). Osmotic stress reduces water-use efficiency while ionic stress interrupts a plant's ionic balance (Ait-El-Mokhtar *et al.*, 2022b). By accelerating the release of O_2, oxidative stress hinders plant cell development and metabolism (Islam *et al.*, 2015).

However, biofertilizers have been efficiently used in several kinds of studies to increase plant growth and development under high-saline environments (Upadhyay *et al.*, 2012). Anli *et al.* (2020a) postulated that date palm plants inoculated with PGPR accumulate osmolytes such as sugars, which elicit both agro-physiological and developmental changes. The PGPR roles are complemented either directly by the activities of other microbes (AMF), or indirectly via the improvement of soil physicochemical conditions (compost application), which can enhance the established relationships between plants and microbes (Upadhyay *et al.*, 2012). According to Ben-Laouane *et al.* (2020), endophytic microorganisms in plant roots, such as AMF and bacteria, significantly stimulate host defense in response to salinity.

9.3.6 Advantages and disadvantages of biofertilizers

The application of biofertilizers in agriculture presents more advantages than disadvantages. On the advantages side, biofertilizers represent environmentally friendly assets that are cost-efficient, promising sustainability of agriculture and constituting natural renewable resources to be exploited. Furthermore, these beneficial tools have the potential to reduce dependence on chemical fertilizers and pesticides, which reduces the environmental cost. In addition, they have been proven, time and again, to be soil-restoring agents and nutrient providers. On the negative side, among the major disadvantages is a lack of technology to produce these biofertilizers (Raimi *et al.*, 2021).

9.3.7 Impact of biofertilizers on the environment

The increase in world population has led to an overexploitation of natural resources and an alarming degradation of agricultural land worldwide (Gomiero, 2016). The massive use of inorganic fertilizers may increase agricultural production and ensure food security; however, as a result, soils become fragile and lose their organic matter, and water resources may become contaminated. The loss of organic matter decreases soil fertility and increases plant susceptibility to nutritional imbalances, thus affecting crop productivity (Mainville *et al.*, 2006). Therefore, it is essential to develop economically and ecologically profitable, efficient, and innovative farming practices. Integrated agriculture based on the use of efficient biofertilizers has become a potential alternative.

Agricultural practices based on the use of biofertilizers may not only assure food security by giving adequate nutrients for plant growth, but also increase biodiversity and restore degraded soils (Bhardwaj *et al.*, 2014; Schütz *et al.*, 2018). Thus, the utilization of biofertilizers is a cost-effective and environmentally friendly option. The use of organic amendments is critical for the bioconversion of organic wastes, the enhancement of nutrient availability, and the physical, chemical, and biological characteristics of the soil (Gaiotti *et al.*, 2017; Chaichi *et al.*, 2018). PGPR can reduce the use of chemical fertilizers and enhance nutrient availability in rhizosphere soil by fixing N_2 and solubilizing inorganic P and K^+ (Tan *et al.*, 2014; Kaushal, 2019), resulting in their availability and absorption by plants. Furthermore, they can fix nutrients and prevent them from leaching out of the soil profile (Vejan *et al.*, 2016). Intracellular PGPR, which have the capacity for symbiotic nitrogen fixation, may further decrease the usage of commercial N fertilizers with the support of legumes and actinorhizal plants. On the other hand, PGPR can increase the soil quality by secreting EPS, which aggregates soil particles (Naseem *et al.*, 2018). A considerable increase in soil aggregation around roots was reported in the rhizosphere of maize inoculated with an EPS-producing PGPR strain (Awad *et al.*, 2012). The secretion of this substance plays a critical role in toxic-ion exclusion and nutrient uptake by plants (Mahmud *et al.*, 2021).

A previous study reported that AMF applied as natural fertilizers enhanced phosphate fertilizer performance, resulting in a reduction in input and the use of cost-effective fertilizers (Zhang *et al.*, 2016). Another study revealed that these fungi can make P available to plants by solubilizing it through the action of specific enzymes including phosphatase (Singh *et al.*, 2020). Furthermore, AMF can play an important role in soil restoration, structure, and quality. They can directly impact the soil and root system architecture by regulating pH and electrical conductivity and enhancing soil structure, aggregation, and root biomass production (Rillig *et al.*, 2015). Due to their mycelial network's extent and thickness, AMF can access the pores containing microaggregates and form stable macroporous aggregates, which allow air and water infiltration and prevent soil erosion (Soka and Ritchie, 2014). In addition, glycoproteins (glomalin) secreted by AMF into the soil may play a crucial role in this process. Glomalin, a hydrophobic and thermotolerant glycoprotein, is slowly biodegraded by soil microorganisms. Glomalin binding to clay particles allows a strong cohesion, more resistant to biological decomposition, that safeguards soil structural stability (Pal, 2014).

The association between microbial inoculation and the supply of organic amendments has shown great importance in agriculture, improving the proliferation of these microorganisms and the establishment of symbiotic relationships with date palm (Anli *et al.*, 2020a). A recent study of Ait-El-Mokhtar *et al.* (2020a) showed that application of compost and AMF alone or in combination alleviated the salinity impact and improved date palm growth and health (Fig. 9.4). Furthermore, Anli *et al.* (2021) reported that application of compost at low dose (5% w/w) increased the number of vascular bundles (xylem and phloem) and number of sclerenchyma fibers, while the combination of compost with AMF significantly improved the number of sclerenchyma fibers and the endoderm lignification (Fig. 9.5). This type of interaction seems to be very important in restoring degraded soils and increasing crop productivity in sustainable agriculture.

Fig. 9.4. State of date palm growth under nonsaline (0 mM NaCl) and saline (240 mM NaCl) conditions after application of compost and arbuscular mycorrhizal fungi (AMF) alone or in combination, 14 months after germination. (Modified from Ait-El-Mokhtar *et al.*, 2020a.)

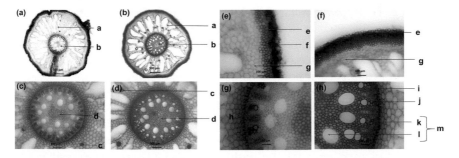

Fig. 9.5. Histological cross-sections of date palm seedling roots after 4 months of growth in the greenhouse. (a) Entire section of control plant; (b) entire section of date palm treated with compost + AMF (*Rhizoglomus irregulare*); (c) central cylinder and part of the control bark; (d) central cylinder and part of the bark of date palm treated with compost + AMF; (e) part of the bark of control; (f) part of the bark of date palm amended with compost + AMF; (g) part of the central cylinder and part of the bark of control; and (h) part of the central cylinder and part of the bark of date palm treated with compost + AMF. a, bark; b, central cylinder; c, sclerenchyma fiber; d, sclerenchyma; e, ectoderm; f, suberoid; g, cortical parenchyma; h, phloem; i, endoderm; j, pericycle; k, protoxylem; l, metaxylem; m, xylem. (Modified from Anli *et al.*, 2021.)

The aforementioned works suggest the capacity of natural biofertilizers to enhance soil structure and increase nutrient uptake, thus minimizing agriculture pollution and fertilizer cost. To ensure modern and sustainable agriculture, the development of biofertilizers will be a most fundamental factor.

9.3.8 Economic feasibility of biofertilizers

Although biofertilizers offer a sustainable solution benefiting both agricultural soils and crop production, their adoption in the agro-market is not yet common, especially in developing countries. Hence, commercialization of good-quality biofertilizers at a reasonable price that does not exceed that of standard fertilizers is a priority. Applied technologies such as fermenters, carrier material for the preparation of biofertilizers, and storage temperatures should seriously be considered (Raimi *et al.*, 2021).

9.4 Conclusions and Prospects

This chapter has highlighted the importance of date palm cultivation in different environments and the positive impact of biofertilizers on growth and vigor. All the research shows that the growth and productivity of date palm can be improved by the application of biofertilizers, especially AMF, PGPR, organic fertilizers, and SWE. It also demonstrates that their use in date palm cultivation can improve welfare in arid and semi-arid areas, where farmers face many constraints such as drought, salinity, and soils poor in organic matter and mineral nutrients. However, the understanding of date palm tolerance mechanisms to abiotic stresses is still superficial. Therefore, further studies are needed from a molecular point of view to understand the adaptation strategies of date palm to extreme conditions. In addition, studies on the application of biostimulants on date palm under metal stress conditions are very scarce.

Despite the fact that biofertilizers have been shown to improve date palm development and production, further studies are needed to address a number of problems: (i) the molecular mechanism(s) responsible for the stimulatory effects; (ii) the practical approach, timing, application rates, and phenological stages to stimulate palm productivity under normal and stress conditions; (iii) overall understanding of the complexity of the date palm–biofertilizer–abiotic stress interaction; and (iv) selection of the appropriate biofertilizers, as effects can vary considerably between date palm species from one ecosystem to another.

References

Abbas, M., Anwar, J., Zafar-ul-Hye, M., Iqbal Khan, R., Saleem, M. *et al.* (2020) Effect of seaweed extract on productivity and quality attributes of four onion cultivars. *Horticulturae* 6(2), 28. DOI: 10.3390/horticulturae6020028.

Abd El-Mageed, T.A., El-Sherif, A.M.A., Abd El-Mageed, S.A. and Abdou, N.M. (2019) A novel compost alleviate drought stress for sugar beet production grown in Cd-contaminated saline soil. *Agricultural Water Management* 226, 105831. DOI: 10.1016/j.agwat.2019.105831.

Abdel Latef, A.A.H., Abu Alhmad, M.F., Kordrostami, M., Abo–Baker, A.-B.A.-E. and Zakir, A. (2020) Inoculation with *Azospirillum lipoferum* or *Azotobacter chroococcum* reinforces maize growth by improving physiological activities under saline conditions. *Journal of Plant Growth Regulation* 39(3), 1293–1306. DOI: 10.1007/s00344-020-10065-9.

Abo-Kora, H.A. and Maie Mohsen, M.A. (2016) Reducing effect of soil salinity through using some strains of nitrogen fixers bacteria and compost on sweet basil plant. *International Journal of PharmTech Research* 9, 187–214.

Adugna, G. (2016) A review on impact of compost on soil properties, water use and crop productivity. *Agricultural Science Research Journal* 4, 93–104.

Ait-El-Mokhtar, M., Laouane, R.B., Anli, M., Boutasknit, A., Wahbi, S. *et al.* (2019) Use of mycorrhizal fungi in improving tolerance of the date palm (*Phoenix dactylifera* L.) seedlings to salt stress. *Scientia Horticulturae* 253, 429–438. DOI: 10.1016/j.scienta.2019.04.066.

Ait-El-Mokhtar, M., Baslam, M., Ben-Laouane, R., Anli, M., Boutasknit, A. *et al.* (2020a) Alleviation of detrimental effects of salt stress on date palm (*Phoenix dactylifera* L.) by the application of arbuscular mycorrhizal fungi and/or compost. *Frontiers in Sustainable Food Systems* 4, 131. DOI: 10.3389/fsufs.2020.00131.

Ait-El-Mokhtar, M., Fakhech, A., Anli, M., Ben-Laouane, R., Boutasknit, A. *et al.* (2020b) Infectivity of the palm groves arbuscular mycorrhizal fungi under arid and semi-arid climate and its edaphic determinants towards efficient ecological restoration. *Rhizosphere* 15, 100220. DOI: 10.1016/j.rhisph.2020.100220.

Ait-El-Mokhtar, M., Boutasknit, A., Ben-Laouane, R., Anli, M., El Amerany, F. *et al.* (2022a) Vulnerability of oasis agriculture to climate change in Morocco. In: Karmaoui, A. (ed.) *Impacts of Climate Change on Agriculture and Aquaculture*. IGI Global, Hershey, Pennsylvania, pp. 76–106.

Ait-El-Mokhtar, M., Fakhech, A., Anli, M., Ben-Laouane, R., Boutasknit, A. *et al.* (2022b) Compost as an eco-friendly alternative to mitigate salt-induced effects on growth, nutritional, physiological and biochemical responses of date palm. *International Journal of Recycling Organic Waste in Agriculture* 11, 85–100.

Ait Rahou, Y., Ait-El-Mokhtar, M., Anli, M., Boutasknit, A., Ben-Laouane, R. *et al.* (2021) Use of mycorrhizal fungi and compost for improving the growth and yield of tomato and its resistance to *Verticillium dahliae*. *Archives of Phytopathology and Plant Protection* 54(13–14), 665–690. DOI: 10.1080/03235408.2020.1854938.

Al-Abdoulhadi, A., Dinar, H.A., Ebert, G. and Büttner, C. (2012) Influence of salinity levels on nutrient content in leaf, stem and root of major date palm (*Phoenix dactylifera* L) cultivars. *International Research Journal of Agricultural Science and Soil Science* 2, 341–346.

Al-Dosary, N., Al-Dobai, S. and Faleiro, J. (2016) Review on the management of red palm weevil *Rhynchophorus ferrugineus* Olivier in date palm *Phoenix dactylifera* L. *Emirates Journal of Food and Agriculture* 28(1), 34–44. DOI: 10.9755/ejfa.2015-10-897.

Ali, O., Ramsubhag, A. and Jayaraman, J. (2021) Biostimulant properties of seaweed extracts in plants: implications towards sustainable crop production. *Plants* 10(3), 531. DOI: 10.3390/plants10030531.

Al-Karaki, G.N. (2006) Nursery inoculation of tomato with arbuscular mycorrhizal fungi and subsequent performance under irrigation with saline water. *Scientia Horticulturae* 109(1), 1–7. DOI: 10.1016/j.scienta.2006.02.019.

Al Kharusi, L., Sunkar, R., Al-Yahyai, R. and Yaish, M.W. (2019) Comparative water relations of two contrasting date palm genotypes under salinity. *International Journal of Agronomy* 2019, 1–16. DOI: 10.1155/2019/4262013.

Al-Khateeb, S.A., Al-Khateeb, A.A., El-Beltagi, H.S. and Sattar, M.N. (2019a) Genotypic variation for drought tolerance in three date palm (*Phoenix dactylifera* L.) cultivars. *Fresenius Environmental Bulletin* 28, 4671–4683.

Al-Khateeb, S.A., Al-Khateeb, A.A., Sattar, M.N., Mohmand, A.S. and El-Beltagi, H.S. (2019b) Assessment of somaclonal variation in salt-adapted and non-adapted regenerated date palm (*Phoenix dactylifera* L). *Fresenius Environmental Bulletin* 28, 3686–3695.

Al-Khayri, J.M., Naik, P.M., Jain, S.M. and Johnson, D.V. (2018) Advances in date palm (*Phoenix dactylifera* L.) breeding. In: Al-Khayri, J.M., Jain, S.M. and Johnson, D.V. (eds) *Advances in Plant Breeding Strategies: Fruits*. Springer, Cham, Switzerland, pp. 727–771.

Allbed, A., Kumar, L. and Aldakheel, Y.Y. (2014) Assessing soil salinity using soil salinity and vegetation indices derived from IKONOS high-spatial resolution imageries: applications in a date palm dominated region. *Geoderma* 230–231, 1–8. DOI: 10.1016/j.geoderma.2014.03.025.

Al-Mulla, L., Bhat, N.R. and Khalil, M. (2013) Salt-tolerance of tissue-cultured date palm cultivars under controlled environment. *International Journal of Biological, Biomolecular, Agricultural, Food and Biotechnological Engineering* 7, 468–471.

Anli, M., Baslam, M., Tahiri, A., Raklami, A., Symanczik, S. *et al.* (2020a) Biofertilizers as strategies to improve photosynthetic apparatus, growth, and drought stress tolerance in the date palm. *Frontiers in Plant Science* 11, 516818. DOI: 10.3389/fpls.2020.516818.

Anli, M., Kaoua, M.E., Ait-el-Mokhtar, M., Boutasknit, A., Ben-Laouane, R. *et al.* (2020b) Seaweed extract application and arbuscular mycorrhizal fungal inoculation: a tool for promoting growth and development of date palm (*Phoenix dactylifera* L.) cv «Boufgous». *South African Journal of Botany* 132, 15–21. DOI: 10.1016/j.sajb.2020.04.004.

Anli, M., Symanczik, S., El Abbassi, A., Ait-El-Mokhtar, M., Boutasknit, A. *et al.* (2021) Use of arbuscular mycorrhizal fungus *Rhizoglomus irregulare* and compost to improve growth and physiological responses of *Phoenix dactylifera* 'Boufgouss. *Plant Biosystems - An International Journal Dealing with All Aspects of Plant Biology* 155(4), 763–771. DOI: 10.1080/11263504.2020.1779848.

Arias, E., Hodder, A. and Oihabi, A. (2016) FAO support to date palm development around the world: 70 years of activity. *Emirates Journal of Food and Agriculture* 28(1), 1–11. DOI: 10.9755/ejfa.2015-10-840.

Awad, N.M., Turky, A.S., Abdelhamid, M.T. and Attia, M. (2012) Ameliorate of environmental salt stress on the growth of *Zea mays* L. plants by exopolysaccharides producing bacteria. *Journal of Applied Sciences Research* 8, 2033–2044.

Badawi, M.A. (2020) Irrigation water saving of date palm tree plantations using soil amendments in UAE. *International Journal of Sustainable Development and Science* 3(2), 1–27. DOI: 10.21608/ijsrsd.2020.120513.

Barnawal, D., Bharti, N., Maji, D., Chanotiya, C.S. and Kalra, A. (2012) 1-Aminocyclopropane-1-carboxylic acid (ACC) deaminase-containing rhizobacteria protect *Ocimum sanctum* plants during waterlogging stress via reduced ethylene generation. *Plant Physiology and Biochemistry* 58, 227–235. DOI: 10.1016/j.plaphy.2012.07.008.

Baslam, M., Qaddoury, A. and Goicoechea, N. (2014) Role of native and exotic mycorrhizal symbiosis to develop morphological, physiological and biochemical responses coping with water drought of date palm, *Phoenix dactylifera*. *Trees* 28(1), 161–172. DOI: 10.1007/s00468-013-0939-0.

Basu, A., Prasad, P., Das, S.N., Kalam, S., Sayyed, R.Z.. *et al.* (2021) Plant growth promoting rhizobacteria (PGPR) as green bioinoculants: recent developments, constraints, and prospects. *Sustainability* 13(3), 1140. DOI: 10.3390/su13031140.

Baumert, V.L., Vasilyeva, N.A., Vladimirov, A.A., Meier, I.C., Kögel-Knabner, I. *et al.* (2018) Root exudates induce soil macroaggregation facilitated by fungi in subsoil. *Frontiers in Environmental Science* 6, 140. DOI: 10.3389/fenvs.2018.00140.

Begum, N., Qin, C., Ahanger, M.A., Raza, S., Khan, M.I. *et al.* (2019) Role of arbuscular mycorrhizal fungi in plant growth regulation: implications in abiotic stress tolerance. *Frontiers in Plant Science* 10, 1068. DOI: 10.3389/fpls.2019.01068.

Begum, N., Ahanger, M.A. and Zhang, L. (2020) AMF inoculation and phosphorus supplementation alleviates drought induced growth and photosynthetic decline in *Nicotiana tabacum* by up-regulating antioxidant metabolism and osmolyte accumulation. *Environmental and Experimental Botany* 176, 104088. DOI: 10.1016/j.envexpbot.2020.104088.

Ben, H., Mahmoud, B., Chaker, R., Rigane, H., Maktouf, S. *et al.* (2019) Change of soil quality based on humic acid with date palm compost incorporation. *International Journal of Recycling of Organic Waste in Agriculture* 8(3), 317–324. DOI: 10.1007/s40093-019-0254-x.

Benidir, M., El Massoudi, S., El Ghadraoui, L., Lazraq, A., Benjelloun, M. *et al.* (2020) Study of nutritional and organoleptic quality of formulated juices from jujube (*Ziziphus lotus* L.) and dates (*Phoenix dactylifera* L.) fruits. *The Scientific World Journal* 2020, 9872185. DOI: 10.1155/2020/9872185.

Ben-Laouane, R., Baslam, M., Ait-El-Mokhtar, M., Anli, M., Boutasknit, A. *et al.* (2020) Potential of native arbuscular mycorrhizal fungi, rhizobia, and/or green compost as alfalfa (*Medicago sativa*) enhancers under salinity. *Microorganisms* 8(11), 1695. DOI: 10.3390/microorganisms8111695.

Benmeddour, Z., Mehinagic, E., Meurlay, D.L. and Louaileche, H. (2013) Phenolic composition and antioxidant capacities of ten Algerian date (*Phoenix dactylifera* L.) cultivars: a comparative study. *Journal of Functional Foods* 5(1), 346–354. DOI: 10.1016/j.jff.2012.11.005.

Berg, G. (2009) Plant-microbe interactions promoting plant growth and health: perspectives for controlled use of microorganisms in agriculture. *Applied Microbiology and Biotechnology* 84(1), 11–18. DOI: 10.1007/s00253-009-2092-7.

Bhardwaj, D., Ansari, M.W., Sahoo, R.K. and Tuteja, N. (2014) Biofertilizers function as key player in sustainable agriculture by improving soil fertility, plant tolerance and crop productivity. *Microbial Cell Factories* 13, 66. DOI: 10.1186/1475-2859-13-66.

Bhattacharyya, P.N. and Jha, D.K. (2012) Plant growth-promoting rhizobacteria (PGPR): emergence in agriculture. *World Journal of Microbiology & Biotechnology* 28(4), 1327–1350. DOI: 10.1007/s11274-011-0979-9.

Chaichi, W., Djazouli, Z., Zebib, B. and Merah, O. (2018) Effect of vermicompost tea on faba bean growth and yield. *Compost Science & Utilization* 26(4), 279–285. DOI: 10.1080/1065657X.2018.1528908.

Chao, C.T. and Krueger, R.R. (2007) The date palm (*Phoenix dactylifera* L.): overview of biology, uses, and cultivation. *HortScience* 42(5), 1077–1082. DOI: 10.21273/HORTSCI.42.5.1077.

Chen, S., Zhao, H., Zou, C., Li, Y., Chen, Y. *et al.* (2017) Combined inoculation with multiple arbuscular mycorrhizal fungi improves growth, nutrient uptake and photosynthesis in cucumber seedlings. *Frontiers in Microbiology* 8, 2516. DOI: 10.3389/fmicb.2017.02516.

Cherif, H., Marasco, R., Rolli, E., Ferjani, R., Fusi, M. *et al.* (2015) Oasis desert farming selects environment-specific date palm root endophytic communities and cultivable bacteria that promote resistance to drought. *Environmental Microbiology Reports* 7(4), 668–678. DOI: 10.1111/1758-2229.12304.

Coutinho, E.S., Barbosa, M., Beiroz, W., Mescolotti, D.L.C., Bonfim, J.A. *et al.* (2019) Soil constraints for arbuscular mycorrhizal fungi spore community in degraded sites of rupestrian grassland: implications for restoration. *European Journal of Soil Biology* 90, 51–57. DOI: 10.1016/j.ejsobi.2018.12.003.

Cui, Y.-N., Li, X.-T., Yuan, J.-Z., Wang, F.-Z., Guo, H. *et al.* (2020) Chloride is beneficial for growth of the xerophyte *Pugionium cornutum* by enhancing osmotic adjustment capacity under salt and drought stresses. *Journal of Experimental Botany* 71(14), 4215–4231. DOI: 10.1093/jxb/eraa158.

Debode, J., Ebrahimi, N., D'Hose, T., Cremelie, P., Viaene, N. *et al.* (2020) Has compost with biochar added during the process added value over biochar or compost to increase disease suppression? *Applied Soil Ecology* 153, 103571. DOI: 10.1016/j.apsoil.2020.103571.

Del Buono, D. (2021) Can biostimulants be used to mitigate the effect of anthropogenic climate change on agriculture? It is time to respond. *Science of the Total Environment* 751, 141763. DOI: 10.1016/j.scitotenv.2020.141763.

Duo, L.A., Liu, C.X. and Zhao, S.L. (2018) Alleviation of drought stress in turfgrass by the combined application of nano-compost and microbes from compost. *Russian Journal of Plant Physiology* 65(3), 419–426. DOI: 10.1134/S102144371803010X.

Eden, M., Gerke, H.H. and Houot, S. (2017) Organic waste recycling in agriculture and related effects on soil water retention and plant available water: a review. *Agronomy for Sustainable Development* 37(2), 11. DOI: 10.1007/s13593-017-0419-9.

El Hadrami, A. and Al-Khayri, J.M. (2012) Socioeconomic and traditional importance of date palm. *Emirates Journal of Food and Agriculture* 24, 371–385.

Elias, D.M.O., Rowe, R.L., Pereira, M.G., Stott, A.W., Barnes, C.J. *et al.* (2017) Functional differences in the microbial processing of recent assimilates under two contrasting perennial bioenergy plantations. *Soil Biology and Biochemistry* 114, 248–262. DOI: 10.1016/j.soilbio.2017.07.026.

El-Juhany, L.I. (2010) Degradation of date palm trees and date production in Arab countries: causes and potential rehabilitation. *Australian Journal of Basic and Applied Sciences* 4, 3998–4010.

El Rabey, H.A., Al-Malki, A.L., Abulnaja, K.O. and Rohde, W. (2015) Proteome analysis for understanding abiotic stress (salinity and drought) tolerance in date palm (*Phoenix dactylifera* L.). *International Journal of Genomics* 2015, 407165. DOI: 10.1155/2015/407165.

El Rabey, H.A., Al-Malki, A.L. and Abulnaja, K.O. (2016) Proteome analysis of date palm (*Phoenix dactylifera* L.) under severe drought and salt stress. *International Journal of Genomics* 2016, 7840759. DOI: 10.1155/2016/7840759.

El-Sharabasy, S. and Ragab, A.A. (2009) Effect of phytohormones produced by PGPR (rhizobium) strains on *in-vitro* propagation of date palm (*Phoenix dactylifera* L.)

Zaghloul cultivar. *Egyptian Journal of Agricultural Research* 87(1), 267–276. DOI: 10.21608/ejar.2009.193166.

El-Sharabasy, S.F., Orf, H.O.M., Abotaleb, H.H., Abdel-Galeil, L.M. and Saber, T.Y. (2018) Effect of plant growth promoting rhizobacteria (PGPR) on growth and leaf chemical composition of date palm plants cv. Bartamuda under salinity stress. *Middle East Journal of Agriculture Research* 7, 618–624.

Fasusi, O.A., Cruz, C. and Babalola, O.O. (2021) Agricultural sustainability: microbial biofertilizers in rhizosphere management. *Agriculture* 11(2), 163. DOI: 10.3390/agriculture11020163.

Ferjani, R., Marasco, R., Rolli, E., Cherif, H., Cherif, A. *et al.* (2015) The date palm tree rhizosphere is a niche for plant growth promoting bacteria in the oasis ecosystem. *BioMed Research International* 2015, 153851. DOI: 10.1155/2015/153851.

Gaiotti, F., Marcuzzo, P., Belfiore, N., Lovat, L., Fornasier, F. *et al.* (2017) Influence of compost addition on soil properties, root growth and vine performances of *Vitis vinifera* cv Cabernet sauvignon. *Scientia Horticulturae* 225, 88–95. DOI: 10.1016/j.scienta.2017.06.052.

Gałązka, A., Niedźwiecki, J., Grządziel, J. and Gawryjołek, K. (2020) Evaluation of changes in glomalin-related soil proteins (GRSP) content, microbial diversity and physical properties depending on the type of soil as the important biotic determinants of soil quality. *Agronomy* 10, 1279. DOI: 10.3390/agronomy10091279.

Ghadbane, M., Medjekal, S., Benderradji, L., Belhadj, H. and Daoud, H. (2021) Assessment of arbuscular mycorrhizal fungi status and rhizobium on date palm (*Phoenix dactylifera* L.) cultivated in a Pb contaminated soil. In: Ksibi, M., Ghorbal, A., Chakraborty, S., Chaminé, H.I., Barbieri, M. *et al.* (eds) *Recent Advances in Environmental Science from the Euro-Mediterranean and Surrounding Regions (2nd Edition). Proceedings of 2nd Euro-Mediterranean Conference for Environmental Integration (EMCEI-2), Tunisia 2019.* Springer, Cham, Switzerland, pp. 703–707.

Ghirardo, A., Nosenko, T., Kreuzwieser, J., Winkler, J.B., Kruse, J. *et al.* (2021) Protein expression plasticity contributes to heat and drought tolerance of date palm. *Oecologia* 197, 903–919. DOI: 10.1007/s00442-021-04907-w.

Goldshtein, E., Cohen, Y., Hetzroni, A., Cohen, Y. and Soroker, V. (2020) The spatiotemporal dynamics and range expansion of the red palm weevil in Israel. *Journal of Pest Science* 93, 691–702. DOI: 10.1007/s10340-019-01176-8.

Gomiero, T. (2016) Soil degradation, land scarcity and food security: reviewing a complex challenge. *Sustainability* 8, 281. DOI: 10.3390/su8030281.

Harkousse, O., Slimani, A., Jadrane, I., Aitboulahsen, M., Mazri, M.A. *et al.* (2021) Role of local biofertilizer in enhancing the oxidative stress defence systems of date palm seedling (*Phoenix dactylifera*) against abiotic stress. *Applied and Environmental Soil Science* 2021, 1–13. DOI: 10.1155/2021/6628544.

Hazzouri, K.M., Flowers, J.M., Nelson, D., Lemansour, A., Masmoudi, K. *et al.* (2020) Prospects for the study and improvement of abiotic stress tolerance in date palms in the post-genomics era. *Frontiers in Plant Science* 11, 293. DOI: 10.3389/fpls.2020.00293.

Hirich, A. and Jacobsen, S. (2014) Deficit irrigation and organic compost improve growth and yield of quinoa and pea. *Journal of Agronomy and Crop Science* 200(5), 390–398. DOI: 10.1111/jac.12073.

Hussain, M.I., Farooq, M. and Syed, Q.A. (2020) Nutritional and biological characteristics of the date palm fruit (*Phoenix dactylifera* L.) – a review. *Food Bioscience* 34, 100509. DOI: 10.1016/j.fbio.2019.100509.

Ibrahim, M.A. and Hamzah, H.A. (2019) Effect of the harvest date, storage period and package on the storability of date palm (*Phoenix dactylifera* L. cv. Taberzal) fruits. *Plant Archives* 19, 226–231.

Ibrahim, M.I.M., Ali, H.S., Sahab, A.F. and Al-Khalifa, A.R.S. (2013a) Co-occurrence of fungi, aflatoxins, ochratoxins A and fumonsins in date palm fruits of Saudi Arabia. *Journal of Applied Sciences Research* 9, 1449–1456.

Ibrahim, M.M., El-Beshbeshy, R.T., Kamh, N.R. and Abou-Amer, A.I. (2013b) Effect of NPK and biofertilizer on date palm trees grown in Siwa Oasis, Egypt. *Soil Use and Management* 29(3), 315–321. DOI: 10.1111/sum.12042.

Ilyas, N., Mazhar, R., Yasmin, H., Khan, W., Iqbal, S. *et al.* (2020) Rhizobacteria isolated from saline soil induce systemic tolerance in wheat (*Triticum aestivum* L.) against salinity stress. *Agronomy* 10(7), 989. DOI: 10.3390/agronomy10070989.

Islam, F., Yasmeen, T., Ali, S., Ali, B., Farooq, M.A. *et al.* (2015) Priming-induced antioxidative responses in two wheat cultivars under saline stress. *Acta Physiologiae Plantarum* 37(8), 153. DOI: 10.1007/s11738-015-1897-5.

Ju, W., Liu, L., Fang, L., Cui, Y., Duan, C. *et al.* (2019) Impact of co-inoculation with plant-growth-promoting rhizobacteria and rhizobium on the biochemical responses of alfalfa-soil system in copper contaminated soil. *Ecotoxicology and Environmental Safety* 167, 218–226. DOI: 10.1016/j.ecoenv.2018.10.016.

Kadhim, S.A. and Kadhim, D.H. (2021) Geographical distribution of land use problems in date palm cultivation in Baghdad Governorate. *Al-Adab Journal* 137, 495–526.

Kanwal, S., Ilyas, N., Batool, N. and Arshad, M. (2017) Amelioration of drought stress in wheat by combined application of PGPR, compost, and mineral fertilizer. *Journal of Plant Nutrition* 40(9), 1250–1260. DOI: 10.1080/01904167.2016.1263322.

Karbout, N., Mlih, R., Latifa, D., Bol, R., Moussa, M. *et al.* (2021) Farm manure and bentonite clay amendments enhance the date palm morphology and yield. *Arabian Journal of Geosciences* 14(9), 818. DOI: 10.1007/s12517-021-07160-w.

Kaushal, M. (2019) Microbes in cahoots with plants: MIST to hit the jackpot of agricultural productivity during drought. *International Journal of Molecular Sciences* 20(7), 1769. DOI: 10.3390/ijms20071769.

Khaliel, A.S., Arabia, S. and Words, K. (1994) The effect of Mg on Ca, K and P content of date palm seedlings under mycorrhizal and non-mycorrhizal conditions. *Mycoscience* 35(3), 213–217. DOI: 10.1007/BF02268440.

Khalifa, N. and Yousef, L.F. (2015) A short report on changes of quality indicators for a sandy textured soil after treatment with biochar produced from fronds of date palm. *Energy Procedia* 74, 960–965. DOI: 10.1016/j.egypro.2015.07.729.

Khan, N., Bano, A., Rahman, M.A., Guo, J., Kang, Z. *et al.* (2019) Comparative physiological and metabolic analysis reveals a complex mechanism involved in drought tolerance in chickpea (*Cicer arietinum* L.) induced by PGPR and PGRs. *Scientific Reports* 9, 2097. DOI: 10.1038/s41598-019-38702-8.

Khan, W., Rayirath, U.P., Subramanian, S., Jithesh, M.N., Rayorath, P. *et al.* (2009) Seaweed extracts as biostimulants of plant growth and development. *Journal of Plant Growth Regulation* 28(4), 386–399. DOI: 10.1007/s00344-009-9103-x.

Liu, J.F., Arend, M., Yang, W.J., Schaub, M., Ni, Y.Y. *et al.* (2017) Effects of drought on leaf carbon source and growth of European beech are modulated by soil type. *Scientific Reports* 7(1), 42462. DOI: 10.1038/srep42462.

Mahmoudi, H., Hosseininia, G., Azadi, H. and Fatemi, M. (2008) Enhancing date palm processing, marketing and pest control through organic culture. *Journal of Organic Systems* 3, 29–39.

Mahmud, A.A., Upadhyay, S.K., Srivastava, A.K. and Bhojiya, A.A. (2021) Biofertilizers: a nexus between soil fertility and crop productivity under abiotic stress. *Current Research in Environmental Sustainability* 3, 100063. DOI: 10.1016/j.crsust.2021.100063.

Mainville, N., Webb, J., Lucotte, M., Davidson, R., Betancourt, O. *et al.* (2006) Decrease of soil fertility and release of mercury following deforestation in the Andean Amazon, Napo River Valley, Ecuador. *Science of the Total Environment* 368(1), 88–98. DOI: 10.1016/j.scitotenv.2005.09.064.

Meddich, A. and Boumezzough, A. (2017) First detection of *Potosia opaca* larva attacks on *Phoenix dactylifera* and *P. canariensis* in Morocco: focus on pests control strategies and soil quality of prospected palm groves. *Journal of Entomology and Zoology Studies* 5, 984–991.

Meddich, A., Jaiti, F., Bourzik, W., Asli, A.E. and Hafidi, M. (2015) Use of mycorrhizal fungi as a strategy for improving the drought tolerance in date palm (*Phoenix dactylifera*). *Scientia Horticulturae* 192, 468–474. DOI: 10.1016/j.scienta.2015.06.024.

Meena, H. and Meena, R.S. (2017) Assessment of sowing environments and bio-regulators as adaptation choice for clusterbean productivity in response to current climatic scenario. *Bangladesh Journal of Botany* 46, 241–244.

Mihi, A., Tarai, N. and Chenchouni, H. (2017) Can palm date plantations and oasification be used as a proxy to fight sustainably against desertification and sand encroachment in hot drylands? *Ecological Indicators* 105, 365–375. DOI: 10.1016/j.ecolind.2017.11.027.

Mirzaee, M.R., Tajali, H. and Javadmosavi, S.A. (2014) *Thielaviopsis paradoxa* causing neck bending disease of date palm in Iran. *Journal of Plant Pathology* 96, S4.113–S4.131.

Mouloud, G., Samir, M., Laid, B. and Daoud, H. (2017) Isolation and characterization of rhizospheric *Streptomyces* spp. for the biocontrol of *Fusarium* wilt (bayoud) disease of date palm (*Phoenix dactylifera* L.). *Journal of Scientific Agriculture* 1, 132. DOI: 10.25081/jsa.2017.v1.46.

Nadeem, S.M., Ahmad, M., Zahir, Z.A., Javaid, A. and Ashraf, M. (2014) The role of mycorrhizae and plant growth promoting rhizobacteria (PGPR) in improving crop productivity under stressful environments. *Biotechnology Advances* 32(2), 429–448. DOI: 10.1016/j.biotechadv.2013.12.005.

Naseem, H., Ahsan, M., Shahid, M.A. and Khan, N. (2018) Exopolysaccharides producing rhizobacteria and their role in plant growth and drought tolerance. *Journal of Basic Microbiology* 58(12), 1009–1022. DOI: 10.1002/jobm.201800309.

Niazi, S., Khan, I.M., Rasheed, S., Niazi, F., Shoaib, M. *et al.* (2017) An overview: date palm seed coffee, a functional beverage. *International Journal of Public Health and Health Systems* 2, 18–25.

Pačuta, V., Rašovský, M., Michalska-Klimczak, B. and Wyszyński, Z. (2021) Grain yield and quality traits of durum wheat (*Triticum durum* Desf.) treated with seaweed- and humic acid-based biostimulants. *Agronomy* 11, 1270.

Pal, A. (2014) Role of glomalin in improving soil fertility: a review. *International Journal of Plant & Soil Science* 3, 1112–1129. DOI: 10.9734/IJPSS/2014/11281.

Patankar, H.V., Al-Harrasi, I., Al-Yahyai, R. and Yaish, M.W. (2018) Identification of candidate genes involved in the salt tolerance of date palm (*Phoenix dactylifera* L.) based on a yeast functional bioassay. *DNA and Cell Biology* 37(6), 524–534. DOI: 10.1089/dna.2018.4159.

Puglia, D., Pezzolla, D., Gigliotti, G., Torre, L., Bartucca, M.L. *et al.* (2021) The opportunity of valorizing agricultural waste, through its conversion into biostimulants, biofertilizers, and biopolymers. *Sustainability* 13(5), 2710. DOI: 10.3390/su13052710.

Quiroga, G., Erice, G., Aroca, R., Chaumont, F. and Ruiz-Lozano, J.M. (2017) Enhanced drought stress tolerance by the arbuscular mycorrhizal symbiosis in a drought-sensitive maize cultivar is related to a broader and differential regulation of host plant aquaporins than in a drought-tolerant cultivar. *Frontiers in Plant Science* 8, 1056. DOI: 10.3389/fpls.2017.01056.

Raimi, A., Roopnarain, A. and Adeleke, R. (2021) Biofertilizer production in Africa: current status, factors impeding adoption and strategies for success. *Scientific African* 11, e00694. DOI: 10.1016/j.sciaf.2021.e00694.

Raza, A., Razzaq, A., Mehmood, S.S., Zou, X., Zhang, X. *et al.* (2019) Impact of climate change on crops adaptation and strategies to tackle its outcome: a review. *Plants* 8(2), 34. DOI: 10.3390/plants8020034.

Ren, H., Huang, B., Fernández-García, V., Miesel, J., Yan, L. *et al.* (2020) Biochar and rhizobacteria amendments improve several soil properties and bacterial diversity. *Microorganisms* 8(4), 502. DOI: 10.3390/microorganisms8040502.

Ribeiro, C.M. and Cardoso, E.J.B.N. (2012) Isolation, selection and characterization of root-associated growth promoting bacteria in Brazil pine (*Araucaria angustifolia*). *Microbiological Research* 167(2), 69–78. DOI: 10.1016/j.micres.2011.03.003.

Rillig, M.C., Aguilar-Trigueros, C.A., Bergmann, J., Verbruggen, E., Veresoglou, S.D. *et al.* (2015) Plant root and mycorrhizal fungal traits for understanding soil aggregation. *New Phytologist* 205, 1385–1388. DOI: 10.1111/nph.13045.

Rouphael, Y. and Colla, G. (2020) Editorial: biostimulants in agriculture. *Frontiers in Plant Science* 11, 40. DOI: 10.3389/fpls.2020.00040.

Saeed, E.E., Sham, A., El-Tarabily, K., Elsamen, F.A., Iratni, R. *et al.* (2016) Chemical control of black scorch disease on date palm caused by the fungal pathogen *Thielaviopsis punctulata* in United Arab Emirates. *Plant Disease* 100, 2370–2376.

Salama, M., El-Samak, A., El-Morsy, A. and Aly, K. (2012) The beneficial effect of minimizing mineral nitrogen fertilization on Sewy date palm trees by using organic and biofertilizers. *Journal of Plant Production* 3(9), 2411–2424. DOI: 10.21608/jpp.2012.84988.

Saleh, A.A., Sharafaddin, A.H., El-Komy, M.H., Ibrahim, Y.E., Hamad, Y.K. *et al.* (2017) *Fusarium* species associated with date palm in Saudi Arabia. *European Journal of Plant Pathology* 148(2), 367–377. DOI: 10.1007/s10658-016-1095-3.

Sandhya, V., Ali, S.Z., Grover, M., Reddy, G. and Venkateswarlu, B. (2010) Effect of plant growth promoting *Pseudomonas* spp. on compatible solutes, antioxidant status and plant growth of maize under drought stress. *Plant Growth Regulation* 62(1), 21–30. DOI: 10.1007/s10725-010-9479-4.

Schütz, L., Gattinger, A., Meier, M., Müller, A., Boller, T. *et al.* (2018) Improving crop yield and nutrient use efficiency via biofertilization – a global meta-analysis. *Frontiers in Plant Science* 8, 2204.

Shareef, H.J., Abdi, G. and Fahad, S. (2020) Change in photosynthetic pigments of date palm offshoots under abiotic stress factors. *Folia Oecologica* 47(1), 45–51. DOI: 10.2478/foecol-2020-0006.

Sharma, S.B., Sayyed, R.Z., Trivedi, M.H. and Gobi, T.A. (2013) Phosphate solubilizing microbes: sustainable approach for managing phosphorus deficiency in agricultural soils. *SpringerPlus* 2, 587.

Siddiq, M. and Greiby, I. (2013) Overview of date fruit production, postharvest handling, processing, and nutrition. In: Siddiq, M., Aleid, S.M. and Kader, A.A. (eds) *Dates: Postharvest Science, Processing Technology and Health Benefits*. Wiley, Chichester, UK, pp. 1–28.

Singh, B., Upadhyay, A.K., Al-Tawaha, T.W., Al-Tawaha, A.R. and Sirajuddin, S.N. (2020) Biofertilizer as a tool for soil fertility management in changing climate. *IOP Conference Series: Earth Environmental Science* 492(1), 012158. DOI: 10.1088/1755-1315/492/1/012158.

Soka, G. and Ritchie, M. (2014) Arbuscular mycorrhizal symbiosis and ecosystem processes: prospects for future research in tropical soils. *Open Journal of Ecology* 04(01), 11–22. DOI: 10.4236/oje.2014.41002.

Sulaiman, M., Purayil, F.T., Krishankumar, S., Kurup, S.S. and Pessarakli, M. (2021) Accumulation of toxic elements in soil and date palm (*Phoenix dactylifera* L.) through fertilizer application . *Journal of Plant Nutrition* 44(7), 958–969. DOI: 10.1080/01904167.2020.1866604.

Symanczik, S., Lehmann, M.F., Wiemken, A., Boller, T. and Courty, P.E. (2018) Effects of two contrasted arbuscular mycorrhizal fungal isolates on nutrient uptake by *Sorghum bicolor* under drought. *Mycorrhiza* 28, 779–785.

Taha, F.H. and Abood, M.R. (2018) Influence of some organic fertilizers on date palm cv. Barhi. *Iraqi Journal of Agricultural Sciences* 49(4), 632–638. DOI: 10.36103/ijas.v49i4.72.

Tan, K.Z., Radziah, O., Halimi, M.S., Khairuddin, A.R., Habib, S.H. *et al.* (2014) Isolation and characterization of rhizobia and plant growth-promoting rhizobacteria and their effects on growth of rice seedlings. *American Journal of Agricultural and Biological Sciences* 9(3), 342–360. DOI: 10.3844/ajabssp.2014.342.360.

Thwaini, Q.S., Abed, I.A. and Alrawi, A.A. (2021) Evaluation of the efficiency of date seeds as a carrier of PGPR inoculants under different storage temperature. *IOP Conference Series: Earth Environmental Science* 761(1), 012021. DOI: 10.1088/1755-1315/761/1/012021.

Toubali, S., Tahiri, A., Anli, M., Symanczik, S., Boutasknit, A. *et al.* (2020) Physiological and biochemical behaviors of date palm vitroplants treated with microbial consortia and compost in response to salt stress. *Applied Sciences* 10(23), 8665. DOI: 10.3390/app10238665.

Ullah, U., Ashraf, M., Shahzad, S.M., Siddiqui, A.R., Piracha, M.A. *et al.* (2016) Growth behavior of tomato (*Solanum lycopersicum* L.) under drought stress in the presence of silicon and plant growth promoting rhizobacteria. *Soil & Environment* 35, 65–75.

Upadhyay, S.K., Singh, J.S., Saxena, A.K. and Singh, D.P. (2012) Impact of PGPR inoculation on growth and antioxidant status of wheat under saline conditions. *Plant Biology* 14(4), 605–611. DOI: 10.1111/j.1438-8677.2011.00533.x.

Vafa, Z.N., Sohrabi, Y., Sayyed, R.Z., Luh Suriani, N. and Datta, R. (2021) Effects of the combinations of rhizobacteria, mycorrhizae, and seaweed, and supplementary irrigation on growth and yield in wheat cultivars. *Plants* 10(4), 811. DOI: 10.3390/plants10040811.

Valencia-Cantero, E., Hernández-Calderón, E., Velázquez-Becerra, C., López-Meza, J.E., Alfaro-Cuevas, R. *et al.* (2007) Role of dissimilatory fermentative iron-reducing bacteria in Fe uptake by common bean (*Phaseolus vulgaris* L.) plants grown in alkaline soil. *Plant and Soil* 291(1–2), 263–273. DOI: 10.1007/s11104-007-9191-y.

Van Oosten, M.J., Pepe, O., De Pascale, S., Silletti, S. and Maggio, A. (2017) The role of biostimulants and bioeffectors as alleviators of abiotic stress in crop plants. *Chemical and Biological Technologies in Agriculture* 4(1), 5. DOI: 10.1186/s40538-017-0089-5.

Vejan, P., Abdullah, R., Khadiran, T., Ismail, S. and Nasrulhaq Boyce, A. (2016) Role of plant growth promoting rhizobacteria in agricultural sustainability-a review. *Molecules* 21(5), 573. DOI: 10.3390/molecules21050573.

Xie, W., Hao, Z., Zhou, X., Jiang, X., Xu, L. *et al.* (2018) Arbuscular mycorrhiza facilitates the accumulation of glycyrrhizin and liquiritin in *Glycyrrhiza uralensis* under drought stress. *Mycorrhiza* 28(3), 285–300. DOI: 10.1007/s00572-018-0827-y.

Yaish, M.W. and Kumar, P.P. (2015) Salt tolerance research in date palm tree (*Phoenix dactylifera* L.), past, present, and future perspectives. *Frontiers in Plant Science* 6, 348.

Yakhin, O.I., Lubyanov, A.A., Yakhin, I.A. and Brown, P.H. (2017) Biostimulants in plant science: a global perspective. *Frontiers in Plant Science* 7, 2049. DOI: 10.3389/fpls.2016.02049.

Yaseen, R., Zafar-ul-Hye, M. and Hussain, M. (2019) Integrated application of ACC-deaminase containing plant growth promoting rhizobacteria and biogas slurry improves the growth and productivity of wheat under drought stress. *International Journal of Agriculture and Biology* 21, 869–878.

Yu, Y.Y., Li, S.M., Qiu, J.P., Li, J.G., Luo, Y.M. *et al.* (2019) Combination of agricultural waste compost and biofertilizer improves yield and enhances the sustainability of a pepper field. *Journal of Plant Nutrition and Soil Science* 182(4), 560–569. DOI: 10.1002/jpln.201800223.

Zhang, S., Wang, L., Ma, F., Zhang, X. and Fu, D. (2016) Arbuscular mycorrhiza improved phosphorus efficiency in paddy fields. *Ecological Engineering* 95, 64–72. DOI: 10.1016/j.ecoleng.2016.06.029.

Zhang, Y., Gao, X., Shen, Z., Zhu, C., Jiao, Z. *et al.* (2019) Pre-colonization of PGPR triggers rhizosphere microbiota succession associated with crop yield enhancement. *Plant and Soil* 439(1–2), 553–567. DOI: 10.1007/s11104-019-04055-4.

Zhen, J., Lazarovitch, N. and Tripler, E. (2020) Effects of fruit load intensity and irrigation level on fruit quality, water productivity and net profits of date palms. *Agricultural Water Management* 241, 106385. DOI: 10.1016/j.agwat.2020.106385.

Zougari-Elwedi, B., Hdiouch, A., Boughalleb, N. and Namsi, A. (2019) Responses of date palm seedling to co-inoculation with phosphate solubilizing bacteria and mycorrhizal arbuscular fungi. *International Journal of Environment, Agriculture and Biotechnology* 4(2), 581–588. DOI: 10.22161/ijeab/4.2.43.

Pest and Disease Management in Date Palm

10

Rashad Rasool Khan ⓘ, Imran Ul Haq ⓘ and Summar Abbas Naqvi* ⓘ

University of Agriculture, Faisalabad, Pakistan

Abstract

Date palm is one of the most nutritious fruits with enormous economic importance, but its growth, vigor, and productivity are decreasing due to many biotic and abiotic stresses. Among these factors biotic stresses, i.e., pests and diseases, are the major threat to date palm trees throughout the world. Date palms are attacked by several insect pests under field conditions. Although few of them cause serious economic losses to the crop, the stem-boring beetles and weevils wreak havoc, causing significant damage to the plants and resulting in death. Dates are also attacked by several mites and insect pests while in storage. Similarly, the physiological and pathological problems have a negative impact on potential yield of date palms. Among entomological and pathological factors, red palm weevil, black beetle, longhorn beetle, dubas bug, and fungal diseases cause catastrophic effects on yield and quality of date crops. These stresses occur because of lack of knowledge, poor pre- and post-care of offshoots, and poor management in the orchard. This chapter provides information on important pre- and postharvest pest and diseases of the date palm tree, leaf, and fruit, and suggests that integrated management of these pests and diseases is the most acceptable and efficient strategy; however, chemical controls are also suggested where insect and disease control failures are frequently experienced.

Keywords: Artificial intelligence, Date palm, Decline, *Fusarium*, Red palm weevil, Stress

10.1 Introduction

Biodiversity of date palm is currently at risk due to various reasons such as replacement of elite/superior cultivars by nondescript types and attacks by pests and diseases. Similarly, the natural date palm diversity presently

*Corresponding author: summar.naqvi@uaf.edu.pk

© CAB International 2023. *Date Palm* (eds J.M. Al-Khayri *et al.*)
DOI: 10.1079/9781800620209.0010

in its natural zones is threatened by desertification, deforestation for fuel and timber, urbanization, and lack of proper documentation. Many local selections of date palm with good quality are nearly extinct due to a lack of interest in conserving this diverse genetic material.

In recent years, the red palm weevil (RPW) (*Rhynchophorus ferrugineus*) has been the most destructive insect of palm plantations throughout the world (Bertone *et al.*, 2010), and the Food and Agriculture Organization of the United Nations (FAO) designated it as a category-1 pest on date palm in the Middle East. Losses in global production of dates have been estimated at 30% due to plant diseases and pests since the mid-1980s (FAO, 2018). The annual loss in the Gulf region of the Middle East due to eradication of severely infested palms has been estimated to range from US$1.74 million to US$8.69 million at 1 and 5% infestation, respectively, over the same period (El-Sabea *et al.*, 2009).

Out of the various diseases, sudden death/decline is a very detrimental disease of date palm. It is a pathological problem, mainly caused by fungi (*Fusarium solani*), and the prevalence of this sudden death is increasing day by day. Its symptoms are similar to those of bayoud disease in Saudi Arabia (Maitlo *et al.*, 2009). Many *Fusarium* species have been reported as date palm (*Phoenix dactylifera* L.) pathogens worldwide. The most serious one is *Fusarium oxysporum* f. sp. *albedinis* that causes *Fusarium* wilt of date palm trees (Zaid and Arias-Jiménez, 2002). *F. oxysporum* f. sp. *albedinis* fungus was responsible for the loss of more than 15,000,000 trees in Morocco and Algeria between 1870 and 1970 (Djerbi, 1982). *Fusarium proliferatum* has been reported as a date palm pathogen in many date-palm-growing countries, e.g., Saudi Arabia (Abdalla *et al.*, 2000), Spain (Armengol *et al.*, 2005), Canary Islands (Hernández-Hernández *et al.*, 2010), Iran (Mansoori, 2012), Iraq (Hameed, 2012), Israel (Cohen *et al.*, 2010), and the USA (Munoz and Wang, 2011). Similarly, isolates of *F. solani* have been recovered from date palm trees in different countries such as Iran (Mansoori and Kord, 2006), Iraq (Al-Yasiri *et al.*, 2010), Oman (Al-Sadi *et al.*, 2012), Pakistan (Maitlo *et al.*, 2014), and Egypt (Alkahtani *et al.*, 2011).

Various cultural, biological, and chemical methods have been adopted for the management of pests and diseases of date palm. In the recent past, different transgenic approaches based on RNA interference (RNAi) were developed against the most devastating pests and diseases of the date palm.

10.2   Important Pre- and Postharvest Pests

Like many other crops, date palms are attacked by a variety of insect pests. A significant number of insect pests attack the foliage as well as the stem of the date palm, besides the green fruit on the plant. However, certain insects also damage the date fruits during storage. As many as 112 species of insects and mites are associated with date palms worldwide, including

22 species attacking dates while in storage (El-Shafie, 2012). Only ten arthropod pests are reported to cause serious damage to the date palms (El-Shafie, 2012). These major insect pests belong to Coleoptera (RPW, rhinoceros beetle, and longhorn beetle), Hemiptera (dubas bug (DB)), and Lepidoptera (lesser date moth (LDM)) (El-Shafie *et al.*, 2017). This chapter also covers some arthropod pests damaging dates during storage.

The coleopterous insects, especially RPW and wood-boring beetles (rhinoceros and longhorn), damage the stem through boring and feeding on the cellulose part. Severe infestation weakens the date palm and hinders quality production of date fruits. The plants show stunted growth and start to wither; if the pest infestation remains unchecked, the severely infested plant is destroyed completely. DB is a hemipteran insect and is a phloem feeder that sucks the cell sap from date palm leaves. While sucking the plant sap, DB releases honeydew that serves as a medium for the growth of sooty mold. Necrotic spots and sooty mold hinder photosynthetic activity of the leaves and hence plant growth is affected negatively. LDM larvae, in turn, feed on the green fruit of date palms and negatively affect the quality and quantity of the fruit harvest (Gassouma, 2004; Al-Zadjali *et al.*, 2006; Wakil *et al.*, 2015a, b; El-Shafie *et al.*, 2017).

Several insects also damage dates while in storage and affect the quality of marketable date fruits. Farmers largely rely on pesticide application for pest management in date palm orchards; however, the environmental and health hazards of pesticides convince them to implement integrated pest management (IPM). El-Shafie (2012) listed 45 species of predators and parasitoids associated with the insect pest complex in date palms worldwide. The latest IPM interventions using mating disruption, sterile insect release technique (SIT), or release of insects with dominant lethal gene (RIDL) are species-specific, eco-friendly, and hence are gaining more attention from scientists (Wakil *et al.*, 2015a). Attract and kill strategies involving semiochemicals are most widely adapted against RPW in the Arabian Peninsula and other countries (Faleiro, 2006). This chapter outlines the nature and extent of damage caused by some major date palm pests, their distribution and biology, and enlightens the major interventions involved in their integrated management.

10.2.1   Red palm weevil (RPW)

10.2.1.1   Distribution and host range

RPW, *R. ferrugineus* (Olivier) (family Curculionidae, order Coleoptera), is native to tropical Asia (Roda *et al.*, 2011); however, it has spread globally, including to the Middle East, North Africa, the Mediterranean, and parts of Central America and the Caribbean, as shown in Fig. 10.1 (Dembilio and Jaques, 2015; Ashry *et al.*, 2020). The true *R. ferrugineus* is endemic to the northern and western regions of continental South-east Asia, Sri Lanka,

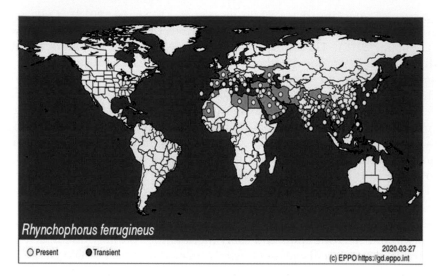

Fig. 10.1. Red palm weevil distribution worldwide. (From EPPO, 2020, used with permission.)

and the Philippines according to cytochrome oxidase I (COI) haplotype data and is responsible for practically all invasive populations globally (Rugman-Jones *et al.*, 2013). The pest causes severe damage to commercial and ornamental palm trees and its severe infestation has devastated numerous palm orchards in many countries (El-Mergawy and Al-Ajlan, 2011; Rach *et al.*, 2013). Nearly 40 palm species from 23 genera are known RPW hosts globally including date palm and coconut (Faleiro, 2006).

10.2.1.2   Life history and damage

The immature stage (grubs) causes severe damage to date palms (Fig. 10.2). All types of palm trees support all growth stages of RPW. The adult female weevil excavates holes with its snout and eggs are laid individually in separate holes (Dembilio and Jaques, 2015; Al-Dosary *et al.*, 2016). A single female lays around 300 or more eggs that hatch into conical legless larvae within 2–5 days and these grubs grow inside the palm trunk for 1–3 months (Al-Dosary *et al.*, 2016). Younger larvae feed on the palm tissues and move inward to the soft heartwood, whereas the fully grown, brown-headed, yellowish-white larva possesses strong chitinized mandibles to bore into the palm trunks (Al-Dosary *et al.*, 2016). Depending on the host plant or diet, the number of larval instars varies from 13 to 17 (Dembilio and Jaques, 2015) and the fully grown larva forms a cocoon (pupal stage) that lasts for 14–21 days before transforming into a reddish-brown adult with black spots on its thorax and a long proboscis. Adult male weevils are characterized by a tuft of bristles on the dorsal tip of their snout and both sexes can survive

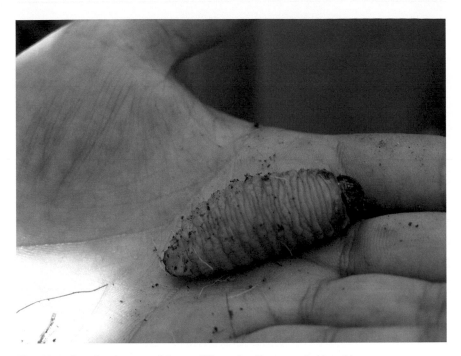

Fig. 10.2. A red palm weevil larva. (Photo by Summar A. Naqvi.)

for an average of 98 days (Al-Dosary *et al.*, 2016). Adult weevils stay on the same plant until complete consumption of the meristem, resulting in death of the palm or its offshoots (Figs 10.3 and 10.4). Afterward, adults disperse from the dead host in search of new plants (Dembilio and Jaques, 2015).

10.2.1.3 Control

DETECTION, TRAPPING, AND MONITORING.   Early detection of RPW and monitoring its activity can help to significantly protect palms against pest infestation and area-wide spreading. Early detection of the pest remains a challenge even to apply curative or preventive methods for RPW management. Use of an optical-fiber distributed acoustic sensor (DAS) enabled detection of feeding sounds produced by 12-day-old larvae in infested trees and was effective to provide 24/7 real-time monitoring of more than 1000 palm trees (Ashry *et al.*, 2020). Using food-baited pheromone (ferrugineol) traps for mass trapping of adults plays a significant role in RPW management (Faleiro, 2006; Al-Dosary *et al.*, 2016). The pheromone lure needs to be replaced (1–3 months) depending on the trap's exposure to sunlight and the lure used. A single pheromone trap per hectare is installed in low-infested orchards, which may reach to 4–10 traps/ha in the case of severe infestation (Al-Dosary *et al.*, 2016).

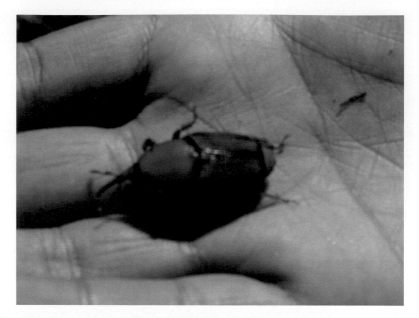

Fig. 10.3. A red palm weevil adult. (Photo by Summar A. Naqvi.)

Fig. 10.4. The red palm weevil and its infestation in a date palm. (Photo by Summar A. Naqvi.)

CULTURAL CONTROL.   Field sanitation and removal of RPW breeding sites play a pivotal role in its successful management. Escaping RPW infestations from incompletely burnt plants suggest avoiding the burning of greener plants; however, severely infested palm trees can be destroyed through tree choppers or shredders. Offshoot pruning or leaf shaving must be followed by an insecticide treatment of the cut or injured palm surface to avoid egg laying by the hovering females (Dembilio and Jaques, 2015).

BIOLOGICAL CONTROL. Certain reports enumerate more than 50 natural enemies including fungi, bacteria, viruses, nematodes, yeasts, mites, insects, and vertebrates attacking RPW species (Dembilio and Jaques, 2015; Al-Dosary *et al.*, 2016). Some reports from the Middle East indicate three phoretic mite species associated with RPW collected from date palm orchards of the United Arab Emirates (Al-Deeb *et al.*, 2011). Entomopathogenic nematodes (*Steinernema carpocapsae*), bacteria (*Bacillus* spp.), and fungi (*Beauveria bassiana*) are reported to demonstrate efficient control of RPW (Dembilio *et al.*, 2010; Qayyum *et al.*, 2020).

CHEMICAL CONTROL. To control the RPW infestation and its spread to non-infested date palm plantations, chemical pesticides are sprayed as preventive measures. Precautionary application of insecticides to palms through stem injection in the early stage of attack is known to prevent infestation of such palms (Al-Dosary *et al.*, 2016). Neonicotinoids (imidacloprid and thiamethoxam), avermectin (abamectin), and organophosphates (chlorpyrifos and phosmet) are suggested against RPW and can be applied as spray on the canopy, injected into the trunk, as a soil drench, or by dipping the offshoots before transplantation (Al-Shawaf *et al.*, 2013; Dembilio and Jaques, 2015). Fumigation of palms or their parts with phosphine (PH_3) and methyl bromide (CH_3Br) before transportation is also important as a quarantine control to avoid infestation in the plants and the potential threat of spread to noninfested areas (Llácer and Jacas, 2010).

TRANSGENIC CONTROL. The *Vg* gene codes for the primary yolk protein precursor vitellogenin (Vg), which is essential for all oviparous animals, including insects, to reproduce successfully. Vg is produced in fat body cells in female insects, which is subsequently translated and released into the hemolymph, before being sequestered by oocytes by endocytosis through a specific receptor termed the Vg receptor (VgR). The silencing of the *RfVg* gene suggests that RNAi technology might be a viable alternative to traditional coleopteran pest management, particularly for *R. ferrugineus* (Rasool *et al.*, 2021).

10.2.2 Rhinoceros beetle

10.2.2.1 Distribution and host range

Oryctes elegans and *Oryctes agamemnon* (Coleoptera: Scarabaeidae: Dynastinae), also known as rhinoceros or dynastid beetles (Figs 10.5 and 10.6), are important economic pests of date palms that damage the fruit stalks and bunches by boring into them (Soltani *et al.*, 2008; El-Shafie, 2012;

Fig. 10.5. A rhinoceros beetle or black beetle. (Photo by Summar A. Naqvi.)

Bedford *et al.*, 2015). *O. elegans* is reported as abundant in the Gulf countries as well as Iraq, Iran, and northern Pakistan, whereas *O. agamemnon* is
mostly found in the Arabian Peninsula (Bedford *et al.*, 2015; Khalaf, 2018).

10.2.2.2 Life history and damage

Adult dynastids bore tunnels into the rachis, leaf bases, fruit stalks, and
bunches (Bedford *et al.*, 2015), while no adults of *O. agamemnon* were
reported attack inflorescences or green leaves in Tunisia (Soltani *et al.*, 2008).
Younger date palms (10–20 years) of short statured varieties having fragile
textured leaves are usually preferred by dynastid beetles (Khalaf, 2018).

Male *O. elegans* excrete an aggregation pheromone that attracts both
sexes. Under the set conditions of 28–30°C, larvae emerge from eggs after
10 days of incubation and undergo three instar stages in 91 days, which
later transform into adults after passing through a prepupal (7 days) and
a pupal (31 days) stage. The adults survive for about 4 months and start
laying eggs 2–3 weeks after mating (Khalaf, 2018). *O. agamemnon* produces
one generation in a year and adult activity booms in June–July and ends
in September. Females lay about 100 white eggs that turn glossy brown
and hatch in 2–3 weeks. Grubs are found from late June, are fully grown
in September (Fig. 10.7), and after overwintering, pupate for 2–3 weeks
(Blumberg, 2008).

Fig. 10.6. Rhinoceros beetle or black beetle and its infestation. (Photo by Summar A. Naqvi.)

10.2.2.3 Control

TRAPPING AND MONITORING. Light traps can be used for mass trapping and monitoring of beetles. Light traps with a source of white light from a mercury vapor bulb and equipped with battery-storage-based solar panels are used successfully; however, the efficiency of light traps can be affected during nights with brighter moonlight (Bedford *et al.*, 2015). *Oryctes* activity is also monitored through pheromone trapping in the Middle East (Khalaf, 2018); however, pheromone trapping is available only for *O. elegans* and fresh date palm tissue is added to enhance its efficacy (Bedford *et al.*, 2015).

CULTURAL CONTROL. Pruning and removal of older leaves is very helpful for physical collection and destruction of larvae (Ali and Hama, 2016). This also eliminates sites for beetles to hide and oviposit in the crevices and for larvae to develop and bore into the leaves.

QUARANTINE CONTROL. The odor of organic manure or animal compost can attract dynastid females and the heaps can serve as good breeding sites (Ehsine *et al.*, 2009). Importing organic fertilizer from countries infested with dynastid beetles can also serve as a source of new infection. Proper management of organic manures in date palm orchards and quarantine checks for imported manures can help prevent plant infestations (Bedford *et al.*, 2015).

Fig. 10.7. Rhinoceros beetle or black beetle larvae. (Photo by Summar A. Naqvi.)

BIOLOGICAL CONTROL. The entomopathogenic fungus (*Metarhizium anisopliae*) has shown promising results to kill dynastid beetles under laboratory conditions. Entomopathogenic nematodes (*Steinernema* spp. and *Rhabditis* spp.) are also isolated from dynastid grubs and are potential candidates for biological control of dynastids (Bedford *et al.*, 2015).

CHEMICAL CONTROL. A proper spraying of palm inflorescences with contact or stomach insecticides can provide effective control of adult beetles hovering for settling and oviposition. In orchards, where insecticides are applied for DB or LDM control, the hovering beetle can also be effectively controlled (Bedford *et al.*, 2015). Contact and persistent insecticides, when applied to the injured parts of palms after pruning practices, can help in avoiding oviposition by beetles. Stem injection or irrigation with neonicotinoids (imidacloprid and thiamethoxam) can also provide satisfactory control (Khalaf, 2018).

10.2.3 Longhorn beetle

10.2.3.1 Distribution and host range

The longhorn beetle, *Jebusa hammerschmidt* (Coleoptera: Cerambycidae), also known as the longhorn date palm stem borer or the date palm stem

Fig. 10.8. Longhorn beetle larva and its infestation in date palm. (Photo by Summar A. Naqvi.)

borer (DPB), is widely distributed in the native regions of date palm (Sama *et al.*, 2010). Unlike RPW, its distribution is not large, and the pest is of serious concern in severe infestations (Al-Deeb and Khalaf, 2015; Khalaf, 2018). Reports describe its prevalence in the Gulf countries, Iraq, Iran, Egypt, Jordan, Algeria, and India (Khalaf, 2018). *J. hammerschmidt* is reported to attack date palms mainly and is spread to noninfested regions through the transportation of infested trees or offshoots (Al-Deeb and Khalaf, 2015; Khalaf, 2018).

10.2.3.2 *Life history and damage*

The reddish-brown adult longhorn beetle possesses long, 12-segmented antennae almost equal in length to the body (female) or longer than the body (male), well-developed oval eyes, and long slender legs. The elongate, cylindrical, and robust body of adult beetles is covered with short pubescence (El-Shafie, 2015). The pest damages the plant by boring into the trunk and feeding on plant tissues, leading to leaf or trunk breakage and exposing the injuries to plant pathogens (Al-Jaboory, 2007). Females lay about 30–150 white and elongated eggs individually on fibers, bases of leaves, or in small crevices in the tree trunk (El-Shafie, 2015). Small grubs hatch out of the eggs after 2 weeks of incubation, bore into the tree trunk, and spend a larval (Figs 10.8 and 10.9) span of about 3 months; however, it can reach up to 10 months when hibernating in winter (Al-Deeb and

Fig. 10.9. A longhorn beetle larva. (Photo by Summar A. Naqvi.)

Khalaf, 2015; Khalaf, 2018). The pupal stage lasts for about 3 weeks and the adults come out by making circular exit holes. In the northern hemisphere, adults are mostly observed in early May and after mating the females start laying eggs (Khalaf, 2018).

10.2.3.3 Control

TRAPPING AND MONITORING.   Light traps can be used for mass trapping and monitoring of beetles. Light traps with a source of white light and equipped with battery-storage-based solar panels are used successfully (Al-Deeb and Khalaf, 2015).

CULTURAL CONTROL.   Older and dried leaf bases, leaf thorns, and dried bunch stalks should be removed to avoid hiding beetles and prevent egg laying. Properly pruned plants provide fewer chances of harboring beetles compared to unpruned trees (Khalaf, 2018). Use of balanced fertilization and adequate irrigation, along with other farm practices to keep healthy and vigorous date palms, also help to avoid pest attack (Al-Deeb and Khalaf, 2015).

BIOLOGICAL CONTROL.   The entomopathogenic fungus *B. bassiana* has shown tremendous potential (72.7% mortality) when tested against *J. hammerschmidt* grubs (Fayyadh *et al.*, 2013). Entomopathogenic nematodes (*Steinernema* spp.), a parasitic fly (*Megaselia* sp.), and a poxviridae virus were also reported to affect the adult and grubs of *J. hammerschmidt* (Al-Jaboory, 2007).

CHEMICAL CONTROL. Like for other concealed borers, pesticides may provide incomplete control of longhorn borer (El-Shafie, 2015). Systemic insecticides as recommended for RPW control are commonly used for longhorn borer management; however, the application requires keen attention as the grubs may be absent even if exit holes are present in leaves (Al-Deeb and Khalaf, 2015). Control results for surface spraying may vary as grubs live inside the trunk and may not be exposed to the pesticide, although stem injection may provide some satisfactory control of the beetle (Al-Deeb and Khalaf, 2015).

10.2.4 Dubas bug (DB)

10.2.4.1 Distribution and host range

The Old-World date bug, also commonly known as DB, *Ommatissus lybicus* de Bergevin (Hemiptera: Tropiduchidae), is a key sucking insect pest of the date palm, causing serious damage to date palms by affecting their growth and resulting in yield reduction. DB is an oligophagous insect that is known to complete its life cycle on the date palm *P. dactylifera*, and it has shown high specificity to date palms. DB has been reported as a major pest of date palm in the Middle East and North Africa, in Iraq, Iran, Oman, Pakistan (Shah *et al.*, 2013), the United Arab Emirates, and Yemen. Its presence is recorded in Algeria, Bahrain, Egypt, Jordan, Kuwait, Libya, Palestine, Saudi Arabia, Sudan, and Tunisia (El-Shafie, 2012; El-Shafie *et al.*, 2017). According to scientists studying the effects of climate change on *O. lybicus* infestation, northern Oman is on the verge of becoming infested and may see severe pest influxes in the future due to its ecological suitability (Shabani *et al.*, 2018).

10.2.4.2 Life history and damage

Adults are yellowish-brown to greenish insects with a powerful chitinous saw-like tool around the ovipositor of female that is used to bore into plant tissues. Females are typically longer than males, measuring 5–6 mm vs 3–3.5 mm for males, and lay light green elongate eggs that change color from yellowish-white to bright yellow before hatching. After hatching from the eggs in 39 days at optimal temperature (27.5°C), the nymphs undergo four life stages with five instars and take 84 days to reach adulthood (Fig. 10.10). The first generation of this bivoltine species extends from February to May in the northern hemisphere, and is called the spring generation, whereas the second one extends from August to November and is called the autumn generation (El-Shafie, 2015). The two generations vary from one country to another depending on the weather variations.

Dubas bug is an oligophagous insect that is known to complete its life cycle on date palms, *P. dactylifera*, and causes significant direct damage to

Fig. 10.10. Dubas bug nymph (a) and adult (b). (Photos by Rashad Rasool Khan.)

palms through feeding, honeydew production, or damage associated with oviposition (Al-Khatri, 2018). It attacks the leaves of date palm trees by sucking the nutrient fluid sap, resulting in direct damage. As the nymphs' and adults' feeding continues, honeydew secreted by the insects accumulates on the leaf surface and over time leads to the growth of black sooty mold that affects the process of photosynthesis. In addition, the presence of a large amount of honeydew on the leaflets and the accumulation of dust on the honeydew lead to light green or yellowish-green leaflets (El-Shafie, 2015). Moreover, females oviposit eggs into any green or soft date palm tissue including the fruit bunch (except the fruits). This makes necrosis scar symptoms on the leaf tissue where the female lays its eggs. It was reported that extremely heavy populations may cause 50% economic losses and lead to the death of some palms (Shah *et al.*, 2013).

10.2.4.3 Control

CULTURAL CONTROL.   Denser plant population, higher humidity levels, and greater shading provide a more suitable environment for DB population establishment (Howard *et al.*, 2001) as well as hindering the plant protection practices performed by farmers. Provision of appropriate space between palms and canopies should be ensured by farmers through silvicultural practices such as amputation of excess suckers and offshoots and clipping of older leaves (Ali and Hama, 2016). In Panjgur (Pakistan), Shah *et al.* (2013) endorsed the removal of the lower two or three leaves before hatching of the pest's eggs to avoid high pest incidence and crop losses.

BIOLOGICAL CONTROL.   There is no evidence of organized or well accomplished biocontrol against DB; however, several reports describe the coexistence of predators and parasitoids along with DB. Several biocontrol agents like *Chrysoperla carnea*, *Aprostocetus* spp., *Cheilomenes sexmaculata*, and *Runcina* spp. were listed on different DB life stages in Oman while surveying various pest-infested date palm orchards (Al-Khatri, 2018). *Pseudoligosita babylonica* was reported as an egg parasitoid of DB in different date-palm-growing areas (Al-Khatri, 2004, 2018), where it showed more than 70% parasitism in some locations (Al-Khatri, 2017). Other natural enemies coexisting with DB in the field, including mantids, lacewings, and coccinellids, play a significant role in natural suppression of its increasing populations (Ali and Hama, 2016). Besides entomophagous insects, a predacious mite, *Anystis agilis* (Banks), was reported to ravenously eat DB nymphs in Iraq (Al-Jaboory, 2007). Additionally, fungal suspensions of *Trichoderma harzianum* and *Trichoderma viride* displayed 68.5 and 65.8% mortality of nymphal instars, respectively, after 72 hour of application (Sfih *et al.*, 2009).

CHEMICAL CONTROL. Date palm orchards with heavy pest infestation are generally sprayed with chemical insecticides; however, in denser orchards, uniform coverage is ineffectual with ground spraying (Wakil *et al.*, 2015b). Many reports describe the efficacies and toxicities of several pesticides from organophosphate, pyrethroid, nicotinoid, and botanical classes against DB (Abbaszadeh, 2014; Al-Jamali and Kareem, 2014; Bashomaila *et al.*, 2014; Mahmoudi *et al.*, 2014; Salim *et al.*, 2017). The Ministry of Agriculture, Fisheries Wealth and Water Resources, Oman has tested many pyrethroid and organophosphate pesticides to control this pest through aerial and ground spraying. Pesticide aerial sprays included deltamethrin, fenitrothion, etofenprox, malathion, dichlorvos, and combinations of fenitrothion and esfenvalerate, whereas deltamethrin, phenthoate, esfenvalerate, and dichlorvos were sprayed in ground applications (Al-Khatri, 2004; MAF, 2008, 2009, 2018). During 1999 and 2006, about 400 t of insecticides was sprayed through aerial application in Oman (Anonymous, 2010).

10.2.5 Lesser date moth (LDM)

10.2.5.1 Distribution and host range

The LDM, *Batrachedra amydraula* Meyrick (Lepidoptera: Batrachedridae), also known as Hummeira (Al-Jorany *et al.*, 2015), is reported as a serious pest attacking almost all varieties of date palm in many countries, especially in Asia and Africa (Perring *et al.*, 2015; Ali and Hama, 2016; Ali, 2018). The pest is highly specific and restricted to date palms only; however, its severe infestations are reported in African and Asian date-palm-growing countries (Ali and Hama, 2016; Ali, 2018).

10.2.5.2 Life history and damage

LDM larvae start to damage inflorescences and continue feeding on ensuing fruit stages. Resultantly infested fruits dry, turn red, and drop, causing serious yield losses (Ali and Hama, 2016). Ecological study reports describe three overlapping generations in a year. Adults emerge from the overwintering generation, feed on pollen, and female moths oviposit eggs individually on petioles and developing fruits (Kakar *et al.*, 2010; Ali and Hama, 2016). LDM larvae damage dates for a shorter period of time (March–August), with the first generation feeding on new hababouk dates from late March to late April in the northern hemisphere, while the second and third generations appear in May and June, respectively (Blumberg, 2008; Ali and Hama, 2016). LDM larvae burrow into fruit through the cap, feed on the fruit contents leaving behind the fruit rind, and may cause 50–70% fruit yield losses (Blumberg, 2008; Ali, 2018). The LDM is also reported to damage dates in storage (Shayesteh *et al.*, 2010).

10.2.5.3   Control

TRAPPING AND MONITORING.   Monitoring of LDM can be done through light traps fitted with a 150 W mercury bulb (Kakar *et al.*, 2010); however, pheromone traps can also be used for monitoring as well as management purposes (Ali and Hama, 2016). Pheromones traps equipped with insecticides can provide more effective pest management.

CULTURAL CONTROL.   Farm sanitation practices for removal and destruction of fallen infested fruits and removal of bunches remaining on the tree after harvest can significantly reduce pest reinfestation (Perring *et al.*, 2015; Ali and Hama, 2016). Bunch covering with a light cloth or plastic net after pollination helps avoid oviposition by female moths and reduces fruit damage (Perring *et al.*, 2015; Ali and Hama, 2016; Ali, 2018).

BIOLOGICAL CONTROL.   One to two releases of the egg parasitoid, *Trichogramma* spp. (300–500 individuals/tree), backed with five pairs of larval parasitoids, *Bracon* spp., has shown potential for effective LDM control. Releases are recommended against the first generation as inoculation (Ali and Hama, 2016). Recently, a *Goniozus* species found on *B. amydraula* in Oman has been dubbed *Goniozus omani* (Polaszek *et al.*, 2019). Dust or spray formulations of *Bacillus thuringiensis* at 3 g/L water (6–7L/tree) are used to control LDM based on data of monitoring traps (Ali and Hama, 2016).

CHEMICAL CONTROL.   Early-season application of synthetic insecticides at the time of pollination can effectively control LDM infestations. Various formulations of Matrixine Plus® (abamectin 5% w/w; oxymatrine 2.4% w/w) at 0.6 ml/L, Matrixine® (oxymatrine 2.4% w/w) at 1 and 1.5 ml/L, Fytomax PM® (azadirechtin 0.1% w/w) at 5 and 10 ml/L, and Decis® (deltamethrin 2.5% w/w) at 1 ml/L were evaluated under field conditions against LDM. Matrixine at 1.5 ml/L and Decis at 1 ml/L provided 92.3% mortality after 14 days of pesticides spraying (MAF, 2018).

10.2.6   Pests of stored dates

10.2.6.1   Distribution and host range

Vertebrate and invertebrate pests cause significant postharvest losses in dates, and these losses due to insect infestation, birds, and fermentation are quite high in most of the Middle East (Abo-El-Saad and El-Shafie, 2013). In Oman, a survey reported 6.7–22.7% infestation in date palm fields and stored dates in various cities (Al-Zadjali *et al.*, 2006). Moths (Lepidoptera), beetles and weevils (Coleoptera), and some noninsect pests (mites, mollusks, rats, birds, bats) are reported to cause significant losses in dates while in storage (Abo-El-Saad and El-Shafie, 2013). Among the coleopterans, insects from the families Nitidulidae, Silvanidae, Cucujidae, Tenebrionidae, Dermestidae, and Scarabidae are listed as pests of stored dates, whereas those from the lepidopterans include members from the families Pyralidae and Phyticidae (Abo-El-Saad and El-Shafie, 2013; Burks *et al.*, 2015; El-Shafie *et al.*, 2017).

10.2.6.2   Life history and damage

Coleopterous and lepidopterous insects possess variable life history parameters and are mostly dependent on the available climatic conditions. Like insect pests of most other stored products, insect pests of stored dates also flourish due to certain factors like pre-storage infestation of dates (on bunches), infestation during transit (transportation in already infested containers), improper storage conditions (high temperature and humidity), and high moisture contents of dates prior to or during storage (Abo-El-Saad and El-Shafie, 2013).

10.2.6.3   Control

Some activities like elimination of initial infestation, proper sanitation of storage facilities, fumigation of storage houses and stored dates, and proper drying of dates are suggested prior to storage for avoiding pest infestations and damage in stored dates.

MONITORING AND DETECTION.   Different sampling method observations are performed to monitor the stored dates and storage structures for pest infestations. Certain insects and their stages are observed visually in the storage structure, prior to date storage. Similarly, dates are also observed for any pest infestation or insect remnants (insect body parts, frass, or excreta) before storage. Various traps (light, sticky, pitfall, refuge, food bait, or pheromone) can be used to detect and manage different storage pest infestations (Abo-El-Saad and El-Shafie, 2013; Burks *et al.*, 2015).

BIOLOGICAL CONTROL.   The egg parasitoid, *Trichogramma* spp., and the larval parasitoid, *Bracon* spp., can be used for control of various lepidopterous pests of stored dates (Abo-El-Saad and El-Shafie, 2013; Burks *et al.*, 2015).

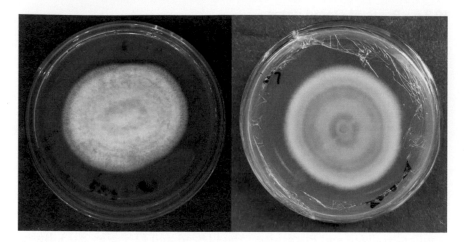

Fig. 10.11. Growth of *Fusarium oxysporum* on potato dextrose agar medium: front view (left) and back view (right). (Photos by Imran Ul Haq.)

CHEMICAL CONTROL. Surface treatment of the storage facility with residual pyrethroid (cyfluthrin) or neonicotinoid (chlorfenapyr) insecticide, insecticide aerosol spraying, and structural fumigation (sulfuryl fluoride) are suggested prior to date storage (Arthur and Subramanyam, 2012; Burks *et al.*, 2015).

10.3   Important Pre- and Postharvest Diseases

10.3.1   Bayoud disease

Bayoud disease is the principal fungal enemy of date palm (Essarioui and Sedra, 2017). It is caused by the fungus *F. oxysporum* f. sp. *albedinis* and may result in yield losses of up to 75% and damage to the whole palm (Killian and Maire, 1930; Toutain, 1965). Bayoud was first recorded around 1870 in Morocco (Foex and Vanssiere, 1919). This highly damaging disease destroyed more than 3 million date palms in Algeria and 10 million in Morocco in a century (Toutain and Louvet, 1974).

10.3.1.1   Aetiology

Bayoud disease is tracheomycosis that occurs due to an attack of *F. oxysporum* f. sp. *albedinis* (Killian and Maire, 1930), as shown in Figs 10.11, 10.12 and 10.13. It has been described by many scientists. Louvet *et al.* (1970) reported the same fungal pathogen causing bayoud disease. *Fusarium* attacks through the roots of the palm and colonizes the vascular system, causing wilt and ultimately leading to tree death (Essarioui *et al.*, 2018).

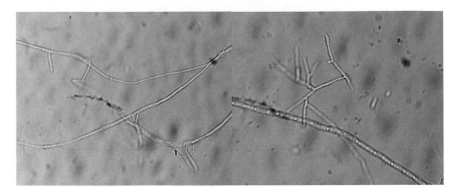

Fig. 10.12. Phaloides of *Fusarium oxysporum* viewed under a compound microscope. (Images by Imran Ul Haq.)

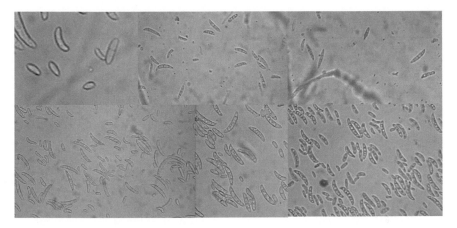

Fig. 10.13. Single-celled, oval-shaped microconidia and fusiform macroconidia (three to seven septations) of *Fusarium oxysporum* viewed under a compound microscope. (Images by Imran Ul Haq.)

Previously, morphological parameters were used for the identification of *F. oxysporum*. Molecular markers based on the translation elongation factor gene proved an informative locus for differentiating the various *Fusarium* species. Phylogenetic studies separate the f. sp. *albedinis* and f. sp. *palmarum* (Elliott *et al.*, 2010). The informative loci ITS and LSU were also used for diversity studies of *Fusarium* (O'Donnell *et al.*, 2008).

10.3.1.2 Symptoms

Bayoud is the significant yield-reducing parameter in date palm production. The disease progression, symptoms, and strategies to manage disease have been well described. Mature palms, young palms, and suckers are affected by bayoud. The typical sign may appear as gray discoloration on leaves of the central portion of the crown, and then withering occurs. Withering progresses from lower to apical portion, then leaf stalk and whole pinnae withering may occur within a few weeks. At first, half of the leaf becomes affected, then symptoms appear at the other side of the leaf and the fungus ultimately damages the whole leaf. On splitting, affected tissues show the sign of the fungus. Reddish-brown discoloration of affected leaves occurs. As it is a systemic problem, the symptoms are also seen in the vascular system of the whole tree. Death of *Fusarium*-affected palms at any growth stage mainly depends on cultivated variety and environmental conditions. The infected roots become reddish-brown in color and die.

10.3.1.3   Disease cycle, epidemiological parameters, and infection cycle

F. oxysporum f. sp. *albedinis* overwinters as chlamydospores or sclerotia on the dead infected roots and in soil. Invasion of *Fusarium* to the vascular tissues occurs through palm roots (de la Perriere and Benkhalifa, 1991). After entry of the pathogenic fungus into the vascular system, it starts to colonize the stem (Ghaemi *et al.*, 2011). Irrigation plays a significant role in dissemination of the fungus from infected to healthy palms. The planting of infected offshoots is one of the primary causes of disease dispersal. Symptoms development after invasion highly depends on cultural practices (Freeman and Maymon, 2000). Heavy irrigation can accelerate the development of bayoud disease (Benzohra *et al.*, 2017). Intercropping of fodders and vegetables (Djerbi *et al.*, 1985) causes an increase in the biological activity of the soil, saprophytic growth of *F. oxysporum*, and may act as symptomless carriers causing inoculum buildup and dissemination of this fungal pathogen (de la Perriere and Benkhalifa, 1991).

10.3.1.4   Control

GENETIC CONTROL. The cultivation of genetically resistant varieties is highly effective and may be the only strategy for protecting date palm plantations from bayoud disease (Sedra and Djerbi, 1986; Saaidi, 1992; El Modafar, 2010; Megateli and Berdja, 2015). Scientists have reported that genetic control is the most reliable and effective against this disease (Saleh *et al.*, 2015). Several resistant varieties have been developed from natural plantations (Djerbi *et al.*, 1989; Sedra, 2003, 2005, 2011). Recently, a new resistant line of date palm was reported that produces good-quality dates (Boumedjout, 2010).

REGULATORY MANAGEMENT. As bayoud is disseminated through contaminated plant debris, infested soils, and infected suckers, regulatory control can restrict the dispersal of the disease. Different countries have already passed such legislation, including Egypt, Iraq, Algeria, Mauritania, Saudi Arabia, Tunisia, and the USA.

CULTURAL CONTROL.   Root grafting is also a cause of disease spread. A trench around the palm can avoid root contact; in this way, the palm can be saved from bayoud for many years (Djerbi, 1983). The spread of infection could be reduced by controlling the irrigation rate, especially in hot regions (Dubost and Kellou, 1974). Proper phytosanitary practices and avoiding growing alternate hosts as intercrop are helpful approaches to prevent the spread of disease. Eradication of infected palms at the stage of early disease detection is also helpful in control. Use of CH_3Br is effective for good soil sterilization because it possesses a strong penetration capacity into the soil (Frederix and Den Brader, 1989).

BIOLOGICAL CONTROL.   Biological control of pathogens by using antagonistic microorganisms to suppress plant diseases has proven to be a promising alternative management strategy (El Hadrami *et al.*, 2011). The efficacy of several biocontrol agents has been evaluated, including fungal and bacterial antagonists, to protect plants against several plant pathogens (Arfaoui *et al.*, 2007; El Hassni *et al.*, 2007). The antagonistic microorganisms may exhibit different modes of action such as induced resistance, competition, direct parasitism, and antibiosis (Compant *et al.*, 2005). The biological agents can also enhance the defense mechanism of plants as an effective alternative to using pesticides against plant diseases (Dehne, 1982). Several scientists have reported bacteria that promote plant growth, such as *Pseudomonas* spp. and *Bacillus* spp. (El Hassni *et al.*, 2007). Nonreactive soil hinders the development of bayoud disease effectively (Sedra and Maslouhy, 1995). The amendment of date palm soils (to enhance the suppressive activity of soils) by bacteria such as *Bacillus* (Chakroune *et al.*, 2008) and *Actinomycetes* (Amir and Sabaou, 1983; Maslouhy, 1989), as well as fungi including species of nonpathogenic *Fusarium* (Sneh, 1998), *Aspergillus*, and *Penicillium* (Chakroune *et al.*, 2008), has proven an effective means for biological suppression of pathogenic *Fusarium* inoculum in soil (Ouhdouch *et al.*, 1996).

CHEMICAL CONTROL.   Identification of the primary source of infection is crucial in chemical control. Before planting, the pathogen overwintering in soil is treated with CH_3Br. Fungicides such as prochloraz, bromuconazole, benomyl, thiophanate methyl, and carbendazim through drip irrigation are effective in controlling soilborne *F. oxysporum* in many field crops. Hence, chemigation through flood irrigation, drenching, and drip irrigation, at the early stages of disease development, could prove effective (Benzohra *et al.*, 2015).

10.3.2   Decline

The decline disease has many similarities to bayoud regarding pathogens, symptoms, and other aspects. Several *Fusarium* species have been reported

Fig. 10.14. Symptoms of decline in date palm: (a) *Fusarium* symptoms on rachis; (b) infection on cross-section of rachis. (Photos by Imran Ul Haq.)

to be involved in decline. Symptoms appear on all parts of the tree like the trunk, roots, and leaves (Figs 10.14 and 10.15). The symptoms are more like wilt and dieback.

10.3.2.1   Aetiology

The complex of *Fusarium* species is involved in decline, including *F. proliferatum* (Abdalla *et al.*, 2000), *F. solani*, and *Fusarium monliforme* (Rashed and Abdel Hafeez, 2001). Molecular diagnosis also has been done using primers of internal transcribed spacer (ITS) regions, β-tubulin gene, and translational elongation factor 1-α (TEF1-α) (Alwahshi *et al.*, 2019).

10.3.2.2   Symptoms

Other symptoms of decline are almost like bayoud disease, as discussed in Section 10.3.1.2 above. Yellow streaks appear on the rachis and the palm gradually dries out (Fig. 10.14). Necrosis occurs on the fruit stalk, and new stalks become stunted (Sarhan, 2001).

10.3.2.3   Dissemination of pathogen

Pathogens causing decline are disseminated to a healthy field by the movement of animals, humans, irrigation water, soil, infested plants parts, and wind.

10.3.2.4   Control

CULTURAL CONTROL.   Pathogen spread in decline occurs through grafting, irrigation, and injuries during field operations. Hence, utmost care while selecting grafts, optimum irrigation, especially in the summer, and care during farm operations are crucial management strategies.

Fig. 10.15. Symptoms of decline in date palm: (a, b) wilted leaves; (c) affected tree. (Photos by Imran Ul Haq.)

BIOLOGICAL CONTROL. Biological control is preferred in controlling disease as pathogens rarely develop resistance against biocontrol agents. For decline of date palm, *T. harzianum* and *Trichoderma longibrachiatum* are effective antagonists. Among endophytic fungi, *Ulocladium chartarum* significantly reduces disease spread (Nishad and Ahmed, 2020).

CHEMICAL CONTROL. Chemical control is indispensable when decline symptoms start to appear in the field. Bavistin, thiophanate methyl, fosetyl-aluminum, and mancozeb can effectively control the pathogen and consequent symptoms. If chemicals are applied at early stages, plants recover and produce fruit the next season (Markhand *et al.*, 2013).

Fig. 10.16. The appearance of fruiting bodies (sori) on leaflets of date palm. (Photos by Imran Ul Haq.)

10.3.3   Graphiola leaf spot

Graphiola leaf spot of *P. dactylifera* is also known as false smut. *Graphiola phoenicis* is the causal agent of the Graphiola leaf spot of the date palm. The disease is found in all countries growing date palm as a cash crop such as Mali, Senegal, Algeria, Qatar, Egypt, Libya, Morocco, Niger, Mauritania, the USA, and Argentina (Sedra, 2003, 2011; El-Deeb *et al.*, 2007; El-Gariani *et al.*, 2007). The primary host of this pathogen is the date palm tree. In semi-arid and arid regions, the intensity of *Graphiola* is very high. It is observed that if the infection occurs in a severe form, it can cause a reduction in tree growth (Abbas and Abdulla, 2004). This disease is widely present in humid areas. This false smut can also occur in other species of the *Phoenix* genus such as *Phoenix canariensis* and *Phoenix sylvestris.* Environmental conditions play a critical role in disease severity and incidence.

10.3.3.1   Aetiology

The fruiting bodies (sori) of *Graphiola* are 0.5–1.2 mm in diameter and contain hard walls that are dark in color. After the maturation of fruiting bodies (sori), the ostiole develops from each sorus (Fig. 10.16). The 2.5–3 μm, smooth, thick-walled spermacia, spherical to elliptical, can be observed on the filaments. These filaments help in the dispersal of the pathogen (Sepúlveda *et al.*, 2017). The superficial appearance of Graphiola leaf spot is similar to mealybug or scale insect infestation, but when closely observed, the former shows the emergence of yellow powdery spores on white filaments (Abbas and Abdulla, 2004).

10.3.3.2 Symptoms

The initial symptoms appear on the lower and upper sides of the leaf as tiny, yellow-brown spots or pustules that later turn black (Mani *et al.*, 2015). Younger plants show necrotic lesions on the leaflets of the palm. The symptoms of Graphiola leaf spot are difficult to observe as compared to the signs. Both symptoms and signs are present on the older leaves of the date palm, while in the case of severe infection, younger leaves are also infected. The fruiting structures (sori) emerge from the leaf surface and can easily be observed; their diameter is less than 1.6 mm, and these sori (fruiting bodies) are black. At the maturity of sori, they produce yellow spores that further form short and thread-like structures. These thread-like structures (filaments) are the sites from which spores are dispersed. After spore dispersion, these sori look like black craters or black cups. The number of sori that appear indicates the severity of the disease (Elliott, 2018).

10.3.3.3 Diagnosis and perpetuation

Infection takes place in small spots on both the upper and lower sides of the date palm leaf. The fungus forms fruiting structures (i.e. black-colored sori) from the infected leaf portion. After maturity, sori rupture releasing hyaline and spherical spores that serve as the primary inoculum. Insects, birds, and wind are main sources of spore dissemination.

This fungus has a very long latent period (time from infection to emergence of the first spore) of 10–11 months, compared to other leaf pathogens where the latent period is measured in weeks. When disease symptoms are observed in the field, the fungus started the infection almost a year ago.

10.3.3.4 Epidemiology

For the establishment of this fungus on the palm, the favourable temperature range is 6–23°C. The germination and penetration of spores into the leaf require moisture for at least 10–12 hour (Mani *et al.*, 2015). The exact conditions required for the development of Graphiola leaf spot are not well known, but this disease is mainly observed in humid regions.

10.3.3.5 Control

CULTURAL CONTROL. The disease appears in humid conditions; therefore, it is necessary to minimize the humidity. Proper spacing is an essential strategy to manage the disease. At the time of irrigation, water should be applied below the canopy of the date palm. To minimize the dew period, trees should be irrigated before sunrise (early morning) because dew is already present at this time. It is recommended that the infested and older leaves are removed to reduce disease inoculum. Pruning is not recommended if the plant suffers from nutrient deficiency such as potassium deficiency. GENETIC CONTROL. The use of genetically tolerant lines or varieties is also effective. The date palm varieties susceptible to Graphiola leaf spot such as Zahidi, Maktoom, Khisab, and Ashrasi should not be grown. In contrast, genetically tolerant varieties can be cultivated, such as Khastawi, Tadala, Iteema, and Jouzi (Nixon, 1957; Sinha *et al.*, 1970). CHEMICAL CONTROL. Chemical control of the fungus employs chemicals such as four or five applications of maneb at 15 day intervals after sporulation. Bordeaux mixture is also effective to control Graphiola leaf spot of date palm.

10.3.4 Pestalotia leaf spot

Pestalotia causes bud rot, leaf spot, and blight in date palm trees. The fungus attacks date palm leaves and affects the whole leaf from top to bottom. It is an opportunistic fungus that can be isolated from the healthy tissues of date palm. In establishing an infection in plant tissues, the fungus requires a wound to penetrate, and these wounds are created by other pathogens. *Pestalotia* can attack guava, apple, date palm, pomegranate, mango, and avocado.

10.3.4.1 Symptoms

The disease can develop on the rachis, or be confined only to leaflets, or may occur on both. Small spots appear on the leaf that are brownish black in the beginning. These spots do not progress under unfavorable conditions. These spots increase in number and size upon favourable conditions and then collapse, resulting in rachis or leaf blight. As the disease progresses, the color of spots becomes gray with black margins. On the petiole, spots merge to form lesions. After destroying petiole tissues in the plant the fungus then attacks the vascular tissue. The pathogen affects the growing point when it is spread to the apical meristem tissues. The effect of this fungus has been seen on both mature and juvenile trees of palm. The disease might not be very destructive to mature date palms if it is restricted to causing leaf spots. The juvenile tree may be severely affected by the fungus because of the absence of a trunk and few leaves.

10.3.4.2 Control

CULTURAL CONTROL. As the fungus is disseminated through spores dispersed by water (irrigation and rain splashes), it is essential to maintain proper sanitation and adopt water management practices. Water management is done by lessening the duration of time for which the leaf is exposed to high humidity. For this purpose, irrigation should be done early in the morning, and overhead irrigation should be avoided. Leaf wetness can also be decreased by proper ventilation. Proper plant-to-plant and row-to-row distances should be maintained to keep sufficient air circulation and proper penetration of sunlight. Proper nutrients should be applied to trees because chlorosis occurs in leaves due to nutrient deficiency, leading to wound development, which helps the pathogen enter leaf tissues. It is recommended to use broad-spectrum fungicides as a foliar spray. The diseased leaves should be pruned before applying fungicides to reduce the disease inoculum.

CHEMICAL CONTROL. Application of fungicide will control the spread of disease to new or uninfected leaves. Broad-spectrum fungicides such as mancozeb, sulfur, and captan are effective when sprayed before disease spread.

10.3.5 *Botryodiplodia theobromae* rot

10.3.5.1 Aetiology

Botryodiplodia theobromae (syn. *Lasiodiplodia theobromae*) is an important plant pathogen that is widespread in many parts of the world. It is a vigorous and aggressive pathogen that causes different diseases on many hosts, i.e. root rots, gummosis, fruit rot, dieback, witches' broom disease, blights, leaf spot, and stem necrosis. It infects a large number of host plants such as date palm, guava, avocado, apple, grapevine, mango, and peach, as well as some pomes and other stone fruit trees (Abdl-Megid and Gafer, 1966; Darvas and Kotze, 1987; Latham and Dozier, 1989; Barakat *et al.*, 1990; Aly *et al.*, 2002; Baiuomy *et al.*, 2003). *B. theobromae* is a notorious plant pathogen reported from 500 hosts and mostly restricted to 40°N to 40°S of the equator. It has been reported in date palm from date-palm-growing regions worldwide. This fungus is most prevalent in tropical and subtropical regions of the world.

Colonies of *Botryodiplodia* are fluffy, have grayish sepia, mouse gray to black in color, with abundant aerial mycelium; the back side of the colony is dark and somber to black. Simple or compound pycnidia, stromatic, often aggregated, frequently setose, ostiolate, are up to 5 mm wide. Simple, hyaline conidiophores are often septate, cylindrical, and rarely branched. Simple, hyaline conidiogenous cells are cylindrical to pyriform. Hyaline conidia are unicellular when immature, subovoid to ellipsoid, thick-walled, granulose; mature conidia have one septation, cinnamon to fawn color, and sometimes longitudinal striations, having $20–30\,\mu m \times 10–15\,\mu m$ size.

10.3.5.2 Symptoms

A yellowish-brown lesion that extends on the rachis is a characteristic symptom of the disease. Leaves turn yellowish-white. After disease progression, lesions turn brown, and the terminal bud may be infected. Other symptoms include wilting and accidental desiccation of the youngest leaves. The youngest leaves are easily pulled by hand because their base is soft rotted.

The base of the youngest leaves dies, and blackish mycelial growth appears on it. At times, mature leaves remain uninfected for many weeks after the loss of the bud. In addition to this, the pathogen attacks the date palm's trunk and disintegrates its fibers, leading to charcoal soft rot. On the infected area, deep cracks and constrictions appear. Symptoms in the form of necrosis appear on a topmost portion of the palm, crown, and terminal bud and offshoot of date palm, and yellowish-brown streaks may appear on leaves, resulting in loss of the offshoot bud.

10.3.5.3 Disease cycle

Growth rate and disease severity are mainly dependent on climatic conditions, especially temperature and relative humidity. The most favourable range of temperature is 30–35°C; 30°C is the optimum temperature for growth. Fungal growth ceases at 40°C (Saha *et al.*, 2008). The fungus requires 70% relative humidity for its growth. The fungus overwinters in the form of pycnidia externally on the diseased wood. Conidia produced in pycnidia are striated, dark brown, and two-celled.

The conidia are disseminated by rain and wind. After the landing of conidia on wounded wood, the disease develops. The conidia germinate in the wood of diseased plants and infect the vascular system. Canker formation starts around the infection point. Complete deterioration of vascular tissues leads to necrosis and dieback. Sometimes, on the outside of cankers, pseudothecia are formed that produce ascospores. Ascospores spread in the same way as conidia and infect the host through wounds. Upon unfavorable conditions, *B. theobromae* forms chlamydospores and specialized hyphae. Upon favourable conditions, chlamydospores germinate and cause a new infection. The specialized hyphae act as the source of infection (Ogundana, 1983).

10.3.5.4 Control

CULTURAL CONTROL.   Hot water treatment to eradicate the pathogen is a proven efficient method to control the disease.

CHEMICAL CONTROL.   Chemical management of fungal pathogens has proven most effective. The use of thiabendazole and mancozeb, individually or combined, has been evaluated *in vitro* and proven effective (da Silva Pereira *et al.*, 2012). Other fungicides, for example, dicarboximides, phthalamides, oximinoacetates, and triazoles, are very effective in controlling this fungus. Washing of infected parts and treatment with thiabendazole can reduce the level of inoculum.

10.3.6 Fruit rot

Some bacteria (*Acetobacter*) and fungi are the causal agents of date palm tree fruit rot. For example, the fungi that cause fruit rot include *Alternaria alternata, Aspergillus fumigatus, Aspergillus japonicus, Aspergillus niger, Aspergillus flavus, Fusarium* spp., *Penicillium* spp., *Botryodiplodia* spp., *Ceratostomella* spp., *Cladosporium* spp., and *Thielaviopsis paradoxa.* The problem of fruit rot in a date palm tree is present throughout the world. In storage, fruit rot can cause significant loss in the rainy season. At the khalal and maturation stage, high humidity and rain are of great concern due to losses (Carpenter and Elmer, 1978). Environmental conditions such as humidity and rainfall play a significant role in fruit rot. Losses of 10–15% have been recorded due to fruit rot disease (Darley and Wilbur, 1955; Calcat, 1959; Djerbi *et al.,* 1989).

10.3.6.1 Symptoms

There are two types of symptoms on fruit: some show soft water- or oil-soaked spots that are translucent and different in size; others show brown, rusty spots covering almost one side of the fruit. The size of these spots increases with time, and they often collapse and form an area with a dark-chocolate margin that is grayish or light creamy color in the center. The center of the spots is brown. In these centers, spores appear in pustules form. After the enlargement of soft spots, they give a rotting appearance. This leads to rupturing of epidermal tissues, resulting in water loss, and the fruit becomes mummified.

Poor postharvest handling and high humidity facilitate the development of disease. The fungal spores are dispersed through wind and rain.

10.3.6.2 Control

CULTURAL CONTROL. Adoption of preventive measures can be helpful in the management of the disease. Good ventilation in the field may reduce the moisture level, which minimizes humidity. At the khalal stage, fruit branches should be covered with paper bags to minimize fruit wetting. The avoidance of bird and insect attacks and fruit injuries may be helpful in the management of the disease.

CHEMICAL CONTROL. A dusting of 5% w/w ferbam, malathion, and 50% w/w sulfur can manage the disease (Djerbi, 1983).

10.4 Genome Editing for Management of Stresses

Tissue culture, breeding (marker-assisted), and DNA fingerprinting biotechnological approaches have previously been used in date palm but could not improve quality and yield significantly. However, it may be possible to make modifications in the genome of date palm to provide resistance against biotic and abiotic stresses. Precise genetic modification (genome editing) could be an effective supplementary tool for improving crops. Several genome editing techniques including zinc finger nucleases (ZFNs) (Shukla *et al.*, 2009; Zhang and Voytas, 2011; de Pater *et al.*, 2013), transcription activator-like effector nucleases (TALENs) (Christian *et al.*, 2013; Wendt *et al.*, 2013), and clustered regularly interspaced short palindromic repeats (CRISPR)/CRISPR-associated protein 9 (Cas9) have been efficiently used for different crops including fruit trees. Due to competency and simplicity, CRISPR/Cas9-based management approaches have a high success rate. CRISPR/Cas9 has been successful as a genome editing tool for resistance against different diseases such as citrus canker (Peng *et al.*, 2017), powdery mildew (Zhang *et al.*, 2017), rice blast (Wang *et al.*, 2017), phytophthora (Fang and Tyler, 2016), and plant viruses (Khatodia *et al.*, 2017). Date palm is a diploid tree ($2n=36$). The average genome size is 670 Mb (Al-Mssallem *et al.*, 2013) and consists of 18 chromosomes, which is why extensive attention is needed to determine the efficacy of CRISPR/Cas9. In date palm, mutations have been used to create genetic variation and selection against bayoud disease (Jain *et al.*, 2011); however, such mutagenesis has a severe drawback (Braatz *et al.*, 2017). The use of the CRISPR/Cas9 system for specific mutagenesis enables scientists to understand gene expression and target multiple loci in the date palm genome, and can also be employed to explore secondary metabolites, target pathogen effectors, and improve qualitative and quantitative traits. This approach is useful to target date palm pathogens through gene drives.

10.5   Artificial Intelligence in Date Palm Pathogen Management

Artificial intelligence (AI) has proven itself as an outstanding, precise, and robust technology. This technology helps to trace a problem and propose actions to solve that problem. Similarly, geoinformatics tools and technologies – including Geographic Information Systems (GIS), Global Positioning System (GPS), remote sensing (RS), modelling, web apps, sensor technology, image processing by high-powered computer systems, geotagging, drones – can be used to collect, archive, analyse, and visualize pests and diseases efficiently and economically for integrated pest and disease management (Biradar *et al.*, 2017). Previously, intelligent systems were developed by using PHP as the web programming language. Evaluation of the diagnostic process showed the expert system running well in diagnosing the date palm diseases. It could be concluded that the AI system can better help farmers diagnose date palm diseases early and works as an assistant tool to produce a better quality of date palm product. After nearly 30 years of development, its application has spread to cover various fields of agriculture in areas such as disease diagnosis and pest control, water scheduling, soil preparation, weed management, etc. With the recent agricultural growth, farmers require advice or expert knowledge to take decisions on different aspects in their domain to get a high yield. In the agricultural domain, knowledge and expertise in making decisions exists, but the major problem is making it available to a large number of growers. Expert systems introduce a powerful tool to provide efficient and effective advice to growers anytime and anywhere. The primary goal of this AI system is as an expert system to diagnose date palm fungal diseases. The domain of the system is limited to the following diseases: inflorescence rot, Fusarium wilt, Graphiola leaf spot, black scorch, and Diplodia disease. The main objectives are to provide a fast solution for the user to detect early date palm diseases. For future work, the domain of an agricultural system will be extended to accommodate new types of date palm diseases (Al-Ahmar, 2009).

10.6   Conclusions and Prospects

Date palm is an economically important fruit crop and holds a considerable position in trade and economies around the globe. Entomological, physiological, and pathological problems have had a negative impact on the potential yield of date palms. Among pathological factors, fungal diseases cause catastrophic effects on yield and quality of date crop. Fungal diseases are the major cause of losses and significantly reduce profitability. Despite research work focusing on the identification of pests and diseases and their cultural and chemical control, RNAi-based pest management can be a futuristic approach. Furthermore, there is a dire need to apply AI in early detection of pests and diseases on date palm and for the use of drones in their control.

References

Abbas, E.H. and Abdulla, A.S. (2004) First report of false smut disease caused by *Graphiola phoenicis* on date palm trees in Qatar. *Plant Pathology* 53(6), 815. DOI: 10.1111/j.1365-3059.2004.01084.x.

Abbaszadeh, G. (2014) Efficacy of different insecticides against dubas bug, *Ommatissus lybicus* (Deberg) Aschae and Wilson. *Annals of Plant Protection Sciences* 22(2), 240–243.

Abdalla, M.Y., Al-Rokibah, A., Moretti, A. and Mulè, G. (2000) Pathogenicity of toxigenic *Fusarium proliferatum* from date palm in Saudi Arabia. *Plant Disease* 84(3), 321–324. DOI: 10.1094/PDIS.2000.84.3.321.

Abdl-Megid, K. and Gafer, K. (1966) Diplodia leaf stalk and off-shoots disease of date palm. *Journal of Microbiology of the United Arab Republic* 1(2), 201–206.

Abo-El-Saad, M. and El-Shafie, H. (2013) Insect pests of stored dates and their management. In: Siddiq, M., Aleid, S.M. and Kader, A.A. (eds) *Dates: Postharvest*

Science, Processing Technology and Health Benefits. Wiley, Chichester, UK, pp. 81–104.

Al-Ahmar, M.A. (2009) An object-oriented expert system for diagnosis of fungal diseases of date palm. *International Journal of Soft Computing* 4(5), 201–207.

Al-Deeb, M.A. and Khalaf, M.Z. (2015) Longhorn stem borer and frond borer of date palm. In: Wakil, W., Faleiro, J.R. and Miller, T.A. (eds) *Sustainable Pest Management in Date Palm: Current Status and Emerging Challenges.* Springer, Cham, Switzerland, pp. 63–72.

Al-Deeb, M.A., Muzaffar, S.B., Abuagla, A.M. and Sharif, E.M. (2011) Distribution and abundance of phoretic mites (Astigmata, Mesostigmata) on *Rhynchophorus ferrugineus* (Coleoptera: Curculionidae). *Florida Entomologist* 94(4), 748–755. DOI: 10.1653/024.094.0403.

Al-Dosary, N.M., Al-Dobai, S. and Faleiro, J.R. (2016) Review on the management of red palm weevil *Rhynchophorus ferrugineus* Olivier in date palm *Phoenix dactylifera* L. *Emirates Journal of Food and Agriculture* 28(1), 34–44. DOI: 10.9755/ejfa.2015-10-897.

Ali, A.-S.A. (2018) IPM of lesser date moth. In: El Boushi, M. and Faleiro, J.R. (eds) *Date Palm Pests and Diseases, Integrated Management Guide.* International Center for Agricultural Research in the Dry Areas (ICARDA), Beirut, pp. 105–113.

Ali, A.-S.A. and Hama, N.N. (2016) Integrated management for major date palm pests in Iraq. *Emirates Journal of Food and Agriculture* 28(1), 24–33. DOI: 10.9755/ejfa.2016-01-032.

Al-Jaboory, I.J. (2007) Survey and identification of the Biotic factors in the date palm environment and its application for designing IPM-program of date palm pests in Iraq. *University of Aden Journal of Natural and Applied Sciences* 11(3), 423–457.

Al-Jamali, N.A. and Kareem, T.A. (2014) Evaluation the efficiency of spinetoram 12SC against dubas bug on the date palm *Ommatissus binotatus lybicus* (Homoptera: Tropiduchidae). In: *Proceedings of the 5th International Date Palm Conference, Abu Dhabi, United Arab Emirates,* Khalifa International Date Palm Award, Abu Dhabi, March 16–18, 2014, pp. 467–470.

Al-Jorany, R.S., Al-Jboory, I.J. and Hassan, N. (2015) Evaluation of the sex pheromone efficiency of the lesser date moth, *Batrachedra amydraula* Meyrick (Lepidoptera: Batrachedridae). *Journal of Life Sciences* 9(5), 242–247.

Alkahtani, M., El-Naggar, M.A., Omer, S.A., Abdel-Kareem, E.M. and Ammar, M.I. (2011) Effect of toxic *Fusarium moniliforme* on some biochemical component of some date palm cultivars. *American Journal of Food Technology* 6(9), 730–741. DOI: 10.3923/ajft.2011.730.741.

Al-Khatri, S.A. (2004) Date palm pests and their control. In: Peña, J.E. (ed.) *Proceedings of Date Palm Regional Workshop on Ecosystem-Based IPM for Date Palm in Gulf Countries,* Al-Ain, United Arab Emirates, March 28–30, 2004, pp. 28–30.

Al-Khatri, S.A. (2017) Efficiency of some insecticides against lesser date moth, *Batrachedra amydraula.* In: *1st International Conference on 'Integrated Protection of Date Palms,'* Al Manama, Kingdom of Bahrain, March 13–14, 2017.

Al-Khatri, S.A. (2018) IPM of dubas bug. In: Bouhssini, M.E. and Faleiro, J.R. (eds) *Date Palm Pests and Diseases, Integrated Management Guide.* International Center for Agricultural Research in the Dry Areas (ICARDA), Beirut, pp. 94–104.

Al-Mssallem, I.S., Hu, S., Zhang, X., Lin, Q., Liu, W. *et al.* (2013) Genome sequence of the date palm *Phoenix dactylifera* L. *Nature Communications* 4, 2274. DOI: 10.1038/ncomms3274.

Al-Sadi, A.M., Al-Jabri, A.H., Al-Mazroui, S.S. and AlMahmooli, I.H. (2012) Characterization and pathogenicity of fungi and oomycetes associated with root diseases of date palms in Oman. *Crop Protection* 37, 1–6.

Al-Shawaf, A.M., Al-Shagag, A., Al-Bagshi, M., Al-Saroj, S., Al-Bather, S. *et al.* (2013) A quarantine protocol against red palm weevil *Rhynchophorus ferrugineus* (Olivier) (Coleptera: Curculiondae) in date palm. *Journal of Plant Protection Research* 53(4), 409–415. DOI: 10.2478/jppr-2013-0061.

Alwahshi, K.J., Saeed, E.E., Sham, A., Alblooshi, A.A., Alblooshi, M.M. *et al.* (2019) Molecular identification and disease management of date palm sudden decline syndrome in the United Arab Emirates. *International Journal of Molecular Sciences* 20(4), 923. DOI: 10.3390/ijms20040923.

Al-Yasiri, I.I., Saad, N.A., Nasser, A.R., Hassan, S.A. and Zaid, K.M. (2010) The relationship between the fungus *Fusarium solani* and some pathological phenomena on date palm trees and the effectiveness of some systemic fungicide for their control. *Acta Horticulturae* 882, 505–514. DOI: 10.17660/ActaHortic.2010.882.57.

Aly, A.Z., Tohamy, M.R.A., Atia, M.M.M., El-Shimy, H. and Kamhawy, M.A.M. (2002) Grapevine twigs tip die-back disease in Egypt. *Egyptian Journal of Phytopathology* 30, 45–56.

Al-Zadjali, T.S., Abd-Allah, F.F. and El-Haidari, H.S. (2006) Insect pests attacking date palms and dates in Sultanate of Oman. *Egyptian Journal of Agricultural Research* 84(1), 51–59. DOI: 10.21608/ejar.2006.228947.

Amir, H. and Sabaou, N. (1983) Le palmier dattier et la fusariose. XII – Antagonisme dans le sol de deux actinomycétes vis-à-vis de *Fusarium oxysporum* f. sp. *albedinis* responsable du Bayoud. *Mémoires de La Société d'Histoire Naturelle de l'Afrique Du Nord* 13, 47–60.

Anonymous (2010) *IPM of Dubas Research Program.* The Research Council, Sultanate of Oman.

Arfaoui, A., El Hadrami, A., Mabrouk, Y., Sifi, B., Boudabous, A. *et al.* (2007) Treatment of chickpea with *Rhizobium* isolates enhances the expression of phenylpropanoid defense-related genes in response to infection by *Fusarium oxysporum* f. sp. *ciceris. Plant Physiology and Biochemistry* 45(6–7), 470–479.

Armengol, J., Moretti, A., Perrone, G., Vicent, A., Bengoechea, J.A. *et al.* (2005) Identification, incidence and characterization of *Fusarium proliferatumon* ornamental palms in Spain. *European Journal of Plant Pathology* 112(2), 123–131. DOI: 10.1007/s10658-005-2552-6.

Arthur, F.H. and Subramanyam, B. (2012) Chemical control in stored products. In: Hagstrum, D.W., Phillips, T.W. and Cuperus, G.W. (eds) *Stored Product Protection.* Kansas State University, Manhattan, Kansas, pp. 95–100.

Ashry, I., Mao, Y., Al-Fehaid, Y., Al-Shawaf, A., Al-Bagshi, M. *et al.* (2020) Early detection of red palm weevil using distributed optical sensor. *Scientific Reports* 10(1), 3155. DOI: 10.1038/s41598-020-60171-7.

Baiuomy, M.A.M., Kamhawy, M.A., El-Shemy, H.A. and Moustafa, M.M. (2003) Root-rot of guava and it's control in Egypt. *Zagazig Journal of Agricultural Research* 30, 801–817.

Barakat, F.M., Abdel Salam, A.M., Abada, K.A.A. and Korra, A.K.M. (1990) Pathogenicity of fungi associate with die-back of peach and some pome and stone fruit trees. In: *Proceedings of the Sixth Congress of Phytopathology*, Cairo, Egypt, June 10–14, 1990, pp. 311–323.

Bashomaila, S.M., Al-Jboory, I.J. and Madi, A.O. (2014) Field efficacy of bio-rational pesticide fytomax N against dubas bug, *Ommatissus lybicus* de Berg (Homoptera: Tropiduchidae) in autumn and spring generation. In: *Proceedings of the 5th International Date Palm Conference, Al-Ain, United Arab Emirates*, Khalifa International Date Palm Award, Abu Dhabi, March 16–18, 2014, pp. 387–392.

Bedford, G.O., Al-Deeb, M.A., Khalaf, M.Z., Mohammadpour, K. and Soltani, R. (2015) Dynastid beetle pests. In: Wakil, W., Romeno Faleiro, J. and Miller, T.A. (eds) *Sustainable Pest Management in Date Palm: Current Status and Emerging Challenges*. Springer, Cham, Switzerland, pp. 73–108.

Benzohra, I.E., Megateli, M. and Berdja, R. (2015) Bayoud disease of date palm in Algeria: history, epidemiology and integrated disease management. *African Journal of Biotechnology* 14(7), 542–550.

Benzohra, I.E., Megateli, M., Elayachi, B.A., Zekraoui, M., Djillali, K. *et al.* (2017) Integrated management of Bayoud disease on date palm (*Phoenix dactylifera* L.) caused by *Fusarium oxysporum* f. sp. *albedinis* in Algeria. *Journal Algérien Des Régions Arides* 14, 93–100.

Bertone, C., Michalak, P.S. and Roda, A. (2010) New pest response guidelines, red palm weevil (*Rhynchophorus ferrugineus*). Available at: www.aphis.usda.gov/import_export/plants/ (accessed 1 March 2011).

Biradar, C., Assi, L., Omer, K., El-Shama, K., El-Bouhssini, M. *et al.* (2017) *Geo-ICTs for Mapping and Managing of Pests and Diseases Risks in Date Palm*. Global Forum for Innovation in Agriculture (GFIA), Abu Dhabi.

Blumberg, D. (2008) Review: date palm arthropod pests and their management in Israel. *Phytoparasitica* 36(5), 411–448. DOI: 10.1007/BF03020290.

Boumedjout, H. (2010) Morocco markets Bayoud-resistant strains of date palm. www.natureasia.com/en/nmiddleeast/article/10.1038/nmiddleeast.2010.220 DOI: 10.1038/nmiddleeast.2010.220. (accessed 8 December 2022).

Braatz, J., Harloff, H.-J., Mascher, M., Stein, N., Himmelbach, A. *et al.* (2017) CRISPR-Cas9 targeted mutagenesis leads to simultaneous modification of different homoeologous gene copies in polyploid oilseed rape (*Brassica napus*). *Plant Physiology* 174(2), 935–942. DOI: 10.1104/pp.17.00426.

Burks, C.S., Yasin, M., El-Shafie, H.A. and Wakil, W. (2015) Pests of stored dates. In: Wakil, W., Faleiro, J.R. and Miller, T.A. (eds) *Sustainable Pest Management in Date Palm: Current Status and Emerging Challenges*. Springer, Cham, Switzerland, pp. 237–286.

Calcat, A. (1959) Diseases and pests of date palm in the Sahara and North Africa. *FAO Plant Protection Bulletin* 8, 5–10.

Carpenter, J.P. and Elmer, H.S. (1978) *Pests and Diseases of Date Palm*. Agriculture Handbook No. 523. US Department of Agriculture, Washington, DC.

Chakroune, K., Bouakka, M., Lahlali, R. and Hakkou, A. (2008) Suppressive effect of mature compost of date palm by-products on *Fusarium oxysporum* f. sp. *albedinis*. *Plant Pathology Journal* 69(6), 521–529. DOI: 10.3923/ppj.2008.148.154.

Christian, M., Qi, Y., Zhang, Y. and Voytas, D.F. (2013) Targeted mutagenesis of Arabidopsis thaliana using engineered TAL effector nucleases. *G3 Genes Genomes Genetics* 3(10), 1697–1705. DOI: 10.1534/g3.113.007104.

Cohen, Y., Freeman, S., Zveibil, A., Ben Zvi, R., Nakache, Y. *et al.* (2010) Reevaluation of factors affecting bunch drop in date palm. *HortScience* 45(6), 887–893. DOI: 10.21273/HORTSCI.45.6.887.

Compant, S., Duffy, B., Nowak, J., Clément, C. and Barka, E.A. (2005) Use of plant growth-promoting bacteria for biocontrol of plant diseases: principles, mechanisms of action, and future prospects. *Applied and Environmental Microbiology* 71(9), 4951–4959.

Darley, E.F. and Wilbur, W.D. (1955) Results of experiment on control of fruit spoilage of Deglet Noor and Saidy dates in California, 1935–1954. *Date Growers' Institute Annual Report* 32, 14–15.

Darvas, J.M. and Kotze, J.M. (1987) Fungi associated with pre- and post-harvest diseases of avocado fruit at Westfalia Estate, South Africa. *Phytophylactica* 19, 83–85.

da Silva Pereira, A.V., Martins, R.B., Michereff, S.J., da Silva, M.B. and Câmara, M.P.S. (2012) Sensitivity of *Lasiodiplodia theobromae* from Brazilian papaya orchards to MBC and DMI fungicides. *European Journal of Plant Pathology* 132(4), 489–498. DOI: 10.1007/s10658-011-9891-2.

Dehne, H.W. (1982) Interactions between vesicular arbuscular mycorrhizal fungi and plant pathogens. *Phytopathologische Zeitung* 72, 1115–1119.

de la Perriere, R.A.B. and Benkhalifa, A. (1991) Progression of *Fusarium* wilt of the date palm in Algeria. *Drought* 2, 119–128.

Dembilio, O., Llácer, E., Martínez de Altube, M.D.M. and Jacas, J.A. (2010) Field efficacy of imidacloprid and *Steinernema carpocapsae* in a chitosan formulation against the red palm weevil *Rhynchophorus ferrugineus* (Coleoptera: Curculionidae) in *Phoenix canariensis*. *Pest Management Science* 66(4), 365–370. DOI: 10.1002/ps.1882.

Dembilio, Ó. and Jaques, J.A. (2015) Biology and management of red palm weevil. In: Wakil, W., Faleiro, J.R. and Miller, T.A. (eds) *Sustainable Pest Management in Date Palm: Current Status and Emerging Challenges.* Springer, Cham, Switzerland, pp. 13–36.

de Pater, S., Pinas, J.E., Hooykaas, P.J.J. and van der Zaal, B.J. (2013) ZFN-mediated gene targeting of the Arabidopsis protoporphyrinogen oxidase gene through *Agrobacterium*-mediated floral dip transformation. *Plant Biotechnology Journal* 11(4), 510–515. DOI: 10.1111/pbi.12040.

Djerbi, M. (1982) Bayoudh disease in North Africa, history, distribution, diagnosis and control. *Date Palm Journal* 1, 153–197.

Djerbi, M. (1983) Diseases of the date palm (*Phoenix dactylifera*) regional project for palm and seedlings with an hypoaggressive *Fusarium oxysporum* isolate. *Journal of Phytopathology* 152(3), 182–189.

Djerbi, M., Sedra, M.H., El Idrissi Ammari, M.A., Assari, K. and Chadli, F. (1985) Caractéristiques culturales et identification du *Fusarium oxysporum* f. sp. *albedinis*, agent causal du Bayoud. *Annales de l'Institut National de La Recherche Agronomique de Tunisie* 58, 1–8.

Djerbi, M., Aouad, L., Filali, H., Saaidi, M., Chtioui, A. *et al.* (1989) Preliminary results of selection of high-quality Bayoud-resistant clones among natural date palm population in Morocco. In: *Proceedings of the Second Symposium on the Date Palm in Saudi Arabia, Al-Hassa, Saudi Arabia,* King Faisal University, Al-Hassa, Saudi Arabia, March 3–6, 1986, pp. 383–399.

Dubost, D. and Kellou, R. (1974) Organisation de la recherche et de la lutte contre le bayoud en Algérie. *Bulletin d'Agronomie Saharienne* 1, 5–13.

Ehsine, M., Belkadhi, M.S. and Chaieb, M. (2009) Bio-ecologic observations on rhinoceros beetle *Oryctes agamemnon* (Burmeister 1847) on the palm dates oasis of Rjim Maatoug in southwestern Tunisia. *Journal of Arid Land Studies* 19, 379–382.

El-Deeb, H.M., Lashin, S.M. and Arab, Y.A. (2007) Distribution and pathogenesis of date palm fungi in Egypt. *Acta Horticulturae* 736, 421–429. DOI: 10.17660/ ActaHortic.2007.736.39.

El-Gariani, N.K., El-Rayani, A.M. and Edongali, E.A. (2007) Distribution of phytopathogenic fungi on the coastal region of Libya and their relationships with date cultivars. *Acta Horticulturae* 736, 449–455. DOI: 10.17660/ ActaHortic.2007.736.42.

El Hadrami, A., Adam, L.R. and Daayf, F. (2011) Biocontrol treatments confer protection against *Verticillium dahliae* infection of potato by inducing antimicrobial metabolites. *Molecular Plant-Microbe Interactions* 24(3), 328–335. DOI: 10.1094/ MPMI-04-10-0098.

El Hassni, M., El Hadrami, A., Daayf, F., Chérif, M., Barka, E.A. *et al.* (2007) Biological control of bayoud disease in date palm: selection of microorganisms inhibiting the causal agent and inducing defense reactions. *Environmental and Experimental Botany* 59(2), 224–234. DOI: 10.1016/j. envexpbot.2005.12.008.

Elliott, M.L. (2018) *Graphiola Leaf Spot (False Smut) of Palm.* Publication #PP-216. UF/ IFAS Extension, University of Florida, Gainesville, Florida. DOI: 10.32473/ edis-pp140-2006.

Elliott, M.L., Des Jardin, E.A., O'Donnell, K., Geiser, D.M., Harrison, N.A. *et al.* (2010) *Fusarium oxysporum* f. sp. *palmarum,* a novel forma specialis causing a lethal disease of *Syagrus romanzoffiana* and *Washingtonia robusta* in Florida. *Plant Disease* 94(1), 31–38. DOI: 10.1094/PDIS-94-1-0031.

El-Mergawy, R.A.A.M. and Al-Ajlan, A.M. (2011) Red palm weevil, *Rhynchophorus ferrugineus* (Olivier): economic importance, biology, biogeography and integrated pest management. *Journal of Agricultural Science and Technology A* 1, 1-23.

El Modafar, C. (2010) Mechanisms of date palm resistance to Bayoud disease: current state of knowledge and research prospects. *Physiological and Molecular Plant Pathology* 74(5–6), 287–294. DOI: 10.1016/j.pmpp.2010.06.008.

El-Sabea, A.M.R., Faleiro, J.R. and Abo-El-Saad, M.M. (2009) The threat of red palm weevil *Rhynchophorus ferrugineus* to date plantations of the Gulf region in the Middle-East: an economic perspective. *Outlooks on Pest Management* 20(3), 131–134. DOI: 10.1564/20jun11.

El-Shafie, H.A.F. (2012) Review: list of arthropod pests and their natural enemies identified worldwide on date palm, *Phoenix dactylifera* L. *Agriculture and Biology Journal of North America* 3(13), 516–524. DOI: 10.5251/abjna.2012.3.12.516.524.

El-Shafie, H.A.F. (2015) Biology, ecology and management of the longhorn date palm stem borer *Jebusaea hammerschmidti* (Coleoptera: Cerambycidae). *Outlooks on Pest Management* 26(1), 20–23. DOI: 10.1564/v26_feb_06.

El-Shafie, H.A.F., Abdel-Banat, B.M.A. and Al-Hajhoj, M.R. (2017) Arthropod pests of date palm and their management. *CAB Reviews* 2017(49), 1–18. DOI: 10.1079/PAVSNNR201712049.

EPPO (2020) EPPO Global Database. *Rhynchophorus ferrugineus* (RHYCFE). Distribution. European and Mediterranean Plant Protection Organization

(EPPO), Paris. Available at: gd.eppo.int/taxon/RHYCFE/distribution (accessed 27 March 2020).

Essarioui, A. and Sedra, M.H. (2017) Lutte contre la maladie du bayoud par solarisation et fumigation du sol. Une expérimentation dans les palmeraies du Maroc. *Cahiers Agricultures* 26(4), 45010. DOI: 10.1051/cagri/2017043.

Essarioui, A., Ben-Amar, H., Khoulassa, S., Meziani, R., Amamou, A. *et al.* (2018) Gestion du bayoud du palmier dattier dans les oasis marocaines. *Revue Marocaine Des Sciences Agronomiques et Vétérinaires* 6(4), 537–543.

Faleiro, J.R. (2006) A review of the issues and management of the red palm weevil *Rhynchophorus ferrugineus* (Coleoptera: Rhynchophoridae) in coconut and date palm during the last one hundred years. *International Journal of Tropical Insect Science* 26(3), 135–154.

Fang, Y. and Tyler, B.M. (2016) Efficient disruption and replacement of an effector gene in the oomycete *Phytophthora sojae* using CRISPR/Cas9. *Molecular Plant Pathology* 17(1), 127–139. DOI: 10.1111/mpp.12318.

FAO (2018) FAO renews support to date palm production. Food and Agriculture Organization of the United Nations, Rome. Available at: www.fao.org/news-room/detail/FAO-renews-support-to-date-palm-production/en (accessed 19 March 2018).

Fayyadh, M.A., Abdalqader, A.Z., Fadel, A.A., Muhammed, E.A., Manea, A.A. *et al.* (2013) Formulation of biopesticide from *Beauveria bassiana* as part of biological control of date palm stem borer (*Jebusaea hammerschmidti*). *Advances in Agriculture & Botanics* 5(2), 84–90.

Foex, E. and Vanssiere, P. (1919) Les maladies du dattier au Maroc. *Journal d'Agriculture Tropicale* 162, 336–339.

Frederix, M.J.J. and Den Brader, K. (1989) Résultats des essais de désinfection des sols contenant des échantillons de *Fusarium oxysporum* f.sp. *albedinis*. FAO/PNUD/RAB/88/024. Ghardaia, Algérie.

Freeman, S. and Maymon, M. (2000) Reliable detection of the fungal pathogen *Fusarium oxysporum* f. sp. *albedinis*, causal agent of bayoud disease of date palm, using molecular techniques. *Phytoparasitica* 28(4), 341–348. DOI: 10.1007/BF02981829.

Gassouma, M.S. (2004) Pests of the date palm (*Phoenix dactylifera*). In: Moustafa, A.T. (ed.) *Proceedings of the Regional Workshop on Date Palm Development in the Gulf Cooperation Countries of the Arabian Peninsula, Abu Dhabi, United Arab Emirates*, CAB International, Wallingford, UK, May 29–31, 2004, pp. 29–31.

Ghaemi, A., Rahimi, A. and Banihashemi, Z. (2011) Effects of water stress and *Fusarium oxysporum* f. sp. *lycopersici* on growth (leaf area, plant height, shoot dry matter) and shoot nitrogen content of tomatoes under greenhouse conditions. *Iran Agricultural Research* 29, 51–62.

Hameed, M.A. (2012) Inflorescence rot disease of date palm caused by *Fusarium proliferatum* in southern Iraq. *African Journal of Biotechnology* 11(35), 8616–8612.

Hernández-Hernández, J., Espino, A., Rodríguez-Rodríguez, J.M., Pérez-Sierra, A., León, M. *et al.* (2010) Survey of diseases caused by *Fusarium* spp. on palm trees in the Canary Islands. *Phytopathologia Mediterranea* 49(1), 84–88.

Howard, F.W., Giblin-Davis, R.M., Moore, D. and Abad, R.G. (2001) *Insects on Palms.* CAB International, Wallingford, UK.

Jain, S.M., Al-Khayri, J.M. and Johnson, D.V. (eds) (2011) *Date Palm Biotechnology.* Springer, Dordrecht, The Netherlands.

Kakar, M.K., Nizamani, S.M., Rustamani, M.A. and Khuhro, R.D. (2010) Periodical lesser date moth infestation on intact and dropped fruits. *Sarhad Journal of Agriculture* 26(3), 393–396.

Khalaf, M.Z. (2018) IPM of date palm borers. In: El Bouhssini, M. and Faleiro, J.R. (eds) *Date Palm Pests and Diseases, Integrated Management Guide*. International Center for Agricultural Research in the Dry Areas (ICARDA), Beirut, pp. 75–93.

Khatodia, S., Bhatotia, K. and Tuteja, N. (2017) Development of CRISPR/Cas9 mediated virus resistance in agriculturally important crops. *Bioengineered* 8(3), 274–279. DOI: 10.1080/21655979.2017.1297347.

Killian, C. and Maire, R. (1930) Le bayoud, maladie du dallier. (Abstract in Review of Applied Mycology 10, 99–100), *Bulletin de La Société d'Histoire Naturelle de l'Afrique Du Nord* 21, 89–101.

Latham, A.J. and Dozier, W.A. Jr. (1989) First rot of an apple rot caused by *Botryodiplodia theobromae* in the United States. *Plant Disease* 73(12), 1020. DOI: 10.1094/PD-73-1020A.

Llácer, E. and Jacas, J.A. (2010) Efficacy of phosphine as a fumigant against *Rhynchophorus ferrugineus* (Coleoptera: Curculionidae) in palms. *Spanish Journal of Agricultural Research* 8(3), 775–779. DOI: 10.5424/sjar/2010083-1278.

Louvet, J., Bulit, J., Toutain, G. and Rieuf, P. (1970) Bayoud, fusarium wilt of the date palm, symptoms and nature of the disease, means of control. *Al Awamia* 35, 161–182.

MAF (2008) Efficacy of two insecticides against dubas bug *Ommatissus lybicus* de Bergevin during spring generation 2008. In: *Annual Report 2008*. Directorate General of Agriculture and Livestock Research, Ministry of Agriculture and Fisheries, Muscat.

MAF (2009) Efficacy of some insecticides against Dubas bug *Ommatissus lybicus* de Bergevin during spring generation 2009. In: *Annual Report 2009*. Directorate General of Agriculture and Livestock Research, Ministry of Agriculture and Fisheries, Muscat.

MAF (2018) Efficiency of some insecticides against lesser date moth. In: *Annual Report 2018*. Directorate General of Agriculture and Livestock Research, Ministry of Agriculture and Fisheries, Muscat.

Mahmoudi, M., Sahragard, A., Pezhman, H. and Ghadamyari, M. (2014) Efficacy of biorational insecticides against dubas bug, *Ommatissus lybicus* (Hem.: Tropiduchidae) in a date palm orchard and evaluation of kaolin and mineral oil in the laboratory. *Journal of the Entomological Society of Iran* 33(4), 1–10.

Maitlo, W., Markhand, G., Soad, A.A., Pathan, M. and Lodhi, A. (2009) Comprehensive study on pathogen causing sudden decline disease of date palm (*Phoenix dactylifera* L.). In: Markhand, G.S. and Soad, A.A. (eds) *Proceedings of the First International Dates Seminar, Khairpur, Sindh, Pakistan*, Date Palm Research Institute, Shah Abdul Latif University, Khairpur, Pakistan, July 28, 2009, pp. 71–76.

Maitlo, W.A., Markhand, G.S., Abul-Soad, A.A., Lodhi, A.M. and Jatoi, M.A. (2014) Fungi associated with sudden decline disease of date palm (*Phoenix dactylifera* L.) and its incidence at Khairpur, Pakistan. *Pakistan Journal of Phytopathology* 26(1), 67–73.

Mani, J.K., Sharma, S.K., Singh, R. and Shekar, C. (2015) Role of weather parameters on graphiola leaf spot disease incidence of date palm (*Phoenix dactylifera* L.) in South Haryana. *Journal of Agrometeorology* 14(Sp. Iss. 2012), 117–121.

Mansoori, B. (2012) *Fusarium proliferatum* induces gum in xylem vessels as the cause of date bunch fading in Iran. *Journal of Agricultural Science and Technology* 14(5), 1133–1140.

Mansoori, B. and Kord, M. (2006) Yellow death: a disease of date palm in Iran caused by *Fusarium solani*. *Journal of Phytopathology* 154(2), 125–127.

Markhand, G.S., Abul-Soad, A. and Jatoi, M. (2013) Chemical control of sudden decline disease of date palm (*Phoenix dactylifera* L.) in Sindh, Pakistan. *Pakistan Journal of Botany* 45, 7–11.

Maslouhy, A. (1989) Contribution à l'étude *in vitro* et *in situ* des antagonistes de *Fusarium oxysporum* f. sp. *albedinis*, agent causal du Bayoud. Thèse de troisième cycle, Université Cadi Ayyad, Marrakech.

Megateli, M. and Berdja, R. (2015) Bayoud disease of date palm in Algeria: history, epidemiology and integrated disease management. *African Journal of Biotechnology* 14(7), 542–550.

Munoz, A. and Wang, S. (2011) Detection of *Fusarium oxysporum* f. sp. *canariensis* and *F. proliferatum* from palms in southern Nevada. *Phytopathology* 101, S125–S125.

Nishad, R. and Ahmed, T.A. (2020) Survey and identification of date palm pathogens and indigenous biocontrol agents. *Plant Disease* 104(9), 2498–2508. DOI: 10.1094/PDIS-12-19-2556-RE.

Nixon, R.W. (1957) Differences among varieties of the date palm in tolerance to *Graphiola* leaf spot. *Plant Disease Reporter* 41, 1026–1028.

O'Donnell, K., Sutton, D.A., Fothergill, A., McCarthy, D., Rinaldi, M.G. *et al.* (2008) Molecular phylogenetic diversity, multilocus haplotype nomenclature, and *in vitro* antifungal resistance within the *Fusarium solani* species complex. *Journal of Clinical Microbiology* 46, 2477–2490.

Ogundana, S.K. (1983) Life cycle of *Botryodiplodia theobromae*, a soft rot pathogen of yam. *Journal of Phytopathology* 106(3), 204–213.

Ouhdouch, Y., Boussaid, A. and Finance, C. (1996) A *Kitasatosporia* strain with non-polyenic activity against the agent of date palm vascular wilt. *Actinomycetes* 7, 1–3.

Peng, A., Chen, S., Lei, T., Xu, L., He, Y. *et al.* (2017) Engineering canker-resistant plants through CRISPR/Cas9-targeted editing of the susceptibility gene CsLOB1 promoter in citrus. *Plant Biotechnology Journal* 15(12), 1509–1519.

Perring, T.M., El-Shafie, H.A. and Wakil, W. (2015) Carob moth, lesser date moth, and raisin moth. In: Wakil, W., Faleiro, J.R. and Miller, T.A. (eds) *Sustainable Pest Management in Date Palm: Current Status and Emerging Challenges*. Springer, Cham, Switzerland, pp. 109–167.

Polaszek, A., Almandhari, T., Fusu, L., Al-Khatri, S.A.H., Al Naabi, S. *et al.* (2019) *Goniozus omanensis* (Hymenoptera: Bethylidae) an important parasitoid of the lesser date moth *Batrachedra amydraulameyrick* (Lepidoptera: Batrachedridae) in Oman. *PloS ONE* 14(12), e0223761. DOI: 10.1371/journal.pone.0223761.

Qayyum, M.A., Saleem, M.A., Saeed, S., Wakil, W., Ishtiaq, M. *et al.* (2020) Integration of entomopathogenic fungi and eco-friendly insecticides for management of red palm weevil, *Rhynchophorus ferrugineus* (Olivier). *Saudi Journal of Biological Sciences* 27, 1811–1817.

Rach, M.M., Gomis, H.M., Granado, O.L., Malumbres, M.P., Campoy, A.M. *et al.* (2013) On the design of a bioacoustic sensor for the early detection of the red palm weevil. *Sensors* 13(2), 1706–1729.

Rashed, M.F. and Abdel Hafeez, N.E. (2001) Decline of date palm trees in Egypt. In: Al-Badawy, A.A. (ed.) *Proceedings of the Second International Conference on Date Palms, Al-Ain, United Arab Emirates,* UAE University, Al-Ain, United Arab Emirates, March 25–27, 2001, pp. 25–27.

Rasool, K.G., Mehmood, K., Tufail, M., Husain, M., Alwaneen, W.S. *et al.* (2021) Silencing of vitellogenin gene contributes to the promise of controlling red palm weevil, *Rhynchophorus ferrugineus* (Olivier). *Scientific Reports* 11(1), 21695.

Roda, A., Kairo, M., Damian, T., Franken, F., Heidweiller, K. *et al.* (2011) Red palm weevil (*Rhynchophorus ferrugineus*), an invasive pest recently found in the Caribbean that threatens the region. *EPPO Bulletin* 41(2), 116–121.

Rugman-Jones, P.F., Hoddle, C.D., Hoddle, M.S. and Stouthamer, R. (2013) The lesser of two weevils: molecular-genetics of pest palm weevil populations confirm *Rhynchophorus vulneratus* (Panzer 1798) as a valid species distinct from *R. ferrugineus* (Olivier 1790), and reveal the global extent of both. *PLoS ONE* 8, e78379.

Saaidi, M. (1992) Field behavior of 32 date palm cultivars towards bayoud: a 25 years survey. *Agronomie* 12, 359–370.

Saha, A., Mandal, P., Dasgupta, S. and Saha, D. (2008) Influence of culture media and environmental factors on mycelial growth and sporulation of *Lasiodiplodia theobromae* (Pat.) Griffon and Maubl. *Journal of Environmental Biology* 29(3), 407–410.

Saleh, A.A., El-Komy, M.H., Eranthodi, A., Hamoud, A.S. and Molan, Y.Y. (2015) Variation in a molecular marker for resistance of Saudi date palm germplasm to *Fusarium oxysporum* f. sp. *albedinis,* the causal agent of bayoud disease. *European Journal Plant Pathology* 143, 507–514.

Salim, H.A., Zewain, Q.K., Hassan, K.A. and Salman, M.M. (2017) Effectiveness of Talstar and Decis insecticides against dubas bug, *Ommatissus binotatus* f. sp. *lybicus* (Homoptera: Tropiduchidae) at Diyala Governorate, Iraq. *Journal of Genetic and Environmental Resources Conservation* 5(1), 24–27.

Sama, G., Buse, J., Orbach, E., Friedman, A.L.L., Rittner, O. *et al.* (2010) A new catalogue of the Cerambycidae (Coleoptera) of Israel with notes on their distribution and host plants. *Munis Entomology & Zoology* 5(1), 1–55.

Sarhan, A.R.T. (2001) A study on the fungi causing decline of date palm trees in middle of Iraq. In: Al-Badawy, A.A. (ed.) *Proceedings of the Second International Conference on Date Palms, Al-Ain, United Arab Emirates,* UAE University, Al-Ain, United Arab Emirates, March 25–27, 2001, pp. 424–429.

Sedra, M.H. (2003) *Le Palmier Dattier Base de La Mise En Valeur Des Oasis Au Maroc. Techniques Phoénicicoles et Création d'oasis.* INRA-Éditions, Rabat.

Sedra, M.H. (2005) Caractérisation des clones sélectionnés du palmier dattier et prometteurs pour combattre la maladie du Bayoud. In: Boulanouar, B. and Kradi, C. (eds) *Actes du Symposium International sur le Développement Durable des Systèmes Oasiens, Erfoud, Maroc,* INRA-Éditions, Rabat, March 8–10, 2005, pp. 72–79.

Sedra, M.H. (2011) Molecular markers for genetic diversity and bayoud disease resistance in date palm. In: Jain, S.M., Al-Khayri, J.M. and Johnson, D.V. (eds) *Date Palm Biotechnology.* Springer, Dordrecht, The Netherlands, pp. 533–550.

Sedra, M.H. and Djerbi, M. (1986) Comparative study of morphological characteristics and pathogenicity of two *Fusarium oxysporum* causing respectively
the vascular wilt disease of date palm (Bayoud) and Canary Island palm. In:
*Proceedings of the Second Symposium on the Date Palm in Saudi Arabia, Al-Hassa,
Saudi Arabia*, King Faisal University, Al-Hassa, Saudi Arabia, March 3–6, 1986,
pp. 359–365.

Sedra, M.H. and Maslouhy, M.A. (1995) La fusariose vasculaire du palmier dattier
(Bayoud). II. Action inhibitrice des filtrats de culture de six microorganisms
antagonistes isolés des sols de la palmeraie de marrakech sur le développement *in vitro* de *Fusarium oxysporum* f. sp. *albedinis*. *Al Awamia* 90, 1–8.

Sepúlveda, G., Arismendi, M., Huanca-Mamani, W., Cárdenas-Ninasivincha, S.,
Salvatierra, R. *et al.* (2017) Presencia del falso carbon (*Graphiola phoenicis*
(Moug. ex Fr) Poit.) sobre Palma de canarias (*Phoenix canariensis*) en Isla de
Pascua, Chile. *Ciencia e Investigación Agraria* 44(3), 307–311.

Sfih, M.S., Mehid, M.R. and Al-Pnayan, M.A. (2009) Effect of some fungi and
plant extracts in *Ommatissus binotatus* var. *lybicus* de Berg. (Tropiduchidae:
Homoptera). *Basrah Journal for Date Palm Research* 8(2), 117–133.

Shabani, F., Kumar, L. and Al Shidi, R.H.S. (2018) Impacts of climate change
on infestations of dubas bug (*Ommatissus lybicus* Bergevin) on date palms in
Oman. *PeerJ* 6, e5545.

Shah, A., Mohsin, A.U., Hafeez, Z., Naeem, M. and Haq, M.I.U. (2013) Eggs distribution behaviour of dubas bug (*Ommatissus lybicus*: Homoptera: Tropiduchidae)
in relation to seasons and some physico-morphic characters of date palm
leaves. *Journal of Insect Behavior* 26(3), 371–386.

Shayesteh, N., Marouf, A. and Amir-Maafi, M. (2010) Some biological characteristics of the *Batrachedra amydraula* Meyrick (Lepidoptera: Batrachedridae) on
main varieties of dry and semi-dry date palm of Iran. *Julius-Kühn-Archiv* 425,
151–155.

Shukla, V.K., Doyon, Y., Miller, J.C., DeKelver, R.C., Moehle, E.A. *et al.* (2009) Precise
genome modification in the crop species *Zea mays* using zinc-finger nucleases.
Nature 459(7245), 437–441.

Sinha, M.K., Singh, R. and Jeyarajan, R. (1970) Graphiola leaf spot on date palm
(*Phoenix dactylifera* L.) susceptibility of date varieties effect on chlorophyll
content. *Plant Disease Reporter* 54, 617–619.

Sneh, B. (1998) Use of non-pathogenic or hypovirulent fungal strains to protect
plants against closely related fungal pathogens. *Biotechnology Advances* 16, 1–32.
DOI: 10.1016/S0734-9750(97)00044-X.

Soltani, R., Ikbel, C. and Hamouda, M.H.B. (2008) Descriptive study of damage
caused by the rhinoceros beetle, *Oryctes agamemnon*, and its influence on date
palm oases of Rjim Maatoug, Tunisia. *Journal of Insect Science* 8, 57.

Toutain, G. (1965) Note on the epidemiology of bayoud in North Africa. *Al Awamia*
15, 37–45.

Toutain, G. and Louvet, J. (1974) Fight against the bayoud. IV. Orientations of the
fight in Morocco. *Al Awamia* 53, 114–162.

Wakil, W., Faleiro, J.R. and Miller, T.A. (2015a) *Sustainable Pest Management in Date
Palm: Current Status and Emerging Challenges*. Springer, Cham, Switzerland.

Wakil, W., Faleiro, J.R., Miller, T.A., Bedford, G.O. and Krueger, R.R. (2015b)
Date palm production and pest management challenges. In: Wakil, W.,
Romeno Faleiro, J. and Miller, T.A. (eds) *Sustainable Pest Management in Date*

Palm: Current Status and Emerging Challenges. Springer, Cham, Switzerland, pp. 1–11.

Wang, M., Mao, Y., Lu, Y., Tao, X. and Zhu, J.-K. (2017) Multiplex gene editing in rice using the CRISPR-cpf1 system. *Molecular Plant* 10, 1011–1013.

Wendt, T., Holm, P.B., Starker, C.G., Christian, M., Voytas, D.F. *et al.* (2013) TAL effector nucleases induce mutations at a pre-selected location in the genome of primary barley transformants. *Plant Molecular Biology* 83(3), 279–285. DOI: 10.1007/s11103-013-0078-4.

Zaid, A. and Arias-Jiménez, E.J. (eds) (2002) *Date Palm Cultivation.* FAO Plant Production and Protection Paper No. 156 Rev. 1. Food and Agricultural Organizationof The United Nations, Rome.

Zhang, F. and Voytas, D.F. (2011) Targeted mutagenesis in *Arabidopsis* using zinc-finger nucleases. In: Birchler, J.A. (ed.) *Plant Chromosome Engineering: Methods and Protocols.* Humana Press, Totowa, New Jersey, pp. 167–177.

Zhang, Y., Bai, Y., Wu, G., Zou, S., Chen, Y. *et al.* (2017) Simultaneous modification of three homoeologs of taedr1 by genome editing enhances powdery mildew resistance in wheat. *Plant Journal* 91, 714–724.

Organic Date Production

11

Glenn C. Wright* (ID)

University of Arizona, Yuma, Arizona, USA

Abstract

Dates have been grown organically for thousands of years. Only relatively recently have certification procedures been developed to guide organic date producers. These procedures vary according to the certification authority but require at the outset the development and submission of an organic plan that describes the organic date orchard and the procedures that will be undertaken to achieve organic status. Once the decision has been made to grow organically, the planting, fertilization, cover crop establishment, and pest control steps must all be organic. Each certification authority has a list of allowed and prohibited substances that must be followed to remain organically certified. Organic procedures must continue during harvest and in the packinghouse. Fruit quality of an organic date does not appear to be inferior to a conventionally grown date, and some feel that organic dates have superior fruit quality. The economics of growing organic dates have not been studied; however, there is no doubt that the price received by the grower for organic dates is superior to that for conventional dates, which may overcome the higher cost of organic production. It is likely that organic dates will command an increasing portion of the overall date market.

Keywords: Certification, Fruit quality, History, Harvesting, Packing, Production practices

11.1   Introduction

Organic fruits and vegetables are one of the largest and fastest-growing segments of the global organic food market (Mordor Intelligence, 2019). The global organic fruits and vegetables market was valued at over US$30 billion in 2019 and is estimated to reach more than US$55 billion by 2027, a compound annual growth rate of 7.9% (Research and Markets, 2020). Some of the factors leading to the growth that were noted in the report

---

*gwright@arizona.edu

© CAB International 2023. *Date Palm* (eds J.M. Al-Khayri *et al.*)
DOI: 10.1079/9781800620209.0011

include governmental support, increasing demand for innovative food and beverages, urbanization, ready-to-eat products, growing disposable income, and increasing concern about health-related issues in developing countries. North America is the largest market for organic fruits and vegetables, followed by Europe and Asia-Pacific (Mordor Intelligence, 2019).

With the arrival of the COVID-19 pandemic, there was a surge in organic produce demand to US$56.5 billion in 2020 (Melhim, 2021). Factors behind the surge included a desire by consumers to mitigate the risk from the food they consume, disease-related lockdowns that kept people at home and restaurants closed, pandemic relief payments that led to higher disposable income, and online shopping. While demand for organic produce may slow, according to Melhim (2021), a preference for organic produce and for cooking at home will keep organic demand above pre-pandemic levels.

Imports will make up an increasingly substantial portion of the organic produce consumed in North America and Europe. According to Melhim (2021), the number of non-US certified organic operations increased by 37% to reach about 17,000 operations in just over 150 countries. About half of those operations supplied fruit and vegetables. Midway through 2021, organic fresh produce shipments accounted for 46% of the total food shipments, more than three times the average for 2018 through 2020.

From the standpoint of dates, it is clear that consumers will purchase organic dates because they are associated with a healthier and safer product. Consumers are also more familiar with organic dates, and the prices of organic dates are competitive with those of conventionally grown dates. Nevertheless, in a recent study in Arizona, consumers were willing to pay a premium for Medjool dates that were labeled as pesticide-free (Grebitus and Hughner, 2021). Therefore, date producers need the knowledge of organic production to meet the increased demand for organic dates.

The objective of this chapter is to provide an overview of organic date production. Dates were grown organically for thousands of years before conventional fertilizers and pesticides were available. Within this chapter, the organic certification process and the plan that governs the transition from conventional to organic production are described. Included are summaries of planting, soil fertility, and pest control practices that are allowed in the orchard under organic certification. Methods for preventing contamination of organic dates prior to and during harvest are presented, as is a description of the organic packinghouse. Finally, there is a brief discussion of the fruit quality of organic dates and the economics of organic date farming.

11.2   History and Description of Organic Date Palm Production

Date palms have long been associated with organic farming practices, such as application of manure and use of cover crops, as there were no alternatives for many years. Surprisingly, those practices led to yields similar to what we would expect today. In ancient Babylon, dates in the Tigris/ Euphrates Delta plain had yields as high as 105 kg per palm (Scheil, 1913), were artificially pollinated, were planted from offshoots, and were abundantly irrigated using canals and aqueducts. No doubt, farmers employed cover crops in the orchards, such as barley, wheat, millet, or emmer, and grew vegetables such as onions, garlic, and cucumbers in the shade of the palm trees (van der Crabben, 2021). Use of manure, however, was unknown (van der Crabben, 2021), and this practice appears to have been done first in Europe as early as 5900 to 2400 BCE (Bogaard *et al.*, 2013). However, the practice of manuring the soil eventually become commonplace in Mesopotamia. By 1903, Arab farmers manured the silty soil of the Shatt-al Arab delta with cow manure only; horse manure was regarded as too likely to burn the plants (Fairchild, 1903). Presumably, the manure was mixed with the soil at planting, and for subsequent applications, manure was used to fill a trench around each palm. Fairchild (1903) also noted that lucerne (alfalfa) (*Medicago sativa* L.) was occasionally used as a cover crop, but there was no use of other legumes.

Similarly, the date was important in the Nile Valley, beginning from 3000 BCE (Danthine, 1937). One can still see dates grown organically in Egypt; farmers use manure, while cover crops such as lucerne are found on the orchard floor. Date culture was also found in northwest Africa, always in the river valleys and oases, where sufficient water was available. In the early 1900s, Harry Simon, an American businessman, visited an Algerian oasis to inspect the dates. There, he noted, the palms had been planted in stony soil. Each palm was planted in a 2.5 m × 2.5 m × 2.5 m hole dug through the sandstone. The holes were filled to a depth of 1.5 m with a mixture of soil and manure. As the palms grew, the holes were gradually filled with the same mix (Simon, 1978). In the USA, the use of manure and cover crops was standard practice for many years, beginning in the early 1900s. Some growers continue to reject conventional fertilization methods; organic methods are still the preferred fertilization method in many US orchards, and it seems as if organic farming of dates will be increasingly preferred in the USA and wherever else it can be practiced.

11.3 The Organic Certification Process

The governmental authorities responsible for organic regulation and certification are generally the ministries or departments within the government that are responsible for regulating agricultural production. These include the Ministry of Agriculture and Fisheries (Ministère de l'Agriculture et de la Pêche Maritime) in Morocco and the US Department of Agriculture (USDA) in the USA. Sometimes, the agency is a regional body, such as the Gulf Standards Organization for the Gulf Cooperation Council (GCC) countries or the European Union for their members. These regional bodies allow each member state to establish an authority, such as a ministry of agriculture, to regulate organic agriculture. For information regarding individual countries, see Global Organic Trade Guide (2022). However, those bodies do not always conduct the actual certification. Instead, the agencies have given private firms the authority to certify organic farmers according to the regulations in place in that country. For example, in Egypt, there is Organic Egypt (Organic Egypt, 2020); in Europe, there is EcoCert (Groupe EcoCert, 2022); and in the USA there is the California Certified Organic Farmers (CCOF) (California Certified Organic Farmers, 2020). All these firms provide certification and auditing for organic farming. Some of these firms provide certification and auditing for several countries. Primus AuditingOps (Primus AuditingOps, 2015) provides organic certification and auditing for farmers in the USA, Canada, Europe, Japan, and Mexico. These companies are not limited to organic certification; they will also audit for food safety, and for good agricultural practices programs such as GlobalGAP (FoodPLUS GmbH, 2022).

The steps needed to become organically certified vary according to the country, but there are many similarities. It is important to note that for the first three years, the organic date orchard will be designated as "in transition". Practically, this means that organic certification requires that crops do not receive any synthetic chemicals including fertilizers or pesticides for three years prior to the harvest of the crops. Meanwhile, the producer cannot sell the crop as "organic" and receive the higher prices that typically accompany organic dates.

The first step is the submission of the organic plan (called the Organic System Plan in the USA). This document guides the producer through the required standards and helps to describe the organic practices. If using a private firm, this step includes an application fee. Before submission, a private firm will have a certification specialist review the plan to determine if anything is missing or needs further explanation. Then an onsite inspection is arranged, and the annual certification fee is determined based on the plan. The purpose of the inspection is to verify that the information in the plan is happening in practice. The inspector will report findings to the private firm, where it will be reviewed for compliance and accuracy. If completed successfully, the organic farm will be certified "in transition" for the first three years and fully certified in the fourth year (California Certified Organic Farmers, 2020). This inspection will reoccur annually, at minimum, to update the firm of plan changes, new labels, or other changes and to ensure compliance.

11.4   Writing the Organic Plan

There are some typical components of a USDA Organic System Plan. They are available in a document that can be found at https://www.ams.

usda.gov/sites/default/files/media/FINAL%20Streamlined%20OSP.doc (accessed 30 November 2022). The actual components for other nations' plans will vary according to the regulating body.

A separate plan should be submitted for each parcel that is not adjacent to another one, and/or has a distinct usage history. Briefly, the plan should contain:

- A description and address of the production location.
- A map of the location.
- A description of the boundaries of the location and buffer zones that protect the organic orchard from conventional farming lands nearby.
- A description of the production practices and management history of the orchard for three years before the first organic harvest. This should include a description of the varieties, soil fertility and pest management practices, and materials.
- A list of the application dates of prohibited materials used prior to organic certification.
- Any documentation of land management history, including current organic certificate, pesticide use reports, and records of brand names, formulations, manufacturers, and application dates of all materials used for three years prior to the harvest of an organic crop.
- Other management plans and certifications that are being pursued.
- Marketing and sales methods.
- A description of the monitoring methods and timing for soil and crop observation.
- A description of the recordkeeping system.

Regarding compost and manure, there needs to be a description of:

- compost and manure management; and
- how the recordkeeping system demonstrates compliant practices related to manure and compost.

There will need to be a description of the natural resources of the operation, including a description of:

- the soil, water, woodlands, wetlands, and wildlife;
- the practices used to maintain or improve these natural resources; and
- how the monitoring methods verify and the recordkeeping system documents that the natural resources practices are effectively implemented.

Information should be included on the organic status of the date offshoots or tissue culture-derived plants on the location including if the planting stock is or is not certified organic. There should be purchase records if applicable. If offshoots are taken from conventional trees, they are not considered to be organic until the end of the three-year transition period.

There should be a list of recurrent or potential pest, disease, and weed problems, including:

- Arthropods, vertebrates, other pests, diseases, and weeds.
- Preventive practices and strategies that make up the weed, pest, and disease management plan.
- A description of how the monitoring methods and the recordkeeping system demonstrate compliant practices.
- A list of all substances used or planned to be used for pest management. The list requires the product name and formulation, the manufacturer, the intended use, any restrictions or preventive practices used, and third-party verification that the material is allowed.
- A description of how the recordkeeping system demonstrates compliant practices related to sourcing and use of materials.

Also included should be an audit trail, which is a description of how the recordkeeping system links to the next component to create a continuous audit trail to track the organic date from its production site; input substances, planting stock; farm management practices and crop rotation; harvest, postharvest handling, storage, and transport; and final sale or release from custody.

There should be a list of names, brands, and labels under which the dates will be marketed, along with a color sample of the labels. Also included should be:

- any other printed, visual, written, or graphic material that will be used to identify the product as organic; and
- a description of the lot numbering system for nonretail packaging.

There should be a risk assessment to prevent contamination or comingling of the organic dates with conventional dates. This assessment should include:

- an identification of potential risks, such as soil and water contamination by prohibited materials, or residues, plant nutrients, heavy metals, pathogens, or comingling between organic and nonorganic product; and
- a list of facilities used for postharvest handling storage, packing, or processing that are owned by the applicant.

11.5   Planting the Organic Date Orchard

Date fruit that are marketed as organic need to be grown from offshoots or tissue culture plants that also conform to organic standards (European Union, 2022). Therefore, those offshoots must be collected from an organic orchard, and the tissue culture plant must be grown in a certified organic nursery. If not, the newly planted orchard must go through the three-year transition period before being certified as organic. Any soil amendment added at planting time must also conform to the rules of the organic program.

11.6 Soil Fertility and Cover Crops in Organic Date Orchards

Soil fertility in the organic date orchard must be maintained and improved using manures, cover crops, and biostimulants.

11.6.1 Manure

As mentioned previously, manure, often known as "composted manure", has been applied to date palms for thousands of years. Manures provide plant nutrients, improve the water-holding capacity of soils, and provide cation exchange sites to help maintain nutrients in the soil for plant uptake. These characteristics are particularly important in sandy soils in which dates are often cultivated.

The USDA National Organic Program has several regulations that apply to manures and manure-based composts (USDA-AMS, 2022a). Raw manure is not allowed on land where the crop is for human consumption. Composted manures may be applied to the date orchard if there is an initial carbon/nitrogen ratio of between 25:1 and 40:1 and temperatures between 55 and 77°C were sustained for 3 days using a vessel or static aerated pile system, or for 15 days using a windrow composting system. For the compost in the windrow system, the materials must be turned at least five times followed by an adequate curing period.

If compost is made onsite, adequate records must be maintained to show that the compost was produced according to the above regulations. If purchased, the grower should have documentation showing that the compost has been produced in a way that meets the above regulations. If stored or handled, manure must not contaminate covered produce, food-contact surfaces, areas used for growing, harvesting, holding, and packing, and water sources and distribution systems. Sewage sludge is not allowed in organic production. For more information, see ATTRA Sustainable Agriculture (2015).

Traditionally, manure is broadcast by hand or by a manure spreader across the entire orchard floor, banded along the tree rows, injected into the soil, or applied around the trunks of individual trees. Applications are easier in flood or basin irrigated orchards, where irrigation lines do not impede movement of the spreading or incorporating equipment. Organic growers in the USA apply up to 50 t/ha to mature dates.

The most common sources of manure used for date palms include cow, chicken, sheep, and goat. Composition of various manures is found in Table 11.1.

Table 11.1. Mineral composition of various manures. (Adapted from Mitchell, 2008; Crop Fertility Services, 2022; Sawyer, 2009. The nutrient content of many more organic amendments can be found in Mitchell, 2008.)

Type	Approximate composition (%)					
	N	P (P_2O_5)	K (K_2O)	Ca	Mg	S
Cattle	1.5	1.5	1.2	1.1	0.3	[a]
Sheep	0.6	0.3	0.2	[a]	[a]	[a]
Young non-laying chicken with litter	3.0	3.0	2.0	1.8	0.4	0.3
Caged laying hen	1.5	1.3	0.5	6	0.4	0.3
Hen litter	1.8	2.8	1.4	[a]	[a]	[a]
	Approximate composition (kg/L)					
Chicken – pelleted	4–5	3–4	2–3	[a]	[a]	[a]
Chicken – liquid	6.8–7.5	3.5–4.0	4.2–6.2	[a]	[a]	[a]

[a]Present in undetermined amounts.

Manure contains both inorganic ammonium nitrogen and organic nitrogen. Ammonium nitrogen is present as urine and may account for up to 50% of the total nitrogen (Ketterings *et al.*, 2005), while feces is organic nitrogen and is converted to plant-usable nitrogen quite slowly. To gain the advantage of the ammonium nitrogen, manure should always be incorporated into the soil wherever possible, otherwise it will volatilize, especially on high-pH soils. If manure is injected into the soil, 100% of the ammonium nitrogen will be retained for plant use. However, if it is incorporated within one day of application, 35% of the ammonium will be lost via volatilization, and if incorporation occurs within 5 days, 83% of the ammonium nitrogen will be lost (Ketterings *et al.*, 2005). Growers with pressurized irrigation systems will sometimes temporarily remove the irrigation lines so that manure can be incorporated, then the lines will be reinstalled.

Organic nitrogen must be mineralized before it is available to the date palm. Mineralization can take 3 to 5 years, and the rate of mineralization can depend on the type of amendment, the application rate, the temperature, the incorporation depth, and other factors. A model has been built to quantify nitrogen mineralization (Geisseler *et al.*, 2021), and a handy guide for the producer can be found at Geisseler Lab (2021).

11.6.2 Cover crops

Cover crops may be used in date palm orchards where stands of the cover crop can be established. Practically, this means that the use of cover crops is limited to areas where flood or basin irrigation is practiced. Soils with date palms under drip or microsprinkler irrigation, in areas with limited rainfall, are unlikely to be wet enough for enough time for a seed stand to become established. Additionally, cover crops will compete with dates, especially when the palms are newly planted, for water and nutrients.

Where a cover crop can be established, they provide numerous benefits, including reduced tillage and soil compaction, increased organic matter and water-holding capacity, recycled nutrients, N_2 fixation, lower soil temperatures, habitat for beneficial insects and microbes, and weed suppression (Hargrove and Fry, 1987). An extensive study was conducted over several years to determine which cover crops were the most suitable for conditions in the deserts of Southern California (Abdul-Baki *et al.*, 2007). A drought-tolerant barley (*Hordeum vulgare* cv. Seco) and vetch (*Vicia villosa* cv. Lana) were superior cool-season cover crops. Sudan grass (*Sorghum* × *drummondii*), cowpea (*Vigna unguiculata*), *Sesbania cannabina*, and forage soybeans (*Glycine max*) were superior summer-season cover crops. Orchards where these cover crops were established needed no additional fertilization during the experimental trial. However, because of the potential for damaging the cover crops during normal orchard farming operations, the authors suggest establishing the cover crops only in every other basin, while maintaining the alternates clear using tillage to facilitate traffic movement through the orchard. Dried pruned leaves could be chopped and incorporated in the bare basins, and the authors suggest supplementing the tree with about 1 kg N fertilizer to avoid N depletion and speed the decomposition of the leaf tissue. This would need to be organic-N to avoid breaking certification.

11.6.3   Biostimulants

Biostimulants are products containing naturally occurring substances and microbes that are used to stimulate plant growth, enhance resistance to plant pests, and reduce abiotic stress. Ingredients might include carbon complexes, organic acids such as humic and fulvic acids, microalgae, and beneficial bacteria, such as *Bacillus amyloliquefaciens* and *Bacillus subtilis*.

Recently, there have been many scientific works showing growth improvements following biostimulant application to date palms. Fulvic acid and biofertilization have been shown to promote growth of tissue culture-produced date palms during acclimation (Abd-El Kareim *et al.*, 2014), while addition of humic acid to the rooting media improved survivability of tissue culture-produced plants in the acclimation stage (Abdel-Gelil *et al.*, 2009). More recently, it was shown that compost, a mycorrhizal consortium, and a *Pantoea agglomerans* strain significantly improved growth of potted dates in the greenhouse (El Kinany *et al.*, 2022). A book entitled *Biostimulants for Sustainable Agriculture in Oasis Ecosystem Towards Improving Date Palm Tolerance to Biotic and Abiotic Stress* has recently been released by the General Secretariat of Khalifa International Award for Date Palm and Agricultural Innovation (Meddich, 2021).

There are many proprietary biostimulants available from agricultural suppliers, usually containing multiple ingredients. These compounds are usually liquids and may be injected in pressurized irrigation systems, sprayed on the orchard floor, or carried into the orchard in flood irrigation waters. Thus, these liquids are preferred by growers whose pressurized system limits their ability to apply manure.

11.7   Managing Weeds, Insects, and Diseases in Organic Date Orchards

Pesticides approved for organic farming are typically natural toxins with a few synthetic compounds added to the list. Nevertheless, there are far fewer pesticides allowed than could be used in conventional farming. In the USA, the National Organic Program is the decision-making group that approves pesticides for organic use. Approved pesticides for other nations are made by their respective organic farming authorities. For the USA, a short list of the major substances allowed can be found at AgDaily (2019) and a more comprehensive list is located at Code of Federal Regulations (2022). Of special note is neem oil, which can be used as a contact insecticide or as a fungicide.

11.7.1 Weeds

Weed control in many organic farming operations is by tillage in between the trees or by use of hand labor around the trunks (Fig. 11.1), or by animals such as cows or goats. Herbicides that are organic have one or more of the following ingredients: acetic acid, citric acid, D-limonene (citrus oil), clove oil or clove-leaf oil, cinnamon oil, lemon grass oil, eugenol, 2-phenethyl propionate, sodium lauryl sulfate, ammonium nonanoate, or pelargonic acid + fatty acids (Shaffer, 2019).

11.7.2 Insects

There are several insects of date palms of worldwide importance. Some, but not all, may be controlled organically. Not all insect pests are listed here. For additional information on insect pests of dates, see Zaid *et al.* (2002), El-Shafie (2012), and , Chapter 10, this volume).

11.7.2.1 Palm weevils

The red palm weevil (*Rhynchophorus ferrugineus*) is one of the most dangerous pests of palms around the globe. It is widespread in the eastern hemisphere in most date-producing regions. Unfortunately, there is no effective organic control; therefore, where red palm weevil is endemic, organic date production is quite difficult. The weevil eggs hatch and the larvae burrow into the trunk of the palm, moving upward toward the apical meristem, consuming the soft tissue. Near the top of the palm, the larvae pupate in cocoons of palm fibers. Later, the adults emerge and continue feeding on the meristem until it is consumed and completely dead (Dembilio and Jaques, 2015). Because of the seriousness of this pest, there has been significant research directed toward its control. Management of the weevil that is consistent with organic farming practices includes trapping and monitoring with pheromone traps. Research directed toward biological control includes work on parasitic mites, entomopathogenic viruses, nematodes, bacteria, and fungi. Some of these control methods have been commercialized. Extensive details can be found in Dembilio and Jaques (2015).

The South American palm weevil (*Rhynchophorus palmarum*) (Fig. 11.2) is native to parts of Mexico, South America, and the Caribbean (Hoddle, 2022). It was found in Tijuana, Mexico in 2010, and has been moving northward since, with weevil populations firmly established in the USA in 2014. Currently, it prefers Canary Island date palms (*Phoenix canariensis*),

Fig. 11.1. Organic date orchard near Yuma, Arizona, USA. Note tilled row middles. Weeds around trunks are controlled by hand. (Photo by Glenn Wright.)

but there is no indication that it would not feed upon date palms (*Phoenix dactylifera*). To date, there have been no finds of living South American palm weevils in date palms in the USA. Biology of this weevil is like that of the red palm weevil, and symptoms are identical. A pheromone trap has been developed to monitor the weevil. Preliminary research in Brazil has

Fig. 11.2. South American palm weevil. (From Hoddle, 2022. Photo by Mike Lewis. Used with permission.)

indicated that a parasitic fly in the genus *Billaea* aggressively attacks South American palm weevil larvae. However, there has been no commercial development from this work to date. For more information, see Hoddle (2022).

11.7.2.2 Lepidopteran pests of dates

The carob moth (*Ectomyelois* = *Apomyelois ceratoniae*), the lesser date moth (*Batrachedra amydraula*), and the raisin moth (*Cadra* = *Ephestia figulilella*) are three primary pests of dates whose biologies are timed with the seasonal occurrence of the date fruit (Perring *et al.*, 2015). Larvae of all three pests feed on the developing fruit, causing scarring, in the case of the lesser date moth, or interior damage, which may also be an entry point for secondary pathogens such as *Aspergillus*.

For the carob moth (Fig. 11.3), cultural control can be quite effective in reducing populations of the insects. These practices can include removing waste dates from the floor of the date orchard (Carpenter and Elmer, 1978) by disking the orchard floor. Other cultural and physical control methods include removing abscised date fruit that are stuck in the bunches with a cleaning tool, cutting the center strands of the date bunches to facilitate the drop of abscised fruit, and the use of bags with a mesh size of 1.7 mm or smaller (Nay *et al.*, 2006; Nay and Perring, 2009; Zirari and Laaziza Ichir, 2010). Other control methods that may be used include pheromone traps

Fig. 11.3. Carob moth larva. (Photo by Justin Nay. Used with permission.)

for mating disruption (Mafra-Neto *et al.*, 2013). No parasitoids, predators, or entomopathogenic bacteria have been commercialized against carob moth (Perring *et al.*, 2015).

Cultural practices that can reduce populations of lesser date moth include covering the flowers with paper, covering the date bunches with cloth, thinning the fruit, inserting rings in the bunches to separate the fruit, and removing fallen fruit from the orchard floor (Harhash *et al.*, 2003). A sex pheromone has been identified (Levi-Zada *et al.*, 2011), but there is no evidence that it has been commercialized for use by the date producer. There are no predators or parasites being used for management of lesser date moth (Perring *et al.*, 2015), but *Bacillus thuringiensis*, spinosad, horticultural oil, an herbal source containing extract from *Sophora flavescens*, *Azadirachta indica*, *Melia azedarach*, and emamectin benzoate controlled lesser date moth in a Saudi Arabian field study (Habib and Essaadi, 2007).

For the raisin moth, a principal strategy is to clean up and destroy infected fruit from the orchard floor (Perring *et al.*, 2015). Early harvesting of the fruit and netting the bunches will also reduce the infestation (Kehat *et al.*, 1969).

11.7.2.3   Hemipteran pests of dates

There are many hemipteran pests of dates. Some of the most notable include the date dubas bug (*Ommatissus lybicus*), the Parlatoria date scale (*Parlatoria blanchardii*), and the red date scale (*Phoenicococcus marlatti*). The pink hibiscus mealybug (*Maconellicoccus hirsutus*) is a recent pest of dates in California.

The date dubas bug is found in throughout the date-growing areas of the Middle East, North Africa, southeast Russia, Iran, and Spain (Perring *et al.*, 2015). Both adults and nymphs feed on the sap of the leaflets, leaf midribs, fruit stalks and fruits, and will produce honeydew. Feeding weakens the leaves, causing them to become chlorotic and die. Honeydew can also accumulate and encourage the growth of sooty mold (El-Shafie *et al.*, 2015). Maintenance of sufficient space between trees and removal of excess leaves and offshoots will aid in the control as the insects will become sun scorched and desiccate (Talhouk, 1983). There is no commercial biological control of dubas bug (El-Shafie *et al.*, 2015).

Parlatoria scale is found wherever dates are grown in the eastern hemisphere, but it was eradicated from the USA by the 1930s. The scales feed on the date palm leaves, fruit stalks, and fruit. Parlatoria scale injects toxic saliva leading to discolored areas where transpiration is reduced, nutrients depleted, and photosynthesis hindered, leading to reduced productivity (Benassy, 1990). Cultural control includes pruning infested leaves. There are also a host of biological control agents, including beetles and mites (El-Shafie *et al.*, 2015). Insecticidal soaps, horticultural oils, and neem oils are effective against scales in general, if there is thorough coverage.

Red date scale is found wherever date palms are grown. Heavy feeding weakens the young palms, leading to death (Moustafa, 2012). Several predators appear to feed on red date scale (El-Shafie *et al.*, 2015), but none are commercialized. Chemical control of this scale is similar to that of Parlatoria scale.

Pink hibiscus mealybug is a relative newcomer to the date-growing areas of the USA, although it has previously been reported in Saudi Arabia (Talhouk, 1983). Feeding damage by pink hibiscus mealybug results in deformed fruits, leaves, and offshoots, stunted plant growth, and eventual plant death. There does not appear to be any specific research on chemical control for this insect; however, organic insecticides such as neem oil, horticultural oil, and insecticidal soap may be effective with sufficient coverage. A biological control program utilizing *Anagyrus callidus* (Triapitsyn *et al.*, 2019) as a parasitoid has been established in Southern California.

11.7.2.4   Mite pests of dates

There are various mites that damage date palms. In California, the Banks grass mite (*Oligonychus pratensis*) is the only mite pest of dates in North America. Elsewhere, the Old-World date mite (*Oligonychus afrasiaticus*) is a major pest in the Middle East and North Africa. In California, *O. pratensis* damages the cultivar Deglet Noor almost exclusively (Negm *et al.*, 2015). Damage is typically scarring and webbing on the fruit. In Tunisia, a European strain of *Neoseiulus californicus*, a predatory phytoseiid mite, is commercially available (Othman *et al.*, 2001). Several organic pesticides have been evaluated against Old-World date mite and been shown to be effective. Also, neem oil and horticultural oils are labeled to control mites.

Spraying the bunches with water to destroy webbing might also be useful (Negm *et al.*, 2015).

11.7.3   Diseases

There are several diseases of date palms of worldwide importance. Some diseases of fungal or phytoplasma origin may be controlled organically. Not all diseases are listed here. For additional information, see Zaid *et al.* (2002) and , Chapter 10, this volume).

11.7.3.1   Bayoud disease

Bayoud disease, caused by *Fusarium oxysporum* f. sp. *albedinis*, is arguably the most serious disease of date palms, having been first reported in 1870 in Morocco (Benzohra *et al.*, 2015). The fungus lives in the soil and will attack the palm tree at its base. From there it moves through the xylem vessels until it reaches the growing point, which is ultimately killed. There is no chemical cure for the disease. Horticultural practices that encourage disease development include irrigation and the use of cover crops, such as lucerne, which require additional irrigation. There are no cultural controls that reduce the incidence of bayoud other than removal of infected trees, including the roots, and cessation of irrigation. Various governments have taken action to prevent the spread of bayoud disease. These include quarantine measures against all offshoots, plant material, manure, and soil originating from bayoud-infected countries or regions, quarantine measures against the import of lucerne seeds from infected countries or regions (Zaid *et al.*, 2002), and prohibitions against the use of offshoot planting material within bayoud-infested countries.

11.7.3.2   Black mold

Black mold (*Aspergillus niger*) is sometimes found inside ripe dates that have become moistened by rain before harvest. Stand thinning, fruit thinning, and insertion of metal rings to spread the strands will improve aeration in the fruit bunch and reduce the incidence of this disease. Also, careful fruit picking that does not remove the calyx will limit the entry points through which the fungus can gain access to the interior of the fruit.

11.7.3.3   Black scorch

Black scorch, caused by *Thielaviopsis punctulata* or *Thielaviopsis paradoxa* (de Beer *et al.*, 2014), is found in most date-growing areas worldwide (Fig. 11.4). Both species are soilborne wound pathogens that can affect all parts of the date trees at all ages and over a wide geographical area (Saeed *et al.*, 2016). Horticultural disease control practices that are consistent with

Fig. 11.4. Crown of a date palm afflicted with black scorch. Note contorted, yellowed, blackened, and dying fronds. (Photo by Glenn Wright.)

organic farming include avoidance of wounds, such as mechanical damage or wounding due to insect infestation (Chase and Broschat, 1993). Severely affected trees should be removed from the orchard. Less severe infestations can be controlled with *Trichoderma* or *Chaetomium* spp. (Soytong *et al.*, 2005; Sánchez *et al.*, 2007).

11.7.3.4 Diplodia

Diplodia (*Diplodia phoenicum*) is a fungus that attacks the leaves and terminal buds of date palms, particularly the offshoots (Zaid *et al.*, 2002). This disease is a wound pathogen, so avoidance of wounds, tool disinfestation, good field sanitation, and use of copper-based fungicides is suggested.

11.7.4.5 Graphiola leaf spot and Khamedj disease

Graphiola leaf spot (*Graphiola phoenicis*) and Khamedj disease (*Mauginiella scaettae*) may be controlled with copper-based fungicides (Zaid *et al.*, 2002). Good field sanitation is also necessary to reduce incidences of the disease.

11.7.4.6 Palm yellows (lethal yellowing)

Palm yellows (lethal yellowing of palms) is a phytoplasma-caused disease that infects several species of palm trees. The disease, caused by *Candidatus* Phytoplasma palmae, is transmitted via insect vectors. Injection of tetracycline has been shown to cause remission of palm yellows symptoms in young date palms (Montasser *et al.*, 2012).

11.8 Allowed and Prohibited Substances in Organic Farming

In general, crops grown under organic farming specifications must not be genetically modified organisms (GMOs) and must not have been exposed to ionizing radiation. Furthermore, the use of artificial fertilizers, herbicides, and pesticides is limited, as are plant growth regulators and plant hormones that have been derived from synthetic substances. The USDA has a list of permitted substances for organic farming at Code of Federal Regulations (2022). Also, pesticide label databases, such as CDMS (http://www.cdms.net/ (accessed 30 November 2022)) or Agrian (https://www.agrian.com (accessed 30 November 2022)), allow the user to identify products that can be used in organic production. These lists have been developed by other countries as well according to their national guidelines.

Some agreements have been established between nations that recognize their respective organic programs. These agreements recognize that the organic standards and oversight systems of the two partners are comparable and verifiable. The USA has such agreements with Canada, the European Union, Japan, South Korea, Switzerland, Taiwan, and the UK. In the case of New Zealand and Israel, the USA accredits certifying agents who are authorized to certify individual farms that meet US organic standards.

11.9 Preventing Contamination of Organic Dates Prior to and During Harvest

As part of the organic plan, there is a risk assessment of how to prevent contamination or comingling of the organic date with the conventional crop. Potential contamination might occur via soil and water,

Fig. 11.5. Dates drying in the sun. (Photo by Glenn Wright.)

contamination by prohibited materials, or residues, plant nutrients, heavy metals, pathogens, or comingling between organic and nonorganic product.

Fruit should be grown and harvested in such a way that there is no contact between the fruit and the soil or water. Growers should ensure that organic fruit are not inadvertently sprayed with pesticides or with unapproved materials. Inferior quality rotting dates should be removed from the bunches and from the orchard floor.

Organic dates should be covered to the greatest extent possible. These coverings should include the bunch bag, coverings on the harvest trays or boxes during transport, and coverings of any carton in which the dates are stored. Sun drying of dates (Fig. 11.5) presents some risks, not only because of the potential for contamination with dust that may have heavy metals or carry manure, but also in the case of spray drift. Operations involving organic dates should be separated in space and time from conventional operations. Organic and conventional dates should not be sprayed, irrigated, or harvested at the same time or in the same place. A method of tracking should be developed so that the two types of dates do not become comingled in the field. Fields should be well marked as organic so that there are fewer chances for errors.

11.10   The Organic Packinghouse

As in the field, conventional and organic dates should not be comingled in the packinghouse. USDA National Organic Program regulations require a full cleanup of packinghouse equipment if organic dates are comingled or if they share the same surface. Therefore, it is best to have a dedicated organic line (Fig. 11.6), or if only a single line is available, have the organic dates pass through the packing process earlier in the day than the conventional ones, and clean the packing line at the end of the day.

Peracetic acid (CH_3CO_3H), commonly used to disinfect equipment, and is allowable in the USA for organic crop production to disinfect equipment, seed, and asexually propagated planting material. It is also permitted in hydrogen peroxide (H_2O_2) formulations used as algicides, disinfectants, and cleaners at a concentration of no more than 6% v/v as indicated on the pesticide product label (Code of Federal Regulations, 2022).

As previously mentioned, there is a risk with sun-dried dates because of the potential for contamination. Therefore, artificial drying in the packinghouse is preferred for organic dates. It is worth noting that for Medjool dates, timely bunch enclosing with bags of appropriate mesh size before harvest, followed by artificial drying exposure to 53°C as a "kill step", and subsequent freezing at −15 to −18°C eliminates the need for postharvest fumigation for insects in the packinghouse.

One organic alternative to conventional fumigants, carbon dioxide, has been developed for postharvest insect control (Pure & Eco India, 2022). The treatment works by pumping carbon dioxide gas into a sealed space to dramatically lower the oxygen percentage in the atmosphere. The insects are deprived of oxygen and start to die off as a result. Another option may be anisole, a volatile compound extracted from anise seed. Research showed that in a 16 h test in the laboratory, anisole killed 100% of the treated weevils, thrips, and beetles (Yang and Liu, 2021). Further research will be needed to prove the effectiveness of anisole, then this compound may be approved for organic production.

11.11   Fruit Quality of Organic Dates

The author conducted an informal survey of date producers and packinghouse personnel as to whether organically grown date fruit had better quality than conventional-grown fruit, and the responses were mixed. Some said that there was no difference, while others believed that the fruit quality of organic fruit was better. It is possible that organic practices, such as the use of manures, cover crops, and biostimulants to improve root and tree growth and improve soil structure, positively affect fruit size and the degree of skin separation, the two factors affecting fruit grade. Also, absence of insect larvae, which can be achieved using nonconventional methods, certainly improves fruit quality.

Fig. 11.6. Organic packing line at Yuma, Arizona, USA. (Photo by Glenn Wright.)

In the USA, there has been little legitimate research on the effects of organic production on the quality of fruits and vegetables. Partisans of the position that organic fruit is superior and partisans who believe that there is no difference can find work to support their positions. One of the most cited studies did find that:

Fig. 11.7. Organic Medjool dates produced in the USA. (Photo by Glenn Wright.)

organic crops contained significantly more vitamin C, iron, magnesium, and phosphorus and significantly less nitrates than conventional crops. There were nonsignificant trends showing less protein but of a better quality and a higher content of nutritionally significant minerals with lower amounts of some heavy metals in organic crops compared to conventional ones.

(Worthington, 2004)

11.12 Economics and Cost-Effectiveness of Organic Date Farming

There have been no official studies of the economics of organic versus conventional date growing. Growing costs for organic dates versus conventional dates might be higher because of the cost of organic fertilizers and pesticides, yet this difference could be reduced as the cost of the petroleum-based raw materials necessary for conventional fertilizers increases. Also, where there is little pest pressure, the costs of growing organic dates will likely be similar to those for conventional dates. Date producers in the USA are increasingly producing organic dates rather than conventional dates (Fig. 11.7). This is because organic date growing in an area where there are no pests and diseases that require conventional growing methods requires few additional costs, and the benefits can be seen in the price return. For example, in March 2022, a

5 kg box of conventionally grown large Medjool dates at the Los Angeles terminal market had a price of US$26.50–27.50. Meanwhile, a 5 kg box of organic large Medjool dates had a price of US$53.00. This difference of wholesale price at the terminal market, according to the present author's informal survey, translates to 10 to 15% more return to the grower, and justifies the increasing movement to organic date production (USDA-AMS, 2022b). Some in the industry feel that almost 90% of the US and Mexican Medjool date crop will be farmed organically within 5 to 10 years because of this price premium.

11.13 Conclusions and Prospects

For producers of high-quality dates, growing organic is becoming increasingly popular. Where organic fertilizers and pesticides are available, and where serious date pests are not a threat or where the pest complex can be controlled by organic pesticides, or cultural or biological control, the numbers of growers transitioning to organic date production will likely increase. Where it is difficult to find organic fertilizers and pesticides and the pest complex cannot be mitigated using organic pesticides, or cultural or biological control, organic production is not likely to gain traction.

Future research should focus on the testing of current practices, fertilizers, and pest control methods for organic production. Furthermore, additional organic fertilization products should be developed so that organic growers can have more options. Efforts should also be directed toward increasing the number of pest control options for organic growers. In particular, these might include the development of organic methods for red palm weevil and South American palm weevil control.

Since customers who can already buy a high-value date may be ready to pay a price premium for organic dates, the economics of growing organic dates will likely make sense. Based on these factors, it seems clear that organic dates will claim an increased share of the date market.

References

Abd-El Kareim, A.H.I., El-Sayed, G. and Afifi, M. (2014) Fulvic acid and biofertilization as a tool for promoting growth of date palm plants during acclimatization stage. *Journal of Biochemistry and Environmental Science* 9, 385–404.

Abdel-Gelil, L.M., Zaid, E. and Madboly, E.A. (2009) Effect of humic acid on date palm plantlets during rooting and acclimitization stages. In: *Proceedings of the 4th Conference on Recent Technologies for Agriculture, Faculty of Agriculture*, Cairo University, Giza, Egypt, November 3–5, 2009. Available at: https://www.researchgate.net/publication/326776212_EFFECT_OF_HUMIC_ACID_ON_DATE_PALM_PLANTLETS_DURING_ROOTING_AND_ACCLIMATIZATION_STAGES (accessed 1 December 2022).

Abdul-Baki, A., Aslan, S., Linderman, R., Cobb, S. and Davis, A. (2007) *Soil, Water and Nutritional Management of Date Orchards in the Coachella Valley and Bard*, 4th edn. California Date Commission, Palm Desert, California.

AgDaily (2019) The list of organic pesticides approved by the USDA. Available at: www.agdaily.com/technology/the-list-of-pesticides-approved-for-organic-prod uction/ (accessed 22 September 2021).

ATTRA Sustainable Agriculture (2015) Tipsheet: manure in organic production systems. Available at: www.ams.usda.gov/sites/default/files/media/Manure% 20in%20Organic%20Production%20Systems_FINAL.pdf (accessed 2 October 2021).

Benassy, C. (1990) Date palm. In: Rosen, D. (ed.) *Armored Scale Insects, Their Biology, Natural Enemies and Control*, Vol. 4B. World Crop Pests, Elsevier Academic Press, Amsterdam, pp. 585–591.

Benzohra, I.E., Megateli, M. and Berdja, R. (2015) Bayoud disease of date palm in Algeria: history, epidemiology and integrated disease management. *African Journal of Biotechnology* 14, 542–550.

Bogaard, A., Fraser, R., Heaton, T.H.E., Wallace, M., Vaiglova, P. *et al.* (2013) Crop manuring and intensive land management by Europe's first farmers. *Proceedings of the National Academy of Sciences of the United States of America* 110(31), 12589–12594. DOI: 10.1073/pnas.1305918110.

California Certified Organic Farmers (2020) CCOF homepage. Available at: www.c cof.org/ (accessed 30 September 2021).

Carpenter, J.B. and Elmer, H.B. (1978) In: *Pests and Diseases of Date Palm*. Agriculture Handbook no.527, US Department of Agriculture, Washington, DC.

Chase, A.R. and Broschat, T.K. (1993) *Diseases and Disorders of Ornamental Palms*. American Phytopathological Society, St. Paul, Minnesota.

Code of Federal Regulations (2022) The National List of Allowed and Prohibited Substances. Available at: www.ecfr.gov/compare/2022-02-28/to/2022-02-27/t itle-7/subtitle-B/chapter-I/subchapter-M/part-205/subpart-G/subject-group -ECFR0ebc5d139b750cd/section-205.601 (accessed 11 March 2022).

Crop Fertility Services (2022) What is the analysis of pelleted chicken manure? Available at: www.cropfertilityservices.com/analysis-pelleted-chicken-manure /#:~:text=A%20typical%20chicken%20manure%20pellets%20analysis%20Pel letized%20chicken,and%20roughly%202-3%25%20%28again%2C%20per%2 0ton%29%20Potassium%20range (accessed 15 January 2022).

Danthine, H. (1937) *Le Palmier Dattier et Les Arbres Sacres Dans l'iconographie de l'Asie Occidentale Ancienne*. P. Geuthner, Paris.

de Beer, Z.W., Duong, T.A., Barnes, I., Wingfield, B.D. and Wingfield, M.J. (2014) Redefining *Ceratocystis* and allied genera. *Studies in Mycology* 79, 187–219. DOI: 10.1016/j.simyco.2014.10.001.

Dembilio, O. and Jaques, J.A. (2015) Biology and management of red palm weevil. In: Wakil, W., Romeno Faleiro, J. and Miller, T. (eds) *Sustainable Pest Management in Date Palm: Current Status and Emerging Challenges*. Springer, Cham, Switzerland, pp. 13–36.

El Kinany, S., El Hilali, R., Achbani, E.H., Haggoud, A. and Bouamri, R. (2022) Enhancement of date palm growth through the use of organic fertilizer and microbial agents. *Journal of Soil Science and Plant Nutrition* 22(2), 1468–1477. DOI: 10.1007/s42729-021-00746-z.

El-Shafie, H. (2012) Review: list of arthropod pests and their natural enemies identified worldwide on date palm, *Phoenix dactylifera* L. *Agriculture and Biology Journal of North America* 3(13), 516–524. DOI: 10.5251/abjna.2012.3.12.516.524.

El-Shafie, H.A.F., Peña, J.E. and Khalaf, M.Z. (2015) Major hemipteran pests. In: Wakil, W., Romeno Faleiro, J. and Miller, T. (eds) *Sustainable Pest Management in Date Palm: Current Status and Emerging Challenges.* Springer, Cham, Switzerland, pp. 169–204.

European Union (2022) Organic production and products. Available at: https://ec.europa.eu/info/food-farming-fisheries/farming/organic-farming/organic -production-and-products_en#organicseeddatabase (accessed 6 March 2022).

Fairchild, D.G. (1903) *Persian Gulf Dates and Their Introduction into America.* US Government Printing Office, Washington, DC. DOI: 10.5962/bhl.title.65122.

FoodPLUS GmbH (2022) What We Do. Available at: www.globalgap.org/uk_en/w hat-we-do/ (accessed 30 November 2022).

Geisseler Lab (2021) UC Davis, University of California. Nutrient Management. Available at: http://geisseler.ucdavis.edu/Amendment_Calculator.html (accessed 16 January 2022).

Geisseler, D., Smith, R., Cahn, M. and Muramoto, J. (2021) Nitrogen mineralization from organic fertilizers and composts: Literature survey and model fitting. *Journal of Environmental Quality* 50(6), 1325–1338. DOI: 10.1002/jeq2.20295.

Global Organic Trade Guide (2022) Country list. Available at: https://globalorgan ictrade.com/country-list (accessed 30 November 2022).

Grebitus, C. and Hughner, R.S. (2021) Consumer demand and preferences for Medjool dates grown in Arizona. *Arizona Food Industry Journal* (June), 7.

Groupe EcoCert (2022) EcoCert homepage. Available at: www.ecocert.com/en (accessed 3 February 2022).

Habib, D.M. and Essaadi, S.H. (2007) Biocontrol of the lesser date moth *Batrachedra amydraula* Meyrik (Cosmopteridae = Batrachedridae) on date palm trees. *Acta Horticulturae* 736, 391–397. DOI: 10.17660/ActaHortic.2007.736.35.

Hargrove, W.L. and Fry, W. (1987) The need for legume cover crops in conservation tillage production. In: Power, J. (ed.) *The Role of Legumes in Conservation Tillage Systems.* Soil and Water Conservation Society, Athens, Georgia, pp. 1–4.

Harhash, M., Mourad, A.K. and Hammad, S.M. (2003) Integrated crop management of the lesser date moth *Batrachedra amydraula* Meyr. (Lepidoptera: Cosmopteridae) infesting some date-palm varieties in Egypt. *Communications in Agricultural and Applied Biological Sciences* 68, 209–221.

Hoddle, M. (2022) Biology and management of South American palm weevil, *Rhynchophorus palmarum* (L.) (Coleoptera: Curculionidae), in California. Available at: https://biocontrol.ucr.edu/south-american-palm-weevil#:~:text= South%20American%20palm%20weevil%20%28SAPW%29%2C%20Rhynch ophorus%20palmarum%20%28L.%29,%29%20in%20Tijuana%20Baja%20C alifornia%20Mexico%20in%202010 (accessed 11 March 2022).

Kehat, M., Blumberg, D. and Greenberg, S. (1969) Experiments on the control of raisin moth *Cadra figulilella* Gregs, (Phyticidae Pyralidae), on dates in Israel. *Israeli Journal of Agricultural Research* 19, 121–128.

Ketterings, Q.M., Albrecht, G., Czymmek, K. and Bossard, S. (2005) *Nitrogen Credits from Manure.* Cornell University Cooperative Extension, Ithaca, New York.

Khan, R.R., Haq, I.U. and Naqvi, S.A. (2023) Pest and disease management in date palm. In: Al-Khayri, J.M., Jain, S.M., Johnson, D.V. and Krueger, R.R. (eds) *Date Palm.* CAB International, Wallingford, UK, pp. 297–338.

Levi-Zada, A., Fefer, D., Anshelevitch, L., Litovsky, A., Bengtsson, M. *et al.* (2011) Identification of the sex pheromone of the lesser date moth, *Batrachedra amydraula,* using sequential SPME auto-sampling. *Tetrahedron Letters* 52(35), 4550–4553. DOI: 10.1016/j.tetlet.2011.06.091.

Mafra-Neto, A., de Lame, F.M., Fettig, C.J., Munson, A.S., Perring, T.M. *et al.* (2013) Manipulation of insect behavior with Specialized Pheromone and Lure Application Technology (SPLAT®). In: Beck, J.J., Coats, J.R., Duke, S.O. and Koivonunen, M.E. (eds) *Pest Management with Natural Products.* American Chemical Society, Washington, DC, pp. 31–58.

Meddich, A. (2021) *Biostimulants for Sustainable Agriculture in Oasis Ecosystem Towards Improving Date Palm Tolerance to Biotic and Abiotic Stress.* General Secretariat of Khalifa International Award for Date Palm and Agricultural Innovation, Abu Dhabi.

Melhim, A. (2021) The state of North American organic produce: a closer look at post-pandemic trends and emerging issues. Available at: https://research.rab obank.com/far/en/sectors/fresh-produce/the-state-of-north-american-organ ic-produce.html (accessed 15 April 2022).

Mitchell, C.C. (2008) *Nutrient Content of Fertilizer Materials.* Alabama Cooperative Extension System, Auburn, Alabama.

Montasser, M.S., Hanif, A.M., Al_Awadhi, H.A. and Suleman, P. (2012) Tetracycline therapy against phytoplasma causing yellowing disease of date palms. *FASEB Journal* 26(S1), 800.1. DOI: 10.1096/fasebj.26.1_supplement.800.1.

Mordor Intelligence (2019) Organic fruits and vegetables market size – segmented by type (fruits, vegetables) and geography – growth, trends, and forecast (2019–2024). Available at: www.mordorintelligence.com/industry-reports/or ganic-fruits-and-vegetables-market (accessed 18 March 2022).

Moustafa, M. (2012) Host plant, distribution and natural enemies of the red date scale insect, *Phoenicococcus marlatti* (Hemiptera: Phoenicococcidae) and its infestation status in Egypt. *Journal of Basic & Applied Zoology* 65(1), 4–8. DOI: 10.1016/j.jobaz.2012.02.001.

Nay, J.E. and Perring, T.M. (2009) Effect of center cut strand thinning on fruit abscission and *Ectomyelois ceratoniae* (Lepidoptera: Pyralidae) infestation in California date gardens. *Journal of Economic Entomology* 102(3), 948–953. DOI: 10.1603/029.102.0313.

Nay, J.E., Boyd, E.A. and Perring, T.M. (2006) Reduction of carob moth in 'Deglet Noor' dates using a bunch cleaning tool. *Crop Protection* 25(8), 758–765. DOI: 10.1016/j.cropro.2005.10.010.

Negm, M.W., De Moraes, G.J. and Perring, T.M. (2015) Mite pests of date palms. In: Wakil, W., Romeno Faleiro, J. and Miller, T. (eds) *Sustainable Pest Management in Date Palm: Current Status and Emerging Challenges.* Springer, Cham, Switzerland, pp. 347–389.

Organic Egypt (2020) Organic Egypt homepage. Available at: https://organicegyp t.org/ (accessed 2 February 2022).

Othman, K., Rhouma, A., Belhadj, R., Alimi, E., Fallah, H. *et al.* (2001) Lutte biologique contre un acarien ravageur des dattes: essai d'utilisation de

Neoseiulus californicus contre Oligonychus afrasiaticus dans les palmeraies du Djerid (Sud tunisien). *Phytoma* 540, 30–31.

Perring, T.M., El-Shafie, H.A.F. and Wakil, W. (2015) Carob moth, lesser date moth, and raisin moth. In: Wakil, W., Romeno Faleiro, J. and Miller, T. (eds) *Sustainable Pest Management in Date Palm: Current Status and Emerging Challenges.* Springer, Cham, Switzerland, pp. 109–167.

Primus AuditingOps (2015) Primus AuditingOps homepage. Available at: https:// primusauditingops.com/ (accessed 30 September 2021).

Pure & Eco India (2022) Carbon dioxide fumigation of organic commodities – the safest way to export. Available at: https://pureecoindia.in/carbon-dioxide-fum igation-of-organic-commodities-a-safe-way-to-export/ (accessed 30 November 2022).

Research and Markets (2020) Organic fruits and vegetables market by product type, form and end user, distribution channel: global opportunity analysis and industry forecast 2020–2027. Available at: www.researchandmarkets.c om/reports/5306388/organic-fruits-and-vegetables-market-by-product?ut m_source=BW&utm_medium=PressRelease&utm_code=hm33vp &utm_campaign=1516676+-+Global+Organic+Fruits+%26+Vegetable s+Market+2020-2027%3a+A+%2455%2b+Billion+Opp (accessed 18 March 2022).

Saeed, E.E., Sham, A., El-Tarabily, K., Abu Elsamen, F., Iratni, R. *et al.* (2016) Chemical control of black scorch disease on date palm caused by the fungal pathogen *Thielaviopsis punctulata* in United Arab Emirates. *Plant Disease* 100(12), 2370–2376. DOI: 10.1094/PDIS-05-16-0645-RE.

Sánchez, V., Rebolledo, O., Picaso, R.M., Cárdenas, E., Córdova, J. *et al.* (2007) *In vitro* antagonism of *Thielaviopsis paradoxa* by *Trichoderma longibrachiatum. Mycopathologia* 163(1), 49–58. DOI: 10.1007/s11046-006-0085-y.

Sawyer, J. (2009) *What Are Average Manure Nutrient Analysis Values?* Iowa State University Extension, Ames, Iowa.

Scheil, V. (1913) De l'exploitation des dattiers dans l'ancienne babylonie. *Revue d'Assyriologie et d'archelogie Orientale* 10, 1–9.

Shaffer, G. (2019) Organic herbicides. Available at: https://extension.sdstate.edu/ organic-herbicides (accessed 13 November 2021).

Simon, H. (1978) *The Date Palm: Bread of the Desert.* Dodd, Mead and Company, New York.

Soytong, K., Pongak, W. and Kasiolarn, H. (2005) Biological control of *Thielaviopsis* bud rot of *Hyophorbe lagenicaulis* in the field. *Journal of Agricultural Technology* 1, 235–245.

Talhouk, A.S. (1983) The present status of date palm pests in Saudi Arabia. In: Makki, Y.M. (ed.) *Proceedings of the First Symposium on Date Palm in Saudi Arabia, Al-Hassa, Saudi Arabia,* King Faisal University, Al-Hassa, Saudi Arabia, March 23–25, 1982, pp. 432–438.

Triapitsyn, S.V., Andreason, S.A., Dominguez, C. and Perring, T.M. (2019) On the origin of *Anagyrus callidus* (Hymenoptera: Encyrtidae), a parasitoid of pink hibiscus mealybug *Maconellicoccus hirsutus* (Hemiptera: Pseudococcidae). *Zootaxa* 4671, 283–289. DOI: 10.11646/zootaxa.4671.2.9.

USDA-AMS (2022a) NOP Handbook: Guidance & Instructions for Accredited Certifying Agents & Certified Operations. Available at: www.ams.usda.gov/rule s-regulations/organic/handbook (accessed 15 January 2022).

USDA-AMS (2022b) Specialty Crops Terminal Markets Standard Reports. Available at: www.ams.usda.gov/market-news/fruit-and-vegetable-terminal-markets-standard-reports (accessed 11 March 2022).

van der Crabben, J. (2021) Agriculture in the Fertile Crescent and Mesopotamia. Available at: www.worldhistory.org/article/9/agriculture-in-the-fertile-crescent–mesopotamia/ (accessed 2 February 2022).

Worthington, V. (2004) Nutritional quality of organic versus conventional fruits, vegetables, and grains. *Journal of Alternative and Complementary Medicine* 7, 161–173.

Yang, X. and Liu, Y.B. (2021) Anisole is an environmentally friendly fumigant for postharvest pest control. *Journal of Stored Products Research* 93, 101842. DOI: 10.1016/j.jspr.2021.101842.

Zaid, A., de Wet, P.F., Djerbi, M. and Oihabi, A. (2002) Diseases and pests of date palm. In: Zaid, A. and Arias-Jiménez, E. (eds) *Date Palm Cultivation*. FAO Plant Production and Protection Paper no.156 Rev. 1, Food and Agriculture Organization of the United Nations, Rome, pp. 227–282.

Zirari, A. and Laaziza Ichir, L. (2010) Effect of different kinds of bunch coverings on date palm fruit (*Phoenix dactylifera* L.) moth's infestation rate. *Acta Horticulturae* 882, 1009–1014. DOI: 10.17660/ActaHortic.2010.882.117.

12

Agroecological Practices on Traditional Date Farms

Rashid Al-Yahyai* ⓘ, Khalid Al-Hashmi and Rhonda Janke ⓘ

Sultan Qaboos University, Al-Khoud, Oman

Abstract

Agroecology is a set of agricultural practices that considers the natural environment in crop cultivation. The Food and Agriculture Organization of the United Nations defines it as the science and practice of applying ecological concepts and principles to manage interactions between plants, animals, humans, and the environment for food security and nutrition. Sustainable agroecological systems in agricultural production aim to create balanced natural resource utilization while maintaining adequate and profitable food systems to ensure food security and thriving farming communities in line with the United Nations' Sustainable Development Goals (SDGs). Date is one of the most suitable crops for achieving sustainable agroecological farming systems, particularly in harsh agroclimatic regions of the world. This chapter covers the basic concepts of sustainable agroecological systems and links to the SDGs, utilizing date palm as a model crop. Although literature concerning agroecology in general, and on date palm particularly, is lacking and there are plenty of gaps to be filled, this chapter highlights the importance of both the practice and the science of agroecology in date production. Case studies and examples from Oman and around the world are presented in the hope that this will drive further research on this topic and the adoption of sustainable agroecological elements in agriculture and crop production.

Keywords: Agricultural practices, Agroecology, Food security, Food systems, Sustainability

12.1 Introduction

Recent studies have shown a direct impact of changing farming practices on the biodiversity of fruit crop orchards and plantations. The sustainability of biological resources is threatened by overexploitation. Destruction

*Corresponding author: alyahyai@squ.edu.om; alyahyai@gmail.com

© CAB International 2023. *Date Palm* (eds J.M. Al-Khayri *et al.*)
DOI: 10.1079/9781800620209.0012

of plant communities caused by cutting, burning, weeding, cleaning, over-fertilizing, overwatering, and pesticide pollution leads to reductions in the biological resources and thus the productivity of farmlands. Scientists estimate that there are about 1.4 million species on earth, including 750,000 insects, 41,000 vertebrates, and 360,000 plants. Some 80,000 plant species are edible, but only 150 of them are currently grown on a large scale and 30 are considered industrial food crops (Mika, 2004).

Biodiversity in vascular plants depends on geographical latitude, abundance, and habitat heterogeneity. Biodiversity in arid and semi-arid regions is essential for world flora as the species growing in these conditions often tolerate abiotic stress conditions such as drought and heat. The flora of Oman, for example, is estimated at 1300 plant species including native crop plants such as date palms with around 250 cultivars (Al-Yahyai and Al-Khanjary, 2008). Some of the richest plant communities are autogenic and seminatural plant communities created by natural forces under favourable conditions, particularly those representing traditional farming systems where inputs are organic and diversity ensures food supply throughout the year. When they are destroyed by modern farming practices, the resulting anthropogenic communities are dominated by a very limited set of species. However, the choice does not have to be binary, either natural ecosystems or industrial agriculture. There are many ways to use ecological principles in the design and functioning of diverse agricultural systems to achieve the benefits of natural system ecology and yet produce food, feed, and/or fiber for human communities.

The discipline of ecology investigates the kind of interactions there are within a plant community and between different plant communities; how biodiversity develops; whether it is stable; whether it recovers after being disturbed; and whether the patterns of plant communities are similar to those in place before the disturbance occurred (Mika, 2004). Date palm cultivation has historically contributed to conservation and agrobiodiversity because it has long followed sustainable practices within agroecological systems.

12.2 Agroecology and Sustainable Development

Agroecology is a recent term that is generally used to describe agricultural practices that consider environmental aspects in addition to economic aspects of crop cultivation. The Food and Agriculture Organization of the United Nations (FAO) defines agroecology as the science and practice of applying ecological concepts and principles to manage interactions between plants, animals, humans, and the environment for food security and nutrition (FAO, 2022). This holistic approach to farming considers all aspects of ecological systems and social concepts in the management of the food-producing system, including sustainable agriculture.

The objective is to optimize food production and enhance food security and food sovereignty while preserving natural resources and environmental components, including animals, birds, insects, soil, water, and traditional farming practices. Therefore, the adoption of agroecological systems to produce sustainable food is at the heart of the United Nations' Sustainable Development Goals (SDGs) as it serves multiple global developmental objectives while maintaining and considering the local culture and people where the approach is applied (FAO, 2022). Thus, the design and practice of agroecology in date palm agriculture can address one or more of the following SDGs:

1. Zero hunger, food security, nutrition, and health.
2. Poverty alleviation.
3. Climate change resilience.
4. Biodiversity.
5. Youth engagement.
6. Gender self-determination.
7. Human rights.

SDGs, according to FAO (2022), can be attained through ten approaches that are essential components of agroecological systems and include:

1. *Biological diversity:* Biological diversity is key to conserving, protecting, and enhancing natural resources including crop diversity while producing adequate and nutritious food through sustainable farming techniques.
2. *Participation:* Knowledge sharing may lead to better agricultural innovations that can address local challenges.
3. *Synergistic food system:* Enhancing key functions through the food-supply chain and supporting food production in multiple ecosystems.
4. *Efficiency:* Agroecological practices lead to higher production using fewer resources by optimizing the inputs.
5. *Recycling:* Agricultural production with lower economic inputs and environmental footprints.
6. *Resilience:* Human, community, and ecosystem resilience is vital to sustainable agricultural food systems.
7. *Human and social values:* Local values lead to protecting and improving rural livelihoods and encourage the equity and social well-being that are essential for sustainable food and agricultural systems.
8. *Traditional knowledge:* Local culture and food knowledge support healthy, diversified, and culturally appropriate diets, thus agroecology contributes to food security and nutrition while maintaining the health of ecosystems.
9. *Responsible governance:* Responsible and effective governance mechanisms are essential for sustainable agricultural food systems that can be implemented at the local and global levels.

10. *Circular and solidarity economy:* Reconnecting producers and consumers provides innovative solutions for living within our planetary boundaries while ensuring the social foundation for inclusive and sustainable development.

Again, these are in alignment and can be achieved through the practice of agroecological farming methods. The details of how this is accomplished are illustrated in the rest of this chapter. These elements of sustainable development through the use of agroecology have been adopted by the FAO and are being implemented across the globe to provide the foundation for achieving the SDGs. It is therefore essential to consider the role of the agroecology of date palm production as one of the most widely cultivated crops across the globe, particularly in arid and semi-arid regions. The sustainability of date palm production stems from the fact that this particular crop has been in cultivation following the traditional system of agroecology for millennia, and it is worth noting the main components that led to its sustainability.

12.3 Date Palm Agroecological Farming Systems

Date palm plantations across its cultivation regions, primarily the Middle East and North Africa (MENA), follow traditional agroecological practices that consider all the environmental variables and the availability of the dominant scarce natural resources. These cultivation systems have many aspects of communality that have made date palm cultivation one of the longest in human history. The following sections offer details about some date palm sustainable cultivation practices.

12.3.1 Historical and traditional systems of cultivation

Traditionally date palm is grown in desert or mountain oases (Al-Yahyai, 2007; Jaradat, 2011; Al-Yahyai and Manickavasagan, 2012) in smaller concentrations or in large-scale plantations near riverbanks (such as in Egypt and Sudan). Planting date palms close to each other and in multilayered forest-style plantations provides an ample habitat for both crop diversity and natural ecological components, such as animals and pollinators (de Grenade, 2013). The understory of the date palm plantation in traditional agroecological systems is normally occupied by larger fruit crops such as bananas, citrus, and mango, among others. Below that and serving as cover crop, fodder crops and cereal crops such as lucerne (alfalfa), wheat, and barley, are planted. In many instances, the planted crops are diverse and include leguminous crops that provide nitrogen to the soil. Nutrients from crop residues and livestock waste lead to the recycling of nutrients and the addition of organic matter, which enhances soil fertility over time.

12.3.2 Date palm cultivar diversity

One key element of agroecological farming systems is the in-farm conservation of crop and plant biodiversity. Traditional date palm plantations are among the most genetically diverse in the world. This is particularly true in relation to modern plantations that focus solely on single cultivars on expansive monocultural lands.

Date palms have been grown for centuries and farmers have secured a wide range of varieties that can produce year-round fruits in their three consumption states: khalal, rutab, and tamar. Hundreds of date palm cultivars are currently in cultivation across the world, including male varieties that are a source of pollen for female varieties, with different varieties being more common in different countries (Al-Yahyai and Al-Khanjary, 2008; Jaradat, 2011; Haider *et al.*, 2015; Abul-Soad *et al.*, 2017; Faci, 2019; Świąder *et al.*, 2020) (Table 12.1).

New genetic resources for date palm can be made available through gene banks and the development of varieties using biotechnological and molecular tools. Recent research is exploring the genetic characteristics of existing cultivars with tolerance to specific environmental stresses such as salinity, heat, and drought (Al Kharusi *et al.*, 2017, 2019). Preservation and expansion of date palm genetic resources are essential for sustainable agricultural food production. The utilization of crop wild relatives (CWRs) has helped many crops evolve and develop through breeding programs over decades. Wild relatives of date palm have long been thought of as nonexistent; however, a recent study has reported the presence of wild date palms in Oman (Gros-Balthazard *et al.*, 2017). Therefore, further consideration of wild date palms and molecular identification of the unique characteristics of these palms is essential for further development of date palm cultivars.

12.3.3 Species diversity and traditional multispecies cropping systems

In addition to the diverse date palm varieties discussed above, ecologically sound date plantations also provide a pool of diverse fruit crops, including citrus, mango, banana, and papaya. Diversity is one of the basic elements of agroecology, and this is achieved by the multiple layers of crops being planted in traditional farming systems of date palm. This has ensured year-round availability of dates and food from other plants. Crop biodiversity is complemented with the natural plant and animal biodiversity that cohabit the date palm plantations. Several wild plant species grow in these plantations and fulfil nutrition and medical purposes. Al-Hashmi and Al-Yahyai (2020) reported a total of 51 medicinal and aromatic plants from 16 different plant species and 13 families in a traditional date palm farm in Oman (Table 12.2).

Table 12.1. Most common cultivated varieties of date palm in major date-producing countries. (Adapted from Świąder *et al.*, 2020.)

Country	Varieties of cultivated dates
Egypt	Amhat, Hayany, Samany, Siwi, Zoghloul
Saudi Arabia	Ajwa, Al-Barakah, Al-Qaseem, Berhi, Gur, Helwet El-Goof, Hiladi, Hulwa, Khalasah, Khasab, Majnaz, Mishriq, Miskani, Nabbut Ghrain, Nabtat Seyf, Rothanat, Ruzeiz, Sag'ai, Sebakat Al-Riazh, Sahal, Sellaj, Shashi, Sokkary, Tanjeeb, Tayyar, Thamani, Umelkhashab, Um Rahim, Zamil, Zaghloul
Iran	Allmehtari, Barhi, Dayri, Estamaran, Gantar, Halawi, Kabkab, Khassui, Khazravi, Mazafati, Mordarsang, Piarom, Pyarom, Rabbi, Sayer, Shahani, Shakkar, Sowaidani, Zahedi
Algeria	Deglet Nour, Iteema, Thoory
Iraq	Amir Hajj, Barhi, Dayri, Halawy, Khadrawi, Khastawi Maktoom, Sayer, Zahidi
Pakistan	Basra, Dhakki, Gulistan, Hsaini, Kajur, Khadrawi, Mobini, Mozafati, Obaidullah, Sabzo, Shakri, Zaidi
Sudan	Abid Rahim, Barakawi, Bentamoda, Birier, Gondaila, Jawa, Khatieb, Kulma Suda, Medina, Mishriq, Mishriq Wad, Mishriq Wad Lagi, Zughloul
Oman	Fard, Khalas, Khasab, Naghal, Khunaizi, Bu Narejah, Mabsli, Um Sella, Maan
United Arab Emirates	Berhi, Bomaan, Khalas, Lolo, Fard
Tunisia	Ammari, Angou, Arichti, Bejjou, Bisr Helou, Bouhattam, Brance de dates, Deglet Nour, Eguiwa, Ftimi, Garn ghazel, Gounda, Gousbi, Hamraya, Hissa, Kenta, Kentichi, Ksebba, Korkobbi, Lagou, Lemsi, Mattata, Mermella, Rouchdi, Touzerzayet

The inclusion of farm animals and poultry, and the need to grow food and fodder crops underneath the date palms, have further increased the diversity of agroecosystems in date palm farms (Fig. 12.1). This contrasts with plantations where a single or very few varieties are planted on hundreds or thousands of hectares (Fig. 12.2). In Oman, about 60% of the farmed land area is on farms with a crop/livestock mix, 39% is on farms with crops only, and less than 1% of land is used to raise livestock (MAF, 2015).

This shift from species-rich date palm farms in traditional settings to modern farms has led to genetic erosion of date palm cultivars as a few superior cultivars dominate the latter and there is the potential loss of hundreds of other varieties due to lack of interest and profitability in their cultivation. Other factors that have eroded the biodiversity of date palm

Table 12.2. List of plants found in date palm plantations and their known uses in Oman. (From Al-Hashmi and Al-Yahyai, 2020.)

Governorate	Farm location	Scientific name	Location			Common uses[a]
			Under palm	Under other trees	Open space/ border	
Al Dakhlia	Samail, Fankh	*Myrtus communis* L.		×		A, U
		Lawsonia inermis L.				A
		Ficus sycomorus L.			×	M
		Forsskaolea tenacissima L.			×	M
		Datura metel L.			×	M
		Chrozophora oblongifolia L.		×		M
		Launaea mucronata (Forssk.) Muschl.	×	×	×	M
		Ricinus communis L.			×	U
		Aerva javanica L.			×	U
Al Dakhlia	Bidbid, Fanja	*L. mucronata* (Forssk.) Muschl.	×		×	M
		Calotropis procera (Aiton) W.T. Aiton			×	M
		M. communis L.			×	A, U
Al Dakhlia	Samail, Daser	*L. mucronata* (Forssk.) Muschl.	×			M
		L. inermis L.			×	A
		Abutilon fruticosum L.		×		M
		F. sycomorus L.			×	M
		C. procera (Aiton) W.T. Aiton			×	M

Continued

Table 12.2. Continued

Governorate	Farm location	Scientific name	Location			Common uses[a]
			Under palm	Under other trees	Open space/ border	
Al Batinah	Rustaq, Grief	*Lycium shawii* L.	x	x	x	M
		F. tenacissima L.			x	M
		D. metel L.	x		x	M
		L. mucronata (Forssk.) Muschl.	x		x	M
		R. communis L.	x		x	U
		A. javanica L.			x	U
		C. procera (Aiton) W.T. Aiton			x	M
		L. inermis L.			x	A
Al Batinah	Rustaq, Dhuhli	*Ammi majus* L.	x		x	M
		F. sycomorus L.			x	M
		D. metel L.	x		x	M
		L. mucronata (Forssk.) Muschl.	x		x	M
		L. inermis L.			x	A

Continued

Table 12.2. Continued

| Governorate | Farm location | Scientific name | Location | | | Common uses[a] |
			Under palm	Under other trees	Open space/ border	
Al Batinah	Rustaq, Sarnah	*F. tenacissima* L.			×	M
		D. metel L.	×	×	×	
		L. mucronata (Forssk.) Muschl.	×			M
		C. procera (Aiton) W.T. Aiton			×	M
		L. inermis L.			×	
		M. communis L.			×	A, U
		F. sycomorus L.			×	
Muscat	Bowshar, Hamam	*F. sycomorus* L.			×	M
		L. mucronata (Forssk.) Muschl.	×	×		M
		F. tenacissima L.			×	M
		Ficus cordata subsp. *salicifolia* L.			×	M
		Asphodelus fistulosus L.	×			
Muscat	Bowshar, Sunub	*D. metel* L.			×	M
		F. cordata subsp. *salicifolia* L.			×	M
		A. fistulosus L.	×			M
		Lunaea mucronata (Forssk.) Muschl.	×			
		F. sycomorus L.			×	

Continued

Table 12.2. Continued

| Governorate | Farm location | Scientific name | Location | | | Common uses[a] |
			Under palm	Under other trees	Open space/ border	
Muscat	Bowshar, Bowshar	*F. tenacissima* L.			×	M
		Pennisetum setaceum L.			×	U
		A. majus L.	×	×		
		L. mucronata (Forssk.) Muschl.	×	×	×	M

[a]A, aromatic; M, medicinal; U, utility. Presence of the plant is indicated by ×.

Fig. 12.1. Agroecological date palm plantations are biologically diverse and resource-conserving sustainable systems of production. (Photo by Rashid Al-Yahyai.)

plantations are loss of plants due to desertification, salinity, pests, diseases, and climate change, the removal of mature palms for landscaping, and non-replacement of old palms with similar cultivars (Jaradat, 2011; Abul-Soad *et al.*, 2017).

Fig. 12.2. Examples of agroecological production that have been investigated in date palm plantations of Oman. (a) Traditionally managed date palm farm. (b) Modern date palm farm. (c) Native wild and crop biodiversity in a traditional date palm farm. (d) Exotic plant species introduced to a traditional date palm farm. (Photos by Rashid Al-Yahyai.)

Genetic improvement is an integral part of the sustainability and biodiversity components of the oasis agroecological systems. However, few controlled studies on crop breeding within these oases have been reported. Crop genetic resource diversity in oasis agroecological systems developed from the farmers, often living in communities surrounded by wild and domesticated plants and animals, sharing knowledge, exchanging seeds and plant materials, and continuously improving valuable traits in different crops (Brush, 2007). The large number of date palm cultivars is an indication of the large extent of date palm breeding and selection in these traditional systems. Furthermore, beside female cultivars, growers in the oases systems have also selected a range of male date palm cultivars for the purpose of pollination that will ultimately increase the yield and quality of dates (Al-Yahyai and Khan, 2015; Haider *et al.*, 2015). Genetic outcrossing potential is another method for crop improvement in diversity-rich oases (de Grenade, 2013). Nevertheless, this millennia-old selection, breeding, and crop development system is currently under threat, creating an immediate need for conservation approaches such as participatory agrobiodiversity conservation (Peano *et al.*, 2021).

Due to the nature of the growth habit of the monocotyledonous date palm, the plantations in agroecological farming systems provide shading for a second and third storey of plants (de Grenade, 2013). This is largely due to the microclimate habitat that date palms provide, which allows for a range of cultivated and naturally growing plants to thrive. Furthermore, the date palm roots provide a microenvironment for a range of beneficial and pathogenic fungi (Al-Sadi *et al.*, 2012) and bacteria (Cherif *et al.*, 2015). The latter study found that date palm roots shape endophytic growth-promoting fungi and bacteria under drought conditions, thus contributing an essential ecological service to the entire oasis ecosystem.

A recent meta-analysis of intercropping systems confirms their yield and other advantages and classifies the mechanisms as either resource acquisition or resource conversion efficiency (Stomph *et al.*, 2020). These advantages are due to differential resource acquisition in time and space in multispecies systems. Resources include light, water, and nutrients. The yield advantages due to intercropping are often quantified as the 'land equivalent ratio' (LER) or the amount of land required for the same yield of each crop grown as monoculture. Worldwide LERs are in the range of 1.2 to 1.3 for annual crops (Yu *et al.*, 2015). For perennial crops, and mixtures of perennials and annuals, values would likely be higher. The yield advantage is because multispecies cropping systems imitate natural ecosystems, which are made up of plant communities with functionally diverse species (Franco *et al.*, 2015). The theory behind these cropping systems can be traced to ecologists such as Odum (1968) and Vandermeer (2010).

In addition, intercropping reduced disease incidence in 79% of papers reviewed, decreased insect pest abundance in 68% of experiments reviewed, and decreased weed pressure in 86% of published experiments (Stomph *et al.*, 2020). The mechanisms for decreased pest pressure of all types have been attributed to modification of microclimate, reduction of host density, and reduced vector dispersal. Ultimately, intercropping might be the most effective way to achieve a sustainable intensive crop production while reducing losses from pests and diseases (Brooker *et al.*, 2014).

12.3.4   Utilization of natural resources – water

The most essential elements of crop production are the resources required to maintain production, such as water, soil, fertility, and pest control. Traditional date palm cultivation is dependent on balanced ecological components that are essential for the yield and quality of dates. In addition, conservation of these resources ensures that other elements of livelihood are maintained, including water security (quality and quantity), other farm components like livestock, a diverse food supply from other crops and wild plants within the farm, and the social and cultural aspects of the village or community where date palm is cultivated. These considerations have always been priorities in the unique mountain and desert oases of Oman.

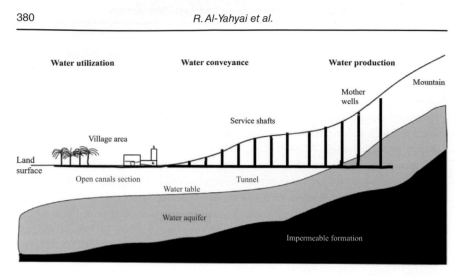

Fig. 12.3. Sketch of daudi falaj. (Used with permission from Al-Ghafri, 2018.)

Irrigation water for date palms in the oasis system comes from multiple sources that include two primary water-supplying techniques: wells and aflaj. Aflaj, also referred to as qanat in other parts of the date palm cultivation world, refers to delivering water to the plantings through canals that are either above or under the ground (Al-Ghafri, 2018). The type of aflaj system is based on the sources of water, which include:

- Daudi or eddi falaj: The water is channeled to farms through villages from an underground aquifer (Fig. 12.3). The water of this type of aflaj has domestic uses such as for drinking and bathing and cleaning, among others.
- Aini: This type of aflaj means that the water source is ain, which means spring. The spring water flows to the surface from underground and is channeled to farms. There are several springs in Oman, some of which do not dry out despite long periods of drought. Water temperature also varies by spring.
- Ghaili: This refers to small dams that are placed in wadis so that when water flows for longer periods of time after a rainy season, the water is channeled to farms and used for irrigation (Fig. 12.4). Ghaili aflaj is a form of rainwater harvesting and is subject to discontinuity during drought years.

Wells are also an important complement to the aflaj system of irrigation and when adequate water resources are available, wells alone can provide the needed irrigation water for date palm farms. Water in traditional farms is extracted from a well using animal power. Some wells have different water extraction techniques utilizing diesel or electric engines that have been developed in recent decades. Currently, farmers with wells are encouraged to adopt modern irrigation systems (MIG) that consist of plastic tubing

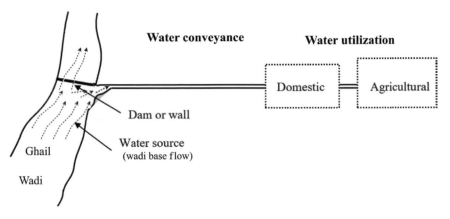

Fig. 12.4. Sketch of ghaili falaj. (Used with permission from Al-Ghafri, 2018.)

with either drip lines or bubblers at each tree, set on timers to deliver a set amount of water at predetermined intervals. These systems are more efficient in delivering water and reducing water loss due to leaching, but do not provide water to the understory species that are planted within a traditional date palm agroecosystem.

To maximize the utilization of the scarce water resources in these arid and semi-arid regions of the world where date palm is commonly cultivated, traditional farming techniques have resorted to flood irrigation. This zero-energy system is dependent on the water delivered through gravity and channels from the water sources. The flood irrigation system is highly efficient when considering the multitude of benefits it provides to multistory farms. It irrigates not only date palms, but also other second-layer fruit crops as well as fodder and grain crops in the third layer of the traditional date palm farm. It also provides a microclimate for other important ecological components including animals, birds, beneficial insects, pollinators, and soil biota. Therefore, this traditional water delivery technique that lasted for centuries must be viewed from the perspective of the holistic benefits it provides and not merely on the date palm fruit yield alone.

In Oman, the percentage of land growing dates that is flood irrigated is 88%, as compared to MIG on 12% of cultivated land. This suggests that a large portion of dates in Oman are still grown in traditional systems. This contrasts with vegetable crops in Oman, where it is reported that 81% of land is irrigated from wells (MIG) and only 19% is flood irrigated. Forage crops report 55% MIG and 45% flood irrigation systems (MAF, 2015). Date palm farms in Oman also tend to be mostly smaller farms. By area farmed, 58% of the land area in date palm farms is farms less than 5 feddan (2.1 ha). By farm number, 85% of date farms are smaller than 2.1 ha, and 66% of farms are in the smallest size class of less than 1 feddan or 0.42 ha (MAF, 2015) (Table 12.3).

Table 12.3. Date palm farms in Oman by farm size. (Used with permission from MAF, 2015.)

Farm size (feddan)		Cropped area (feddan)	Number of farms	% Cropped area	% Number of farms
Min.	Max.				
0.01	0.99	16,841	77,959	29.32	65.91
1	1.99	5,965	9,914	10.39	8.38
2	4.99	10,557	12,868	18.38	10.88
5	9.99	9,839	9,074	17.13	7.67
10	19.99	8,004	6000	13.94	5.07
20	29.99	2,413	1342	4.20	1.13
30	39.99	950	453	1.65	0.38
40	49.99	476	208	0.83	0.18
50	99.99	909	274	1.58	0.23
100	199.99	367	113	0.64	0.10
200	Or more	1,109	70	1.93	0.06
Total		57,430	118,275	100.00	100.00

Because date palms are irrigated year-round, they use more cubic meters of water per year ($9257\,m^3$/feddan) than vegetable and field crops (between 1831 and $6254\,m^3$/feddan), which are irrigated only during their growing season (MAF, 1993). In Oman, 38% of the total irrigation water applied is used to irrigate the 7.6 million date palms, and so research has focused on ways to water them more efficiently (Al-Mulla and Al-Gheilani, 2016).

One suggestion is to plant higher-yielding cultivars to increase water-use efficiency, but as discussed earlier, this would be at the cost of losing some traditional heritage varieties. Deficit irrigation has been tested in a number of trials, but generally finds lower yields with lower water levels, even in controlled experiments (Rahnama *et al.*, 2012; Mattar *et al.*, 2021). In addition, treated wastewater can be used combined with deficit irrigation. When normal or excessive water is applied, the trees did not suffer yield loss, but at deficit levels, they did (Mattar *et al.*, 2021). Water with higher salt levels has also been compared, but at salinity levels of 3.94 dS/m yields were decreased at all irrigation levels, especially with deficit irrigation (Mattar *et al.*, 2021). Deficit irrigation can result in higher fruit quality, as measured by total or reducing sugars content (Al-Yahyai and Al-Kharusi, 2012).

12.3.5   Soil fertility management

Because agroecological systems implement an integrated and sustainable approach to food production at the local level, including traditional and cultural practices, the sources of other essential components of date palm production are also within the farm. Traditionally, many date palm growers also raised animals, particularly poultry, sheep, goat, cow, and camels, either alone or in combination. The animals provide protein from eggs, meat, and/or milk to the community/family diet, and they also provide fertilizers that enrich the soil with organic matter. Certain animals, such as oxen and donkeys, traditionally provided power to extract the water from wells and plow the soil.

Until now, much of the fertilizers being applied to traditional date palm farms in date-growing countries such as Saudi Arabia, Morocco, and Oman came from animal manure (Al-Yahyai and Khan, 2015; Almadini *et al.*, 2021; Ou-Zine *et al.*, 2021). Long-term manure applications are responsible for increased soil organic matter levels in some traditional agriculture fields in Oman (Buerkert *et al.*, 2005). The Ministry of Agriculture's guidance on date palm fertilization in Oman includes recommendations for applying a combination of manures and soluble fertilizers at planting and annually to bearing trees (MAF, 1993). Indeed, several research studies specifically on date palm fertilization report advantages to applying either manure or compost alone or in combination with fertilizers, as contrasted to no fertility applications or fertilizers alone (Marzouk and Kassem, 2011; Elsadig *et al.*, 2017). One study with the date variety Barhy found that nitrogen sources composed of 50% fertilizer + 50% compost resulted in the best yield, but the best fruit quality was obtained with 100% compost (El-Sayed *et al.*, 2019). Farm manure combined with bentonite clay resulted in increased yields for date palms in Tunisia (Karbout *et al.*, 2019). In addition to better fruit quality from using compost, the understory crops like wheat will benefit from the addition of organic matter. Higher levels of protein and zinc in wheat were found to be associated with higher levels of soil organic matter (Wood *et al.*, 2018).

Organic amendments (manures and composts) not only provide nutrients but also improve water infiltration and water-holding capacity (Janke *et al.*, 2022). Increasing evidence suggests that soils that receive organic amendments also foster higher levels of soil microbial diversity (Gosling *et al.*, 2006), which can help reduce disease incidence. A study with compost made from date palm waste showed significant antagonistic activity against Fusarium wilt in date palms in Morocco (Mohamed *et al.*, 2020).

Organically amended soils also benefit from arbuscular mycorrhizal fungi (AMF), which form symbiotic relationships with date palm root systems as well as with other plants (Oehl *et al.*, 2004). AMF have been identified in arid environments, including on native plants and in traditional date palm agroecosystems in Oman, and new species have been identified (Al-Yahya'ei *et al.*, 2011). A total of 25 AMF morphospecies were identified in that study in Oman. Species richness and spore abundance were significantly higher on both oases (traditional) and experiment station date palms as compared to wild desert plants, with the infection potential slightly higher using AMF from traditional date palm farms.

AMF had previously been thought to help plants mainly with water and phosphorus uptake, but AMF can also improve date palm seedling growth without salt as well as increase salt stress tolerance (Toubali *et al.*, 2020). AMF are also known to increase salt tolerance in other crop plants, including wheat (Talaat and Shawky, 2014). Infection with AMF can also inhibit *Fusarium* infection of crop roots (Al Hmoud and Al Momany, 2015).

In addition to livestock waste, there are other overlooked sources of soil carbon on traditional farms. Date palms produce between 20 and 40 kg of dry leaves per year that are removed by pruning (Milh *et al.*, 2015). Currently, pruned leaves are disposed of by dumping or burning, but on-farm composting or biochar production would preserve this carbon and allow it to be reapplied to the soil. Biochar made from date palm leaves is beneficial to soil, especially with low-temperature pyrolysis (Usman *et al.*, 2015). Low-temperature biochar made from date palm waste has also been found to reduce heavy metal mobility in soil in addition to reducing carbon dioxide contributions to the environment (Al-Wabel *et al.*, 2019). Date palm biochar increased soil water retention up to 20%, increased the cation exchange capacity, and lowered the sodium adsorption ratio in sandy soils from the United Arab Emirates (Khalifa and Yousef, 2015).

In summary, a review of organic matter amendments in date palms concludes that traditional farming methods and other farming practices aligned with agroecological principles can successfully improve soil water balance, reduce salinization effects, and supply nutrients (Milh *et al.*, 2015).

12.3.6 Pest management and control – insects, weeds, and disease

Several pests and diseases affect date palm growth and production. Traditional control methods have been primarily focused on cultivation practices, such as rotation of understory crops, enhanced biodiversity of crops, providing proper environment for parasites and predators (such as ants and birds), utilization of deterrent plants that are grown within and on the border of farms, use of ash and smoke of certain plants, extraction and application of biopesticides from plants known to have pesticidal properties, and other techniques that combine traditional knowledge of wild plants and skills of pest and disease management techniques. Plant extracts of commonly found plants in date plantations, such as castor bean, datura, and oleander, are effective in controlling major pests such as red palm weevil (Ali *et al.*, 2020). Intercropping in date palm plantations has been found to reduce pest populations by increasing biodiversity and ecological stability, including the abundance of pests' natural enemies (Latifian, 2017).

The abundance of natural predators is higher in intercropped polyculture systems as compared to monocultures (Liu *et al.*, 2018). In date palms in Oman, more beneficial natural enemies of dubas bug were found in unsprayed plots of trees as compared to sprayed, and the natural enemies were more abundant in the understory crops than in the date palm leaves (Al-Ansari, 2018). Surprisingly, the dubas bug or pest populations were higher in the sprayed plots.

12.3.7 Socio-economic aspects of traditional date palm agroecosystems

Farmers have adopted and developed several cultural practices to maintain productive date palms for decades in a sustainable, nutritious, and profitable way. Some of these techniques have not been studied well or addressed in the literature. Examples of these techniques include:

- building terraces in steep mountain slopes to cultivate date palms;
- using wood columns to support date palms that are unstable or bent because of wind;
- development of a range of traditional crafts, such as ropes and baskets, that are used to climb and harvest dates;
- long-term storage of pollination strands; and

Fig. 12.5. Traditional harvesting (a) of dates at khalal stage for the purpose of cooking (*tabseel*) (b). (Photos by Rashid Al-Yahyai.)

- utilization of all parts of date palm in house construction, household items, boat making, etc.

Social customs and traditions are an integral part of date palm cultivation in the traditional agroecosystems and one of the elements of agroecology. Traditional farming practices support social connections and group work from planting to harvesting to marketing and consumption. In Oman, the majority of farm products are for family consumption (88% of farms) and only 0.55% are for industry or export. The remainder (9.5%) are marketed within Oman (MAF, 2015).

Date palm culture in different countries is full of social activities, such as harvesting, where all gather to celebrate harvest, with women playing a major role in this practice (Fig. 12.5a). Another key activity worth noting and constituting a major date consumption stage in many countries is the cooking of khalal dates in a process that is called *tabseel* in Oman (Fig. 12.5b). In this process, dates at the khalal (or besir) stage are boiled in large containers of water and then sun-dried, packed, and shipped to date processing factories. The sale of date palm fruit is another social activity that takes place at the khalal stage when they are offered for sale in a local auction, called *t'tanat*. Once harvested, dates are packed at the tamar stage. This is also ceremonial as the local people gather dates to store them in clay or palm baskets in a process known as *kanaz*. Several dishes are also made from dates, from the date syrup (called *debs*) to traditional delicacies and meals that are shared within the farming communities. Furthermore, date palms in agroecosystems provide other by-products that are also communal in nature and used for construction, household utensils, furniture, and other products (see Salomón-Torres, 2023, Chapter 17, this volume). Other aspects of the socioeconomics of date palm cultivation include the exchange and sale of irrigation water shares from aflaj and wells among growers (Al-Ghafri, 2018).

Growing dates in an oasis-type farming system can be profitable. In Saudi Arabia, a survey of 30 farms determined the fixed cost, variable cost, and gross and net income for an average plantation size of 2.5 ha (Al-Abbad *et al.*, 2011). Using a selling price of 4.0 SR (Saudi riyal)/kg (US$1.07/kg), the farmers realized a 25% return on investment over their fixed and variable costs. Their sales were to the local market (40%), factories (37%), and used on the farm (23%). The majority of the farmers noted that they are selling to known customers, who come back to purchase from the farm based on the quality of the product. This survey did not mention if the plantations were monoculture or polyculture plantings, but the tree density was 130 per hectare, which would be considered as medium to low density.

In Oman, even though date is the most widely grown crop with 8 million trees, the marketing of the fruit lags behind the production (Abdul-Razak, 2010). Statistics from 2007 showed that 3.4% of the crop was sold as export, 51.5% was consumed in-country by humans, 20.0% was used for animal feed, but 25.7% of the crop was deemed as 'surplus'. Recent government investment in processing facilities and date product development should help to improve this statistic and increase the use of these surplus dates. The profitability of dates, as with any crop, is based only partly on production, and the rest on access to markets and price. Quality control in the processing and marketing steps also plays a key role in the determination of profitability (Alsughayir, 2013).

In theory, it should be possible to compare the profitability of polyculture traditional date farms with more modern, monoculture farms, geared toward just date production and sales. However, these direct comparisons are difficult to find, and even if they did exist would have little meaning, since most of the traditional systems are not part of the market economy but grown for home use. The diversity of crops grown in the traditional system is also an important part of the nonmarket economy, which, as mentioned above, represents 88% of the farms in Oman.

12.3.8 Integration of livestock in date palm agroecosystems

Farming systems that combine or integrate animal husbandry and crop production vary across regions and agroecological zones. Animals play a significant role in crop production as well as being a source of nutrients in many parts of the world. Asian growers incorporate animals into farming operations (Ruthenberg, 1971; McDowell and Hildebrand, 1980). The practice is similar to that in agroecological date production systems. For example, buffalo in rice farms provide (i) traction for cultivating fields and (ii) milk and meat that are consumed domestically or sold in markets. Animal manure is another by-product associated with this integrated system, where cattle, fowl (primarily chickens and ducks), and swine are also commonly raised on rice farms. Livestock in date palm plantations may also include camels, sheep, and goats, as a source of meat and milk, and cattle, donkeys, and horses as service animals. Animals either freely graze in date palm farms or they are kept in a corner of the farm and fed from pasture crops, such as lucerne and grass, grown under the date palms (Fig. 12.6).

Dates are nutritious and are integrated into animal diets in many parts of the world. Studies have shown that integrating dates into animal diets leads to increased animal weight and other desirable characteristics that benefit the farmers who integrate animals into the date production in agro-ecological systems. A study found that feeding of grazing goats using a date supplement that included energy-rich by-products of fish and date palm increased their daily organic matter intake and thus covered their requirements for optimum growth and production (Dickhoefer *et al.*, 2011). The dates that are fed to animals are usually the dates that prematurely fall to the ground, cultivars that are not marketed for human consumption, or poor-quality dates of small size or discolored fruits. Nevertheless, several growers grow dates specifically for animal feed. The share of animal feed can reach a quarter (25%) of all date production in a country like Oman, where agroecological systems of production are dominant (Al-Yahyai and Khan, 2015).

12.4   Case Studies on the Agroecology of Date Palm Farms

Research on agroecology in general and especially on date palm is very limited. This is largely due to the misconception that traditional agroecological systems are backward and not profitable. Thus, governments in many date-producing countries have initiated projects that aimed to convert the existing traditional means of date growing into more industrial forms, which has led to failure as the growers were not consulted and have not learned to live with modern powered equipment and cultivation systems. It is important that agroecological systems are kept, and this can only be achieved by studying the advantage of these systems on socio-economic well-being in rural communities as well as on global climate change mitigation and adaptation and conservation of biodiversity and the environment.

12.4.1   Agroecology of date palm plantations in Oman

One case study of the agroecology of date palm compared traditional date palm farms in three regions of Oman to a modern farm (Al-Hashmi and Al-Yahyai, 2020). As indigenous knowledge, traditional farming practices, and local landraces continue to disappear due to the passing of the elders, there is a pressing need to study the traditional means of cultivation that is largely built around date palm as the main crop in the country. Ten transects were selected for the study, each of which was 1 feddan (*c.*4200 m^2),

Fig. 12.6. An integrated date palm–animal production system is an essential component of agroecology. Farm animals including poultry (a), camels (b), and goats (c) are examples of the animals used in this system of production. (Photos by Rashid Al-Yahyai.)

and data were gathered about the number of date palms, the cultivars, the understory permanent crops grown, and other plants of economic importance. The data showed a large diversity in fruit and other perennial crops, medicinal plants, and wild trees on traditional farms compared to a typical modern farm. There was also a diverse number of date palm cultivars.

Few local owners were working on their farms, and they relied instead on expatriate laborers who introduced changes in the understory crops and local plants to befit their needs and customs. Modern date farms, characterized by monoculture, modern irrigation, and other management practices, were lacking cultivar and crop biodiversity. These modern practices may in the long term affect the agroecology of entire villages if applied in isolated oases. The study also included a survey of plants in the traditional farms that follow agroecological systems and found great biodiversity in functional wild plants that served medicinal and nutritional needs of the date growers. Additionally, among the most common plants found in date palm plantations were ones used for cosmetic purposes, such as henna (*Lawsonia inermis*) and myrtle (*Myrtus communis*), illustrating the extent to which date palm plantations serve as forests for many other plant and animal species with a range of functional values.

Three traditional date palm plantations were surveyed from three locations in Oman (governorates of Batinah South, Dakhlia, and Muscat), and one modern farm at Sultan Qaboos University (SQU), following the most productive practices, was selected as a control (as exemplified in Figs 12.1 and 12.2). The area selected per plantation was 1 feddan, as over 70% of date palm farms are less than 1 feddan. Plantations were randomly selected, and representative of the practices being followed in each location. The results showed that traditional farms following agroecological principles in three governorates of Oman (Batinah, Dakhlia, and Muscat) had much higher biodiversity compared to the modern farm (SQU) (Table 12.4) and had higher numbers of date palm cultivars (Table 12.5).

Although traditional farms are biodiverse in plant species as well as date palm cultivars compared to modern farms, subtle changes are taking place. The traditional crop and plant species that are associated with traditional farms including cosmetic and medicinal plants, such as henna (*L. inermis*), are being replaced by exotic plants. Plants used as spices, such as Indian bay leaves (*Cinnamomum tamala*), or for their medicinal properties, such as neem (*Azadirachta indica*), are increasingly common on farms where expatriate laborers have replaced the local people in managing the date palm farms.

The data from the case study indicated that modern farm models that opted for a single superior cultivar are very low in agrobiodiversity. The loss in biodiversity has a broader impact on the availability to produce a variety of other crops and plants that are utilized for medicinal and other purposes. Improving current practices of cultivation to enhance yield and

Table 12.4. Diversity of date palm plantations in four locations of Oman.

Biodiversity index[a]	Dakhlia	Batinah	Muscat	SQU
Number of taxa (S)	14	7	19	1
Total number of individuals (N)	428	320	616	72
Dominance (D)	0.45	0.82	0.43	1.00
Shannon (H)	1.18	0.43	1.42	0.00
Simpson ($1 - D$)	0.55	0.18	0.57	0.00
Evenness (eH/S)	0.23	0.22	0.22	1.00
Menhinick	0.68	0.39	0.77	0.12
Margalef	2.15	1.04	2.80	0.00
Equitability (J)	0.45	0.22	0.48	0.00
Fisher's alpha	2.78	1.26	3.71	0.16
Berger–Parker	0.62	0.90	0.63	1.00

[a]See Hammer *et al*. (2001) for details.

Table 12.5. Diversity of date palm cultivars in four locations of Oman.

Biodiversity index[a]	Dakhlia	Batinah	Muscat	SQU
Number of taxa (S)	16	17	15	2
Total number of individuals (N)	266	289	387	72
Dominance (D)	0.2	0.11	0.19	0.92
Shannon (H)	2.05	2.42	1.88	0.17
Simpson ($1 - D$)	0.80	0.89	0.81	0.08
Evenness (eH/S)	0.48	0.66	0.44	0.59
Menhinick	0.98	1.00	0.76	0.24
Margalef	2.69	2.82	2.35	0.23
Equitability (J)	0.74	0.86	0.69	0.25
Fisher's alpha	3.74	3.95	3.10	0.38
Berger–Parker	0.37	0.20	0.30	0.96

[a]See Hammer *et al*. (2001) for details.

fruit quality should consider the overall impact on the date palm plantation's ecological benefits, including loss of bird and wildlife habitat and sources of nectar for honeybees, loss of medicinal and traditional non-crop plants, loss of fodder that supplements farmer income particularly in rural communities, and loss of year-round food availability.

12.4.2   Other case studies

The role of date palms in agroecological oasis systems in Baja California Sur, Mexico, was studied by de Grenade (2013). Date palms were introduced to the oases on the Baja California peninsula by Jesuit missionaries

Fig. 12.7. Seedling date palm groves in the valley of San Ignacio, Baja California peninsula, Mexico. (Photo from de Grenade, 2013.)

and later became naturalized, becoming an ecological keystone species. Date palms grown from seedlings have been densely planted and have partially replaced the native fan palms of *Washingtonia* spp. The date palms provided overstory cover for diverse smaller crop trees and shrubs grown underneath and altered the oasis microclimate, forming complex agroecosystems. These relatively recent date palm oases, compared to millennia-old ones in the MENA region, have provided an agroecological system where food from dates is obtained for the local communities, dates feed the oasis's local and migratory fauna, and date palm by-products are utilized for building and craft materials. This case study further highlights the importance of the date palm as a keystone species in food security and ecological restoration in challenging environments (Fig. 12.7).

Another case study on date palm agroecological systems on another *Phoenix* palm species that is commonly cultivated in Asia, i.e. *Phoenix sylvestris*, highlighted a similar agroforestry practice in the region of Jashore in Bangladesh. The findings from that study carried out by Mondol *et al.* (2021) indicated that the most important feature of the palm-based agroforestry system was to generate income and diversified products for the rural farmers. This approach can be replicated utilizing *Phoenix dactylifera* in regions with abundant water resources. Unsurprisingly, traditional farming practices are eroding rapidly across the globe and thus implementing the elements of agroecology described previously is an essential and urgent step in sustaining the world's food security needs.

12.5 Conclusions and Prospects

Agroecology has recently been presented as a means of adapting to and mitigating climate change, preserving biodiversity and the environment, and ensuring local communities' food security. The date palm has been in cultivation for millennia thanks to sustainable practices that aligned with the principles of agroecological production systems. The sustainable production of date has many facets, which have been presented in this chapter.

If we revisit the United Nations' SDGs, we find that traditional agriculture and preservation of the integrity of the agroecosystems for date production have value as a food production system and to society. Biological diversity is conserved, and even enhanced, in traditional date palm multispecies systems. Efficiency and recycling of nutrients within the farming system are enhanced through the production of multiple crops and crop–livestock integration, which results in a circular economy. This leads to greater resilience since multiple genotypes of dates and multiple food and fodder species are planted in the same plots. Traditional knowledge is utilized to preserve these systems, which supports local human and social values and community and family participation. This synergistic food system would not be possible without responsible governance at the local, regional, and national levels.

That said, there are also ways we can improve on the foundation of traditional agriculture and enhance what is there now. Research shows that soil quality is enhanced with the addition of soil carbon, and in addition to livestock waste, date palm fronds could be chopped and used as a mulch, like compost, or made into biochar at the local level. Long-term research at experiment stations and on farms could document the benefits realized by using these methods, including disease suppression and crop salinity tolerance. Recycled water, including treated wastewater, could be considered for areas where understory crops are not consumed by humans, especially on farms that are now using well water and MIG.

The adoption of the principles of agroecology in date production is essential for sustainable date production to meet the increasing demand for food and to preserve the limited natural resources necessary to continue crop cultivation. Multistorey production and animal integration provide additional layers of food security and sustainable food production for rural communities. It is highly recommended that these practices that have lasted for centuries or millennia are further studied, and the elements of agroecology are applied in modern date production systems.

References

Abdul-Razak, N.A. (2010) Economics of date palm agriculture in the Sultanate of Oman, current situation and prospects. *Acta Horticulturae* 882, 137–146. DOI: 10.17660/ActaHortic.2010.882.15.

Abul-Soad, A.A., Jain, S.M. and Jatoi, M.A. (2017) Biodiversity and conservation of date palm. In: Ahuja, M.R. and Jain, S.M. (eds) *Biodiversity and Conservation of Woody Plants*, Vol. 17. Sustainable Development and Biodiversity, Springer, Cham, Switzerland, pp. 313–353. DOI: 10.1007/978-3-319-66426-2_12.

Al-Abbad, A., Al Jamal, M., Al Elaiw, Z., Al Shreed, F. and Belaifa, H. (2011) A study on the economic feasibility of date palm cultivation in the Al Hassa oasis of Saudi Arabia. *Journal of Development and Agricultural Economics* 3(9), 463–468.

Al-Ansari, S.M. (2018) The effect of pesticide pressure and other factors on the dubas bug and non-target arthropods in date palm orchards of Oman. MSc. thesis, Sultan Qaboos University, Al-Khoud, Oman.

Al-Ghafri, A. (2018) Overview about the Aflaj of Oman. In: *Proceeding of the International Symposium of Khattaras and Aflaj, Erachidiya, Morocco, 9 October 2018.* Available at: www.researchgate.net/publication/328560516_Overview_about_the_Aflaj_of_Oman (accessed 6 December 2022).

Al-Hashmi, K. and Al-Yahyai, R. (2020) Biodiversity of medicinal and aromatic plants in modern and traditional date palm farms in Northern and central Oman. *Acta Horticulturae* 1299, 415–420. DOI: 10.17660/ActaHortic.2020.1299.62.

Al Hmoud, G. and Al Momany, A. (2015) Effect of four mycorrhizal products on *Fusarium* root rot on different vegetable crops. *Journal of Plant Pathology & Microbiology* 06(2), 255. DOI: 10.4172/2157-7471.1000255.

Ali, M., Mohanny, K., Mohamed, G. and Allam, R. (2020) Insecticidal potential of some plant extracts in nano and normal form on immatures stages of red palm weevil *Rhynchophorus ferrugineus. SVU-International Journal of Agricultural Sciences* 2(2), 306–315. DOI: 10.21608/svuijas.2020.43031.1037.

Al Kharusi, L., Assaha, D.V.M., Al-Yahyai, R. and Yaish, M.W. (2017) Screening of date palm (*Phoenix dactylifera* L.) cultivars for salinity tolerance. *Forests* 8(4), 136. DOI: 10.3390/f8040136.

Al Kharusi, L., Sunkar, R., Al-Yahyai, R. and Yaish, M.W. (2019) Comparative water relations of two contrasting date palm genotypes under salinity. *International Journal of Agronomy* 2019, 1–16. DOI: 10.1155/2019/4262013.

Almadini, A.M., Ismail, A.I.H. and Ameen, F.A. (2021) Assessment of farmers practices to date palm soil fertilization and its impact on productivity at Al-Hassa oasis of KSA. *Saudi Journal of Biological Sciences* 28(2), 1451–1458. DOI: 10.1016/j.sjbs.2020.11.084.

Al-Mulla, Y.A. and Al-Gheilani, H.M. (2016) Increasing water productivity enhances water saving for date palm cultivation in Oman. *Journal of Agricultural and Marine Sciences* 22(1), 87. DOI: 10.24200/jams.vol22iss1pp87-91.

Al-Sadi, A.M., Al-Jabri, A.H., Al-Mazroui, S.S. and Al-Mahmooli, I.H. (2012) Characterization and pathogenicity of fungi and oomycetes associated with root diseases of date palms in Oman. *Crop Protection* 37, 1–6. DOI: 10.1016/j.cropro.2012.02.011.

Alsughayir, A. (2013) The impact of quality practices on productivity and profitability in the Saudi Arabian dried date industry. *American Journal of Business and Management* 2, 340–346. Available at: https://worldscholars.org/index.php/ajbm/article/view/465 (accessed 30 November 2022).

Al-Wabel, M.I., Usman, A.R.A., Al-Farraj, A.S., Ok, Y.S., Abduljabbar, A. *et al.* (2019) Date palm waste biochars alter a soil respiration, microbial biomass carbon, and heavy metal mobility in contaminated mined soil. *Environmental Geochemistry and Health* 41(4), 1705–1722. DOI: 10.1007/s10653-017-9955-0.

Al-Yahya'ei, M.N., Oehl, F., Vallino, M., Lumini, E., Redecker, D. *et al.* (2011) Unique arbuscular mycorrhizal fungal communities uncovered in date palm plantations and surrounding desert habitats of Southern Arabia. *Mycorrhiza* 21(3), 195–209. DOI: 10.1007/s00572-010-0323-5.

Al-Yahyai, R. (2007) Improvement of date palm production in the Sultanate of Oman. *Acta Horticulturae* 736, 337–343. DOI: 10.17660/ActaHortic.2007.736.32.

Al-Yahyai, R. and Al-Khanjary, S. (2008) Biodiversity of date palm in the Sultanate of Oman. *African Journal of Agricultural Research* 3(6), 389–395.

Al-Yahyai, R. and Al-Kharusi, L. (2012) Sub-optimal irrigation affects chemical quality attributes of dates during fruit development. *African of Agricultural Research* 7(10), 1498–1503. DOI: 10.5897/AJAR11.1553.

Al-Yahyai, R. and Khan, M.M. (2015) Date palm status and perspective in Oman. In: Al-Khayri, J.M., Jain, S.M. and Johnson, D.V. (eds) *Date Palm Genetic Resources and Utilization*, Vol. 2. *Asia and Europe*. Springer, Dordrecht, The Netherlands, pp. 207–240. DOI: 10.1007/978-94-017-9707-8_6.

Al-Yahyai, R. and Manickavasagan, A. (2012) An overview of date palm production. In: Manickavasagan, A., Essa, M.M. and Sukumar, E. (eds) *Dates: Production, Processing, Food, and Medicinal Values*. CRC Press, Boca Raton, Florida, pp. 3–12.

Brooker, R.W., Bennett, A.E., Cong, W.-F., Daniell, T.J., George, T.S. *et al.* (2014) Improving intercropping: a synthesis of research in agronomy, plant physiology and ecology. *The New Phytologist* 206(1), 107–117. DOI: 10.1111/nph.13132.

Brush, S.B. (2007) Farmers' rights and protection of traditional agricultural knowledge. *World Development* 35(9), 1499–1514. DOI: 10.1016/j.worlddev.2006.05.018.

Buerkert, A., Nagieb, M., Siebert, S., Khan, I. and Al-Maskri, A. (2005) Nutrient cycling and field-based partial nutrient balances in two mountain oases of Oman. *Field Crops Research* 94(2–3), 149–164. DOI: 10.1016/j.fcr.2004.12.003.

Cherif, H., Marasco, R., Rolli, E., Ferjani, R., Fusi, M. *et al.* (2015) Oasis desert farming selects environment-specific date palm root endophytic communities and cultivable bacteria that promote resistance to drought. *Environmental Microbiology Reports* 7(4), 668–678. DOI: 10.1111/1758-2229.12304.

de Grenade, R. (2013) Date palm as a keystone species in Baja California peninsula, Mexico oases. *Journal of Arid Environments* 94, 59–67. DOI: 10.1016/j.jaridenv.2013.02.008.

Dickhoefer, U., Mahgoub, O. and Schlecht, E. (2011) Adjusting homestead feeding to requirements and nutrient intake of grazing goats on semi-arid, subtropical highland pastures. *Animal* 5(3), 471–482. DOI: 10.1017/S1751731110001783.

Elsadig, E.H., Aljuburi, H.J., Elamin, A.H.B. and Gafar, M.O. (2017) Impact of organic manure and combination of N P K S, on yield, fruit quality and fruit mineral content of Khenazi date palm (*Phoenix dactylifera* L.) cultivar. *Journal of Scientific Agriculture* 1, 335. DOI: 10.25081/jsa.2017.v1.848.

El-Sayed, M.A., Abdalla, A.S., El-Hameed, M.M.A. and El-Naggar, H.M.F. (2019) Effect of using plant compost and EM as partial replacement of inorganic N fertilizer on fruiting of Barhy date palms. *New York Science Journal* 12(1), 16–27. https://doi.org/10.7537/marsnys120219.03

Faci, M. (2019) Typology and varietal biodiversity of date palm farms in the North-East of Algerian Sahara. *Journal of Taibah University for Science* 13(1), 764–771. DOI: 10.1080/16583655.2019.1633006.

FAO (2022) Agroecology Knowledge Hub. Food and Agriculture Organization of the United Nations, Rome. Available at: www.fao.org/agroecology/overview/en/ (accessed 24 January 2022).

Franco, J.G., King, S.R., Masabni, J.G. and Volder, A. (2015) Plant functional diversity improves short-term yields in a low-input intercropping system. *Agriculture, Ecosystems & Environment* 203, 1–10. DOI: 10.1016/j.agee.2015.01.018.

Gosling, P., Hodge, A., Goodlass, G. and Bending, G.D. (2006) Arbuscular mycorrhizal fungi and organic farming. *Agriculture, Ecosystems & Environment* 113(1–4), 17–35. DOI: 10.1016/j.agee.2005.09.009.

Gros-Balthazard, M., Galimberti, M., Kousathanas, A., Newton, C., Ivorra, S. *et al.* (2017) The discovery of wild date palms in Oman reveals a complex domestication history involving centers in the Middle East and Africa. *Current Biology* 27(14), 2211–2218. DOI: 10.1016/j.cub.2017.06.045.

Haider, M.S., Khan, I.A., Jaskani, M.J., Naqvi, S.A., Hameed, M. *et al.* (2015) Assessment of morphological attributes of date palm accessions of diverse agro-ecological origin. *Pakistan Journal of Botany* 47(3), 1143–1151.

Hammer, Ø., Harper, D.A.T. and Ryan, P.D. (2001) PAST: Paleontological statistics software package for education and data analysis. *Palaeontologia Electronica* 4(1). Available at: http://palaeo-electronica.org/2001_1/past/issue1_01.htm (accessed 6 December 2022).

Janke, R., Blackburn, D., Khan, M., Al Busaidi, A., Al Busaidi, W. *et al.* (2022) A 25 year history of the use of organic soil amendments in Oman: a review. *Journal of Agricultural and Marine Sciences* 27(1), 41–49. Available at: www.researchgate.net/publication/356191556_A_25_Year_History_of_the_use_of_Organic_Soil_Amendments_in_Oman_A_review (accessed 6 December 2022).

Jaradat, A.A. (2011) Biodiversity of date palm. Soils, plant growth and crop production. In: UNESCO-EOLSS (ed.) *Encyclopedia of Life Support Systems: Land Use, Land Cover and Soil Sciences.* EOLSS Publishers, Oxford, pp. 1–31.

Karbout, N., Mlih, R., Latifa, D., Bol, R., Moussa, M. *et al.* (2019) Farm manure and bentonite clay amendments enhance the date palm morphology and yield. *Arabian Journal of Geosciences* 14(9), 818. DOI: 10.1007/s12517-021-07160-w.

Khalifa, N. and Yousef, L.F. (2015) A short report on changes of quality indicators for a sandy textured soil after treatment with biochar produced from fronds of date palm. *Energy Procedia* 74, 960–965. DOI: 10.1016/j.egypro.2015.07.729.

Latifian, M. (2017) Integrated pest management of date palm fruit pests: a review. *Journal of Entomology* 14(3), 112–121. DOI: 10.3923/je.2017.112.121.

Liu, J.L., Ren, W., Zhao, W.Z. and Li, F.R. (2018) Cropping systems alter the biodiversity of ground- and soil-dwelling herbivorous and predatory arthropods in a desert agroecosystem: implications for pest biocontrol. *Agriculture, Ecosystems & Environment* 266, 109–121. DOI: 10.1016/j.agee.2018.07.023.

MAF (1993) *North Batinah Integrated Study Vol. 1, 2, 3, 4. Land Resources, Soil Survey and Land Classification Project, Sultanate of Oman.* Ministry of Agriculture and Fisheries, Muscat.

MAF (2015) *Agricultural Census 2012/2013.* Ministry of Agriculture and Fisheries, Muscat.

Marzouk, H.A. and Kassem, H.A. (2011) Improving fruit quality, nutritional value and yield of Zaghloul dates by the application of organic and/or mineral fertilizers. *Scientia Horticulturae* 127(3), 249–254. DOI: 10.1016/j.scienta.2010.10.005.

Mattar, M.A., Soliman, S.S. and Al-Obeed, R.S. (2021) Effects of various quantities of three irrigation water types on yield and fruit quality of 'Succary' date palm. *Agronomy* 11(4), 796. DOI: 10.3390/agronomy11040796.

McDowell, R.E. and Hildebrand, P.E. (1980) *Integrating and Animal Production: Making the Most of Resources Available to Small in Developing Countries.* Rockefeller Foundation, New York.

Mika, A. (2004) The importance of biodiversity in natural environment and in fruit plantations. *Journal of Fruit and Ornamental Plant Research* 12, 11–21.

Milh, R., Bol, R., Amelung, W. and Brahim, N. (2015) Soil organic matter amendments in date palm groves of the Middle Eastern and North African region: a mini-review. *Journal of Arid Land* 8(1), 77–92. DOI: 10.1007/s40333-015-0054-8.

Mohamed, O.-Z., Yassine, B., Hilali Rania, E., El Hassan, A., Abdellatif, H. *et al.* (2020) Evaluation of compost quality and bioprotection potential against *Fusarium* wilt of date palm. *Waste Management* 113, 12–19. DOI: 10.1016/j.wasman.2020.05.035.

Mondol, M.A., Alam, N.E.K. and Islam, K.K. (2021) Contribution of traditional date palm (*Phoenix sylvestris*) agroforestry in income generation and livelihood improvements: a case of Jashore district, Bangladesh. *International Journal of Environment, Agriculture and Biotechnology* 6(1), 261–269. DOI: 10.22161/ijeab.61.33.

Odum, E.P. (1968) *Fundamentals of Ecology,* 2nd edn. Saunders, London.

Oehl, F., Sieverding, E., Mäder, P., Dubois, D., Ineichen, K. *et al.* (2004) Impact of long-term conventional and organic farming on the diversity of arbuscular mycorrhizal fungi. *Oecologia* 138(4), 574–583. DOI: 10.1007/s00442-003-1458-2.

Ou-Zine, M., Symanczik, S., Rachidi, F., Fagroud, M., Aziz, L. *et al.* (2021) Effect of organic amendmenton soil fertility, mineral nutrition, and yield of Majhoul date palm cultivar in Drâa-Tafilalet region, Morocco. *Journal of Soil Science and Plant Nutrition* 21(2), 1745–1758. DOI: 10.1007/s42729-021-00476-2.

Peano, C., Caron, S., Mahfoudhi, M., Zammel, K., Zaidi, H. *et al.* (2021) A participatory agrobiodiversity conservation approach in the oases: community actions for the promotion of sustainable development in fragile areas. *Diversity* 13(6), 253. DOI: 10.3390/d13060253.

Rahnama, A.A., Agdol, M.A. and Moehebi, H. (2012) Study of the different irrigation and fertilization levels effects on fruit set and yield of tissue cultured Barhee date palms. *International Journal of Agriculture and Crop Sciences* 4, 1666–1671. Available at: www.researchgate.net/publication/234163000_Study_of_the_different_irrigation_and_fertilization_levelseffects_on_fruit_set_and_yield_of_tissue_culturedBarhee_date_palms (accessed 30 November 2022).

Ruthenberg, H. (1971) *Farming Systems in the Tropics.* Clarendon, Oxford.

Salomón-Torres, R. (2023) Nonfood products and uses of date palm. In: Al-Khayri, J.M., Jain, S.M., Johnson, D.V. and Krueger, R.R. (eds) *Date Palm.* CAB International, Wallingford, UK, pp. 546–579.

Stomph, T., Dordas, C., Baranger, A., de Rijk, J., Dong, B. *et al.* (2020) Designing intercrops for high yield, yield stability and efficient use of resources: are there principles? *Advances in Agronomy* 160, 1–50. Available at: https://doi.org/10.1016/bs.agron.2019.10.002

Świąder, K., Balek, K. and Sleten Hosoglu, M. (2020) Varieties of date palm fruits (*Phoenix dactylifera* L.), their characteristics and cultivation. *Technological Progress in Food Processing* 1, 173–179.

Talaat, N.B. and Shawky, B.T. (2014) Protective effects of arbuscular mycorrhizal fungi on wheat (*Triticum aestivum* L.) plants exposed to salinity. *Environmental and Experimental Botany* 98, 20–31. DOI: 10.1016/j.envexpbot.2013.10.005.

Toubali, S., Tahiri, A., Anli, M., Symanczik, S., Boutasknit, A. *et al.* (2020) Physiological and biochemical behaviors of date palm vitroplants treated with microbial consortia and compost in response to salt stress. *Applied Sciences* 10(23), 8665. DOI: 10.3390/app10238665.

Usman, A.R.A., Abduljabbar, A., Vithanage, M., Ok, Y.S., Ahmad, M. *et al.* (2015) Biochar production from date palm waste: Charring temperature induced changes in composition and surface chemistry. *Journal of Analytical and Applied Pyrolysis* 115, 392–400. DOI: 10.1016/j.jaap.2015.08.016.

Vandermeer, J. (2010) *The Ecology of Agroecosystems.* Bartlett and Jones, Sudbury, Massachusetts.

Wood, S.A., Tirfessa, D. and Baudron, F. (2018) Soil organic matter underlies crop nutritional quality and productivity in smallholder agriculture. *Agriculture, Ecosystems & Environment* 266, 100–108. DOI: 10.1016/j.agee.2018.07.025.

Yu, Y., Stomph, T.-J., Makowski, D. and van der Werf, W. (2015) Temporal niche differentiation increases the land equivalent ratio of annual intercrops: a meta-analysis. *Field Crops Research* 184, 133–144. DOI: 10.1016/j.fcr.2015.09.010.

Date Harvest

13

Mohamed Marouf Aribi*

Higher National School of Biotechnology – Taoufik Khaznadar, Constantine, Algeria

Abstract

The cultivation of date palm is one of the main agricultural activities in the Saharan and pre-Saharan zones and it is the keystone species of the oasis ecosystems. The problems of marketing dates are the result of several constraints, chiefly the unsatisfactory presentation of fruit due to the traditional methods of harvesting, storage and packaging, conservation difficulties, and the lack of proper treatment of dates both before and after harvest. This chapter is intended primarily for agricultural service officials and those responsible for improvement plans concerning the various aspects of the date fruit's economy, culture, industry, and trade. It is well illustrated, especially for readers in countries that have only recently established modern processing and packaging facilities. Farmers in these countries still need to refine how to harvest, handle, pack, and transport dates so that the packer can accept them. Therefore, major emphasis here is placed on the on-the-spot operations and other activities of the farmer. Packers have come to expect the delivery of dates covered with sand, infested with insects, crushed, or otherwise with defects that would remove almost any commercial value. This chapter shows the farmer that the various operations that contribute to satisfactory marketing must begin at the palm grove itself, outlining methods and processes and identifying the machines and equipment that are of interest in modern processing and packaging facilities. Strict hygiene and proper handling are essential everywhere if, together, farmers, packers, and traders are to achieve good economic results. Indeed, in the phoenicicultural regions, fresh dates give rise daily to important local gathering, handling, and marketing activities.

Keywords: Dates, Handling, Harvest, Packaging, Palm

13.1   Introduction

Date fruit has been gaining research attention because both the pulp and seed are utilized as food and feed since they contain numerous nutrients

*mar-bio-tp@live.fr

© CAB International 2023. *Date Palm* (eds J.M. Al-Khayri *et al.*)
DOI: 10.1079/9781800620209.0013

such as protein, sugar, lipid, vitamins, magnesium, iron, potassium, and a significant amount of calcium (Bouhlali *et al.*, 2017; Al Juhaimi *et al.*, 2018). In desert areas, dates represent a providential fruit, one of the pillars of the oasis economy, providing a large portion of food needs. Their success over an extended period of time is explained by the nutritional qualities of these fruits, which are particularly rich in sugars and are suitable for sustaining long-term physical effort.

On a dry matter (DM) basis, date fruit contains 44–88% carbohydrates, 0.2–0.4% fat, 6.4–11.5% fiber, minerals, and vitamins, and a high protein concentration (2.3–5.6%) compared to other widely grown fruits (Elleuch *et al.*, 2008). In addition, recent studies have shown that date fruit is an excellent source of phenolic compounds and thus has an elevated antioxidant capacity (Ghiaba *et al.*, 2013). Dates are eaten fresh, dry, or in various processed forms (Yahia and Kader, 2011). Fresh dates are generally marketed at three stages of growth and development (khalal, rutab, tamar) and can be eaten as soft, semidried, or dried fruits, depending on their water content at harvest.

The decision to harvest at a specific stage depends on cultivar characteristics, especially soluble tannin levels, climatic conditions, and market demand (Glasner *et al.*, 2002). Time of harvest is based on the date fruit's appearance and texture (related to moisture and sugar content). Proper timing of harvest reduces the incidence and severity of cracking or splitting of the date skin, excessive dehydration, insect infestation, and attack by microorganisms (Kader and Hussein, 2009). Losses during harvesting, postharvest handling, and marketing are high due to the incidence of physical and physiological disorders, pathological diseases, and insect infestation.

Deglet Nour is the commercial variety that occupies most of the international date fruit trade (Ashraf and Hamidi-Esfahani, 2011; Yahia and Kader, 2011). The economic and commercial importance of this fruit is paramount. The export market for standard dates from Algeria and Tunisia, the export of which is exclusively reserved for the Deglet Nour variety, is part of a declining trend (Chebbi and Gil, 2002).

The worldwide production of date fruit was recorded to be 3.43 million t in 1990, which was harvested from about 625,000 ha; over the past three decades, the demand from global markets has increased reaching 8.52 million t from 1,092,000 ha (Sarraf *et al.*, 2021). Egypt, Iran, and Saudi Arabia have always been the largest producers of dates (FAO, 2018). Due to the importance and growing acceptance of date fruits, there is a need to discuss the advantages and disadvantages of past and current fruit processing techniques. Fig. 13.1 shows the important factors for the handling and storage of date fruits that make the date trade profitable for its producers and satisfactory to consumers (Sarraf *et al.*, 2021), as described in this chapter.

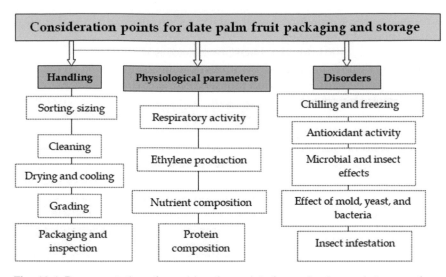

Fig. 13.1. Representation of consideration points for packaging and storage of date fruits. (Adapted from Sarraf *et al.*, 2021.)

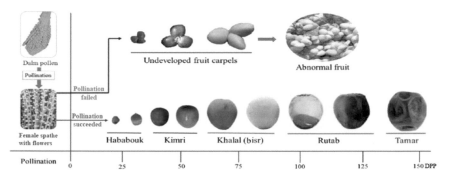

Fig. 13.2. The five growth stages of a date fruit by days post pollination (DPP). (From Ali-Dinar *et al.*, 2021, used under Creative Commons CC BY license.)

13.2 Fruit Developmental Stages

When the date has grown a little and taken on an apple-green hue, it is called tchimri in Iraq. The development of the date at the green or kimri stage is divided into two phases (Fig. 13.2).

The first is characterized by a rapid increase in weight and volume, an accumulation of reducing sugars, a slow but increasing increase in total sugars and solids, an extremely high real acidity, and a high humidity level although slightly lower than that of the next phase. The characteristics of the second phase are a slower increase in weight and volume, a significant drop in the rate of accumulation of reducing sugars, a considerable slowdown in the formation of total sugars, a slight decrease in real acidity, and an increase in humidity.

The transition to the next stage, or khalal, is marked by a change in color of the skin of the fruit, which turns from green to yellow, to chrome, to yellow speckled with red (often more on one side than the other), pink, or scarlet. At this stage, the weight gain is increasingly slow (toward the end, the weight may even decrease), the accumulation of reducing sugars is low, the proportions of sucrose, total sugars, and solids increase rapidly, the actual acidity and the moisture content decrease. It should be noted that it is at the khalal stage that dates, both with sucrose and with reducing sugars, accumulate the maximum amount of sugar in the form of sucrose.

The next stage of development is called rutab (wet). The date becomes more-or-less translucent, its skin turns from yellow, chrome, or scarlet to an almost black-brown or green in some varieties (e.g., in the varieties Khadar, Khadrawi, Khadrai, and Khadouuri, all these names meaning *green*), and the date becomes soft. This stage is often referred to as the ripening stage and the date is considered ripe when completely soft. However, the notion of maturity is relative, with some consumers applying it at the khalal stage, others at the tamar stage. It has been said, regarding the Barhi date with reducing sugars, that all the sucrose accumulated at the previous stage (khalal) is inverted.

The final stage of date ripening is called tamar. The fruit has then lost a lot of water and the ratio of sugar to remaining water is high enough to prevent fermentation. In the so-called "soft" varieties, the pulp is first soft and then becomes increasingly firm while remaining supple. In most varieties the skin adheres to the pulp and becomes wrinkled as the latter diminishes in volume; in some cases, however, the very fragile skin cracks when the pulp is reduced and thus leaves exposed fragments of sticky flesh that attract insects or allow grains of sand to adhere. The color of the epidermis and pulp gradually darkens. Soft dates in the tamar stage keep perfectly for a year at room temperature if they remain loose without compression. The dry, or hard, date that does not go through the rutab stage or soften at the tamar stage has about the same water content as the soft date at this last stage, but it has a much firmer texture. In most varieties the date fruit softens from top to bottom. When it is the top that ripens first, there is a difference in composition between this soft end and the still hard base. In the case of some varieties, desiccation occurs at this stage. The dates in question usually have a hard, rounded, shiny, buff-colored base and a pointed, wrinkled, dark-colored top. A stoppage of development is also sometimes observed in fruits that normally pass through the rutab to arrive at the tamar stage. This is especially the case during seasons when the wind is dry and hot instead of hot and humid. The phenomenon is found in the Hallawi variety of Basra. The age of the tree also affects the dryness of the fruit. The dates of very young or very old palm trees are drier and have a harder base than those of trees in the full period of production. Dates classified as having reducing sugars resemble, in this state, the fruits of the ordinary dry varieties because they contain a certain proportion of sucrose.

13.3 Harvesting Technology

The methods of picking are of direct interest to the packer: the packer has no cleaning work to do if the dates are picked properly, while he must remove the grains of sand and other impurities if the fruits are dropped on the ground. Moreover, the essential processing operations depend to a large extent on the time of the picking, the way it is done, and the care it is given. The poor condition in which dates arrive at the packing company when they have not been properly chosen is the source of many difficulties in some countries and greatly increases operating costs. Therefore, the agricultural advisory services concerned must systematically encourage the use of new techniques designed to collect directly the fruits harvested so that they do not touch the ground at any time and to avoid significant losses during the harvest.

Dates are harvested in July to August at the khalal stage or in September to December at the rutab and tamar stages in the northern hemisphere production regions; proper timing of harvest reduces the incidence and severity of cracking or splitting of dates, excessive dehydration, insect infestation, and attack by microorganisms (Yahia *et al.*, 2014). As ripening of dates is progressive on the bunch, some fruits can become overripe while others are still at the khalal or rutab stage; selective picking of individual dates or strands is often practiced for good quality at prime maturity (Yahia *et al.*, 2014). When this approach is adopted, multiple pickings are made before harvesting of all fruits is completed (Yahia *et al.*, 2014). Hand harvesting can account for as much as 45% of the operational costs, and therefore efforts have been made to develop mechanical harvesting methods to conduct the harvesting more conveniently and faster than traditional methods (Al-Suhaibani *et al.*, 1991; Ibrahim *et al.*, 2007). Some trials have been conducted on Deglet Nour dates in the Coachella Valley, California, using platforms built on extensible towers, enabling the picker to move from one palm to the other (Brown, 1982). If the tree is low, it is not difficult to reach the bunches, but in general it is necessary to climb the tree or use a ladder to reach them.

The most common way is to climb barefoot along the trunk, but it is painful, may cause falls, and does not allow tools and containers to be easily carried to the top. If the tree is very high, recesses used to wedge the feet can be cut along the trunk. Sometimes the gatherer climbs using only hands and feet, carrying a rope with which he attaches himself once he reaches the top. At other times, he uses a ring rope to facilitate the ascent (Fig. 13.3).

Fig. 13.3. Simple model of climbing rope and hatchet for waist, used in Biskra, Algeria. (Photos by Mohamed Marouf Aribi.)

Fig. 13.4. The grower shakes the bunch (Deglet Nour dates), knocking out the rutab and tamar fruits, Biskra, Algeria. (Photos by Mohamed Marouf Aribi.)

The planter picks the rutab dates one by one. Dates picked and placed in a container one by one are not likely to get dirty, but this process is laborious. They are a little more likely to drop when the picker puts them in his shirt in the stomach area.

The dates do not all ripen at the same time on the bunch. The difference in the times of arrival at maturity varies according to the climate: the warmer it is the shorter the interval. In marginal growing areas, it may take up to 2 months between the appearance of the first and last rutab dates on the same tree. It is then often necessary to repeat the hand picking to be able to ship to the packer batches of nearly uniform maturity. If the bunch is shaken, the rutab and tamar dates fall while the khalal remain attached to the pedicel (Fig. 13.4). To avoid contamination of date fruits, it is useful to put a container under the bunch, but it often happens that, to take less trouble, the grower shakes the bunches in which most fruits are tamar

Fig. 13.5. Cutting of whole date bunch, Biskra, Algeria. (Photo by Mohamed Marouf Aribi.)

without collecting them in a basket. Dirt then agglomerates especially on the rutab stage fruits, which are sticky.

A later development was the use of a hydraulic crane built on a truck, the crane having a basket used by the picker to reach the top of the tree; the picker cuts off the whole bunch and places it in the basket, which is lowered down by the crane to a shaker-trailer for shaking (Yahia *et al.*, 2014). After shaking, fruits fall into the bulk bins placed beneath the shaker-trailer; the bulk bins are then lifted on to trucks to be transported to the packinghouse (Yahia *et al.*, 2014).

Usually, a cursory triage is done during the harvest. Cull fruit is thrown into a special basket to serve as livestock feed. The khalal are put in another basket for fresh consumption by the farmer and his family, and the rutab in a third basket. The rutab dates are then spread in the sun to complete maturation up to the tamar stage. When the dates are delivered in bunches to the packer, as is the case for a large part of the Deglet Nour harvest in North Africa, there is hardly any sorting in the palm grove.

The removal of the entire bunch is the most widely used method of collection in the Saharan regions. A small sickle with a very fine sawtooth blade is usually used to cut the stem (Fig. 13.5).

If the dates are tamar, and therefore moderately hard and not sticky, one usually places the whole bunch, at once, either on the ground or on mats or canvases. In some areas of the Sahara, such as in Tunisia and Biskra (Algeria), men use the rope from the top of the tree to the ground to lower the bunch hand over hand. No other instrument is used other than the

Fig. 13.6. Descent of the bunch on a rope, Biskra, Algeria. (Photos by Mohamed Marouf Aribi.)

sickle and the dates arrive on the ground without being soiled or damaged (Fig. 13.6).

Most dates are harvested at the fully ripe rutab (light brown and soft) and tamar (dark brown and soft, semidry, or dry) stages when they have much greater levels of sugars, lower contents of moisture and tannins (disappearance of astringency) and are softer than the khalal stage dates (Yahia *et al.*, 2014).

Khalal dates are sometimes marketed on branches or bunches. Entire bunches are harvested when the dates are fully yellow and lowered to ground level, then hung on a carrier for transportation to the packing-house or the market (Yahia *et al.*, 2014).

Dates need to be dehydrated to the optimal moisture content for preserving their quality during subsequent handling and storage (Yahia, 2009). If ambient conditions allow, dehydration can be done using solar energy by spreading the dates on trays that will be exposed to the sun or under plastic tunnels until drying is completed to the desired moisture level (Yahia, 2009).

Similarly, ambient air drying can be done inside plastic greenhouses with good air circulation; drying in plastic houses, which can be constructed at a reasonable cost, protects the dates from dust, birds, rodents, and other damaging factors (Yahia *et al.*, 2014).

Processing dates by blanching in water at 96°C and subsequent dehydration at 60°C for 18–20 h results in good-quality dehydrated dates as compared to those without heat treatment (Kulkarni *et al.*, 2008). Studies recommend using steam for dehydration to improve resistance to microbial pathogens (Kader, 2003).

Hydrocooling is one of the cooling methods of date fruits at the khalal stage that requires 10–20 min at near 0°C temperature (Elansari, 2008). Also, tearing the skin of date fruits, due to the breaking of cell walls, can be facilitated in low temperatures and above 20% w/w moisture (Glasner

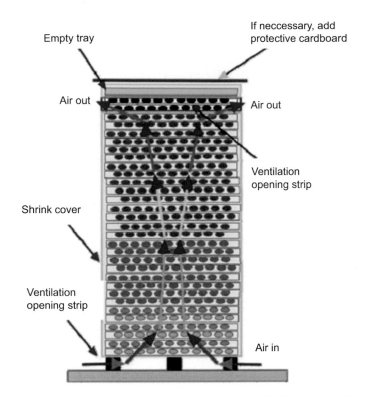

Empty tray

If neccessary, add protective cardboard

Air out

Air out

Ventilation opening strip

Shrink cover

Ventilation opening strip

Air in

Fig. 13.7. Schematic representation of the sun dryer for Medjool dates. The pallet contains trays of dates covered by shrink film with a strip at the top and the bottom to permit ventilation. (From Navarro, 2006.)

et al., 2002). Some dry date fruits with 20% w/w or lower moisture can be stored at −18, 0, 4, and 20°C for more than 1 year, 1 year, 8 months, and 1 month, respectively (Siddiq and Greiby, 2014). Conventional sun drying methods consist of exposing trays containing dates to direct solar irradiation; in this method (Fig. 13.7), the dates are exposed to occasional strong winds that cause dust to deposit on the dates, which reduces their appearance and quality because of the difficulty in separating the dust by washing them (Navarro, 2006).

13.4   Field Sorting and Grading

Dates arriving from the farm may be contaminated with dirt, dust, and sand particles, plant/field debris, and chemical products and should be cleaned to remove these particles, which stick to the date skin. Cleaning can be achieved by blowing air on the fruits and brushing the dates softly to avoid damage to the fruit skin or by washing the fruits with running water (Yahia *et al.*, 2014).

Spray jets can be used for soft dates instead of washers; washing with sanitizers is important to remove soil and debris and for water disinfection to avoid cross-contamination between clean and contaminated product (Yahia *et al.*, 2014).

Most sanitizing solutions achieve higher microbial reductions immediately after washing compared to water washing; however, after storage, epiphytic microorganisms grow rapidly, reaching similar levels (Yahia *et al.*, 2014). Date processors rely on wash-water sanitizers to reduce microbial counts to maintain quality and extend the shelf life of the end product (Gil *et al.*, 2009).

Dates are sorted to remove culls and to separate them into uniform sizes; sorting can be carried out manually or mechanically in crates or on moving belts (Yahia *et al.*, 2014). Discarded fruits consist of dates with defects and abnormalities such as parthenocarpic (unpollinated) fruits, immature or overripe fruits, fruits mechanically damaged during harvesting or on the palm, fruits damaged by birds or insects, and fruits with physiological disorders or diseases (Yahia *et al.*, 2014). During this step, workers sort and remove dates with any indication of infestation as well as other particles and damaged dates (Naturland, 2002).

Dates are eaten both fresh and as processed products, most commonly as dried and dehydrated dates (Chonhenchob *et al.*, 2012). The fruit is increasingly receiving attention for consumption as it is known to provide a wide range of functional and nutritional values. Several studies have shown that dates contain an important source of bioactive compounds with potential health benefits (Barreveld, 1993).

Packaging plays a significant role in managing the effective and sustainable supply chain. The common date supply chain consists of delivery of harvested dates to the collecting facility, packinghouse, or distribution center for preparing, sorting, and packing the fruits for destination channels (wholesale, retail, processing plants) (Chonhenchob *et al.*, 2012). Packaging protects dates from physical damage, moisture absorption or loss, and insect reinfestation during subsequent storage and handling steps.

Some dates are marketed in 15 lb flats of fiberboard or wood (6.8 kg, approximately), others in 5 or 10 lb cartons (about 2.3 and 4.5 kg, respectively) (Yahia *et al.*, 2014). Packages include transparent film bags and trays overwrapped with film; round fiberboard cans with metal tops and bottoms, containing 0.5–1 kg, are also used, as commonly are rigid transparent plastic containers with a capacity of 0.2–0.3 kg (Yahia *et al.*, 2014). Small consumer packages are used too, such as bags containing about 50–60 g of dates (Ait-Oubahou and Yahia, 1999). Data such as weight, country of origin, quality, and date of expiry should appear on the package labels (Glasner *et al.*, 2002).

Additionally, nutritional labeling, already required on the retail packages by many date-importing countries, should be added on all retail packages, including those intended for local markets (Yahia *et al.*, 2014).

13.5 Operations

Harvested dates usually arrive at the packer in boxes delivered by truck; therefore, a dock of suitable height must be provided. As the first quality check is done at the time of delivery, the dock must have sufficient space to empty the crates on trays. The agent responsible for receiving dates must be well acquainted with the qualities and must at the same time have tact, firmness, and great honesty in dealing with the producers. The value of the enterprise's output depends largely on it. Even for an experienced professional, it is useful to randomly take, from time to time, a hundred dates from different areas of a crate and to classify them quickly by quality, according to local uses.

To sample dates, an automatic system is used, consisting of a hopper whose lower part is provided with a slot on one-tenth of its width, thus allowing passage of a tenth of the dates, which fall into a container placed below. The sample is then sorted by hand. If one out of every ten or 20 crates is moved to the sampler, depending on the size of the delivery, a representative sample is obtained that does not take a lot of time to analyze. The dates are intimately mixed by gravity, about a tenth is taken, and the process is repeated until a 50-ounce (about 1400 g) sample remains. A new sampling is conducted at the time of delivery to the packing plant, by random sampling of several boxes equal to the square root of the total number.

The harvested fruit is transferred into large plastic or wooden bins for transport to the packinghouse (Fig. 13.8); large wooden, plastic, or cardboard cases of various sizes are also used, as well as baskets, sacks (for very dry fruit), and trays, focusing on the need to prevent damage to the fruit (especially to soft and sensitive fruit) (Yahia *et al.*, 2014) (Fig. 13.9).

The fruit must be transported carefully in the early hours of the morning to avoid the heat; refrigeration during transport is advisable (Yahia *et al.*, 2014).

Those varieties marketed on the bunches must not be shaken during transportation to prevent the fruit from falling off the bunches; speedy transport will also prevent infection by pests that attack the fruit during the postharvest period (Yahia *et al.*, 2014).

13.6 Artificial Maturation

Farahnaky and Afshari-Jouybari (2011) and Afshari-Jouybari *et al.* (2013) confirm that the date is a climacteric fruit by observing an increase in

Fig. 13.8. Transport of dates on bunches by truck from the palm grove to the packinghouse, Biskra, Algeria. (Photos by Mohamed Marouf Aribi.)

Fig. 13.9. Packing crate used for date bunches in Algeria. (Photo by Mohamed Marouf Aribi.)

ethylene production during ripening. On the contrary, other authors consider the date not to be climacteric (Yahia and Kader, 2011).

For dates stored at 20°C, carbon dioxide (CO_2) production is less than 25 ml/kg per h at the khalal stage, and less than 5 ml/kg per h at the rutab and tamar stages (Kader and Hussein, 2009). Under the same storage conditions, the same authors found ethylene production values of less than 0.5

µl/kg per h at the khalal stage, and less than 0.1 µl/kg per h at the rutab and tamar stages.

Marouf and Khali (2019) studied the effect of polyethylene terephthalate (PET) packaging on the postharvest physiology of Deglet Nour dates under different storage conditions (22 ± 1°C, 75–80% relative humidity (RH) compared to 10 ± 2°C, 85–90% RH) in Algeria. They used an experimental setup consisting of a box with dimensions of 7.5 cm × 15.0 cm × 7.5 cm. They found that PET packaging limited the respiratory intensity during storage, and they concluded that the combination of thermization–PET packaging and cold storage (10 ± 2°C) is an excellent storage method. It ensures an optimal physiological basis for dates by limiting their respiratory activity. Values of respiration rate are needed for an optimum design of polymeric packages when packaging takes place in a modified atmosphere (Jemni *et al.*, 2016b).

In the same study, dates showed typical non-climacteric fruit respiration after harvest. According to the classification of tropical and subtropical fruits reported by Gross *et al.* (2002) and Paull and Duarte (2011), dates are non-climacteric fruits, with very low respiration rate and low ethylene production, and therefore have very low metabolic activity. Serrano *et al.* (2001) reported that in the early maturation stages of cv. Negros dates grown in Spain, a small peak of ethylene production was detected followed by a peak of respiratory activity, suggesting that the date is a climacteric fruit.

The sensory and nutritional qualities of date fruit are closely linked to its degree of maturity (Baliga *et al.*, 2011). The maturity stages of dates use Iraqi terminology (Al-Shahib and Marshall, 2003; Saafi *et al.*, 2009; Baliga *et al.*, 2011): hababouk (stage I), kimri (stage II), khalal (stage III), rutab (stage IV), and tamar (stage V).

The drying and artificial maturation of dates are two intimately related processes. Heating accelerates the maturation process if it has been carried out naturally up to a certain stage of fruit development (Fig. 13.10). In palm groves, not all dates ripen at the same time on the bunch. For technological and economic reasons, unripe diets are harvested, and maturation is artificially conducted. This technique has the advantage of reducing the duration of maturation and preventing damage caused by infestation due to rain or ripe fruit dropping from the bunches. Fruits eaten at the soft stage (rutab) should therefore be picked at different intervals. This is possible when only a few trees are harvested whose dates are reserved for family feeding. Maturation is an ongoing process, so any triage is necessarily somewhat arbitrary.

There are three stages marked by signs visible to the naked eye and corresponding to the beginning of significant changes in the appearance and composition of the fruit. These indices are change of skin color from green to yellow or red (kimri, tchimri, or gamag stage); softening of the date apex; and extension of softening to the whole fruit – or more precisely,

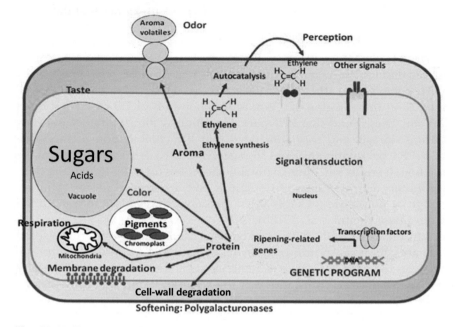

Fig. 13.10. Representation of the series of metabolic changes that occur in dates during the ripening process. (From El-Hadrami and Al-Khayri, 2012.)

since some varieties never become soft over the entire fruit surface, to the whole part that normally softens. There is no generally accepted definition of "mature date". Deglet Nour is not considered mature until the end of the rutab stage.

Dates had to be handled in certain ways from the beginning of crop domestication, and the same was probably true of figs, apricots, and grapes. In current use, "processing" refers to the different operations to which dates are subjected after harvest to alter composition, color, shape, size, or consistency. It produces a marketable product that preserves most of the fruit.

The operations included in postharvest treatment are drying, hydration, heating, refrigeration, chemical preservation, coloring, pressing, and crushing. The objectives of the treatment are to: (i) hasten maturation; (ii) ripen dates that would not ripen properly on the tree; (iii) transform the khalal into a well-preserved product; (iv) permit the sale of rutab and soft tamar in distant markets; (v) soften hard dates; (vi) improve the appearance, consistency, and shelf life of tamar dates; and (vii) destroy insects, mold, yeast, bacteria, etc.

Deglet Nour matures more unevenly and matures over a longer period than most other varieties. On the other hand, the packer requires that it be a little soft, almost ripe. It is therefore more convenient to harvest when most of the dates have started to soften and to continue maturing in sheds where the bunches can be monitored, and the good dates detached to pack. Dates that are in an intermediate stage between the khalal and rutab, with a water content of 35–40% w/w, reach the tamar stage (where they are quite ripe and soft) in a few weeks, while rutab with a water content of 25–28% w/w ripen in a few days. However, many Deglet Nour subjected to this almost natural maturation mode lose their amber color and must be sold as a second choice.

The main enzymes involved in the maturation of dates at the tamar stage have been studied by Hasegawa *et al.* (1972). They found in a previous study (Hasegawa *et al.*, 1969) on different batches of Deglet Nour dates at the tamar stage with a greater or lesser degree of drying (75–85% DM) that polygalacturonase (PG) activity had a value of 0.9–1.3 IU/g fresh fruit. In addition, if the dates have high water content (50% w/w for example) and if the maturation process is too long, they may ferment before ripening. Drummond (1924) and Tate and Hilgeman (1958) recommend a rather high temperature, and Nixon (1959) a rather low temperature. In general, dates ripen better when the humidity is high, but they may reach an excessive moisture content.

Research published so far does not provide a uniform indication of the best temperature and RH for each stage of maturity and each water content of the main varieties normally ripened by artificial means. All we can say to the new packer in the trade is that he must equip his maturation chambers to be able to vary the temperature and humidity at will and to search by trial and error for the best possible result. An additional stage of maturation can be distinguished by the fact that some dates are eaten rutab, at a time when they are "botanically ripe", while others are pushed to the tamar stage where they are kept for a long time. Deglet Nour harvest should occur at the end of the rutab stage and therefore Deglet Nour does not need post-maturation. There is little post-ripening in the packing plants because the Deglet Nour dates must be at the end of the rutab stage, while most of the other varieties that arrive at the conditioner are already at the tamar stage.

Artificial maturation can also be carried out using chemical or physical agents. For fruits with a climacteric peak, this operation occurs when the respiration rate increases. For dates, artificial maturation can only occur if the fruit has already reached a definite degree of maturation, called the "critical moment" by Vinson (1924). The critical moment precedes veraison, the first dark and soft spot that is generally observed at the top of the date and that announces the passage from the khalal to the rutab stage (Dowson, 1982), by a few days. This passage is usually marked by a significant accumulation of sugars. In Deglet Nour, the critical moment begins when the soft spot reaches half the distance between the base and the top of the fruit (Al-Ogaidi and Aref, 1985). Below this point, thermal or chemical stimulation has no effect. On the other hand, this action can kill or stimulate the protoplasm of tissue cells to release previously insoluble or inactive diastases. Vinson (1924) tested about 100 chemicals for the artificial maturation of dates and noted the notable effect of acetic acid vapor.

Temperature plays a decisive role in the maturation of dates. It increases the rate of chemical reactions by two to three times for each 10°C rise according to Van't Hoff's law (Al-Ogaidi and Aref, 1985). However, the correct temperature depends on several factors: date variety, initial water content, sugar and solids contents, air velocity, and RH. Rygg (1975) reports that in the USA, less ripe fruits require a higher temperature and RH than more ripe fruits. For Deglet Nour, it is recommended not to exceed a temperature of 35°C to preserve its scent and characteristic blonde color.

Kahramanoglu and Usanmazm (2019) studied hastening the ripening process of date fruits cv. Medjool by some preharvest and postharvest applications. Three doses (500, 750, and 1000 ppm) of ethephon, two controls, and apple vinegar (4–5% acidity; g/100 g malic acid) were used in the preharvest studies with or without bunch bagging (perforated black polyethylene bags). Postharvest experiments were conducted with visually mature firm (khalal) fruits of the cv. Medjool that were subjected to: (i) dipping into distilled water; (ii) dipping into ethephon at 500 ppm; (iii) dipping into ethephon at 750 ppm; (iv) dipping into apple vinegar (4–5% acidity; g/100 g malic acid); (v) dipping into grape vinegar (4–5% acidity; g/100 g malic acid); and (vi) freezing at −18°C for 3 days. They found that bunch bagging enhanced fruit ripening on the tree, increased fruits' total soluble solids (TSS), and reduced titratable acidity (TA) at both the tamar and khalal stages. Preharvest application of 1000 ppm ethephon (with bunch bagging) was found to have the highest ratio of tamar (37%) and rutab (46%) fruits whereas preharvest apple vinegar application had only a slight effect on fruit ripening (12% tamar) as compared with control (8% tamar). Kader and Hussein (2009) reported that acetic acid, which is the primary acid in apple and grape vinegar, is used postharvest to enhance ripening of date fruits. Awad (2007) also noted that ethephon and abscisic acid combination applied postharvest significantly enhanced fruit ripening of cv. Helali.

Mohammed *et al.* (2021) designed a study to find a precise combination of temperature and RH to artificially ripen unripe Bisr fruits in controlled-environment chambers (Fig. 13.11); the Bisr fruits (cv. Khalas) were placed immediately after harvesting in the treatment chambers of the system with three set-point temperatures (45, 50, 55°C) and six set-point RHs (30, 35, 40, 45, 50, 55%) until ripening; the optimal treatment combination for artificial ripening of Bisr fruits of cv. Khalas was 50°C and 50% RH.

In the same study, the authors found that the combination of 50°C and 50% RH provided good fruit size, color, firmness, TSS, pH, and sugars content, a reduction in fruit weight loss, and optimum fruit ripening time. Low temperature and RH delayed the ripening process, deteriorated fruit quality, and caused more weight loss (Mohammed *et al.*, 2021).

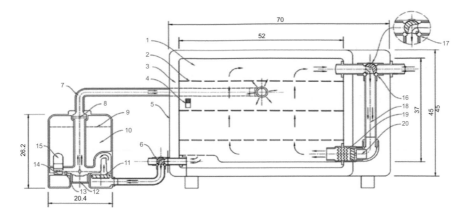

Fig. 13.11. Schematic diagram of an artificial ripening system. 1, Processing chamber; 2, sample shelves; 3, glass wool; 4, temperature and relative humidity sensor; 5, outer cover; 6, electrical three-way valve; 7, mist duct; 8, plastic mesh; 9, upper limit of water level; 10, water tank; 11, direct current air fan; 12, ultrasonic transducer; 13, water level sensor; 14, water flow regulator; 15, filter; 16, electrical three-way valve; 17, the electrical valve is in the position of air intake from the outside; 18, air duct; 19, nickel-chrome heat coil; 20, direct current air fan (all dimensions in cm). (From Mohammed *et al.*, 2021, used under Creative Commons CC BY license.)

Saleem *et al.* (2004) studied the influence of hot water treatment on the ripening/curing of Dhakki dates. The time of dipping in hot water was fixed at 5 min and temperature of the hot water varied at 35, 70, and 93°C. The treated samples were then allowed to ripen/cure for 72 h under air circulated in a cabinet dehydrator adjusted at 38–40°C. They found that the treatment with 70°C water temperature performed better than that with 35 and 93°C, resulting in 55% product yield of acceptable quality. The yield of improved quality product was further increased to 70% on the optimization of the treatment time to 3 min. The ripening of Dhakki dates does not require the fruits to stay on the tree beyond the fully mature Doka (equivalent to rutab) stage for want of development and hence saves at least 2 weeks' hang-on period.

Dipping in hot water is a technique reserved for dry dates and cannot be used for fruit destined for export.

In modern packinghouses, prematurely harvested dates are ripened in controlled atmospheres, the degrees of temperature and humidity varying with the nature of the cultivar (Barreveld, 1993).

Ripening of immature Medjool dates using solar energy in shrink film-covered pallets was demonstrated as a practical solution (Navarro, 2006). This type of maturation is based on the use of solar energy by exposing the pallets to solar heat in the open. Calculations showed that the energy required for field maturation is cumulative and predictable.

Fig. 13.12. Two models of shrink film-covered pallets for ripening Medjool dates using solar energy. (From Navarro, 2006.)

In field conditions, covering the pallets with shrink film may cause slight condensation. The author recommends that a ventilation window must be left open to prevent excessive accumulation of moisture at the top layers of the shrink film-covered pallet (Fig. 13.12).

13.7   Physiological Disorders

Significant quantities of dates are lost annually in the Arab world; indeed, Besbes *et al.* (2009) gave a rate of 30% of the total production of Tunisian dates being thrown away each year. This is mainly due to alteration of the quality of these fruits leading to an unattractive appearance to the consumer. Loss of date fruit quality is the result of physiological deterioration occurring with increased moisture content and storage temperature (Yahia, 2004). These cause changes in fruit texture, structure, and flavor (Jeantet *et al.*, 2006). The alteration of the date is generally caused by physiological phenomena continuing within the date after harvest as well as by its treatment and manipulation. This usually leads to the occurrence of browning along with some other forms of deterioration.

Fruit darkening (beyond that normally desired) can be due to both enzymatic and nonenzymatic browning, which increases with higher moisture content and higher temperatures; enzymatic browning can be inhibited at low oxygen concentrations and low temperatures (Lobo *et al.*, 2014). For dates harvested and marketed at the khalal stage, avoiding fruit

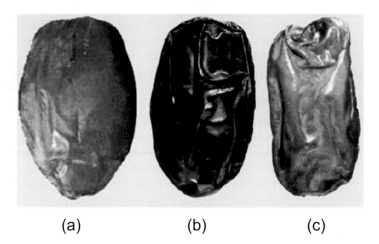

(a)   (b)   (c)

Fig. 13.13. Medjool date: (a) immediately after ripening, the skin does not adhere to the mesocarp, and it is about to be separated; (b) a non-skin separating fruit; (c) a skin separating example. (From Lustig *et al.*, 2014.)

damage/bruising is critical in minimizing enzymatic browning (Lobo *et al.*, 2014).

Oxidative browning reactions generally occur more rapidly at high than low temperatures (Sidhu, 2006). Belarbi *et al.*, 2001aBelarbi *et al.*, 2001b studied the drying effects of temperature ranging from 60 to 80°C on two determinant criteria of Deglet Nour date quality: light color and soft texture. They assumed that the heat treatment should be carried out under controlled conditions to avoid nonenzymatic browning, while being sufficient to inactivate the enzymes involved in the enzymatic browning.

Skin separation or puffiness develops during ripening of soft date cultivars, which vary in their susceptibility (Fig. 13.13); high temperature and high humidity at a stage before the beginning of ripening may predispose the dates to skin separation (skin becomes dry, hard, brittle, and separates from the flesh) (Lobo *et al.*, 2014).

When skin extension during the khalal stage does not exceed the elastic limit, the skin remains tight with a high modulus of elasticity (E) and adheres to the shrinking mesocarp without separation. These considerations may explain the difference between "skin separated" and "non-skin separated" fruits. Varying skin strains that cause creep can only occur during the khalal stage growth of the fruit when it is still supplied with water from the tree and has turgor pressure. They stop in the rutab stage when the water supply from the tree has been finally interrupted, the turgor pressure disappears, and the mesocarp continues to shrink (Lustig *et al.*, 2014).

Storage at recommended temperatures minimizes skin disorders, which occur mainly in cultivars in which glucose and fructose are the main sugars. Sugaring, if not too severe, can be reduced by gentle heating of the affected dates (Rygg, 1975; Yahia, 2004). These spots can be minimized at sufficiently low storage temperatures (Choehom *et al.*, 2004).

Saltveit (2000) showed that low nonfreezing temperatures lead to at least long-term enrichment of tissue with reducing sugars, insoluble pectins, citric acids, pyruvic acid, acetaldehyde, ethanol and ketone acids, chlorogenic acid and similar phenolic compounds, abscisic acid, etc. Total sugars reach their minimum at 0°C storage conditions.

On the other hand, cold causes a decrease in malic acid and total TA, soluble pectins, ascorbic acid, etc. According to the same author, the activity of phenylalanine ammonia-lyase (PAL), hydroxycinnamyl CoA, and hydroxycinnamyl transferase quinate would increase at cold temperatures, while the activity of peroxidase and ascorbate oxidase would decrease (Saltveit, 2000). According to Kacperska and Kubacka-Zebalska (1989), cold influences the activity of PAL in plants. Thus, a temperature of 2°C results in a marked increase in the activity of PAL observed after 2 days, subsequently the specific activity of PAL was almost constant while the total activity (amount of active enzymes) increased by 30% for 6 days.

It is common to observe browning phenomena in fleshy plant organs (fruits and vegetables), particularly during postharvest handling, storage, and technological transformations (crushing, thawing). The appearance of brown pigments, the color of which is superimposed on the natural color, leads to significant changes in organoleptic and nutritional quali-ties (Tirilly and Bourgeois, 1999). The browning phenomenon covers a generally very complex set of reactions. In terms of basic mechanisms, these reactions can be classified into two main groups: enzymatic versus nonenzymatic browning processes (Macheix *et al.*, 1990).

Date enzymes generally belong to three groups: pectic enzymes (Myhara *et al.*, 2000), oxidase enzymes, and carboxylase enzymes. Reynes (1997) considers that five enzymes are involved in the maturation and browning of dates: invertase, polyphenol oxidase (PPO), peroxidase, pectin methylesterase, and PG.

Browning occurs in two stages: an enzymatic stage catalyzes the forma-tion of quinones; then, a nonenzymatic stage results in the formation of melanins (insoluble brown polymers) by polymerization and condensa-tion of quinones (Rivas and Whitaker, 1973). These enzymes are located primarily in the membranes of the cell organelles (mitochondria and chloroplasts), and their phenolic substrates are in the vacuoles. When the cell is damaged, the enzyme and substrates may come into contact, leading to rapid oxidation of phenols (Chazarra *et al.*, 2001; Chisari *et al.*, 2007).

Enzymatic browning occurs in plants that are rich in phenolic compounds, which are naturally present in dates (Albagnac *et al.*, 2002). Enzymatic browning poses significant color problems with some fruits and vegetables, especially when the tissues of these plants are diseased or have been damaged by bruising during handling operations or by certain processing treatments, as is often the case with dates (Al-Bekr, 1972).

Dates are rich in total phenolic compounds (Mansouri *et al.*, 2005; Al-Farsi *et al.*, 2007). Deglet Nour is the richest variety with approximately 493.5 mg gallic acid equivalents (GAE)/100 g (Besbes *et al.*, 2009). Date PPO activity is optimal at pH 4.5–6.5 (Hasegawa and Maier, 1980). Browning activity is intensified with increased water content and temperature (Yahia and Kader, 2011). PPO is relatively resistant to heat treatment at temperatures below 63°C (Reynes, 1997). According to Belarbi *et al.* (2003), the activity of this enzyme is reduced by more than 50% after a heat treatment of 70°C for 10 min.

Due to their composition (reducing sugars, amino acids, proteins, phenols, and minerals, etc.), dates are also altered by nonenzymatic browning. Nonenzymatic browning refers to a very complex set of reactions leading to the formation of brown or black "melanoid" pigments, unfavorable in the date, and often also to undesirable changes in the odor, flavor, and nutritional quality (Cheftel *et al.*, 1979; Chavéron, 1999). Nonenzymatic browning occurs during technological treatments or storage of dates. It is accelerated by heat (Cheftel *et al.*, 1979).

As much as these compounds are sought in certain foods to develop desirable tastes and appearances, they are considered a negative quality factor in other foods such as fruits and vegetables (Machiels and Istasse, 2002). The Maillard reaction affects not only the color of the date, but also its flavor, taste, and nutritional quality (Van Boekel, 2001).

The Maillard reaction can also occur at low temperatures of 5–45°C (Bulut and Kilic, 2009; Özhan *et al.*, 2010). Maier and Schiller (1961), cited by Hazbavi *et al.* (2013), reported that color degradation of Deglet Nour dates is caused by both enzyme and nonenzymatic pathways at 28 and 40°C.

The two main forms of nonenzymatic browning are caramelization and the Maillard reaction. Both vary in reaction rate as a function of water activity. The Maillard reaction is responsible for the production of the flavor when foods are cooked. It is a chemical reaction that takes place between the amine group of a free amino acid and the carbonyl group of a reducing sugar, usually with the addition of heat. Melanoidins are brown, high-molecular-weight heterogeneous polymers that are formed when sugars and amino acids combine through the Maillard reaction at high temperatures and low water activity.

13.8 Biotic Problems (Diseases, Insects) and Their Control

13.8.1 Diseases

In some countries or areas with low heat accumulation, dates that have not been eaten as fresh and hard must be eaten as soon as they have softened, since the heat is not strong enough to dry them before fermentation begins. On the other hand, in warmer regions, the fruit dries out to take the consistency of a prune. At this stage, dates are no longer perishable and can be packed for export worldwide.

The mold *Aspergillus parasiticus* is able to penetrate the intact date fruit tissue and can produce aflatoxin in 10 days at 28°C in all stages of maturity except the tamar stage, which does not support mold growth (Ahmed *et al.*, 1997) due to low water content. However, extracts from all four stages of date fruit were able to support growth of this mold for aflatoxin production (Ahmed *et al.*, 1997), and the amounts of aflatoxin produced increased with ripeness.

Apart from sizable quantities of dates being consumed at perishable immature stages (khalal and rutab), the majority of date fruits are consumed in the dry tamar stage with moisture content of less than 20% w/w (Sabbri *et al.*, 1982).

The richness of Deglet Nour dates in invertase causes the inversion of sucrose; the transformation of sucrose into glucose and fructose by invertase may result in a decrease in the relative moisture equilibrium of the date and a change in its quality and natural flavor. Other types of alterations have been noted in the literature, including sugar spotting or sugar stains, which affect all varieties and are characterized by the formation of granular sugar deposits just below the skin and in the flesh of the fruit (Jarrah and Benjamin, 1982).

Susceptibility of date fruit to postharvest diseases increases after harvest, during ripening, and over prolonged or inadequate storage because of physiological changes in the fruit favoring pathogen development (Yahia *et al.*, 2014). *Zygosaccharomyces* spp. yeasts are more tolerant of high sugar content than others found in dates; yeast-infected dates develop an alcoholic odor (become fermented) (Rygg, 1975; Lobo *et al.*, 2014; Yahia *et al.*, 2014). Lactic acid bacteria are present only at the rutab stage in some varieties (Tafti and Fooladi, 2005). Microbial spoilage can also be caused by molds and bacteria.

High humidity increases perishability by encouraging the activity of yeasts (Hardenburg *et al.*, 1990). *Acetobacter* bacteria may convert the alcohol into acetic acid (vinegar), thereby resulting in souring of dates (due to accumulation of ethanol and/or acetic acid) with moisture content above 25% w/w when kept at temperatures above 20°C, and its severity increases with duration and storage temperature (Lobo *et al.*, 2014).

Growth of *Aspergillus flavus* on dates can result in aflatoxin contamination, which makes them unsafe for human consumption and unmarketable (Yahia, 2004; Tafti and Fooladi, 2005; Yahia and Kader, 2011).

Current knowledge suggests that refrigeration is the best way to control pathogens in stored date fruits; microbial pathogens decrease when the fruits are stored at temperatures below 5°C as compared to 25°C (Abdul Aly *et al.*, 2018).

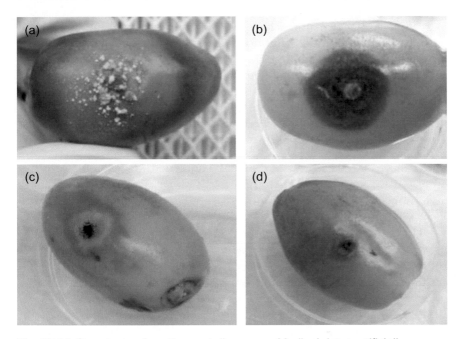

Fig. 13.14. Symptoms of postharvest disease on Medjool dates artificially inoculated with *Penicillium expansum* (a), *Alternaria alternata* (b), *Aspergillus* sp., (c) or *Cladosporium cladosporioides* (d) and incubated at 20°C for 10 days. (From Palou *et al.*, 2016.)

The most common fungal genera responsible for decay spoilage losses because of their pathogenicity are *Alternaria, Aspergillus, Cladosporium, Fusarium, Rhizopus,* and *Penicillium* (Nasser, 2017; Al-Mutarrafi *et al.*, 2019; Quaglia *et al.*, 2020); all may grow on high-moisture dates, especially when harvested following rain or a period of high humidity (Lobo *et al.*, 2014). The incidence and etiology of postharvest diseases affecting fresh date fruit in the palm grove of Elche (Spain) were determined under local environmental conditions; the most important causal agents of disease were *Penicillium expansum, Alternaria alternata, Cladosporium cladosporioides,* and a black aspergillus species belonging to the *Aspergillus niger* clade. These fungi were identified by macroscopic and microscopic morphology and/or DNA amplification and sequencing (Palou *et al.*, 2016) (Fig. 13.14).

Several methods can be adopted to control disease development (Yahia *et al.*, 2014):

- Dry the dates to 20% w/w moisture or lower to greatly reduce incidence of molds and yeasts.
- Maintain recommended temperature and RH ranges throughout the handling system. In date-producing countries, it is common to consume fruit at the khalal and rutab stages during the harvesting season. Dates

at these stages of maturity have high moisture content and are prone to rapid deterioration. Therefore, for optimum shelf life, such fruits should be handled and marketed using low-temperature controlled storage just like other perishable commodities.
- Avoid temperature fluctuations to prevent moisture condensation on the dates, which may encourage growth of decay-causing microorganisms.
- Use adequate sanitation procedures in the packinghouse and storage rooms to reduce potential sources of microbial contamination.

Date fruits of some cultivars at the khalal and rutab stages of maturity are preferred by consumers (Al-Mulhim and Osman, 1986); however, being high in moisture, they are the most susceptible to contamination (Aidoo et al., 1996).

The most common molds found in date fruit belong to the genera *Aspergillus, Penicillium,* and *Cladosporium* (Gherbawy, 2001; Ragab et al., 2001; Shenasi et al., 2002b). *A. niger, A. flavus,* and *Aspergillus fumigatus* are the species of *Aspergillus* most often reported in stored dates (Shenasi et al., 2002b; Atia, 2011). They are present at high frequencies: 39–83%, 40–49%, and 26–40%, respectively (Al-Sheikh, 2009; Colman et al., 2012). Species of the genus *Alternaria* may also be present on the date fruit: *A. alternata* (22–62%) and *Alternaria chlamydospora* (24%) (Al-Sheikh, 2009). All these species can produce mycotoxins, posing a health hazard to consumers (Logrieco et al., 2003; Patriarca et al., 2007). Yeasts that affect dates are the most tolerant of high sugar contents such as *Zygosaccharomyces* spp. and *Hansenula* spp. (Ait-Oubahou and Yahia, 1999). The formation of pockets of gas under the skin, white aggregates of cells, bleached pulp, and alcoholic odors are characteristics of yeast-infected dates (Yahia and Kader, 2011).

It has been reported that acetic acid bacteria (Kader and Hussein, 2009) and lactic acid bacteria (Shenasi et al., 2002a) may be responsible for acidification. This acidification is amplified in dates harvested at the rutab stage, especially in soft varieties.

The development of fungal microflora depends on the temperature, the humidity of the storage air, the water activity of the fruit, and the initial microbial load at harvest. Apart from contamination resulting from noncompliance with good harvest practices, the initial fungal load of the date depends on the stage and physiological state of the harvest. The pH of dates is usually slightly acidic or acidic and ranges from 5 to 6. It is harmful to bacteria but suitable for fungal flora development (Reynes, 1997).

Shenasi et al. (2002a) reported a relatively high fungal load (in terms of colony-forming units (CFU)) in cv. Fard at the first kimri maturity stage ($1.7 \log_{10} CFU/g$) that increased strongly at the second rutab stage ($5.7 \log_{10} CFU/g$), then decreased significantly at the final tamar stage of maturity ($1.9 \log_{10} CFU/g$).

13.8.2 Insects

Insects can cause damage to dates at different developmental stages (Carpenter and Elmer, 1978; Dowson, 1982; Ait-Oubahou and Yahia, 1999; Marouf *et al.*, 2013; Jemni *et al.*, 2014; Wakil *et al.*, 2015):

- *Oligonychus afrasiaticus* McGregor and *Oligonychus pratensis* Banks are mites that cause a disorder known as Bou Faroua disorder, which affects fruit at the hababouk stage.
- *Coccotrypes dactyliperda* (date stone beetle) has the same consequences, with the fruit dropping at the immature green stage.
- *Parlatoria blanchardi* (date palm scale) attacks the fruit while still green and forms white filaments around the fruit that reduce photosynthesis and the fruits do not reach maturity.
- *Ectomyelois ceratoniae* Zeller (date carob moth) is another lepidopteran that is widely distributed in different producing areas of date culture and causes significant postharvest losses in stored dates.
- *Batrachedra amydraula* Meyr (lesser date moth), *Carpophilus hemipterus* (dried-fruit beetle), *Carpophilus mutilatus* (confused sap beetle), *Urophorus humeralis* (pineapple beetle), and *Haptoncus luteolus* (pineapple sap beetle) can cause serious damage to dates on the bunch or after harvest.
- *Vespa orientalis* (Oriental hornet), *Cadra figulilella* (raisin moth), *Arenipses sabella* (greater date moth), and *Tyrophagus lintneri* Osborn (mushroom mite) can infest stored dates.
- *Ephestia cautella* Walker (fig moth) is an important postharvest pest in some growing regions that can attack dates in the orchard, packinghouses, or stores (Ahmed *et al.*, 1994). Dates at the kimri, khalal, and rutab stages are not attacked by this insect, only fruits at the tamar stage.
- *Oryzaephilus surinamensis* L. (sawtooth grain beetle) is a serious insect pest of stored dates in some regions.

Soft and semisoft dates are more susceptible to insect infestation than dry dates (Bouka *et al.*, 2001; Idder *et al.*, 2009). Insect infestation and damage to dates is one of the main causes of postharvest losses in quality and quantity (Kader and Hussein, 2009). The most encountered insect pests include *E. ceratoniae* Zeller (Fig. 13.15), *Ectomyelois decolor* Zeller, and *Ephestia calidella* Guenée which are Lepidoptera, and *O. surinamensis* L. (Coleoptera) and *O. afrasiaticus* (Ben-Lalli, 2010).

Ben-Lalli's (2010) observations, made in the palm groves of Socobio-Sud in the region of Biskra (southeastern Algeria), show that more than 30% of organic dates of the Deglet Nour variety are infested by pests, particularly Lepidoptera. Insect pests degrade stored dates and cause weight loss and a decrease in the commercial value of the fruit. Pests include *Myeloîs phoenicie* and *O. afrasiaticus* (Ferry *et al.*, 1998). In the USA, the

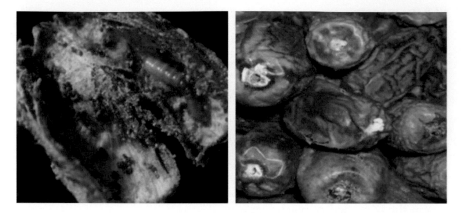

Fig. 13.15. Damage caused by *Ectomyelois ceratoniae* on Deglet Nour date. (From Mehaoua, 2014.)

infestation rate varies from 10 to 40% on the Deglet Nour variety (Nay and Perring, 2006).

13.8.3 Control methods

Fumigation by methyl bromide (CH_3Br), irradiation, microwaves, heat treatment, modified atmosphere packaging (MAP), ozonation, and biological control can be used to control insects in dates (Paull and Armstrong, 1994; Yahia, 2009; Yahia and Kader, 2011).

13.8.3.1 Fumigation

Fumigation is preferable to all processes because it is quick, inexpensive, and easy to use (Aegerter and Folwell, 2000). It does not require further drying of dates and does not cause browning, but its effect on quality criteria is not always clarified. It can be applied under vacuum or at atmospheric pressure. It is currently the most used technique in the major date-producing countries.

Dowson and Aten (1963) report that in the dried fruit industry, fumigation is the treatment by which insects introduced into fruit are subjected to the action, in a confined space, of a toxic gas to destroy them. Dusting and spraying on the tree, or the application of insecticides around piles of crates containing fruit, cannot be considered as fumigation. Fumigation in the palm grove is sparse due to the lack of competent applicators (Dowson and Aten, 1963). It is different from what is done in the packinghouse because it is not intended to destroy the same types of insects.

CH_3Br at 30 g/m^3 (30 ppm) for 12–24 h at temperatures above 16°C is very effective for insect disinfestation (Yahia and Kader, 2011). Teisseire (1959) and Munier (1973) report that CH_3Br is used at doses of 20 g/m^3

at atmospheric pressure for 13 to 15 h and 80 g/m^3 for 1 h after a previous vacuum. In Iraq, Hussain (1974) reported the use of a 1 kg dose per 41.6 m^3 in a confined space for 24 h. Ahmed *et al.* (1982) compared CH$_3$Br fumigation and irradiation of Zahidi dates and reported that both techniques were efficient for disinfestation during the first period of storage (25 days), but reinfestation of dates occurred during storage leading to detection of live insects.

In view of the toxicity of these products to humans, the regulations of each country fix the doses not to be exceeded. According to Hassouna *et al.* (1996), the doses used are not dangerous for humans if aerations are carried out for several days. Hilton and Banks (1997) reported that 82% of the CH$_3$Br applied was released 24 h after exposure.

In France, the decree of January 25, 1971, allows a maximum CH$_3$Br residue level on and in fruits, vegetables, and cereals of 0.1 mg/kg. In the USA, its use is prohibited, while it is 80 g/m^3 in Tunisia (Reynes, 1997).

In Algeria fumigation with CH$_3$Br remains the only chemical treatment applied to all date conditioning units. The recommended dose is 80 g/m^3, but no specific constraints associated with the use of this product (other than the standard) have been identified at the industrial level.

CH$_3$Br is one of the major substances currently threatening the ozone layer around the earth. Once sprayed, this product, which is very stable, reaches the upper atmosphere where it damages the ozone layer. In 1995, industrialized countries agreed to phase out the production and use of CH$_3$Br to comply with the Montreal Protocol on Substances that Deplete the Ozone Layer (Gan *et al.*, 2001).

13.8.3.2 Irradiation

Eman *et al.* (1994) investigated the possibility of substituting irradiation for currently used CH$_3$Br. A treatment of 300 krad seemed very effective for disinfestation and to prevent the development of aflatoxins. However, the high cost of this technique, the risk of accident, and the reluctance of consumers to accept irradiated products can hinder the development of this process.

Ionizing radiation at doses below 1 kGy (a level currently approved for use in fruits and vegetables) has potential for effective insect disinfestation without negative effects on the quality of dates (Ahmed, 1981; Al-Taweel *et al.*, 1993).

Ahmed *et al.* (1982) found that a dose of 0.86 kGy was adequate for the disinfestation of polyethylene-wrapped small date packages, causing complete inhibition of adult emergence of both *Ephestia* and *Oryzaephilus*. Al-Taweel *et al.* (1990) reported that a dose of 0.44 kGy for 30 min was sufficient to disinfest dates and no live insects could be detected after a storage period of 185 days. Azelmat *et al.* (2005) found that 0.3 kGy was the minimum needed to prevent damage from feeding and prevent adult emergence, and 0.45 kGy was required to kill the fourth instar of *Plodia interpunctella* (Huber) (Lepidoptera: Pyralidae). Studies by El-Sayed and Baeshin (1983), Grecz *et al.* (1989), and Al-Khahtani *et al.* (1998) showed that panelists could not discriminate between control and 0.2 to 6.0 kGy irradiated dates. Aleid *et al.* (2013) showed that sensory quality was affected at doses >3 kGy. Ionizing radiation is also an effective disinfestation method that has negligible effects on fruit quality (Çopur and Tamer, 2014).

13.8.3.3   Microwaves

Microwaves have been used to heat products by converting electromagnetic energy to heat energy and are a potential quarantine treatment method to control some insect pests (Ooi *et al.*, 2002; Zouba *et al.*, 2009). Nevertheless, work on disinfection by microwaves is rare and their industrial use is practically nonexistent.

Reynes (1997) reported that applying a 65°C microwave treatment for 2 min destroyed *E. ceratoniae* eggs and larvae. Wahbah (2003) reported that microwave radiation at 2540 MHz for 19 to 22 s was sufficient to cause 50% mortality of the two species *O. surinamensis* and *Tribolium castaneum*, respectively. Al-Azab (2007) indicated that 20 s at 2450 MHz was necessary to produce 100% mortality for all *E. cautella* stages, with sensitivity being pupae > adults > eggs > larvae.

13.8.3.4   Heat treatment

In view of the disadvantages of other food and date preservation techniques, in particular residues of CH_3Br, the use of heat disinfestation, mainly for Lepidoptera, has become much more widespread (Johnson *et al.*, 2003; Wang *et al.*, 2004).

Research studies have been conducted on the effects of temperature on ripe date fruits (tamar) for storage, transportation, and postharvest insect and disease control purposes (Marouf and Khali, 2020). Heat treatment of dates at 60–70°C for 2 h killed 100% of both the fig moth and the sawtooth grain beetle but resulted in a shiny appearance or glazing of the fruit (Hussain, 1974).

Heated air at 50–55°C for 2–4 h (from the time the fruit temperature reaches 50°C or higher) is effective for insect disinfestation (Navarro, 2006), but the use of higher temperatures is not recommended because it makes the color of the dates darker. Hussein *et al.* (1989) reported that boiling water is more efficient in controlling insect infestation of dates than exposure to hot air at 70°C.

Al-Azawi (1985) reported that 100% mortality could be achieved for the various life stages (eggs, larvae, adults) of *C. hemipterus* when exposed to temperatures of 40–60°C. To determine the most thermally resistant life stage, Wang *et al.* (2004) selected nine combinations: 48°C for 2 min, 48°C for 5 min, 48°C for 10 min, 50°C for 2 min, 50°C for 3 min, 50°C for 5 min, 52°C for 1 min, 52°C for 2 min, and 52°C for 5 min. These authors showed that all larval forms and eggs were destroyed by the treatments 50°C for 5 min and 52°C for 2 min.

Exposing dates to temperatures of 65–80°C for 30 min to 4 h at high humidity controls insects (Yahia, 2004); however, this approach is not always very efficient for controlling insects in dates with high moisture content because high temperatures for prolonged periods may cause darkening, the appearance of a dull color, and loss of flavor. Rafaeli *et al.* (2006) described an effective, short-duration, and inexpensive method using a postharvest heating container; they found that the optimum temperature regime for maximum escape of beetles from the fruit was 55°C for 2.5 h attained at a rate of 1.8°C/min.

13.8.3.5 Modified atmosphere packaging (MAP) and controlled atmosphere packaging (CAP)

Some fresh vegetable products have been the subject of numerous works for their preservation in so-called "modified" atmospheres (also called MAP = modified atmosphere packaging) or "controlled" atmospheres (also known as CAP = controlled atmosphere packaging) (Kader, 1992; Varoquaux and Nguyen-The, 1993). This work has led to the proposal of theoretical models for predicting fresh product respiratory rates as a function of CO_2 and oxygen (O_2) concentrations (Lee *et al.*, 1991; Fonseca *et al.*, 2002; Murray *et al.*, 2007).

The rare studies on packaging and dates (Awad, 2007) have focused on increasing the maturation of dates in bunches and not as postharvest conservation atmospheres.

The use of modified or controlled atmospheres for products of plant origin is often associated with low-temperature storage. Most of these studies, combining various percentages of CO_2 and O_2, showed positive effects (Choehom *et al.*, 2004; Hertog *et al.*, 2004; Nguyen *et al.*, 2004; Gómez and Artés, 2005; Escalona *et al.*, 2006; Serrano *et al.*, 2006; Shen *et al.*, 2006). The combination of heat treatment in modified or controlled atmospheres is beginning to generate increasing interest and recent studies are reported (Murray *et al.*, 2007).

High CO_2 levels in MAP inhibit the growth of many spoilage microorganisms of fresh food (Sandhya, 2010; Homayouni *et al.*, 2015; Jemni *et al.*, 2016a). The combined use of CO_2 and fumigation provides better control of the pests *O. surinamensis* and *Tribolium confusum* in stored dates without causing undesirable quality changes (El-Mohandes, 2010). Presence of O_2 promotes several deteriorative reactions in foods including fat and pigment oxidation and browning reactions (Sandhya, 2010). Packing infested dates in polyethylene bags with 80–90% vacuum resulted in 100% mortality of date pests after 2 days (Hussain, 1974).

A 4 h exposure at 2.8% v/v O_2 at 26°C caused over 80% of the initial nitidulid beetle populations to emigrate from the infested dried dates (Navarro *et al.*, 1998a, Navarro *et al.*, 1998b). Al-Azab (2007) used a mixture of modified atmosphere (65% CO_2, 15% nitrogen (N_2), 20% O_2; v/v) and found that exposure for 24 h at 34°C and 65% RH caused 100% mortality of the adults of *E. cautella*. El-Mohandes (2009) found that 100% mortality of the adults of *O. surinamensis* and *T. confusum* was achieved after 36 h exposure to a CO_2 concentration of 75% v/v at 25°C and 55% RH.

13.8.3.6   Ozonation

Niakousari *et al.* (2010) exposed dates contaminated with all life stages (adults, larvae, eggs) of Indian meal moth (*P. interpunctella*) and sawtooth grain beetle (*O. surinamensis*) to gaseous ozone (O_3) (600, 1200, 2000, 4000 ppm) for 1–2 h. The insecticidal effect of O_3 is due to its ability to diffuse through biological cell membranes and its high oxidation potential (Jemni *et al.*, 2015).

13.8.3.7   Biological control

Due to the desire to present a healthy product to the demanding consumer, especially in the external market, some countries have found other alternatives or additions to chemical treatments. Biological control is the control of infestation by natural enemies (predators and parasites). Studies report the results of biological control of *Ectomyelois ceratoniae* (Doumandji-Mitiche, 1989; Dhouibi and Jarraya, 1996). Of all the parasites, only *Phanerotoma* gave encouraging results, but this was difficult to generalize. Parasitism due to *Phanerotoma planifrons* and *Phanerotoma flavitestacea* was more important (10–35% parasitism for export dates and 25% for discarded dates at the time of sorting) (Biliotti and Daumal, 1969). Some biological methods for control of the insect pests of stored dates, such as the sterile insect technique, cytoplasmic incompatibility, and the use of parasites, have been tried (Ahmed, 1981; Ahmed *et al.*, 1982, 1994), but none commercially.

13.9   Storage Conditions and Their Influence

Storage of fruit at temperatures below freezing results in the destruction of fruit tissue (Barrett *et al.*, 2005). Indeed, it was reported by Mikki and Al-Taisan (1993) and Hui *et al.* (2006) that freezing dates at the rutab stage for 6 months at −18 ± 2°C led to an increase in their water content, reducing sugars, and pH as well as a decrease in tannin content. As a result, the development of a sweet taste and a decrease in astringency were remarked and the dates became brown (Goneum *et al.*, 1993). Refrigeration can be used as an alternative to temporarily limit enzymatic browning (Barrett *et al.*, 2005).

Cold allows long-term storage of dates (Estanove, 1990). It limits weight loss, defers browning, and stabilizes acidity and organoleptic characteristics. Dates are very resistant to low temperature, which thus can be used to significantly reduce insect infestation (Yahia, 2004; Yahia and Kader, 2011). Temperatures below 13°C will prevent feeding damage and reproduction, and temperatures of 5°C or lower are effective in controlling different developmental stages of insects (Barreveld, 1993). Fig moth larvae could live for 85 days at 2–6°C, but storage at 0°C can result in total mortality of larvae of the fig moth and adults of the grain beetle after 15 and 27 days, respectively (Hussain, 1974). Thus, packed fumigated dates could be kept free of infestation at 4°C for as long as 1 year (Hussain, 1974). Although the literature is very prolific on chilling injury, particularly for tropical and subtropical fruits (Lafuente *et al.*, 2003; Trakulnaleumsai *et al.*, 2006), such reports relating to dates remain few.

13.10   Grading Standards

Yahia (2004) reported that the skin of good-quality dates should be smooth, golden brown, orange, green, or dark brown depending on the variety; and the texture soft and syrupy, or firm and dry, depending on the cultivar.

Dowson (1982), on his part, gave further criteria for high-quality fresh dates, citing the need for adequate color and size, with thick flesh; free of dirt, sand, and leaf particles as well as of damage caused by birds, insects, rodents, fungi, and molds; in addition to the formation of sugar crystals.

The quality levels of dates are based on the uniformity of color/size and the absence of defects or damage caused by handling after the harvest and storage of the fruits, such as discoloration of the flesh, rupture of the skin, fruit deformation, scarring, sunburn, rot, browning, fermentation, improper maturation, mechanical damage, dirt, or any other foreign matter.

13.10.1   Classification according to date quality criteria

13.10.1.1   Maturity stage

Some dates from marginal growing areas such as the Libyan coast are called "fresh dates" (at khalal stage). They have an optimal size, weight, and sugar content. They are slightly astringent, hard, fibrous, and bright yellow. Other so-called "dry" dates (at tamar stage) have a minimum size and weight, as is the case with the majority of current varieties.

13.10.1.2 Sugar composition

This is the main compositional criterion. A distinction is made between sucrose dates (dominant sucrose) and reductive sugar dates (dominant glucose and fructose).

The first category includes, for example, the Deglet Nour and Mech Degla varieties, and the second category includes the Ghars varieties. Knowledge of the sugar/water ratio makes it possible to define the stability of the fruit during conservation and the quality of the cultivar (Reynes, 1997).

The relative contents of sucrose and reducing sugars, in part, characterize a variety and influence the texture of the date (Coggins and Knapp, 1969).

Cook and Furr (1953) and Munier (1973) proposed a classification based on r, the sugar/water ratio, distinguishing three categories of dates: soft dates, $r < 3.5$; semisoft dates, $2.5 < r < 3.5$; and dry dates, $r > 3.5$ (Cook and Furr, 1953).

Hussein et al. (1976) and Barreveld (1993) classified the Saudi dates into three categories:

- Soft dates, not containing sucrose and with a water content of about 30% w/w.
- Semidry dates, with a water content of 20 to 30% w/w and high sucrose content, such as Deglet Nour.
- Dates known as dry, containing less than 20% w/w water and that generally have an equivalent amount of reducing sugars and sucrose. This classification is particularly prevalent in Algeria.

Munier (1973) considers the ratio R, stone weight/date weight, to be another characteristic that is a function of date variety, ecological factors, and growing conditions. This ratio is 8–12% and 11–12% for the Algerian varieties Deglet Nour and Ghars, respectively.

13.10.2 Commercial quality control of whole dates

The UNECE Standard DDP-08, 2010 (UNECE, 2010) applies to naturally occurring or processed, whole uncut dates of varieties derived from *Phoenix dactylifera* L. intended to be delivered to consumers (Table 13.1). It does not apply to dates for processing industries, pressed dates, or frozen dates.

The purpose of the standard is to define the qualities to be presented by dates at the stage of dispatch after packaging. In all categories, considering the specific provisions laid down for each category and the tolerances allowed, dates must be:

- Whole and healthy.
- Not including products affected by decay or alteration such as to render them unfit for consumption.

Table 13.1. Quality standards for dates. (From UNECE, 2010; FAO/WHO, 2019.)

Characteristic	UNECE Standard DDP-08, 2010			FAO/WHO Codex Alimentarius Standard for Dates (CXS 143-1985)
Moisture content (w/w)				
Cane sugar varieties[a]	<26%			Max. 26%
Invert sugar varieties[a]	<30%			Max. 26%
Deglet Nour variety[a], natural	<30%			Max. 26%
Permitted food additives				Glycerol, sorbitol
Class	Extra – I – II			
Minimum calibration and tolerance	4.75 g			4.75 g
	Max. 10% with lower weight			Max. 5% with lower weight
Quality tolerance (%)	Extra	I	II	
Overall tolerance	5	10	20	1
Sour/rotten/musty fruit	0	0	1	6
Contaminated dead insects	3[b]	5[b]	8[b]	0
Contaminated live insects	0	0	0	6
Damaged/immature	2	4	6	7
Stained	3	5	7	
Mineral impurities (g/kg)				
Processed dates	1	1	1	1
Dates in the natural state	2	2	2	1
Branches				
Minimum length	10 cm			

Continued

Table 13.1. Continued

Characteristic	UNECE Standard DDP-08, 2010	FAO/WHO Codex Alimentarius Standard for Dates (CXS 143-1985)
Fruits/10 cm long	4	
% Authorized for detached fruit	10% by weight	
Class concerned	Extra and Class I	
Marking		
Identification	Packer and/or shipper	Packer and/or shipper and/or importer, etc.
Nature of the product	Date/variety name/in diets or branches (in dry matter: DM)	Date or date coated with glucose syrup
Origin of the product	Country name	Country of origin
Business characterization	Class/net unit weight	Net weight/consumption cutoff date
Disinsectization	—	—

[a]The word "variety" is used generically to indicate dates rich in sucrose (cane sugar, which dates are called dry), dates with invert sugar, rich in glucose and fructose (which dates are called soft); the variety Deglet Nour containing invert sugars and sucrose is called semisoft.

[b]Germany/Poland/Switzerland/UK wish to maintain the tolerances: Extra 2%, Class I 4%, and Class II 6%.

- Ripe, fleshy, and supple.
- Clean.
- Free from living insects or any living parasite or even their visible traces.
- Free from mold, foreign odor, or flavor, or from fermentation.
- Of specified moisture content. For dates of the Deglet Nour varieties, in the natural state, the maximum moisture content is set at 30% w/w.

Dates are classified into three categories as follows:

- *Extra class.* Dates in this category must be of superior quality. They must have the typical shape, development, and coloring of the variety, they must have a color ranging from amber to brown and an abundant, oily, or semi-oily and unctuous flesh, the epicarp must be translucent and, depending on the variety, adherent to the flesh. They must be free from defects except for very slight superficial alterations provided they do not affect the general appearance of the product, its quality, its preservation, or its presentation in the packaging.
- *Class I.* Dates in this category must be of good quality. They must have characteristics typical of the variety. The flesh must be sufficiently abundant, fat or half fat, considering the variety. They may contain the following slight defects, if they do not affect the general appearance of the product, its quality, its preservation, or its presentation in the packaging: a slight defect of the epicarp not affecting the pulp, a slight defect in shape or development, slight wrinkles.
- *Class II.* This category includes dates that cannot be classified in the higher categories but correspond to the minimum characteristics defined above. They may contain the following defects provided they retain their essential characteristics of general appearance, quality, preservation, and presentation: defects of the epicarp not affecting the pulp, defects of shape or development, defects of coloring. The size is determined by the minimum weight of the fruits; whatever the variety, the weight of the whole dates is fixed at 4.75 g. Quality and size tolerances are allowed in each package for products that do not meet the specified quality requirements; minimum impurity quality tolerance and size tolerances (not more than 1 g/kg). However, for natural whole dates, not more than 2 g/kg. For all categories 10% of dates may have a unit weight less than 4.75 g, but none less than 4 g.

13.11 Innovative Technologies

Given the disadvantages associated with traditional harvesting methods, several scientists have proposed effective methods and techniques to limit losses and damage and ensure the harvesting of fruits and vegetables in the best possible conditions.

Numerous studies are being conducted for the purpose to perform intelligent and precise operations with quick sorting of date fruits to improve grading efficiency (Siddiq and Greiby, 2014).

This then highlights the importance of innovation in agriculture, including technological advancements like sensors, robots, Global Positioning Satellite (GPS) technology, and the Internet of Things (IoT) (Rose and Chilvers, 2018). Visible–near infrared (Vis-NIR) spectroscopy is known to have a long history of use for analyzing fruit color and the quality of fresh products (De Jager and Roelofs, 1996). Several companies produce packing lines with multiple cameras and sensors, integrated with Vis-NIR spectrometric probes to predict different quality parameters of the fruits (Walsh *et al.*, 2020).

Kondo (2010) reported that automation is highly applicable in orange grading, eggplant grading, and leek preprocessing. Alavi (2012) reported that the Mamdani fuzzy inference system can be used to provide decision making for the classification of cv. Mozafati date fruits.

A machine vision device was designed to evaluate date quality by reflective NIR imaging through analyzing two-dimensional images, which improved grading accuracy (Lee *et al.*, 2008).

Al Ohali (2011) designed and tested a prototype computer vision-based date grading/sorting system using a defined set of fruit external quality features. The computer-mediated fruit quality assessment and sorting system has two subsystems: a computer vision system and a fruit handling system. The computer vision system captures the image of the underlying fruit and transmits it to an image processor. The recognizer performs the quality assessments and classifies the underlying fruit into prespecified quality classes and directs the sorter to direct the fruit to the appropriate bin. The system classifies dates into three quality categories (grades 1, 2, 3) defined by experts. The system can sort dates with 80% accuracy (Al Ohali, 2011).

Mechanization of date harvest aims at addressing the cost and scarcity of specialized labor, the hazard and the burden of the palm climbing operation, and in practice is the only option where date palms are culti-vated in large, specialized plantations (Nourani *et al.*, 2021). Date harvest mechanization can only be partially achieved and is based on facilitating machines, ladders, lifters, or platforms that allow the harvesters to reach up to the fruit bunches level (Garbati Pegna *et al.*, 2012; Nourani *et al.*, 2017; Bonechi *et al.*, 2018).

Experiments were carried out by Nourani *et al.* (2021) from November 11 to 16, 2018 with two different electromechanical shakers on date palms of the Mech Degla and Deglet Nour varieties grown at the Bio-resources Station El Outaya (34°55′44.9″N, 5°39′00.1″E), located 12 km north of CRSTRA (Scientific and Technical Research Center on Arid Regions) in Biskra (Algeria). The aim of the experiment, focused on the harvesting capacity in terms of mass per unit time of two different vibrating harvesting

Fig. 13.16. Holly (left) and Alice (right) heads. These two models of olive electromechanical harvesting head, both equipped with oscillating combs and produced by Campagnola Srl, Zola Predosa, Italy, were tested by Nourani *et al.* (2021). Both heads, which differ for the beating system, are provided with their own electric motor, and can be carried at the end of a telescopic aluminum pole, extensible up to 2.2 m. (From Nourani *et al.*, 2021.)

heads on two date varieties, was to verify the actual possibility of using this method for harvesting dates and hence to provide basic information for understanding its dynamics (Nourani *et al.*, 2021) (Fig. 13.16).

These tests showed the possibility of vibration harvesting as a viable technique for harvesting dates that are not susceptible to damage when dropping on the ground or that are destined for prompt processing (Fig. 13.17), hence making further investigation and development of this technique worthwhile (Nourani *et al.*, 2021).

Altaheri *et al.* (2019) have presented a comprehensive dataset for date fruits that can be used by the research community for multiple tasks including automated harvesting, visual yield estimation, and classification tasks. The dataset contains images of date fruit bunches of different date varieties, captured at different pre-maturity and maturity stages (Fig. 13.18). These images cover multiple sets of variations such as multiscale images, variable illumination, and different bagging states. Research on automated date fruit harvesting is limited as there is no public dataset for date fruits to aid in this.

In Altaheri *et al.*'s (2019) study, 152 Barhi date bunches belonging to 13 palms were weighed after harvesting, and their images were captured in front of white graph paper (Fig. 13.19). They marked the whole bunches in the palm, captured their images from different angles before and during harvesting, recorded a 360° video for each palm, and registered

Fig. 13.17. Preliminary tests with Holly head. Vibrating the lower section of the bunch first, with a downward movement, and proceeding with the section above and so on, in sequence, gives the best results, since most fruits drop vertically on the ground; while, if starting from the top, a large amount of fruit is projected far away, especially from the lower sections, due to the oscillation of the bunch. (From Nourani *et al.*, 2021.)

their characteristics (height, trunk circumference, total yield, number of bunches).

The dataset also has variations in bunch compactness and the angle of view; including these variables is important for building an effective and reliable machine vision system for weight and yield estimation (Altaheri *et al.*, 2019). This dataset can help in advancing research and automating date palm agricultural applications, including robotic harvesting, fruit detection and classification, maturity analysis, and weight/yield estimation.

The machine developed by Faisal *et al.* (2020) allows harvesting whole date clusters from the ground, without need for the operator to climb up the tree, hence avoiding fatigue and risks consequent to this operation,

(a)   **(b)**

Fig. 13.18. Camera position setup for captured videos; the orientation of the camera depends on the palm height. (a) Camera orientation of 45° for a 3 m high palm tree. (b) Real example for palm-3 (B3.K.BW); the palm height is 2.85 m, so the camera orientation was less than 45°. (From Altaheri *et al.*, 2019.)

(a)   (b)

Fig. 13.19. (a) A sample bunch of Barhi dates captured from the front. (b) The dimensions of the graph paper, showing the distances between the centers of the reference shapes. (From Altaheri *et al.*, 2019.)

and can represent an interesting device for farmers who face a lack of skilled labor.

In same study, Faisal *et al.* (2020) proposed an intelligent harvesting decision system (IHDS) based on date fruit maturity level (Fig. 13.20). The decision system used computer vision and deep learning (DL) techniques to detect seven different maturity stages/levels of date fruit (immature stage 1, immature stage 2, pre-khalal, khalal, khalal with rutab, pre-tamar, and tamar).

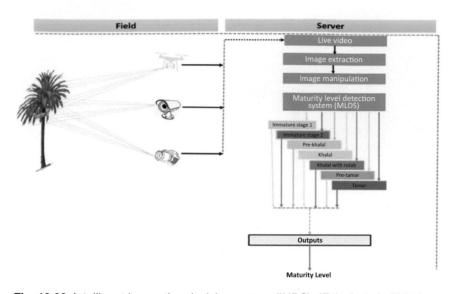

Fig. 13.20. Intelligent harvesting decision system (IHDS). (Faisal *et al.*, 2020.)

In the IHDS, Faisal *et al.* (2020) developed six different DL systems. Each one produced different accuracy levels in terms of the seven maturity stages; the IHDS used datasets that had been collected by the Center of Smart Robotics Research. The maximum performance metrics of the proposed IHDS were 99.4, 99.7, and 99.7% for accuracy, sensitivity (recall), and precision, respectively.

Moreover, an efficient machine vision framework for date fruit harvesting robots was proposed by Altaheri *et al.* (2019); the framework consists of three classification models used to classify date fruit images in real time according to their type, maturity, and harvesting decision. In the classification models, deep convolutional neural networks are utilized with transfer learning and fine-tuning on pretrained models; the dataset has a large degree of variations that reflects the challenges in the date orchard environment including variations in angles, scales, illumination conditions, and date bunches covered by bags (Altaheri *et al.*, 2019). The results of classification models achieve accuracies of 99.01, 97.25, and 98.59% with classification times of 20.6, 20.7, and 35.9 ms for the type, maturity, and harvesting decision classification tasks, respectively.

In Algeria, date harvesting practice is still conducted manually by climbing up the tree, but this operation is expensive and there is a lack of skilled workers, especially because of the increasing number of palms due to the new plantings in recent years. Algerian farmers cannot afford expensive machinery such as motorized elevators, so the problem has been addressed by developing a manual aid that allows this operation to be conducted easily and safely (Nourani *et al.*, 2017).

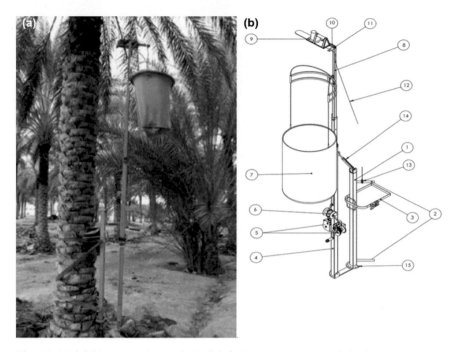

Fig. 13.21. (a) Harvester in the field. (b) Schematic diagram of the harvester. 1, Stabilizing platform; 2, feet; 3, lashing belt; 4, telescopic mast; 5, winches; 6, reel; 7, basket; 8, basket support; 9, electric saw; 10, arm to support the saw; 11, draw wire; 12, return spring; 13, switch; 14, adjuster; 15, mast fixing mechanism. (From Nourani *et al.*, 2017.)

Nourani *et al.* (2017) developed a harvesting machine with four main components (stabilizing platform, lifting device, lowering device, and cutting device) that can be carried manually and used to cut and lower whole clusters without having to climb the palm (Fig. 13.21), thus avoiding fatigue and the risks associated with tree climbing.

Trials conducted with this device showed that a cluster at a height of 6.5 m can be picked and lowered in about 3 min after the gear has been set. The time needed for harvesting and its cost make this harvester an affordable alternative to manual harvesting, and thus an interesting device for farmers who lack skilled laborers. The machine represents a link between manual and mechanical motorized harvesting.

13.12 Conclusions and Prospects

The date palm cultivation sector faces many difficulties such as the lack of production know-how and postharvest technologies, the lack of economies of scale, and the lack of links with research and development. According to experts, nearly 10% of the fruit rots on trees due to lack of pickers. In addition, local price volatility, weak market information system, poor infrastructure, and a high percentage of postharvest losses pose significant challenges for the sector.

The harvest of dates is based on manual labor, representing a real financial burden for farmers. To remedy this, it is very important to innovate and develop automated and intelligent harvesting tools. This will allow farmers to save money, improve crop quality under the best possible conditions, and save labor. Another point is that the manual harvest of dates is heavy, physical, and difficult work.

The transition to automated selective harvesting promises to create more skilled jobs focusing on machine operation and sorting. Selective harvesting refers to picking only those portions of the crop that have reached maturity, without damaging the remainder of the crop for later harvest or to be left in the ground. For the moment, this technique is exclusively done by hand. For years, the modernization of agriculture has made it possible to double production levels by reducing hardship. This was accompanied using large machines and potentially harmful products. The increasing environmental constraint has also encouraged the development of new production methods, such as organic farming or precision farming. These require an increased workforce. In this context, robotics can provide solutions, allowing repetitive tasks to be carried out accurately and to intervene in complex operations such as harvesting dates without exposing individuals to harm. Intelligent, automated harvesting reduces costs and improves crops.

References

Abdul Aly, A., Al Abid, A., Alshwakir, A., Hassan, R., Al Fahaid, Y. *et al.* (2018) Study of the effect of storage temperature on microbial stored dates under vacuum. In: *Sixth International Date Palm Conference, Abu Dhabi, United Arab Emirates,* March 19–21, 2018. Available at: https://repo.mel.cgiar.org/handle/20.500. 11766/9751 (accessed 2 December 2022).

Aegerter, A.F. and Folwell, R.J. (2000) Economic aspects of alternatives to methyl bromide in the post-harvest and quarantine treatment of selected fresh fruits. *Crop Protection* 19, 161–168.

Afshari-Jouybari, H., Farahnaky, A. and Moosavi-Nasab, S. (2013) The use of acetic acid, sodium chloride solutions, and incubation to accelerate the ripening of 'Mazafati' date. *International Journal of Fruit Science* 14, 95–106.

Ahmed, I.A., Ahmed, A.W.K. and Robinson, R.K. (1997) Susceptibility of date fruits (*Phoenix dactylifera*) to aflatoxin production. *Journal of the Science of Food and Agriculture* 74, 64–68.

Ahmed, M.S.H. (1981) Investigations on insect disinfestation on dried dates using gamma radiation. *Date Palm Journal* 1, 107–116.

Ahmed, M.S.H., Al-Hakkak, Z.S., Ali, S.R., Kadhum, A.A., Hassan, I.A. *et al.* (1982) Disinfestation of commercially packed dates, zahdi variety, by ionizing radiation. *Date Palm Journal* 1, 249–273.

Ahmed, M.S.H., Al-Taweel, A.A. and Hameed, A.A. (1994) Reduction of *Ephestia cautella* (Lepidoptera: Pyralidae) infestation rate of stored dates by releasing cytoplasmically incompatible males. *International Journal of Tropical Insect Science* 15(1), 25–29. DOI: 10.1017/S1742758400016726.

Aidoo, K.E., Tester, R.F., Morrison, J.E. and MacFarlane, D. (1996) The composition and microbial quality of pre-packed dates purchased in Greater Glasgow. *International Journal of Food Science & Technology* 31(5), 433–438. DOI: 10.1046/j.1365-2621.1996.00360.x.

Ait-Oubahou, A. and Yahia, E.M. (1999) Post-harvest handling of dates. *Postharvest News and Information* 10(6), 67–74.

Alavi, N. (2012) Date grading using rule-based fuzzy inference system. *Journal of Agricultural Technology* 8, 1243–1254.

Al-Azab, A.M. (2007) Alternative approaches to methyl bromide for controlling *Ephestia cautella* (Walker) (Lepidoptera: Pyralidae). MSc. thesis, King Faisal University, Alahsa, Saudi Arabia.

Al-Azawi, F. (1985) The effect of high temperatures on the dried beetle *Carpophilus hemipterus* L. A pest of stored dates in Iraq. *Date Palm Journal* 3(1), 327–336.

Albagnac, G., Varoquaux, P. and Montigaud, J.C. (2002) *Technologie de transformation des fruits.* Tec&Doc/Lavoisier, Paris.

Al-Bekr, A.J. (1972) *The Date Palm.* Al-Ani Press, Baghdad.

Aleid, S.M., Dolan, K.D., Siddiq, M., Jeong, S. and Marks, B. (2013) Effect of low-energy X-ray irradiation on physical, chemical, textural and sensory properties of dates. *International Journal of Food Science & Technology* 48(7), 1453–1459. DOI: 10.1111/ijfs.12112.

Al-Farsi, M., Alasalvar, C., Al-Abid, M., Al-Shoaily, K., Al-Amry, M., and Al-Rawahy, F. (2007) Compositional and functional characteristics of dates, syrups, and their byproducts. *Food Chemistry* 104, 943–947.

Ali-Dinar, H., Mohammed, M. and Munir, M. (2021) Effects of pollination interventions, plant age and source on hormonal patterns and fruit set of date palm (*Phoenix dactylifera* L.). *Horticulturae* 7, 427. DOI: 10.3390/horticulturae7110427.

Al Juhaimi, F., Ozcan, M.M., Adiamo, O.Q., Alsawmahi, O.N., Ghafoor, K. *et al.* (2018) Effect of date varieties on physico-chemical properties, fatty acid composition, tocopherol contents, and phenolic compounds of some date seed and oils. *Journal of Food Processing and Preservation* 42(4), e13584. DOI: 10.1111/jfpp.13584.

Al-Khahtani, H.A., Abu-Tarboush, H.M., Aldryhim, Y.N., Ahmad, M.A., Bajaber, A.S. *et al.* (1998) Irradiation of dates: insect disinfestations, microbial and chemical assessments, and use of thermo-luminescence technique. In: Afifi, M.A.R. and Al-Badawy, A.A. (eds) *Proceedings the First International Conference on Date Palms, Al-Ain, United Arab, 8–10 March 1998.* UAE University, Al-Ain, United Arab Emirates, pp. 126–148.

Al-Mulhim, F.N. and Osman, G.E. (1986) Household date consumption patterns in Al-Hassa: a cross-sectional analysis. In: *Proceedings of the Second Symposium on the Date Palm in Saudi Arabia, Al-Hassa, Saudi Arabia,* King Faisal University, Al-Hassa, Saudi Arabia, March 3–6, 1986, pp. 513–522 (Vol. I).

Al-Mutarrafi, M., Elsharawy, N.T., Al-Ayafi, A., Almatrafi, A. and Abdelkader, H. (2019) Molecular identification of some fungi associated with soft dates (*Phoenix dactylifera* L.) in Saudi Arabia. *Advancement in Medicinal Plant Research* 7(4), 97–106. DOI: 10.30918/AMPR.74.19.038.

Al-Ogaidi, H.K.H. and Aref, A.A. (1985) *Industrialisation Des Dattes et Produits Cellulosiques Du Palmier Dattier.* Ed. U.A.L.A, Baghdad (in Arabic).

Al Ohali, Y. (2011) Computer vision based date fruit grading system: design and implementation. *Journal of King Saud University - Computer and Information Sciences* 23(1), 29–36. DOI: 10.1016/j.jksuci.2010.03.003.

Al-Shahib, W. and Marshall, R.J. (2003) The fruit of the date palm: its possible use as the best food for the future? *International Journal of Food Sciences and Nutrition* 54(4), 247–259. DOI: 10.1080/09637480120091982.

Al-Sheikh, H. (2009) Date-palm fruit spoilage and seed-borne fungi of Saudi Arabia. *Research Journal of Microbiology* 4(5), 208–213. DOI: 10.3923/jm.2009.208.213.

Al-Suhaibani, S.A., Babier, A.S., Kilgour, J. and Blackmore, B.S. (1991) Field tests of the KSU date palm machine. *Journal of Agricultural Engineering Research* 51, 179–190. DOI: 10.1016/0021-8634(92)80036-R.

Altaheri, H., Alsulaiman, M., Muhammad, G., Amin, S.U., Bencherif, M. *et al.* (2019) Date fruit dataset for intelligent harvesting. *Data in Brief* 26, 104514. DOI: 10.1016/j.dib.2019.104514.

Al-Taweel, A.A., Hameed, A.A., Ahmed, M.S.H. and Ali, M.A. (1990) Disinfestation of packed dates by gamma-radiation using a suitable food irradiation facility. *International Journal of Radiation Applications and Instrumentation. Part C. Radiation Physics and Chemistry* 36(6), 825–828. DOI: 10.1016/1359-0197(90)90186-L.

Al-Taweel, A.A., Ahmed, M.S.H., Shawkit, M.A. and Nasser, M.J. (1993) Effects of sublethal doses of gamma radiation on the mating ability and spermatophore transfer of *Ephestia cautella* (Lepidoptera: Pyralidae). *International Journal of Tropical Insect Science* 14(1), 7–10. DOI: 10.1017/S1742758400013321.

Ashraf, Z. and Hamidi-Esfahani, Z. (2011) Date and date processing: a review. *Food Reviews International* 27(2), 101–133. DOI: 10.1080/87559129.2010.535231.

Atia, M.M.M. (2011) Efficiency of physical treatments and essential oils in controlling fungi associated with some stored date palm fruits. *Australian Journal of Basic and Applied Sciences* 5, 1572–1580.

Awad, M.A. (2007) Increasing the rate of ripening of date palm fruit (*Phoenix dactylifera* L.) cv. Helali by preharvest and postharvest treatments. *Postharvest Biology and Technology* 43(1), 121–127. DOI: 10.1016/j.postharvbio.2006.08.006.

Azelmat, K., Sayah, F., Mouhib, M., Ghailani, N. and ElGarrouj, D. (2005) Effects of gamma irradiation on fourth-instar *Plodia interpunctella* (Hubner) (Lepidoptera: Pyralidae). *Journal of Stored Products Research* 41(4), 423–431. DOI: 10.1016/j.jspr.2004.05.003.

Baliga, M.S., Baliga, B.R.V., Kandathil, S.M., Bhat, H.P. and Vayalil, P.K. (2011) A review of the chemistry and pharmacology of the date fruits (*Phoenix dactylifera* L.). *Food Research International* 44(7), 1812–1822. DOI: 10.1016/j.foodres.2010.07.004.

Barrett, D.M., Somogyi, L.P. and Ramaswamy, H.S. (2005) *Processing Fruits: Science and Technology*, 2nd edn. CRC Press, Boca Raton, Florida.

Barreveld, W.H. (1993) *Date Palm Products.* FAO Agricultural Services Bulletin no.101. Food and Agriculture Organization of the United Nations, Rome. Available at: www.fao.org/docrep/t0681E/t0681e00.htm (accessed 2 December 2022).

Belarbi, A., Aymard, C.H. and Hebert, J.P. (2001a) Evolution of Deglet-Noor date quality on it heat treatments (color and texture). In: Al-Badawy, A.A. (ed.), *Proceedings of the Second International Conference on Date Palms, Al-Ain, United Arab Emirates*, UAE University, Al-Ain, United Arab Emirates, March 25–27, 2001.

Belarbi, A., Aymard, C.H. and Hebert, J.P. (2001b) Points of caution in studying heat inactivation of enzymes, exemplified by the polyphenoloxidase from the Deglet-Nour date (*Phoenix dactylifera* L.). In: Al-Badawy, A.A. (ed.), *Proceedings of the Second International Conference on Date Palms, Al-Ain, United Arab Emirates*, UAE University, Al-Ain, United Arab Emirates, March 25–27, 2001.

Belarbi, A., Aymard, C. and Hebert, J. (2003) Deglet-Nour date polyphenoloxidase activity is very thermolabile and exhibits peculiarities in its thermal inactivation kinetics. *Food Biotechnology* 17, 193–202.

Ben-Lalli, A. (2010) Étude et modélisation de la désinfestation de la datte: analyse du couplage entre transferts thermiques et mortalité des insectes. Thèse de doctorat, Université des sciences et techniques de Montpellier 2, Montpellier, France.

Besbes, S., Drira, L., Blecker, C., Deroanne, C. and Attia, H. (2009) Adding value to hard date (*Phoenix dactylifera* L.): compositional, functional and sensory characteristics of date jam. *Food Chemistry* 112, 406–411.

Biliotti, E. and Daumal, J. (1969) Biologie de *Phanerotoma flavitestacea* Fisher (Hyménoptéra, Braconidae) mise au point d'un élevage permanent en vue de la lutte biologique contre *Ectomyelois ceratoniae* Zeller. *Annales de Zoologie Ecologie Animale* 1, 379–394.

Bonechi, F., Garbati Pegna, F. and Bonaiuti, E. (2018) Performance evaluation of an offroad light aerial platform for date palm cultivation. In: *Sixth International Date Palm Conference, Abu Dhabi, United Arab Emirates*, March 19–21, 2018. Available at: https://repo.mel.cgiar.org/handle/20.500.11766/8121 (accessed 2 December 2022).

Bouhlali, E.D.T., Ramchoun, M., Alem, C., Ghafoor, K., Ennassir, J. *et al.* (2017) Functional composition and antioxidant activities of eight Moroccan date fruit varieties (*Phoenix dactylifera* L.). *Journal of the Saudi Society of Agricultural Sciences* 16(3), 257–264.

Bouka, H., Chemseddine, M., Abbassi, M. and Brun, J. (2001) La pyrale des dattes dans la région de Tafilalet au Sud-Est du Maroc. *Fruits* 56, 189–196.

Brown, E.K. (1982) Date production mechanization in the USA. In: Makki, Y.M. (ed.) *Proceedings of the First Symposium on the Date Palm in Saudi Arabia, Al-Hassa, Saudi Arabia*, King Faisal University, Al-Hassa, Saudi Arabia, March 23–25, pp. 14–24.

Bulut, L. and Kilic, M. (2009) Kinetics of hydroxymethylfurfural accumulation and color change in honey during storage in relation to moisture content. *Journal of Food Processing and Preservation* 33, 22–32.

Carpenter, J.B. and Elmer, H.S. (1978) *Pests and Diseases of the Date Palm*. Agricultural Handbook no.527. US Department of Agriculture, Washington, DC.

Chavéron, H. (1999) *Introduction à la toxicologie nutritionnelle*. Tec&Doc/Lavoisier, Paris, pp. 149–175.

Chazarra, S., García-Carmona, F. and Cabanes, J. (2001) Evidence for a tetrameric form of iceberg lettuce (*Lactuca sativa* L.) polyphenol oxidase: purification and characterization. *Journal of Agricultural and Food Chemistry* 49(10), 4870–4875. DOI: 10.1021/jf0100301.

Chebbi, H.E. and Gil, J.M. (2002) Position compétitive des exportations tunisiennes de dattes sur le marché européen: une analyse shift-share. *New Medit* 1, 40–47.

Cheftel, J.C., Cheftel, H. and Besançon, P. (1979) *Introduction à La Biochimie et à La Technologie Des Aliments, Volume 1 et 2.* Tec&Doc/Lavoisier, Paris.

Chisari, M., Barbagallo, R.N. and Spagna, G. (2007) Characterization and role of polyphenol oxidase and peroxidase in browning of fresh-cut melon. *Journal of Agricultural and Food Chemistry* 56(1), 132–138. DOI: 10.1021/jf0721491.

Choehom, R., Ketsa, S. and van Doorn, W.G. (2004) Senescent spotting of banana peel is inhibited by modified atmosphere packaging. *Postharvest Biology and Technology* 31(2), 167–175. DOI: 10.1016/j.postharvbio.2003.07.001.

Chonhenchob, V., Chinsirikul, W. and Singh, S.P. (2012) Current and innovative packaging technologies for tropical and subtropical fruits. In: Siddiq, M. (ed.) *Tropical and Subtropical Fruits: Postharvest Physiology, Processing and Packaging.* Wiley, Ames, Iowa, pp. 115–134.

Coggins, C.W.J. and Knapp, J.F.C. (1969) Growth development and softening of the Deglet Nour date fruit. *Date Growers' Institute Annual Report* 46, 11–14.

Colman, S., Spencer, T., Ghamba, P. and Colman, E. (2012) Isolation and identification of fungal species from dried date palm (*Phoenix dactylifera*) fruits sold in Maiduguri metropolis. *African Journal of Biotechnology* 11, 12063–12066.

Cook, J.A. and Furr, J. (1953) Kinds and relative amounts of sugars and their relation to texture of some American grown date varieties. *Proceedings of the American Society for Horticultural Science* 61, 722–727.

Çopur, Ö.U. and Tamer, C.E. (2014) Fruit processing. In: Malik, A., Erginkaya, Z., Ahmad, S. and Erten, H. (eds) *Food Processing: Strategies for Quality Assessment.* Food Engineering Series, Springer, New York, pp. 9–35.

De Jager, A. and Roelofs, F. (1996) Prediction of optimum harvest date of Jonagold. In: *Determination and Prediction of Optimum Harvest Date of Apples and Pears. Action 94.* COST, Brussels, pp. 21–31.

Dhouibi, M.H. and Jarraya, A. (1996) *Le Ver Des Dattes, Carob Moth:* Ectomyelois Ceratoniae. Document Inrat et GID, Tunis.

Doumandji-Mitiche, B. (1989) Les parasites de la datte dans les oasis Algériennes et particulièrement les cas d'*Ectomyeloïs ceratoniae* Zeller. In: *Actes du Séminaire Maghrébin sur la Phoéniciculture*, El-Oued, Algeria, December 18–21, 1989.

Dowson, V.H.W. (1982) *Date Production and Protection.* FAO Plant Production and Protection Paper no.35. Food and Agriculture Organization of the United Nations, Rome.

Dowson, V.H.W. and Aten, A. (1963) Composition et maturation. In: *Récolte et Conditionnement Des Dattes.* FAO Progrès et mise en valeur agriculture no. 72. Food and Agriculture Organization of the United Nations, Rome.

Drummond, B. (1924) Artificial maturation of dates and utilization of cull dates by methods of semi-maturation. *Annual Report Date Growers' Institute* 1, 27–28.

Elansari, A.M. (2008) Hydrocooling rates of Barhee dates at the Khalal stage. *Postharvest Biology and Technology* 48, 402–407.

El-Hadrami, A. and Al-Khayri, J.M. (2012) Socioeconomic and traditional importance of date palm. *Emirates Journal of Food and Agriculture* 24, 371–385.

Elleuch, M., Besbes, S., Roiseux, O., Blecker, C., Deroanne, C. *et al.* (2008) Date flesh: Chemical composition and characteristics of the dietary fibre. *Food Chemistry* 111(3), 676–682. DOI: 10.1016/j.foodchem.2008.04.036.

El-Mohandes, M. (2009) Modified atmospheres and/or phosphine fumigation for controlling post-harvest dates pests. In: *Report of the Regional Experts Group Meeting on Applications of Methyl Bromide Alternatives in the Dates Sector, Al-Khobar, Saudi Arabia,* United Nations Environment Program, Regional Office for West Asia, December 13–16, 2009, pp. 18–19.

El-Mohandes, M.A. (2010) Methyl bromide alternatives for dates disinfestations. *Acta Horticulturae* 882, 555–562. DOI: 10.17660/ActaHortic.2010.882.62.

El-Sayed, S. and Baeshin, N.A. (1983) Feasibility of disinfestations of date fruits produced in Saudi Arabia by gamma irradiation. In: Makki, Y.M. (ed.) *Proceedings of the First Symposium on the Date Palm in Saudi Arabia, Al-Hassa, Saudi Arabia,* King Faisal University, Al-Hassa, Saudi Arabia, March 23–25, 1982, pp. 342–350.

Eman, O.A., Farag, S.E.A. and Hamad, A.I. (1994) Comparative study between fumigation and irradiation of semi-dry date fruits. *Food/Nahrung* 38, 612–620.

Escalona, V.H., Aguayo, E. and Artés, F. (2006) Modified atmosphere packaging improved quality of kohlrabi stems. *LWT – Food Science and Technology* 40, 397–403.

Estanove, P. (1990) Note technique: valorisation de la datte. In: Les systèmes agricoles oasiens. *Options Méditerranéennes, Série A,* 11, 301–318.

Faisal, M., Mansour, A., Mohammed, A. and Mohamed, A.M. (2020) IHDS: intelligent harvesting decision system for date fruit based on maturity stage using deep learning and computer vision. *IEEE Access* 8, 167985–167997.

FAO (2018) FAOSTAT database. Crop statistics. Available at: www.fao.org/faostat/en/#compare (accessed 15 November 2021).

FAO/WHO (2019) *Codex Alimentarius Standard for Dates CXS 143-1985. Adopted in 1985. Amended in 2019.* Food and Agriculture Organization of the United Nations, Rome and World Health Organization, Geneva,Switzerland. Available at: www.fao.org/fao-who-codexalimentarius/sh-proxy/fr/?lnk=1&url=https%253A%252F%252Fworkspace.fao.org%252Fsites%252Fcodex%252FStandards%252FCXS%2B143-1985%252FCXS_143e.pdf (accessed 1 December 2022).

Farahnaky, A. and Afshari-Jouybari, H. (2011) Physicochemical changes in Mazafati date fruits incubated in hot acetic acid for accelerated ripening to prevent diseases and decay. *Scientia Horticulturae* 127(3), 313–317. DOI: 10.1016/j.scienta.2010.10.019.

Ferry, M., Bouguedoura, N. and El Hadrami, I. (1998) Patrimoine génétique et techniques de propagation in vitro pour le développement de la culture de palmier dattier. *Science et Changements Planétaires/Sécheresse* 2, 139–146.

Fonseca, S.C., Oliveira, F.A.R. and Brecht, J.K. (2002) Modelling respiration rate of fresh fruits and vegetables for modified atmosphere packages: a review. *Journal of Food Engineering* 52(2), 99–119. DOI: 10.1016/S0260-8774(01)00106-6.

Gan, J., Megonnell, N.E. and Yates, S.R. (2001) Adsorption and catalytic decomposition of methyl bromide and methyl iodide on activated carbons. *Atmospheric Environment* 35(5), 941–947. DOI: 10.1016/S1352-2310(00)00339-3.

Garbati Pegna, F., Battaglia, M. and Bergesio C. (2012) *Italian Machinery and Equipment for Date Palm Field Operations.* University of Florence, Florence, Italy.

Gherbawy, Y. (2001) Use of RAPD-PCR to characterize *Eurotium* strains isolated from date fruits. *Cytologia* 66(4), 349–356. DOI: 10.1508/cytologia.66.349.

Ghiaba, Z., Yousfi, M., Hadjadj, M., Saidi, M. and Dakmouche, M. (2013) Study of antioxidant properties of five Algerian date (*Phoenix dactylifera* L) cultivars by cyclic voltammetric technique. *International Journal of Electrochemical Science* 9, 909–920.

Gil, M.I., Selma, M.V., López-Gálvez, F. and Allende, A. (2009) Fresh-cut product sanitation and wash water disinfection: problems and solutions. *International Journal of Food Microbiology* 134, 37–45.

Glasner, B., Botes, A., Zaid, A. and Emmens, J. (2002) Date harvesting, packing-house management and marketing aspects. In: Zaid, A. and Arias-Jiménez, E.J. (eds) *Date Palm Cultivation*. FAO Plant Production and Protection Paper no.156 Rev. 1, Food and Agriculture Organization of the United Nations, Rome, pp. 237–267.

Gómez, P.A. and Artés, F. (2005) Improved keeping quality of minimally fresh processed celery sticks by modified atmosphere packaging. *LWT - Food Science and Technology* 38(4), 323–329. DOI: 10.1016/j.lwt.2004.06.014.

Goneum, S.I., El-Samahy, S.K., Ibrahim, S.S., El-Fadeel, M.G.A. and Mohammed, S.M. (1993) Compositional changes in the date fruits during ripening by freezing. In: *Program and Abstracts of the Third Symposium on the Date Palm in Saudi Arabia, King Faisal University, Al-Hassa, Saudi Arabia*, King Faisal University, Al-Hassa, Saudi Arabia, abstract no.I-14, January 17–20, 1993.

Grecz, N., Al-Harithy, R., El-Mojaddidi, M.A. and Rahma, S. (1989) Radiation inactivation of microorganisms on dates from Riyadh and Alahsa areas. In: *Proceedings of the Second Symposium on the Date Palm in Saudi Arabia, Al-Hassa, Saudi Arabia*, King Faisal University, Al-Hassa, Saudi Arabia, March 3–6, 1986.

Gross, K., Wang, C.Y. and Saltveit, M. (2002) *The Commercial Storage of Fruits, Vegetables, and Florist and Nursery Crops*. Agriculture Handbook no.66. US Department of Agriculture, Agricultural Research Service, Washington, DC.

Hardenburg, R.E., Watada, A.E. and Wang, C.Y. (1990) *The Commercial Storage of Fruits, Vegetables, and Florist and Nursery Stocks*. Agriculture Handbook no.66, reprinted edn. US Department of Agriculture, Agricultural Research Service, Washington, DC.

Hasegawa, S. and Maier, V.P. (1980) Polyphenol oxidase of dates. *Journal of Agricultural and Food Chemistry* 28(5), 891–893. DOI: 10.1021/jf60231a009.

Hasegawa, S., Maier, V.P., Kaszycki, H.P. and Crawford, J.K. (1969) Polygalacturonase content of dates and its relation to maturity and softness. *Journal of Food Science* 34(6), 527–529. DOI: 10.1111/j.1365-2621.1969. tb12078.x.

Hasegawa, S., Smolensky, D.C. and Maier, V.P. (1972) Hydrolytic enzymes in dates and their application in the softening of tough dates and sugar wall dates. *Date Grower's Institute Annual Report* 49, 6–8.

Hassouna, M., Ghrir, R., Mahjoub, A. and Hamdi, S. (1996) Influence de la fumigation au bromure de méthyle sur la composition chimique des dattes tunisiennes. *Fruits* 49(3), 197–204.

Hazbavi, E., Khoshtaghaza, M.H., Mostaan, A. and Banakar, A. (2013) Effect of storage duration on some physical properties of date palm (cv. Stamaran). *Journal of the Saudi Society of Agricultural Sciences* 14(2), 140–146. DOI: 10.1016/j. jssas.2013.10.001.

Hertog, M.L.A.T.M., Nicholson, S.E. and Jeffery, P.B. (2004) The effect of modified atmospheres on the rate of firmness change of 'Hayward' kiwifruit. *Postharvest Biology and Technology* 31(3), 251–261. DOI: 10.1016/j. postharvbio.2003.09.005.

Hilton, S.J. and Banks, H.J. (1997) Methyl bromide sorption and residues on sultanas and raisins. *Journal of Stored Products Research* 33(3), 231–249. DOI: 10.1016/S0022-474X(97)00003-9.

Homayouni, A., Azizi, A., Keshtiban, A.K., Amini, A. and Eslami, A. (2015) Date canning: a new approach for the long time preservation of date. *Journal of Food Science and Technology* 52(4), 1872–1880. DOI: 10.1007/s13197-014-1291-0.

Hui, Y.H., Barte, J., Pilar-Cano, M., Gusek, T.W., Sidhu, J.S. *et al.* (2006) *Handbook of Fruit and Fruit Processing.* Blackwell, Ames, Iowa.

Hussain, A.A. (1974) *Date Palms and Dates with Their Pests in Iraq.* Mosul University Press, Mosul, Iraq.

Hussein, F., Moustafa, S., Le Sanuirala, F. and Zeid, A. (1976) Studies on physical and chemical characteristics of eighteen date cultivars grown in Saudi Arabia. *Indian Journal of Horticulture* 33, 107–113.

Hussein, F., Souial, G.F., Khalifa, A.S., Gaefar, S.I. and Mousa, I.A. (1989) Nutritional value of some Egyptian soft date cultivars (protein and amino acids). In: *Proceedings of the Second Symposium on the Date Palm in Saudi Arabia, Al-Hassa, Saudi Arabia,* King Faisal University, Al-Hassa, Saudi Arabia, March 3–6, 1986.

Ibrahim, A.A., Ibrahim, H.R. and Abdul-Rasool, N. (2007) Development and testing of a shaker-system for the selective harvest of date fruit. *Acta Horticulturae* 736, 199–203. DOI: 10.17660/ActaHortic.2007.736.17.

Idder, M.A., Idder-Ighili, H., Saggou, H. and Pintureau, B. (2009) Taux d'infestation et morphologie de la pyrale des dattes *Ectomyelois ceratoniae* (Zeller) sur différentes variétés du palmier dattier (*Phoenix dactylifera* L.). *Cahiers Agricultures* 18, 63–71.

Jarrah, A.Z. and Benjamin, N.D. (1982) Activity of polyphenol oxidase and pectin esterase during different stages of growth and development of Khadrawi fruit in Iraq. *Date Palm Journal* 1(2), 5–18.

Jeantet, R., Croguennec, T., Schuck, P. and Brulé, G. (2006) *Sciences des aliments.* Tec&Doc/Lavoisier, Paris.

Jemni, M., Chniti, S., Harbaoui, K., Ferchichi, A. and Artés, F. (2016a) Partial vacuum and active modified atmosphere packaging for keeping overall quality of dates. *Journal of New Sciences* 29, 1656–1665.

Jemni, M., Gómez, P.A., Souza, M., Chaira, N., Ferchichi, A. *et al.* (2014) Combined effect of UV-C, ozone and electrolyzed water for keeping overall quality of date palm. *LWT - Food Science and Technology* 59(2), 649–655. DOI: 10.1016/j.lwt.2014.07.016.

Jemni, M., Otón, M., Souza, M., Dhouibi, M., Ferchichi, A. and Artés, F. (2015) Ozone gas greatly reduced the survival of carob moth larvae in stored date palm fruit. *Journal of New Sciences* 16, 567–573.

Jemni, M., Ramírez, J.G., Otón, M., Artés-Hernández, F., Harbaoui, K. *et al.* (2016b) Passive modified atmosphere packaging and chilling storage for keeping overall quality of dates. *Acta Horticulturae* 1194, 673–680. DOI: 10.17660/ActaHortic.2018.1194.96.

Johnson, J.A., Wang, S. and Tang, J. (2003) Thermal death kinetics of fifth instar *Plodia interpunctella. Journal of Economic Entomology* 96, 519–524.

Kacperska, A. and Kubacka-Zebalska, M. (1989) Formation of stress ethylene depends both on ACC synthesis and on the activity of free radical-generating system. *Physiologia Plantarum* 77(2), 231–237. DOI: 10.1111/j.1399-3054.1989.tb04974.x.

Kader, A.A. (1992) Postharvest biology and technology: an overview. In: Kader, A.A. (ed.) *Postharvest Technology of Horticultural Crops*, 2nd edn. University of California, Division of Agriculture and Natural Resources, Oakland, California, pp. 15–20.

Kader, A.A. (2003) A perspective on postharvest horticulture (1978-2003). *HortScience* 38(5), 1004–1008. DOI: 10.21273/HORTSCI.38.5.1004.

Kader, A.A. and Hussein, A.M. (2009) Harvesting and postharvest handling of dates. In: ICARDA (ed.) *Project on the Development of Sustainable Date Palm Production Systems in the GCC Countries of the Arabian Peninsula.* International Center for Agricultural Research in the Dry Areas (ICARDA), Aleppo, Syria.

Kahramanoglu, I. and Usanmazm, S. (2019) Preharvest and postharvest treatments for increasing the rate of ripening of date palm fruit (*Phoenix dactylifera* L.) cv. Medjool. *Progress in Nutrition* 21(1), 215–224.

Kondo, N. (2010) Automation on fruit and vegetable grading system and food traceability. *Trends in Food Science & Technology* 21(3), 145–152. DOI: 10.1016/j.tifs.2009.09.002.

Kulkarni, S.G., Vijayanand, P., Aksha, M., Reena, P. and Ramana, K.V.R. (2008) Effect of dehydration on the quality and storage stability of immature dates (*Pheonix dactylifera*). *LWT - Food Science and Technology* 41(2), 278–283. DOI: 10.1016/j.lwt.2007.02.023.

Lafuente, M.T., Zacarias, L., Martínez-Téllez, M.A., Sanchez-Ballesta, M.T. and Granell, A. (2003) Phenylalanine ammonia-lyase and ethylene in relation to chilling injury as affected by fruit age in citrus. *Postharvest Biology and Technology* 29(3), 309–318. DOI: 10.1016/S0925-5214(03)00047-4.

Lee, D.-J., Schoenberger, R., Archibald, J. and McCollum, S. (2008) Development of a machine vision system for automatic date grading using digital reflective near-infrared imaging. *Journal of Food Engineering* 86(3), 388–398. DOI: 10.1016/j.jfoodeng.2007.10.021.

Lee, D.S., Haggar, P.E., Lee, J. and Yam, K.M. (1991) Model for fresh produce respiration in modified atmosphere based on principles of enzyme kinetics. *Journal of Food Science* 56(6), 1580–1585.

Lobo, G.M., Yahia, E.M. and Kader, A.A. (2014) Biology and postharvest physiology of date fruit. In: Siddiq, M., Aleid, S.M. and Kader, A.A. (eds) *Dates: Postharvest Science, Processing Technology and Health Benefits*, 1st edn. Wiley, Chichester, UK, pp. 57–80.

Logrieco, A., Bottalico, A., Mulé, G., Moretti, A. and Perrone, G. (2003) Epidemiology of toxigenic fungi and their associated mycotoxins for some Mediterranean crops. *European Journal of Plant Pathology* 109(7), 645–667. DOI: 10.1023/A:1026033021542.

Lustig, I., Bernstein, Z. and Gophen, M. (2014) Skin separation in Majhul fruits. *International Journal of Plant Research* 4(1), 29–35.

Macheix, J.J., Fleuriet, A. and Billot, J. (1990) *Fruit Phenolics*, 1st edn. CRC Press, Boca Raton, Florida, pp. 295–322.

Machiels, D. and Istasse, L. (2002) La réaction de Maillard: importance et applications en chimie des aliments. *Annales de Médecine Vétérinaire* 146, 347–352.

Maier, V.P. and Schiller, F.H. (1961) Studies on domestic dates. III. Effect of temperature on some chemical changes associated with deterioration. *Journal of Food Science* 26(5), 529–534. DOI: 10.1111/j.1365-2621.1961.tb00401.x.

Mansouri, A., Embarek, G., Kokkalou, E. and Kefalas, P. (2005) Phenolic profile and antioxidant activity of the Algerian ripe date palm fruit (*Phoenix dactylifera*). *Food Chemistry* 89(3), 411–420. DOI: 10.1016/j.foodchem.2004.02.051.

Marouf, A., Amir-Maafi, M. and Shayesteh, N. (2013) Two-sex life table analysis of population characteristics of almond moth, *Cadra cautella* (Lepidoptera: Pyralidae) on dry and semi-dry date palm varieties. *Journal of Crop Protection* 2, 171–181.

Marouf, A.M. and Khali, M. (2019) Characterization of postharvest physiology of thermised dates (*Phoenix dactylifera* L.), packed under a modified atmosphere (MAP). *AgroBiologia* 9(2), 1528–1542.

Marouf, A.M. and Khali, M. (2020) Study the effect of postharvest heat treatment on infestation rate of fruit date palm (*Phoenix dactylifera* L.) cultivars grown in Algeria. *Journal of Nutritional Science and Healthy Diet* 1(1), 37–40.

Mehaoua, M.S. (2014) Abondance saisonnière de la pyrale des dattes (*Ectomyelois ceratoniae* Zeller., 1839), bioécologie, comportement et essai de lutte. Thèse de doctorat, Université Mohamed Khider Biskra, Biskra, Algérie.

Mikki, M.S. and Al-Taisan, S.M. (1993) Physico-chemical changes associated with freezing storage of date cultivars at their rutab stage of maturity. In: *Program and Abstracts of the Third Symposium on the Date Palm in Saudi Arabia, King Faisal University, Al-Hassa, Saudi Arabia*, King Faisal University, Al-Hassa, Saudi Arabia, abstract no.I-11, January 17–20, 1993.

Mohammed, M., Sallam, A., Alqahtani, N. and Munir, M. (2021) The combined effects of precision-controlled temperature and relative humidity on artificial ripening and quality of date fruit. *Foods* 10(11), 2636. DOI: 10.3390/foods10112636.

Munier, P. (1973) La date. In: *Le Palmier-Dattier*. G.-P. Maisonneuve et LaRose, Paris.

Murray, R., Lucangeli, C., Polenta, G. and Budde, C. (2007) Combined pre-storage heat treatment and controlled atmosphere storage reduced internal breakdown of 'Flavorcrest' peach. *Postharvest Biology and Technology* 44(2), 116–121. DOI: 10.1016/j.postharvbio.2006.11.013.

Myhara, R.M., Al-Alawi, A., Karkalas, J. and Taylor, M.S. (2000) Sensory and textural changes in maturing Omani dates. *Journal of the Science of Food and Agriculture* 80(15), 2181–2185. DOI: 10.1002/1097-0010(200012)80:15<2181::AID-JSFA765>3.0.CO;2-C.

Nasser, L. (2017) Fungal contamination and invertase activity in dates and date products in Saudi Arabia. *American Journal of Food Technology* 12, 295–300.

Naturland (2002) *Organic Farming in the Tropics and Subtropics. Exemplary Description of 20 Crops: Date Palm*. Naturland e.V, Graefelfing, Germany.

Navarro, S., Donahaye, E., Rindner, M. and Azirieli, A. (1998a) Disinfestations of nitidulid beetles from dried fruits by modified atmospheres. In: *Proceedings of the Annual International Research Conference on Methyl Bromide Alternatives and Emission Reductions, Orlando, Florida, December 1998*, pp. 681–683.

Navarro, S., Donahaye, E., Rindner, M. and Azrieli, A. (1998b) Storage of dried fruits under controlled atmospheres for quality preservation and control of nitidulid beetles. *Acta Horticulturae* 480, 221–226. DOI: 10.17660/ActaHortic.1998.480.38.

Navarro, S. (2006) Postharvest treatment of dates. *Stewart Postharvest Review* 2(6), 1–10. DOI: 10.2212/spr.2006.6.1.

Nay, J.E. and Perring, T.M. (2006) Effect of fruit moisture content on mortality, development, and fitness of the carob moth (Lepidoptera: Pyralidae). *Environmental Entomology* 35(2), 237–244. DOI: 10.1603/0046-225X-35.2.237.

Nguyen, T.B.T., Ketsa, S. and van Doorn, W.G. (2004) Effect of modified atmosphere packaging on chilling-induced peel browning in banana. *Postharvest Biology and Technology* 31(3), 313–317. DOI: 10.1016/j.postharvbio.2003.09.006.

Niakousari, M., Erjaee, Z. and Javadian, S. (2010) Fumigation characteristics of ozone in postharvest treatment of Kabkab dates (*Phoenix dactylifera* L.) against selected insect infestation. *Journal of Food Protection* 73(4), 763–768. DOI: 10.4315/0362-028x-73.4.763.

Nixon, R.W. (1959) *Growing Dates in the United States*, Agriculture Information Bulletin No. 207. US Department of Agriculture, Washington, DC.

Nourani, A., Kaci, F., Garbati Pegna, F. and Kadri, A. (2017) Design of a portable dates cluster harvesting machine. *Agricultural Mechanization in Asia, Africa and Latin America* 48(1), 18–21.

Nourani, A., Francesco, G.P. and Angelo, R. (2021) Mechanically assisted harvesting of dry and semi-dry dates of average to low quality. *Journal of Agriculture and Environment for International Development* 115(1), 85–96. https://doi.org/10.12895/jaeid.20211.1436

Ooi, P.A.C., Winotai, A. and Peña, J.E. (2002) Pests of minor tropical fruits. In: Peña, J.E., Sharp, J.L. and Wysoki, M. (eds) *Tropical Fruit Pests and Pollinators: Biology, Economic Importance, Natural Enemies and Control*. CAB International, Wallingford, UK, pp. 315–330.

Özhan, B., Karadeniz, F. and Erge, H.S. (2010) Effect of storage on nonenzymatic browning reactions in carob pekmez. *International Journal of Food Science & Technology* 45(4), 751–757. DOI: 10.1111/j.1365-2621.2010.02190.x.

Palou, L., Rosales, R., Taberner, V. and Vilella-Espla, J. (2016) Incidence and etiology of post-harvest diseases of fresh fruit of date palm (*Phoenix dactylifera* L.) in the grove of Elx (Spain). *Phytopathologia Mediterranea* 55(3), 391–400. DOI: 10.14601/Phytopathol_Mediterr-17819.

Patriarca, A., Azcarate, M.P., Terminiello, L. and Fernández Pinto, V. (2007) Mycotoxin production by *Alternaria* strains isolated from Argentinean wheat. *International Journal of Food Microbiology* 119(3), 219–222. DOI: 10.1016/j.ijfoodmicro.2007.07.055.

Paull, R.E. and Armstrong, J.W. (eds) (1994) *Insect Pests and Fresh Horticultural Products: Treatments and Responses*. CAB International, Wallingford, UK.

Paull, R.E. and Duarte, O. (eds) (2011) Postharvest technology. In: *Tropical Fruits*, 2nd Edition., Vol. 1. CAB International, Wallingford, UK, pp. 101–122.

Quaglia, M., Santinelli, M., Sulyok, M., Onofri, A., Covarelli, L. *et al.* (2020) *Aspergillus, Penicillium* and *Cladosporium* species associated with dried date fruits collected in the Perugia (Umbria, Central Italy) market. *International Journal of Food Microbiology* 322, 108585. DOI: 10.1016/j.ijfoodmicro.2020.108585.

Rafaeli, A., Kostukovsky, M. and Carmeli, D. (2006) Successful disinfestations of sap-beetle contaminations from organically grown dates using heat treatment: a case study. *Phytoparasitica* 34, 204–212.

Ragab, W., Ramadan, B. and Abdel-Sater, M. (2001) Mycoflora and aflatoxins associated with Saidy dates as affected by technological processes. In: Al-Badawy, A.A. (ed.) *Proceedings of the Second International Conference on Date Palms, Al-Ain,*

United Arab Emirates, UAE University, Al-Ain, United Arab Emirates, March 25–27, 2001, pp. 409–421.

Reynes, M. (1997) Influence d'une technique de désinfestation par micro-ondes sur les critères de qualité physico-chimiques et biochimiques de la datte. Thèse de doctorat, Institut national polytechnique de Lorraine, Nancy, France.

Rivas, N.J. and Whitaker, J.R. (1973) Purification and some properties of two polyphenol oxidases from Bartlett pears. *Plant Physiology* 52, 501–507.

Rose, D.C. and Chilvers, J. (2018) Agriculture 4.0: broadening responsible innovation in an era of smart farming. *Frontiers in Sustainable Food Systems* 2, 87.

Rygg, G.L. (1975) *Date Development, Handling and Packing in the United States.* Agriculture Handbook no.482. US Department of Agriculture, Agricultural Research Service, Washington, DC.

Saafi, E.B., El Arem, A., Issaoui, M., Hammami, M. and Achour, L. (2009) Phenolic content and antioxidant activity of four date palm (*Phoenix dactylifera* L.) fruit varieties grown in Tunisia. *International Journal of Food Science & Technology* 44, 2314–2319.

Sabbri, M.M., Makki, Y.M. and Salehuddin, A.H. (1982) Study on dates consumers preference in different regions of the Kingdom of Saudi Arabia. In: Makki, Y.M. (ed.) *Proceedings of the First Symposium on the Date Palm in Saudi Arabia, Al-Hassa, Saudi Arabia*, King Faisal University, Al-Hassa, Saudi Arabia, March 23–25, 1982, pp. 23–25.

Saleem, S.A., Baloch, A.K., Baloch, M.K., Saddozai, A.A. and Ghafoor, A. (2004) Influence of hot water on ripening/curing of Dhakki dates. *Pakistan Journal of Biological Sciences* 7(12), 2034–2038.

Saltveit, M.E. (2000) Wound induced in phenolic metabolism and tissue browning are altered by heat shock. *Postharvest Biology and Technology* 21, 61–69.

Sandhya (2010) Modified atmosphere packaging of fresh produce: current status and future needs. *LWT – Food Science and Technology* 43, 381–392.

Sarraf, M., Jemni, M., Kahramanoğlu, I., Artés, F., Shahkoomahally, S. *et al.* (2021) Commercial techniques for preserving date palm (*Phoenix dactylifera*) fruit quality and safety: a review. *Saudi Journal of Biological Sciences* 28(8), 4408–4420. DOI: 10.1016/j.sjbs.2021.04.035.

Serrano, M., Pretel, M.T., Botella, M.A. and Amoros, A. (2001) Physicochemical changes during date ripening related to ethylene production. *Food Science and Technology International* 7, 31–36.

Serrano, M., Martinez-Romero, M., Guillén, F., Castillo, S. and Valero, D. (2006) Maintenance of broccoli quality and functional properties during cold storage as affected by modified atmosphere packaging. *Postharvest Biology and Technology* 39(1), 61–68.

Shen, Q., Kong, F. and Wang, Q. (2006) Effect of modified atmosphere packaging on the browning and lignification of bamboo shoots. *Journal of Food Engineering* 77(2), 348–354. DOI: 10.1016/j.jfoodeng.2005.06.041.

Siddiq, M. and Greiby, I. (2014) Overview of date fruit production, postharvest handling, processing, and nutrition. In: Siddiq, M., Aleid, S.M. and Kader, A.A. (eds) *Dates: Postharvest Science, Processing Technology and Health Benefits*, 1st edn. Wiley, Chichester, UK, pp. 1–28. DOI: 10.1002/9781118292419.

Shenasi, M., Aidoo, K.E. and Candlish, A.A.G. (2002a) Microflora of date fruits and production of aflatoxins at various stages of maturation. *International Journal of Food Microbiology* 79, 113–119.

Shenasi, M., Candlish, A.A.G. and Aidoo, K.E. (2002b) The production of aflatoxins in fresh date fruits and under simulated storage conditions. *Journal of the Science of Food and Agriculture* 82(8), 848–853. DOI: 10.1002/jsfa.1118.

Sidhu, J.S. (2006) Date fruits production and processing. In: Hui, Y.H. (ed.) *Handbook of Fruits and Fruit Processing*. Blackwell, Oxford, pp. 391–419.

Tafti, A.G. and Fooladi, M.H. (2005) Microbial contamination on date fruits. In: *Book of Abstracts, First International Symposium and Festival on Date Palm, Bandar Abass, Iran*, Ministry of Jihad-e-Agriculture, Bandar Abbas, Iran, November 20–21, 2005, p. 10.

Tate, H.F. and Hilgeman, R.H. (1958) *Dates in Arizona*, Extension Circular No.165. Arizona Agricultural College, Tucson, Arizona.

Teisseire, P. (1959) Aperçu sur les ennemis et maladies du palmier dattier en Algérie et au Sahara et les moyens mis en œuvre pour le combattre. In: *Rapport de la première réunion technique internationale FAO sur la production et le traitement des dattes, tenue à Tripoli (Libye) 5–11 décembre 1959*, Food and Agriculture Organization of the United Nations.

Tirilly, Y. and Bourgeois, C.L. (1999) *Technologie Des Légumes*. Tec&Doc/Lavoisier, Paris, pp. 216–293.

Trakulnaleumsai, C., Ketsa, S. and van Doorn, W.G. (2006) Temperature effects on peel spotting in 'Sucrier' banana fruit. *Postharvest Biology and Technology* 39(3), 285–290. DOI: 10.1016/j.postharvbio.2005.10.015.

UNECE (2010) *UNECE Standard DDP-08 Concerning the Marketing and Commercial Quality Control of Dates, 2010 edition*. United Nations Economic Commission for Europe, Geneva, Switzerland. Available at: https://unece.org/fileadmin/DAM/trade/agr/standard/dry/dry_e/08Dates_e.pdf (accessed 1 December 2022).

Van Boekel, M.A.J.S. (2001) Kinetic aspects of the maillard reaction: a critical review. *Food/Nahrung* 45, 150–159.

Varoquaux, P. and Nguyen-The, C. (1993) *Les Atmosphères Modifiées de La Quatrième Gamme: Technologie Alimentaire, Les Hautes Pressions*. IFN Dossier, Paris.

Vinson, A.E. (1924) The chemistry of the date. *Date Growers' Institute Report* 1, 11–12.

Wahbah, T.F. (2003) Control of some dried fruit pests. Master thesis, University, Alexandria, Egypt.

Wakil, W., Romeno Faleiro, J. and Miller, T.A. (2015) *Sustainable Pest Management in Date Palm: Current Status and Emerging Challenges*. Sustainability in Plant and Crop Protection. Springer, Cham, Switzerland. DOI: 10.1007/978-3-319-24397-9.

Walsh, K.B., Blasco, J., Zude-Sasse, M. and Sun, X. (2020) Visible-NIR 'point' spectroscopy in postharvest fruit and vegetable assessment: the science behind three decades of commercial use. *Postharvest Biology and Technology* 168, 111246. DOI: 10.1016/j.postharvbio.2020.111246.

Wang, S., Yin, X., Tang, J. and Hansen, J.D. (2004) Thermal resistance of different life stages of codling moth (Lepidoptera: Torticidae). *Journal of Stored Products Research* 40, 565–574.

Yahia, E.M. (2004) Date. In: Gross, K., Wang, C.Y. and Saltveit, M. (eds) *The Commercial Storage of Fruits, Vegetables, and Florist and Nursery Crops*. Agriculture Handbook No. 66. US Department of Agriculture, Agricultural Research Service, Washington, DC, pp. 311–314.

Yahia, E.M. (ed.) (2009) *Modified and Controlled Atmospheres for the Storage, Transportation, and Packaging of Horticultural Commodities.* CRC Press, Boca Raton, Florida.

Yahia, E.M. and Kader, A.A. (2011) Date (*Phoenix dactylifera* L.). In: Yahia, E.M. (ed.) *Postharvest Biology and Technology of Tropical and Subtropical Fruits*, Vol. 3. *Coconato Mango.* Woodhead Publishing Series in Food Science, Technology and Nutrition, Woodhead Publishing, Cambridge, pp. 41–79.

Yahia, E.M., Lobo, M.G. and Kader, A.A. (2014) Harvesting and postharvest technology of dates. In: Siddiq, M., Aleid, S.M. and Kader, A.A. (eds) *Dates: Postharvest Science, Processing Technology and Health Benefits*, 1st edn. Wiley, Chichester, UK, pp. 105–135.

Zouba, A., Khoualdia, O., Diaferia, A., Rosito, V., Bouabidi, H. *et al.* (2009) Microwave treatment for post-harvest control of the date moth *Ectomyelois ceratoniae. Tunisian Journal of Plant Protection* 4, 173–184.

Postharvest Handling of Dates

14

Ibrahim E. Greiby[1]* and Mohamed Abusaa Fennir[2]

[1]Department of Food Science & Technology, Faculty of Agriculture, University of Tripoli, Tripoli, Libya; [2]Department of Agricultural Engineering, Faculty of Agriculture, University of Tripoli, Tripoli, Libya

Abstract

The date palm is an important component in arid and semiarid ecosystems. Its value goes beyond agriculture, as it also has social, cultural, and religious importance. Lately, significant increases in cultivation and production have been recorded. Dates are invaluable fruits: tasty, rich in minerals, antioxidants, and carbohydrates in the simple forms of glucose and fructose. However, they are susceptible to severe losses due to a wide range of pre- and postharvest factors. This chapter deals with conditions and factors affecting date fruit quality at the postharvest level. Cultivars are classified as soft, semidry, and dry. Factors and conditions influencing these classifications are discussed. The palatable stages of date fruit development, i.e., khalal, rutab, and tamar, are discussed in relation to ripening, properties, and nutrient contents. Handling operations, from the farm until the product reaches the consumer, are explained, taking account of fruit properties and requirements for maintaining fruit quality. Important operations such as field packaging, receiving at the warehouse facility, disinfecting, sorting, hydration, dehydration, packing, cooling, and storage are reviewed, taking account of the latest methods, technologies, and research findings as well as traditional and grower experiences. Physiological, pathological, and physical disorders affecting fruit quality and methods for alleviating their negative effects are also presented, as are quality assurance, food safety, and marketing strategies. Innovative and emerging technologies in sorting, grading, and packing are also discussed. Lastly, better utilization of cull dates as economically viable products is discussed.

Keywords: Dates, Handling, Postharvest, Quality, Storage conditions

---

*Corresponding author: i.greiby@uot.edu.ly

© CAB International 2023. *Date Palm* (eds J.M. Al-Khayri *et al.*)
DOI: 10.1079/9781800620209.0014

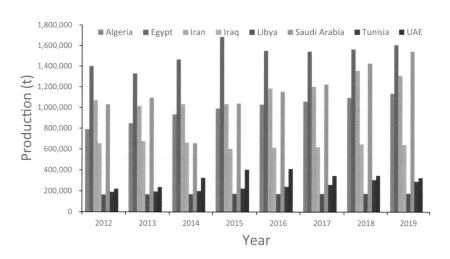

Fig. 14.1. Production of dates in leading countries of production, 2012–2019. UAE, United Arab Emirates. (Data from FAO, 2021.)

14.1 Introduction

Date cultivation goes back in history as far as 4000 BC in the Middle East, specifically in what is now known as Iraq; from there offshoots and seeds were taken and planted wherever climate permitted, thus spreading the area of date production. Nowadays, dates are grown in diverse climatic conditions, in about 100 countries, ranging from very hot and dry to relatively mild; also, the palm is grown for date production and for ornamental purposes. Throughout the long history of propagation and breeding, cultivars have been created, but the exact number existing worldwide is unknown (Sidhu, 2012). However, an approximate estimation is 3000 cultivars (Arias-Jiménez and Zaid, 2002; Salomon-Torres *et al.*, 2021). More than 400 cultivars have been reported in Saudi Arabia (Abdul-Hamid *et al.*, 2020) and Libya (Racchi *et al.*, 2014), and 600 cultivars in Iraq (Khierallah *et al.*, 2015), yet the number of economically valuable cultivars is considerably less than that. It has been estimated that more than 100 million date palm trees exist worldwide (Jonoobi *et al.*, 2019); in 2020 about 1.24 million ha were cultivated with date palms around the world (FAO, 2022). In fact, date production has witnessed a rapid increase in the past three decades, reaching 9.07 million t in 2019, up from 8.4 million t in 2017. Fig. 14.1 shows date production in some countries. The best growing countries are located between latitudes of 24 and 34°N (Zaid and de Wet, 2002). Egypt, Iran, and Saudi Arabia have always been large producers of dates (FAO, 2021). Most date cultivation takes place in the Middle East, extending to the Indian subcontinent and along the Nile River Valley to the Sahara of North Africa. Cultivation practices range from small orchards to vast plantations.

In Saharan oases, the date palm is the main pillar of the ecosystem, and people rely on it for their income and consider it a vital part of their well-being and heritage. In North African oases, the social status of a family is very much related to the number of date palms it owns. The date palm and its fruit are highly revered by desert dwellers for their environmental, economic, nutritional, medicinal, religious, cultural, and architectural values. Dates are a stable and nutritious food, containing about 70% w/w sugar, in addition to minerals, vitamins, phenols, flavonoids, and fiber (Yahia *et al.*, 2014). Dates are consumed fresh, dried, and processed; with milk, grains, and meat they make a viable and healthy diet. For Muslims, dates are essential on the break-fasting table in the month of Ramadan; dates are easy to digest and give quick energy (Mallah *et al.*, 2017). Moreover, dates have great value in other religions, particularly in Christianity and Judaism (Ali *et al.*, 2012). Lately, remarkable attention has been given to the date palm; cultivation now takes place in new regions such as Namibia, Thailand, India, China, and Australia. Investment in dates has become successful and new technologies have been developed for harvesting, handling, packing, and storing. In the last decade, the world has witnessed significant trade in dates; they have been introduced in supermarket chains all over the world. Standards and specifications have been issued, such as those found in ISO and Codex Alimentarius (Yahia and Kader, 2011). Nonetheless, great effort is still needed in improving quality, reducing losses, extending shelf life, and better utilizing cull fruits. This chapter focuses on postharvest issues related to quality and reports recent research investigations and scientific findings in this area.

14.2 Factors Influencing Postharvest Quality

14.2.1 Pre- and postharvest factors

In postharvest technology, it is well known that quality starts from the field. For dates in particular, irrigation, fertilization, disease control, pollination, thinning, and bunch covering (bagging) near ripening are the main preharvest factors influencing quality and are immensely important. However, postharvest factors, including the harvesting process itself, can also influence quality. Attention and care during bunch removal reduces mechanical damage. Harvesting at the proper stage of ripening leads to good eating and storage qualities, harvesting unripe or partially ripe (spotted dates) bunches results in less sweetness and flavor, and delaying harvesting of soft dates leads to rotting, dropping, and souring. In semidry and dry cultivars delayed harvest results in cracking, shrinkage, stiffness, and severe dryness. Nonetheless, harvesting at full ripeness and selective harvesting (picking) is preferred by many growers for the advantages of entering the market early and getting higher prices, extra weight, and quick sales. Harvesting partially ripe (spotted dates) bunches and keeping dates frozen is a common practice for the Deglet Nour cultivar in some countries, such as Libya, since it prevents dryness damage caused by severe heatwaves; however, shelf life after thawing is highly affected. Harvesting of soft cultivars at the khalal stage for the purpose of induced ripening has shown promise (Fennir and Morghem, 2019). Harvesting of soft cultivars as fresh dates at the rutab stage is carried out by selective picking of the ripe fruits; the process is carried out every 2–3 days and extends for 2–3 weeks. In mild and humid conditions, failing to do so leads to souring, rotting, and fruit falling from the bunch (Fennir and Morghem, 2019). However, for semidry and dry cultivars, the same harvesting practice may be followed. Bunches are left to become fully rutab or left to dry out further and become tamar. Generally, low moisture content (10–25% w/w) is needed for uncooled and long storage duration, as dates become less perishable and less vulnerable to microorganisms at those moisture contents (Navarro, 2006).

14.2.2   Losses and their sources

Dates are subject to losses for several reasons. In most producing countries, losses are estimated at about 30% (El-Juhany, 2010; Suhail *et al.*, 2020). Sources of loss are grouped into five main groups as follows.

14.2.2.1   Cultivar-related losses

Cultivar-related losses are caused by the fruit's physiological, morphological, and physical properties, and chemical composition. These may cause damage in shape uniformity, color lightness or darkness, mechanical properties such as softness and fruit hardness related to physicochemical properties (Ahmed and Ramaswamy, 2005), in addition to taste and flavor disorders. Also, some cultivars are more resistant to diseases than others, while others are more susceptible to heat, rain, or moisture. For instance, in Libya, Deglet Nour dates are affected by heat more than Tagyat, Saedi, and Abel under the same conditions. Fig. 14.2 shows cv. Degelt Nour dates collected at harvesting time from a farm located in the City of Houn. Heat damage can be clearly seen, with fruits severely shrunken.

14.2.2.2   Field and weather conditions (environmental conditions)

Field and weather conditions (environmental conditions) are external factors that influence some properties related to storage and stability of dry stored dates, such as sorption characteristics (Chukwu, 2010). Also, unusual high temperatures cause severe dryness even before the fruit reaches the ripening stage. Such phenomena are common in North African Saharan oases, in Jufra and Gadames of Libya, Shat Aljareed of Tunis, and Wergla and southern oases in Algeria. On the other hand, high relative humidity (RH) in Mediterranean coastal regions leads to spread of diseases and

Fig. 14.2. Heat-damaged Deglet Nour dates. (Photo by M. Fennir.)

souring of bunches at ripening, while severe heat and dryness cause fruit shrinkage and loss of texture in soft cultivars that are sold at the khalal stage such as Hellawi, Hurra, and Lemsi. Seasonal rain, such as monsoon rains in southern Pakistan and sub-Saharan countries (Abul-Soad, 2011), and severe flooding may cause uneven ripening and changes in fruit size, shape, taste, and flavor.

14.2.2.3 *Diseases and pests*

Diseases and pests are important sources of loss; pests comprise both verte-brates and invertebrates. El-Shafie (2012) reported 54 species of mites and insects that attack date palms around the world. These may cause severe damage to the palm itself and to its fruits. Infested dates carry insect eggs from the field to the packing shed or storage area that develop into insects when storage conditions permit. Also, stored dates may be infested with fungi and yeasts, especially in storage conditions with high RH and dates with high water content.

14.2.2.4 *Human practices*

Human practices, mainly in postharvest handling chains, may cause severe damage (Suhail *et al.*, 2020). Dates are quite sensitive to pressure espe-cially at the khalal and rutab stages. Rough handling and overfilling of harvest containers lead to substantial mechanical damage such as bruising, shape deformation, cracking, skin removal, seed emerging from flesh, and lumping. Using improper packages and overfilling large boxes cause pressure damage, date adhesiveness, syrupiness, and syrup crystallization. High stacking of boxes also reduces cooling efficiency, causing warm and unventilated areas.

14.2.2.5 Storage and handling conditions

Storage and handling conditions, mainly temperature and RH, are important in keeping good-quality dates. These factors are very much related to
each other: if no water is added to the storage area, a temperature increase
lowers RH, while a temperature decrease raises it. This process is related to
water vapor pressure in the air. Dates absorb and lose moisture in response
to changes in RH (Boubekri *et al.*, 2010): low RH causes dryness of soft
dates, whereas high RH leads to wetting of dry dates and increases their
water content. High temperature during the handling, marketing, and
storage of dates at the khalal and rutab stages is the main cause of losses.
High temperature leads to high respiration rates, microbial growth, and
severe weight loss (Fennir *et al.*, 2014; Zamir *et al.*, 2018). RH is also important during handling of cultivars palatable at khalal, with high temperature
causing shriveling and dryness. Generally, dates are not cold sensitive, thus
handling and storage at low temperature is beneficial. Low temperature
near 0°C along with modified atmosphere (MA) conditions have proven
to be effective in keeping soft cultivars consumed at khalal, such as Burhi,
Hurra, and Hellawi cultivars (Al-Redhaiman, 2004; El-Rayes, 2009; Fennir
et al., 2017). Carbon dioxide (CO_2) from 5 to 20% v/v with 5% v/v oxygen
(O_2) was found effective for extending the shelf life of Berhi for up to
26 weeks and Hurra and Hellawi for 12 weeks. Similar conditions were
reported as effective in preserving some soft cultivars at the rutab stage
(Dehghan-Shoar *et al.*, 2008; Fennir *et al.*, 2021).

14.2.3 Cultivation and cultivars in relation to quality

Dates are mostly grown in subtropical climates. For good yield, high heat
levels between flowering and fruiting are needed and low or no rain
during ripening is preferred. Depending on climate, date cultivars are
divided into three main groups: soft, semidry, and dry. Soft dates require
2000 heat units (temperature above a base of 10°C) and are grown in mild
regions, generally along and near the southern coast of the Mediterranean
Sea below latitude 35°N and in areas with similar climatic conditions elsewhere. Semidry cultivars need about 2750 heat units; they are grown in
warmer locations, commonly around latitude 29°N. Dry cultivars are grown
in hotter and drier desert areas with heat units greater than 2750 (Fennir
and Morgham, 2016; Wright, 2016). Generally, soft dates contain >30%
w/w moisture, semidry dates contain 20–30% w/w moisture, and dry dates
contain <20% w/w (Ghnimi *et al.*, 2018). It is a well-known fact that soft
cultivars requiring less heat do well in warmer regions, but cultivars requiring high heat units do not bear quality dates and fail to ripen in cooler
regions. Also, regions with seasonal rain during ripening are unsuitable
for date cultivation; such conditions lead to rotting and prevent ripening.

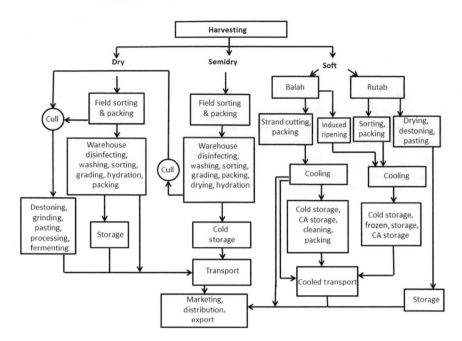

Fig. 14.3. Possible handling steps for dry, semidry, and soft types of dates.

Cultivar properties are very important in palatability and market value; some cultivars are superior in their quality attributes, such as sweetness, flavor, size, color, mastication, and flesh/seed ratio (Kader and Hussein, 2009). Additionally, some nutritional and medicinal advantages are significant (Al-Farsi and Lee, 2008). Other cultivar advantages are similarly important, such as early and even ripening, storability, tolerance to weather conditions and handling, and suitability for processing, mainly making soft paste. In general, properties of date fruits are very much related to environmental and genetic interactions (Ghnimi *et al.*, 2018), i.e., variations in the quality and properties of a cultivar can be observed under different environmental and growing conditions.

14.3 Postharvest Handling

Postharvest operations of date production start with harvesting, followed by several operations until reaching the market destination. Fig. 14.3 shows the layout of date production flow from harvesting until marketing. Harvesting at optimum ripening stage is important in quality, whether at the khalal stage for a few cultivars or at the rutab or tamar stage for the vast majority. For cultivars consumed khalal, ripening is indicated by time, color, and taste, extending for about 3 weeks depending on weather conditions with hot conditions reducing the harvesting period significantly. For cultivars consumed at rutab and tamar, ripening is indicated by color change, normally starting with a few fruits on the bunch and increasing until completely ripe, usually taking 3–4 weeks depending on weather conditions. Generally, quality and consumer needs at the destination market must be taken into consideration. Local markets or farm sale points require minimum or no packing, cooling, wrapping, and labeling compared with distant and export markets. The harvest operation itself is critical. No or minimum mechanical injuries are preferred, also attention and care must be given to all stages of field handling such as filling, emptying, and handling boxes.

From a physiological point of view, date fruit is climacteric; ripening occurs after detachment from the tree. Dates can also respond to artificial and induced ripening. Several studies have investigated methods of ripening using chemical treatments such as sodium chloride (NaCl) and acetic acid (CH_3COOH) (Farahnaky *et al.*, 2009; Markhand *et al.*, 2014; Jahromi, 2015). Also, physical treatments such as heating, freezing, and controlled atmosphere (CA) reportedly induce or accelerate ripening (Shamim *et al.*, 2013; Fennir and Morghem, 2019). Despite the advantages of induced ripening such as reducing losses, avoiding rain damage, reducing harvesting losses, and early marketing, it is blamed for negatively affecting some quality attributes, shelf life, and storability. However, in North Africa detaching bunches of Bronsi, Taboni, and Zaghlol soft cultivars at advanced balah stage and hanging them indoors to ripen is a common practice. Generally, induced ripening is still very much practiced in traditional ways with very limited market potential. From a quality point of view, induced ripening can be done whenever natural ripening is difficult.

14.3.1 Development stages of date fruit

The date fruit is a berry with a single seed (kernel); after pollination it has five main development stages, i.e., hababouk, kimri, khalal (also called balah or bisr in some countries), rutab, and tamar (Kader and Hussein, 2009; Siddiq and Greiby, 2014). However, some references report only the last four stages due to their importance (Yahia *et al.*, 2014). It takes 24–32 weeks from flowering to reach ripening, the four main stages are:

1. Kimri, also known as the green stage, is reached after about 19 weeks from pollination. During this period, the fruit enlarges, and is bitter, astringent, and unpalatable in all cultivars. At this stage thinning is recommended for the purpose of obtaining large size fruit.
2. Khalal (balah or bisr) is reached after nearly 25 weeks from pollination and lasts for about 4 weeks. Depending on cultivar, fruits become yellow, greenish-yellow, or light to dark red (crimson). Generally, the fruit obtains full size, with less moisture, high sugar, and depending on cultivar, low or no astringency. At this stage some cultivars, such as

Hellawi, Lemsi, and Hurra of North Africa, Zaglol of Egypt, and Semany, Hayani, Khaseb, and Berhi of the Arabian Gulf, become sweet and ready for fresh consumption (El-Rayes, 2009; Kassem *et al.*, 2011; Fennir *et al.*, 2017).

3. Rutab is reached after about 4 weeks of the khalal stage. This is known as the soft ripe stage. Depending on cultivar, fruit color can be light honey, yellow greenish, and light to dark brown. The fruit becomes flavorful, chewy, sweet, and soft with moisture content below 50% w/w.

4. Tamar occurs in hot dry regions, if bunches are left on the tree for about 4 weeks after achieving rutab. Dates become fully ripe and dry, with moisture content reduced below 25% w/w. They are sweet, chewy, with colors ranging from light to dark brown. In some regions, some cultivars become rigid and dry.

14.3.2 Handling

In this section, focus is given to handling dates in the three stages of khalal, rutab, and tamar. It is quite evident that handling conditions greatly vary.

14.3.2.1 Khalal dates

For cultivars palatable at khalal, astringency diminishes while the fruit is still hard, high in moisture content, sweet, and crunchy. In North Africa, harvesting takes place in early autumn, lasting from 2 to 4 weeks depending on weather conditions, and high temperature accelerates achievement of the rutab stage. Other factors may influence ripening, such as irrigation, type of soil, vegetation, and age of the tree. Reaching rutab sometimes is not preferable; for instance, Hellawi and Lemsi are less appealing and deteriorate quickly at the rutab stage. Generally, shelf life lasts a few days under normal conditions; cold storage with high moisture content extends shelf life for a few weeks. CA storage with CO_2 content up to 20% v/v is quite effective for the Burhi cultivar with the fruit retaining good eating quality for up to 26 weeks (Al-Redhaiman, 2004; El-Rayes, 2009). However, for other cultivars such as Hurra and Hellawi, CA conditions affected color properties (Fennir *et al.*, 2017). Other treatments such as calcium chloride ($CaCl_2$) and glycerol ($C_3H_8O_3$) have shown potential for extending khalal of Khesab cultivar (Farag and Al-Masri, 1999). Despite the good response of some cultivars to CA and chemical treatments, their applications in storage and packaging in fresh khalal dates are still limited.

14.3.2.2 Rutab dates

The rutab stage is known as the main ripening stage. The fruit becomes soft and sweet, water content is reduced to about 50% w/w, and the color

Fig. 14.4. Semidry dates at rutab stage stored frozen. (Photo by M. Fennir.)

becomes dark, ranging from light honey to dark brown depending on cultivar. Harvesting of rutab dates is selective, since dates in the bunch ripen unequally. In North Africa, harvesting is known as shaking. A wide basket is placed under the bunch, and it is shaken, the ripe dates fall into the basket. This process is repeated every 2–3 days. For soft cultivars, fruits are delicate and perishable, thus cooling and good packing are necessary. Additionally, shelf life is quite short under uncooled conditions; produce is subjected to rotting and souring. On the other hand, because of the lower moisture content and the nature of the fruit, the rutab stage of dry and semidry cultivars has a longer shelf life, is less perishable, and tolerant to handling. Storage of soft cultivars under freezing conditions or CA conditions is successful; shelf life can be extended for months. Semidry cultivars, on the other hand, can be kept for few months at <5°C and for a longer time at −18°C. Fig. 14.4 shows 25 kg storage boxes inside a cold room (at −18°C) in Jufra region of Libya.

14.3.2.3 Tamar dry dates

Dry dates are stable and can be stored for an extended period of time under uncooled conditions, benefiting from low moisture content and water activity. In dry regions, the common practice is allowing bunches after rutab stage to dry out on the tree and become tamar, and to reach moisture content <20% (wet basis, w.b.). Texture-wise, moisture content is related to fruit softness, the lower the moisture content the harder the dates, yet dates

Fig. 14.5. Handling and marketing of dry dates in Cairo, Egypt. (Photos by M. Fennir.)

are preferred chewy. The greatest advantages of reaching tamar are long shelf life and low handling cost. In hot areas, traditional sun drying is the most common. For instance, in southern Algeria sun drying is performed in the field or on the house roof for 2–3 weeks (Manaa *et al.*, 2013). After drying, dates are collected and kept at room temperature. In Gulf States, dates vacuum packaged and irradiated at 1 kGy have shown extended shelf life at room temperature (Ramadan *et al.*, 2016). Nonetheless, in rainy areas such as Punjab province of Pakistan, mechanical and sun drying are mostly applied (Hussain *et al.*, 2015). Dry dates are either vacuum packaged in 1–5 kg bulk or handled bulk in 30–40 kg jute bags. In North Africa dry dates are generally processed to date syrup and vinegar or exported south to nomadic tribes in the great Sahara in northern Chad, Niger, and Mali. In urban communities, dry dates can be hydrated and softened by soaking with minimum loss in quality (Boubekri *et al.*, 2010); also steaming is another effective method. Fig. 14.5 shows handling and marketing of dry dates in the public market in Cairo, Egypt; practices are quite similar elsewhere in North Africa.

14.3.3   Sorting and grading

Sorting is an important step to assure quality, raise product value, and meet health and quality standards. It is worth mentioning that sometimes the fruit is delivered to a warehouse still attached to the strands. For small operations, manual sorting using trained workers with tables or small conveyor lines is sufficient. In advanced and large operations, semiautomatic and fully automatic sorting lines with machine vision technology have become available in recent years. Sorting eliminates damaged, unripe, overdry, and infested dates. Grading and/or cleaning are the second important processes. Normally dates are graded using different systems; some are local while others are regional and international. Local grading such as Fancy, Good, Average; A, B, and C; first, second, and third; and so on exist (Kader and Hussein, 2009). The US Standards for Grading of Dates was issued in 1949 by the US Department of Agriculture; it classified dates into six grades based on four factors: color, character (well developed, well fleshed, and soft), uniformity of size, and absence of defects (USDA, 1955). According to UNECE Standard DDP-08 (UNECE, 2010), dates must be intact, healthy, fully ripe, clean, fleshy, and elastic; free from insects at all stages, molds, fermentation, odors, and off-flavors; moisture content must be not more than 26% (w.b.) for cultivars with high sucrose content such as Deglet Nour and about 30% (w.b.) for cultivars with high fructose content such as Berhi. However, in some countries local systems of grading are used. In North Africa, for instance, Deglet Nour dates are commonly classified as grade A, B, C, and common or dry. In Pakistan, date fruits are graded as Extra Class, Category A, Category B, Good Quality, Fair Quality, and Industrial Grade (Phulpoto *et al.*, 2012). Other systems also exist, among which is the Codex Alimentarius Standard for dates (Yahia and Kader, 2011). Nonetheless, whatever system is used mainly deals with properties such as color, size uniformity, moisture content (dryness), defects, and infestation. Several attempts have been made to use automatic and advanced and computerized sorting systems for distinguishing cultivars, moisture content, physical properties, and color; some methods are being applied while others are at research and development stages (Pourdarbani *et al.*, 2015).

14.3.4   Disinfestation

14.3.4.1   Insects

Dates are susceptible to infestation by insects at all stages of development. El-Shafie (2012) listed 112 mite and insect species associated with date palm diseases; they affect the tree and the dates. Insects such as dubas bug (*Ommatissus lybicus*), the lesser date palm moth (*Batrachedra amydraula*), the longhorn stem borer (*Apriona germarii*), the bunch borer (*Castnia daedalus*), the frond borer (*Phonapate frontalis*), and the Arabian rhinoceros beetle (*Oryctes agamemnon arabicus*) are common sources of losses in palm tree and dates (Khalaf *et al.*, 2013; Al-Saffar and Augul, 2020). In the field, the stone beetle (*Coccotrypes dactyliperda*) is one of many kinds of insects that cause losses in dates by infecting dates at the green stage, resulting early fruit drop at an immature stage (Al-Saffar and Augul, 2020). Similarly, the date palm scale (*Parlatoria blanchardi*) infests green fruits and forms white filaments around fruits, leading to reduction of photosynthesis and the fruits do not reach maturity. Generally, infestation is carried from the field for some insects such as the date carob moth (*Ectomyelois ceratoniae*), a common pest that causes significant postharvest losses. Additionally, the lesser date moth (*B. amydraula*), the dried-fruit beetle (*Carpophilus hemipterus*), the pineapple beetle (*Urophorus humeralis*), the confused sap beetle (*Carpophilus mutilatus*), and the pineapple sap beetle (*Haptoncus luteolus*) are common insects that infest dates in the field and at storage and handling. Other insects infest dates only at the postharvest stage and are responsible for high losses, such as the raisin moth (*Cadra figulilella*), the greater date moth (*Aphomia sabella* Hampson), and the Oriental hornet (*Vespa orientalis*) (Bachrouch *et al.*, 2010; Sarraf *et al.*, 2021). Generally, well-implemented field protection procedures throughout product development stages and pre-storage disinfecting eliminate most postharvest infestation (El-Shafie, 2012).

14.3.4.2   Insect disinfestation treatments

Warehouses in date-producing countries are mostly owned by companies, government, large producers, or farmer associations. Upon receiving, each batch must be weighed, coded, and inspected for meeting facility standards. Since dates may carry insect infestations from the field, generally the first step is fumigation against common pests such as beetles and moths. Previously, methyl bromide (CH_3Br) was a powerful disinfecting treatment when dosed at 15 g/m^3 for 12–24 h at 15–16°C (El-Mohandes, 2010). However, due to its impact on the ozone layer, it has been prohibited worldwide effective 2015 by the Montreal protocol. Nowadays, several alternatives exist, which can be classified as physical, irradiation, and chemical.

PHYSICAL TREATMENTS.   High and low temperature can be effective treatments. A temperature of 55°C for 2 h led to high mortality of Nitidulidae (sap) beetles (Finkelman *et al.*, 2006). Dipping in boiling water has also been reported as a more efficient treatment than hot air at 70°C (Hussein *et al.*, 1989). A low temperature of –18°C was found effective in eliminating eggs and larvae of carob moth in addition to preserving other fruit quality attributes for up to a year (Lallouche *et al.*, 2017). Application of the same temperature for 48 h is considered an effective treatment for eliminating all life stages of insects in stored commodities (Yahia *et al.*, 2014). However, it has been reported that high temperature results in color darkening and the use of water dipping leads to loss of sugar and significant changes in other quality attributes (Hussein *et al.*, 1989; Hazbavi *et al.*, 2015).

IRRADIATION, RADIO FREQUENCY, AND ULTRAVIOLET (UV). Such methods have also been studied for their effectiveness as anti-pest treatments. The use of gamma radiation at 0.7 kGy was found effective in eliminating *Cadra cautella* and *Oryzaephilus surinamensis* (Ahmed, 1991; Khalid *et al.*, 2017). Subjecting dates of the semidry cultivar Sewi to a radio frequency of 27.12 MHz for 6 min was found effective in eliminating larvae, pupae, and adults of the dried-fruit beetle (*C. hemipterus*) with no significant effect on quality (Pegna *et al.*, 2017). Although the technique is currently at the research stage, it is quite promising in terms of time, fewer negative effects on fruits, and good effectiveness in eliminating infestations (Rosi *et al.*, 2017). Ultraviolet (UV) light, specifically UV-C with wavelength in the 240–260 nm range, is considered a promising antimicrobial technique to preserve food and maintain quality. The radiation retards microbial growth by inducing the formation of pyrimidine dimers and stopping microbial cell growth and is considered an alternative to sodium hypochlorite (NaOCl) for partially processed fruits and vegetables (Artés-Hernández *et al.*, 2009). For dates, using UV-C light at 6 kJ/m^2 on Deglet Nour eliminated microbial growth at 20°C (Jemni *et al.*, 2014). Microwaves at 2450 MHz were also applied on dates infested with adults of *Tribolium castaneum* (Herbst) and *O. surinamensis* (L.) as well as larvae of *T. castaneum*, treatment at 800 W and exposure time of 30 s being effective in eliminating both stages with no effect on quality attributes of dates (Manickavasagan *et al.*, 2013).

ALTERNATIVE CHEMICALS. Several alternative compounds to CH_3Br are used; some are applied alone while others are mixed with other agents. Phosphine (PH_3) is applied as aluminum or magnesium phosphide (AlP or Mg_3P_2) at 3 g/m^3 dose with recommended exposure time of 10 to 12 days with temperature above 10°C (EPPO, 2012). Also, it was applied at 2% w/w combined with 98% w/w CO_2 to reduce exposure time to a few hours (Dhouibi *et al.*, 2015). PH_3 combined with CO_2 achieved 100% mortality of *Plodia interpunctella* and *O. surinamensis* (El-Shafie, 2020). Although PH_3 is considered a good alternative to CH_3Br, its effect is slow, and some insects have developed resistance to it (Navarro, 2006). Ethyl formate ($C_3H_6O_2$) at a dose of 114.6 g/m^3 was found effective for eliminating all stages of carob moth (Bassi *et al.*, 2015). Other compounds such as sulfuryl fluoride (SO_2F_2) and carbonyl sulfide (COS) have been found effective for disinfesting other agricultural commodities, yet their application in dates is very much limited (Small, 2009).

OZONE (O_3). The use of ozone (O_3) at >2000 ppm for 2 h was found effective in killing larvae and adults of Indian meal moth (*P. interpunctella*) and sawtooth grain beetle (*O. surinamensis*) in infested dates, while the eggs showed tolerance to the treatment up to 4000 ppm (Niakousari *et al.*, 2010).

CARBON DIOXIDE (CO_2) AND MODIFIED ATMOSPHERE (MA). MA methods have received great attention due to their safety, effectiveness, ease of use, and low cost. The effectiveness of high CO_2 concentrations alone or combined with PH_3 or heat treatment has been investigated for killing insects at all stages. At a large scale, a CO_2 concentration of 60–80% v/v was maintained in a 151 m³ room for 4.5 months, leading to effective insect control and maintaining fruit quality attributes, with effects comparable to storing dates at −18°C (Navarro *et al.*, 2001). High CO_2 concentration at normal pressure was also investigated at the laboratory scale for its effectiveness on insects at all stages (El-Shafie, 2020; Fennir *et al.*, 2020), and at high pressure, its efficiency against several pests was also reported (Riudavets *et al.*, 2010). Among other treatments, CO_2 and CA applications for disinfecting dates may be considered as the least expensive, the safest, and the most effective in maintaining quality attributes.

14.3.5 Cleaning

In the field, dates are subjected to sand, dust, and other debris. Washing of bunches with pressurized water before ripening is a common practice in some regions, generally made near the end of the khalal stage; also, it can be applied when marketing of fresh dates is targeted. It removes dust and sand; and can also be done on ripe and dry bunches during hot conditions and in the middle of the day. Although washing in the field is effective, fruit surface washing at the packing facility is still needed, and high-pressure air at receiving is commonly used. Washing using clean water and surface drying is also done at packing facilities. Water is generally sprayed on dates on moving conveyors followed by subjecting the dates to hot and dry air to remove surface water. Additionally, surface cleaning and shining is done by passing a single layer of dates through rotating brushes. It is a common operation in high-quality dates such as Madjool and Deglet Nour. Use of an edible coating has also been reported to extend the shelf life of fresh dates in addition to improving physical and chemical properties (Abu-Shama *et al.*, 2020) and appearance (Rahemi *et al.*, 2019). Surface cleaning and manual shining using brushes is a common practice in date export facilities in Tunisia.

14.3.6 Cooling

Cold storage is an essential operation in the handling of dates to maintain storage quality attributes and extend shelf life. Facility considerations are durability, airtightness, and accessibility to handling, loading, and unloading. Inside the facility, temperature, RH, gas composition, and hygiene measures are essential. Dates are stored based on their moisture content. High moisture content dates need low temperatures, with freezing preferable, while dry dates at suitable moisture content are generally stored in cool conditions, cold and dry being preferable. Cooling benefits are to prevent color changes, prohibit syrupiness and sugar spots, and reduce chances of disease incidence and insect infestation. In addition, cold storage alleviates spoilage and minimizes flavor, texture, and quality losses. To prevent water loss and delay entering the rutab stage, fruits at the khalal stage should be stored near 0°C and 85 to 95% RH, whereas tamar can be stored for 6–12 months at 0°C. Some semisoft dates like Deglet Nour and Halawi can be stored longer than soft dates such as Madjool and Barhi. Date fruits can be stored for long periods at temperatures below −15.7°C. Dry dates with moisture content ≤20% w/w can be stored for >1 year at −18°C, for 1 year at 0°C, for 8 months at 4°C, and for only 1 month at 20°C (Siddiq and Greiby, 2014). In order to extend the shelf life of khalal dates and maintain their quality during distribution, precooling using an ice–water mixture was found to be very effective; however, it requires decontamination of the water and removal of excess surface moisture from the cooled dates before packing and transportation. Cooling of dates before transportation or storage to 0–10°C and 65–75% RH is recommended (Elansari, 2008). A study on two date cultivars (Khalas and Sukkary) showed that cold storage (3, 6, 9, 12 months at 5°C) and packing type (carton and basket) could significantly affect physical and chemical qualities (moisture, total soluble solids, tannins, pH, water activity, firmness, color, and fruit weight, length, and width). Color properties of Sukkary date were significantly affected by packing type compared to Khalas. Lightness and yellowness increased significantly during the storage period; however, a significant difference between lightness, redness, and yellowness was recognized during storage time, while redness decreased significantly at same situations (Aleid *et al.*, 2014). Generally, dates are stored frozen at about −18°C, cooled at 5°C, and at normal temperature. When dates at the rutab stage are stored frozen, color and other fruit quality properties are retained; for longer storage durations, freezing is always recommended (Kader and Hussein, 2009).

14.3.7   Dehydration and hydration

Moisture content of dates depends on cultivar, prevailing weather during harvesting, and delaying or advancing bunch removal. Generally, for cultivars palatable at a fully ripe stage, harvesting is made at the rutab or tamar stage. However, it is a well-known fact that all the dates in a bunch do not ripen at once; it takes a few weeks for all dates in a bunch to ripen, hence variation in moisture content among dates is inevitable. For even moisture content, dehydration and hydration are commonly applied. The purpose of dehydration and hydration is to improve the quality of the fruit, producing identical fruit with appropriate moisture content to extend stability during storage and marketing.

14.3.7.1   Dehydration

Dates at high moisture content are subject to severe losses and spoilage by a wide range of bacteria and fungi. Losses in quality may be changes in color, taste, appearance, and acceptance, and spoilage leads to health concerns.

Drying is perhaps the oldest, the least expensive, and the most common method for preservation of dates (Kader and Hussein, 2009). Additionally, dried dates have more benefits, greater market potential, and are easier to handle than fresh dates (Saikiran *et al.*, 2018). The drying operation should reduce moisture content to the desired level in addition to maintaining quality attributes such as color, tissue softness, and taste.

Dates need to be dehydrated to the optimal moisture content (23–25% w/w) for preserving quality during handling and storage operations. Dehydration is aimed at achieving a sugar/water ratio around 2 for soft dates, greater than 2 for dry dates, and lower than 2 for very soft dates. Ratios are considered an indicator of date quality in storage. Overdrying to below 20% w/w moisture should be avoided to keep dates texturally soft (Yahia *et al.*, 2014). From an engineering perspective, dehydration is a heat and moisture transfer process aimed at reducing moisture content to the required value. Solar drying is the oldest; it is carried out by placing dates on a mat exposed to sunlight and left until the desired drying level is achieved. However, this method is blamed for contamination and losses by insects, birds, and rodents. Advanced solar drying techniques are available using facilities similar to a greenhouse structure and, in addition, specially designed equipment such as solar drying tunnels and convective hot air-drying equipment are also reported (Behfar and Hamdami, 2012; Saikiran *et al.*, 2018). Other methods such as vacuum drying of dates have been reported as well (Amellal and Benamara, 2008; Sahari *et al.*, 2008). More advanced technology applying microwaves at a temperature range of 60–80°C has been reported to reduce drying time and maintain the nutritional value of dates (Izli, 2017). In general, whatever drying method is selected must be efficient, inexpensive, reliable, low cost, suitable for the local community from a technical perspective, have low energy consumption, and be environmentally friendly.

14.3.7.2 Hydration

Dates, if picked ripe and not overdried, do not require hydration. However, moisture must be increased by hydration in areas with high temperatures and low RH, and some cultivars, such as Amri and Zahidi, have a dry, hard texture when ripe. To hydrate dates, the fruit is saturated with water or steam at the appropriate temperature to generate optimal conditions for enzymatic activity (to make the fruit softer). This softening often goes along with a rise in moisture to a level that exposes the fruits to microbiological elements (at moisture content over 20% w/w and equilibrium moisture content (EMC) over 65% w/w), the proper hydration procedure depends on how long the dates have been exposed to these conditions (Arias-Jiménez and Zaid, 2002). In Libya, for instance, dates produced below latitude 24°N are harvested very dry and require hydration. This is achieved by several methods; among which dipping in

Fig. 14.6. Effect of soaking on size and appearance of Deglet dates affected by severe hot production conditions. (Photo by M. Fennir.)

30°C water for 6 h followed by air flow with intermediate heat at 60°C to remove excess moisture was found the best to retain quality of overdried Algerian Deglet Nour (Boubekri *et al.*, 2010). Also, dipping in hot water or using warm air at 60 to 65°C and 100% RH for 4–8 h was found effective (Yahia and Kader, 2011). Hydration of overdried Libyan Deglet by soaking in water followed by convective air drying was investigated; the effect of this treatment is shown in Fig. 14.6 (Fennir *et al.*, 2021, unpublished results). Hydration changes the dried dates into plump and glossy dates. Generally, forced air improves uniformity of temperature and RH throughout the hydration room.

14.3.8 Packaging

Packaging is mainly based on experience and skill rather than applying conceptual and postharvest engineering principles. It is perhaps one of the most effective postharvest operations; its purposes are protecting fruits from all sources of damage and infestation in addition to facilitating cooling, handling, attracting the consumer, and supporting marketing (Kader, 2002). The package must provide good protection from mechanical damage and pressure, reduce moisture loss or gain, facilitate cooling, promote the produce, be attractive to the consumer, contain any necessary nutritional information, and facilitate handling and marketing by barcoding. Additionally, for distant markets, it should include information on the exporting, importing, and distributing companies, country of origin, and other information required by laws and regulations. Generally, packaging of dates is still very traditional compared to other fruits and vegetables. Cardboard and plastic boxes, bags, and sacks are used, with size and shape varying greatly. Packing by hand and fully automatic operations both exist;

Fig. 14.7. Date pack sizes and types at the local market in Sharjah, United Arab Emirates. (Photo by M. Fennir.)

both must be carried out indoors in hygienic conditions protected from dust and wind. Fresh dates must be packed in special boxes, with attention given to facilitating cooling and ensuring fruit protection. Handling with care and using good-quality packaging material reduces mechanical damage and preserves quality. Lately, emerging techniques in packaging, coating, and wrapping employing nanotechnology appear promising, although their applications in dates are very much limited (Sarraf *et al.*, 2021).

Date packages are available in several sizes, the packed weight whether in kilograms or pounds is mainly related to a country's local system. In countries using the imperial system of units in their marketing chains, such as the USA, dates must be packed accordingly. Some standard packing sizes used in the USA are 5, 10, and 15 lb (Ait-Oubahou and Yahia, 1999). In countries using the metric system, pack sizes ranging from 1 to 10 kg are common depending on the quality of the dates and the targeted market. Dates are also marketed locally and shipped in bulk large boxes (25 kg). Fig. 14.7 shows several box sizes exhibited at a local market.

Generally, common materials used worldwide for packing dates have been tested under laboratory conditions, and approvals have been obtained. Materials must fulfill quality and health standards known as the quality index (QI), which measures the overall quality of the packaged product; results show that the highest QI value reached is 0.5 (Guedri *et al.*, 2016).

14.4 Physical and Physiological Disorders and Control Measures

Some physical and physiological disorders are very much related to field conditions and practices, such as sunburn and skin disorders due to infestation, mechanical damage, or high RH related to irrigation or rainfall. Most of these are alleviated by removing the damaged fruit during sorting. Common physiological disorders are white nose, black nose, and freezing damage. White nose appears after dry weather during the early rutab period and causes rapid fruit ripening. Black nose is a shrinking and blackening of the fruit tip caused by humid weather at the khalal stage. Although it rarely happens in warm areas, freezing damage happens by water freezing and crystallizing inside the fruits (Chao and Krueger, 2007; Sarraf *et al.*, 2021). Dates are also subjected to postharvest disorders, mainly due to handling and surrounding conditions in the packing facility. Bruises caused by pressure and rough handling lead to changes in color and shape; these are considered as physiological and physical factors (Lobo *et al.*, 2014). Additionally, dates are subject to disorders caused by insect pests. Most of these were discussed previously in Section 14.3.4.1, but some disorders are related to taste such as souring caused by yeast. Souring of ripe dates generally occurs in soft dates grown in mild and humid regions.

14.5 Biological Disorders and Control Measures

Undesirable alcoholic flavor and bad taste and odor are the main disorders caused by biological agents, generally due to fermentation by yeast and by mold growth. Temperature and high moisture content (above 20°C and above 25% w/w, respectively) generally trigger microbial growth; they negatively affect the storability and shelf life of date fruits (Kader, 2007; Sarraf *et al.*, 2021). Several fungi genera have high pathogenicity and are responsible for date spoilage at the postharvest level. The most common are *Alternaria, Aspergillus, Cladosporium, Fusarium, Rhizopus,* and *Penicillium* (Al-Bulushi *et al.*, 2017; Sarraf *et al.*, 2021). Cooling to near 0°C and the use of MA is effective in preventing microbial growth. However, O_2 depletion to the level that anaerobic respiration occurs must be prevented. Several antimicrobial agents, such as thionins and defensins, have shown great potential as antimicrobial agents in food preservation (Hintz *et al.*, 2015); however, their applications in preserving dates have not been reported.

14.6 Postharvest Fruit Quality Evaluation

In general, due to consumer awareness of quality factors of foodstuffs in general and fruit in particular, the focus on fruit quality has increased significantly. Customers have higher expectations of fruit qualities such as ripeness, firmness, taste, and flavor. For the assessment of date fruit quality, various technologies have been employed, including spectroscopic techniques (visible–near infrared (Vis-NIR) spectroscopy, multispectral imaging, hyperspectral imaging), acoustic techniques, machine vision, and electronic noses. Quality attributes targeted for assessment with these methods include soluble solids content, acidity, firmness, dry matter, vitamin C, polyphenols, and pigment contents (Wang *et al.*, 2015). The quality of date fruit is mainly dependent on appropriate postharvest handling and processing. Several nutritious products can be obtained from dates. In a review study conducted by Idowu *et al.* (2020) on the therapeutic properties of date fruits, seeds, and by-products, the date can be considered a nutritive, promising medicinal fruit. Dates can also be used in a low-cost, natural diet in vulnerable communities where diseases and food-related illnesses are common. Scientifically, date products can be potentially synergistic in the development of health-improving products for therapeutic purposes.

14.7 Processing and By-products

Generally, processing operations target local and export markets. High-quality dates can be processed into pitted packed dates, date paste, and coated dates. Overdried and usually raw materials derived from low-quality dates, with a low percentage of sugar, but not rotten or fermented, are also used. Pitted pressed dates, date paste, date syrup (sometimes called *dibs* or *rub*), and alcoholic drinks are considered the most common by-products derived from date fruits; other products are ground pitted dates for making dough and diced dates (Arias-Jiménez and Zaid, 2002). Date syrup (as a by-product) was produced from the Deglet Nour cultivar in a pilot-scale operation through a three-stage process, consisting of discontinuous solid/liquid extraction, clarification, and evaporation. This yielded 0.7 kg syrup/kg fruit and 70% Brix, and the date syrup color was characterized by pigments (color parameters *L*, *a*, and *b*) in date syrup; however, sucrose-substituted cakes showed a higher value of both total phenolic compounds and antioxidant activity, showing the potential of using date syrup as an efficient sweetener (Lajnef *et al.*, 2021). A study on second-grade date (date juice) and lemon by-products showed a possibility to produce low-calorie jellies using different sugar contents and pH values. The most valued jellies were those with the lowest sugar content, with a minor preference for that with a pH of 3.5 and with no significant differences found between scores for the other sensory characteristics (Masmoudi *et al.*, 2010). Other parts of the date palm (date palm leaflets waste) can be utilized. For example, composite materials can be made using the fibers from palm leaves as the reinforcement and expanded polystyrene (EPS) waste dissolved in gasoline as the matrix. Several combinations of reinforcement sizes (0.1–0.315,

0.315–0.5, and 0.5–1 mm) and fiber/matrix weight ratios (70, 75, and 80 wt%) were considered to investigate the properties of the leaflets–polystyrene composite (LPC) (Masri *et al.*, 2018). Date palm wastes (DPW) have attracted specific attention as a source of renewable energy in Tunisia (Bensidhom *et al.*, 2018).

14.8 Quality Assurance and Safety

Dates are noted for their inherent nutritional and functional properties, being an excellent source of simple carbohydrates, mainly glucose and fructose. However, quantities of these components differ among cultivars and depend on the maturity stage. The rutab stage has less total sugar than the tamar stage (Al-Farsi and Lee, 2008; Al-Mssallem *et al.*, 2019). Quality is normally determined by physical properties of the fruits, such as color, shape, texture, and size, and sensory attributes such as flavor and physical appearance; however, the nutritional content is evaluated with reference to their chemical properties. Color varies significantly according to the maturity stage cultivar (Amoro *et al.*, 2015; Abdul-Hamid *et al.*, 2020). Date fruit chemical composition can differ depending on cultivar, soil conditions, agronomic practices, and the ripening stage. The composition of date varieties is shown in Table 14.1. The main sugars found in date flesh are fructose, glucose, and sucrose. The more complete reduction of sugars in some cultivars suggests the presence of a more noticeable invertase activity, which would considerably reduce its content in sucrose. Dates have a fairly low content of protein, ranging from 0.46 to 4.73 g/100 g dry matter (Borchani *et al.*, 2010; Abdul Rahman, 2015; Jemni *et al.*, 2019).

14.9 Innovative Technologies

Assurance of food safety and security and the necessity of supplying products to consumers in a short time and with high quality highlight the need for automation technologies such as sensors, robots, and Global Positioning Satellite (GPS) technology. In a study investigating the role of agricultural extension facilities in improving date production, about 72% of the respondents were conscious of up-to-date technologies but only 43% of them had been educated by extension workers; it was concluded that most of the respondents (70%) were not satisfied with extension workers (Ullah *et al.*, 2014). Information about the products to guarantee traceability, including grower name, harvest time, agrichemicals applied, etc., can be reported automatically, and this information is nowadays being highly requested by consumers (Sarraf *et al.*, 2021). In addition to scientific studies, industry is also working on automation in agriculture and processing. There are now several companies in the market that produce packing lines with multiple cameras and sensors equipped with Vis-NIR

Table 14.1. Composition of the date pulp (g/100 g dry weight). (From Borchani *et al.*, 2010; Abdul Rahman, 2015; Jemni *et al.*, 2019.)

Variety	Dry matter	Total sugars	Ash	Protein	Fat
Alligh	82.94 ± 0.70	84.59 ± 0.18	2.18 ± 0.22	1.22 ±0.03	0.56 ± 0.19
Deglet Nour	86.42 ± 0.75	88.02 ± 0.60	1.78 ± 0.10	1.71 ± 0.08	0.40 ± 0.11
Bajo	86.88 ± 0.59	79.90 ± 0.31	1.73 ± 0.04	1.28 ± 0.08	0.11 ± 0.04
Boufeggous	88.70 ± 0.68	86.72 ± 0.95	1.58 ± 0.05	1.51 ± 0.16	0.14 ± 0.00
Goundi	90.57 ± 0.37	84.79 ± 0.91	1.85 ± 0.03	2.85 ± 0.20	0.35 ± 0.21
Ikhouat	87.97 ± 0.40	78.86 ± 0.33	2.59 ± 0.52	0.66 ± 0.03	0.07 ± 0.00
Kenta	88.22 ± 0.79	85.11 ± 0.46	1.75 ± 0.02	0.90 ± 0.02	0.06 ± 0.01
Kentichi	87.29 ± 0.18	77.44 ± 0.26	1.74 ± 0.05	0.46 ± 0.01	0.11 ± 0.04
Lagou	73.10 ± 0.60	77.31 ± 0.15	2.08 ± 0.02	1.83 ± 0.05	0.25 ± 0.00
Touzerzaillet	70.66 ± 0.38	78.58 ± 0.77	2.11 ± 0.19	1.49 ± 0.05	0.57 ± 0.04
Tranja	87.85 ± 0.55	83.95 ± 0.35	2.23 ± 0.09	2.42 ± 0.85	0.14 ± 0.07
Ajwa	77.20 ± 0.10	74.30 ± 0.20	3.43 ± 0.01	2.91 ± 0.02	0.47 ± 0.01
Shalaby	84.80 ± 0.20	75.90 ± 0.50	3.39 ± 0.01	4.73 ± 0.01	0.33 ± 0.05
Khodari	80.50 ± 0.10	79.40 ± 0.30	3.42 ± 0.04	3.42 ± 0.03	0.18 ± 0.04
Anabarah	70.50 ± 0.20	78.40 ± 0.20	2.33 ± 0.01	3.49 ± 0.01	0.51 ± 0.04
Sukkari	78.80 ± 0.10	78.50 ± 0.10	2.37 ± 0.05	2.76 ± 0.01	0.52 ± 0.01
Suqaey	85.50 ± 0.10	79.70 ± 0.20	2.29 ± 0.03	2.73 ± 0.04	0.41 ± 0.05
Safawy	76.40 ± 0.30	75.30 ± 0.10	1.68 ± 0.01	2.48 ± 0.02	0.12 ± 0.03
Burni	75.60 ± 0.10	81.40 ± 0.04	2.02 ± 0.01	2.50 ± 0.04	0.67 ± 0.01
Labanah	98.50 ± 0.10	71.20 ± 0.10	3.94 ± 0.02	3.87 ± 0.05	0.72 ± 0.02
Mabroom	78.70 ± 0.10	76.40 ± 0.07	2.79 ± 0.05	1.72 ± 0.05	0.27 ± 0.01

Values are presented as mean ± standard error.

spectrometric probes to assess different quality parameters of the fruits (Walsh *et al.*, 2020). A study on date fruit classification for robotic harvesting (based on difference in size, shape, color, and texture properties of various fruits) concluded that a deep convolutional neural network (CNN) could accomplish robust date fruit classification without the preprocessing of images to remove background noise or enhance illumination. The best precisions obtained by the proposed date fruit classification models were about 99.01, 97.25, and 98.59% with classification times of 20.6, 20.7, and 35.9 ms for the type, maturity, and harvesting decision classification tasks, respectively. Future work to improve the dataset by including testing images captured from different date plantations and decrease the confusion in the maturity detection of date fruit will be conducted (Altaheri *et al.*, 2019).

14.10 Conclusions and Prospects

In the last two decades, cultivation of dates has witnessed a rapid increase; high demand for quality dates strongly exists. Generally, dates are subjected to substantial losses due to several factors. Some are less controllable as they result from preharvest conditions, while others can be managed to some extent, as they are postharvest factors. Harvesting and handling practices are quite important for quality dates, as are other factors such as sorting, grading, cooling, hydration, dehydration, and packaging. Significant advancements have been achieved in improving postharvest operations, leading to reduced losses with quality maintained. Additionally, effective methods employing MA in storage and packaging are available. Quality assurance and quality guidelines and standards have been introduced. Nonetheless, research and development efforts are still very much needed. These must focus on further improvement of storage systems, packaging materials, disinfection using safe advanced methods, quality assurance, and analysis. Further effort is still needed for enhancing quality through employing methods and conditions that maintain color, firmness, freshness, and overall nutritional quality. Logistical methods are necessary for improving supply and cold storage chains, as well as modernized marketing strategies and planning. Also, investment and research are important in applying promising handling technologies such as robotics, machine vision, and low-cost analysis and quality assurance tools. Lastly, better utilization of low-quality dates to produce economically viable products such as antioxidants, fibers, phenols, sugars, alcohols, vinegars, date syrups, and date beverages, and their use in special and medicinal diets, needs to be improved.

References

Abdul-Hamid, N.A., Mustaffer, N.H., Maulidiani, M., Mediani, A., Ismail, I.S. *et al.* (2020) Quality evaluation of the physical properties, phytochemicals, biological activities and proximate analysis of nine Saudi date palm fruit varieties. *Journal of the Saudi Society of Agricultural Sciences* 19(2), 151–160. DOI: 10.1016/j.jssas.2018.08.004.

Abdul Rahman, A.E. (2015) Nutritional composition of fruit of 10 date palm (*Phoenix dactylifera* L.) cultivars grown in Saudi Arabia. *Journal of Taibah University for Science* 9(1), 75–79. DOI: 10.1016/j.jtusci.2014.07.002.

Abul-Soad, A.A. (2011) *Date Palm in Pakistan, Current Status and Prospective.* US Agency for International Development (USAID) Firms Project Report. Office of the Economic Growth and Agriculture, USAID Pakistan.

Abu-Shama, H.S., Abou-Zaid, F.O.F. and El-Sayed, E.Z. (2020) Effect of using edible coatings on fruit quality of Barhi date cultivar. *Scientia Horticulturae* 265, 109262. DOI: 10.1016/j.scienta.2020.109262.

Ahmed, J. and Ramaswamy, H.S. (2005) Physico-chemical properties of commercial date pastes (*Phoenix dactylifera*). *Journal of Food Engineering* 76(3), 348–352. DOI: 10.1016/j.jfoodeng.2005.05.033.

Ahmed, M.S.H. (1991) Irradiation disinfestation and packaging of dates. In: *Insect Disinfestation of Food and Agricultural Products by Irradiation.* International Atomic Energy Agency (IAEA), Vienna, pp. 7–26.

Ait-Oubahou, A. and Yahia, M.E. (1999) Postharvest handling of dates. *Postharvest News and Information* 10(6), 67–74.

Al-Bulushi, I.M., Bani-Uraba, M.S., Guizani, N.S., Al-Khusaibi, M.K. and Al-Sadi, A.M. (2017) Illumina MiSeq sequencing analysis of fungal diversity in stored dates. *BMC Microbiology* 17(1), 72. DOI: 10.1186/s12866-017-0985-7.

Aleid, S.M., Elansari, A.M., Zhen-Xing, T. and Sallam, A.A. (2014) Effect of cold-storage and packing type on Khalas and Sukkary dates quality. *Advance Journal of Food Science and Technology* 6(5), 603–608. DOI: 10.19026/ajfst.6.82.

Al-Farsi, M.A. and Lee, C.Y. (2008) Nutritional and functional properties of dates: a review. *Critical Reviews in Food Science and Nutrition* 48(10), 877–887. DOI: 10.1080/10408390701724264.

Ali, A., Waly, M., Essa, M.M. and Devarajan, S. (2012) Nutritional and medicinal value of date fruit. In: Manickavasagan, A., Essa, M. and Sukumar, E. (eds) *Dates: Production, Processing, Food, and Medicinal Values.* CRC Press, Boca Raton, Florida, pp. 361–375.

Al-Mssallem, M.Q., Alqurashi, R.M. and Al-Khayri, J.M. (2019) Bioactive compounds of date palm (*Phoenix dactylifera* L.). In: Murthy, H. and Bapat, V. (eds) *Bioactive Compounds in Underutilized Fruits and Nuts.* Reference Series in Phytochemistry, Springer, Cham, Switzerland, pp. 91–105.

Al-Redhaiman, K.N. (2004) Modified atmosphere improves storage ability, controls decay, and maintains quality and antioxidant contents of Barhi date fruits. *Journal of Food, Agriculture and Environment* 2(2), 25–32.

Al-Saffar, H.H. and Augul, R.S. (2020) Survey of the insect pests from some orchards in the middle of Iraq. *Plant Archives* 20, 4119–4125.

Altaheri, H., Alsulaiman, M. and Muhammad, G. (2019) Date fruit classification for robotic harvesting in a natural environment using deeplearning. *IEEE Access* 7, 117115–117133. DOI: 10.1109/ACCESS.2019.2936536.

Amellal, H. and Benamara, S. (2008) Vacuum drying of commondate pulp cubes. *Drying Technology* 26(3), 378–382. DOI: 10.1080/07373930801898232.

Amoro, S.A., Pretel, M., Almansa, M.S. and Botella, M. (2015) Antioxidant and nutritional properties of date fruit from Elche grove as affected by maturation and phenotypic variability of date palm. *Food Science and Technology International* 15(1), 65–72. DOI: 10.1177/1082013208102758.

Arias-Jiménez, E.J. and Zaid, A. (2002) Date harvesting, packinghouse management and marketing aspects. In: Zaid, A. and Arias-Jiménez, E.J. (eds) *Date Palm Cultivation*. FAO Plant Production and Protection Paper No. 156 Rev. 1, Food and Agriculture Organization of the United Nations, Rome. Available at: www.fao.org/3/y4360e/y4360e0d.htm#bm13 (accessed 4 December 2022).

Artés-Hernández, F., Escalona, V.H., Robles, P.A., Martínez-Hernández, G.B. and Artés, F. (2009) Effect of UV-C radiation on quality of minimally processed spinach leaves. *Journal of the Science of Food and Agriculture* 89(3), 414–421. DOI: 10.1002/jsfa.3460.

Bachrouch, O., Mediouni-Ben Jemâa, J., Wissem, A.W., Talou, T., Marzouk, B. *et al.* (2010) Composition and insecticidal activity of essential oil from *Pistacia lentiscus* L. against *Ectomyelois ceratoniae* Zeller and *Ephestia kuehniella* Zeller (Lepidoptera: Pyralidae). *Journal of Stored Products Research* 46(4), 242–247. DOI: 10.1016/j.jspr.2010.07.001.

Bassi, H., Bellagha, S., Lebdi, K.G., Bikoba, V. and Mitcham, E.J. (2015) Ethyl formate fumigation of dry and semidry date fruits: experimental kinetics, modeling, and lethal effecton carob moth. *Journal of Economic Entomology* 108(3), 993–999. DOI: 10.1093/jee/tov032.

Behfar, S. and Hamdami, N. (2012) Mass transfer modeling in date palm (*Phoenix dactylifera*) during hot air convective drying. *Annals of Biological Research* 3(11), 4993–5000.

Bensidhom, G., Trabelsi, A.H., Sghairoun, M., Alper, K. and Ismail Trabelsi, I. (2018) Pyrolysis of Tunisian date palm residues for the production and characterization of bio-oil, bio-char and syngas. In: Kallel, A., Ksibi, M., Ben Dhia, H. and Khélifi, N. (eds) *Recent Advances in Environmental Science from the Euro-Mediterranean and Surrounding Regions. EMCEI 2017.* Advances in Science, Technology & Innovation, Springer, Cham, Switzerland, pp. 1561–1563. DOI: 10.1007/978-3-319-70548-4_453.

Borchani, C., Besbes, S., Blecker, C., Masmoudi, M., Baati, R. and Attia, H. (2010) Chemical properties of 11 date cultivars and their corresponding fiber extracts. *African Journal of Biotechnology* 9(26), 4096–4105.

Boubekri, A., Benmoussa, H., Courtois, F. and Bonazzi, C. (2010) Softening of over-dried 'Deglet Nour' dates to obtain high-standard fruits: impact of rehydration and drying processes on quality criteria. *Drying Technology* 28(2), 222–231. DOI: 10.1080/07373930903526764.

Chao, C.T. and Krueger, R.R. (2007) The date palm (*Phoenix dactylifera* L.): overview of biology, uses, and cultivation. *HortScience* 42(5), 1077–1082. DOI: 10.21273/HORTSCI.42.5.1077.

Chukwu, O. (2010) Moisture-sorption study of dried date fruits. *AU Journal of Technology* 13(3), 175–180.

Dehghan-Shoar, Z., Hamidi-Esfahani, Z. and Abbasi, S. (2008) Effect of temperature and modified atmosphere on quality preservation of Sayer date fruits (*Phoenix dactylifera* L.). *Journal of Food Processing and Preservation* 34(2), 323–334. DOI: 10.1111/j.1745-4549.2008.00349.x.

Dhouibi, M.H., Lagha, A. and Bensalem, A. (2015) Palm dates fumigation in Tunisia: efficiency of phosphine and CO_2 mixtures, at different temperatures, as an alternative to methyl bromide. *International Journal of Agriculture Innovations and Research* 3(60), 2319–1473.

Elansari, A.M. (2008) Hydrocooling rates of Barhee dates at the Khalal stage. *Postharvest Biology and Technology* 48(3), 402–407. DOI: 10.1016/j.postharvbio.2007.11.003.

El-Juhany, L.I. (2010) Degradation of date palm trees and date production in Arab countries: causes and potential rehabilitation. *Australian Journal of Basic and Applied Sciences* 4(8), 3998–4010.

El-Mohandes, M.A. (2010) Methyl bromide alternatives for dates disinfestations. *Acta Horticulturae* 882, 555–562. DOI: 10.17660/ActaHortic.2010.882.62.

El-Rayes, D.A. (2009) Effect of carbon dioxide-enriched atmosphere during cold storage on limiting antioxidant losses and maintaining quality of 'Barhy' date fruits. *Meteorology, Environment and Arid Land Agriculture Sciences* 20(1), 3–22. DOI: 10.4197/met.20-1.1.

El-Shafie, H.A.F. (2012) Review: list of arthropod pests and their natural enemies identified worldwide on date palm, *Phoenix dactylifera* L. *Agriculture and Biology Journal of North America* 3(13), 516–524. DOI: 10.5251/abjna.2012.3.12.516.524.

El-Shafie, W.K.M. (2020) Comparison between using phosphine and/or carbon dioxide for controlling *Plodia interpunctella* and *Oryzaephilus surinamensis* in stored date fruits. *Middle East Journal of Applied Sciences* 10(4), 657–664.

EPPO (2012) Phosphine fumigation of stored products to control stored product insects in general. *Bulletin OEPP/EPPO Bulletin* 42(3), 498–500. DOI: 10.1111/epp.2622.

FAO (2021) FAOSTAT. Crop statistics. Food and Agriculture Organization of the United Nations, Rome. Available at: www.fao.org/faostat/en/#data/QCL (accessed 5 November 2021).

FAO (2022) FAOSTAT. Food and Agriculture Organization of the United Nations, Rome. Available at: www.fao.org/faostat/en/#data/QCL (accessed 20 March 2022).

Farag, K. and Al-Masri, H. (1999) Extending the Khalil stage of Khesab dates using new method of packaging and modified calcium formation. *Emirates Journal of Food and Agriculture* 11(1), 21. DOI: 10.9755/ejfa.v11i1.4944.

Farahnaky, A., Askari, H., Bakhtiyari, M. and Majzoobi, M. (2009) Accelerated ripening of Kabkab dates using sodium chloride and acetic acid solutions. *Iran Agricultural Research* 27(1), 99–112.

Fennir, M.A. and Morgham, M.T. (2016) Effects of controlled atmosphere conditions on storability of Libyan 'Hurra' soft date cultivar. *Journal of Advanced Agricultural Technologies* 3(3), 202–206.

Fennir, M.A. and Morghem, M.T. (2019) Response of two Libyan soft date cultivars to induced ripening under controlled atmosphere conditions. *Scientific Journal of Agricultural Sciences* 1(2), 1–9.

Fennir, M.A., Morgham, M.T. and Raheel, S.E. (2014) Respiration rates of ten Libyan date cultivars (*Phoenix dactylifera*) measured at Balah stage. In: *2014 2nd International Conference on Sustainable Environment and Agriculture. IPCBEE Vol 76*, IACSIT Press, pp. 31–36.

Fennir, M.A., Raheel, S.S. and Morghem, M.T. (2017) Response of Libyan soft date cultivars 'Hellawi' and 'Hurra' at 'Balah' stage to controlled atmosphere conditions. *The Libyan Journal of Agriculture* 22(1), 61–73.

Fennir, M.A., Morghem, M.T., Rheel, S.S. and Bdewi, A.B. (2020) Effectiveness of high levels of carbon dioxide and modified atmosphere conditions on insect infestation development in three dry date cultivars. *The Libyan Journal of Agriculture* 25(3), 82–92.

Fennir, M.A., Bdewi, A.S. and Morghem, M.T. (2021) The potential of using modified atmosphere and cooling for preserving 'Bronsi' and 'Taboni' soft dates at Rutab stage. *The Libyan Journal of Agriculture* 26(1), 76–87.

Finkelman, S., Navarro, S., Rindner, M. and Dias, R. (2006) Use of heat for disinfestation and control of insects in dates: laboratory and field trials. *Phytoparasitica* 34(1), 37–48. DOI: 10.1007/BF02981337.

Ghnimi, S., Al-Shibli, M., Al-Yammahi, H.R., Al-Dhaheri, A., Al-Jaberi, F. *et al.* (2018) Reducing sugars, organic acids, size, color, and texture of 21 Emirati date fruit varieties (*Phoenix dactylifera* L.). *NFS Journal* 12, 1–10. DOI: 10.1016/j.nfs.2018.04.002.

Guedri, W., Jaouadi, M. and Msahli, S. (2016) New approach for modeling the quality of the bagging date using desirability functions. *Textile Research Journal* 86(19), 2106–2116. DOI: 10.1177/0040517515621126.

Hazbavi, I., Khoshtaghaza, M.H., Mostaan, A. and Banakar, A. (2015) Effect of postharvest hot-water and heat treatment on quality of date palm (cv. Stamaran). *Journal of the Saudi Society of Agricultural Sciences* 14(2), 153–159. DOI: 10.1016/j.jssas.2013.10.003.

Hintz, T., Matthews, K.K. and Di, R. (2015) The use of plant antimicrobial compounds for food preservation. *BioMed Research International* 2015, 246264. DOI: 10.1155/2015/246264.

Hussain, I., Ahmad, S., Amjad, M. and Ahmed, R. (2015) Effect of modified sun drying techniques on fruit quality characters of dates harvested at rutab stage. *Journal of Agricultural Research* 52(3), 415–423.

Hussein, F., Souial, G.F., Khalifa, A.S., Gaefar, S.I. and Mousa, I.A. (1989) Nutritional value of some Egyptian soft date cultivars (protein and amino acids). In: *Proceedings of the Second Symposium on the Date Palm in Saudi Arabia*, King Faisal University, Al-Hassa, Saudi Arabia, March 3–6, 1986, pp. 170–171.

Idowu, A.T., Igiehon, O.O., Adekoya, A.E. and Idowu, S. (2020) Dates palm fruits: a review of their nutritional components, bioactivities and functional food applications. *AIMS Agriculture and Food* 5(4), 734–755. DOI: 10.3934/agrfood.2020.4.734.

Izli, G. (2017) Total phenolics, antioxidant capacity, colour and drying characteristics of date fruit dried with different methods. *Food Science and Technology (Campinas)* 37(1), 139–147. DOI: 10.1590/1678-457x.14516.

Jahromi, A.A. (2015) Field early and even ripening of date palm fruits. *Indian Journal of Fundamental and Applied Life Sciences* 5(2), 68–73.

Jemni, M., Sousa, M., Otón, M., Chaira, N. *et al.* (2014) UV-C treatments for keeping the microbial and sensory quality of palm dates throughout shelf life. In: Actes du 4ème Meeting International sur l'Aridoculture et Cultures Oasisennes: «Gestiondes Ressources et Applications Biotechnologiques en Aridoculture et Cultures Oasisennes: perspectives pour développement durable des zones arides», Djerba (Tunisie), 17–19 décembre 2013. *Revue Régions Arides* 35(3/2014), 545–550.

Jemni, M., Chniti, S. and Soliman, S.S. (2019) Date (*Phoenix dactylifera* L.) seed oil. In: Ramadan, M.F. (ed.) *Fruit Oils: Chemistry and Functionality.* Springer, Cham, Switzerland, pp. 815–829. DOI: 10.1007/978-3-030-12473-1.

Jonoobi, M., Shafie, M., Shirmohammadli, Y., Ashori, A., Zarea-Hosseinabadi, H. *et al.* (2019) A review on date palm tree: properties, characterization and its potential applications. *Journal of Renewable Materials* 7(11), 1055–1075. DOI: 10.32604/jrm.2019.08188.

Kader, A.A. (2002) Postharvest biology and technology: an overview. In: *Postharvest Technology of Horticultural Crops*, 3rd edn. Publication no.3311. University of California Agriculture and Natural Resources, Oakland, California, pp. 15–20.

Kader, A.A. (2007) *Recommendations for Maintaining Postharvest Quality.* Department of Plant Science, University of California, Davis, Davis, California.

Kader, A.A. and Hussein, A.M. (2009) *Harvesting and Postharvest Handling of Dates.* International Center for Agricultural Research in the Dry Areas (ICARDA), Aleppo, Syria.

Kassem, H.A., Omar, A.K.H. and Ahmed, M.A. (2011) Response of Zaghloul date palm productivity, ripening and quality to different polyethylene bagging treatments. *American-Eurasian Journal of Agriculture and Environmental Science* 11(5), 616–621.

Khalaf, M., Al-Rubeae, H.F., Al-Taweel, A.A. and Naher, F.H. (2013) First record of Arabian rhinoceros bettle, *Oryctes agamemnon arabicus* Fairmaire on date palm trees in Iraq. *Agriculture and Biology Journal of North America* 4(3), 349–351. DOI: 10.5251/abjna.2013.4.3.349.351.

Khalid, H., Hayder, M.S.H. and Maher, T.S. (2017) The optimal irradiation of Iraqi dates fruit by gamma radiation for disinfestation purposes. *Advances in Physics Theories and Applications* 17, 50–57.

Khierallah, H.S.M., Bader, S.M., Ibrahim, K.M. and Al-Jboory, I. (2015) Date palm status and perspective in Iraq. In: Al-Khayri, J.M., Jain, S.M. and Johnson, D.V. (eds) *Date Palm Genetic Resources and Utilization*, Vol. 2. *Asia and Europe.* Springer, Dordrecht, The Netherlands, pp. 97–140.

Lajnef, I., Khemiri, S., Ben Yahmed, N., Chouaibi, M. and Smaali, I. (2021) Straightforward extraction of date palm syrup from (*Phoenix dactylifera* L.) byproducts: application as sucrose substitute in sponge cake formulation. *Journal of Food Measurement and Characterization* 15(5), 3942–3952. DOI: 10.1007/s11694-021-00970-2.

Lallouche, A., Kolodyaznaya, V., Boulkrane, M.S. and Baranenko, D. (2017) Low temperature refrigeration as an alternative anti-pest treatment of dates. *Environmental and Climate Technologies* 20(1), 24–35. DOI: 10.1515/rtuect-2017-0008.

Lobo, M.G., Yahia, E.M. and Kader, A.A. (2014) Biology and postharvest physiology of date fruit. In: Siddiq, M., Aleid, S.M. and Kader, A.A. (eds) *Dates: Postharvest Science, Processing Technology and Health Benefits*, 1st edn. Wiley, Chichester, UK, pp. 105–135.

Mallah, N.A., Sahito, H.A., Kousar, T., Kubar, W.A., Jatoi, F.A. *et al.* (2017) Varietal analyze of chemical composition moisture, ash and sugar of date palm fruits. *Journal of Advanced Botany and Zoology* 5(1), 1–5.

Manaa, S., Younsi, M. and Moummi, N. (2013) Study of methods for drying dates; review the traditional drying methods in the region of Touat Wilaya of Adrar–Algeria. *Energy Procedia* 36, 521–524. DOI: 10.1016/j.egypro.2013.07.060.

Manickavasagan, A., Alahakoon, P.M.K., Al-Busaidi, T.K., Al-Adawi, S., Al-Wahaibi, A.K. *et al.* (2013) Disinfestation of stored dates using microwave energy. *Journal of Stored Products Research* 55, 1–5. DOI: 10.1016/j.jspr.2013.05.005.

Markhand, S.G., Parveen, Z., Abul-Soad, A.A., Jatoi, M.A. and Saleem, S.A. (2014) Accelerated ripening of var. Aseel dates fruit using sodium chloride and acetic acid solutions. In: *Proceedings of the Fifth International Date Palm Conference, Abu Dhabi, United Arab Emirates, 16–18 March*, Khalifa International Date Palm Award, pp. 363–370.

Masmoudi, M., Besbes, S., Blecker, C. and Attia, H. (2010) Preparation and characterization of jellies with reduced sugar content from date (*Phoenix dactylifera* L.) and lemon (*Citrus limon* L.) by-products. *Fruits* 65(1), 21–29. DOI: 10.1051/fruits/2009038.

Masri, T., Ounis, H., Sedira, L., Kaci, A. and Benchabane, A. (2018) Characterization of new composite material based on date palm leaflets and expanded polystyrene wastes. *Construction Building Materials* 164, 410–418. DOI: 10.1016/j.conbuildmat.2017.12.197.

Navarro, S. (2006) Postharvest treatment of dates. *Stewart Postharvest Review* 2(1), 1–9.

Navarro, S., Donahaye, J.E., Rindner, M. and Azrieli, A. (2001) Storage of dates under carbon dioxide atmosphere for quality preservation. In: Donahaye, E.J., Navarro, S. and Leesch, J.G. (eds) *Proceedings of the 6th International Conference on Controlled Atmosphere and Fumigation in Stored Products, Fresno, California, 29 October–3 November 2001*, Executive Printing Services, Clovis, California, pp. 231–238.

Niakousari, M., Erjaee, Z. and Javadian, S. (2010) Fumigation characteristics of ozone in postharvest treatment of Kabkab dates (*Phoenix dactylifera* L.) against selected insect infestation. *Journal of Food Protection* 73(4), 763–768. DOI: 10.4315/0362-028x-73.4.763.

Pegna, F.G., Sacchetti, P., Canuti, V., Trapani, S., Bergesio, C. *et al.* (2017) Radio frequency irradiation treatment of dates in a single layer to control *Carpophilus hemipterus*. *Biosystems Engineering* 155, 1–11. DOI: 10.1016/j.biosystemseng.2016.11.011.

Phulpoto, N.N., Shah, A.B. and Shaikh, F.M. (2012) Challenges faced by rural-women in dates processing industry in Khairpur Mirs. *Australian Journal of Business and Management Research* 2(1), 64–69. DOI: 10.52283/NSWRCA.AJBMR.20120201A09.

Pourdarbani, R., Ghassemzadeh, H.R., Seyedarabi, H., Nahandi, F.Z. and Vahed, M.M. (2015) Study on an automatic sorting system for date fruits. *Journal of the Saudi Society of Agricultural Sciences* 14(1), 83–90. DOI: 10.1016/j.jssas.2013.08.006.

Racchi, M.L., Bove, A., Turchi, A., Bashir, G., Battaglia, M. *et al.* (2014) Genetic characterization of Libyan date palm resources by microsatellite markers. *3 Biotech* 4(1), 21–32. DOI: 10.1007/s13205-013-0116-6.

Rahemi, M., Roustai, F. and Sedaghat, S. (2019) Use of edible coatings to preserve date fruits (*Phoenix dactylofera* L.). *Journal of Packaging Technology and Research* 4(1), 79–84. DOI: 10.1007/s41783-019-00078-5.

Ramadan, B.R., El-Rify, M.N.A., Abd El-Hamid, A.A. and Abd El-Majeed, M.H. (2016) Effect of some treatments on chemical composition and quality properties of Saidy date fruit (*Phoenix dactylifera* L.) during storage. *Assiut Journal of Agricultural Sciences* 47(5), 107–124. DOI: 10.21608/ajas.2016.2052.

Riudavets, J., Castañé, C., Alomar, O., Pons, M.J. and Gabarra, R. (2010) The use of carbon dioxide at high pressure to control nine stored-product pests. *Journal of Stored Products Research* 46(4), 228–233. DOI: 10.1016/j.jspr.2010.05.005.

Rosi, M.C., Garbati Pegna, F., Nencioni, A., Guidi, R., Bicego, M. *et al.* (2017) Emigration effects induced by radio frequency treatment to dates infested by *Carpophilus hemipterus*. *Insects* 10(9), 273. DOI: 10.3390/insects10090273.

Sahari, M.A., Hamidi-Esfehani, Z. and Samadlui, H. (2008) Optimization of vacuum drying characteristics of date powder. *Drying Technology* 26(6), 793–797. DOI: 10.1080/07373930802046476.

Saikiran, K.C.S., Reddy, N.S., Lavanya, M. and Venkatachalapathy, N. (2018) Different drying methods for preservation of dates: a review. *Current Journal of Applied Science and Technology* 29(5), 1–10. DOI: 10.9734/CJAST/2018/41678.

Salomon-Torres, R., Valdez-Salas, B. and Norzagaray-Plasencia, S. (2021) Date palm: source of food, sweets and beverages. In: Al-Khayri, J.M., Mohan, J.S. and Johnson, D.V. (eds) *The Date Palm Genome*. Vol. 2. *Omics and Molecular Breeding*. Compendium of Plant Genomes. Springer, Cham, Switzerland, pp. 3–26.

Sarraf, M., Jemni, M., Kahramanoğlu, I., Artés, F., Shahkoomahally, S. *et al.* (2021) Commercial techniques for preserving date palm (*Phoenix dactylifera*) fruit quality and safety: a review. *Saudi Journal of Biological Sciences* 28(8), 4408–4420. DOI: 10.1016/j.sjbs.2021.04.035.

Shamim, F., Ali, M.A., Asghar, M., Din A., Babu, I. and Yasmin, Z. (2013) Controlled ripening of date palm fruit and impact on quality during postharvest storage. *Extensive Journal of Applied Sciences* 1–2, 53–57.

Siddiq, M. and Greiby, I. (2014) Overview of date fruit production, postharvest handling, processing, and nutrition. In: Siddiq, M., Aleid, S.M. and Kader, A.A. (eds) *Dates: Postharvest Science, Processing Technology and Health Benefits*, 1st edn. Wiley, Chichester, UK, pp. 1–28.

Sidhu, J.S. (2012) Date fruit production and processing. In: Sinha, N.K., Sidhu, J.S., Barta, J., Wu, J.S.B. and Cano, M.P. (eds) *Handbook of Fruits and Fruit Processing*, 2nd edn. Wiley, Chichester, UK, pp. 629–651.

Small, G.J. (2009) Evaluation of the impact of sulfuryl fluoride fumigation and heat treatment on stored-product insect populations in UK flour mills. *International Pest Control* 51(1–2), 43–46.

Suhail, M., Durrani, Y., Hashmi, M.S., Muhammad, A., Ali, S.A. *et al.* (2020) Postharvest losses of Dhakki dates during supply chain in Pakistan. *Fresenius Environmental Bulletin* 29(1), 299–309.

Ullah, R., Kalim Ullah, K. and Khan, M.Z. (2014) Extension services and technology adoption of date palm (*Phoenix dactylifera* L.) in District Dera Ismail Khan. *Pakistan Journal of Agricultural Research* 27(2), 160–166.

UNECE (2010) *UNECE Standard DDP-08 Concerning the Marketing and Commercial Quality Control of Dates, 2010 Edition*. United Nations Economic Commission for Europe, Geneva, Switzerland. Available at: https://unece.org/fileadmin/DAM/trade/agr/standard/dry/dry_e/08Dates_e.pdf (accessed September 2021).

USDA (1955) *United States Standards for Grades of Dates*. US Department of Agriculture, Agricultural Marketing Service, Washington, DC. Available at: www.ams.usda.gov/sites/default/files/media/Date_Standard%5B1%5D.pdf (accessed December 2022).

Walsh, K.B., Blasco, J., Zude-Sasse, M. and Sun, X. (2020) Visible-NIR 'point' spectroscopy in postharvest fruit and vegetable assessment: the science behind three decades of commercial use. *Postharvest Biology and Technology* 168, 111246. DOI: 10.1016/j.postharvbio.2020.111246.

Wang, H., Peng, J., Xie, C., Bao, Y. and He, Y. (2015) Fruit quality evaluation using spectroscopy technology: a review. *Sensors* 15, 11889–11927.

Wright, G.C. (2016) The commercial date industry in the United States and Mexico. *HortScience* 51(11), 1333–1338. DOI: 10.21273/HORTSCI11043-16.

Yahia, E.M. and Kader, A.A. (2011) Date (*Phoenix dactylifera* L.). In: Yahia, E.M. (ed.) *Postharvest Biology and Technology of Tropical and Subtropical Fruits*, Vol. 3. *Coconato Mango*. Woodhead Publishing Series in Food Science,Technology and Nutrition, Woodhead Publishing, Cambridge, pp. 41–81.

Yahia, E.M., Lobo, M.G. and Kader, A.A. (2014) Harvesting and postharvest technology of dates. In: Siddiq, M., Aleid, S.M. and Kader, A.A. (eds) *Dates: Postharvest Science, Processing Technology and Health Benefits*, 1st edn. Wiley, Chichester, UK, pp. 105–135.

Zaid, A. and de Wet, P.F. (2002) Origin, geographical distribution and nutritional values of date palm. In: Zaid, A. and Arias-Jiménez, E.J. (eds) *Date Palm Cultivation*. FAO Plant Production and Protection Paper No. 156 Rev. 1, Food and Agriculture Organization of the United Nations, Rome. Available at: https://www.fao.org/3/y4360e/y4360e06.htm (accessed 4 December 2022).

Zamir, R., Nazmul Islam, A.B.M., Rahman, A., Ahmed, S. and Omar Faruque, M. (2018) Microbiological quality assessment of popular fresh date samples available in local outlets of dhaka city, Bangladesh. *International Journal of Food Science* 2018, 7840296. DOI: 10.1155/2018/7840296.

Date Food Products

15

Salah Mohammed Aleid* ⓘ

King Faisal University, Al-Ahsa, Saudi Arabia

Abstract

Date is a major crop significantly contributing to human nutrition. Several challenges face date processing such as insect infestation during storage, postharvest cooling, color- and size-based sorting, and insufficient processing and packaging technologies. Topics discussed in this chapter include postharvest handling and the date fruit's chemical and microbial nature. Ready-to-use and semifinished date-derived products with commercial importance and nutritional and health benefits are described in detail. Control of microbial spoilage in date syrup and date paste is highlighted. Enrichment of fermented and unfermented dairy milk and probiotic yoghurt with date syrup is elaborated in detail. This chapter also highlights the date as a natural alternative to white sugar in bakery products and its effect on dough rheological properties and improvements in texture, color, and nutritional values. Date fruits' fractions such as dietary fibers, pectin, and sugars that exhibit some functional properties in the food industry are also discussed. Presented as well are ways of utilizing date by-products containing date seeds and press cake. Comprehensive information about quality control and microbiological safety measures, as well as recent findings from research and development activities, also provides an overview of major recent aspects of date food products. Future scientific research for improving the manufacturing process in date processing is suggested.

Keywords: By-products, Date fruit, Industry, Processing, Quality, Value-added products

15.1   Introduction

Date fruits, the main economic product of the date palm (*Phoenix dactylifera* L.), have good food stand-alone value or as an ingredient in processed foods. They can undergo processing, be eaten fresh, or incorporated in value-added bakery and cereal products, dairy-based products, candies, jellies, jams, juices, pastes, powder (date sugar), and syrup. Field and postharvest fruit losses are high, and processing methods for date products and by-products need improvement (Erskine *et al.*, 2003). Surplus date

---

*seid@kfu.edu.sa

© CAB International 2023. *Date Palm* (eds J.M. Al-Khayri *et al.*)
DOI: 10.1079/9781800620209.0015

fruits can be utilized as a substrate in industrial fermentation processes yielding primary and secondary metabolites such as vinegar, bioethanol, organic acids, and antibiotics, or yielding biomass-based products such as baker's yeasts, single-cell proteins, and biofertilizers (Aleid, 2013). This chapter addresses date processed products as well as basic hygiene and quality control procedures to protect consumers. Recently, there has been more emphasis on dates as a super fruit, delivering exceptional health and nutritional benefits.

15.2 Date Fruits as Raw Materials

The chemical composition of date fruit is of prime importance in the utilization of dates for food-based end uses. Date fruit is an excellent source of nutritional substances. The chemical composition of dates includes carbohydrates, dietary fiber, proteins, fats, minerals, vitamins, enzymes, phenolic acid, and carotenoids, all of which are directly linked to nutritional and health benefits for consumers. A few studies have also confirmed the therapeutic effects of dates and their efficacy and protective effects against many diseases (Ibrahim *et al.*, 2021). Even though several studies have been done on date-based products, such as confectionary, soft drinks, milk shakes, ice cream, and bakery goods, more research and development activities are needed for ingredient and process optimization on a commercial level.

15.3 Date Fruit Chemical Nature

Date fruit consists of three parts: flesh with a thin skin, pit, and cap (calyx). Carbohydrates in date fruits, such as soluble sugars (glucose, fructose, sucrose) and dietary fiber (cellulose, hemicelluloses, pectin, fructans), are the most important components (Kamal-Eldin *et al.*, 2012; Oladzad *et al.*, 2021). The main components of date fruits from previous analysis in King Faisal University research laboratories are given in Table 15.1. Water activity (a_w) in the range of 0.6–0.65 is needed to maintain the quality and chemical stability of the fruit. Polygalacturonase and pectinesterase enzymes contribute to softness while cellulose causes changes in texture. Polyphenol oxidase is responsible for biological changes to polyphenols, including tanning. At least 15 minerals, of which potassium, phosphorus, and iron are present in high amounts, increase the alkalinity of dates. Phenolic components, vitamins C and E, carotenoids, and flavonoids contribute to antioxidant activity. Volatiles such as esters, alcohols, lactones, aldehydes, and ketones possess aromas and flavors, contributing to the desired sensory characteristics. Fibers are chemically bound to the insoluble proteins of the date flesh, and are mainly composed of cellulose, hemicellulose, lignin, and lignocellulose. During the ripening process, these substances are

Table 15.1. Chemical composition of date fruit.

Macronutrient	Content (g/100 g)
Water	10–20
Sugar	65–75
Protein	1–2.2
Fat	0.3–2.0
Pectin	1–4
Crude fiber	2–4.1

Micronutrient	Content (mg/100 g)
K	1090
Ca	71
Mg	62
P	60
Fe	5.8
Zn	1.5
Polyphenols	180–255
Vitamin B_6	0.24

gradually broken down by enzymes into soluble compounds, rendering the fruit more tender and softer (Aleid and Al-Saikhan, 2019). The predominant essential amino acids in fresh and dried dates include lysine and leucine, whereas the nonessential amino acids are proline, glutamic acid, aspartic acid, and glycine (Al-Farsi and Lee, 2008).

The physicochemical properties of dietary fiber extracted from date flesh show some functional properties in the food industry, e.g., high water-holding capacity, high oil-holding capacity, emulsifying, pseudoplasticity behavior of their suspensions, and gel formation. Dietary fiber from dates can be used as an ingredient in food products (dairy, soup, meat, bakery products, jam) to modify textural properties, avoid syneresis, and stabilize high-fat food and emulsions (Elleuch *et al.*, 2008). Regarding texture, dates are classified into three groups as soft, semidry, or dry dates (Fig. 15.1).

In date fruits, pectin is the major noncellulosic cell-wall component (Alhinai *et al.*, 2021). Pectin is defined as a polymer containing galacturonic acid units (at least 65%). The acid groups commonly combine with methanol as a methyl ester. Date texture depends on the amount of pectin methylation: the lower the amount of methylation, the softer the date. Tamar stage date is the softest (39% methylation) and kimri stage date is the hardest (72% methylation) (Rohani, 1988; Myhara *et al.*, 2008). Moisture and fiber content also play a role in determining whether a date is soft, semidry, or dry (Biglari *et al.*, 2008).

Fig. 15.1. Mature semidry date fruits of Suqei variety. (Photo by Mohammed S. Aleid.)

The color of dates is due to the pigments produced by browning reactions during ripening, processing, and storage (Khali and Selselet-Attou, 2007). These brown pigments can be produced by three possible mechanisms: browning of sugars, enzymatic oxidation of polyphenols, and oxidative browning of tannins. The oxidative browning reactions occur more rapidly at higher temperatures than at low temperatures. The enzymatic browning of polyphenols and tannins is faster than sugar-browning reactions (Sidhu, 2006). As date loses its green color and becomes red or yellow, tannins are preserved in cells and turn into insoluble particles. A thin layer of tannin under the date's skin is responsible for the astringency of date at the kimri stage. As date ripens, tannins decrease quickly from the highest value of 1.8–2.5% (w/w) at the kimri stage to about 0.4% (w/w) at the tamar stage of maturity. It is believed that tannins also play a role in the darkening of date after harvest (Hashempoor, 1999).

15.4 Date Fruit Microbial Nature

The microflora of date fruit depends on the maturation stage. Microbial counts are high at the first stage of maturation (kimri) and increase sharply at the rutab stage, then decrease significantly at the final dried tamar stage of maturation. Yeasts and molds are generally considered to be spoilage organisms of date fruit and counts (colony-forming units (CFU)) range from 1.7 to 7.2 $\log_{10}$ CFU/g at the rutab stage. Yeasts and mold counts are higher than bacterial counts in dates. Lactic acid bacteria are present only at the rutab stage in some varieties (Golshan and Fooladi, 2005). Moreover, Kader (2013) reported that microbial spoilage of dates at the tamar stage can be caused by yeasts, molds, and bacteria, and that spoilage can be controlled by drying the dates to 20% (w/w) moisture or lower and by maintaining the recommended temperature and relative humidity ranges throughout the handling process. Date fruit moisture content is more than 15–20% (w/w). Its pH is about 5.5–6.5, which is suitable for microorganism activity, and it is attacked by osmophile and xerophile bacteria, yeasts, and molds due to high dry matter content. If the total microbial count of date is very high, it is necessary to find suitable methods to decrease microbial count of date and its products (Iranmanesh, 2000). Date fruit extract at concentrations of 5–20% w/v exhibited antimicrobial properties and caused distortion, weakening, and eventual cell death by cell-wall lysis in the yeast *Candida albicans* (Shraideh *et al.*, 1998). Flavonoids, which are a major chemical constituent of date fruit, are reported to possess antifungal activity (Jassim and Naji, 2010).

15.5 Industrial Uses of Dates

Date fruits can be used to manufacture semifinished and ready-to-use food-based products such as date juice, date syrup, date paste, and liquid sugar (Ashraf and Hamidi-Esfahani, 2011). Moreover, the food industry incorporates such semifinished products in a variety of value-added products such as dairy-, cereal-, bakery-, and confectionary-based final products. The functional constituents in date fruits including pectin, dietary fiber, and sugars acts as gelling agents (El Fouhil *et al.*, 2011; Di Cagno *et al.*, 2017; Qadir *et al.*, 2020). The fruit has a distinctive flavor and aroma. Because date is a good source of sugar, it can be utilized as a raw material in the fermentation process. This utilization is relatively inexpensive and does not require any special treatment such as acid hydrolysis, steam explosion, or enzymatic treatment. The waste products of the date industry have great application in bioprocess technologies to produce several value-added products (Chandrasekaran and Bahkali, 2013).

15.6 Economic Feasibility for Industrial Uses

15.6.1 Postharvest losses

Gustavsson *et al.* (2011) defined postharvest loss as the decrease of edible food throughout the supply chain that leads to a decrease of edible food for human consumption, while waste refers to the loss at the end of the food chain, i.e., in retail and final consumption. Kitinoja (2016) summarized the causes of food loss and waste as follows: (i) absence or weakness of quality standards; (ii) surplus in supply of food; (iii) lack of appropriate storage facilities; and (iv) deficiency in packing material. Another study (Kader and Hussein, 2009) conducted on harvest and postharvest handling of dates to explore the causes of losses showed that the main causes of high postharvest losses were fermentation, insect infestation, birds, and mechanical damage. Insect infestations of dates likely begin in the field based on certain organic volatile compounds in mature-stage dates that play a fundamental role in insect attraction. The major insect pests attack dates either on the date palm tree, during sun drying of fruits in the field, while awaiting transport, and in storehouses, factories, and packing houses (El-Habbab *et al.*, 2017). Avoidance of mixing fallen dates on the ground with clean ones is advised. In addition, there are certain techniques to protect stored dates, such as fumigation in storehouses by aluminum phosphide or ethyl formate, modified atmosphere storage, microwave technology, ozonation, and biological control (Abo-El-Saad and El-Shafie, 2013).

El-Habbab *et al.* (2017) reported that the actual losses of dates at the processing and retail levels were caused by pest infestation resulting in fruit damage that cannot be repaired; therefore, date damage includes loss of weight and nutrients, reduced quality grade, lower market value, and/or contamination. The leading causes of postharvest losses in quality and quantity are insect infestation and damage caused by insects feeding directly on the dates. To decrease or eliminate insects forced air can be used. On the other hand, yeast, molds, and bacteria cause microbial waste. Conducting proper fumigation for stored dates to reduce losses during storage should be practiced.

15.6.2 Feasible aspects about processed dates

About 90% of the world's dates are consumed as a fresh product, with the remaining 10% going into food products (Aleid, 2009). Rigorous scientific research is needed into the use of new technology for the date industry to decrease the fluctuation of date prices, achieve stability, avoid losses, and use surplus fruits to manufacture value-added products. The date processing sector depends mainly on packaged dates with an underutilization of the full capacity of the processing lines. Most factories prefer to simply package dates because it is an easy process, generates quick profits, and has low risk compared to processing dates into more value-added products (Elsabea, 2012). There are several cost and quality factors in the date processing and packaging industry that need to be considered as major essential factors affecting the economic feasibility of dates. These factors include fumigation systems against fruit pests after harvesting and during storage, dehydration or rehydration systems, cold storage capabilities, grading systems conforming to international standards, advanced cleaning and packaging technologies, and refrigerated transportation.

15.7   Date Value-Added Products

15.7.1   Date packaging

After harvesting, dates are sorted manually by workers who remove damaged and infested dates as well as other foreign matter. Date processing involves removing the cap and in some cases the pits of the dates for better marketing and reduced transfer costs (Barreveld, 1993). Removing date pits may be done by crushing and sieving the fruits or, more sophisticatedly, by mechanical pitting of the fruit (Mahmoudi *et al.*, 2008). The next stage involves fumigation and sterilization of the fruit to prevent pest damage. Major techniques to prevent insect infestation are fumigation, heat treatment, cold storage, and irradiation, of which fumigation is the most common technology. Heat treatment and cold storage are beneficial when applied to dates for other reasons, and irradiation is an effective but not common technology. Methyl bromide was applied to dates for many years. However, it was banned in the USA and other developed countries in 2005. Other alternative fumigants are carbon disulfide, phostoxin, and ethylene oxide (Glasner *et al.*, 2002). At the next stage, dates are graded according to size, color, and moisture, then washing is usually done by automatic machines with water sprays to remove dust or other foreign materials using clean water. After washing, dates are subjected to hot air blown to remove the additional moisture from the date's surface. After that, heat treatment of 60–65°C as a partial pasteurization is applied to limit the activity of microorganisms, enzymes, and insects. Safe moisture content for storage of dates is between 24 and 25% (w/w). An a_w value in the range of 0.6–0.65 is needed to maintain quality and chemical stability (Aleid, 2013). A flow diagram for the commercial date packaging process can be seen in Fig. 15.2.

15.7.2   Modified atmosphere packaging (MAP) for loose dates

In modified atmosphere packaging (MAP) systems, the gas mixture surrounding fruits in the package is changed from the ambient composition. Elevated concentrations of carbon dioxide (CO_2) and reduced levels of oxygen (O_2) inside the package have benefits including reduced respiration, ethylene production and sensitivity to ethylene, reduced softening, and reduced decay (Kader *et al.*, 1989). Active MAP introduces the desired gas mixture into the package prior to sealing, thereby accelerating the process of achieving an equilibrium atmosphere (Zagory and Kader, 1988). It has long been recognized that increases in temperature during shipping, handling, or retailing MAP packages could cause a decrease in package O_2

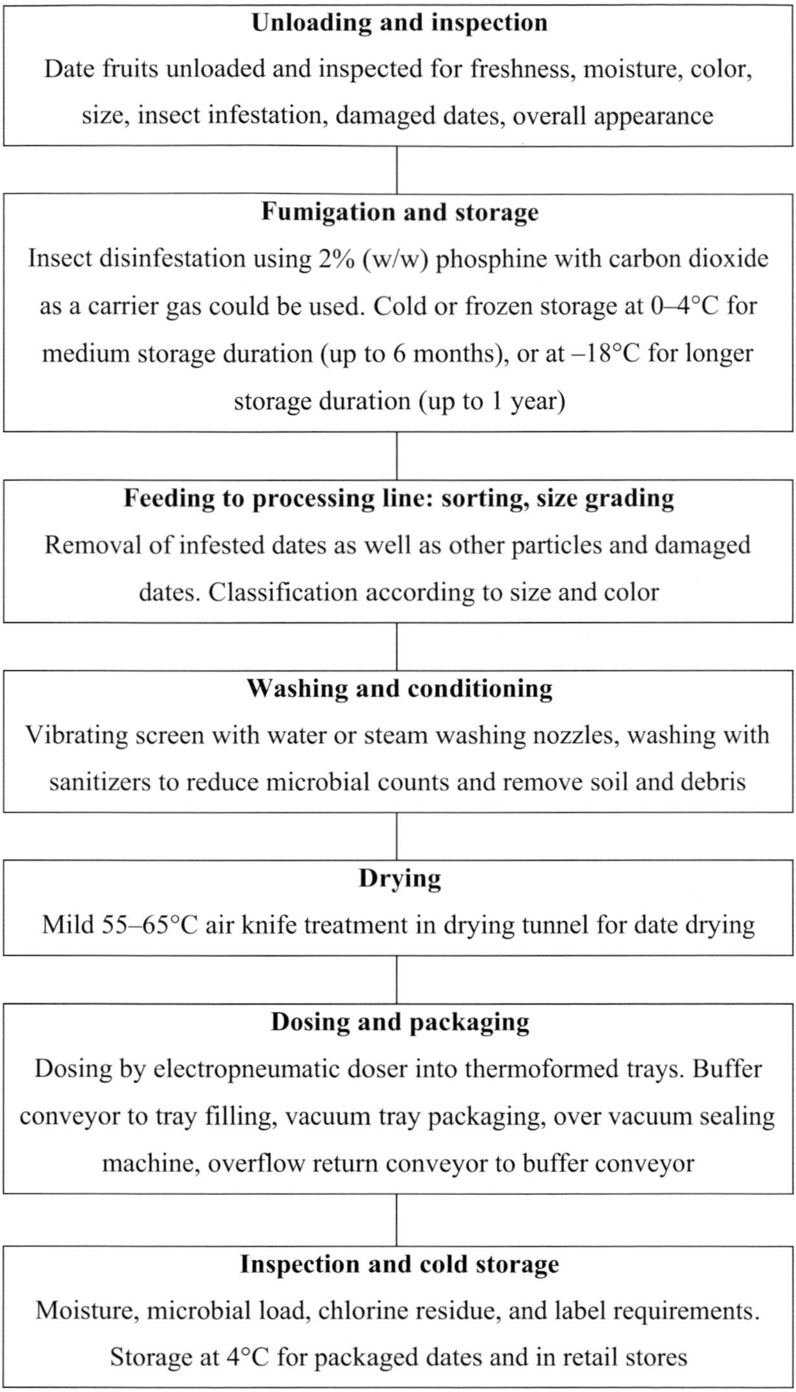

Unloading and inspection

Date fruits unloaded and inspected for freshness, moisture, color, size, insect infestation, damaged dates, overall appearance

Fumigation and storage

Insect disinfestation using 2% (w/w) phosphine with carbon dioxide as a carrier gas could be used. Cold or frozen storage at 0–4°C for medium storage duration (up to 6 months), or at –18°C for longer storage duration (up to 1 year)

Feeding to processing line: sorting, size grading

Removal of infested dates as well as other particles and damaged dates. Classification according to size and color

Washing and conditioning

Vibrating screen with water or steam washing nozzles, washing with sanitizers to reduce microbial counts and remove soil and debris

Drying

Mild 55–65°C air knife treatment in drying tunnel for date drying

Dosing and packaging

Dosing by electropneumatic doser into thermoformed trays. Buffer conveyor to tray filling, vacuum tray packaging, over vacuum sealing machine, overflow return conveyor to buffer conveyor

Inspection and cold storage

Moisture, microbial load, chlorine residue, and label requirements. Storage at 4°C for packaged dates and in retail stores

Fig. 15.2. Flow diagram for the commercial date packaging process.

levels because respiration tends to increase more than permeation of O_2 through polymeric films (Yahia *et al.*, 2004).

MAP techniques preserve the initial date quality against yeast and mold proliferation, eliminate insects, and reduce dehydration. Vacuum packaging and MAP usually decrease date dehydration during storage. Dates stuffed with almond paste that received a 10% (v/v) gas mixture of 20% (w/w) CO_2 and 80% (w/w) nitrogen (N_2) and were stored under 20°C in small amorphous polyethylene terephthalate wrappings, exhibited a shelf life of 6.6 months compared to 4.2 months observed in the case of a normal air sealing. In another case for Deglet Nour natural dates at the tamar stage stored at less than 20°C, the application of partial vacuum packaging increased the shelf life from 3.8 months, obtained with a simple sealing, to 9 months (Achour *et al.*, 2003).

15.7.3   Modified atmosphere packaging (MAP) for fresh dates (bisr)

The usage of MAP in improving the keeping quality of fresh Ghur and Khenaizy date cultivars preserved at 1°C was assessed. A gas mix of 20% (w/w) CO_2 and 80% (w/w) N_2 and a high-barrier semipermeable film allowed moisture and gas diffusion, creating an equilibrium state that postponed maturation advancement with a significant reduction in weight loss and color darkening, with firmness maintained for fresh Ghur dates during 30 days of storage. Khenaizy fresh dates could only be kept for 20 days in MAP storage with changes in the overall color domain to dark glossy brown/cinnamon colors instead of red (Fig. 15.3). The alteration in freshness seems to be a function of modified atmosphere and a_w. A positive strong linear correlation between a_w and total soluble solids (TSS) was obtained for all MAP refrigeration times. The degree of the relationship between a_w and TSS appeared to be influenced by the moisture content (Aleid and Al-Saikhan, 2019).

Weight loss, flesh firmness, TSS, a_w, acidity and appearance of Barhi cultivar dates at the khalal stage in response to vacuum and MAP were studied (Mortazavi *et al.*, 2007). Fruits in MAP treatment had insignificant weight loss, the lowest percentage of rutab fruits, the highest a_w, and only small changes in other parameters tested. However, in the vacuum packaging, weight loss and the number of crumbled fruits were the least, but a large portion of the fruit had changed to rutab stage and fruit firmness was significantly reduced (Mortazavi *et al.*, 2007). Studies have shown that dates exposed to 20% (w/w) CO_2 under modified atmosphere conditions improved during the storage period, maintained their TSS content and total tannins, showed retarded degradation of caffeoylshikimic acid, and maintained best fruit color, firmness, and proper eating quality (Al-Redhaiman, 2004). According to Baloch *et al.* (2006), for dates stored under controlled atmosphere using pure O_2, N_2, and air, O_2 accelerates and N_2 retards the darkening as compared to air. Dates stored under N_2

Control   20 day

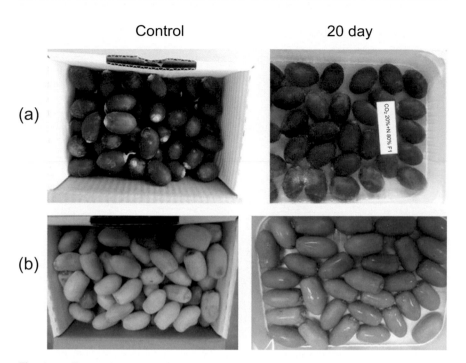

Fig. 15.3. Fresh Khenaizy (a) and Ghur (b) dates stored under control conditions and for 20 days at 1°C under modified atmosphere packaging (MAP). (Photos by Mohammed S. Aleid.)

appeared normal in color and flavor and were resistant to deterioration at the end of the storage period, whereas those under O_2 appeared dark brown with an odor of burnt sugar. Thus, controlled atmosphere containing pure N_2 and no O_2 gives the best quality of dates during storage (Baloch *et al.*, 2006, Baloch *et al.*, 2007).

15.7.4   Date paste

Processing of date fruits into date paste is a technique to preserve the fruit. Date paste comes from the steamed and minced date fruit pulp and has a smooth thick texture. Date paste is a natural sweetener that can be used as a healthy alternative to sugar in the confectionary and bakery industries (Siddiq *et al.*, 2013). Date paste can be used in the baking industry as a filling in pastries and biscuits. Moreover, it can be used as an ingredient in cereals, breads, cakes, cookies, and ice cream. Date paste has been utilized as a filler and a substitute for sugar in many food formulations. The confectionary industry has utilized date paste as one of its major ingredients (Alhamdan and Hassan, 1999).

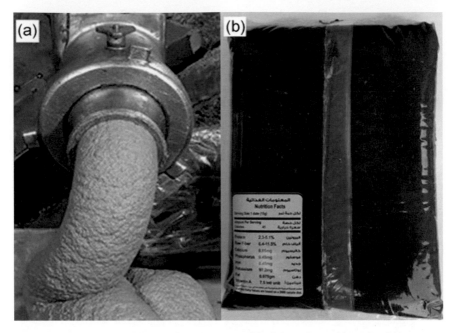

Fig. 15.4. Date paste: (a) out of mixing extruder; (b) packaged in polyethylene bags. (Photos by Mohammed S. Aleid.)

Date seed removal is the main process in date paste processing. The pitting is done in factories either by crushing the fruits or, with more sophistication, by piercing the seeds out, leaving the fruits whole. The calyces may also be mechanically removed. In processing date paste, clean-pitted dates are mixed or minced with simultaneous addition of calculated amounts of steam. Yousif *et al.* (1991a) reported that the optimum steaming time is 3–5 min. A moisture content of 23% (w/w) (0.06 a_w) for date paste could be considered as the lower safe limit for microbial spoilage. Hysteresis is apparent in the date paste isotherm and over the entire relative humidity range. The extruded date paste is usually packaged in high- or low-density polyethylene or polypropylene packages (Fig. 15.4).

Date paste prepared from the Ruzeiz cultivar has a lower a_w value (0.41) and high levels (78% w/w) of invert sugar, is a good source of minerals and trace elements, and has a high level of dietary fiber (7% w/w) (Yousif *et al.*, 1991a). Sánchez-Zapata *et al.* (2011) reported that date paste of the Medjoul cultivar had a high sugar content (53% w/w), especially reducing sugars (fructose and glucose), total and insoluble dietary fiber content of 7 and 4% (w/w), respectively, and natural antioxidants (polyphenol content of 225 mg gallic acid equivalents (GAE)/100 g). The low moisture and the natural organic acids in date paste were two important positive attributes for its storage and potential manufacturing uses.

Aleid (2009) reported that the extraction procedure of date paste from the Khalas cultivar gave a yield of 90% (w/w) as a finished product. The date paste had 11.2% (w/w) moisture, 88.8% (w/w) dry matter, 83.2% (w/w) TSS, and 2.13% (w/w) protein. The pH value of the date paste was 5.05 while the titratable acidity (expressed as a percentage of citric acid) was 0.385%. The date paste obtained had 80% (w/w) total sugars (41% fructose, 38% glucose, and 1% sucrose). The total solids were related mainly to the high sugar concentration in the paste. Ahmed and Ramaswamy (2006) reported that color is an important sensory quality attribute of date paste because it is usually the first property observed by consumers. As a result, minimizing color losses during processing and storage is of primary concern to the processor. Color varies among varieties and different maturity stages of dates. The initial green color converts to yellow then to golden or reddish-brown. Moreover, date paste's color changes due to processing conditions. The dryness and hardness of date paste were reported to be the main problem facing the producers of date paste. Experiments were carried out by Al-Abid *et al.* (2007) to overcome the solidification of date paste. Exposure to steam for 10 min was recommended based on the assessment of moisture, pH, compression force, color, a_w, and sensory evaluation.

15.7.5   Date syrup

15.7.5.1   Date syrup extraction and chemical composition

In date syrup production, the fruits go through juice extraction, filtration, and concentration. Pretreatment involves improving penetration of water by crushing the dates by means of special rollers. Juice extraction can be carried out by a batch or a continuous process in which the production line can work at a steady state continuously (Barreveld, 1993). Using ultrasonic waves in syrup extraction also results in reduced microbial count and extraction time (Entezari *et al.*, 2004). The quantity of crude fiber is important in date processing. For instance, higher amounts of fiber and moisture result in lower production of liquid date sugar (Hashempoor, 1999). During date syrup production and storage, pectin can gel, which results in lower yields of liquid sugar. This effect is eliminated by pretreatment of the extracted juice with pectinase, a pectin-decomposing enzyme (Fallahi, 1996).

Water to fruit weight ratio is a very important parameter in solids extraction, for technical and economic reasons (Ramadan, 1998). The pulp–water at ratio 1:3 gives a high recovery (72% w/w) of TSS (fresh basis). Several technologies have been employed for clarification of date juice, such as juice pretreatment by boiling, precipitating, or hydrolyzing colloids by enzyme, removing insoluble foreign matters by means of filtration or centrifugation, decolorization using activated carbon or ion exchange, removing the minerals using chemicals or ion exchange, and removing high-molecular-weight compounds by foaming (Barreveld, 1993). Cake washing for recycling some of the soluble solids is a part of the two-stage extraction system, the main advantage of which is using the added water twice in a countercurrent system. Two-stage extraction systems are of great importance for saving the taste-producing substances during refining, and should be considered in all processes (Barreveld, 1993). After refining, juice with soluble solids content of 20–25% (w/w) is obtained, which needs to be concentrated. The most common soluble solids content of date syrup is 75% (w/w) (Al-Farsi, 2003).

Aleid *et al.* (1999, 2000) extracted date syrup by assembling pressure, heating, and filtering apparatuses and investigated the effect of this method of extraction on the chemical composition of date syrup. They found that treating the date–water mixture (1:2.5) at 103 kPa for 10 min gave the best TSS recovery. The extracted date syrup had 13.85% (w/w) moisture. Moisture content highly influences the storage stability of syrups, due to its favoring undesirable microbial fermentation (Aleid *et al.*, 2000). The extracted date syrup had 84% (w/w) total sugars on a dry weight basis (44% fructose, 39% glucose, 1% sucrose). The high sugar concentration represented the main portion of the total solids in the date syrup. The monosaccharides (glucose, fructose) were the prevailing sugars whereas sucrose was limited. In detail, fructose concentration was greater than glucose concentration (Table 15.2). Aleid (2006) analyzed date syrup and proved that fructose and glucose sugars were a large part.

El-Sharnouby *et al.* (2009) prepared date syrup from the Ruzeiz date, a soft variety, at different water/date ratios (2:1, 2.5:1, and 3:1). Pectinase and cellulase were used to obtain the maximum TSS extraction. The extraction rate of sugars increased as the water/flesh ratio increased. Also, the use of pectinase/cellulase gave the highest recovery of TSS (65.6 to 70.7% w/w) compared with control (50.5 to 56.3% w/w). The resultant date syrups were evaluated for their physicochemical characteristics and compared to cane syrup (molasses). Results of the organoleptic evaluation proved that date syrup is considered more desirable than cane syrup. Results indicated the possibility of employing pectinase/cellulase to produce concentrated date syrup from date fruits for use in food product development.

15.7.5.2 Date syrup-based carbonated beverages

Soft drinks usually contain water mixed with other ingredients to deliver specific sensorial attributes. Beverage preference is motivated by the sensations it provides (Redondo *et al.*, 2014). The industrial production of soft drinks involves several stages starting with water treatment, followed by adding simple sugar syrup that highly contributes to the taste. Moreover, several ingredients can be added to the simple sugar syrup to form the final sugar-mix concentrate containing, for example, glucose or fructose, aromatic extracts, single or multiple vitamins and minerals, and other additives such as acidulants, colorants, and preservatives. At a final stage,

Table 15.2. Chemical composition of date syrup from Ruzeiz cultivar, dry basis. (From S.I. Al-Jendan, Saudi Food and Drug Authority, Inspection Lab, Eastern Provence, Saudi Arabia, 2021, personal communication.)

Macronutrient	Content (g/100 g)
Moisture	13.8
Total solids	86.2
Total sugars	84
Fructose	44
Glucose	39
Sucrose	1
Ash	1.8
Protein	1.8
Pectin[a]	0.5
Acetic acid	0.5
Butyric acid	0.2

Micronutrient	Content (mg/100 g)
K	940
Na	620
P	160
Ca	72
Mg	45
Fe	0.8
Zn	0.5
Polyphenols[b]	2043

[a]As calcium pectate.
[b]As gallic acid equivalents in the concentrated, thick Khalas date syrup (75°Brix).

the manufactured beverage is prepared as a mixture of the final sugar-mix concentrate, the treated water, and CO_2 gas (in carbonated beverages). Water is considered the main constituent, representing near 87–92% of the drink, and the sweetener usually accounts for about 8–12% of the beverage by weight. An additional common part in nearly all beverages is the acidulant (specific organic or inorganic acid) donating sour taste to the soft drink. The balance between sweetness and sourness is the basic typical taste profile of all flavored soft drinks (Shachman, 2005).

Carbonated date-based soft drinks are prepared by blending diluted date syrup with 0.3% (w/w) citric acid. Flavor substances such as cola, apple, and ginger can be used in addition to 0.06% phosphoric acid that is added to cola drinks based on the final weight of the soft drink. The mixture is pasteurized at 85°C for 15 min and chilled to 4°C. After cooling,

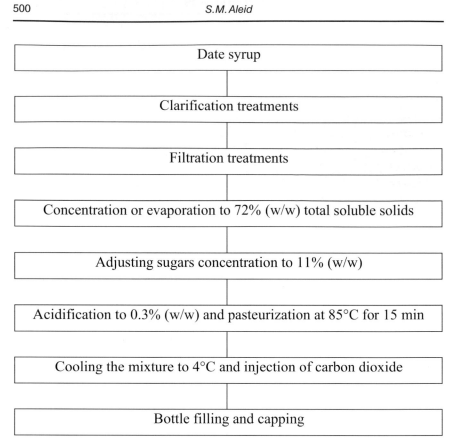

Fig. 15.5. Flow chart for production of carbonated soft drinks from date syrup.

the mixture is injected with CO_2 gas and dispensed into glass bottles to volume before sealing. The standard specification illustrates that total sugar ratio ranges from 8 to 14% (w/w) in the final product. The amount of CO_2 in soft drinks ranges from 2 to 4.5, measured as volume of gas per volume of liquid, or as pressure inside the bottle (psi). The volume of CO_2 gas significantly affects soft drink quality. A flow chart for production of carbonated soft drinks from dates is presented in Fig. 15.5.

15.7.6   Date-based cereal products

The texture of date fruit can be described as dense and tacky, which easily facilitates binding and mixing with complementary ingredients such as wheat flour. Therefore, date fruit can be used in many applications during product development to improve binding capacity (Ibrahim *et al.*, 2021). Date is a natural alternative to white sugar for bakery products. The replacement of sucrose by date paste in bakery products would improve their nutritional value by increasing levels of both minerals and vitamins.

Fig. 15.6. Bread roll stuffed with date paste. (Photos by Mohammed S. Aleid.)

Most of the date sugars are invert sugar resulting in increasing softness when used in bread and cookie production. Sugar is the basic source of energy that yeast converts into CO_2 during the early stages of fermentation and dough proofing. The concentration of sugar used in dough depends on the type of the product and desired crust characteristics, improvement in texture of the crumb, retention of the moisture in the crumb, and adding to the nutritional value of the bread. Sugar is added to provide a pleasant flavor and to develop a desired crust color (Salhlstrum *et al.*, 2004). In baked goods, the addition of date paste helps to improve dough rheological properties, retard the development of staleness, and prolong shelf life. Moreover, date paste can be used as a filling agent in various baked confectionaries as can be seen from Fig. 15.6.

15.7.6.1 Bread fortified with dates

Aleid (2009) investigated the incorporation of date paste of the Khalas variety into Arabic bread as a sugar source at 10, 20, and 30% based on flour weight. Flours with 80 and 95% extraction rate were used. Lower extraction rate indicates that less of the bran and germ remain in the flour after milling. The baked loaves were tested for internal and external quality characteristics. Quality results from baking tests showed a positive relationship between the total bread quality and the addition of date paste. Addition of date paste up to 20% (w/w) of the dough weight of Arabic bread was recommended (Fig. 15.7). Date paste can be dried and milled into powder. Such a process will extend the shelf life and storage cost. Ahmed and Ramaswamy (2006) reported that the use of 4–8% (w/w) date paste in bread formulation resulted in significant improvements in the dough rheological properties, delayed gelatinization, improved gas production and retention, extended the shelf life, improved the crumb

Fig. 15.7. Arabic bread fortified with date paste. (Photo by Mohammed S. Aleid.) The dough was made of wheat flour, water optimum, salt 1.5% (w/w), dry yeast 0.5% (w/w), date paste 20% (w/w), ascorbic acid 100 ppm. After mixing for 4 min, the dough was fermented in a proofing cabinet (30°C, 90–95% relative humidity for 1 h); divided into 100 g balls; fermented for 30 min; sheeted to 7 mm thickness; then allowed to relax for 30 min under the same conditions; and baked at 450 ± 2°C for 1 min.

and crust characteristics, and retarded development of staleness, which is a process when starch molecules crystallize.

The effect of substituting date palm fruit pulp meal (DPFPM) for granulated sugar in bread production was evaluated by Obiegbuna *et al.* (2013). Date palm fruit pulp was oven dried (46°C, 8 h) and milled. DPFPM at 0, 25, 50, 75 and 100% was substituted for granulated sugar in a bread recipe and bread was produced using a straight dough bulk fermentation method. Proofing ability of the dough, bread loaf weight, volume, and specific volume and oven spring were evaluated. Oven spring in bread baking refers to the final burst of rising just after a loaf is put in the oven and before the crust hardens.

The use of DPFPM as a granulated sugar substitute in bread production improved the nutritional value of bread. The overall organoleptic quality was not affected even at 100% replacement. The physical property of the bread, judged mostly by the loaf specific volume that relates loaf weight to its volume, decreased with increasing replacement of granulated sugar. As the decrease was not significant at 75% replacement with DPFPM, but was at 100% replacement, replacement of granulated sugar with a proportion of the date fruit pulp should be encouraged. Using a high proportion may delay the incidence of mold growth on the loaves but may encourage

Table 15.3. Cookies recipe. (From Alsenaien *et al.*, 2015, used under the Creative Commons Attribution License (CC-BY).)

Ingredient	Amount (g)
Flour	100
Sugar	60
Nonfat dried milk	3
Shortening	30
Water[a]	6.75
Ammonium chloride	0.50
Sodium bicarbonate	0.80
Sodium chloride	0.45

[a]Based on 14% moisture basis of flour.

greater growth under prolonged storage. The study confirmed that date palm fruit could be considered an ideal food that provides a wide range of essential nutrients (increased protein, fiber, ash, vitamin, and minerals, and low in fat) with many potential health benefits, as reflected in bread substituted with date palm fruit flour (Awofadeju *et al.*, 2021).

Nwanekezi *et al.* (2015) evaluated the effect of the substitution of date palm fruit pulp for sucrose on bread quality. The 5% (w/w) sucrose in the formula was replaced with date palm fruit pulp at various ratios (100:0, 75:25, 50:50, and 25:75). Proximate analysis revealed that the protein, moisture, ash, crude fiber, and fat contents increased with increasing level of the date palm fruit pulp. However, there was a decrease in the carbohydrate content from 45.39 to 35.13% (w/w) as the level of date palm fruit pulp increased. The addition of date palm fruit pulp had no effect on loaf volume. The sensory analysis revealed that all the loaf samples were acceptable.

15.7.6.2 Cookies and biscuits fortified with dates

Cookies are usually prepared from soft wheat flour. A recipe for making cookies is shown in Table 15.3. After mixing, the cookie dough was rolled to 0.6 mm thickness, cut into cookies of 6.5 mm diameter, and baked at 205°C.

Alsenaien *et al.* (2015) studied the effects of date powder and date syrup, as a sugar substitution, on the physical properties and sensory attributes of cookies. Date powder is usually made when the fruits are sorted, pitted, dried, ground, and sieved to produce a fine powder. An increase in firmness and moisture content of cookies was obtained with the use of date ingredients. The diameter and spread ratio of cookies showed a decrease with increasing level of date powder or date syrup. Partial replacement of sugar with date powder or date syrup produced cookies with a darker and redder color. Sensory evaluation results indicated that cookies with acceptable preference could be supplemented with sucrose substitution of up to 50% with date powder and up to 75% with date syrup. Date-based cookies using date powder or date syrup could be produced at a commercial scale. Yousif *et al.* (1991b) found that the use of date paste in cookies resulted in a higher spread ratio, which increased with substitution of up to 20%. It prevented the crystallization of sucrose in cookies during the cooling period immediately after taking the cookies out of the oven.

Kenawi *et al.* (2016) developed hard sweet biscuits by partial replacement of sucrose with 10, 20, 30, 40, and 50% date powder. Biscuits supplemented with 30% date powder were most acceptable due to improvements in general appearance, taste, texture, and crust color, as compared to the control. A decrease of the specific volume and density and an increase in the hardness were noticed with increased levels of date powder supplementation. An increase was found in the moisture content, starch, ash, fiber, and mineral contents, compared to the control. Sulieman *et al.* (2011) used date powder at 5 and 10% replacement levels of wheat flour for the production of biscuits. The sensory evaluation of the different biscuit samples revealed no significant differences among biscuits made from different blends of wheat flour and date powder. However, panelists gave higher scores to the 5% date powder (cultivar Agwa) than the biscuit made from other blends.

15.7.6.3   Cakes fortified with dates

Date press cake (DPC) is a by-product of date fruit syrup manufacturing that has remained underutilized in the date industry. The addition of date fiber press cake (DFPC) from the date syrup manufacturing process in chocolate cake produced a positive impact on health, storage, and sensory properties. DFPC was dried, ground, and replaced wheat at rates of 5, 10, 15 and 20% (w/w). Both ash and fiber contents increased significantly to reach maximum values of 3.3 and 3.2% (w/w), respectively. The volume of cake decreased with the increase of DFPC percentage. When replaced for wheat at a rate of 15% (w/w), DFPC gave superior sensory properties for freshness and overall acceptance and appearance. Bacterial examination of frozen and refrigerated samples showed that bacterial growth decreased with increased DFPC level during frozen storage for 28 days and during refrigeration for 21 days (Abass *et al.*, 2017).

A high-quality, low-calorie cake was produced, enriched with DFPC and date syrup, with high quality and superior sensory properties. The cake produced from wheat flour and DFPC gave an overall acceptance of 8.7 out of 9 on a hedonic scale. The best acceptance was for the cake with the addition of 6% DFPC when 50% of the added sugar was substituted with date syrup, as can be seen from Fig. 15.8.

(a) (b)

Fig. 15.8. Yellow cake with (a) 100% sucrose as sugar source and (b) 50% sucrose + 50% date syrup + 6% date syrup press cake sugar source. (Photo courtesy of S.H. Alabad.)

15.7.7 Date-based dairy products

Date syrup could be utilized as a natural sweetener in the formulation of new dairy food systems to improve taste, smoothness, and nutritional properties. Newly developed fermented milk products enriched with date syrup are described below.

Alhamdan *et al.* (2021) investigated the nutritional, microbial, and sensory quality attributes of a fermented milk (laban) drink flavored with date syrup during cold storage at 4°C for 7 days. Date syrup was added to laban in specific proportions (2.5, 5, 7.5, 10, 12.5, and 15% of date syrup per total weight of flavored laban) and an appropriate percentage (12.5%, 74°Brix) was selected based on the sensory preference of panelists. The results indicated that flavoring laban with date syrup affected the physicochemical, nutritional, microbial, and sensory quality attributes of the product in different ways. Incorporation of date syrup in the fresh laban drink significantly increased the pH, ash, protein, total solids, sugars, and magnesium. However, acidity, fat, casein, lactose, calcium, total microbial count, and total yeast and mold counts were decreased. During storage, acidity, ash, and microbial load progressively increased, while fat, casein, total solids, and sugars showed a simultaneous reduction as the storage period progressed. The panelists preferred the freshly prepared flavored laban drink compared to the stored one, which is not surprising. After 7 days of storage, the flavored laban drink was more acceptable than an unflavored one. The findings of this research will promote fortifying dairy products with dates to create highly nutritious drinks without the addition of artificial additives, refined sweeteners, and preservatives, which would be accepted by consumers.

Hariri *et al.* (2020) characterized milk–date mixtures used to produce steamed yoghurt. Dates at 9, 20, and 30% based on milk weight were mixed with milk at 30 and 45°C for 2.5 h. The addition of dates to milk increased the levels of density, viscosity, dry matter, proteins, pectin, ash, and mineral salts. Stability of the quantity of lactose and a slight decrease in total fat were observed. The percentage of sugars increased gradually and stabilized at a value of 6% (w/w) compared to the control (4.5% w/w). All samples of milk were of satisfactory hygienic quality. Milk–date mixtures were appreciated by panelists for their color, taste, and flavor. The mixture prepared at 30°C remained stable after 19 h while the mixture at 45°C became coagulated after 7 h. The emulsifying power increased gradually with the concentration of dates used and the foaming power of the control milk was the highest compared with the mixtures, being 26.8% (w/w) at the beginning and then decreasing to 16.6% (w/w) after 45 min. Yoghurts prepared with milk–date mixtures were noticeable for their good biochemical, microbiological, and sensory characteristics. The new milk–date system is very rich in nutritional elements and able to produce steamed yoghurts of good quality and high acceptability to consumers.

Abdollahzadeh *et al.* (2018) investigated the effect of date extract (DE) addition on the microbiological, physicochemical, rheological, and sensory characteristics of probiotic fermented milk. DE was added to milk at the level of 0–12 g/100 ml; the mixtures were then fermented with *Lactobacillus acidophilus* La-5. The initial probiotic concentrations ranged between 8.2 and 8.8 $\log_{10}$ CFU/g. Although the highest DE concentration led to a significant count reduction (from 8.2 to 6.4 $\log_{10}$ CFU/g), the probiotic concentration was above 6 $\log_{10}$ CFU/g in all samples during 14 days of storage at 4°C. All treatments had similar ash, fat, and protein contents. As the DE concentration increased, total solids (from 9.9 to 16.9 g/100 g) and ferric reducing antioxidant power (from 4.0 to 22.2 mg ascorbic acid equivalent/100 g) increased. Samples with higher DE concentration showed lower pH and syneresis, and higher acidity. DE addition slightly increased the viscosity from 53.5 to 70.0 mPa/s. In accordance with the power law model, shear-thinning behavior was observed at all concentrations. DE-added products had lower L^* (lightness) and higher a^* (redness) color values. Fortification with DE did not adversely affect the sensory acceptability. DE, therefore, seems to be a suitable candidate to improve the nutritional quality of probiotic dairy foods.

El-Nagga and Abd El–Tawab (2012) prepared fermented yoghurt enriched with date syrup. Fresh standardized buffalo milk (3.0% fat) was heated to 90°C/15 min, then concentrated date syrup was added based on milk weight at different levels (1, 2, 3%) at 50°C, the mixture rapidly cooled to 42°C, and divided into eight equal portions. Each portion was separately inoculated with starter cultures as follows: (i) a yoghurt, inoculated with 2% active growing culture (mixed 1:1 *Streptococcus thermophilus* and *Lactobacillus delbruekii* subsp. *bulgaricus*), or (ii) a bio-yoghurt, inoculated with 6% active starter cultures of *Bifidobacterium bifidum*, *L. acidophilus*, and *S. thermophilus* (2:1:2). The organoleptic properties of the resultant fermented milk made with/or without addition of date syrup revealed that yoghurt with 2% date syrup had the highest scores for flavor, appearance, and body texture compared to those of the control; this may be due to increased gel firmness and improved body texture. Furthermore, the average total score points decreased gradually with prolongation of the storage period. The changes in total, *S. thermophilus* lactobacilli, and *B. bifidum* counts of the resultant fermented milk products with 2% date syrup demonstrated that viable counts of total bacteria and *S. thermophilus* lactobacilli decreased gradually with the increase of storage period. This decline in viable bacterial count was likely due to the developing acidity of the resultant products (Abd El-Tawab, 2009). Moreover, the bio-yoghurt with 2% date syrup attained the greatest population of *B. bifidum* at the end of storage as compared to the control. Furthermore, yeasts, molds, and *Escherichia coli* were completely absent in either the fresh or stored samples. This finding may be due to the production of antimicrobial agents by bifidobacteria, which suppress the growth of yeasts (Mohamed, 2002).

A new drink was produced by the addition of date syrup to milk in concentrations of 5, 8, and 10% based on milk weight. It was concluded that by increasing the amount of date syrup in the formulation of the drink, characteristics such as dry matter, specific gravity, and turbidity were increased. Color measurement reported that the color parameters L^* (lightness), a^* (redness), and b^* (yellowness) of the samples showed significant differences and using date syrup in formulation of the drink caused changes in the color of the samples. By increasing the amount of date syrup, the acceptability of the samples was increased, with the highest score belonging to the sample including 10% date syrup (Ardali *et al.*, 2014). The difference in color between fresh milk and that enriched with date syrup can be seen in Fig. 15.9. This product represents a two-phase system in which the insoluble or immiscible dispersed phase (date syrup) is distributed through the continuous phase (milk).

15.7.8 Date-based confectionary products

15.7.8.1 Dates in candy and toffee making

Plain date bars can be prepared from date fruit, almonds, coconut, groundnuts, and pistachios, coated with chocolate, and fortified with sesame, skim milk powder, and oat flakes (Sidhu and Al-Hooti, 2005). Date candy was prepared using date paste with roasted groundnuts, and coconut. The combination of date paste and nuts in a 60:40 ratio with chocolate coating resulted in good sensory properties. Zeeshan *et al.* (2017) developed candy from Dhakki dates picked at the khalal stage. The candy was prepared by steeping the Dhakki dates in a sugar syrup at 0, 20, 40, 60, or 70% TSS for 24 h. The syrup was drained and dried at 60°C to ≤16% moisture content. Physicochemical and sensory characteristics such as moisture, pH, TSS, color, flavor, texture, and overall acceptability were studied for a total period of 6 months. Among them, the best treatment was identified based on overall acceptability. Candy prepared from the 60% sugar

Fig. 15.9. (a) Pasteurized fresh milk enriched with date syrup and (b) plain fresh milk. (Photo courtesy of Shaheed M. Alshikhsaleh.)

syrup proved to be best, but the candy prepared from the 40% sugar syrup was equally as good. The least acceptable was the candy of 0% followed by 70% sugar syrups. Sensory properties, moisture, and pH decreased while TSS increased during 6 months of storage. Candy packed in high-density polyethylene bags could be kept safely for up to 6 months.

Khapre and Shah (2016) presented a research work intended to standardize the recipe for the preparation of toffee by incorporating soft date pulp as a nutritional substitute for sugar. Uniformly ripened soft dates were selected and washed with water, then blanched in hot water at

80°C for 1 to 2 min. The seed present in the central portion was scooped out. The peripheral pulpy portion was cut and mixed in a grinder. The extracted pulp was concentrated to about a third of its original volume and then sugar, glucose syrup, fat, and skim milk powder were added. The whole mixture was cooked up to a concentration of 80–85°Brix. The standard recipe for toffee obtained was soft date pulp 1 kg, sugar 550 g, liquid glucose 150 g, fat 150 g, and skim milk powder 150 g, at 86°C cooking temperature. The soft date toffee had good sensory properties and was chemically analyzed as containing moisture 7.2% (w/w), ash 2.2% (w/w), crude protein 4.5% (w/w), crude fat 8% (w/w), TSS 86°Brix, pH 5.9, and acidity 0.30% (w/w).

Date candies were prepared by Al-Rawahi *et al.* (2005) as follows: dry dates were pitted and made into a paste using water and rose water. This was followed by adding and thoroughly mixing melted butter. Powdered toasted pistachios, cashew nuts, and almonds were incorporated in the formulation. Finally, coconut, cardamon, and cinnamon powder were included in the mixture. The paste was then made into balls 2 cm in diameter. These were rolled in coconut powder. Some preparations were further treated by cooking at 50°C for 5 min. Products were aged up to 13 weeks, with the results demonstrating that the textural quality remained stable throughout the storage period. Texture profile analysis (TPA) attributes of hardness, firmness, brittleness, and adhesiveness were evaluated. Over the years, research on the textural properties of food materials developed a protocol that identified certain parameters used to control quality. One school of thought argues that these can be divided into the primary characteristics of firmness, hardness, fracturability (brittleness), and adhesiveness, and the secondary (or derived) properties of springiness and cohesiveness (Peleg, 1983). TPA has been utilized widely by academics and industrialists to record and manipulate the structural behavior of processed foodstuffs to improve their quality (Kasapis *et al.*, 2004).

The cooked versions of the candies developed gained higher values of hardness due to some water loss incurred while cooking. The texture of the date candy was stable over a period of 3 months, which is the maximum time reported by retailers for purchase of the product. Clearly, the overall structural relaxation was maintained at ambient temperature. Structural relaxation in the glassy state of amorphous food components during isothermal storage/aging is also known as physical aging (Syamaladevi *et al.*, 2015). Furthermore, no significant difference was noticed between the two candies in terms of these quality criteria. The TPA attributes (plasticity and adhesiveness) of the paste to the palate were fully investigated via a trained sensory panel that allowed alignment of the consumer's perception to instrumental textural attributes. The connection achieved between instrumental textural attributes and sensory responses should provide immediate feedback on the effect of ingredients on consumer expectations (Al-Rawahi *et al.*, 2006).

15.7.8.2 Dates in jam making

Dates contain high sugar contents that are suitable for jam making (Sidhu, 2006). A sugar/date pulp ratio of 55:45 was used for jam making with 65% (w/w) sugar content, 1% (w/w) pectin, and a pH of 3.0–3.2 (Sidhu and Al-Hooti, 2005). For jelly making, a date juice/sugar ratio of 1:1 was used, and the finished product had a TSS content of 73°Brix and pH of 3.57 (Sidhu and Al-Hooti, 2005). Besbes *et al.* (2009) used second-grade dates with a hard texture (from Tunisian cultivars Allig and Kentichi) with sugar concentration of ~73.3–89.5 g/100 g dry matter in jam production. The corresponding jams were characterized in terms of chemical composition, physical properties (texture and water retention capacities), and sensory properties. Results showed a significant effect of date variety on the composition and physical characteristics of the date jams. Allig jam was richer in reducing sugars and was characterized by its greater firmness and water retention capacity. Allig and Kentichi jams presented high overall acceptability. Results from this work revealed essential information that could promote the commercialization of date jam.

Yousif *et al.* (1987) conducted research on the possibility of using date paste as a replacement for caramel or sugar paste in preparing candy bars. Processing conditions, nutritive value, and sensory properties of the prepared date bars as well as their storability were evaluated. The results indicated that the prepared date bars, either plain or chocolate coated, had good acceptability, possessed high nutritional value, and could be stored for more than 5 months under refrigeration (5°C) without affecting their qualities.

15.8   Utilization of Industrial Date Waste

Qualitatively, not all cultivated date fruits meet commercial standards, and tonnes of date fruit that are immature or of poor appearance but with no reduced nutritional value are treated as date by-products, used for animal feed, or discarded during processing by the relevant industries. Together with the rejected fruit, date seed, which is the inedible part, results in an environmental disposal problem and economic loss (Najjar *et al.*, 2020). Fructose is an important sweetener in beverages and the food industry that can be extracted from date fruit waste. Utilizing date fruit waste to extract fructose is an economical approach because it is waste in the first place (Faiad *et al.*, 2022).

15.8.1   Date seeds

Date seeds have been shown to be carbon-rich, have a high fatty acid content (such as oleic acid), and contain lignin, cellulose, hemicelluloses, and proteins (Mahdi *et al.*, 2018). Date seeds are composed of 3.1–7.1%

(w/w) moisture, 2.3–6.4% (w/w) protein, 5.0–13.2% (w/w) fat, 0.9–1.8% (w/w) ash, and 22.5–80.2% (w/w) dietary fiber. The good nutritional value of date seeds is based on their dietary fiber content, which makes them suitable for the preparation of fiber-based foods and dietary supplements (Al-Farsi *et al.*, 2007). Date seeds have also been used for the manufacturing of a caffeine-free coffee (Habib and Ibrahim, 2009).

Zidan and Abdel Samea (2019) evaluated the nutritional, physical, and sensory properties of biscuits prepared using different levels of date seed powder (DSP). Blends of biscuits were prepared using wheat flour at 72% extraction rate as a control and those in which wheat flour was substituted with 5, 10, and 15% (w/w) of DSP. Nutritional value, physical properties, and sensory characteristics differed with changing levels of DSP. The study concluded that utilization of DSP improved the nutritional value of biscuits. Biscuit color values decreased with increasing DSP, but with no significant differences in color, appearance, and overall acceptability. Biscuits prepared with 5% (w/w) DSP showed the highest texture scores, which decreased with increasing DSP. There were no significant differences between all prepared biscuits in texture values.

Alqattan *et al.* (2020) utilized DSP as a novel fat replacement fiber source in processed cheese blocks. Four replacement levels (0, 5, 10, 15, 20% w/w) were used. DSP improved cheese fiber content and textural properties, and stabilized hardness, adhesiveness, and springiness. Cheese microstructure showed less numerous fat globules and had smooth and homogeneous protein embedded and distributed uniformly compared to the control. Fat replacement at 5% (w/w) with DSP recorded the closest rated sensorial evaluation compared with the control in all criteria and acceptance scores.

15.8.2 Date press cake

DPC is the primary by-product of the date syrup and juice industries that remains after extraction. Enormous amounts of DPC are produced daily as date juicing results in approximately 17–28% DPC by total volume. Discarded DPC has potential as an abundant, low-cost, and green precursor for a variety of bioprocesses (Oladzad *et al.*, 2021). Depending on the method used for juice extraction, date flesh with the remaining sugar and with or without pits is usually left after extraction. Since DPC is wet (about 70% moisture), bulky (about 30% of date weight), and easily deteriorates, it creates a disposal problem (Barreveld, 1993). Most of the protein exists in DPC, and most of the fat exists in date pits. Date pits and DPC are rich sources of dietary fiber and date pits contain the highest levels of phenolic compounds and antioxidant activity. DPC is used for animal feed and in microbial conversions to single-cell proteins by fermentation processes (Al-Farsi *et al.*, 2007). Date by-products containing date pit and DPC can be used for producing alcohol and animal feed. As the date pit contains dietary fiber and phenolic compounds, it can be used in functional foods (Ashraf and Hamidi-Esfahani, 2011).

15.9   Quality Control and Safety Regulations

In recent years, consumers have become more exigent and demand high quality and convenient food products with natural flavors and taste, free from additives and preservatives. Therefore, the challenge for the fruit industry is to develop such products, while considering quality and safety aspects along with consumer acceptance (Tylewicz *et al.*, 2019). Processed date product quality is determined by the quality of the raw materials utilized and the efficiency and care taken during handling, processing, storage, and distribution. It is important to establish quality standards and to ensure their fulfillment with a quality assurance program at the processing facility. Quality evaluation of date-based processed products includes determination of quality value attributes such as titratable acidity or pH, flavor, odor, color, size, shape and symmetry, maturity/ character or soluble/total solids, texture, viscosity or consistency by subjective or instrumental means, and identification of defects such as softness, over-ripeness, or under-ripeness, insect damage, mold, and enzymes responsible for flavor, odor, color, or textural changes (Barrett, 1996).

Silva and Abud (2017) addressed the importance of quality aspects and food industry legislation. Among the tools available to create these relations are good manufacturing practices (GMPs), quality management (ISO series), total quality management, and hazard analysis and critical control points (HACCP). Hence, GMP compliance is considered a minimum procedure for obtaining safe food products. The GMP involves all parameters and control operations used within the food industry including personal hygiene and training aspects, facility, as well as production flowchart, and pest and quality control programs. The control periodicity must be established in the eight standard operating procedures generally required for a food processing industry: (i) hygiene of premises, machines, and tools; (ii) water quality with periodic analyses; (iii) hygiene and health of food handlers; (iv) waste management; (v) preventive maintenance and calibration of equipment; (vi) integrated management of vector and urban pests; (vii) selection and reception of raw materials, packaging, and ingredients; and (viii) food gathering program for products withdrawn from the market due to the food expiration date being exceeded. Moreover, HACCP provides guidance on how to identify biological, chemical, and physical hazards in a particular food processing line and how to control them at the critical control points throughout production.

15.10   Conclusion and Prospects

There is a need for the development of innovative novel products with ensured nutritional quality and safety meeting consumer expectations through the application of various emerging, unconventional technologies such as pulsed electric field, pulsed light, ultrasound, high pressure, and microwave drying, which will increase the stability of date-based products while preserving their taste, texture, and overall acceptability. Currently, there is insufficient usage of date by-products, which are limited to animal feed. Research into date by-products has not been a true reflection of the importance and potential of this crop. Using fermentation technology to produce value-added products such as ethanol, baker's yeast, single-cell protein, citric acid, and vinegar from low-quality date fruits needs to be more explored. Like any fruit industry, date processors face obstacles negatively affecting the date manufacturing process. Technical measures such as applying methods to control microbial contamination through applying GMP in the manufacturing process and adapting quality standards would ensure the quality of the final products. More scientific research to improve manufacturing processes and procedures is essential. For example, the effectiveness of new fumigants such as fogging with hydrogen peroxide for pest control in dates and processing facilities needs to be evaluated in terms of the permeability of dry fog and its effect on increasing the shelf life and promoting the rapid elimination of pests, as well as evaluating its advantages of environmental friendliness due to the resulting nontoxic decomposition products and its broad spectrum of action. Evaluating the application of active nanocomposite films as an antimicrobial packaging to control undesirable microorganisms in date processed products and investing product quality and safety need further studies.

References

Abass, W.F., Yaseen, H.A. and Al-Shaibani, A.M. (2017) Effect of utilization of date fiber from date syrup manufacture on organoleptic properties of chocolate cake and extending the storage life of the product. *International Journal of Science and Nature* 8(3), 662–666.

Abd El-Tawab, Y.A. (2009) A study on yoghurt and yoghurt derivatives. PhD thesis, Al Azhar University, Cairo, pp. 99–101.

Abdollahzadeh, S.M., Zahedani, M.R., Rahmdel, S., Hemmati, F. and Mazloomi, S.M. (2018) Development of *Lactobacillus acidophilus*-fermented milk fortified with date extract. *LWT– Food Science and Technology* 98, 577–582. DOI: 10.1016/j.lwt.2018.09.042.

Abo-El-Saad, M. and El-Shafie, H.A.F. (2013) Insect pests of stored dates and their management. In: Siddiq, M., Aleid, S.M. and Kader, A.A. (eds) *Dates: Postharvest Science, Processing Technology and Health Benefits*. Wiley, Chichester, UK, pp. 81–104. DOI: 10.1002/9781118292419.

Achour, M., Ben Amara, S., Ben Salem, N., Jebali, A. and Hamdi, M. (2003) Effet de différents conditionnements sous vide ou sous atmosphère modifiée sur la conservation de dattes Deglet Nour en Tunisie. *Fruits* 58(4), 205–212. DOI: 10.1051/fruits:2003008.

Ahmed, J. and Ramaswamy, H.S. (2006) Physico-chemical properties of commercial date pastes (*Phoenix dactylifera*). *Journal of Food Engineering* 76(3), 348–352. DOI: 10.1016/j.jfoodeng.2005.05.033.

Al-Abid, M., Al-Shoaily, K., Al-Amry, M. and Al-Rawahy, F. (2007) Maintaining the soft consistency of date paste. *Acta Horticulturae* 36(736), 523–530. DOI: 10.17660/ActaHortic.2007.736.51.

Aleid, S.M. (2006) Chromatographic separation of fructose from date syrup. *International Journal of Food Sciences and Nutrition* 57(1–2), 83–96. DOI: 10.1080/09637480600658286.

Aleid, S.M. (2009) *Utilization of Date Paste in Double Layered Arabic Bread.* Final Technical Report, Research Project #6009. King Faisal University, Alahsa, Saudi Arabia.

Aleid, M. (2013) Date fruit processing and processed products. In: Siddiq, M., Aleid, S.M. and Kader, A.A. (eds) *Dates: Postharvest Science, Processing Technology and Health Benefits.* Wiley, Chichester, UK, pp. 171–202. DOI: 10.1002/9781118292419.

Aleid, S.M. and Al-Saikhan, M.S. (2019) Quality, texture and color properties of fresh 'Ghur' and 'Khenaizy' dates cold stored with modified atmosphere packaging. *Fresenius Environmental Bulletin* 28(11A), 8882–8888.

Aleid, S.M., El-Shaarawy, M.L., Mesallam, A.S. and Al-Jendan, S.I. (1999) Chemical composition and nutritional value of some sugar and date syrups. *Minufiya Journal of Agricultural Research* 24(2), 577–587.

Aleid, S.M., El-Shaarawy, M.L., Mesallam, A.S. and Al-Jendan, S.I. (2000) Influence of storage temperature on the quality of some sugar and date syrups. *Scientific Journal of King Faisal University* 1(1), 69–78.

Al-Farsi, M.A. (2003) Clarification of date juice. *International Journal of Food Science and Technology* 38(3), 241–245. DOI: 10.1046/j.1365-2621.2003.00669.x.

Al-Farsi, M.A. and Lee, C.Y. (2008) Nutritional and functional properties of dates: a review. *Critical Reviews in Food Science and Nutrition* 48(10), 877–887. DOI: 10.1080/10408390701724264.

Al-Farsi, M.A., Alasalvar, C., Al-Abid, M., Al-Shoaily, K., Al-Amry, M. *et al.* (2007) Compositional and functional characteristics of dates, syrups, and their by-products. *Food Chemistry* 104(3), 943–947. DOI: 10.1016/j. foodchem.2006.12.051.

Alhamdan, A.M. and Hassan, B.A. (1999) Water sorption isotherms of date pastes as influenced by date cultivar and storage temperature. In: Afifi, M.A.R. and Al-Badawy, A.A. (eds) *Proceedings of the First International Conference on Date Palms, Al-Ain, United Arab Emirates*, United Arab Emirates University, Al-Ain, United Arab Emirates, March 8–10, pp. 111–125.

Alhamdan, A.M., Al Juhaimi, F.Y., Hassan, B.H., Ehmed, K.A. and Mohamed Ahmed, I.A. (2021) Physicochemical, microbiological, and sensorial quality attributes of a fermented milk drink (laban) fortified with date syrup (dibs) during cold storage. *Foods* 10(12), 3157. DOI: 10.3390/foods10123157.

Alhinai, T.Z.S., Vreeburg, R.A.M., Mackay, C.L., Murray, L., Sadler, I.H. *et al.* (2021) Fruit softening: evidence for pectate lyase action *in vivo* in date (*Phoenix dactylifera*) and rosaceous fruit cell walls. *Annals of Botany* 128(5), 511–525. DOI: 10.1093/aob/mcab072.

Alqattan, A.M., Alqahtani, N.K., Aleid, S.M. and Alnemr, T.M. (2020) Effects of date pit powder inclusion on chemical composition, microstructure, rheological properties, and sensory evaluation of processed cheese block. *American Journal of Food and Nutrition* 8(3), 69–77. DOI: 10.12691/ajfn-8-3-3.

Al-Rawahi, A.S., Kasapis, S. and Al-Bulushi, I.M. (2005) Development of a date confectionery: Part 1. Relating formulation to instrumental texture. *International Journal of Food Properties* 8(3), 457–468. DOI: 10.1080/10942910500267521.

Al-Rawahi, A.S., Kasapis, S., Al-Maamari, S. and Al-Saadi, A.M. (2006) Development of a date confectionery: Part 2. Relating instrumental texture to sensory evaluation. *International Journal of Food Properties* 9(3), 365–375. DOI: 10.1080/10942910600741482.

Al-Redhaiman, K.N. (2004) Modified atmosphere extends storage period and maintains quality of Birhi date fruits. *Acta Horticulturae* 682, 979–986. DOI: 10.17660/ActaHortic.2005.682.127.

Alsenaien, W.A., Alamer, R.A., Tang, Z.-X., Albahrani, S.A., Al-Ghannam, M.A. *et al.* (2015) Substitution of sugar with dates powder and dates syrup in cookies making. *Advance Journal of Food Science and Technology* 8(1), 8–13. DOI: 10.19026/ajfst.8.1455.

Ardali, F.R., Rahimi, E., Tahery, S., Shariati, M.A. and Ali, M. (2014) Production of a new drink by using date syrup and milk. *Journal of Food Biosciences and Technology* 4(2), 67–72.

Ashraf, Z. and Hamidi-Esfahani, Z. (2011) Date and date processing: a review. *Food Reviews International* 27(2), 101–133. DOI: 10.1080/87559129.2010.535231.

Awofadeju, O.F.J., Awe, A.B., Adewumi, O.J. and Adewumi Adeyemo, E. (2021) Influence of substituting sucrose with date palm fruit flour on the nutritional and organoleptic properties of bread. *Croatian Journal of Food Science and Technology* 13(1), 1–6. DOI: 10.17508/CJFST.2021.13.1.01.

Baloch, M.K., Saleem, S.A., Baloch, A.K. and Baloch, W.A. (2006) Impact of controlled atmosphere on the stability of Dhakki dates. *LWT - Food Science and Technology* 39(6), 671–676. DOI: 10.1016/j.lwt.2005.04.009.

Baloch, A.K., Baloch, W.A., Baloch, M.K. and Saleem, S.A. (2007) Shelf stability of Dhakki dates as influenced by water activity and headspace atmosphere. *Acta Horticulturae* 736, 575–586. DOI: 10.17660/ActaHortic.2007.736.57.

Barrett, D.M. (1996) Quality assurance for processed fruit and vegetable products. *Perishables Handling Newsletter* 85, 21–23.

Barreveld, W.H. (1993) *Date Palm Products*. FAO Agricultural Services Bulletin No. 101. Food and Agriculture Organization of the United Nations, Rome.

Besbes, S., Drira, L., Blecker, C., Deroanne, C. and Attia, H. (2009) Adding value to hard date (*Phoenix dactylifera* L.): compositional, functional and sensory characteristics of date jam. *Food Chemistry* 112(2), 406–411. DOI: 10.1016/j.foodchem.2008.05.093.

Biglari, F., AlKarkhi, A.F.M. and Easa, A.M. (2008) Antioxidant activity and phenolic content of various date palm (*Phoenix dactylifera*) fruits from Iran. *Food Chemistry* 107(4), 1636–1641. DOI: 10.1016/j.foodchem.2007.10.033.

Chandrasekaran, M. and Bahkali, A.H. (2013) Valorization of date palm (*Phoenix dactylifera*) fruit processing by-products and wastes using bioprocess technology

- Review. *Saudi Journal of Biological Sciences* 20(2), 105–120. DOI: 10.1016/j. sjbs.2012.12.004.

Di Cagno, R., Filannino, P., Cavoski, I., Lanera, A., Mamdouh, B.M. *et al.* (2017) Bioprocessing technology to exploit organic palm date (*Phoenix dactylifera* L. cultivar Siwi) fruit as a functional dietary supplement. *Journal of Functional Foods* 31, 9–19. DOI: 10.1016/j.jff.2017.01.033.

El Fouhil, A.F., Ahmed, A.M., Darwish, H.H., Atteya, M. and Al-Roalle, A.H. (2011) An extract from date seeds having a hypoglycemic effect. Is it safe to use? *Saudi Medical Journal* 32(8), 791–796.

El-Habbab, M.S., Al-Mulhim, F.N., Aleid, S.M., Abo El-Saad, M., Aljassas, F.M. *et al.* (2017) Assessment of post-harvest loss and waste for date palms in the kingdom of Saudi Arabia. *International Journal of Environmental and Agriculture Research* 3(6), 01–11. DOI: 10.25125/agriculture-journal-IJOEAR-MAY-2017-7.

Elleuch, M., Besbes, S., Roiseux, O., Blecker, C., Deroanne, C. *et al.* (2008) Date flesh: chemical compositionand characteristics of the dietary fiber. *Food Chemistry* 111(3), 676–682. DOI: 10.1016/j.foodchem.2008.04.036.

El-Nagga, E.A. and Abd El–Tawab, Y.A. (2012) Compositional characteristics of date syrup extracted by different methods in some fermented dairy products. *Annals of Agricultural Sciences* 57(1), 29–36. DOI: 10.1016/j.aoas.2012.03.007.

Elsabea, A.M.R. (2012) An economic study of processing problems for the main important varieties of dates in Saudi Arabia. *Annals of Agricultural Sciences* 57(2), 153–159. DOI: 10.1016/j.aoas.2012.08.009.

El-Sharnouby, G.A., Aleid, S.M. and Al-Otaibi, M.M. (2009) Utilization of enzymes in the production of liquid sugar from dates. *African Journal of Biochemistry Research* 3(3), 41–47.

Entezari, M.H., Nazary, H.S. and Khodaparast, H.M.H. (2004) The direct effect of ultrasound on the extraction of date syrup and its micro-organisms. *Ultrasonics Sonochemistry* 11(6), 379–384. DOI: 10.1016/j.ultsonch.2003.10.005.

Erskine, W., Moustafa, A.T., Osman, A.E., Lashine, Z., Nejatian, A. *et al.* (2003) Date palm in the GCC countries of the Arabian Peninsula. Available at: www.researc hgate.net/publication/267714856_Date_Palm_in_the_GCC_countries_of_th e_Arabian_Peninsula (accessed 16 December 2022).

Faiad, A., Alsmari, M., Ahmed, M.M.Z., Bouazizi, M.L., Alzahrani, B. *et al.* (2022) Date palm tree waste recycling: treatment and processing for potential engi- neering applications. *Sustainability* 14(3), 1134. DOI: 10.3390/su14031134.

Fallahi, M. (1996) *Growth, Treatment, and Packaging of Date.* Barsava Publications, Tehran, Iran.

Glasner, B., Botes, A., Zaid, A. and Emmens, J. (2002) Date harvesting, packing- house management and marketing aspects. In: Zaid, A. and Arias-Jiménez, E.J. (eds) *Date Palm Cultivation.* FAO Plant Production and Protection Paper no.156 Rev. 1, Food and Agriculture Organization of the United Nations, Rome, pp. 237–267.

Golshan, A. and Fooladi, M.H. (2005) Microbial contamination on date fruits. In: *Book of Abstracts, First International Symposium and Festival on Date Palm, Bandar Abass, Iran,* Ministry of Jihad-e-Agriculture, Bandar Abbas, Iran, November 20–21, 2005, p. 10.

Gustavsson, J., Cederberg, C., Sonesson, U., Van Otterdijk, R. and Meybeck, A. (eds) (2011) *Global Food Losses and Food Waste: Extent, Causes and Prevention. Study Conducted For the International Congress SAVE FOOD! At Interpack2011,*

Düsseldorf, Germany. Food and Agriculture Organization of the United Nations, Rome.

Habib, H.M. and Ibrahim, W.H. (2009) Nutritional quality evaluation of eighteen date pit varieties. *International Journal of Food Sciences and Nutrition* 60 Suppl 1, 99–111. DOI: 10.1080/09637480802314639.

Hariri, A., Ouis, N., Ibri, K., Bouhadi, D. and Benatouche, Z. (2020) Technological characteristics of fermented milk product manufactured by milk–dates mixtures. *Acta Agriculturae Serbica* 25(49), 27–35.

Hashempoor, M. (1999) *Date Treasure.* Agricultural Education Publication, Tehran.

Ibrahim, S.A., Ayad, A.A., Williams, L.L., Ayivi, R.D., Gyawali, R. *et al.* (2021) Date fruit: a review of the chemical and nutritional compounds, functional effects and food application in nutrition bars for athletes. *International Journal of Food Science & Technology* 56(4), 1503–1513. DOI: 10.1111/ijfs.14783.

Iranmanesh, S.M. (2000) *Introduction on Practical Technology of Date Producing, Preservation, Processing, Packaging and Export (Storing, Packing, By-product and Exporting of Date).* Almahdi Publication Organization, Iran, pp. 125–158.

Jassim, S.A.A. and Naji, M.A. (2010) *In vitro* evaluation of the antiviral activity of an extract of date palm (*Phoenix dactylifera* L.) pits on a Pseudomonas phage. *Evidence- Based Complementary and Alternative Medicine* 7, 816839.

Kader, A.A. (2013) Postharvest technology of horticultural crops – an overview from farm to fork. *Ethiopian Journal of Applied Science and Technology* (Special Issue No. 1), 1–8.

Kader, A.A. and Hussein, A.M. (2009) *Harvesting and Postharvest Handling of Dates.* International Center for Agricultural Research in the Dry Areas (ICARDA), Aleppo, Syria.

Kader, A.A., Zagory, D. and Kerbel, E.L. (1989) Modified atmosphere packaging of fruits and vegetables. *Critical Reviews in Food Science and Nutrition* 28(1), 1–30. DOI: 10.1080/10408398909527490.

Kamal-Eldin, A., Hashim, I.B. and Mohamed, I.O. (2012) Processing and utilization of palm date fruits for edible applications. *Recent Patents on Food, Nutrition & Agriculture* 4(1), 78–86. DOI: 10.2174/2212798411204010078.

Kasapis, S., Al-Oufi, H.S., Al-Maamari, S., Al-Bulushi, I.M. and Goddard, S. (2004) Scientific and technological aspects of fish product development. Part I: Handshaking instrumental texture with consumer preference in burgers. *International Journal of Food Properties* 7(3), 449–462. DOI: 10.1081/JFP-200032935.

Kenawi, M.A., El Sokkary, F.A., Kenawi, M.N., Assous, M.T. and Abd El Galil, Z.A. (2016) Chemical, physical and sensory evaluation of biscuit supplemented with date powder. *Minia Journal of Agricultural Research and Development* 36(2), 215–227.

Khali, M. and Selselet-Attou, G. (2007) Effect of heat treatment on polyphenol oxidase and peroxidase activities in Algerian stored dates. *African Journal of Biotechnology* 6, 790–794.

Khapre, A.P. and Shah, U.A. (2016) Standardization of technology for development of soft date (*Pheonix dactylifera*) toffee as a nutritional enrichment of confectionery. *Asian Journal of Dairy and Food Research* 35(4), 335–337. DOI: 10.18805/ajdfr.v35i4.6636.

Kitinoja, L. (2016) *Innovative Approaches to Food Loss and Waste Issues.* Frontier Issues Brief submitted to the Brookings Institution's Ending Rural Hunger project. The Postharvest Education Foundation, La Pine, Oregon.

Mahdi, Z., Yu, Q.J. and El Hanandeh, A. (2018) Removal of lead(II) from aqueous solution using date seed-derived biochar: batch and column studies. *Applied Water Science* 8(6), 181. DOI: 10.1007/s13201-018-0829-0.

Mahmoudi, H., Hosseininia, G., Azadi, H. and Fatemi, M. (2008) Enhancing date palm processing, marketing and pest control through organic culture. *Journal of Organic Systems* 3, 29–39.

Mohamed, M.A.A. (2002) Studies on cultured milk beverages. PhD thesis, Al-Azhar University, Cairo, pp. 167–170.

Mortazavi, S.M.H., Arzani, K. and Barzegar, M. (2007) Effect of vacuum and modified atmosphere packaging on the postharvest quality and shelf life of date fruits in khalal stage. *Acta Horticulturae* 736, 471–477. DOI: 10.17660/ActaHortic.2007.736.45.

Myhara, R.M., Al-Alawi, A., Karkalas, J. and Taylor, M.S. (2008) Sensory and textural changes in maturing Omani dates. *Journal of the Science of Food and Agriculture* 80, 2181–2185.

Najjar, Z., Stathopoulos, C. and Chockchaisawasdee, S. (2020) Utilization of date by-products in the food industry. *Emirates Journal of Food and Agriculture* 32(11), 808. DOI: 10.9755/ejfa.2020.v32.i11.2192.

Nwanekezi, E.C., Ekwe, C.C. and Agbugba, R.U. (2015) Effect of substitution of sucrose with date palm (*Phoenix dactylifera*) fruit on quality of bread. *Journal of Food Processing Technology* 6, 484.

Obiegbuna, J.E., Akubor, P.I., Ishiwu, C.N. and Ndife, J. (2013) Effect of substituting sugar with date palm pulp meal on the physicochemical, organoleptic and storage properties of bread. *African Journal of Food Science* 7(6), 113–119.

Oladzad, S., Fallah, N., Mahboubi, A., Afsham, N. and Taherzadeh, M.J. (2021) Date fruit processing waste and approaches to its valorization: a review. *Bioresource Technology* 340, 125625. DOI: 10.1016/j.biortech.2021.125625.

Peleg, M. (1983) The semantics of rheology and texture. *Food Technology* 11, 54–61.

Qadir, A., Shakeel, F., Ali, A. and Faiyazuddin, M. (2020) Phytotherapeutic potential and pharmaceutical impact of *Phoenix dactylifera* (date palm): current research and future prospects. *Journal of Food Science and Technology* 57(4), 1191–1204. DOI: 10.1007/s13197-019-04096-8.

Ramadan, B.R. (1998) Preparation and evaluation of Egyptian date syrup. In: Afifi, M.A.R. and Al-Badawy, A.A. (eds) *Proceedings of the First International Conference on Date Palms. Al-Ain, United Arab Emirates,* United Arab Emirates University, Al-Ain, United Arab Emirates, March 8–10, 1998, pp. 86–98.

Redondo, N., Gómez-Martínez, S. and Marcos, A. (2014) Sensory attributes of soft drinks and their influence on consumers' preferences. *Food & Function* 5(8), 1686–1694. DOI: 10.1039/c4fo00181h.

Rohani, A. (1988) *Date Palm.* Tehran University Publication Center, Tehran.

Salhlstrum, S., Park, W. and Shelton, D.R. (2004) Factors influencing yeast formation and effect LMW sugars and yeast fermentation on health bread quality. *Cereal Chemistry Journal* 81, 328–335. DOI: 10.1094/CCHEM.2004.81.3.328.

Sánchez-Zapata, E., Fernández-López, J., Peñaranda, M., Fuentes-Zaragoza, E., Sendra, E. *et al.* (2011) Technological properties of date paste obtained from date by-products and its effect on the quality of a cooked meat product. *Food Research International* 44(7), 2401–2407. DOI: 10.1016/j.foodres.2010.04.034.

Shachman, M. (2005) *The Soft Drinks Companion: A Technical Handbook for the Beverage Industry.* CRC Press, Boca Raton, Florida.

Shraideh, Z.A., Abu-Elteen, K.H. and Sallal, A.K. (1998) Ultrastructural effects of date extract on *Candida albicans*. *Mycopathologia* 142, 119–123.

Siddiq, M., Aleid, S.M. and Kader, A.A. (eds) (2013) *Dates: Postharvest Science, Processing Technology and Health Benefits*. Chichester, UK.

Sidhu, J.S. and Al-Hooti, S.N. (2005) Functional foods from date fruits. In: Shi, J., Ho, C.-T. and Shahidi, F. (eds) *Asian Functional Foods*. CRC Press, Boca Raton, Florida, pp. 491–524.

Sidhu, J.S. (2006) Date fruits production and processing. In: Hui, Y.H. (ed.) *Handbook of Fruit and Fruit Processing*. Blackwell Publishing, Ames, Iowa, pp. 391–419. DOI: 10.1002/9780470277737.

Silva, C.E.F. and Abud, A.K.S. (2017) Tropical fruit pulps: processing, product standardization and main control parameters for quality assurance. *Brazilian Archives of Biology and Technology* 60, 1–19.

Sulieman, A.E., Masaad, M.K. and Ali, O.A. (2011) Effect of partial substitution of wheat flour with date powder on biscuit quality. *Gezira Journal of Agricultural Science* 9(2), 9–16.

Syamaladevi, R.M., Barbosa-Cánovas, G.V., Schmidt, S.J. and Sablani, S.S. (2015) Molecular weight effects on enthalpy relaxation and fragility of amorphous carbohydrates. In: Gutiérrez-López, G., Alamilla-Beltrán, L., del Pilar Buera, M., Welti-Chanes, J., Parada-Arias, E. *et al.* (eds) *Water Stress in Biological, Chemical, Pharmaceutical and Food Systems*. Food Engineering Series, Springer, New York, pp. 161–174.

Tylewicz, U., Tappi, S., Nowacka, M. and Wiktor, A. (2019) Safety, quality, and processing of fruits and vegetables. *Foods* 8(11), 569–571. DOI: 10.3390/foods8110569.

Yahia, E.M., Barry-Ryan, C. and Dris, R. (2004) Treatments and techniques to minimise the postharvest losses of perishable food crops. In: Dris, R. and Jain, S.M. (eds) *Production Practices and Quality Assessment of Food Crops*, Vol. 4. *Postharvest Treatment and Technology*, Springer, Dordrecht, The Netherlands, pp. 95–133.

Yousif, A.K., Abdelmasseh, M., Yousif, M.E. and Saeed, B.T. (1987) Use of date paste in the processing of nutritious candy bars for athletes. *Archivos Latinoamericanos* 34, 241–247.

Yousif, A.K., Morton, I.D. and Mustafa, A.I. (1991a) Processing, evaluation and water relation of date paste. *Tropical Science* 31, 147–158.

Yousif, A.K., Morton, I.D. and Mustafa, A.I. (1991b) Functionality of date paste in bread making. *Cereal Chemistry* 68, 43–47.

Zagory, D. and Kader, A.A. (1988) Modified atmosphere packaging of fresh produce. *Food Technology* 42, 70–77.

Zeeshan, M., Saleem, S.A., Ayub, M., Shah, M. and Jan, Z. (2017) Physicochemical and sensory evaluation of Dhakki dates candy. *Journal of Food Processing & Technology* 8(3), 663–666.

Zidan, N.S. and Abdel Samea, R.R. (2019) Nutritional and sensory evaluation of biscuit prepared using palm date kernels and olive seeds powders. *Journal of Specific Education and Technology* 5, 322–339.

Health Benefits and Nutraceutical Properties of Dates

16

Neeru Bhatt[1], Lyutha Al-Subhi[2], Ayah R. Hilles[3] and Mostafa I. Waly[1]*

[1]Global Science Heritage, Toronto, Ontario, Canada; [2]Sultan Qaboos University, Al-Khoud, Oman; [3]International Islamic University Malaysia, Kuala Lumpur, Malaysia

Abstract

The date palm (*Phoenix dactylifera* L.) is a member of the family *Arecaceae*. It is cultivated for its sweet and nutritious fruits, consumed as a staple food by millions of people in many countries, especially in Southwest Asia and North Africa. It is one of the oldest cultivated fruit crops in the world. Dates have been used for both dietary and pharmaceuticals purposes. Date fruits are a source of carbohydrates such as sucrose, fructose, maltose, dextrose, as well as dietary fiber, and are also rich in micronutrients, vitamins and minerals. Despite dates being sugar-packed, many date varieties have a low glycemic index (GI) and do not stimulate any metabolic and inflammatory markers associated with chronic diseases. Dates are an excellent source of chemical compounds like phenolic acids, tannins, flavonoids, phytosterols, and carotenoids that act as potential therapeutic agents against several diseases, including cancer and heart disease. Ayurvedic practitioners also use date fruit as a remedy for several diseases. The wide bioactive profile of dates makes them an excellent option for use as nutraceuticals. This chapter addresses these health benefits and the nutraceutical composition of dates, along with their potential utilization.

Keywords: Chemical composition, Date, Health benefits, Nutraceuticals, Phytochemicals

*Corresponding author: mostafa@squ.edu.om

© CAB International 2023. *Date Palm* (eds J.M. Al-Khayri *et al.*)
DOI: 10.1079/9781800620209.0016

16.1   Introduction

Food-based approaches are encouraged worldwide to combat health-related issues. Various terms like 'nutraceuticals', 'dietary supplements', 'functional foods', 'medical foods', and 'dietary supplements' are used by herbal therapists and nutritionists. These terms may create confusion when used interchangeably. The term 'nutraceutical' was coined by Stephen Defelice, chairman of the Foundation for Innovation in Medicine, in 1989. It is derived by combining the terms 'nutrition' and 'pharmaceutical'. Nutraceuticals are described as foods or parts of foods that provide health benefits and are used for prevention or treatment of disease (Prabu *et al.*, 2012; Srivastava *et al.*, 2015). The date palm (*Phoenix dactylifera* L.) is widely grown in the arid and semi-arid regions of the globe and is considered one of the oldest and most important staple crops in Southwest Asia and North Africa. Dates are also grown in Australia, Mexico, South America, southern Africa, and the USA, especially in southern California and Arizona (Chao and Krueger, 2007; Al-Harrasi *et al.*, 2014; Hazzouri *et al.*, 2015). Fresh dates are harvested from the palm when they are completely ripe, while dry dates will be left out in the sun for a prolonged period of time to be dried. It was observed that phenolic acids increased significantly in the dry dates while carotenoids were lost from the dates after sun drying (Al-Farsi *et al.*, 2005a). Date fruits are high in carbohydrates and low in fats. The fruits are rich in sucrose, fructose, maltose, dextrose, dietary fibers, and various other bioactive components that offer health benefits. This chapter provides detailed information on the nutritional and bioactive components present in date fruits and their nutraceutical properties.

16.2   Nutritional Values of Date Palm Fruits

The potential of foods to offer health benefits and reduce the risk of chronic disease beyond basic nutrition is known as the functional value of foods. Date fruits are known for their excellent taste and nutritional and health benefits. The proximate composition of dates is presented in Table 16.1. Fresh dates are soft, sweet, and easy to digest. Consumption of date fruits in the morning on an empty stomach can reverse the effects of harmful substances to which the individual has been exposed (Hamad, 2014). The high amount of soluble and insoluble fibers present in dates prevents constipation and facilitates bowel movement, thus improving the digestive system (Al-Shahib and Marshall, 2003). High levels of potassium and low levels of sodium help to balance blood pressure and keep the heart and nervous system healthy. The higher iron content in the fruit is helpful in managing anaemia. In some parts of Morocco, the date fruit is being used to treat diabetes and hypertension (Tahraoui *et al.*, 2007). Various polyphenols present in dates extend antimutagenic, antioxidant,

Table 16.1. Proximate composition of date fruits[a]. (From Ahmed *et al.*, 1995; Al-Hooti *et al.*, 1997; Myhara *et al.*, 1999; Al-Farsi *et al.*, 2005a .).

Component	Fresh dates (g/100 g)	Dried dates (g/100 g)
Moisture	42.4	15.2
Protein	1.5	2.14
Fat	0.14	0.38
Ash	1.16	1.67
Carbohydrate	54.9	80.6
Total sugar	43.4	64.1
Glucose	19.4	29.4
Fructose	22.8	30.4
Dietary fiber	7.5	10.2

[a]Values for fresh dates are the average of the varieties included: Naghal, Khunaizy, Khalas, Barhi, Lulu, Fard, Khasab, Bushibal, Gash Gaafar, and Gash Habash. Values for dried dates are the average of all the varieties above plus Deglet Nour, Medjhool, Hallawi, Sayer, Khadrawi, and Zahidi

anticarcinogenic, and anti-inflammatory bioactivities (Maqsood *et al.*, 2020). Date seeds are also packed with bioactive compounds and have traditionally been used for the management of diabetes, liver diseases, and gastrointestinal disorders in the Middle East (Abu-Odeh and Talib, 2021). The Ajwa variety of date has been found to be effective in fighting oxidative stress-mediated complications, even in COVID-19 patients (Anwar *et al.*, 2022). The nutritional value of date fruits is due to a number of nutrients, as outlined below.

16.2.1 Sugars and fibers

Date fruits are rich in easily digestible sugars and a good source of fiber. The main sugars detected in fresh and dried dates are glucose, fructose, and sucrose. The average quantity of total sugar in fresh dates is 43.4 g/100 g, of which about 4.03 g/100 g is sucrose, 19.4 g/100 g is fructose, and 22.8 g/100 g is glucose. The stage of ripening plays a role in the sugar content of the fruit, and sugar content increases as dates move from the rutab to the tamar stage due to moisture content reduction (Barreveld, 1993). In total, the sugar content of dried dates is 64.1 g/100 g. The distribution of monosaccharides in dried dates is 30.4, 29.4, and 11.6 g/100 g for glucose, fructose, and sucrose, respectively (Yousif *et al.*, 1982; Al-Farsi *et al.*, 2005a; Ismail *et al.*, 2006). The reducing sugars glucose and fructose are the highest essential constituents, and they are found in almost equal amounts in dates. The average energy provided by dried and fresh dates is 314 and 213 kcal/100 g (1314 and 891 kJ/100 g), respectively (Al-Farsi *et al.*, 2005a; USDA, 2007). Five dates (approx. 45 g) contain about 115 kcal (481 kJ), nearly all from carbohydrates. Compared to sucrose, the reducing sugar glucose leads to rapid elevation of blood sugar levels due to its ready absorption after digestion (Barreveld, 1993). Total sugar content in two Omani date cvs., Umsellah and Khalas, were reported as 51.37 and 44.78 g glucose equivalents/100 g dates, respectively (Siddiqi *et al.*, 2020).

The majority of date fruit fibers are composed of cellulose and non-starch polysaccharides (NSPs). They contain 8.1–12.7% by weight of total dietary fibers (84–94% are insoluble and 6–16% are soluble dietary fibers). Insoluble dietary fiber plays a significant physiological role in the human body. It can protect from various conditions like diverticular disease and bowel cancer by increasing bulk, which stimulates the propulsive movement of the large intestine (Marlett *et al.*, 2002). Date dietary fiber exhibits both water-holding capacity and oil-holding capacity. Total dietary fibers in Omani dates were 81.17 g/100 g in cv. Umsellah and 67.35 g/100 g in cv. Khalas (Siddiqi *et al.*, 2020). There was an increase in dietary fiber from 7.5 to 8 g/100 g as dates move from fresh to dry during the ripening process. This increase might be due to reduced moisture content or enzymatic breakdown of substances into soluble compounds (Fennema, 1976). The recommended daily intake (RDI) of dietary fiber is about 25 g (Marlett *et al.*, 2002). Dates are a good source of dietary fiber because 100 g of dates can provide about 32% of the RDI. Not only date fruits but also their seeds are an excellent source of dietary fiber. Date seeds contain about 15% of fiber by weight, characterized by a high level of water-insoluble fibers (Mrabet *et al.*, 2019). Date palm fruit products, especially seeds, may be a good alternative for future food development and nutraceuticals as well (Hinkaew *et al.*, 2021).

16.2.2 Micronutrients

Micronutrients are required in very small amounts, but they perform vital and specific functions in the human body. Dates are known for their high amounts of trace elements such as potassium, phosphorus, magnesium, calcium, selenium, and iron (Table 16.2). As dates have low sodium and high potassium levels, they are considered good for people suffering from hypertension (Appel *et al.*, 1997).

It was reported that, per 100 g, date fruit roughly contains 713 mg of potassium, 64.2 mg of magnesium, 0.24 mg of copper, and 0.31 mg of selenium (Mohamed, 2000; Al-Farsi and Lee, 2008; Ismail *et al.*, 2008). Minerals are important constituents of various tissues and cells, such as teeth, bone, haemoglobin, soft tissues, muscle, and nerve cells (O'Dell and Sunde, 1997; Sardesai, 2011). Date seeds are said to be a good source of selenium. Selenium is an integral part of glutathione peroxidase. This antioxidant enzyme is required for maintenance of the body's defense system

Table 16.2. Mineral contents of different varieties of dates at their different maturity stages. (Adopted from Ahmed *et al.*, 1995. Used with permission.)

Date cultivar	Stage of ripening	Mineral (mg/100 g)							
		Ca	K	Mg	Fe	Cu	Na	Mn	Zn
Barhi	Kimri	88	1,163	209	1.1	0.4	29	1.7	0.8
	Khalal	10	796	45	0.9	0.2	204	1.0	0.4
	Rutab	12	799	89	1.4	0.3	209	0.3	0.2
	Tamar	12	855	82	0.3	0.2	75	0.5	0.1
Fard	Kimri	53	1,243	121	1.5	0.5	66	0.7	0.7
	Khalal	18	1,106	97	1.3	1.3	64	0.5	0.6
	Rutab	14	1,414	68	1.2	0.3	282	0.5	0.3
	Tamar	14	914	63	1.2	0.4	141	0.5	0.2
Khulas	Kimri	101	1,101	151	2.2	0.6	52	0.7	0.5
	Khalal	60	789	89	1.6	0.4	83	0.4	0.3
	Rutab	18	588	62	1.4	0.4	212	0.3	0.3
	Tamar	16	630	62	1.7	0.4	82	0.3	0.3

against various infections and to protect body tissues from oxidative stress (Baliga *et al.*, 2011).

Deficiencies of selenium have been associated with infertility in men and women (Mistry *et al.*, 2012). The RDI of selenium is 55 mg, which is available in approximately 28 g of dates, as the selenium content in ten date varieties grown in Saudi Arabia was found to range from 1.48 to 2.96 mg/g (Al-Showiman *et al.*, 1994).

Like minerals, vitamins are essential nutrients for maintaining good health. The concentration of water-soluble vitamins (B-complex and C) is comparatively high in date fruits compared to fat-soluble vitamins. These vitamins act as coenzymes involved in the synthesis of new cells as well as in the metabolism of fat, protein, and carbohydrates. Vitamin C (ascorbic acid) acts as a powerful antioxidant to defend the body from various diseases by protecting tissue from oxidative stress (Whitney and Rolfes, 2007). Vitamins B_2 (riboflavin), B_3 (niacin), B_6, and B_9 (folic acid) are present in moderate concentrations in dry dates. Intake of 100 g of dates can provide 1.4 mg of vitamin B_3 (RDI = 14 mg), while it can supply 0.2 mg of vitamin B_6 (RDI = 1 mg), and 0.05 mg of vitamin B_9 (RDI = 0.4 mg). On average, date fruit per 100 g contains vitamin A (23.85 µg), vitamin B_1 (thiamine) (78.61 µg), vitamin B_2 (116.5 µg), vitamin B_3 (1442 µg), vitamin B_6 (207 µg), vitamin B_9 (53.75 µg), and vitamin C (3900 µg) (USDA, 2007; Al-Farsi and Lee, 2008). Vitamin E or tocopherol was also found in all its isomers (α, β, γ, δ) in date fruits. Yousif *et al.* (1982) observed that date fruits have high vitamin levels, e.g. ascorbic acid (2.4–17.5 mg/100 g), thiamine (0.08–0.13 mg/100 g), and riboflavin (0.13–17.5 mg/100 g).

Table 16.3. Bioactive components present in date fruits.

Phenolic acids	Flavonoids	Sterols	Carotenoids
Caffeic acid	Quercetin	Esterone	Xanthophylls
Chlorogenic acid	Luteolin	Ergosterol	α-Carotene
Syringic acid	Apigenin	Estrogen	Lutein
Ferulic acid	Isoquercitrin	Brassicasterol	β-Carotene
Protocatechuic acid	Rutin	β-Sitosterol	Zeaxanthin
Catechin acid		Stigmasterol	Lycopene
Gallic acid		Campesterol	β-Cryptoxanthin
p-Coumaric acid		Isofucosterol	
Resorcinol			
Vanillic acid			
Salicylic acid			
Ellagic acid			

16.2.3 Amino acids

Amino acids, the structural units of proteins, are involved in almost every body function, including growth and development, healing and repair, digestion, and metabolism. Like other nutrients, the amino acids content of dates is reduced as they attain maturity, which may be due to reduced moisture content (Ishurd *et al.*, 2004). Date fruits are low in protein, which ranges from 2.5 to 6.5 g/100 g (Chaira *et al.*, 2009). Some essential amino acids are found in dates. Glycine, leucine, lysine, aspartic acid, and glutamic acid are predominately found in fresh dates, while proline, leucine, glycine, aspartic acid, and glutamic acid are found in dried date fruits. In a study, 23 different amino acids were detected, some of which are uncommon in other fruits (Al-Farsi *et al.*, 2005a).

16.2.4 Bioactive components

The bioactive components that offer protection against oxidative stress, such as phenolics, carotenoids, anthocyanins, procyanidins, and flavonoids, are abundant in date fruits (Al-Farsi *et al.*, 2005b; Allaith, 2008; Gayathri and Thilagavathi, 2021) (Table 16.3, Fig. 16.1). The phenolic contents of fresh and dried dates are 193.7 and 239.5 mg/100 g, respectively, although they may vary from one variety to another. Oxidative decomposition of phenolics may occur enzymatically (via polyphenol oxidase and glycosidase) or thermally during the drying of date fruits (Shahidi and Naczk, 2003). However, there is an increase in phenolics in some varieties that is possibly due to the breakdown of tannins and the release of phenolic compounds during the drying process (Al-Farsi and Lee, 2008). A wide range

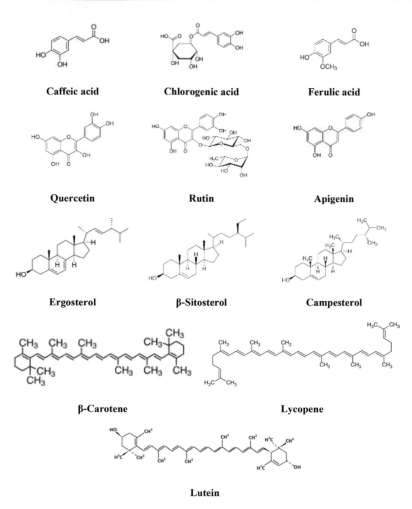

Fig. 16.1. Chemical structures of bioactive compounds present in date fruits.

of bioactive components has been reported in Omani date cvs. Umsellah and Khalas. The total phenolics in cvs. Umsellah and Khalas were 164.22 and 103.85 mg/100 g, respectively, and gallic acid, 35.77 and 27.41 mg/100 g, respectively. Caffeic and syringic acids were greater in Umsellah as compared to Khalas. The *p*-coumaric acid contents in Umsellah and Khalas were 24.94 and 21.69 mg/100 g, respectively (Siddiqi *et al.*, 2020). Al-Farsi *et al.* (2007) found that total phenolic content ranged from 172 to 246 mg gallic acid equivalents (GAE)/100 g in three date varieties grown in Oman. Flavonoids are valued for their anti-inflammatory, antioxidant, antiallergic, hepatoprotective, antithrombic, and antiviral properties (Al-Farsi *et al.*, 2005b; Vayalil, 2012). They have the potential to scavenge free radicals via acting as electron-donating and/or metal ion-chelating agents (Rice-Evans and Burdon, 1993).

Date fruits are a good source of carotenoids, although the concentration is influenced by several factors, such as the drying method, analysis conditions, variety, and maturation stage of dates. Date fruits that are red in color usually have hydrocarbon carotenoids like γ-carotene, neurosporene, and lycopene. Yellow-colored date varieties contain a mixture of carotenol fatty acid esters in addition to other carotenoids (Fennema, 1976). Carotenoids can be degraded or isomerized to form *cis* isomers during the drying process. The rate of carotene degradation or isomerization is dependent on drying temperature and duration (Al-Farsi *et al.*, 2005b; Fiedor and Burda, 2014). Lutein, neoxanthin, and β-carotene are predominately found in dates. The total carotenoid contents in fresh and dried dates are 913 and 973 µg/100 g dates, respectively (Ben-Amotz and Fishier, 1998; Al-Farsi *et al.*, 2005b; Boudries *et al.*, 2007; Fiedor and Burda, 2014).

Anthocyanins are present only in fresh dates, especially red-colored varieties. On average, dates contain about 0.87 mg of anthocyanins per 100 g of dates. Drying has an adverse effect on the anthocyanin content of dates (Shahidi and Naczk, 1995; Al-Farsi *et al.*, 2005b; Al-Farsi and Lee, 2008). Storage conditions, light, agronomic factors, and genetics may affect anthocyanin concentration (Shahidi and Naczk, 2003).

Date seeds are rich in nutritive substances like proteins, fats, oils, dietary fibers, and minerals (Table 16.2) that make them not only important as a food product but also give them purported medicinal value in different ailments. The phenolic compounds and flavonoids make them more attractive with multiple effects on human health with their antioxidant, anti-inflammatory, antimutagenic, nephroprotective, hepatoprotective, antidiabetic, anti-obesity, and anticancer properties (Hossain *et al.*, 2014; Masmoudi-Allouche *et al.*, 2016; Khalid *et al.*, 2017). Earlier studies have also found antitumor activities of water extract of date seed (variety not mentioned) against HCT-15 colon cancer cells (Sundar *et al.*, 2017), and of methanolic extracts of date cvs. Arechti and Korkobbi against HepG2 and HeLa cells (Mansour *et al.*, 2011; Thouri *et al.*, 2019). Besides this, *in vivo* radioprotective and *in vitro* antiproliferative effects of date seed n-hexane extract is reported against human liver carcinoma (HepG2), human lung carcinoma (A549), and human breast adenocarcinoma (MCF-7) cell lines (Al-Sheddi, 2019). Furthermore, oil derived from date palm cv. Deglet Nour has shown antiproliferative activity in HeLa cells (Mansour *et al.*, 2011).

16.3 Date Palm Use in Ayurveda

Ayurveda, an ancient Indian system of treatment, believes in a holistic approach to disease, which involves physical, spiritual, psychological, social, and environmental factors. The medicines prescribed by Ayurveda

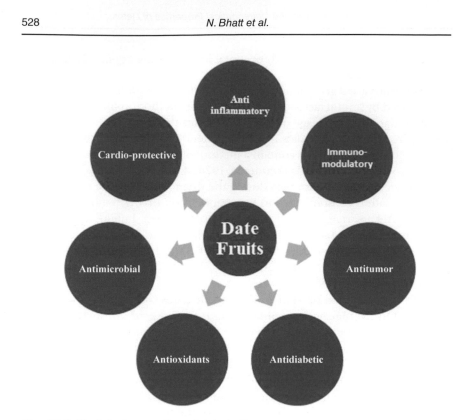

Fig. 16.2. Medicinal properties of date fruits.

physicians are mainly procured from plants, although minerals and metals are also sometimes used. Date fruit is considered a wonder fruit in Ayurveda and valued for its sweet (*madhura*), cooling (*sheetha*), tonic, fattening, and aphrodisiac properties. Ayurveda physicians use date fruit as a remedy for numerous diseases such as teeth and gum problems, urinary tract infections, upper and lower respiratory tract infections, general weakness, sciatica, dehydration, asthma, hiccups, cystitis, tuberculosis, nervous disorders, skin problems, leprosy, gonorrhea, anaemia, decreased sperm count, purpura, oedema, sepsis, cognitive dysfunction, anxiety, psychosis, muscular paralysis, cardiovascular disorders, kidney and liver disorders, and microbial and viral infections (Nadkarni, 1976; Anonymous, 2005). The medical properties of dates are summarized in Fig. 16.2.

16.3.1   Antidiabetic properties

Some date varieties are known for their low glycemic index (GI), like many other fruits. Studies among individuals with non-insulin-dependent diabetes (NIDDM) and gestational diabetes have revealed that dates can be used in low-GI diets (Lock *et al.*, 1988; Famuyiwa *et al.*, 1992) even though the diets provided higher carbohydrate intake (75 g). Date sugars are phenol rich, a potent antioxidant, and strong inhibitors of α-glycosidase enzyme (Ranilla *et al.*, 2008). The high dietary fiber content of dates makes them a low-GI food. Dates are a consolidated source of minerals involved in glucose metabolism and may be potentially beneficial for the prevention of diabetes. These minerals include magnesium, which is essential for energy-dependent transport systems, glycolysis, oxidative energy metabolism, insulin synthesis, secretion, and signaling; chromium, which enhances insulin action; manganese and copper, which can alter glucose homeostasis (Mooradian and Morley, 1987); and selenium, which has insulin-mimetic action (Mooradian and Morley, 1987; Stapleton, 2000). Selenium stimulates glucose uptake and regulates glycolysis, gluconeogenesis, fatty acid synthesis, and the pentose phosphate pathway. The high amounts of phyto-oestrogens may potentially help to maintain normal glucose metabolism in both healthy populations and diabetic patients.

16.3.2   Anti-inflammatory properties

Chronic inflammation gives rise to arthritis, atherosclerosis, and many other life-threatening disease conditions. The bioactive substances in dates may be responsible for anti-inflammatory mechanisms by targeting prostaglandins, which are involved in the late phase of acute inflammation and pain perception. A research report showed that ethyl acetate, methanolic, and water extracts of cv. Ajwa dates inhibited the lipid peroxidation cyclooxygenase enzymes COX-1 and COX2 (Zhang *et al.*, 2013). Another important research study reported that the methanolic extract of date fruit showed a vital role in reducing foot swelling and plasma fibrinogen (Mohamed and Al-Okbi, 2004). Not only date fruits but also date plant parts exhibit a role as anti-inflammatory substances. An animal study showed that date pollen has a potential protective effect via modulation of cytokine expression (Elberry *et al.*, 2011), while another study revealed that the leaves of dates could be considered a good source of natural antioxidant and anti-inflammatory drugs (Eddine, 2013). Dietary antioxidants in dates help to protect the body from neurological diseases, cardiovascular diseases (Abdollahi *et al.*, 2004), and gastric ulcers by minimizing oxidative stress (Al-Qarawi *et al.*, 2005).

16.3.3   Antimicrobial properties

Date fruit possesses strong antimicrobial activity and can prevent several diseases caused by microorganisms. The current pandemic caused by COVID-19 has shaken the world and compelled thinking on alternative approaches to promote immunity. One very recent research study has suggested that date consumption can effectively prevent coronavirus infection (Mafruchati, 2020). Dates are rich in ascorbate, carotenoids, selenium, and other antioxidants that may protect the body from oxidative damage caused by the lymphocyte phagocytosis activity of pests and pathogens. Chronic and recurrent infections and inflammatory conditions like abscesses, boils, and ulcers can also be controlled by regular consumption of date fruits and with external applications of various Ayurvedic preparations made of dates (Anonymous, 2005; Sastry, 2008). The methanol and acetone extracts of date pits inhibited the growth of Gram-positive and Gram-negative bacteria (Ammar *et al.*, 2009; Jassim and Naji, 2010). Another recent study in support of date fruits' effect as an antimicrobial on *Klebsiella pneumonia* and *Escherichia coli* showed a role in reducing the side effects of drugs such as methylprednisolone (Aamir *et al.*, 2013). A commercially available date fruit cultivar (Khadrawi) was tested for antimicrobial properties. The date fruits were found rich in polyphenols, tannins, flavonoids, and flavanols, which exerted potent bacteriostatic effects against Gram-positive and Gram-negative *E. coli* and *Staphylococcus aureus*, respectively (Taleb *et al.*, 2016).

16.3.4 Anticancer properties

Date fruits contain a wide range of phenolics that have proven anticarcinogenic activities (Mitscher *et al.*, 1996; Yamada and Tomita, 1996). Phenolics can inactivate the enzymes that are involved in the formation of malignant tumors at different stages (Uenobe *et al.*, 1997; Kuroda and Inoue, 1988). Caffeic and ferulic acids are the main phenolic acids in dates and inhibit the development of skin tumors because they react with nitriles and prevent the formation of nitrosamines (Kaul and Khanduja, 1998). Quercetin of dates not only suppressed the growth of gastric cancer cells significantly but also blocked the cell cycle progression from G1 to S phase in rats (Yoshida *et al.*, 1990). In another study, Al-Qarawi *et al.* (2005) found that ethanolic and aqueous extracts of dates significantly decreased the severity of gastric ulcers in rats. Date seed extracts may be a possible therapeutic agent against cancer (Khattak *et al.*, 2020). The methanolic extract of Medjool dates had significant antioxidant properties and anticancer activity against MCF-7 cells in an *in vitro* study; this effect may have been derived from phenolic compounds and sugars detected in the extract (Rozila *et al.*, 2019). The antigenotoxicity of date pits is due to its ability to scavenge the alkyl radical or inhibit the aromatase activity of cytochrome P450 or block the reaction between methane diazonium ion and DNA (Diab and Aboul-Ela, 2012). Studies have reported that β-D-glucan from dates has antitumor activity (Ishurd *et al.*, 2002, 2004). The glucans exhibited a dose-dependent anticancer activity with an optimum activity at a dose of 1 mg/kg in tumour in rats (Khan *et al.*, 2011).

16.3.5   Cardioprotective properties

It was recorded that dates prevent cardiovascular diseases, which might be due to inhibition of platelet aggregation and oxidation of low-density lipoprotein. Phenolics may be able to reduce blood pressure due to their anti-inflammatory and antithrombotic effects (Gerritsen *et al.*, 1995; Muldoon and Kritchevsky, 1996). Phenolics are also able to affect type II diabetes by inhibiting the activities of α-glucosidase and α-amylase from increasing blood glucose levels (Andlauer and Fürst, 2003; McCue and Shetty, 2004). Date palm fruit extracts have the potential to mobilize endogenous circulating progenitor cells, which can promote tissue repair following ischaemic injury (Alhaider *et al.*, 2017). Consumption of Ajwa dates prevented the depletion of endogenous antioxidants such as superoxide dismutase (SOD) and catalase (CAT) and inhibited lipid peroxidation in rats (Al-Yahya *et al.*, 2016). Al-Yahya *et al.* (2016) further reported that Ajwa dates have the capability to downregulate the expression of proinflammatory cytokines (interleukin 6 (IL-6), interleukin 10 (IL-10), tumour necrosis factor α (TNFα)) and apoptotic markers and upregulate the anti-apoptotic protein Bcl2, indicating antioxidant, hypolipidemic, anti-inflammatory, and anti-apoptotic potential against myocardial damage. Additionally, date fruit extract administration protected against doxorubicin-induced myocardial toxicity by reducing cardiac injury markers, reestablishing of oxidant/antioxidant parameters, ameliorating lipid profile, and lessening the histopathological changes. These effects may be due to its high contents of flavonoids (anthocyanins, apigenin, isoquercitrin, quercetin, quercitrin, procyanidins, luteolin, rutin) and its lipid-lowering effects (Mubarak *et al.*, 2018).

16.3.6   Immunomodulatory properties

Immunostimulation is a prophylactic or therapeutic concept that aims to stimulate the nonspecific immune system. Dates, due to their high fiber and phenolic contents, can positively impact the immune system. Phenolics derived from dates can suppress hypersensitive immune response due to their anti-allergic immune-modulatory activities. The immunomodulatory activities of phenolics also include anti-inflammatory responses triggered by the suppression of proinflammatory pathways (Ma and Kinneer, 2002).

16.3.7   Date palm in indigenous medicine

Medicine is the third important product secured from plants after food and oxygen. Preindustrial people were entirely dependent on plants for the management of various ailments. This legacy is continuing in many parts of the world and especially in rural areas. Traditionally, date fruits are useful in treating many disease conditions such as sore throat, fever, colds, intestinal troubles, bronchial catarrh, gonorrhea, oedema, cystitis, and liver and abdominal issues (Morton and Dowling, 1987). The fruit serves as a tonic and restorative and is also used as an analgesic to mitigate pain. In addition, it is widely used as an aphrodisiac, a sweetener, a diuretic, and in the treatment of vomiting, vertigo, and unconsciousness. Dates contain a substantial amount of dietary fiber and facilitate the evacuation of the bowels. Dried dates improve cardiovascular health by removing cholesterol from the arteries. They have high calcium content and improve bone health.

Dates can be a good source of many nutrients, including iron. A deficiency of iron can contribute to anaemia, a condition characterized by fatigue, dizziness, brittle nails, and shortness of breath. Fortunately, increasing the intake of iron-rich foods such as dates may help provide relief from anaemia symptoms (Hyder *et al.*, 2002).

16.3.8   Antioxidant properties of date palm cultivars

The major antioxidant activity of dates is due to their phenolic contents. Phenolics impart their activities in two ways: (i) by safeguarding macromolecules such as lipids, nucleic acids, and proteins from oxidative damage by scavenging free radicals (Jakus, 2000; Dröge, 2002; Al-Farsi and Lee, 2008); and (ii) by modulating cell physiology. The capability of any food item to defend living cells from the action of free radicals to prevent degenerative disorders deriving from persistent oxidative stress is considered as antioxidant activity. Antioxidant activities are determined by the presence of biologically active compounds like polyphenols, carotenoids, lignans, glucosinolates, vitamins, and oxalates. Such compounds exhibit a broad spectrum of biological activities such as antiallergic, antiatherogenic, anti-inflammatory, antimicrobial, antioxidant, antithrombotic, cardioprotective, as well as the ability to modify gene expression (Tapiero *et al.*, 2002; Nakamura *et al.*, 2003). The diverse chemical structures, their interactions, biological roles, and different modes of action make it difficult to assess a single and reliable procedure for the evaluation of antioxidant activity. Antioxidant activity increases in dates as they advance to maturity. On average, antioxidant content of dates determined by the oxygen radical absorbance capacity (ORAC) method was 1656 and 1025 µmol Trolox equivalents (TE)/100 g in fresh and dried dates, respectively (Al-Farsi *et al.*, 2005b).

16.4   Factors Affecting Antioxidant Activity in Date Fruits

Various physical and chemical characteristics affect antioxidant levels in date fruits. These factors include genotype variation, fruit maturity stages, preharvest conditions (climate, temperature, and light), cultural practices, and postharvest handling and processing.

16.4.1   Type of solvent used for extraction

Extraction plays a crucial role in evaluating antioxidant activity of date fruits. Extraction is influenced mainly by the type of solvent, i.e., organic solvent (ethanol, methanol, acetone, and/or mixtures) (Velioglu *et al.*, 1998; Liyana-Pathirana and Shahidi, 2006), aqueous extraction (hot or cold), or a mixture of organic solvent and water. Antioxidant activity is also influenced by extraction time and temperature (Ziaedini *et al.*, 2010; Marquez *et al.*, 2014; Wissam *et al.*, 2016), as well as by the chemical composition and physical characteristics of the analysed sample (Luthria, 2006). It has been found that a mixture of methanol–water improves the extraction efficiency since it results in extracts with the highest level of antioxidants (Iqbal *et al.*, 2005; Rufino *et al.*, 2010). However, alkaline hydrolysis, acid hydrolysis, and enzymatic digestion have each also been used as extraction processes (Verma *et al.*, 2009; Bennet *et al.*, 2010; Delgado-Andrade *et al.*, 2010; White et al., 2010; Navarro-González *et al.*, 2011; Royer *et al.*, 2011; Tang *et al.*, 2015). Different solvents show different antioxidant values within the same variety. In Ajwa dates, the aqueous, methanolic, and ethyl acetate extracts resulted in 91, 70, and 88% inhibition, respectively, against lipid peroxidation (Zhang *et al.*, 2017). In another study, the antioxidant activity of extracts from Ajwa date was 74.19 mg GAE/ml based on lipid peroxidation and 2,2-diphenyl-1-picrylhydrazyl (DPPH) (Arshad *et al.*, 2015). Seven date cultivars (Tantebouchte, Biraya, Degla Baidha, Deglet Nour, Ali Ourached, Ghars, Tansine) showed high polyphenol, flavonoid, and flavonol contents and consequently high antioxidant activity; the methanolic extract of Ali Ourached had the highest antioxidant activity (Ali Haimoud *et al.*, 2016).

16.4.2   Antioxidant activity of flavonoid and total phenolic contents

All cultivars of date fruit, whether ripe or unripe, contain phenolic (caffeic acid, ferulic acid, protocatechuic acid, catechin, gallic acid, *p*-coumaric acid, resorcinol, chlorogenic acid, and syringic acid) and flavonoid (quercetin, luteolin, apigenin, isoquercitrin, rutin) compounds and consequently exhibit great antioxidant activity (Al-Farsi *et al.*, 2005b; Gayathri and Thilagavathi, 2021). Therefore, the antioxidant properties of dates can be used as a natural and cost-effective tool in foods as a replacement for synthetic sources. The phenolic concentrations in Omani date cvs. such as Khasab, Khalas, and Fard were reported to be between 217 and 343 mg ferulic acid equivalents/100 g (Al-Farsi *et al.*, 2005b). Six ripe date fruit cultivars (Bouskri, Bousrdon, Bousthammi, Boufgous, Jihl, Majhoul) from Morocco exhibited good antioxidant potential, particularly cvs. Bousrdon

and Jihl (Bouhlali *et al.*, 2016). The antioxidant potential of cv. Bousrdon
was based on phenolic and flavonoid contents, whereas the highest antioxidant activity of cv. Jihl was based on the DPPH scavenging activity and
ferric reducing power (Bouhlali *et al.*, 2016). Saudi date cvs. such as Al
Sagey, Helwat Al Jouf, and Al Sour exhibited high antioxidant capacity
because of their high phenolic (caffeic acid, ferulic acid, protocatechuic
acid, catechin, gallic acid, *p*-coumaric acid, resorcinol, chlorogenic acid,
syringic acid) and flavonoid (quercetin, luteolin, apigenin, isoquercitrin,
rutin) contents. Chemical analysis showed that cv. Al Sagey possessed the
highest phenolic contents and the strongest antioxidant potential. In addition, cvs. Al Sagey and Helwat Al Jouf showed comparable glutathione and
ascorbate redox status, whereas cv. Al Sour had the least glutathione redox
status (Hamad, 2014).

    In another investigation, ten Algerian cultivars (Mech Degla, Deglet
Ziane, Deglet Nour, Thouri, Sebt Mira, Ghazi, Degla Beida, Arechti,
Halwa, Itima) were analysed and found to have potent antioxidant
potential (Benmeddour *et al.*, 2013). The cvs. Ghazi, Arechti, and Sebt
Mira had the most potent antioxidant capacities and the highest phenolic contents. Gallic, ferulic, coumaric, and caffeic acids were found
to be the main phenolics present, whereas isoquercitrin, quercitrin,
rutin, quercetin, and luteolin were the main flavonoids. In addition,
five Algerian cultivars (Deglet Nour, Degla Baidha, Ghars, Tamjhourt,
Tafezauine) were analysed by Zineb *et al.* (2012), who reported that
total phenolics, total flavonoids, and antioxidant activities were high
based on scavenging assays of the DPPH radical and 2,20-azino-bis(3
-ethylbenzothiazoline-6-sulfonate) (ABTS), as well as potassium fer-
ricyanide complex as reducing power assay. Tunisian cvs. Allig, Bejo,
and Deglet Nour also showed strong antioxidant activities, with Allig
being the highest, followed by Bejo and Deglet Nour. The antioxidant
potential of the cultivars correlated positively with their total phenolic
and flavonoid contents. The most widely used antioxidant assays (DPPH
radical scavenging activity, ferric reducing antioxidant power (FRAP)
assay, hydrogen peroxide (H_2O_2) scavenging activity, and metal chelat-
ing activity) were used (Kchaou *et al.*, 2014). However, it was observed
that there was no correlation between the total phenolic content and
the antioxidant activity of the date extracts. Those phenols are not the
only phytochemicals responsible for antioxidant activity in date; there
may be other nonphenolic constituents present that exhibit antioxidant
activity (Khanavi *et al.*, 2009).

16.4.3   Stage of ripening

The ripening stage of dates also influences antioxidant activity. Six
Mauritanian date cultivars (Ahmar dli, Ahmar denga, Bou seker, Tenterguel,
Lemdina, Tijib) at two edible stages of ripening (blah = khalal) and fully

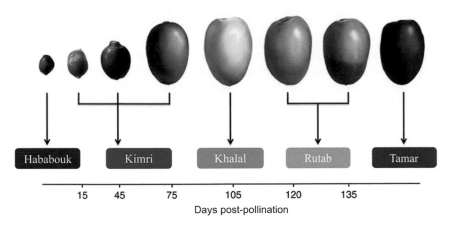

Fig. 16.3. Ripening stages of date fruits. (Modified from Al-Mssallem *et al.*, 2013.)

ripe (tamar) were investigated. The average total phenolics at the blah and tamar stages were recorded as 728.5 and 558.9 mg GAE/100 g dry matter, respectively. Average flavonoid content was 119.6 and 67.3 mg quercetin equivalents/100 g dry matter, respectively, being comparatively higher in the blah stage than the tamar stage in all cultivars. Total phenolics were higher than flavonoids in both the tamar and blah stages (Lemine *et al.*, 2014). Four Tunisian cvs., Gondi, Gasbi, Khalt Dhahbi, and Rtob Ahmar, at three maturation stages, mature firm (besser or khalal), partially ripe (rutab), and ripe (tamar), were analysed for their antioxidant contents and activities. The besser stage showed the highest antioxidant contents (Amira *et al.*, 2012). The phytochemical composition and phenolic profile changes as well as antioxidant and antibacterial properties of five cultivars (Beidh Hmam, Degla, Khalt Ahmar, Rtob, Rtob Hodh) at three distinct maturity stages (besser, rutab, tamar) were reported by Chaira *et al.* (2009). The results showed that the antioxidant contents decreased with increasing fruit ripening, with a corresponding decrease in the antioxidant and antibacterial activities. Caffeic, ferulic, and *p*-coumaric acids were the major acids detected, especially at the besser stage. It was shown that ten Tunisian cultivars (Smiti, Kenta, Bekrari, Mermella, Garn ghzal, Nefzaoui, Baht, Korkobbi, Bouhattam, Rotbi) possess high antioxidant potentials (Chaira *et al.*, 2009) as well as three other Tunisian cultivars (Deglet Nour, Allig, Bejo) (Kchaou *et al.*, 2016). The ripening stages of date fruits are shown in Fig. 16.3.

16.4.4 Postharvest management

Dates are a non-climacteric fruit, with very low respiration rate and low ethylene production, and therefore exhibit very low metabolic activities. The effects of chilling and storage on antibacterial properties and antioxidant capacities of Saudi Arabian cultivars (Mabroom, Safawi, Ajwa) compared to an Iranian cultivar (Mariami) were analysed by Samad *et al.* (2016). The results suggested that storage at 4 and 20°C increased the anthocyanin content of date fruits, whereas chilling for 8 weeks increased the phenolic content. They also showed antibacterial activity against *S. aureus, Bacillus cereus, Serratia marcescens,* and *E. coli.* In Egypt, antioxidant and antimicrobial properties of tamar stage date fruits were evaluated with the results showing high antioxidant contents mostly due to esculetin, tannic acid, gallic acid, itaconic acid, and ferulic acid (Sohaimy *et al.,* 2015).

Not only the date fruit, but also the date seed is rich in antioxidant activity in different cultivars of Omani dates. Seeds from date cvs. Mabseeli, Um-sellah, and Shahal presented high antioxidant activity (580–929 µmol TE/g fresh weight) (Al-Farsi *et al.,* 2007). Dates were found to be the source of the second most potent natural antioxidants out of 28 different fruits extensively studied in China (Guo *et al.,* 2003).

16.5   Conclusion and Prospects

Date fruits are widely grown in the arid and semi-arid regions of the globe and are considered one of the oldest and most important staple crops in Southwest Asia and North Africa. Date fruits are high in carbohydrates and low in fats. The fruits are abundant in monosaccharides such as glucose and fructose, which give a sweetness and unique taste to the fruits. Date sugar is absorbed more gradually than refined sugar due to its high fiber content. Traditionally date fruits are useful in the treatment of several health/disease conditions. Dates are rich in bioactive components, making them an excellent option for use as a nutraceutical. Systematic research is needed to explore the health benefits of date fruit and date seeds in order to develop various value-added food products, supplements, and nutraceuticals. This can open new doors for food and pharmaceutical companies to diversify and provide added value to date palm cultivation.

References

Aamir, J., Kumari, A., Khan, M.N. and Medam, S.K. (2013) Evaluation of the combinational antimicrobial effect of *Annona squamosa* and *Phoenix dactylifera* seeds methanolic extract on standard microbial strains. *International Research Journal of Biological Sciences* 2, 68–73.

Abdollahi, M., Ranjbar, A., Shadnia, S., Nikfar, S. and Rezaiee, A. (2004) Pesticides and oxidative stress: a review. *Medical Science Monitor* 10, RA141–RA147.

Abu-Odeh, A.M. and Talib, W.H. (2021) Middle east medicinal plants in the treatment of diabetes: a review. *Molecules* 26(3), 742. DOI: 10.3390/molecules26030742.

Ahmed, I.A., Ahmed, A.W.K. and Robinson, RK. (1995) Chemical composition of date varieties as influenced by the stage of ripening. *Food Chemistry* 54(3), 305–309. DOI: 10.1016/0308-8146(95)00051-J.

Al-Farsi, M. and Lee, C.Y. (2008) Nutritional and functional properties of dates: a review. *Critical Reviews in Food Science and Nutrition* 48(10), 877–887. DOI: 10.1080/10408390701724264.

Al-Farsi, M., Alasalvar, C., Morris, A., Baron, M. and Shahidi, F. (2005a) Compositional and sensory characteristics of three native sun-dried date (*Phoenix dactylifera* L.) varieties grown in Oman. *Journal of Agricultural and Food Chemistry* 53(19), 7586–7591. DOI: 10.1021/jf050578y.

Al-Farsi, M., Alasalvar, C., Morris, A., Baron, M. and Shahidi, F. (2005b) Comparison of antioxidant activity, anthocyanins, carotenoids, and phenolics of three native fresh and sun-dried date (*Phoenix dactylifera* L.) varieties grown in Oman. *Journal of Agricultural and Food Chemistry* 53(19), 7592–7599. DOI: 10.1021/jf050579q.

Al-Farsi, M., Alasalvar, C., Al-Abid, M., Al-Shoaily, K., Al-Amry, M. *et al.* (2007) Compositional and functional characteristics of dates, syrups, and their by-products. *Food Chemistry* 104(3), 943–947. DOI: 10.1016/j.foodchem.2006.12.051.

Alhaider, I.A., Mohamed, M.E., Ahmed, K.K.M. and Kumar, A.H.S. (2017) Date palm (*Phoenix dactylifera*) fruits as a potential cardioprotective agent: the role of circulating progenitor cells. *Frontiers in Pharmacology* 8, 592. DOI: 10.3389/fphar.2017.00592.

Al-Harrasi, A., Rehman, N.U., Hussain, J., Khan, A.L., Al-Rawahi, A, *et al.* (2014) Nutritional assessment and antioxidant analysis of 22 date palm (*Phoenix dactylifera*) varieties growing in Sultanate of Oman. *Asian Pacific Journal of Tropical Medicine* 7(S1), S591–S598. DOI: 10.1016/S1995-7645(14)60294-7.

Al-Hooti, S., Sidhu, J.S. and Qabazard, H. (1997) Physicochemical characteristics of five date fruit cultivars grown in the United Arab Emirates. *Plant Foods for Human Nutrition* 50(2), 101–113. DOI: 10.1007/BF02436030.

Ali Haimoud, S., Allem, R. and Merouane, A. (2016) Antioxidant and anti-inflammatory properties of widely consumed date palm (*Phoenix dactylifera* L.) fruit varieties in Algerian oases. *Journal of Food Biochemistry* 40(4), 463–471. DOI: 10.1111/jfbc.12227.

Allaith, A.A.A. (2008) Antioxidant activity of Bahraini date palm (*Phoenix dactylifera* L.) fruit of various cultivars . *International Journal of Food Science & Technology* 43(6), 1033–1040. DOI: 10.1111/j.1365-2621.2007.01558.x.

Al-Mssallem, I.S., Hu, S., Zhang, X., Lin, Q., Liu, W. *et al.* (2013) Genome sequence of the date palm *Phoenix dactylifera* L. *Nature Communications* 4, 2274. DOI: 10.1038/ncomms3274.

Al-Qarawi, A.A., Abdel-Rahman, H., Ali, B.H., Mousa, H.M. and El-Mougy, S.A. (2005) The ameliorative effect of dates (*Phoenix dactylifera* L.) on ethanol-induced gastric ulcer in rats. *Journal of Ethnopharmacology* 98(3), 313–317. DOI: 10.1016/j.jep.2005.01.023.

Al-Shahib, W. and Marshall, R.J. (2003) The fruit of the date palm: its possible use as the best food for the future? *International Journal of Food Sciences and Nutrition* 54(4), 247–259. DOI: 10.1080/09637480120091982.

Al-Sheddi, E.S. (2019) Anticancer potential of seed extract and pure compound from *Phoenix dactylifera* on human cancer cell lines. *Pharmacognosy Magazine* 15(63), 494. DOI: 10.4103/pm.pm_623_18.

Al-Showiman, S.S., Al-Tamrah, S.A. and BaOsman, A.A. (1994) Determination of selenium content in dates of some cultivars grown in Saudi Arabia.

International Journal of Food Sciences and Nutrition 45(1), 29–33. DOI: 10.3109/09637489409167014.

Al-Yahya, M., Raish, M., AlSaid, M.S., Ahmad, A., Mothana, R.A. *et al.* (2016) "Ajwa" dates (*Phoenix dactylifera* L.) extract ameliorates isoproterenol-induced cardiomyopathy through downregulation of oxidative, inflammatory and apoptotic molecules in rodent model. *Phytomedicine* 23(11), 1240–1248. DOI: 10.1016/j. phymed.2015.10.019.

Amira, E.A., Behija, S.E., Beligh, M., Lamia, L., Manel, I. *et al.* (2012) Effects of the ripening stage on phenolic profile, phytochemical composition and antioxidant activity of date palm fruit. *Journal of Agricultural and Food Chemistry* 60(44), 10896–10902. DOI: 10.1021/jf302602v.

Ammar, N.M., Lamia, T., Abou, E., Nabil, H.S., Lalita, M.C. *et al.* (2009) Flavonoid constituents and antimicrobial activity of date (*Phoenix dactylifera* L.) seeds growing in Egypt. In: Proceedings of the 4th Conference on Research and Development of Pharmaceutical Industries (Current Challenges). *Medicinal and Aromatic Plant Science and Biotechnology* 3(Special Issue 1), 1–5.

Andlauer, W. and Fürst, P. (2003) Special characteristics of non-nutrient food constituents of plants – phytochemicals. Introductory lecture. *International for Vitamin and Nutrition Research* 73(2), 55–62. DOI: 10.1024/0300-9831.73.2.55.

Anonymous (2005) *The Wealth of India – Raw Materials Volume VII*. Council of Science and Industrial Research, New Delhi.

Anwar, S., Raut, R., Alsahli, M.A., Almatroudi, A., Alfheeaid, H. *et al.* (2022) Role of Ajwa datefruit pulp and seed in the management of diseases through *in vitro* and *in silico* analysis. *Biology* 11(1), 78. DOI: 10.3390/biology11010078.

Appel, L.J., Moore, T.J., Obarzanek, E., Vollmer, W.M., Svetkey, L.P. *et al.* (1997) A clinical trial of the effects of dietary patterns on blood pressure. *New England Journal of Medicine* 336(16), 1117–1124. DOI: 10.1056/ NEJM199704173361601.

Arshad, F.K., Haroon, R., Jelani, S. and Masood, HB. (2015) A relative *in vitro* evaluation of antioxidant potential profile of extracts from pits of *Phoenix dactylifera* L. (Ajwa and Zahedi dates). *International Journal of Advanced Science and Technology* 35, 28–37.

Baliga, M.S., Baliga, B.R.V., Kandathil, S.M., Bhat, H.P. and Vayalil, P.K. (2011) A review of the chemistry and pharmacology of the date fruits (*Phoenix dactylifera* L.). *Food Research International* 44(7), 1812–1822. DOI: 10.1016/j. foodres.2010.07.004.

Barreveld, W. (1993) *Date Palm Products*. Agricultural Services Bulletin no.101. Food and Agriculture Organizationof the United Nations, Rome.

Ben-Amotz, A. and Fishier, R. (1998) Analysis of carotenoids with emphasis on 9-*cis* β-carotene in vegetables and fruits commonly consumed in Israel. *Food Chemistry* 62(4), 515–520. DOI: 10.1016/S0308-8146(97)00196-9.

Benmeddour, Z., Mehinagic, E., Meurlay, D.L. and Louaileche, H. (2013) Phenolic composition and antioxidant capacities of ten Algerian date (*Phoenix dactylifera* L.) cultivars: a comparative study. *Journal of Functional Foods* 5(1), 346–354. DOI: 10.1016/j.jff.2012.11.005.

Bennet, R.N., Shiga, T.M., Hassimoto, N.M.A., Rosa, E.A.S., Lajolo, F.M. *et al.* (2010) Phenolics and antioxidant properties of fruit pulp and cell wall fractions of post-harvest banana (*Musa acuminata* Juss.) cultivars. *Journal of Agricultural and Food Chemistry* 54, 1646–1658.

Boudries, H., Kefalas, P. and Hornero-Mendez, D. (2007) Carotenoid composition of Algerian date varieties (*Phoenix dactylifera*) at different edible maturation stages. *Food Chemistry* 101(4), 1372–1377. DOI: 10.1016/j.foodchem.2006.03.043.

Bouhlali, E.T., Bammou, M., Sellam, K., Benlyas, M., Alem, C. *et al.* (2016) Evaluation of antioxidant, antihemolytic and antibacterial potential of six Moroccan date fruit (*Phoenix dactylifera* L.) varieties. *Journal of King Saud University - Science* 28(2), 136–142. DOI: 10.1016/j.jksus.2016.01.002.

Chaira, N., Smaali, M.I., Martinez-Tomé, M., Mrabet, A., Murcia, M.A. *et al.* (2009) Simple phenolic composition, flavonoid contents and antioxidant capacities in water-methanol extracts of Tunisian common date cultivars (*Phoenix dactylifera* L.) . *International Journal of Food Sciences and Nutrition* 60(sup7), 316–329. DOI: 10.1080/09637480903124333.

Chao, C.C.T. and Krueger, R.R. (2007) The date palm (*Phoenix dactylifera* L.): overview of biology, uses, and cultivation. *HortScience* 42(5), 1077–1082. DOI: 10.21273/HORTSCI.42.5.1077.

Delgado-Andrade, C., Conde-Aguilera, J.A., Haro, A., Pastoriza de la Cueva, S. and Rufián-Henares, J.Á. (2010) A combined procedure to evaluate the global antioxidant response of bread. *Journal of Cereal Science* 52(2), 239–246. DOI: 10.1016/j.jcs.2010.05.013.

Diab, K.A.S. and Aboul-Ela, E.I. (2012) *In vivo* comparative studies on antigenotoxicity of date palm (*Phoenix dactylifera* L.) pits extract against DNA damage induced by *N*-nitroso-*N*-methylurea in mice. *Toxicology International* 19(3), 279–286. DOI: 10.4103/0971-6580.103669.

Dröge, W. (2002) Free radicals in the physiological control of cell function. *Physiological Reviews* 82(1), 47–95. DOI: 10.1152/physrev.00018.2001.

Eddine, L.S. (2013) Antioxidant, anti-inflammatory and diabetes related enzyme inhibition properties of leaves extract from selected varieties of *Phoenix dactylifera* L. *Innovare Journal of Life Sciences* 1, 14–18.

Elberry, A.A., Mufti, S.T., Al-Maghrabi, J.A., Abdel-Sattar, E.A., Ashour, O.M. *et al.* (2011) Anti-inflammatory and antiproliferative activities of date palm pollen (*Phoenix dactylifera*) on experimentally-induced atypical prostatic hyperplasia in rats. *Journal of Inflammation* 23(1), 40. DOI: 10.1186/1476-9255-8-40.

Famuyiwa, O.O., Elhazmi, M.A.F., Aljasser, S.J., Sulimani, R.A., Jayakumar, R.V. *et al.* (1992) A comparison of acute glycemic and insulin-response to dates (*Phoenix dactylifera*) and oral dextrose in diabetic and nondiabetic subjects. *Saudi Medical Journal* 13(5), 397–402.

Fennema, O.R. (1976) *Principles of Food Science. Part I: Food Chemistry.* Marcel Dekker, New York.

Fiedor, J. and Burda, K. (2014) Potential role of carotenoids as antioxidants in human health and disease. *Nutrients* 6(2), 466–488. DOI: 10.3390/nu6020466.

Gayathri, M. and Thilagavathi, S. (2021) Phytochemical composition and antioxidant activity of two varieties of unripe date palm. *International Journal of Current Research and Review* 13(7), 106–111. DOI: 10.31782/IJCRR.2021.13730.

Gerritsen, M.E., Carley, W.W., Ranges, G.E., Shen, C.P., Phan, S.A. *et al.* (1995) Flavonoids inhibit cytokine-induced endothelial cell adhesion protein gene expression. *American Journal of Pathology* 147(2), 278–292.

Guo, C., Yang, J., Wei, J., Li, Y., Xu, J. *et al.* (2003) Antioxidant activities of peel, pulp and seed fractions of common fruits as determined by FRAP assay. *Nutrition Research* 23(12), 1719–1726. DOI: 10.1016/j.nutres.2003.08.005.

Hamad, I. (2014) Phenolic profile and antioxidant activity of Saudi date palm (*Phoenix dactylifera* L.) fruit of various cultivars. *Life Science Journal* 11, 1268–1271.

Hazzouri, K.M., Flowers, J.M., Visser, H.J., Khierallah, H.S., Rosas, U. *et al.* (2015) Whole genome re-sequencing of date palms yields insights into diversification of a fruit tree crop. *Nature Communications* 6, 8824. DOI: 10.1038/ncomms9824.

Hinkaew, J., Aursalung, A., Sahasakul, Y., Tangsuphoom, N. and Suttisansanee, U. (2021) A comparison of the nutritional and biochemical quality of date palm fruits obtained using different planting techniques. *Molecules* 26, 2245. https://doi.org/10.3390/molecules26082245

Hossain, M., Waly, M.I., Singh, V., Sequeira, V. and Rahman, M. (2014) Chemical composition of date-pits and its potential for developing value-added product – a review. *Polish Journal of Food and Nutrition Sciences* 64(4), 215–226. DOI: 10.2478/pjfns-2013-0018.

Hyder, S.M.Z., Persson, L.Å., Chowdhury, A.M.R. and Ekström, E.C. (2002) Do side-effects reduce compliance to iron supplementation? A study of daily- and weekly-dose regimens in pregnancy. *Journal of Health, Population, and Nutrition* 20(2), 75–79.

Iqbal, S., Bhanger, M.I. and Anwar, F. (2005) Antioxidant properties and components of some commercially available varieties of rice bran in Pakistan. *Food Chemistry* 93(2), 265–272. DOI: 10.1016/j.foodchem.2004.09.024.

Ishurd, O., Sun, C., Xiao, P., Ashour, A. and Pan, Y. (2002) A neutral beta-D-glucan from dates of the date palm, *Phoenix dactylifera* L. *Carbohydrate Research* 337(14), 1325–1328. DOI: 10.1016/s0008-6215(02)00138-6.

Ishurd, O., Zgheel, F., Kermagi, A., Flefla, M. and Elmabruk, M. (2004) Antitumor activity of beta-D-glucan from Libyan dates. *Journal of Medicinal Food* 7(2), 252–255. DOI: 10.1089/1096620041224085.

Ismail, B., Henry, J., Haffar, I. and Baalbaki, R. (2006) Date consumption and dietary significance in the United Arab Emirates. *Journal of the Science of Food and Agriculture* 86(8), 1196–1201. DOI: 10.1002/jsfa.2467.

Ismail, B., Haffar, I., Baalbaki, R. and Henry, J. (2008) Physico-chemical characteristics and sensory quality of two date varieties under commercial and industrial storage conditions. *LWT - Food Science and Technology* 41(5), 896–904. DOI: 10.1016/j.lwt.2007.06.009.

Jakus, V. (2000) The role of free radicals, oxidative stress and antioxidant systems in diabetic vascular disease. *Bratislavske Medical Journal* 101(10), 541–551.

Jassim, S.A.A. and Naji, M.A. (2010) *In vitro* evaluation of the antiviral activity of an extract of date palm (*Phoenix dactylifera* L.) pits on a *Pseudomonas* phage. *Evidence-Based Complementary and Alternative Medicine* 7(1), 57–62. DOI: 10.1093/ecam/nem160.

Kaul, A. and Khanduja, K.L. (1998) Polyphenols inhibit promotional phase of tumorigenesis: relevance of superoxide radicals. *Nutrition and Cancer* 32(2), 81–85. DOI: 10.1080/01635589809514723.

Kchaou, W., Abbès, F., Attia, H. and Besbes, S. (2014) *In vitro* antioxidant activities of three selected dates from Tunisia (*Phoenix dactylifera* L.). *Journal of Chemistry* 2014, 1–8. DOI: 10.1155/2014/367681.

Kchaou, W., Abbès, F., Mansour, R.B., Blecker, C., Attia, H. *et al.* (2016) Phenolic profile, antibacterial and cytotoxic properties of second grade date extract from Tunisian cultivars (*Phoenix dactylifera* L.). *Food Chemistry* 194, 1048–1055. DOI: 10.1016/j.foodchem.2015.08.120.

Khalid, S., Khalid, N., Khan, R.S., Ahmed, H. and Ahmad, A. (2017) A review on chemistry and pharmacology of Ajwa date fruit and pit. *Trends in Food Science & Technology* 63, 60–69. DOI: 10.1016/j.tifs.2017.02.009.

Khan, M.A., Chen, H., Tania, M. and Zhang, D. (2011) Anticancer activities of *Nigella sativa* (black cumin). *African Journal of Traditional, Complementary, and Alternative Medicines* 8(5 Suppl), 226–232. DOI: 10.4314/ajtcam.v8i5S.10.

Khanavi, M., Saghari, Z., Mohammadirad, A., Khademi, R., Hadjiakhoondi, A. and Abdollahi, M. (2009) Comparison of antioxidant activity and total phenols of some date varieties. *DARU Journal of Pharmaceutical Sciences* 17, 104–108.

Khattak, M.N.K., Shanableh, A., Hussain, M.I., Khan, A.A., Abdulwahab, M. *et al.* (2020) Anticancer activities of selected Emirati date (*Phoenix dactylifera* L.) varieties pits in human triple negative breast cancer MDA-MB-231 cells. *Saudi Journal of Biological Sciences* 27(12), 3390–3396. DOI: 10.1016/j.sjbs.2020.09.001.

Kuroda, Y. and Inoue, T. (1988) Antimutagenesis by factors affecting DNA repair in bacteria. *Mutation Research* 202(2), 387–391. DOI: 10.1016/0027-5107(88)90200-x.

Lemine, F.M.M., Ahmed, M.V.O.M., Maoulainine, L.B.M., Bouna, Z.E.A.O., Samb, A. *et al.* (2014) Antioxidant activity of various Mauritanian date palm (*Phoenix dactylifera* L.) fruits at two edible ripening stages. *Food Science & Nutrition* 2(6), 700–705. DOI: 10.1002/fsn3.167.

Liyana-Pathirana, C.M. and Shahidi, F. (2006) Antioxidant properties of commercial soft and hard winter wheats (*Triticum aestivum* L.) and their milling fractions. *Journal of the Science of Food and Agriculture* 86(3), 477–485. DOI: 10.1002/jsfa.2374.

Lock, D.R., Bar-Eyal, A., Voet, H. and Madar, Z. (1988) Glycemic indices of various foods given to pregnant diabetic subjects. *Obstetrics and Gynecology* 71(2), 180–183.

Luthria, D.L. (2006) Significance of sample preparation in developing analytical methodologies for accurate estimation of bioactive compounds in functional foods. *Journal of the Science of Food and Agriculture* 86(14), 2266–2272. DOI: 10.1002/jsfa.2666.

Ma, Q. and Kinneer, K. (2002) Chemoprotection by phenolic antioxidants. Inhibition of tumor necrosis factor alpha induction in macrophages. *The Journal of Biological Chemistry* 277(4), 2477–2484. DOI: 10.1074/jbc.M106685200.

Mafruchati, M. (2020) The use of dates against COVID-19, based on effectiveness or religion's believe? Trends and relevance analysis in big data. *Systematic Reviews in Pharmacy* 11(8), 394–399.

Mansour, R.B., Lassoued, S., Dammak, I., Elgaied, A., Besbes, S. *et al.* (2011) Cytotoxicity evaluation and antioxidant activity of date seed oil from 'Deglet-nour Tunisian cultivar' (*Phoenix dactylifera* L.). *Natural Products* 7(1), 16–20.

Maqsood, S., Adiamo, O., Ahmad, M. and Mudgil, P. (2020) Bioactive compounds from date fruit and seed as potential nutraceutical and functional food ingredients. *Food Chemistry* 308, 125522. DOI: 10.1016/j.foodchem.2019.125522.

Marlett, J.A., McBurney, M.I. and Slavin, J.L. (2002) Position of the American Dietetic Association: health implications of dietary fiber. *Journal of the American Dietetic Association* 102(7), 993–1000. DOI: 10.1016/s0002-8223(02)90228-2.

Marquez, A., Perez-Serratosa, M., Varo, M.A. and Merida, J. (2014) Effect of temperature on the anthocyanin extraction and color evolution during controlled dehydration of Tempranillo grapes. *Journal of Agricultural and Food Chemistry* 62(31), 7897–7902. DOI: 10.1021/jf502235b.

Masmoudi-Allouche, F., Touati, S., Mnafgui, K., Gharsallah, N. *et al.* (2016) Phytochemical profile, antioxidant, antibacterial, antidiabetic and anti-obesity activities of fruits and pits from date palm (*Phoenix dactylifera* L.) grown in south of Tunisia. *Journal of Pharmacognosy and Phytochemistry* 5(3), 15–22.

McCue, P.P. and Shetty, K. (2004) Inhibitory effects of rosmarinic acid extracts on porcine pancreatic amylase *in vitro*. *Asia Pacific Journal of Clinical Nutrition* 13(1), 101–106.

Mistry, H.D., Broughton Pipkin, F., Redman, C.W.G. and Poston, L. (2012) Selenium in reproductive health. *American Journal of Obstetrics and Gynecology* 206(1), 21–30. DOI: 10.1016/j.ajog.2011.07.034.

Mitscher, L.A., Telikepalli, H., McGhee, E. and Shankel, D.M. (1996) Natural antimutagenic agents. *Mutation Research* 350(1), 143–152. DOI: 10.1016/0027-5107(95)00099-2.

Mohamed, A. (2000) Trace element levels in some kinds of dates. *Food Chemistry* 70(1), 9–12. DOI: 10.1016/S0308-8146(99)00232-0.

Mohamed, D.A. and Al-Okbi, S.Y. (2004) *In vivo* evaluation of antioxidant and anti-inflammatory activity of different extracts of date fruits in adjuvant arthritis. *Polish Journal of Food and Nutrition Sciences* 54(4), 397–402.

Mooradian, A.D. and Morley, J.E. (1987) Micronutrient status in diabetes mellitus. *American Journal of Clinical Nutrition* 45(5), 877–895.

Morton, J.F. and Dowling, C.F. (1987) Date. In: Morton, J.F. (ed.) *Fruits of Warm Climates*. Distributed by Creative Resource Systems, Miami, Florida/Winterville, North Carolina, pp. 5–11.

Mrabet, A., Hammadi, H., Rodríguez-Gutiérrez, G., Jiménez-Araujo, A. and Sindic, M. (2019) Date palm fruits as a potential source of functional dietary fiber: a review. *Food Science and Technology Research* 25(1), 1–10. DOI: 10.3136/fstr.25.1.

Mubarak, S., Hamid, S.A., Farrag, A.R., Samir, N. and Hussein, J.S. (2018) Cardioprotective effectof date palm against doxorubicin-induced cardiotoxicity. *Asian Journal of Pharmaceutical and Clinical Research* 11(7), 141. DOI: 10.22159/ajpcr.2018.v11i7.24453.

Muldoon, M.F. and Kritchevsky, S.B. (1996) Flavonoids and heart disease. *British Medical Journal* 312(7029), 458–459. DOI: 10.1136/bmj.312.7029.458.

Myhara, R.M., Karkalas, J. and Taylor, M.S. (1999) The composition of maturing Omani dates. *Journal of the Science of Food and Agriculture* 79(11), 1345–1350. DOI: 10.1002/(SICI)1097-0010(199908)79:11<1345::AID-JSFA366>3.0.CO;2-V.

Nadkarni, K.M. (1976) *Indian Materia Medica*, Vol. 1. Popular Prakashan, Bombay, India.

Nakamura, Y., Watanabe, S., Miyake, N., Kohno, H. and Osawa, T. (2003) Dihydrochalcones: evaluation as novel radical scavenging antioxidants. *Journal of Agricultural and Food Chemistry* 51(11), 3309–3312. DOI: 10.1021/jf0341060.

Navarro-González, I., García-Valverde, V., García-Alonso, J. and Periago, M.J. (2011) Chemical profile, functional and antioxidant properties of tomato peel fiber. *Food Research International* 44(5), 1528–1535. DOI: 10.1016/j.foodres.2011.04.005.

O'Dell, B.L. and Sunde, R.A. (1997) *Handbook of Nutritionally Essential Mineral Elements*. CRC Press, Boca Raton, Florida. DOI: 10.1201/9781482273106.

Prabu, S.L., Suriyaprakash, T., Dinesh, K., Suresh, K. and Ragavendran, T. (2012) Nutraceuticals: a review. *Elixir Pharmacy* 46, 8372–8377.

Ranilla, L.G., Kwon, Y.I., Genovese, M.I., Lajolo, F.M. and Shetty, K. (2008) Antidiabetes and antihypertension potential of commonly consumed carbohydrate sweeteners using *in vitro* models. *Journal of Medicinal Food* 11(2), 337–348. DOI: 10.1089/jmf.2007.689.

Rice-Evans, C. and Burdon, R. (1993) Free radical-lipid interactions and their pathological consequences. *Progress in Lipid Research* 32(1), 71–110. DOI: 10.1016/0163-7827(93)90006-i.

Royer, M., Diouf, P.N. and Stevanovic, T. (2011) Polyphenol contents and radical scavenging capacities of red maple (*Acer rubrum* L.) extracts. *Food and Chemical Toxicology* 49(9), 2180–2188. DOI: 10.1016/j.fct.2011.06.003.

Rozila, I., Abdul Manap, N., Ghazali, L., Kamal, N., Abdul Hakeem, W. *et al.* (2019) The antioxidant properties and anticancer effect of Medjool dates (*Phoenix dactylifera* L.) on human breast adenocarcinoma (MCF-7) cells: *in vitro* study. *Frontiers in Pharmacology* 10, 63–65. DOI: 10.3389/conf.fphar.2019.63.00038.

Rufino, M. d. S.M., Alves, R.E., de Brito, E.S., Pérez-Jiménez, J., Saura-Calixto, F. *et al.* (2010) Bioactive compounds and antioxidant capacities of 18 non-traditional tropical fruits from Brazil. *Food Chemistry* 121(4), 996–1002. DOI: 10.1016/j.foodchem.2010.01.037.

Samad, M.A., Hashim, S.H., Simarani, K. and Yaacob, J.S. (2016) Antibacterial properties and effects of fruit chilling and extract storage on antioxidant activity, total phenolic and anthocyanin content of four date palm (*Phoenix dactylifera*) cultivars. *Molecules* 21(4), 419. DOI: 10.3390/molecules21040419.

Sardesai, V. (2011) *Introduction to Clinical Nutrition*, 3rd edn. CRC Press, Boca Raton, Florida. DOI: 10.1201/b16601.

Sastry, J.L.N. (2008) *Illustrated Dravyaguna Vijnana*, 3rd edn., Vol. 2. Chaukhambha Orientalia, Varanasi, India.

Shahidi, F. and Naczk, M. (1995) Antioxidant properties of food phenolics. In: *Food Phenolics: Sources, Chemistry, Effects, Applications*. Technomic Publishing Co, Lancaster, Pennsylvania, pp. 235–277.

Shahidi, F. and Naczk, M. (2003) *Phenolics in Food and Nutraceuticals*, 2nd edn. CRC Press, Boca Raton, Florida. DOI: 10.1201/9780203508732.

Siddiqi, S.A., Rahman, S., Khan, M.M., Rafiq, S., Inayat, A. *et al.* (2020) Potential of dates (*Phoenix dactylifera* L.) as natural antioxidant source and functional food for healthy diet. *The Science of the Total Environment* 748, 141234. DOI: 10.1016/j.scitotenv.2020.141234.

Sohaimy, S.A., Abdelwahab, A.E. and Brennan, C.S. (2015) Phenolic content, antioxidant and antimicrobial activities of Egyptian date palm fruits. *Australian Journal of Basic and Applied Sciences* 9, 141–147.

Srivastava, S., Sharma, P.K. and Kumara, S. (2015) Nutraceuticals: a review. *Journal of Chronotherapy and Drug Delivery* 6, 1–10.

Stapleton, S.R. (2000) Selenium: an insulin-mimetic. *Cellular and Molecular Life Sciences* 57(13–14), 1874–1879. DOI: 10.1007/PL00000669.

Sundar, R.D.V., Segaran, G., Shankar, S., Settu, S. and Ravi, L. (2017) Bioactivity of *Phoenix dactylifera* seed and its phytochemical analysis. *International Journal of Green Pharmacy* 11(2), S292–S297.

Tahraoui, A., El-Hilaly, J., Israili, Z.H. and Lyoussi, B. (2007) Ethnopharmacological survey of plants used in the traditional treatment of hypertension and diabetes in south-eastern Morocco (Errachidia province). *Journal of Ethnopharmacology* 110(1), 105–117. DOI: 10.1016/j.jep.2006.09.011.

Taleb, H., Maddocks, S.E., Morris, R.K. and Kanekanian, A.D. (2016) The antibacterial activity of date syrup polyphenols against *S. aureus* and *E. coli. Frontiers in Microbiology* 7, 198. DOI: 10.3389/fmicb.2016.00198.

Tang, Y., Li, X., Zhang, B., Chen, P.X., Liu, R. *et al.* (2015) Characterizations of phenolics, betanins and antioxidant activities in seeds of three *Chenopodium quinoa* Willd. genotypes. *Food Chemistry* 166, 380–388. DOI: 10.1016/j.foodchem.2014.06.018.

Tapiero, H., Tew, K.D., Ba, G.N. and Mathé, G. (2002) Polyphenols: do they play a role in the prevention of human pathologies? *Biomedicine & Pharmacotherapy* 56(4), 200–207. DOI: 10.1016/s0753-3322(02)00178-6.

Thouri, A., La Barbera, L., Canuti, L., Vegliante, R., Jelled, A. *et al.* (2019) Antiproliferative and apoptosis-inducing effect of common Tunisian date seed (var. Korkobbi and Arechti) phytochemical-rich methanolic extract. *Environmental Science and Pollution Research International* 26(36), 36264–36273. DOI: 10.1007/s11356-019-06606-9.

Uenobe, F., Nakamura, S. and Miyazawa, M. (1997) Antimutagenic effect of resveratrol against Trp-P-1. *Mutation Research* 373(2), 197–200. DOI: 10.1016/s0027-5107(96)00191-1.

USDA (2007) *USDA National Nutrient Database for Standard Reference, Release 21.* US Department of Agriculture, Agricultural Research Service, Beltsville, Maryland.

Vayalil, P.K. (2012) Date fruits (*Phoenix dactylifera* Linn.): an emerging medicinal food. *Critical Reviews in Food Science and Nutrition* 52, 249–271.

Velioglu, Y.S., Mazza, G., Gao, L. and Oomah, B.D. (1998) Antioxidant activity and total phenolics in selected fruits, vegetables, and grain products. *Journal of Agricultural and Food Chemistry* 46(10), 4113–4117. DOI: 10.1021/jf9801973.

Verma, B., Hucl, P. and Chibbar, R.N. (2009) Phenolic acid composition and antioxidant capacity of acid and alkali hydrolysed wheat bran fractions. *Food Chemistry* 116(4), 947–954. DOI: 10.1016/j.foodchem.2009.03.060.

White, B.L., Howard, L.R. and Prior, R.L. (210) Release of bound procyanidins from cranberry pomace by alkaline hydrolysis. *Journal of Agricultural and Food Chemistry* 58, 7572–7579.

Whitney, E. and Rolfes, S.R. (2007) *Understanding Nutrition.* Cengage Learning, Boston, Massachusetts.

Wissam, Z., Ali, A. and Rama, H. (2016) Optimization of extraction conditions for the recovery of phenolic compounds and antioxidants from Syrian olive leaves. *Journal of Pharma and Phytochemistry* 5, 390–394.

Yamada, J. and Tomita, Y. (1996) Antimutagenic activity of caffeic acid and related compounds. *Bioscience, Biotechnology, and Biochemistry* 60(2), 328–329. DOI: 10.1271/bbb.60.328.

Yoshida, M., Sakai, T., Hosokawa, N., Marui, N., Matsumoto, K. *et al.* (1990) The effect of quercetin on cell cycle progression and growth of human gastric cancer cells. *FEBS Letters* 260(1), 10–13. DOI: 10.1016/0014-5793(90)80053-l.

Yousif, A.K., Benjamin, N.D., Kado, A., Alddin, S.M. and Ali, S.M. (1982) Chemical composition of four Iraqi date cultivars. *Date Palm Journal* 1, 285–294.

Zhang, C.R., Aldosari, S.A., Vidyasagar, P.S., Nair, K.M. and Nair, M.G. (2013) Antioxidant and anti-inflammatory assays confirm bioactive compounds in

Ajwa date fruit. *Journal of Agricultural and Food Chemistry* 61(24), 5834–5840. DOI: 10.1021/jf401371v.

Zhang, C.-R., Aldosari, S.A., Vidyasagar, P.S.P.V., Shukla, P. and Nair, M.G. (2017) Health-benefits of date fruits produced in Saudi Arabia based on *in vitro* antioxidant, anti-inflammatory and human tumor cell proliferation inhibitory assays. *Journal of the Saudi Society of Agricultural Sciences* 16(3), 287–293. DOI: 10.1016/j.jssas.2015.09.004.

Ziaedini, A., Jafari, A. and Zakeri, A. (2010) Extraction of antioxidants and caffeine from green tea (*Camelia sinensis*) leaves: kinetics and modeling. *Food Science and Technology International* 16(6), 505–510. DOI: 10.1177/1082013210367567.

Zineb, G., Boukouada, M., Djeridane, A., Saidi, M. and Yousfi, M. (2012) Screening of antioxidant activity and phenolic compounds of various date palm (*Phoenix dactylifera*) fruits from Algeria. *Mediterranean Journal of Nutrition and Metabolism* 5(2), 119–126. DOI: 10.1007/s12349-011-0082-7.

Nonfood Products and Uses of Date Palm

17

Ricardo Salomón-Torres* ⓘ

Universidad Estatal de Sonora, Sonora, Mexico

Abstract

The date palm (*Phoenix dactylifera* L.) is considered a multipurpose tree. The main purpose of its cultivation is to provide a highly nutritious fruit, with great benefits to human health. But other components of this tree, including the trunk, leaves, pulp, and seed, are also exploited for other purposes. The wood of its trunk, in addition to providing fuel, is used to make tables, benches, other types of furniture, flowerpots, small artworks, and shoe soles. The leaves can be used to produce baskets, bags, or boxes, as well as a great diversity of handicrafts and accessories. Likewise, they are used as roof covers and walls in houses, as protective fences, and in the production of biofertilizers and compost. The pulp of the fruit has been used to produce alcohol, which is used to make antibacterial gel. The date seed is capable of producing a high-quality oil that has already been used in some cosmetic and pharmaceutical applications. Finally, the date palm is very attractive when used as an ornamental plant in avenue gardens in cities and in hotel resorts around the world. This chapter provides a broad review of the main nonfood uses and products of the date palm, its influence in religious activities, traditional uses, and some of its potential uses, highlighting the use of its leaves, trunk, fruit pulp, and seed.

Keywords: Date palm, Date seed, Palm leaves, Palm trunk, Palm waste

17.1 Introduction

The date palm (*Phoenix dactylifera* L.) and its fruit have acquired great economic, social, historical, cultural, and religious importance, mainly in the countries of the Middle East and North Africa (MENA). The economic significance of this crop is due to the fact that the sale and export of dates are one of the main sources of income in most of these countries, playing

*ricardo.salomon@ues.mx

© CAB International 2023. *Date Palm* (eds J.M. Al-Khayri *et al.*)
DOI: 10.1079/9781800620209.0017

a strategic role in their economies (Abd Rabou and Radwan, 2017). The socio-economic aspects of this crop are observed in the great demand for labour, machinery, fertilizers, and pesticides as well as in its supply chain for activities such as planting, cultivation, harvesting of fruits, treatment, packaging, transportation of products, and consumption (Hanieh *et al.*, 2020). Its historical influence has been reflected in coins, paintings, and sculptures of ancient civilizations such as the Sumerian, Assyrian, Babylonian, Egyptian, Greek, and Roman, up to the present day (Al-Yahyai and Manickavasagan, 2012; Rivera *et al.*, 2019). Its cultural significance has been present since its domestication, which was governed under the laws relating to the cultivation and sale of dates in the Hammurabi code in ancient Babylon (Haider, 2015), and has endured to the present day, as reflected in the daily consumption of its fruit, due to its high nutritional content, and its use as an ornamental plant in private gardens and public roads (Cohen and Glasner, 2015). The religious activities of the three main monotheistic religions of the world have also been tightly associated with the date palm. The date is closely linked to the diet of the Islamic peoples during the month of Ramadan, while for Judaism, the date is one of the seven species cultivated in the Promised Land for the people of Israel. For Christianity, the date palm is used in festivities such as Easter and Palm Sunday (Schorr *et al.*, 2018).

All these characteristics have recently been recognized by the United Nations Educational, Scientific and Cultural Organization (UNESCO), by accepting as intangible cultural heritage of humanity, the knowledge, skills, traditions, and practices related to this crop in fourteen MENA countries (UNESCO, 2019). The date palm represents one of the main unifying cultural elements in Arab culture, being the main cultural symbol of many communities and individuals in the MENA area. Likewise, in 2020 the Food and Agricultural Organization of the United Nations (FAO) approved the proposal of the Kingdom of Saudi Arabia that the year 2027 be recognized by the United Nations as the International Year of the Date Palm, because this crop has a proven socio-economic importance and for its contribution to the eradication of poverty (Committee on Agriculture, 2020).

The date palm is a source of food and one of the main crops for MENA countries, which makes its fruit a food security crop for that region (Ghnimi *et al.*, 2017). Its intake provides many health benefits, such as the purported prevention of some diseases, common cancers, and other degenerative diseases (Al-Farsi and Lee, 2012). Their high nutritional properties make dates a functional food, as well as being a rich source of energy (Salomón-Torres *et al.*, 2018). These nutritional characteristics have been used by the food industry and food scientists for the development of a wide range of processed products, products with added value, and by-products derived from fruit and seed (Salomón-Torres *et al.*, 2021a).

The date palm, in addition to being a rich nutritional source, is considered a multi-purpose tree because most of its parts (except the roots) can be used for the purpose that best suits them (Barreveld, 1993). For many centuries, it has been the source for the production of a wide variety of products with economic value, mainly food. It has also provided materials for construction in rural areas and for the elaboration of a great diversity of modern handicrafts, fashion accessories, religious objects, fine and artisan furniture, cosmetics, and some unusual products (Johnson, 2016; Elsayegh, 2018; Rivera *et al.*, 2019). The rediscovery of palm by-products and the maximization of their added value have been the objectives of the recent conferences organized by the ByPalma association, which has focused exclusively on palm-derived by-products throughout the world, highlighting their current and potential products, where date palm by-products play a preponderant role (ByPalma, 2021).

According to FAO data, there are more than 100 million date palms in the world (FAO, 2020) that generate thousands of tonnes of waste every year after the fruit is harvested that accumulates on agricultural land as landfill material or is incinerated without any specific use (Jonoobi *et al.*, 2019). However, these wastes have great potential to be used in many applications. These nonfood usable parts of the date palm are the trunk (including bark), leaves (midribs, leaflets, sheath, and spines), and reproductive organs (spathes, fruit stalk, spikelets, pollen, seeds, and poor-quality fruit) (Barreveld, 1993). Except for the trunk, all these parts are considered waste as they are derived from annual pruning.

This chapter provides a broad review of nonfood and economic uses of the exploitable parts of the date palm. Likewise, the influence that the date palm has on some activities in the three main monotheistic religions of the world, on the culture and traditions of various peoples, as ornamentation, and other potential uses are discussed.

17.2   Trunk

The date palm trunk (DPT) is a vertical and cylindrical stem that is covered with the bases of dry petioles from leaves that have been removed. A DPT is available for exploitation at the end of its useful life cycle (El-Mously and Darwish, 2020). Its main use is as wood. Date palm wood is not of high quality; however, it has great resistance, which is very useful as poles, beams, rafters, lintels, girders, pillars, jetties, and light footbridges (Barreveld, 1993). DPTs have also been marketed for decorative purposes, handcrafted, and sometimes open to the buyer's imagination. In Spain, they can be ordered online for a cost of €110.00 per meter length (Space Garden, 2015). Likewise, the wood provided by the trunk is also used as timber or fuel (Chao and Krueger, 2007). The products and applications described below are developed from date palm wood.

17.2.1 Furniture

Due to the fact that the vascular structure of the DPT is thick (Fig. 17.1a), difficult to cut (Fig. 17.1b) and polish, it has only been possible to manufacture rustic furniture such as benches, tables, and chairs (Fig. 17.1c and f), some of which have fine finishes with varnish, paint, or another type of protection for the wood (Visit Elche, 2015). Craftsmen use thinner trunks for table legs and thicker trunks for bench legs. In the Gaza Strip, it is possible to observe how locals have used pieces of DPT as tables in some parks and coastal chalets (Abd Rabou and Radwan, 2017). Likewise, in Orihuela, Spain, the DPT is transformed from agricultural waste to urban furniture (TodoPalmera, 2021).

17.2.2 Building material

DPTs have been used as beams and columns in traditional rural dwellings (El-Mously and Darwish, 2020). They are also used as ornamental columns (Fig. 17.1g) instead of structural columns, for decoration in houses, resorts, restaurants, ranches, and haciendas (Visit Elche, 2015; Abd Rabou and Radwan, 2017). Thin cuts of wood can be used in doors and windows of rustic houses or as fences for farmyard animals (Fig. 17.1h). Also, the trunk can be cut into quarters to be used as beams to support roofs made of palm leaves. Another common use of the trunk is as a structure in the manufacture of beach pergolas made with palm leaves, providing a cooler environment inside (Abd Rabou and Radwan, 2017). As DPT is a sustainable material and an alternative to other types of wood, in Spain there is production of palm wood panels that may be used as thermal and sound insulators in the construction of buildings (Información, 2014). The trunk has also been used in rural areas as rustic footbridges across small rivers and canals.

17.2.3 Pots and containers

A DPT can be transformed into garden furniture. In Elche, Spain, pots, planters, and containers are usually made with DPT fragments (TodoPalmera, 2021) (Fig. 17.2a and b). In MENA countries, it is possible to find DPT pots hanging on or attached to the outside wall of a house (Fig. 17.2c and d). For the manufacture of flowerpots or containers, a cavity is made in the trunk. The depth and width of the hole in the center of the trunk can vary according to its use. Once the hole is made, the inside is sanded to make the walls smooth and the outside to remove all rough sections. Finally, the desired color tone is added, obtaining a beautiful piece of furniture that will give a country theme, cheerful and full of freshness, or create a warm setting on a terrace or garden. In traditional beekeeping in Oman,

Fig. 17.1. Uses of date palm trunk (DPT) for furniture and construction material. (a) Transverse cut of DPT. (b) Cutting of the DPT with specialized equipment to manufacture benches. (c, d) Benches built with DPT. (e, f) Two different shapes of chairs made from a DPT. (g) The DPT as a structure to support a palm leaf roof. (h) Wood cuts from a DPT used as a fence for free-range livestock. (Photo (a) from Wikipedia, https://commons.wikimedia.org/w/index.php?curid=81347533 (accessed 7 December 2022), used under the Creative Commons Attribution-Share Alike 4.0 International license; photos (b, c, d, g) used with permission of Miguel Angel Sánchez Martínez from TodoPalmera Company; photos (e, f, h) courtesy of Mr Baruch Glasner.)

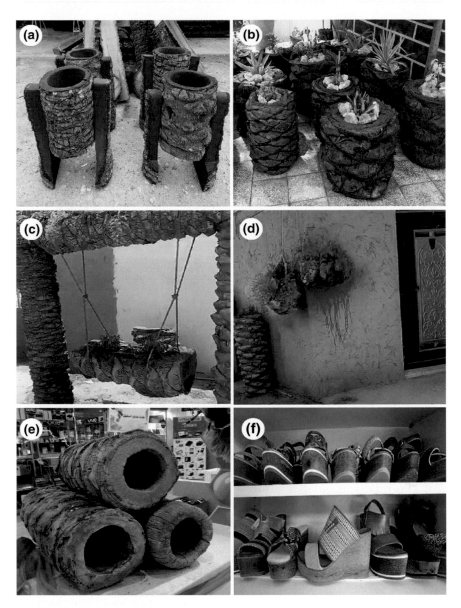

Fig. 17.2. Other uses of the date palm trunk (DPT). (a) Container made with DPT. (b) Rustic pots made from DPT. (c, d) DPT pots hanging or attached to the outside wall. (e) Hollow DPT to be used as honeycomb containers. (f) Women's shoe soles constructed with DPT. (Photos (a, b) used with permission of Miguel Angel Sánchez Martínez from TodoPalmera Company; photos (c, d, e) courtesy of Dr Abdulbasit Oudah Ibrahim; photo (f) reproduced with permission from Mr Manuel Lorenzo Ramón.)

hollow DPTs are used as containers to house hives of honey-producing bees (Fig. 17.2e).

17.2.4 Women's shoe soles

In Spain, an environmental solution is being given to the DPTs that end up being cut down due to damage caused by the red palm weevil (*Rhynchophorus ferrugineus*) or because they end their life cycle. This recycling system consists of cutting the DPT into planks, which are profiled and turned into shape. The material then goes through a chemical process and reinforcements are added to prevent the soft wood from breaking. The result of all this process is to obtain 100% ecological soles for women's shoes (Fig. 17.2f), where the fibers of the trunk give an elegant appearance to the footwear. This material has advantages over other shoe soles, such as producing a woman's shoe with a thick sole and light weight, as well as the comfort it gives the wearer and its reasonable final price (El Mundo, 2016). The company that markets this product is Exclusivas Indupal SL, which works only with DPT of at least 3m in length, having to wait several years for the trunk to dry completely, and then finally taking only 60% of it, corresponding to its densest section.

17.2.5 Art and decoration

In the field of art and decoration, potential uses of the DPT increase. There is the possibility of making almost anything: engravings, small sculptures, and decorative objects (Medjool Lovers, 2021). It is possible to find engravings in small DPTs, with texts in Arabic referring to verses from the Quran related to the Surah of Maryam ('Saint Mary' or 'Virgin Mary') (*Sura* 19, v. 25). This verse is translated in two versions as follows: 'And shake towards thyself the trunk of the palm-tree: It will let fall fresh ripe dates upon thee' and 'Shake the branch of the palm-tree, drawing it towards you, it will cause fresh and ripe dates to fall upon you.' It is also possible to carve the DPT to obtain beautiful sculptures, such as the figure of a camel and the head of a horse (Medjool Lovers, 2021). An example of a decorative object with DPT are pieces of the trunk (obtained from a transverse section) with the Arabic word 'Orjoon', which means the date fruit stalk (date bunch). This word is also used in the Quran in the Surah of Yasin (*Sura* 36, v. 39): "And to the Moon, we decree it phases, until (it is waning and) looks like the old (and withered) lower part of a date-stalk' (*orjoon* in Arabic).

17.3 Leaves

An adult date palm produces 12–15 leaves each year, with approximately 120–240 leaflets on the midrib (El-Mously and Darwish, 2020). The

Table 17.1. Annual estimated production of residues of an adult date palm. (Adapted from Hosseinkhani *et al.*, 2014.)

Residues on each palm			Leaf residues on each palm		
Part	Dry weight (kg)	%	Part	Dry weight (kg)	%
Leaves	18	52.94	Midrib	5.4	30.00
Bunches	6	17.56	Rachis	7.8	43.33
Bunches arm	7	20.56	Leaflet	4.8	26.67
Sheaths	3	8.82			
Total	34	100.00	Total	18.0	100.00

appearance of new date palm leaves (DPLs) will cause the pruning of the same amount of dry DPLs each year, where each leaf has an average dry weight of 2–3 kg (Mallaki and Fatehi, 2014). Likewise, an adult palm produces between 34–40 kg of residues (Hosseinkhani *et al.*, 2014; Mallaki and Fatehi, 2014), of which 53% corresponds to its leaves and the rest to other parts of the palm (Table 17.1) .

According to FAO production data for 2020, there are 1.24 million ha of date palms in the world (FAO, 2020). Assuming each hectare has 156 palms (8 m × 8 m), and that each palm produces an average of 34 kg of waste (Table 17.1), estimated waste production without considering the trunk would be 6.61 million t of raw material that may be used in a wide variety of industries.

Each adult date palm produces approximately 4.8 kg of leaflet and 7.8 kg of rachis dry weight, which represent 14.12 and 22.94% of the total residues of each palm, respectively. With this amount of material, a whole ecological and sustainable industry has been developed that prevents the accumulation of palm waste in the environment and allows the manufacture of a great diversity of handicrafts, furniture, and other products, as described below.

17.3.1 Furniture

The manufacture of chairs, tables, baby beds, small sofas, furniture, decorations, cages, and other objects from DPLs is an industry that has made several localities famous, such as the Qena governorate in Egypt. This activity is carried out with the great skill and talent of its artisans, some of whom are known as 'leaves artists'. They claim that this profession has been inherited from generation to generation in their families since ancient times. The modernization and development of this industry in Qena derived from the commercial competition it faced from plastic products, which had invaded popular markets (Soutalomma, 2020). After the diversification of DPL-derived products, they became very popular in

Upper Egypt. In addition to providing a natural product, DPL-derived products have a greater durability than plastic and have begun to displace plastic items.

The manufacturing process begins with the pruning of the DPL, the separation of the leaflets from the midrib, and the drying for a period of 3–10 days until it reaches total dryness. Then the rachis is cut to specific lengths with perforations according to the type of furniture to be manufactured (Al-Maaref, 2020) (Fig. 17.3a-e). Before starting its assembly and in order to prolong its life and avoid damage mainly by weevils, the rachis is sprayed with a chemical compound for its protection. An artisan can produce 15–20 cages, a bed, two chairs, or up to three small tables in one working day, which can last up to 14 hour. Furniture manufactured in the Gaza Strip with DPLs is characterized by its low cost, beauty, resistance, but in essence it is the reflection of its great Palestinian cultural legacy, which is very attractive for tourism (Abd Rabou and Radwan, 2017). Other types of furniture are also manufactured, such as wardrobes or containers for the exhibition of small merchandise (Fig. 17.3f).

Items in demand are cages for birds and containers for fruits and vegetables, since they are more resistant, safe, and fresh than cardboard, plastic, or wooden boxes. These are widely used for the packaging and sale of a great variety of products (Al Khaleej, 2018). Cages are used for the transport on donkeys, camels, and vehicles of small poultry or domestic birds for sale in markets, as well as for the domestic breeding of small species of birds (Barreveld, 1993).

17.3.2   Handicrafts and accessories

17.3.2.1   Handbags

The production of women's handbags made from leaflets is highly developed in Jordan (SAAF, 2021). With carefully selected leaflets, their manufacturers generate high-quality bags following the ancient weaving traditions, where, by joining and intertwining the leaflets, a bag with a unique, modern, and fashionable style is created, in addition to generating a sustainable and biodegradable product (Fig. 17.4a and b). Likewise, the fashion industry for its spring/summer 2018 and 2019 collections put special emphasis on bags and accessories made of hand-woven leaflets as ecological products, which in combination with other materials set trends in women's fashion (Elsayegh, 2018).

It is also possible to find, in most of the MENA countries, a great diversity of handcrafted handbags of very high quality. These are manufactured with different designs, sizes, colors, and decorations, and their main attraction is their durability and accessible price (Abd Rabou and Radwan, 2017; SAAF, 2021). These handbags are popular in the summer, because women prefer to use leaflet bags rather than bags made of other materials that

Fig. 17.3. Manufacture of furniture and other objects with date palm leaf (DPL). (a) Rustic table built with midribs. (b) Elegant set of table and chairs made with DPL, with fine handcrafted finishes. (c, d, e) Rustic furniture made with DPL in the Egyptian market. (f) Containers made with DPL. (Photos (a, b) courtesy of Dr Abdulbasit Oudah Ibrahim; photo (c) courtesy of Mr Baruch Glasner; photos (d, e, f) courtesy of Mr Khaled M. Alali.)

Fig. 17.4. Manufacture of handicrafts and accessories with date palm leaf (DPL). (a, b) DPL women's bags, handwoven in Jordan. (c) DPL baskets in the market of the Kingdom of Saudi Arabia. (d, e) Small containers to hold dates. (f) A rolled-up floor mat in a market in the Kingdom of Saudi Arabia. (g) An elegant placemat set for a meal. (h) A hat with a fine finish, made with DPL in Jordan. (i, j) Sandals made with DPL. (k) DPL bracelets in two different presentations. (l) A lampshade woven with DPL. (m) Pillow made of DPL. (n) Handcrafted decorative item. (o) DPL-covered book. (p) DPL-covered glass bottle. (Photos (a, b, d, g, h, i, j, k, m, o) reproduced with permission from Dr Nehaya Qasem from Jordan SAAF Facebook Page; photos (c, f) courtesy of Dr Jameel Al-Khayri; photo (e) courtesy of Dr Abdulbasit Oudah Ibrahim; photos (l, n) by Dr Ricardo Salomon; photo (p) from Wikipedia, https://es.wikipedia.org/wiki/Artesan%C3%ADa_de_ la_palma#/media/Archivo:Ampolla_Folrada_de_Pauma.png (accessed 8 December 2022), used under the Creative Commons Attribution-Share Alike 4.0 International license.)

retain heat. Through the use of textile dyes and textile printing techniques applied to leaflets, Elsayegh (2018) proposed the production of a bag woven with leaflets, highlighting combination with recycled materials such as fabrics and leather.

17.3.2.2 Baskets

Handling dried leaflets in woven handicrafts is difficult due to their texture, so it is necessary to immerse them in water until they soften to facilitate their use. For the dyeing of the leaflets, water is boiled in a large container, to which the required dye is added. Leaflets are then added to the container for 5 min, after which they are removed and put in the shade for drying (Alarab, 2016). The small leaflets are used to make smaller baskets and other small and fragile items, while the long and thick leaflets are used in large baskets, items, or tools that require greater resistance for their use. Baskets are easy to prepare, being a famous and traditional industry where women commonly participate in their manufacture (Abd Rabou and Radwan, 2017). There is a great variety of modern and innovative designs for baskets, plates, and large pots made with leaflets. Small and medium-sized baskets are commonly used to hold bread, fruits, vegetables, clothing, and when they have a lid, valuables such as jewelry (Fig. 17.4c-e).

17.3.2.3 Floor mats

Floor mats are a traditional piece of furniture, which are made by hand to give a beautiful look to furnished rooms (Fig. 17.4f). They are also used in rooms to sit or sleep on, to dry products in the sun, or for religious purposes, such as Indian wedding ceremonies or as a prayer mat for Muslims. The fine or rustic finish that the mat has will depend on the thickness and size of the leaflet used, while its color will be according to the use that is given to it. These are manufactured in large quantities in MENA countries due to their high demand, especially in winter (Abd Rabou and Radwan, 2017). Using the same braiding technique as for mats, other products such as small tablecloths and fans can also be made. The tablecloths can be circular or square in shape, sporting a range of colors, with other decorative parts. These are used as a placemat for eating (Fig. 17.4g). Likewise, fans are commonly square in shape and are used for manual heat mitigation. The design for women is usually in beautiful bright colors, while the one used by men is colorless, as is customary in MENA countries.

17.3.2.4 Hats

Hats made of leaflets are ideal to protect the head from sunlight and are ideal accessories for both men and women for trips to the beach, stays in the garden, or for agricultural activities on sunny summer days. Their production is very common in MENA countries, where they can be found from very rustic or classic styles for common jobs to fine fashion hats in prestigious brands (Fig. 17.4h). These latter have beautiful and delicate finishes in different shapes, sizes, and colors, so they can be used at social events (Abd Rabou and Radwan, 2017) or to mark a summer lifestyle trend. It is possible to find them in local markets as well as digital online sales platforms, where their cost will vary according to the quality of the hat.

17.3.2.5   Sandals

The manufacture of leaflet sandals dates back to ancient Egypt, where they were initially used by royalty and priests to protect their feet or to be used in ceremonies (El-Mously and Darwish, 2020). Later they were used by soldiers, farmers, workers, and common people, who received them as part of their salary. These were characterized by the combined use of different sizes of leaflets with small fabrics (Wendrich, 2009). Various designs of sandals, a legacy of ancient Egypt, are parts of exhibits in various museums in Egypt and the UK. Currently, the production of footwear with leaflets continues (Fig. 17.4i), mostly for a female market, with fashionable sandals or with beautiful leaflet decorations on the footwear (Fig. 17.4j). Leaflet sandals are the ideal footwear to walk comfortably in hot climates and/or on the beach (SAAF, 2021).

17.3.2.6   Canes

A cane is an essential item for walking for elderly or physically disabled people. They are also very useful for exploring slopes and in some cases can be used as a defense weapon against aggression (Johnson, 2016). Dried DPL midribs are the raw material for manufacturing lightweight canes in India (Sukumar, 2012). Likewise, canes made from the stem of the date fruit have been observed (Segas, 2021).

17.3.2.7   Accessories

A wide variety of accessories have been developed utilizing leaflets, using the same braiding techniques and with the skilled hands of the artisans. These decorations seek to replace materials such as fabrics, plastics, synthetic fibers, and metals to create accessories with the identity of the country of origin. Within this wide range of accessories, it is possible to find bracelets (Fig. 17.4k), necklaces, rings, picture frames, artistic paintings (some with verses from the Holy Quran), disposable tissue boxes, lampshades (Fig. 17.4l), pillows (Fig. 17.4m), plates, vases, woven flowers, trays (to offer food and drinks), works of art, small pieces of decoration (Fig. 17.4n), brooms, brushes, book covers (Fig. 17.4o), coverings (such as for large glass bottles) (Fig. 17.4p), cup holders, curtains, napkin rings, and

Fig. 17.5. Use of date palm leaf (DPL) as a building material. (a) Interior walls of a house built with DPL. (b) A warehouse built with petioles. (c) A fence built with DPL in Mexico. (d) DPL serving as support for the growth of other crops. (Photo (a) courtesy of Dr Abdulbasit Oudah Ibrahim; photo (b) used with permission of Miguel Angel Sánchez Martínez from TodoPalmera Company; photo (c) by Ricardo Salomon; photo (d) courtesy of Kapil Mohan MSc.)

many others (Barreveld, 1993; Abd Rabou and Radwan, 2017; El-Mously and Darwish, 2020).

17.3.3 Building material

17.3.3.1 Houses

The construction of houses and sheds with DPL is not limited to MENA countries, it also extends to all countries with a presence of date palms. The existence of houses made with this material responds to the need to have a cool and pleasant environment in areas where the climate is very hot and humid, particularly in desert and coastal areas. DPLs offer protection from the sun's rays and dusty winds (El-Mously and Darwish, 2020). They are commonly used as a building material on the roofs and walls of simple houses (Fig. 17.5a). A weakness of this type of construction is its vulnerability to meteorological phenomena such as hurricanes (Khuyut, 2020). However, DPLs can also be mixed with other materials (such as clay paste) to use for walls or ceilings; this offers greater strength and durability and can withstand rain. The construction of this type of housing is not only associated with low-income populations, but also with tourist complexes to give visitors a stay with an ecological and rustic environment. In addition to midribs and leaflets, petioles can also be used as construction material. These are at the lower end of the DPL, are triangular in shape, and they are commonly left attached to the DPT after pruning. Small huts built with petioles have been observed acting as small warehouses (Fig. 17.5b). The manufacture of doors and windows with midribs was common in ancient Egypt. These were tied with ropes and reinforced diagonally with midribs. These traditional doors can still be found in rural areas of Egypt (El-Mously and Darwish, 2020).

17.3.3.2 Fences

DPLs are very often used to build fences that provide protection from the wind and create favourable microclimates for horticulture (El-Juhany, 2010). These are placed like a fence with the petioles touching the ground. The DPLs are joined together with several layers of rope or wire to give the fence shape and direction (Fig. 17.5c). They can also be used to prevent the invasion of private gardens by animals. Fences can be built with DPLs or only with midribs, which, depending on the purpose of their construction, can be very close to each other, not allowing visibility from one side to the other.

17.3.3.3 Fishing boats

The use of small DPL-built fishing boats is a tradition that still exists in the United Arab Emirates. These boats are known by their Arabic name of *shasha*. Around 90% of the material required for their construction comes from the midribs, while the remaining material can be obtained from various timber trees. Construction of these boats requires the labour of a single man for a single day (Johnson, 2016). The *shasha* is designed to sail short distances, since its main function is to be a fishing boat that can carry up to four people.

17.3.3.4 Others

Other uses of midribs that have been reported are as a fiber source for rope making, fishing rods, support stakes for vine plants and other crops (Fig. 17.5d), and for ripening date bunches (Popenoe, 1973). They have also been used to protect young seedlings or offshoots from the heat of the sun (Barreveld, 1993).

17.3.4 Organic fertilizer

To facilitate access to the crown and as part of good orchard management, old DPLs can be pruned after fruit harvest and before pollination. The pruned leaves, when crushed, can become organic fertilizer (Cohen and Glasner, 2015). When used as a natural fertilizer, DPL can improve soil properties. In the Gaza Strip, these residues are used for the manufacture of fertilizer such as humus, which provides nutrients to plants, in addition to being considered a safe alternative to chemical fertilizers (Abd Rabou and Radwan, 2017).

17.3.5 Livestock feed

The composition of the DPL is made up of 54.12% of dry matter, 89.86% of organic matter, 24.48% of crude fiber, 59.10% of neutral detergent fiber, 8.51% of crude protein, 24.69% of hemicellulose, and 16.24% of lignin, among others. These characteristics are considered of low nutritional value as animal feed (Mahrous *et al.*, 2021). However, as it is an abundant, cheap, and highly available material in MENA countries, DPL mixed with other forages is widely used as feed for ruminants. In addition, its high fiber content aids in the digestion of food, is a good source of energy, and contributes to the rates of weight gain in livestock, mainly sheep and goats.

In Kuwait, a comparative study was carried out where two breeds of dairy cattle were given a diet of various forages and ground DPL for 12 weeks (Bahman *et al.*, 1997). The researchers reported that cows that ingested DPL maintained the same level of milk production and composition, as well as weight gain, as cows that ingested barley straw. The study concluded that DPL is an acceptable alternative to barley straw as fodder for dairy cows in Kuwait. Another study carried out in Egypt with lambs concluded that the use of DPL in mixed rations with other forages improved the growth, voluntary intake, digestibility of nutrients, and ruminal fermentation (Mahrous *et al.*, 2021). In the Gaza Strip, a machine has been manufactured to chop, grind, and convert palm waste (DPL, fibers, and seeds) into animal fodder (Abd Rabou and Radwan, 2017).

17.4 Seed

The seed of the fruit of the date palm (DPS) is usually oblong and ventrally grooved, containing a small embryo and a hard endosperm, accounting for 6–18% of the total weight of the fruit, this varying according to the variety and agroclimatic conditions of its culture (Salomón-Torres *et al.*, 2020). Despite the fact that DPS contains high nutritional attributes for humans, it is still considered an agricultural residue in many countries. One of its main exploitation alternatives is the extraction of oil to be used as a food supplement for livestock, poultry, and some fish (Sablani *et al.*, 2008; Mahgoub *et al.*, 2012).

17.4.1   Oil applications

DPS is made up of ~10% lipids and has high contents of phenols, flavonoids, and antioxidants. These characteristics make DPS oil an ideal candidate to replace other vegetable oils, such as maize, palm, and coconut, for the formulation of foods such as mayonnaise (Basuny and AL-Marzooq, 2011) as well as for body creams, shampoos, and shaving cream. A great number of studies have characterized the nutritional, antioxidant, and phytochemical contents of DPS (Shi *et al.*, 2014; Adeosun *et al.*, 2016; Golshan *et al.*, 2017), highlighting the high quality of its oil (García-González *et al.*, 2019) and the wide range of applications it could have, especially in the cosmetics industry (Shi *et al.*, 2014; Alharbi *et al.*, 2021). However, there are few commercial products on the market for which it has been used.

In Israel, three types of shampoo called Date Seed Oil Shampoo were formulated, which in its testing phase had encouraging comments on its quality (Devshony *et al.*, 1992). Currently a by-product derived from DPS (cv. Sahel) is produced in Burkina Faso, which is known as Desert Date Oil. The high antioxidant activity of this oil protects the skin against ultraviolet-B rays and helps to rejuvenate damaged skin. A women's cooperative takes care of its extraction, using a special cold pressing process, so that its nutritional, protective, and structuring properties are not lost (Ecco-Verde, 2021). Likewise, using this oil with other ingredients, a shampoo rich in antioxidant polyphenols was developed to protect fine hair and for hair that requires greater volume and texture (Nature's, 2021).

In the Gaza Strip, DPS is washed and dried in the sun to be ground up and used as an eyeliner. On some occasions, it is also used to produce charcoal (Abd Rabou and Radwan, 2017).

17.4.2   Animal feed

DPSs are used as an alternative nutritional supplement for animals since due to their high content of dietary fiber, phenolic acids, and flavonoids, they contribute to growth stimulation in poultry and some fish and the production of milk and meat in cattle. DPS is very hard and cannot be digested whole by animals; therefore, it must be crushed until it reaches a desired texture and is mixed with the animal's usual diet. A study carried out in the Kingdom of Saudi Arabia showed that up to 20% DPS can be mixed in the diets of lactating goats without producing negative effects on their health or production, making it a good alternative ingredient in ruminant feed (Al-Suwaiegh, 2016). The use of 10% DPS in the diet of broilers improved their weight gain, feed conversion, and growth performance, compared to a maize–soybean meal diet (Al-Farsi and Lee, 2011).

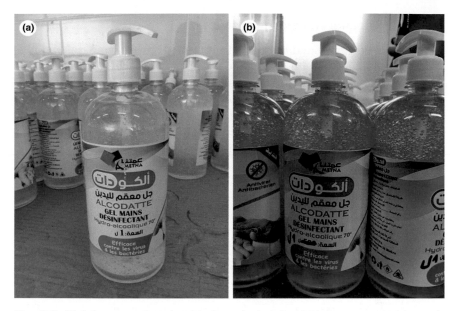

Fig. 17.6. Disinfectant gel made with date alcohol. (a, b) Various presentations of the hand sanitizer gel. (Photos (a, b) used with permission of Khobzi Abdelmadjid from SARL Ametna Company.)

17.5 Pulp

17.5.1 Alcohol production

The fruit of the date palm (DPF), which is known as the date, is an oblong or ellipsoid berry, where the pulp/seed weight ratio varies according to the cultivar (Krueger, 2021). The fermentation of sugars in DPF is a biochemical process in which sugars are converted into ethanol and carbon dioxide by the activity of various enzymes (Salomón-Torres *et al.*, 2021a). The resulting alcohol can be diluted to various percentages of purity and be used for beverages, fuels, and medicinal purposes (Sivakumar, 2002). The alcohol produced during DPF winemaking can become a target product when separated from the fermented liquor (Barreveld, 1993). In December 2021, a factory was inaugurated in Biskra, Algeria, that produces more than 3000 liters of surgical medical alcohol daily, extracted from DPFs not intended for human consumption. With this, this factory will provide more than 15% of the medical alcohol needs in that country (Province, 2021). This product complies with the international standard ISO 22000, is marketed under the name ALCODATTE, and is already used in various public health institutions in Algeria. Likewise, based on this medicinal alcohol, a hand sanitizer gel with a base of 70% natural alcohol was developed (Fig. 17.6a and b).

17.5.2   Bioethanol and biogas production

Recently a study evaluated the cost of bioethanol production, taking DPF residues as a source. It was determined that the production cost for each liter of ethanol was US$0.68, where the raw material cost represents 85.3% of the total production cost (Taghizadeh-Alisaraei *et al.*, 2019). Electricity, heat, and the manufacture of biomaterial would be generated from the ethanol produced. Likewise, the ethanol produced can be used as a supplementary biofuel in motor vehicles, thereby contributing to the reduction of environmental pollution. Another study evaluated biogas production using DPF waste as an energy source using a batch reactor (Lattieff, 2016). The results determined that the solid concentration at 0.15 (w/w) generated the highest biogas yield of 182 l/kg volatile solids with a methanol composition of 63%. This shows that DPF is a suitable source for biogas production.

17.5.3   Pharmaceutic product

A group of researchers developed a study to explore the effect of DPF on skin care. Using a 4% pulp extract in the form of a cream, the facial skin treatment significantly influenced all the parameters analysed, such as skin elasticity, pigmentation, shine, redness, and hydration. The study concluded that the efficiency of the treatment was due to the numerous active principles that make up DPF (Meer *et al.*, 2017).

17.5.4   Animal feed

Commonly, small DPFs or those that do not meet market quality standards are used to develop by-products such as paste, jam, syrup, sugar, beverages, and others (Salomón-Torres *et al.*, 2021a). Due to their high nutritional content, they can also be used as supplementary feed for livestock and poultry. A survey of sheep, goat, and dromedary breeders in the Nefzaoua region of Tunisia showed that 38% of the breeders feed the livestock with DPF without any treatment, 26% soak them with water to improve their tenderness, and 19% mix with straw, grasses, or concentrates (Genin *et al.*, 2004). A group of those surveyed (63%) use this food only during the winter, while 25% use it all year-round. Livestock producers in that region consider this a good and inexpensive food source. A study carried out in Iran evaluated the use of whole DPF waste in the diets of young ostriches (Najafi *et al.*, 2021). The treatments consisted of adding 0, 10, 20 and 30% of DPF to their usual diet. The results of the experiment suggested that the inclusion of dates in feed rations did not negatively affect productive performance. The 10% portion resulted in the best results in the evaluation.

17.6   Religious and Traditional Uses

17.6.1   Religious uses

The date palm has religious values deeply rooted in the three main monotheistic religions of the world, which are Christianity, Judaism, and Islam
(Qasim and Naqvim, 2012). In the sacred literature of these three religions
there are several citations where the date palm and its fruit are highlighted
as part of their history, traditions, and culture.

17.6.1.1   Judaism

For Judaism, the date palm had a great participation in the development of
its culture. In its sacred book, the Tanakh, there are multiple references to
the date palm. In one of them it is indicated how it was part of the decoration of the walls and doors of the Temple of Solomon (1st Kings 6:29–32):
'And he carved all the walls of the house around various figures, cherubs,
palm trees and flower buds, inside and out. The two doors were made of
olive wood; and he carved on them figures of cherubs, palm trees, and
flower buds, and overlaid them with gold; He also covered the cherubs and
the palm trees with gold' (VRV-60). It is also seen as a symbol of strength
and fruitfulness (Psalm 92:12): 'The righteous will flourish like the palm
tree; He will grow like a cedar in Lebanon' (VRV-60). Likewise, the date
is one of the seven species (foods) that played a preponderant role in the
feeding of the ancient Israelites (Deuteronomy 8:8): 'land of wheat and
barley, of vines, fig trees and pomegranates; land of olive trees, oil and
honey (date syrup)' (VRV-60). Sukkot is a Jewish holiday, also known as
the Feast of Huts or the Feast of Tabernacles, which reminds Jews of their
pilgrimage through the desert until they reached the land that flows with
milk and honey. The origin of this commandment is found in the Torah
(Leviticus 23:24–43). For this festival it is necessary to gather four species
(*Arbaat Haminim*), among which a tender and closed DPL known as *lulav*
stands out (Fig. 17.7a). Likewise, the tabernacles used in this feast are
decorated with DPLs (Fig. 17.7b).

17.6.1.2   Christianity

For Christianity, the DPL is associated with the triumphal and prophetic
entry of Christ into the City of Jerusalem on a Sunday, 6 days before the start
of the Passover feast. According to John 12:13: "they took palm branches
and went out to meet him, crying out: Hosanna! Blessed is he who comes
in the name of the Lord, the King of Israel!' (VRV-60). Currently this event
is celebrated as Palm Sunday, thus initiating Holy Week. Palm leaves are
typically distributed at religious services on this day. For this celebration in
Spain, a great diversity of commemorative objects such as flowers, crosses,

Fig. 17.7. Religious uses of date palm. (a) The four species (*Arbaat Haminim*) for the Feast of First Fruits in Israel. (b) Decoration for the Feast of Tabernacles in Israel. (c) Maryam shakes the palm tree (Holy Quran, *Sura* 19, v. 25). (d, e) Various Christian objects made with white palm. (f, g) Christian religious objects alluding to Palm Sunday in Spain. (h) Date palm crown with its leaves tied upwards, in order to obtain the yellow leaves. (Photo (a) from Wikipedia, https://es.wikipedia.org/wiki/Cuatro_especies#/media/Archivo:Arbaat_haminim-new.jpg (accessed 8 December 2022), used under the Creative Commons Attribution-Share Alike 4.0 International license; photo (b) from Wikipedia, https://es.wikipedia.org/wiki/Sucot#/media/Archivo:Suka.jpg (accessed 8 December 2022), used under Creative Commons Attribution 2.0 Generic license; image (c) from Wikipedia, https://es.wikipedia.org/wiki/Cor%C3%A1n#/media/Archivo:Maryam.jpg (accessed 8 December 2022), public domain; photos (d, e, f, g) courtesy of Dr Mohammed Aziz; photo (h) used with permission of Miguel Angel Sánchez Martínez from TodoPalmera Company.)

and other figures are made with yellowish-white braided leaflets (Fig. 17.7d-g). To achieve this color in the leaflets, the DPLs are tied up around the trunk, covering their tips with a wrap (Fig. 17.7h). This way, no light enters them, and the new DPLs will grow yellowish-white. DPLs are also seen as an icon of the promised victory of the soul over the body, according to Revelation 7:9: "After this I looked, and behold a great multitude, which no one could count, from all nations and tribes and peoples and languages, who stood before the throne of the Lamb, dressed in white clothes, and with palms in their hands' (VRV-60).

17.6.1.3   Islam

For Islam, the date plays an essential role in the diet of the Muslim world, as well as to break the fast on the days of Ramadan. The date palm also has deep importance and significance in the Holy Quran (Fig. 17.7c), where it is mentioned 20 times. The verses where the date palm is discussed explain the botanical, physiological, morphological, pharmacological, and nutritional aspects, as well as other aspects. According to the Islamic tradition, 'the tree of life' in the Garden of Eden corresponds to a date palm (Qasim and Naqvim, 2012). One of the best-known references to the date palm is the Surah of Maryam (*Sura* 19 v. 23–36): "And she gave birth next to the trunk of the palm tree. He said: I wish I had died before this disappearing into oblivion! And he called to her from below: Do not be sad, your Lord has placed a stream at your feet. Shake the trunk of the palm tree at you and ripe, fresh dates will fall. Eat and drink, and refresh your eyes. And if you see any human, tell him: I have made a promise of fasting to the Merciful and today I cannot speak to anyone.' Likewise, there is in Islam the traditional prayer bead strand, which sometimes is made with DPSs. This is made up of 99 beads, with marker beads separating them into three sets of 33, as each of three prayers is repeated 33 times (Johnson, 2016).

17.6.2   *Traditional uses*

17.6.2.1   Popular tradition

A popular tradition inherited in some MENA countries is that when the graves of deceased relatives are visited, DPLs are placed around or on the graves. Others carry DPL bouquets on holidays. Likewise, DPLs can be observed in the streets, as a sign indicating that there is a consolation present somewhere very close (Abd Rabou and Radwan, 2017).

17.6.2.2   Burning of the atxes

In Elche, Spain, there is a tradition of the arrival of the Three Wise Men from the East, with which the festive acts of Christmas terminate. This

Fig. 17.8. Some traditional uses of the date palm. (a) Burning of the Atxes in Spain. (b) An altar for the Day of the Dead in Mexico, decorated with date palm leaves. (c) Participant in the world date palm seed toss championship in Spain. (d) Entrance to a residential area in the City of Mexicali, in Mexico. (Photos (a, c) used with permission of Miguel Angel Sánchez Martínez from TodoPalmera Company; photos (b, d) by Ricardo Salomón.)

tradition consists of lighting the way for them, towards the dwelling of the children who are waiting for their gifts, by using a lit torch (Fig. 17.8a). This torch, known by the name of 'Atxes', is made by hand with dried leaflets, rachises, and as combustion material the fibers of the trunk (sheaths), which are tied with esparto rope. The burning of the Atxes takes place on January 6, representing the end of the parade of the Three Wise Men.

17.6.2.3 Day of the dead

The Day of the Dead is a Mexican tradition that is celebrated on November 1 and 2, in which deceased relatives are honored. This cultural heritage is the fusion of some Catholic celebrations with various indigenous customs of pre-Hispanic Mexico. This tradition consists of building a multicolored altar with a photo of the deceased, adorned with fruits, flowers, water, bread, alcoholic beverages, and foods that were liked by the deceased. Its construction is temporary and the elements with which the altar is adorned may vary according to the region of the country. It is common to observe Day of the Dead altars adorned with DPLs in the cities of Mexico where this plant is grown (Fig. 17.8b).

17.6.2.4　World date seed toss championship

In Spain, since 2009 a traditional tournament called 'World Date Seed Toss Championship' has been held. This competition consists of launching the DPS the farthest distance possible, using only the air impulse generated with the force of the lungs (La Vanguardia, 2019). This pleasant activity is part of the date festivities in the Altábix neighbourhood, in the eastern part of the historic palm grove of the city of Elche, which seeks to stimulate the consumption of dates as a highly nutritious fruit. The record to beat in the launch of the seed is the distance of 14.43 m, achieved in 2017 (Fig. 17.8c).

17.7　Others

17.7.1　Fruit bunches

The stem of the bunches is solid, fibrous, flexible, and strong, measures more than 1m in length, and is curved due to the weight of the fruits it supports (Johnson, 2016). Despite being one of the least valued parts of the palm, it is preferred for making ropes and belts by date palm climbers (Barreveld, 1993).

17.7.2　Spines

Spines are modified leaflets located near the base of the DPL. They are very sharp, with lengths of up to 20 cm. They have the function of protecting the fruits and central tender parts of the palm against any predator. Commonly when DPLs are cut, they no longer have their spines due the spines being removed for good access to the fruit bunches. The spines can be used for the manufacture of fish traps, as sewing needles for traditional fabrics, or for other uses where a pointed utensil is required (Barreveld, 1993; El-Mously and Darwish, 2020).

17.7.3　Leaf sheath

The new DPL grows covered with a tender tissue that remains attached to the trunk of the palm when the DPL develops (Barreveld, 1993). It is estimated that each palm can produce around 1 kg of leaf sheath each year, the sheaths having high contents of cellulose, hemicellulose, and lignin (Mahmoud, 2016). This material, in addition to its value as fuel, is used to protect newly planted offshoots, to make nets (for transporting materials), fishing nets, and brushes, as stuffing for cloth bags, to produce many types of ropes and twine, and for other uses (Barreveld, 1993; El-Mously and Darwish, 2020). Recently, this type of palm fiber has been evaluated with very promising results for its use as a filler in polymer composites, as an ecological filter in water treatment, and as a new carrier for the adsorption of invertase in the production of invert sugar (Mahmoud, 2016; Alshammari et al., 2019).

17.7.4 Ornaments

Due to its elegance and aesthetic beauty, the date palm is a preferred ornamental plant, adorning avenues, public parks, green spaces, high-end housing units, shopping centers, and tourist complexes (Fig. 17.8d). They are preferred since they act as windbreaks, can moderate the temperature, raise relative humidity, stabilize soils, and well as combat desertification (Abd Rabou and Radwan, 2017). The price of an adult date palm to be used as an ornamental plant can reach up to US$1500 in Mexico, for instance. Also, in order to harmonize with the environment, cell towers shaped like a date palm have been observed.

17.7.5 Tourism

Recently in North-west Mexico, a tourist event known as 'The Date Tourist Route' has been developing. This consists of a series of fairs, exhibitions, and social events seeking to encourage local consumption of the date and promote the date industry in the region. At these events, gourmet dishes accompanied with dates are offered and a great diversity of gastronomic products based on the DPF are presented (Ortiz-Uribe et al., 2019). On the coast of Dubai, United Arab Emirates, there is a project known as the Palm Islands, which are artificial islands characterized by having the shape of a date palm. A commercial and residential structure is intended for these islands, and it is expected that they will become one of the best tourist destinations in the world. The island, known as The Palm Jumeirah, is currently completed and is mainly a residential and tourist area, while the islands called The Palm Jebel Ali and The Palm Deira are currently under construction. In Spain there is a theme park based on the date palm (Multiaventura, 2021). The Multiadventure Park is a magical environment full of palm trees, which offers its visitors a great diversity of sports-related activities and great fun.

17.7.6 Date palm fragrance

The fragrant scent of date palm is the main ingredient in a perfume presented at the Liwa Date Festival in the United Arab Emirates. It took 2 years to develop this aromatic fragrance, seeking to pay tribute to the cultural heritage and traditions of that country. This date palm aroma was fused with the best aromatic ingredients from France, generating an essence unique in the world of perfumes (National News, 2012).

17.7.7 Traditional herbal medicine

Heart of palm is a food product obtained from the date palm that can be eaten alone, as a salad, or as a cooked vegetable (Krueger, 2021). This can be bought for consumption in some fairs in southern Spain, but its commercialization involves killing the palm. Male palms that do not produce good pollen or poorly productive palms are commonly sacrificed for this purpose. In India a chewing gum from this part of the palm is used to treat diarrhoea and genitourinary ailments. It is diuretic and demulcent (El-Juhany, 2010).

Due to its tannin content, DPF is used medicinally for its purifying and astringent properties in the body. This is ingested as an infusion, syrup, decoction, or paste in treatments for sore throats, bronchial catarrh, and colds. Likewise, it relieves fever, cystitis, gonorrhea, oedema, liver and abdominal problems, as well as counteracting alcohol intoxication (El-Juhany, 2010).

DPS has been used in traditional medicine to alleviate various ailments such as toothache (Barreveld, 1993) and fever (El-Juhany, 2010). For its application it is necessary to grind the seed to a fine powder, so that it becomes an ingredient in a paste. Roasted and ground DPSs are used in eye drops or ointments to stimulate the growth of long eyelashes (El-Juhany, 2010). The intake of powdered DPS is also recommended, since their polyphenols strengthen the antioxidant defense system of the human body (Platat *et al.*, 2019).

A home remedy to treat kidney problems, nervousness, and effervescent blood disorders is the sap of leaflets (El-Juhany, 2010). Likewise, the consumption of leaf sheath infusions is recommended, since they cleanse the body and protect the skin from diseases (Abd Rabou and Radwan, 2017).

Date palm pollen has been used in traditional medicine to treat male infertility, while the consumption of male flowers has been used to improve fertility (Salomón-Torres *et al.*, 2021b). In popular medicine its intake through distilled water, honey, or directly is prescribed for therapeutic and medicinal purposes. It is used to strengthen bones, treat frigidity, inflammatory bowel, and stomach ulcers, help digestion, and stop bleeding (Abd Rabou and Radwan, 2017).

17.8 Some Potential Uses

17.8.1 Wax extraction

Residues from the annual pruning of DPLs could become a potential alternative source of wax with natural extraction. There is currently a high demand for natural waxes, derived from the preferences of the cosmetics and sanitary products markets. It is also compatible with the trend toward the development of ecological and sustainable products of natural origin (Attard *et al.*, 2018). One study showed that DPL had better wax extraction performance compared to other waxes from agricultural residues. A low manufacturing cost in extraction makes it an excellent option for the industrial exploitation of natural waxes (Al Bulushi *et al.*, 2018). Another study reported that the hydrophobic molecules of this wax could potentially be used in detergents, surface coatings, or nutraceuticals (Attard *et al.*, 2018).

17.8.2   Fuel and energy production

The DPL can be used as a natural biofuel, clean burning biomass, or firewood to cook food, among others. A study analysed the characteristics of DPL for its potential use as a raw material in various energy conversion processes, such as pyrolysis, gasification, and torrefaction compared to other biomass types (Sulaiman *et al.*, 2016). Among other parameters, its carbon and fixed carbon content were lower and its volatile matter content very similar to the rest of the biomasses compared. It was concluded that the DPL has the potential to be used as a renewable energy source. Researchers from Iran designed a steam power plant to burn date palm waste as fuel for the simultaneous generation of electrical energy and the desalination of seawater (Mallaki and Fatehi, 2014). Their results revealed that if a biomass burning plant of this type was put into operation, burning of 140,000 t of biomass annually would generate approximately 62 GW/h of electricity and purify 2.27 million t of seawater annually, while also reducing greenhouse gases.

17.8.3   Paper from date palm rachis fiber

The rachis of the date palm offers a pulp yield between 41 and 45%, in addition to offering a satisfactory shine after bleaching and providing a resistance property similar to that of spruce and poplar. These characteristics of the rachis pulp make it an ideal candidate for the manufacture of writing and printing paper (Khristova *et al.*, 2005). In Tunisia, a study was carried out to determine the potential of date palm rachises as a raw material for the manufacture of paper compared to other sources of fiber such as wood, nontimber species of trees, and agricultural waste (Khiari *et al.*, 2011). All pulps and papers were characterized according to their morphological, chemical, and physical properties, concluding that the date palm rachis could be considered a good source of fibers for papermaking.

17.8.4   Date palm for producing wood composites

After its pruning process, the date palm can generate up to 40 kg of waste, among which are petiole, rachis, leaflets, sheath, bunch, pedicels, spathe, and spines. These wastes are very abundant, are sources of natural fibers, and are a renewable raw material (Almi *et al.*, 2015a). A study carried out in Algeria evaluated the physical and mechanical properties of four date palm residues (petiole, rachis, bunch, and sheath) for use in making composites (Almi *et al.*, 2015b). The results revealed that the rachis and petiole showed better properties than the rest of the residues, making them ideal candidates for the integration with natural compounds in various applications, such as materials for the construction, automotive, and furniture industries. However, before use, these fibers must be pretreated to repel moisture. Another study determined that the wood from date palm waste is composed mainly of cellulose (40%), lignin (28%), and hemicellulose (20%), which give it an advantage for use in the development of light, highly resistant materials (Almi *et al.*, 2015a).

17.8.5 Date palm as an absorbent for heavy and toxic metals

Date palm waste can also be used as a filtration medium for industrial wastewater. Date palm fiber filters have the potential to be a commercial technology for the absorption of heavy and toxic metals in wastewater, obtaining acceptable results compared to filters commonly used for these purposes (Shafiq *et al.*, 2018). This material could be used in place of commercial activated carbon for the removal of different contaminants in aqueous solutions (Ahmad *et al.*, 2012).

17.9 Conclusions and Prospects

The date palm has fed humanity since ancient times and has been part of human life, in its economic, social, historical, cultural, and religious activities. In addition to providing highly nutritious and energy-rich food, it has also provided traditional medicines, construction materials, handicrafts, animal feed, fertilizers, and other items that have facilitated daily life. More than just a waste material after pruning, nonfruit portions of the date palm should be considered an essential and valuable raw material, since skilled hands of craftsmen have transformed its leaves into beautiful furniture, appreciated crafts, and a great variety of useful products and accessories. Religious activities would not be complete without the historical role that the date palm plays in them. Likewise, a great number of studies have concluded the great potential that all the discarded parts of this tree have for exploitation in a great variety of industries.

References

Abd Rabou, A.N. and Radwan, E.S. (2017) The current status of the date palm (*Phoenix dactylifera* L.) and its uses in the Gaza Strip, Palestine. *Biodiversitas*

Journal of Biological Diversity 18(3), 1047–1061. DOI: 10.13057/biodiv/d180324.

Adeosun, A.M., Oni, S.O., Ighodaro, O.M., Durosinlorun, O.H. and Oyedele, O.M. (2016) Phytochemical, minerals and free radical scavenging profiles of *Phoenix dactilyfera* L. seed extract. *Journal of Taibah University Medical Sciences* 11(1), 1–6. DOI: 10.1016/j.jtumed.2015.11.006.

Ahmad, T., Danish, M., Rafatullah, M., Ghazali, A., Sulaiman, O. *et al.* (2012) The use of date palm as a potential adsorbent for wastewater treatment: a review. *Environmental Science and Pollution Research International* 19(5), 1464–1484. DOI: 10.1007/s11356-011-0709-8.

Alarab (2016) Wickers palm strands weave the heritage of the Emirates (in Arabic). Available at: https://alarab.co.uk/الخوجدائلخنيلتسنجـتراثلإمارات (accessed 4 December 2021).

Al Bulushi, K., Attard, T.M., North, M. and Hunt, A.J. (2018) Optimisation and economic evaluation of the supercritical carbon dioxide extraction of waxes from waste date palm (*Phoenix dactylifera*) leaves. *Journal of Cleaner Production* 186, 988–996. DOI: 10.1016/j.jclepro.2018.03.117.

Al-Farsi, M.A. and Lee, C.Y. (2011) Usage of date (*Phoenix dactylifera* L.) seeds in human health and animal feed. In: Preedy, V.R., Watson, R.R. and Patel, B. (eds) *Nuts and Seeds in Health and Disease Prevention.* Academic Press, London, pp. 447–452. DOI: 10.1016/B978-0-12-375688-6.10053-2.

Al-Farsi, M.A. and Lee, C.Y. (2012) The functional value of the dates. In: Manickavasagan, A., Essa, M.M. and Sukumar, E. (eds) *Dates: Production, Processing, Food, and Medicinal Values.* CRC Press, Boca Raton, Florida, pp. 351–359.

Alharbi, K.L., Raman, J. and Shin, H.J. (2021) Date fruit and seed in nutricosmetics. *Cosmetics* 8(3), 59. DOI: 10.3390/cosmetics8030059.

Al Khaleej (2018) Palm leaf furniture is a popular industry in southern Egypt (in Arabic). Available at: www.alkhaleej.ae/ملحق/أثاثجريدلنخيليلصناعةرائجة جنوبـمصر (accessed 4 December 2021).

Al-Maaref, A.A. (2020) Industrialists are fighting to survive. Furniture for the poor from palm trees in Qena (in Arabic). Available at: https://m.akhbarelyom.com/news/newdetails/3167429/1/فيديو-وصور—صناعيةـاحراربنـللبقاء. أثاثـلفقراءـمنـجريدةلنخيليلبقنا (accessed 4 December 2021).

Almi, K., Lakel, S., Benchabane, A. and Kriker, A. (2015a) Characterization of date palm wood used as composites reinforcement. *Acta Physica Polonica A* 127(4), 1072–1074. DOI: 10.12693/APhysPolA.127.1072.

Almi, K., Benchabane, A., Lakel, S. and Kriker, A. (2015b) Potential utilization of date palm wood as composite reinforcement. *Journal of Reinforced Plastics and Composites* 34(15), 1231–1240. DOI: 10.1177/0731684415588356.

Alshammari, B.A., Saba, N., Alotaibi, M.D., Alotibi, M.F., Jawaid, M. *et al.* (2019) Evaluation of mechanical, physical, and morphological properties of epoxy composites reinforced with different date palm fillers. *Materials* 12(13), 2145. DOI: 10.3390/ma12132145.

Al-Suwaiegh, S.B. (2016) Effect of feeding date pits on milk production, composition and blood parameters of lactating ardi goats. *Asian-Australasian Journal of Animal Sciences* 29(4), 509–515. DOI: 10.5713/ajas.15.0012.

Al-Yahyai, R. and Manickavasagan, A. (2012) An overview of date palm production. In: Manickavasagan, A., Essa, M.M. and Sukumar, E. (eds) *Dates: Production, Processing, Food, and Medicinal Values.* CRC Press, Boca Raton, Florida, pp. 3–11.

Attard, T.M., Bukhanko, N., Eriksson, D., Arshadi, M., Geladi, P. *et al.* (2018) Supercritical extraction of waxes and lipids from biomass: a valuable first step towards an integrated biorefinery. *Journal of Cleaner Production* 177, 684–698. DOI: 10.1016/j.jclepro.2017.12.155.

Bahman, A.M., Topps, J.H. and Rooke, J.A. (1997) Use of date palm leaves in high concentrate diets for lactating Friesian and Holstein cows. *Journal of Arid Environments* 35(1), 141–146. DOI: 10.1006/jare.1995.0145.

Barreveld, W.H. (1993) *Date Palm Products.* FAO Services Bulletin No. 101, 1st edn. Food and Agriculture Organization of the United Nations, Rome.

Basuny, A.M.M. and AL-Marzooq, M.A. (2011) Production of mayonnaise from date pit oil. *Food and Nutrition Sciences* 02(09), 938–943. DOI: 10.4236/fns.2011.29128.

ByPalma (2021) 2nd World Conference on Byproducts of Palms and their Applications. Available at: bypalma.com/ (accessed 9 December 2021).

Chao, C.T. and Krueger, R.R. (2007) The date palm (*Phoenix dactylifera* L.): overview of biology, uses, and cultivation. *HortScience* 42(5), 1077–1082. DOI: 10.21273/HORTSCI.42.5.1077.

Cohen, Y. and Glasner, B. (2015) Date palm status and perspective in Israel. In: Al-Khayri, J.M., Jain, S.M. and Johnson, D.V. (eds) *Date Palm Genetic Resources and Utilization*, Vol. 2. *Asia and Europe.* Springer, Dordrecht, The Netherlands, pp. 265–298. DOI: 10.1007/978-94-017-9707-8_8.

Committee on Agriculture (2020) Proposal for an international year of date palm. Available at: www.fao.org/3/nd415en/ND415EN.pdf (accessed 15 November 2021).

Devshony, S., Eteshola, E. and Shani, A. (1992) Characteristics and some potential applications of date palm (*Phoenix dactylifera* L.) seeds and seed oil. *Journal of the American Oil Chemists' Society* 69(6), 595–597. DOI: 10.1007/BF02636115.

Ecco-Verde (2021) Organic desert date oil. Available at: www.ecco-verde.com/naje l/organic-desert-date-oil (accessed 22 December 2021).

El-Juhany, L.I. (2010) Degradation of date palm trees and date production in Arab countries: causes and potential rehabilitation. *Australian Journal of Basic and Applied Sciences* 4(8), 3998–4010.

El-Mously, H. and Darwish, E.A. (2020) Date palm byproducts: history of utilization and technical heritage. In: Midani, M., Saba, N. and Alothman, O.Y (eds) *Date Palm Fiber Composites: Processing, Properties and Applications.* Springer, Singapore, pp. 3–71. DOI: 10.1007/978-981-15-9339-0_1.

El Mundo (2016) Un zapato que da una 'segunda vida' a la palmera (in Spanish). Available at: www.elmundo.es/comunidad-valenciana/2016/05/02/572715f7 e2704e0a388b45ed.html (accessed 22 November 2021).

Elsayegh, H. (2018) Using printed palm leaflets in modern crafts according the international fashion trends. In: *1st World Conference on By-Products of Palm Trees and Their Applications. Aswan, Egypt,* Bypalma, December 15–17, 2019, pp. 333–342. Available at: www.mrforum.com/wp-content/uploads/open_access/9781644900178/29.pdf (accessed December 9).

FAO (2020) FAOSTAT. Data. Crops and livestock products. Food and Agriculture Organization of the United Nations, Rome. Available at: www.fao.org/faostat/en/#data/QC (accessed 26 December 2021).

García-González, C., Salomón-Torres, R., Montero-Alpírez, G., Chávez-Velasco, D., Ortiz-Uribe, N. *et al.* (2019) Effect of pollen sources on yield oil extraction and fatty acid profile of the date seed (*Phoenix dactylifera* L.) cultivar Medjool from Mexico. *Grasas y Aceites* 70(3), 315. DOI: 10.3989/gya.0936182.

Genin, D., Kadri, A., Khorchani, T., Sakkal, K., Belgacem, F. *et al.* (2004) Valorisation of date-palm by-products (DPBP) for livestock feeding in southern Tunisia. I – potentialities and traditional utilisation. *Nutrition and Feeding Strategies of Sheep and Goats under Harsh Climates* 226, 221–226.

Ghnimi, S., Umer, S., Karim, A. and Kamal-Eldin, A. (2017) Date fruit (*Phoenix dactylifera* L.): an underutilized food seeking industrial valorization. *NFS Journal* 6, 1–10. DOI: 10.1016/j.nfs.2016.12.001.

Golshan, T., Solaimani, D. and Yasini, A. (2017) Physicochemical properties and applications of date seed and its oil. *International Food Research Journal* 24(4), 1399–1406.

Haider, N. (2015) Date palm status and perspective in Syria. In: Al-Khayri, J.M., Jain, S.M. and Johnson, D.V. (eds) *Date Palm Genetic Resources and Utilization*, Vol. 2. *Asia and Europe*. Springer, Dordrecht, The Netherlands, pp. 387–421. DOI: 10.1007/978-94-017-9707-8_12.

Hanieh, A.A., Hasan, A. and Assi, M. (2020) Date palm trees supply chain and sustainable model. *Journal of Cleaner Production* 258, 120951. DOI: 10.1016/j.jclepro.2020.120951.

Hosseinkhani, H., Euring, M. and Kharazipour, A. (2014) Utilization of date palm (*Phoenix dactylifera* L.) pruning residues as raw material for MDF manufacturing. *Journal of Materials Science Research* 4(1), 46–62. DOI: 10.5539/jmsr.v4n1p46.

Información (2014) Los mil y un usos de la palmera datilera (in Spanish). Available at: www.informacion.es/elche/2014/05/09/mil-usos-palmera-datilera-628095 1.html (accessed 29 November 2021).

Johnson, D.V. (2016) Unusual date palm products: prayer beads, walking sticks and fishing boats. *Emirates Journal of Food and Agriculture* 28(1), 12. DOI: 10.9755/ejfa.2015-11-1021.

Jonoobi, M., Shafie, M., Shirmohammadli, Y., Ashori, A., Zarea-Hosseinabadi, H. *et al.* (2019) A review on date palm tree: properties, characterization and its potential applications. *Journal of Renewable Materials* 7(11), 1055–1075. DOI: 10.32604/jrm.2019.08188.

Khiari, R., Mauret, E., Belgacem, M.N. and Mhemmi, F. (2011) Tunisian date palm rachis used as an alternative source of fibres for papermaking applications. *BioResources* 6(1), 265–281. DOI: 10.15376/biores.6.1.265-281.

Khristova, P., Kordsachia, O. and Khider, T. (2005) Alkaline pulping with additives of date palm rachis and leaves from Sudan. *Bioresource Technology* 96(1), 79–85. DOI: 10.1016/j.biortech.2003.05.005.

Khuyut (2020) Palm fronds in Yemen (in Arabic). Available at: www.khuyut.com/blog/08-21-2020-09-40 (accessed 7 December 2021).

Krueger, R.R. (2021) Date palm (*Phoenix dactylifera* L.) biology and utilization. In: Al-Khayri, J.M., Jain, S.M. and Johnson, D.V. (eds) *The Date Palm Genome*, Vol. 1. *Phylogeny, Biodiversity and Mapping*. Compendium of Plant Genomes. Springer, Cham, Switzerland, pp. 3–28. DOI: 10.1007/978-3-030-73746-7_1.

Lattieff, F.A. (2016) A study of biogas production from date palm fruit wastes. *Journal of Cleaner Production* 139, 1191–1195. DOI: 10.1016/j.jclepro.2016.08.139.

La Vanguardia (2019) Elche celebra el lunes el campeonato mundial de lanzamiento de hueso de dátil (in Spanish). Available at: www.lavanguardia.com/local/valencia/20190425/461859356975/elche-celebra-el-lunes-el-campeonato-mundial-de-lanzamiento-de-hueso-de-datil.html (accessed 23 December 2021).

Mahgoub, O., Kadim, I.T. and Al-Marzooqi, W. (2012) Use of date palm by-products in feeding livestock. In: Manickavasagan, A., Essa, M.M. and Sukumar, E. (eds) *Dates: Production, Processing, Food, and Medicinal Values.* CRC Press, Boca Raton, Florida, pp. 323–336.

Mahmoud, D.A.R. (2016) Utilization of palm wastes for production of invert sugar. *Procedia Environmental Sciences* 34, 104–118. DOI: 10.1016/j.proenv.2016.04.011.

Mahrous, A.A., El-Tahan, A.A.H., Hafez, Y.H., El-Shora, M.A., Olafadehan, O.A. *et al.* (2021) Effect of date palm (*Phoenix dactylifera* L.) leaves on productive performance of growing lambs. *Tropical Animal Health and Production* 53(1), 72. DOI: 10.1007/s11250-020-02493-2.

Mallaki, M. and Fatehi, R. (2014) Design of a biomass power plant for burning date palm waste to cogenerate electricity and distilled water. *Renewable Energy* 63, 286–291. DOI: 10.1016/j.renene.2013.09.036.

Medjool Lovers (2021) Medjool Lovers (in Arabic). Available at: www.facebook.com/groups/2014490635462376/ (accessed 1 December 2021).

Meer, S., Akhtar, N., Mahmood, T. and Igielska-Kalwat, J. (2017) Efficacy of *Phoenix dactylifera* L. (date palm) creams on healthy skin. *Cosmetics* 4(2), 13. DOI: 10.3390/cosmetics4020013.

Multiaventura (2021) Multiaventura Elche (in Spanish). Available at: https://multiaventuraelche.com/ (accessed 17 November 2021).

Najafi, S., Ghasemi, H.A., Hajkhodadadi, I. and Khodaei-Motlagh, M. (2021) Nutritional value of whole date waste and evaluating its application in ostrich diets. *Animal* 15(3), 100165. DOI: 10.1016/j.animal.2020.100165.

National News (2012) Scent of a date palm stirs the emotions. Available at: www.thenationalnews.com/uae/scent-of-a-date-palm-stirs-the-emotions-1.597758 (accessed 25 December 2021).

Nature's (2021) Volumizing shampoo. Available at: www.natures.it/en/prodotti/shampoo-volumizzante/ (accessed 22 December 2021).

Ortiz-Uribe, N., Salomón-Torres, R. and Krueger, R. (2019) Date palm status and perspective in Mexico. *Agriculture* 9(3), 46. DOI: 10.3390/agriculture9030046.

Platat, C., Hillary, S., Tomas-Barberan, F.A., Martinez-Blazquez, J.A., Al-Meqbali, F. *et al.* (2019) Urine metabolites and antioxidant effect after oral intake of date (*Phoenix dactylifera L.*) seeds-based products (powder, bread and extract) by human. *Nutrients* 11(10), 2489. DOI: 10.3390/nu11102489.

Popenoe, P.B. (1973) *The Date Palm.* Field Research Projects, Miami, Florida.

Province (2021) Industry Sector (in Arabic). Available at: http://wilayabiskra.dz/?p=4969 (accessed 23 December 2021).

Qasim, M. and Naqvim, S.A. (2012) Dates: a fruit from heaven. In: Manickavasagan, A., Essa, M.M. and Sukumar, E. (eds) *Dates: Production, Processing, Food, and Medicinal Values.* CRC Press, Boca Raton, Florida, pp. 341–349.

Rivera, D., Obón, C., Alcaraz, F., Laguna, E. and Johnson, D. (2019) Date-palm (*Phoenix*, Arecaceae) iconography in coins from the Mediterranean and

West Asia (485 BC–1189 AD). *Journal of Cultural Heritage* 37, 199–214. DOI: 10.1016/j.culher.2018.10.010.

SAAF (2021) SAAF Jordan. Available at: www.facebook.com/SAAFJORDAN (accessed 2 December 2021).

Sablani, S.S., Shrestha, A.K. and Bhandari, B.R. (2008) A new method of producing date powder granules: physicochemical characteristics of powder. *Journal of Food Engineering* 87(3), 416–421. DOI: 10.1016/j.jfoodeng.2007.12.024.

Salomón-Torres, R., Ortiz-Uribe, N., Valdez-Salas, B., Rosas-González, N., García-González, C. *et al.* (2018) Nutritional assessment, phytochemical composition and antioxidant analysis of the pulp and seed of medjool date grown in Mexico. *PeerJ* 7, e6821. DOI: 10.7717/peerj.6821.

Salomón-Torres, R., Sol-Uribe, J.A., Valdez-Salas, B., García-González, C., Krueger, R. *et al.* (2020) Effect of four pollinating sources on nutritional properties of medjool date (*Phoenix dactylifera* L.) seeds. *Agriculture* 10(2), 45. DOI: 10.3390/agriculture10020045.

Salomón-Torres, R., Valdez-Salas, B. and Norzagaray-Plasencia, S. (2021a) Date palm: Source of foods, sweets and beverages. In: Al-Khayri, J.M., Jain, S.M. and Johnson, D.B. (eds) *The Date Palm Genome. Vol. 2. Omics and Molecular Breeding.* Compendium of Plant Genomes. Springer, Cham, Switzerland, pp. 3–26. https://doi.org/10.1007/978-3-030-73750-4_1

Salomón-Torres, R., Krueger, R., García-Vázquez, J.P., Villa-Angulo, R., Villa-Angulo, C, *et al.* (2021b) Date palm pollen: features, production, extraction and pollination methods. *Agronomy* 11(3), 504. DOI: 10.3390/agronomy11030504.

Schorr, M., Valdez-Salas, B., Salomon-Torres, R., Ortiz-Uribe, N. and Eliezer, A. (2018) The date industry: history, chemistry, processes and products. *The Israel Chemist and Chemical Engineer* 4, 30–35.

Segas (2021) Canes made of unusual materials. Available at: www.canesegas.com/materials/4.date.palm.cane.html (accessed 9 December 2021).

Shafiq, M., Alazba, A.A. and Amin, M.T. (2018) Removal of heavy metals from waste-water using date palm as a biosorbent: a comparative review. *Sains Malaysiana* 47(1), 35–49. DOI: 10.17576/jsm-2018-4701-05.

Shi, L.E., Zheng, W., Aleid, S.M. and Tang, Z.X. (2014) Date pits: chemical composition, nutritional and medicinal values, utilization. *Crop Science* 54(4), 1322–1330. DOI: 10.2135/cropsci2013.05.0296.

Sivakumar, N. (2002) Fermentative products using dates as a substrate. In: Manickavasagan, A., Essa, M.M. and Sukumar, E. (eds) *Dates: Production, Processing, Food, and Medicinal Values.* CRC Press, Boca Raton, Florida, pp. 305–315.

Soutalomma (2020) How did the palm leaf industry turn from beds and benches to chandeliers and antiques? (in Arabic). Available at: www.soutalomma.com/Article/921753/كيف-تحولت-صناعة-جريد-النخيل-من-سراير-ومقاعد-الى-نجف (accessed 4 December 2021).

Space Garden (2015) Troncos de palmera para decoración (in Spanish). Available at: www.spacegarden.eu/es/macetas/272-tronco-de-palmera.html (accessed 30 November 2021).

Sukumar, E. (2012) Dates in indigenous medicines of India. In: Essa, M.M. and Sukumar, E. (eds) *Dates: Production, Processing, Food, and Medicinal Values.* CRC Press, Boca Raton, Florida, pp. 387–396.

Sulaiman, S.A., Bamufleh, H.S., Tamili, S.N.A., Inayat, M. and Naz, M.Y. (2016) Characterization of date palm fronds as a fuel for energy production. *Bulletin of the Chemical Society of Ethiopia* 30(3), 465–472. DOI: 10.4314/bcse.v30i3.15.

Taghizadeh-Alisaraei, A., Motevali, A. and Ghobadian, B. (2019) Ethanol production from date wastes: adapted technologies, challenges, and global potential. *Renewable Energy* 143, 1094–1110. DOI: 10.1016/j.renene.2019.05.048.

TodoPalmera (2021) TodoPalmera (in Spanish). Available at: www.facebook.com/todopalmera/ (accessed 29 November 2021).

UNESCO (2019) Decision of the Intergovernmental Committee: 14.COM 10.B.3. United Nations Educational, Scientific and Cultural Organization, Paris. Available at: ich.unesco.org/en/decisions/14.COM/10.B.3 (accessed November 2021).

Visit Elche (2015) Usos de las palmeras (in Spanish). Available at: www.visitelche.com/usos-de-las-palmeras/ (accessed 11 November 2021).

Wendrich, W.Z. (2009) Basketry. In: Nicholson, P.T. and Shaw, I. (eds) *Ancient Egyptian Materials and Technology*. Cambridge University Press, Cambridge, pp. 254–267.

Economics and Marketing of Dates in Saudi Arabia

18

Roshini Brizmohun[1] ⓘ, Raga Elzaki[2] ⓘ, Ozcan Ozturk[3] ⓘ, Mohammed Al-Mahish[2]* ⓘ and Abda Abdalla Eman[2] ⓘ

[1]University of Mauritius, Reduit, Mauritius; [2]King Faisal University, Al-Ahsa, Saudi Arabia; [3]Hamad Bin Khalifa University, Doha, Qatar

Abstract

Date production is an income source for North African and Asian countries, specifically the Arabian Peninsula, and date serves as an important perennial food that plays a key role in human nutrition and welfare. Indeed, it is considered one of the main sources of food in several countries, particularly in the Arabian Gulf and Middle East. The economic importance of the date palm for producing countries means it is considered a sector with strong potential for employment creation, income generation, and trade. The leading producers of dates in the world are Egypt, Saudi Arabia, Iran, and Algeria, while Saudi Arabia is classified as the largest date-consuming country worldwide. However, there will be a substantial reduction in arable land for date palm cultivation in the future due to climate change, increasing temperature, and dust storms. In addition, climate change can influence the date market directly through negative impacts on storage and distribution, and indirectly through negative impacts on productivity. This chapter discusses the economics of date production globally and in Saudi Arabia specifically. It concisely covers major topics of importance to date globally, including the socioeconomics of date production, date production and consumption, world trade, marketing systems, prices, challenges and opportunities for date marketing, regulations and standards of the date trade, and the economic feasibility of date production. In addition, this chapter contains detailed explanations of macroeconomic information of dates in Saudi Arabia as a case study.

Keywords: Dates market, Dates trade, Economics, Macroeconomic, Socioeconomics

*Corresponding author: malmahish@kfu.edu.sa

© CAB International 2023. *Date Palm* (eds J.M. Al-Khayri *et al.*)
DOI: 10.1079/9781800620209.0018

Fig. 18.1. Date fruit varieties in Saudi Arabia. (Photo by Mohammed Al-Mahish.)

18.1   Introduction

The date palm, *Phoenix dactylifera* L., is widely grown for its edible fruit. Date production has a long history in the Arabian Peninsula as an important fruit crop playing a key role in the welfare, culture, history, environment including agroecosystems, and nutrition of the population within that part of the world (Dhehibi *et al.*, 2018). Fig. 18.1 displays examples of the most important types of date fruit varieties in the Arabian Peninsula, which differ in textures and colors.

The date fruit is known to be highly nutritious and has a spiritual value. Historically, the date palm is considered a traditional crop in the Old World and has been introduced as modern plantations in the USA, Israel, and the southern hemisphere (Botes and Zaid, 2002). The economic importance of the date palm lies in the fact that its fruit is not only consumed but is also considered a sector with a good potential for employment, income generation, and trade in many countries. The fruit is marketed all over the world as fresh fruits or high-value confectionary. In very dry arid areas, the fruit is an important source of subsistence and resilience for local communities, given the date palm's adaptability to harsh conditions including tolerance to high temperatures, salinity, and drought (Frija *et al.*, 2017; Dhehibi *et al.*, 2018).

This chapter reviews the existing literature on the economics of date production and marketing with a focus on Saudi Arabia. Although the crop has a long cultural history in the Middle Eastern region of the world, the economic potential of the date fruit is increasingly viewed by countries as a source for revenue generation. This has led to an accelerated amount of research and strategies by countries with a comparative advantage to exploit date production and marketing with a view to increasing trade to supply an increasing global demand. Over the past years, economic research has focused on developing policies and programs that can improve the efficiency of production and reduce transaction costs to increase the profit margin for traders of date fruits. The market for both fresh dates and dates for confectionaries is growing, inducing countries to increase the efficiency of production and marketing to exploit export market potential.

18.2   Socioeconomics of Date Production

Date production, consumption, and marketing occupy an important place in the economic structure of agriculture in the Middle East and North Africa. The Kingdom of Saudi Arabia is one of the major producers and exporters of dates in the region, hence dates have a significant role in foreign exchange earnings to that country. The National Center for Palms and Dates (2020) indicates that the average annual growth of Saudi Arabia's date exports value and quantity reached 12%. Emphasis is being put on increasing the production and quality of dates produced in Saudi Arabia (El-Habbab *et al.*, 2017). In recent years, the quality and reputation of the date sector in Saudi Arabia have improved, with the country known to be producing premium quality dates. Investment to expand the local sector to satisfy world demand is paying off (Alabdulkader *et al.*, 2017). Dates can be viewed as a strategic crop to increase revenue to Saudi Arabia as it has a comparative advantage in date production, where the comparative advantage in date production means the ability of Saudi Arabia to produce dates at a lower opportunity cost than other countries. Hence, the increasing trend in date production and exports over the period 2000–2020 suggests that Saudi Arabia should focus on accessing external markets competitively.

World consumption of dates increased to 1.1 million t in 2018 (Statista, 2021). The global consumption of table dates amounted to approximately

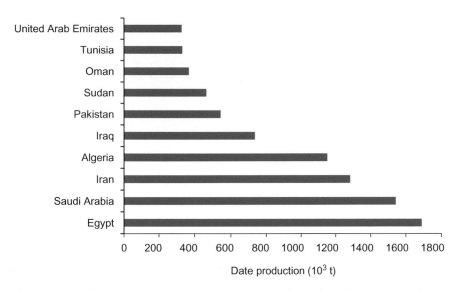

Fig. 18.2. Leading date-producing countries, 2020. (Data from Statista, 2021.)

875,000 t in 2019, down from 927,000 t in 2016. The per capita consumption of dates across countries varies greatly. Oman has the highest per capita consumption of 68 kg/year and Saudi Arabia 34 kg/year, while other countries have a relatively lower per capita consumption (Frija *et al.*, 2017; Dhehibi *et al.*, 2018). Consumption of dates fluctuates in Saudi Arabia due to competition from other fresh fruits and sweets. Moreover, changes in family size and eating habits influence date consumption (Aleid *et al.*, 2015). One potential strategy to increase the consumption of dates is to develop innovative products from primary products and by-products to create new markets.

18.3 World Production and Revenues

The global production of dates as an agricultural industry, which increased from about 1.8 million t in 1961 to 5.4 million t in 2001, reached around 9.45 million t in 2020, with the top ten producing countries in descending order being Egypt, Saudi Arabia, Iran, Algeria, Iraq, Pakistan, Sudan, Oman, Tunisia, and the United Arab Emirates (Fig. 18.2).

Table 18.1 and Fig. 18.2 show that Saudi Arabia is one of the major countries for date fruit production and was ranked second in the world in 2019 in terms of quantity produced (17.34%) according to statistics from the Food and Agriculture Organization of the United Nations (FAO).

The areas suitable for date production are in the arid regions of Southwest Asia, the Middle East, and North Africa. The trends in plantation

Table 18.1. Date production in the top producing countries, 2019. (Data from FAO, 2021.)

Rank	Producer country	Production (t)	Rank	Producer country	Production (t)
1	Egypt	1,603,762	21	Jordan	23,375
2	Saudi Arabia	1,539,756	22	Mauritania	21,926
3	Iran	1,307,908	23	Chad	21,458
4	Algeria	1,136,025	24	Niger	19,769
5	Iraq	639,315	25	Somalia	14,166
6	Pakistan	483,071	26	Albania	14,035
7	Sudan	438,700	27	Bahrain	13,000
8	Oman	372,572	28	Mexico	12,365
9	United Arab Emirates	323,478	29	Palestine	7729
10	Tunisia	288,700	30	Syria	3567
11	Libya	174,850	31	Benin	1454
12	China	172,587	32	Kenya	1100
13	China, mainland	158,642	33	Mali	670
14	Kuwait	105,867	34	Cameroon	638
15	Morocco	101,537	35	Peru	367
16	Yemen	64,375	36	Namibia	354
17	USA	55,700	37	Eswatini	311
18	Israel	43,412	38	Djibouti	118
19	Turkey	41,570	39	Colombia	16
20	Qatar	25,843			

Total production = 9,248,033 t

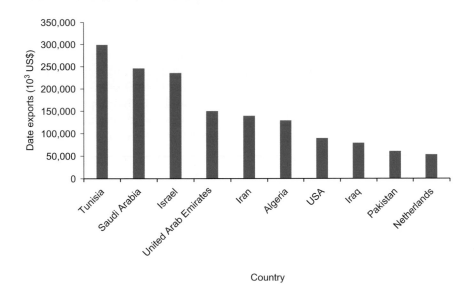

Fig. 18.3. Top date exporters, 2020. (Data from Statista, 2021.)

areas, production, and productivity among countries cultivating dates vary considerably. For instance, Bahrain, Kuwait, and Qatar had very limited areas under cultivation in 2016, with efforts being made to increase the planted areas. Moreover, the highest average yields were recorded in Kuwait (22.03 t/ha), Qatar (11.13 t/ha), and Oman (10.34 t/ha) in 2016 (Dhehibi *et al.*, 2018). Therefore, the productivity for dates within the Middle East varies significantly, with the average yield at the global level estimated at 6 t/ha. The fruit crop is marketed all over the world and is an important subsistence crop in desert regions. According to FAO statistics (FAO, 2019a), global date production covered an area of more than 1 million ha during 1999–2001, and the world total production reached 9.45 million t in 2020 (Statista, 2022). Date cultivation occurs in Asia (648,300 ha), Africa (435,700 ha), Europe (947 ha), and the Americas (7022 ha), with Asia and Africa accounting for, respectively, 55.8 and 43.4% of the total world harvest.

18.4 World Exports and Imports

In 2020, Tunisia was the largest exporter of dates worldwide, with exports amounting to a value of around US$299 million, followed by Saudi Arabia with exports worth around US$246 million, as shown in Fig. 18.3 (Statista, 2021). However, in terms of volume, Iraq is the largest exporter of dates, followed by the United Arab Emirates and Iran. Saudi Arabia is a net exporter and imports very limited quantities (Frija *et al.*, 2017).

18.5 Regulations and Standards of Trade

Quality indices for dates include fruit size, shape, color, texture (chewiness), cleanliness, and freedom from defects and decay-causing pathogens. The Codex Alimentarius Standard for Dates CXS 143–1985 was adopted in 1985 and amended in 2019 to describe the standards that apply to whole dates in pitted and unpitted styles packed ready for consumption (FAO, 2019b). Quality factors in the Codex standard for dates state the following:

- Dates should possess the characteristic color and flavor for the variety, be at the proper stage of ripeness, and be free from live insects, insect eggs, and mites.
- Moisture content of 26 to 30% (wet basis), depending on the variety.
- Minimum fruit size of 4.75 g (unpitted) or 4.0 g (pitted).
- Absence of defects including blemishes, mechanical damage, unripe, unpollinated, embedded dirt or sand, damaged by insects and/or mites, souring, mold, and decay. Dates and their products should be free from objectionable matter and free from microorganisms that represent a hazard to human health.

Al-Abdoulhadi *et al.* (2011) described four texture parameters (hardness, springiness, cohesiveness, and resilience) to standardize the quality norms of cvs. Khalas, Sheshi, and Reziz dates. The underlying reason was to protect the identity of the cultivars and at the same time strengthen the quality norms to boost exports. The main limitations to be overcome to meet the demands of the date industry include seasonality, perishability, quality standards, packaging, and logistics (Jamshed and Ahmad, 2018).

18.6 Economic Feasibility of Date Production

Dates are one of the main sources of food in several countries, particularly in the Arabian Gulf and Middle Eastern countries. Apart from its cultural and religious roles in Muslim traditions, dates are consumed for their many health and nutritional benefits. The date fruit is rich in fiber, potassium, iron, and vitamin B, as well as containing several types of acids, salts, and minerals. In addition, it helps prevent cardiovascular diseases and diabetes. As consumer awareness of such health benefits of dates increases, the demand for and the hence the production of dates have increased over time (FAO, 2021).

Because the date palm tree is well adapted to desert environments with extreme temperatures and water shortages, a significant portion (more than 95%) of the world's dates is produced in the arid regions of West Asia and North Africa. This is at least partially due to that area being the date palm's center of origin and diversity, thus resulting in the cultural values described previously.

The economic feasibility of date production is determined based on a cost–benefit analysis (CBA) approach. If the benefits outweigh the costs of any given project, then investment takes place. On the cost side, there are fixed and variable costs of date production. Although they vary by country, costs include land acquisition and clearing costs, machinery costs, labour expenses, and plant procurement costs. In addition, there are managerial, operational, harvesting, storing, and marketing costs. Once costs are calculated, they then are translated into the price in local currency. On the other hand, to determine the economic feasibility and benefits of date production, certain financial metrics are used. The most common ones include net present value (NPV), internal rate of return (IRR), return on investment (ROI), and payback period (PP) (Nordin *et al.*, 2004).

NPV is the sum of the discounted net future value for a given period. The decision rule for this metric is that if the NPV of a project or investment is positive, it means that the discounted present value of all future cash flows related to that project or investment will be positive and, hence, accepted; otherwise, it will be rejected. IRR is used to estimate the economic profitability of a potential investment. It allows comparing potential rates of annual return over time. Among the investment options, the investment with the highest IRR would be considered the most attractive. ROI measures the number of benefits gained on a particular project relative to the initial investment costs. PP measures the amount of time required to recover the initial cost of investment, being the time when an investment reaches the breakeven point.

In addition to CBA, some studies (Al-Abbad *et al.*, 2011) use SWOT (strengths, weaknesses, opportunities, threats) analysis to measure the economic feasibility and profitability of date production.

18.7 Marketing Systems and Prices

In the Gulf Cooperation Council (GCC) countries, marketing of dates is predominately done at the domestic level except for the United Arab Emirates and Saudi Arabia, which focus on export of dates. The marketing channels for dates are country specific and depend on existing institutions to distribute the products in the domestic market. The two marketing channels for marketing dates locally are through direct traditional marketing to consumers and marketing of dates to factories. In the former case, the dates are not processed (therefore no sorting, grading, steaming, or washing). However, dates are marketed according to the stage of maturity of the fruit, being either fresh or ripe. In the latter scenario, the dates that are delivered to the factories are of good quality under prefixed norms and standards set by the factory. The price of the dates will depend on the quality and quantities supplied to the factory (Dhehibi *et al.*, 2018).

Although Saudi Arabia is recognized as having a competitive date sector, the marketing efficiency of dates can be improved significantly. Date quality and the costs of marketing such as transportation, storage, sorting, grading, and packaging, unavailability of workers, the emergence of a black market, insufficient marketing information systems, and lack of coordination among market components and players are currently obstacles and challenges to be overcome (Alabdulkader *et al.*, 2017). Date dealers can achieve an additional 51% value added (marketing margin) once they reach full economic efficiency in date marketing (Alabdulkader *et al.*, 2017). Therefore, improving the cost and technical efficiencies can enhance the marketing efficiency for dates in Saudi Arabia.

Past research on date prices indicates that the low export price of dates in Saudi Arabia is attributed to the lack of focus on the production of high-quality dates, although more recently this trend is being reversed (El-Habbab *et al.*, 2017).

18.7.1   Other economic uses of dates

Date pits and dates falling from trees prior to maturity are often used as animal feed. This practice is used in both the Arabian Gulf and North African countries such as Tunisia. The potential of date fruits, by-products, and tree wastes as raw material to produce ethanol is being investigated in Saudi Arabia (Aleid *et al.*, 2015). Biomass waste generated from date palm can be used for biofuel generation through the fast pyrolysis technique (Al Yahya *et al.*, 2021). Therefore, the development of a holistic approach to maximize the outputs from date palm cultivation for economic purposes is recommended.

18.8   Challenges and Opportunities for Date Marketing

The literature on the economics and marketing of dates indicates that there is an increasing potential in the global trade of dates. The driving force for the demand of dates pertains to changing consumer preferences for natural, healthy food products. Innovative products derived from dates will most likely be in demand as the high sugar content of dates replaces sweeteners in products such as healthy fruit bars, breakfast cereals, and sugar-free spreads with dates (CBI, 2022). From the producer angle, countries have the potential to increase date production. However, an improved coordination of research, policies, and strategies among countries in the Middle East and North Africa region may contribute to a meaningful increase in global trade. It is to be noted that some countries have developed country-level policies for increasing date production and exports. For instance, the strategy of Saudi Arabia is to increase the export of Saudi dates and become the world's biggest date exporter in 2021 through increasing date exports by 12% per year to reach Saudi riyal (SAR) 1.75 billion and 222,000 t, thus contributing to achieving the Saudi Vision 2030 of raising the non-oil share of exports. Frija *et al.* (2017) showed that Saudi Arabia is the leading producer of dates in the GCC region based on the statistics. Moreover, Saudi Arabia is the most active in terms of exports on the GCC market, with high and increasing export values to the different GCC countries, specifically the United Arab Emirates. Frija *et al.*'s (2017) study determined that the GCC countries together have a strong potential for dominating the date market. A more recent study indicated that Saudi Arabia is very competitive in the global date export market, having a comparative advantage over at least four countries among the top six exporters in three markets (Africa, Asia, Organization of Islamic Countries) but being less competitive in the American and European markets (Alnafissa *et al.*, 2021). That same study demonstrated a possible reversal in the comparative advantage for Saudi dates over time, indicating that efforts should be made to maintain and improve the comparative advantage in the future. One way to achieve a greater market share is to improve coordination among the different trade strategies of the GCC countries through specialization and division of tasks, creating important opportunities for gaining more prominence in the world date market. Moreover, to improve competitiveness in the date industry, Saudi Arabia requires appropriate policies that support date exports. One strategy to demarcate the products from Saudi Arabia is to differentiate the market according to preferences for date varieties, quality requirements, and ability to pay for date products, which is often linked to prices. Therefore, further research that supports the date industry in Saudi Arabia should focus more on date quality, packaging, and selling prices for exporting dates. Other variables to take into consideration during research for a better perspective on the export potential of Saudi dates include trade agreements and quality factors such as pesticide residues (Alnafissa *et al.*, 2021).

Date quality is key to successful marketing of the product. The value attributes that influence consumer perception of quality include date cultivar, taste, texture, size, color, crust cohesion, freshness, no skin fractures, and free from pests and diseases. Product differentiation strategies are now increasingly being used in the date market to attract consumers (Aleid *et al.*, 2015). Packaging of dates can also influence consumer acceptance of dates. In Saudi Arabia, the types of packaging used for dates include paperboard box, plastic box, metal box, and vacuum bag. Cardboard boxes are used for packing loose dates that are cleaned, graded, and weighed. Therefore, focusing on the product marketing mix for differentiation of dates to enhance consumer acceptance is a key consideration to which exporting countries should pay attention.

Understanding the price marketing mix and its impact on consumer demand for dates is the basis for successful strategies to promote date exports. Previous studies on the price elasticities of demand for dates in the Asian market suggest that price plays an important role in the export of dates. The United Arab Emirates can increase its export potential in the international market, with India, Sri Lanka, Pakistan, and Jordan having high price elasticities, indicating that these markets are very sensitive to date import prices (Elashry *et al.*, 2010). Elashry *et al.* (2010) concluded that date exporters and policy makers should consider the impact of date prices when exploring export markets in Asia. The results of their study therefore can be extended to other date-producing countries and indicate that efforts should be made to be competitive in the international market as importing countries tend to be very price sensitive. On the other hand, the prices obtained for dates by producers and exporters need to be high enough for them to remain in business. Therefore, a thorough analysis of the price elasticities of demand for dates in different countries is relevant for exporters to develop appropriate strategies.

The economics of date production and marketing in producing countries should focus on benefiting from economies of scale to reduce costs of production and improve coordination among stakeholders to decrease transaction costs. Saudi Arabia is placing emphasis on introducing labor-saving methods of cultivation and modern irrigation systems; improved packaging; industrialization of food processing; and product diversification (Aleid *et al.*, 2015). More research focused on determining appropriate cultivars and production practices suited for specific regions may also contribute to increasing production capacities within countries.

The date sector in Saudi Arabia has gained much support and development over the last decade. Production difficulties pertaining to the cultivation of dates in Saudi Arabia include water scarcity and salinity, soil erosion and desertification, insect pest infestations and diseases, and insufficient processing and packaging facilities. Leading date producers in Saudi Arabia have invested in world-class facilities and operations. The large farms have made substantial capital expenditures in replacement trees and new trees that are not yet in production are expected to increase production capacity in the future (Aleid *et al.*, 2015). Alnafissa *et al.* (2021) predicted that Saudi Arabia has the potential to increase its exports of dates to international markets over the period 2020 to 2025 since there is a surplus of production over domestic consumption. The model forecasted an increase of more than 3.8-fold in the export potential of dates from Saudi Arabia. Therefore, programs targeting support to date farms and exporters to tap international markets provide a good opportunity for Saudi Arabia.

An improvement in the efficiency of moving the products from producers to consumers may contribute significantly to reducing transaction costs and increasing revenues for marketing channel intermediaries. Alabdulkader *et al.* (2017) proposed that the marketing efficiency of dates from Saudi Arabia requires policies that would improve both technical and cost efficiencies rather than focusing on the traditional concept of market efficiency. Therefore, countries like Saudi Arabia should conduct appropriate research to determine the transaction costs of moving dates from producers to consumers and identify appropriate marketing functions that will reduce costs and hence improve marketing efficiency.

Segmentation, targeting, and positioning are important considerations for successful marketing of products. Jamshed and Ahmad (2018) proposed that countries adopt niche marketing strategies and target marketing to improve the export potential of dates. The same authors recommend the need to minimize competition within market segments and develop new products as alternatives to contemporary food and beverages, given the high nutritional value of date fruits and seeds. Strategic options for date marketing include date paste, date syrup, and date vinegar. The development of premium value-added products from date resources may contribute to alleviate nutrition- and health-related problems in the developed and developing worlds. Therefore, Saudi Arabia needs to explore the potential of developing niche products for high-value markets to increase revenue rather than focusing solely on trading the raw commodity.

The potential for growth of dates in the European market appears to be steadily increasing as consumers seek more healthy and nutritious food. Although Europe is not the largest import market for dates, it is the fastest-growing importing region in the world. Europe increased its world import share from 10% in 2015 to more than 16% in 2019. Total European date imports reached 162,000 t in 2019, for a value of €405 million. The rising interest in healthy snacking and sugar replacement makes dates a very good alternative. Moreover, the increasing number of immigrants in Europe is a driving force leading to increased consumption of dates in Europe. The countries with the highest growth potential for dates include France, Germany, the UK, Spain, Italy, and the Netherlands. Internal European trade consists of simple reexporting of imported dates, but a significant part consists of added-value trade, including operations such as retail packing. European imports from developing countries increased over the last 5 years of the 2010s, from 91,000 t in 2015 to 126,000 t in 2019. The current increase and the expected increase of imported quantities offer good opportunities for suppliers in developing countries (CBI, 2022).

Climate change is expected to influence date production in Saudi Arabia. Research shows that there will be a substantial reduction in areas suitable for date palm cultivation in the Gulf region due to climate change and increasing temperatures (Shabani *et al.*, 2014). It is predicted that the current date palm cultivation in central and southern parts of Saudi Arabia will remain suitable for palm production by 2050 but will witness a significant reduction in climatic suitability for date palm cultivation by 2100 (Allbed *et al.*, 2017). Therefore, decision makers should identify management strategies that address the impacts of climate change on date production.

18.9   Saudi Arabian Dates: A Case Study

Saudi Arabia achieved vast economic growth after the oil revolution and according to World Bank classification, it ranks in the high-income countries list (World Bank, 2020). However, the economy of Saudi Arabia is still dependent on oil as the major source of economic growth.

Saudi Arabia is a desert country with virtually no permanent rivers or lakes and with only limited bursts of rainfall during a brief period each year. This phenomenon hinders the agricultural activities in the country, which are facing many challenges. The most challenging are adverse weather (hot and dry climate) with low precipitation, sandy and salinized soils with low fertility, high incidence of diseases and pests, and water scarcity. Even though the agricultural sector of Saudi Arabia faces many challenges, the country has the largest dynamic projects in the Middle East related to the agricultural sector (date palm, poultry, dairy, and aquaculture) that are being adopted in various regions (Ministry of Environment, Water and Agriculture, 2020a).

The most common crops include seasonal fruits and vegetables, cereals, and fodder. The key driver of crop production in Saudi Arabia is the date palm, which is significantly contributing to agroecosystems in the country.

Dates occupy a special place in the economic structure of Saudi agriculture with respect to production. In addition, the date palm is considered a symbol of life in the desert, because it tolerates high temperatures, drought, and salinity more than many other species of fruit crop. Date palms are grown in the various regions of Saudi Arabia. Fig. 18.4 displays the 12 regions for date production in Saudi Arabia; however, the most important and ancient date-growing regions are Riyadh, Qaseem, Eastern Region, and Madinah.

From Table 18.2, it is notable that Riyadh and Qaseem have the highest percentages of date palm trees, followed by Madinah and Eastern Region. Conversely, Northern Borders, Jazan, and Al-Bahah have the lowest numbers of trees, possibly due to the wetter climates. Elfeky and Elfaki (2019) and Al-Omran *et al.* (2021) indicated that for saving water, improving growth and soil properties, and preventing and controlling pests and diseases, the cultivated farms in the producing regions have adopted two irrigation systems for date production, i.e., drip or surface irrigation, as shown in Table 18.2.

According to the annual statistical report issued by the Ministry of Environment, Water and Agriculture (2020b), the total cultivated area of organic date palms was estimated as 6313.56 ha with a total production of 16,591.52 t in 2020.

The harvested area of the date palms covers an area of 152,705 ha with yield of 100,964 kg/ha through 2020 (FAO, 2019a). The date palm accounted for over 31 million trees with annual date production exceeding 1.5 million t and with more than 400 varieties (Ministry of Environment,

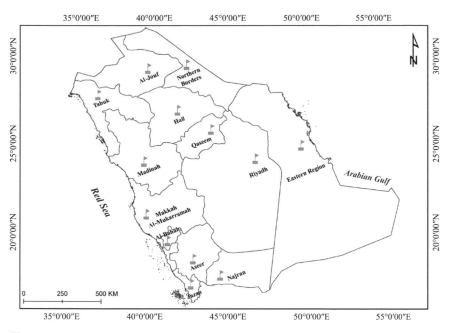

Fig. 18.4. Regions of date production in Saudi Arabia. (Constructed by the authors.)

Water and Agriculture, 2020b). The most popular varieties of dates were Khalas, Sukkari, Hulwa, Ajwa, Ruthana, Segae, Barhi, and Rushodia in 2019 (Table 18.3).

18.9.1 Trend of date production in Saudi Arabia

In recent decades a great deal of attention has been given to date production in Saudi Arabia and application of modern technological developments has contributed to increasing production. The trend of Saudi date production has increased sharply during the period of 1980–2019 (Fig. 18.5). In 2019, the total production of dates was 1,539,000 t, achieving a self-sufficiency rate of 125% (Ministry of Environment, Water and Agriculture, 2020c). The production increased by 28.41% in the period 2015–2019 compared to the period 1980–1984. During the period from 1980 to 1985, it is noted that date production registered the lowest production quantity. The annual growth rates were negative (–2.56%) between 2005 and 2009, due to the financial crisis of 2007 to 2009, whereas the highest growth rate occurred between 1995 and 1999 (30.56%). Likewise, the cumulative growth rate during the period 1980–2019 estimated by a simple linear trend model shows a highly significant rise of 22,574.85 t/year (Table 18.4).

However, recently several constraints have affected date production. The average general losses from pests and diseases are estimated at 12.19%

Table 18.2. Date palms by region, number of trees, and irrigation system in Saudi Arabia, 2019/20 season. (Data from General Authority for Statistics, 2021; National Center for Palms and Dates, 2020.)

Region	Total no. of trees[a]	% of total no. of trees	No. of fruiting trees[a]	% of total no. of fruiting trees	Area by irrigation system[b] (donum[c])	
					Drip	Surface
Riyadh	7,924,947	25.37	6,290,624	24.53	151,657	131,445
Makkah Al-Mukarramah	1,243,909	3.98	1,044,284	4.07	4,566	38,482
Madinah	4,751,040	15.21	3,812,367	14.87	32,161	150,872
Qaseem	7,542,914	24.15	6,187,301	24.13	240,662	57,816
Eastern Region	4,042,524	12.94	3,431,533	13.38	43,678	48,245
Aseer	1,117,738	3.58	983,955	3.84	9,548	33,435
Tabuk	1,012,499	3.24	875,468	3.41	7,042	26,159
Hail	1,973,528	6.32	1,740,970	6.79	29,637	27,353
Northern Borders	24,918	0.08	9,001	0.04	293	235
Jazan	8,822	0.03	3,813	0.01	182	302
Najran	526,333	1.69	391,492	1.53	4,316	9,813
Al-Bahah	80,935	0.26	7,1216	0.28	669	2,441
Al-Jouf	984,048	3.15	798,649	3.11	15,656	6,149
Total	31,234,155	100.00	2,5640,673	100.00	540,068	532,745

[a]Data from National Center for Palms and Dates, 2020
[b]Data from General Authority for Statistics (2021)
[c]"Donum" is a local term for the units used by farmers; 1 donum = 1000 m^2

Table 18.3. Distribution of date palm trees in Saudi Arabia according to cultivar, 2019. (Data from Ministry of Environment, Water and Agriculture, 2020b.)

Cultivar	No. of fruiting trees	Total no. of trees	Cultivar	No. of fruiting trees	Total no. of trees
Khalas	5,886,502	7,476,760	Kassab	13,759	17,966
Yellow Sukkari	3,251,477	4,405,483	Um Raheem	16,604	17,861
Barni	1,571,990	2,019,553	Khesab	15,811	17,634
Sefri	1,632,922	1,900,996	Hatmi	15,865	17,458
Hulwa	1,279,673	1,560,258	Halawi Ahmer	12,479	14,979
Barhi	1,182,802	1,448,849	Um Khashab	12,139	14,474
Ajwa	621,424	817,673	Magafri	11,563	13,304
Segae	587,088	803,712	Sari Alkharj	11,441	13,291
Ruthana	664,333	782,056	Sabaka	7130	11,339
Red Sukkari	507,311	655,218	Um Mignaz	10,030	10,624
Khodry	533,157	651,695	Assela	8839	9599
Sari	544,851	600,224	Um Kbar	6600	7295
Maktomi	426,577	525,267	Halawi Abyed	4578	6632
Ruzeiz	446,641	502,594	Magi	4930	5780
Nabtat Ali	303,303	415,619	Zamli	4983	5516
Nabtat Seif	284,712	340,795	Moakil	4706	5330
Shagra	264,465	308,470	Beraimi	3594	4017
Sullaj	240,067	286,841	Wsaili	3836	4010
Shaishee	197,647	227,965	Tayyar	3309	3536
Beid	184,505	206,685	Khoje	2992	3499
Rushodia	130,759	175,884	Khuylde	3069	3467

Continued

Table 18.3. Continued

Cultivar	No. of fruiting trees	Total no. of trees	Cultivar	No. of fruiting trees	Total no. of trees
Hilalia	122,565	157,288	Tanajeeb	3114	3371
Meneifi	120,906	146,999	Nabtat Sama	2533	3325
Beid Najran	110,986	136,277	Un Qouz	2980	3289
Ghur	97,021	127,632	Offendih	2918	3234
Barni Al Ais	97,299	118,542	Sukarat Yanbou	2877	3217
Wannana	61,389	78,008	Qatarah	2915	3154
Qasbi	60,727	73,951	Querrey	2733	3110
Hshishi	36,189	69,494	Aouinat	2772	2885
Anbara	45,230	63,844	Thawee	1983	2513
Fankha	50,218	58,153	Aeidiya	952	2270
Shalabi	47,238	55,091	Sandi	1928	2010
Khenaizy	35,485	44,509	Hrmosze	1418	1601
Magali	31,741	36,488	Garawya	1258	1473
Shahal	25,424	28,646	Asabe Al Arous	1312	1439
Lnat Massad	24,210	28,187	Marzban	1263	1374
Dekhaini	21,218	25,504	Bint Al Seyed	692	1189
Bakeira	10,324	20,888	Toree	1034	1035
Kassab	13,759	17,966	Other dates	781,834	940,606

Total no. of fruiting trees = 22,735,149   Total no. of trees = 28,570,804

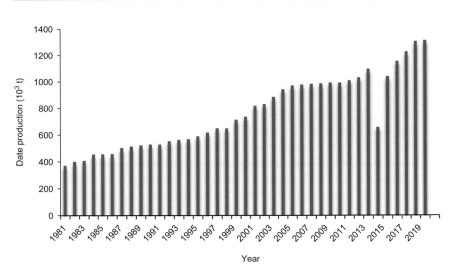

Fig. 18.5. Date production in Saudi Arabia, 1980–2019. (Data from Ministry of Environment, Water and Agriculture, 2020b.)

Table 18.4. Trend of date production and growth rate in Saudi Arabia, 1980–2019. (Authors' calculations.)

Years	Cumulative production (t)	Cumulative growth rate for 5 years (%)
1980–1984	1,974,042	24.18
1985–1989	2,451,436	11.74
1990–1994	2,739,218	17.38
1995–1999	3,215,408	30.86
2000–2004	4,207,652	16.65
2005–2009	4,908,139	−2.56
2010–2014	4,782,698	26.05
2015–2019	6,028,605	–

Y (growth rate) $= -4.44 + 22{,}5740.85 \times$ time, $R = 0.90$, $t = 18.95$ ($P=0.000$), $F_{(1, 38)} = 358.91$, no. of observations $= 40$.

and as much as 7.8% of date production can potentially be lost during marketing (El-Habbab *et al.*, 2017). Additionally, due to the fluctuation associated with climate change, some trees in scattered regions are in danger of drying out because of human activities.

18.9.2 Trend of date consumption in Saudi Arabia

Dates are a major food source and income source for most of the Saudi population and play significant roles in the economy, society, history, and agroecosystems (Jain *et al.*, 2011; Aleid *et al.*, 2015). Dates have been considered a part of Arabian culture for millennia. Dates can be consumed directly as a fresh fruit or be packed and processed in several ways, such as by-products derived from dates (jam, jelly, juice, syrup, fermented beverage). Also, other parts of the date palm tree can be used for various purposes, e.g., leaves for handicrafts.

The date consumption growth rate plays an essential role in forecasting future date production trends. Saudi Arabia occupies the first rank in the world in terms of average per capita consumption of dates according to Al-Shreed *et al.* (2012, Table 3), reaching 32.89 kg/year in 2019. Based on the annual report issued by the National Center for Palms and Dates, the total date production was 1,539,755 t in 2019 with approximately 66% consumed locally (National Center for Palms and Dates, 2020).

The per capita consumption of dates in Saudi Arabia has fluctuated significantly over the years (Table 18.5). Date consumption showed negative growth rates in 1991, 1998, 2008, 2010, 2012, 2014, and 2019. The highest growth rate of date consumption was in 2015 and was mainly linked with high production that year. The minimum per capita consumption in 2014 was 16.99 kg/year. This reduced consumption has affected production of dates adversely, despite all the support offered by the government (Al-Abbad *et al.*, 2011). Furthermore, date consumption is common in Saudi Arabia at various ages, although the group having the highest consumption of dates is youths followed by adults (Al-Mssallem *et al.*, 2019).

18.9.3   International trade

Based on FAO (2021) data regarding the international trade of Saudi dates, the country exported 182,317 t with a value of US$229,833 million and imported 339 t with a value of US$1.425 million in 2019. Furthermore, Saudi Arabia exported 21,7721 t with a value of US$247,198 million and imported 4001 t with a value of US$8.287 million in 2020, resulting in a positive balance of trade for dates in 2019 and 2020, respectively. The National Center for Palms and Dates (2020) indicated that from 2015 to 2020, the increase in cumulative growth rate of exported dates reached 68.87% (in quantity, i.e. tonnes) and 73.14% (in value, i.e. US dollars).

Saudi dates were exported to more than 75 and 107 countries in 2019 and 2020, respectively (National Center for Palms and Dates, 2020). The United Arab Emirates, Kuwait, Yemen, Morocco, India, Indonesia, Oman, and Turkey are the major destinations for Saudi dates.

18.9.4   Dates export and import

A simple estimation of the balance of trade in dates is given by:

Balance of dates trade = Saudi Arabia's exports − Saudi Arabia's imports   (18.1)

Table 18.5. Date consumption in Saudi Arabia, 1990–2019. (Data from FAO, 2021; General Authority for Statistics, 2021.)

Year	Consumption (t)	Growth rate per year (%)[a]	Per capita consumption (kg/year)[a]
1990	511,928	4.51	31.53
1991	510,102	−0.36	30.41
1992	534,225	4.73	30.91
1993	544,999	2.02	30.68
1994	551,140	1.13	30.26
1995	554,938	0.69	29.77
1996	586,092	5.61	30.79
1997	623,929	6.46	32.15
1998	623,230	−0.11	31.50
1999	705,110	13.14	34.92
2000	706,705	0.23	34.20
2001	786,092	11.23	37.08
2002	796,348	1.30	36.52
2003	851,191	6.89	37.90
2004	896,572	5.33	38.76
2005	922,178	2.86	38.72
2006	935,214	1.41	38.17
2007	935,352	0.01	37.14
2008	932,164	−0.34	36.01
2009	945,833	1.47	35.52
2010	922,237	−2.49	33.63
2011	977,336	5.97	34.57
2012	966,551	−1.10	33.15
2013	998,497	3.31	33.23
2014	525,325	−47.39	16.99
2015	919,395	75.01	28.99
2016	1,021,699	11.13	31.49
2017	1,078,250	5.53	32.57
2018	1,141,467	5.86	33.87
2019	1,128,637	−1.12	32.98

[a]Authors' calculations.

From Table 18.6, the balance of dates trade in Saudi Arabia is clearly positive at an increasing rate, which means that Saudi Arabia has created strategies that encourage a trade surplus in the long term.

Table 18.6. Balance of dates trade in Saudi Arabia, 1990–2019. (Data from FAO, 2021.)

Year	Import quantity (t)	Import value (10³ US$)	Export quantity (t)	Export value (10³ US$)	Trade surplus/deficit[a]
1990	4,346	2,588	20,299	13,959	11,371
1991	300	180	18,272	10,671	10,491
1992	160	110	18,428	14,876	14,766
1993	172	85	18,181	25,223	25,138
1994	0	0	16,622	13,604	13,604[b]
1995	0	0	34,323	21,785	21,785[b]
1996	30	35	30,846	21,095	21,060
1997	0	0	25,310	21,133	21,133[b]
1998	82	54	24,852	19,073	19,019
1999	210	150	7,100	5,300	5,150
2000	109	81	28,248	18,320	18,239
2001	86	80	31,881	18,694	18,614
2002	733	239	33,925	24,248	24,009
2003	1,978	741	34,875	24,585	23,844
2004	2,814	1,275	47,535	31,739	30,464
2005	2,788	1,743	51,098	32,456	30,713
2006	2,265	1,234	44,087	36,183	34,949
2007	1,568	894	48,762	40,529	39,635
2008	876	2,223	55,121	57,995	55,772
2009	1,232	1,686	47,059	35,349	33,663
2010	4,053	1,933	73,362	78,126	76,193
2011	47,026	39,111	77,795	86,293	47,182

Continued

Table 18.6. Continued

Year	Import quantity (t)	Import value (10^3 US$)	Export quantity (t)	Export value (10^3 US$)	Trade surplus/deficit[a]
2012	5,783	4,380	70,314	77,989	73,609
2013	3,109	2,441	99,770	103,571	101,130
2014	495	746	131,977	129,670	128,924
2015	1,223	1,384	120,358	136,264	134,880
2016	258	448	131,568	152,296	151,848
2017	637	938	146,579	186,718	185,780
2018	549	1,285	161,941	201,393	200,108
2019	939	1,425	182,317	229,833	228,408

[a]Authors' calculations.
[b]During this year, full self-sufficiency from dates is achieved.

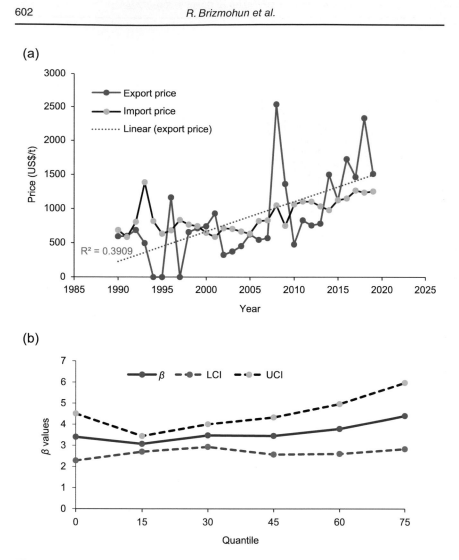

Fig. 18.6. (a) Export and import prices of dates per tonne in Saudi Arabia, 1990–2019. (b) Coefficients of the quantile model evaluating the effect of dates export value (independent variable) on gross domestic product (dependent variable). (β, regression coefficient; LCI, lower confidence interval; UCI, upper confidence interval.) (Authors' calculations.)

18.9.5 Influence of date exports on gross domestic product

The effect of dates export value on the Saudi economy for the period of 1990–2019 is highly significant and has a positive impact on enhancing the country's gross domestic product (GDP) according to linear regression ($r = 0.88$, $P<0.01$). Because of the price volatility for exported dates (Fig. 18.6a), the quantile regression model (Koenker and Bassett, 1978; Taylor, 2005;

Table 18.7. Coefficients of the asymmetric quantile model for Saudi dates: effect of dates export value (independent variable) on gross domestic product (dependent variable). (Authors' calculations.).

Quantile interval (%)	Statistics					95% confidence interval	
	β	SE	t	$P > |t|$	R^2	LCI	UCI
Quantile 0	3.40	0.54	6.25	0.00	0.60	2.28	4.51
Quantile 15	3.07	0.18	16.88	0.00	0.53	2.70	3.44
Quantile 30	3.47	0.26	13.30	0.00	0.56	2.93	4.00
Quantile 45	3.45	0.43	8.03	0.00	0.58	2.57	4.33
Quantile 60	3.78	0.58	6.55	0.00	0.57	2.60	4.96
Quantile 75	4.40	0.77	5.75	0.00	0.53	2.83	5.97

LCI, lower confidence interval; SE, standard error; UCI, upper confidence interval.

Huang, 2012) has been applied. The volatility is estimated by using the asymmetric quantile intervals of 15, 30, 45, 60, 75, and 90% in assessing the dates export value volatility and its respective forecast on the Saudi GDP. The general formula of the quantile regression model for Saudi Arabian dates is as follows:

$$Q\left(\text{GDP}_i^{\text{SA}}|\text{DEV}_i^{\text{SA}}, q\right) = \text{DEV}_i^{\text{SA}}\beta_q\text{Pr}\left(\text{GDP}_i^{\text{SA}} < \text{DEV}_i^{\text{SA}}\beta_q|\text{DEV}_i^{\text{SA}}\right) \quad (18.2)$$
$$= q, \; 0 < q < 1$$

$$\hat{\beta}_q = \arg\min_{\hat{\beta}_q} \sum_i \left\{ q \cdot 1 \left(\text{GDP}_i^{\text{SA}} \geq \text{DEV}_i^{\text{SA}}\hat{\beta}_q\right) \cdot \left|\text{GDP}_i^{\text{SA}} - \text{DEV}_i^{\text{SA}}\hat{\beta}_q\right| \right. $$
$$\left. + (1 - q) \cdot 1 \left(\text{GDP}_i^{\text{SA}} < \text{DEV}_i^{\text{SA}}\hat{\beta}_q\right) \cdot \left[\text{GDP}_i^{\text{SA}} - \text{DEV}_i^{\text{SA}}\hat{\beta}_q\right] \quad (18.3)$$

where Q is the quantile regression parameter for $i = 1, \dots, q$, permitting smoothing of the changes in the quantile over time; superscript SA indicates Saudi Arabia; GDP_i^{SA} is the gross domestic product (dependent variable); DEV_i^{SA} is the vector of dates export value (independent variable); q is the value of the quantile; $\hat{\beta}_q$ is the regression coefficient, which is estimated by minimizing the weight sum of the absolute deviation; and $1(.)$ is an indicator function.

The regression coefficient increases gradually to reach a value of more than 4 as the quantile interval increases (Table 18.7, Fig. 18.6b). The estimation results of this quantile regression model can be interpreted as showing a significant effect of date exports on economic growth. This implies that the expected future GDP would increase significantly at high date export levels.

To achieve this positive impact of the date exports, the Ministry of Environment, Water and Agriculture is collaborating with the National Center for Palms and Dates to facilitate export procedures and open sales outlets outside Saudi Arabia, so that Saudi dates can reach world markets.

The Ministry of Environment, Water and Agriculture aims to increase the exports of Saudi dates and to make Saudi Arabia the world's largest exporter of dates (Ministry of Environment, Water and Agriculture, 2020c).

18.9.6   Factors influencing the date market

The main marketing objectives of the Saudi Arabian date industry are to increase exports, raise domestic consumption, provide marketing services, provide information on the date sector, and strengthen the center's international and local relations (National Center for Palms and Dates, 2020).

Based on data gathered from the agricultural production survey conducted by the General Authority for Statistics (2019), about 85% of date production was sold with the remaining 15% being surplus (Table 18.8). The Riyadh region appears to be the biggest producer of dates and has the greatest date surplus (22.86%). The Eastern Region and Hail showed similar percentages of surplus dates, while Jazan sold most of its production (Table 18.8).

However, the date market faces several challenges and factors influencing the marketing procedures as shown in Fig. 18.7. Factors influencing the date market can be summarized as institutional, consumer behavior, climate change, economic, religious, and social factors.

Institutional factors affecting the date market can be identified as the marketing strategy adopted both nationally and on farm as well as the stage of marketing. Various authors have confirmed that consumer behavior plays an important role in all product and service markets (Liu *et al.*, 2016; Gillani *et al.*, 2021). Four main types of consumer behavior are: habitual buying behavior, variety-seeking behavior, dissonance-reducing buying behavior, and complex buying behavior (Kotler, 2012; Katrodia, 2021). The National Center for Palms and Dates (2020) revealed that the main features influencing the decision to buy dates are color, size, tenderness, shape, packing, flavor, and freshness.

Allbed *et al.* (2017) indicated that decision makers should formulate management strategies that focus on the impact of climate change on agriculture to attain long-term sustainable production of cash crops such as date palm in Saudi Arabia. Climate change can influence the date market directly through negative impacts on storage and distribution, and indirectly through a negative impact on productivity.

The economic factors influencing the date market can be concentrated on price policy and formulation of prices. In general, different policies can be carried out to address the policy objective of supporting the date market, such as export subsidies and facilitating marketing processes. Also, offering suitable credits at low interest rate at the harvesting, storage, and distribution stages can enhance the marketing process.

Likewise, other factors that affect the date market are the religious and social norms. During the fasting month of Ramadan, the date market is

Table 18.8. Total production, total sold production, and supply surplus of dates by producing region in Saudi Arabia, 2019. (Data from General Authority for Statistics, 2019.)

Region	Total production (t)	Sold production (t)	Supply surplus (t)	% of total supply surplus
Riyadh	402,411.6	351,073.0	51,338.60	22.86
Makkah Al-Mukarramah	59,825.5	45,235.0	14,590.50	6.50
Madinah	213,668.3	189,643.8	24,024.50	10.70
Qaseem	372,827.1	345,057.7	27,769.40	12.36
Eastern Region	213,515.9	182,293.0	31,222.90	13.90
Aseer	55,209.7	37,272.0	17,937.70	7.99
Tabuk	50,639.0	42,480.2	8158.80	3.63
Hail	97,390.3	66,041.7	31,348.60	13.96
Northern Borders	401.7	276.8	124.90	0.06
Jazan	131.1	92.3	38.80	0.02
Najran	27,180.3	19,314.9	7865.40	3.50
Al-Bahah	3351.8	2342.6	1009.20	0.45
Al-Jouf	43,203.4	34,045.0	9158.40	4.08
Total	1,539,755.8	1,315,167.9	224,587.70	100.00
% of total production	–	85	15	–

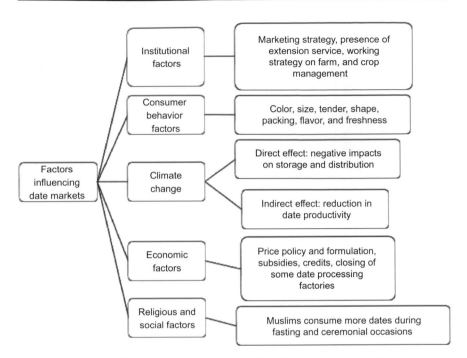

Fig. 18.7. Factors affecting date marketing in Saudi Arabia. (Authors' formulation.)

operating in all Muslim countries, and in particular the Saudi market. Al-Abbad *et al.* (2011) found that the date palm farmers market their products in three different ways: selling dates directly from the farm (23%), selling in the local markets (40%), or selling through factories (37%).

18.10   Strategies to Promote Date Marketing

18.10.1   Definition of marketing and marketing strategy

Agricultural marketing is defined as all activities that carry the agricultural product from the farm to the final consumer. It consists of amenities interrelated to the transportation, planning, regulation, orientation, and handling of agricultural products in a way that gratifies farmers, intermediaries, and consumers (Al-Hazmi, 2021). From this definition, agricultural marketing can be begun before production as represented by market research, planning, and implementing for production (Emam, 2002). Thus, this definition of agricultural marketing is compatible with the term 'supply chain'. Agricultural marketing aims at conducting the marketing functions necessary to transfer goods from production areas to consumption centers.

The broad definition of marketing strategy refers to an organization's integrated design of decisions that stipulate its vital choices regarding goods, marketplaces, marketing activities, and marketing resources in the creation, communication, and/or delivery of products that offer value to customers in exchanges with the organization, thereby enabling the organization to achieve specific objectives (Varadarajan, 2010, 2015). The term illustrates the crucial decisions as behavior patterns in the marketplace; in the proposed definition of marketing strategy, they can be elaborated as an organization's crucial choices, such as those pertaining to marketing activities to perform and the manner of performance of these activities in the chosen markets and market segments, and the allocation of marketing resources among markets, market segments, and marketing activities. E-marketing strategies involve operating current and developing communications and data platforms (e-mail, social media) to convey information between clients and the firm (Sheth and Sharma, 2005). Regarding the objective of marketing management, the objective of traditional marketing is to create demand for the product, i.e., supplier side (Kotler, 1973), while the objective of e-marketing concentrates on the customer perspective, i.e. the manufacturing will begin when the client orders.

18.10.2   Dates in Saudi Arabia

Saudi Arabia has the largest cultivated area for date palms in the world at about 152,710 ha, which is equivalent to about 12.36% of the total worldwide date palm cultivated area, and date production equals 1,541,770 t, equivalent to about 16.31% of global production in 2019 (FAO, 2020). Despite these characteristics of the Saudi date sector at global level, the sector faces many obstacles resulting in low efficiency in the domestic and world markets (Alabdulkader *et al.*, 2017). Such obstacles might be due to many reasons, among them being inefficiencies of marketing functions and services such as transportation, storage, sorting, grading, and packaging, shortage of seasonal labour, lack of inadequate marketing information systems, and lack of coordination among market components and players. These obstacles can be summarized as an ineffective supply chain. Fig. 18.8 illustrates the date supply chain in Saudi Arabia. It shows that the date follows different ways to reach the consumer, from farmer to consumer directly or through traders or from farmer to processer, or to overseas consumers through export.

Many efforts are being made by Saudi Arabia to overcome such obstacles. During the 27th Session of the FAO's Committee on Agriculture (COAG) held in September 2020, the Kingdom requested to establish the International Year of the Date Palm in 2027. The request was approved, with the COAG documenting that the observance of an International Year of the Date Palm by the international community would contribute to raising awareness about climate change (sustainable cultivation of date

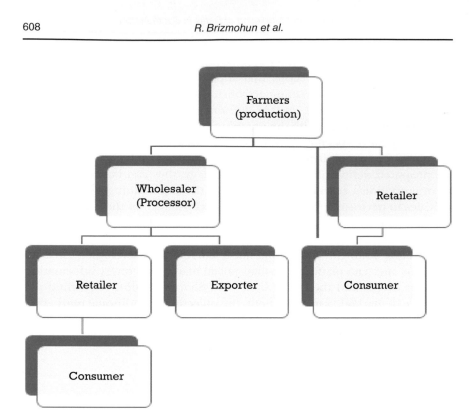

Fig. 18.8. Supply chain of dates in Saudi Arabia. (Constructed by the authors.)

palm under climate change), marketing efficiencies (supply chain), sources of income, and increased investment in agribusiness. Also, the Ministry of Environment, Water and Agriculture (2020c) is operating closely with the National Center for Palms and Dates in Saudi Arabia to simplify export processes and develop new global markets.

18.10.3 Strategies to promote date marketing

In Saudi Arabia, marketing strategies toward dates are inadequate. The marketing strategy has been constructed on a one-way opinion of producer behavior factors, in the context of firm administration (Kotler, 2012). It seems a traditional marketing management. Traditional marketing concentrates on the services and products that firms produce and aims to create a demand for the product; consequently, marketing management has traditionally been viewed as demand management (Kotler, 1973). Also, the marketing mix seems inadequate. The marketing mix comprises price, product, promotion, and place. The date sector faces an absence of marketing strategy due to the absence of marketing departments in the industry. This is attributed to excess market demand relative to supply generating the so-called seller's market, allowing the producer to sell very profitably.

Accordingly, these circumstances represent the main reason for the lack of marketing research (Alabdulkader *et al.*, 2017). The date market is activated annually through date festivals in different regions of Saudi Arabia. These festivals allow direct interactions between sellers and buyers of dates. Regarding the international market, marketing strategies should be designed to shed light on the occupation of Saudi Arabia in the production of dates; and on directing studies to know the demand of global markets, delivering information on the countrywide vital players, and encouraging exports with supportive policies. Also, an efficient campaign promoting the date fruit's health and nutritional advantages should be run during universal gatherings (conferences, festivals, etc.) (Dhehibi *et al.*, 2018).

The National Center for Palms and Dates has formally launched the 'Saudidates' platform, pioneering a suitable and well-organized business-to-business electronic market to conduct date marketing activities between international and national buyers and sellers in the Kingdom. The Saudidates platform was launched to oversee and boost the ability of Saudi Arabia's national exporters to access global markets (Arab News, 2022). The platform was launched under the slogan 'The Homeland of Dates', which shows dates as a national product.

For the future dates marketing strategy, the modern marketing system should be introduced to increase quality assurance and create added value. The system includes sorting, grading, packaging, transporting, storage, and manufacturing (Elsabea, 2012).

18.11   Conclusion and Prospects

Thorough analysis of the economics and marketing of dates and their products from Saudi Arabia indicates a good potential for revenue generation from the date industry. Although much support already exists for the sector, as the development of the industry progresses, flaws and challenges become apparent. Therefore, further research must be geared toward improving the date industry in Saudi Arabia. On the production side, the identification of cultivars that produce date fruit characteristics with potential high demand should be undertaken. The impact of climate change on date production is another area that needs attention. From the cost of production perspective, it appears that improved production techniques utilizing better technologies need to be developed for increased productivity. Research into new product development for better product differentiation based on global trends will surely move the Saudi date industry on the path to higher revenues. Consumer preferences for naturally sweet products are gaining momentum, with innovative products containing dates as an ingredient having increasing potential in the food industry. It is therefore important for Saudi Arabia to focus its research on collaboration with food industries in trendsetting countries.

Studies on the marketing functions required for reducing transaction costs and improving marketing efficiency represent one avenue of research. Along the same line, better coordination among countries involved in date production and marketing in the region will support the improvement of regional marketing efficiencies. Possible multilateral and bilateral trade agreements with importing countries, especially in untapped markets of the European Union and the USA, can give a tremendous boost to the date industry in Saudi Arabia. One strategy for date marketing that is derived from understanding the repackaging strategy undertaken in European countries implies that Saudi Arabia should aim at developing innovative packaging to benefit from value addition. The potential for improved date marketing therefore exists and needs better coordination at the country level for a successful strategy.

References

Al-Abbad, A., Al-Jamal, M., Al-Elaiw, Z., Al-Shreed, F. and Belaifa, H. (2011) A study on the economic feasibility of date palm cultivation in the Al-Hassa Oasis of Saudi Arabia. *Journal of Development and Agricultural Economics* 3(9), 463–468.

Al-Abdoulhadi, I.A., Al-Ali, S., Khurshid, K., Al-Shryda, F., Al-Jabr, A.M. *et al.* (2011) Assessing fruit characteristics to standardize quality norms in date cultivars of Saudi Arabia. *Indian Journal of Science and Technology* 4(10), 1262–1266. DOI: 10.17485/ijst/2011/v4i10.5.

Alabdulkader, A.M., Elhendy, A.M., Al Kahtani, S.H. and Ismail, S.M. (2017) Date marketing efficiency estimation in Saudi Arabia: a two-stage data envelopment analysis approach. *Pakistan Journal of Agricultural Sciences* 54(2), 475–485. DOI: 10.21162/PAKJAS/17.3109.

Aleid, S.M., Al-Khayri, J.M. and Al-Bahrany, A.M. (2015) Date palm status and perspective in Saudi Arabia. In: Al-Khayri, J.M., Jain, S.M. and Johnson, D.V. (eds) *Date Palm Genetic Resources and Utilization*, Vol. 2. *Asia and Europe*. Springer, Dordrecht, The Netherlands, pp. 49–95. DOI: 10.1007/978-94-017-9707-8_3.

Al-Hazmi, N.M. (2021) Obstacles to agricultural marketing of dates crop in date production companies in the Kingdom of Saudi Arabia (an applied study on the production companies of dates in Al-Kharj Governorate). *International Journal of Modern Agriculture* 10(2), 1665–1673.

Allbed, A., Kumar, L. and Shabani, F. (2017) Climate change impacts on date palm cultivation in Saudi Arabia. *The Journal of Agricultural Science* 155(8), 1203–1218. DOI: 10.1017/S0021859617000260.

Al-Mssallem, M.Q., Elmulthum, N.A. and Elzaki, R.M. (2019) Nutritional security of date palm fruit: an empirical analysis for Al-Ahsa region in Saudi Arabia. *Scientific Journal of King Faisal University* 20(2), 47–54.

Alnafissa, M., Ghanem, A., Alamri, Y. and Alagsam, F. (2021) The competitiveness of Saudi dates in global markets and its effect on future exports. *Pakistan Journal of Agricultural Sciences* 58(04), 1115–1122. DOI: 10.21162/PAKJAS/21.741.

Al-Omran, A.M., Al-Khasha, A. and Eslamian, S. (2021) Irrigation water conservation in Saudi Arabia. In: Eslamian, S. and Eslamian, F. (eds) *Handbook of*

Water Harvesting and Conservation: Case Studies and Application Examples. Wiley-Blackwell, Hoboken, New Jersey, pp. 373–383.

Al-Shreed, F., Al-Jamal, M., Al-Abbad, A., Al-Elaiw, Z., Ben Abdallah, A. *et al.* (2012) A study on the export of Saudi Arabian dates in the global markets. *Journal of Development and Agricultural Economics* 4(9), 268–274. DOI: 10.5897/JDAE12.058.

Al Yahya, S., Iqbal, T., Omar, M.M. and Ahmad, M. (2021) Techno-economic analysis of fast pyrolysis of date palm waste for adoption in Saudi Arabia. *Energies* 14(19), 6048. DOI: 10.3390/en14196048.

Arab News (2022) Saudi Arabia launches digital initiative to promote dates globally. Available at: www.arabnews.pk/node/2104191/saudi-arabia (accessed 19 December 2022).

Botes, A. and Zaid, A. (2002) The economic importance of date production and international trade. In: Zaid, A. and Arias-Jiménez, E.J. (eds) *Date Palm Cultivation.* FAO Plant Production and Protection Paper No.156 Rev. 1, Food and Agriculture Organization of the United Nations, Rome. Available at: www.fao.org/3/Y4360E/y4360e07.htm#bm07 (accessed 10 December 2022).

CBI (2022) Entering the European market for dates. Centre for the Promotion of Imports from developing countries, Ministry of Foreign Affairs, The Hague, The Netherlands. Available at: www.cbi.eu/market-information/processed-fruit-vegetables-edible-nuts/dates-0/market-entry (accessed 19 December 2022).

Dhehibi, B., Ben Salah, M. and Frija, A. (2018) Date palm value chain analysis and marketing opportunities for the Gulf Cooperation Council (GCC) countries. In: Kulshreshtha, S.N. (ed.) *Agricultural Economics: Current Issues.* IntechOpen, London. DOI: 10.5772/intechopen.82450.

Elashry, M.K., Gheblawi, M.S. and Sherif, S.A. (2010) Analysis of export demand for United Arab Emirates' dates in world markets. *Bulgarian Journal of Agricultural Science* 16(2), 123–134.

Elfeky, A. and Elfaki, J. (2019) A review: date palm irrigation methods and water resources in the Kingdom of Saudi Arabia. *Journal of Engineering Research and Reports* 9(2), 1–11. DOI: 10.9734/jerr/2019/v9i217012.

El-Habbab, M.S., Al-Mulhim, F., Al-Eid, S., Abo El-Saad, M., Aljassas, F. *et al.* (2017) Assessment of post-harvest loss and waste for date palms in the Kingdom of Saudi Arabia. *International Journal of Environmental and Agriculture Research* 3(6), 01–11. DOI: 10.25125/agriculture-journal-IJOEAR-MAY-2017-7.

Elsabea, A.M.R. (2012) An economic study of processing problems for the main important varieties of dates in Saudi Arabia. *Annals of Agricultural Sciences* 57(2), 153–159. DOI: 10.1016/j.aoas.2012.08.009.

Emam, A.A. (2002) *Agricultural Marketing* (in Arabic). Ministry of Education, Khartoum.

FAO (2019a) FAOSTAT Statistics Database. Food and Agriculture Organization of the United Nations, Rome. Available at: www.fao.org/faostat/en/#data/QCL (accessed 12 December 2020).

FAO (2019b) Codex Standard for Dates. Food and Agriculture Organization of the United Nations, Rome. Available at: www.google.com.sa/url?sa=t&rct=j&q=&esrc=s&source=web&cd=&cad=rja&uact=8&ved=2ahUKEwiigfKcw4X8AhW3UaQEHRP3DcMQFnoECA4QAQ&url=https%3A%2F%2Fwww.fao.org%2Finput%2Fdownload%2Fstandar

ds%2F256%2FCXS_143e.pdf&usg=AOvVaw1Du9vWJZbuNeybSWkvzt_T (accessed 19 December 2022).

FAO (2020) FAOSTAT Statistics Database. Food and Agriculture Organization of the United Nations, Rome. Available at: www.fao.org/faostat/en/#data/QCL (accessed 15 December 2022).

FAO (2021) FAOSTAT Statistics Database. Food and Agriculture Organization of the United Nations, Rome. Available at: www.fao.org/faostat/en/#data/QCL (accessed 24 May 2022).

Frija, A., Dhehibi, B., Salah, M.B. and Aw-Hassan, A. (2017) Competitive advantage of GCC date palm sector in the international market: market shares, revealed comparative advantages, and trade balance indexes. *International Journal of Marketing Studies* 9(6), 1. DOI: 10.5539/ijms.v9n6p1.

General Authority for Statistics (2019) Agricultural production survey bulletin. Available at: www.stats.gov.sa/sites/default/files/AgricultureProduction Surv ey 2019 EN.pdf (accessed 15 December 2022).

General Authority for Statistics (2021) Agriculture Census. Available at: www.stats.g ov.sa/en/22 (accessed 15 December 2022).

Gillani, A., Kutaula, S., Leonidou, L.C. and Christodoulides, P. (2021) The impact of proximity on consumer fair trade engagement and purchasing behavior: the moderating role of empathic concern and hypocrisy. *Journal of Business Ethics* 169(3), 557–577. DOI: 10.1007/s10551-019-04278-6.

Huang, A.Y.H. (2012) Volatility forecasting by quantile regression. *Applied Economics* 44(4), 423–433. DOI: 10.1080/00036846.2010.508727.

Jain, S.M., Al-Khayri, J.M. and Johnson, D.V. (eds) (2011) *Date Palm Biotechnology*. Springer, Dordrecht. DOI: 10.1007/978-94-007-1318-5.

Jamshed, M. and Ahmad, S. (2018) Niche marketing of date palm based food and beverages as health products. *Journal of Economic Cooperation and Development* 39(2), 49–68.

Katrodia, A. (2021) A study of identity consumer purchasing behavior and factors that influence consumer purchase decision: with reference to Durban. *Journal of the Research Society of Pakistan* 58(3), 60–71.

Koenker, R. and Bassett, G. (1978) Regression quantiles. *Econometrica* 46(1), 33. DOI: 10.2307/1913643.

Kotler, P. (1973) The major tasks of marketing management. *Journal of Marketing* 37(4), 42–49. DOI: 10.1177/002224297303700407.

Kotler, P. (2012) *Kotler on Marketing*. The Free Press, New York.

Liu, Z., Wang, Y., Mahmud, J., Akkiraju, R., Schoudt, J. *et al.* (2016) To buy or not to buy? Understanding the role of personality traits in predicting consumer behaviors. In: Spiro, E. and Ahn, Y.Y. (eds) *Social Informatics*, Vol. 10047. SocInfo 2016. Lecture Notes in Computer Science, Springer, Cham, Switzerland, pp. 337–346. DOI: 10.1007/978-3-319-47874-6_24.

Ministry of Environment, Water and Agriculture (2020a) Domestic agricultural production sector achieves high sufficiency in 2020. Available at: www.mewa.g ov.sa/en/MediaCenter/News/Pages/News142010.aspx https://www.mewa.go v.sa/ar/MediaCenter/News/Pages/News1004.aspx (accessed 1 January 2023).

Ministry of Environment, Water and Agriculture, (2020b) Agricultural Book. Available at: www.mewa.gov.sa/ar/InformationCenter/Researchs/Reports/G eneralReports/الكتاب الإحصائي 2020.pdf (accessed 1 January 2023).

Ministry of Environment, Water and Agriculture (2020c) FAO approves Saudi Arabia's proposal to declare 2027 the International Year of Date Palm. Available at: www.mewa.gov.sa/en/MediaCenter/News/Pages/News201220.a spx (accessed 20 March 2022).

National Center for Palms and Dates (2020) Annual Report. Available at: https ://ncpd.org.sa/elnakhel/public/storage/reports/10831847831623051434_ رير قت 2020.4.pdf (accessed 19 December 2022).

Nordin, A.B.A., Simeh, M.A., Amiruddin, M.N., Weng, C.K. and Salam, B.A. (2004) Economic feasibility of organic palm oil production in Malaysia. *Oil Palm Industry Economic Journal* 4(2), 29–38.

Shabani, F., Kumar, L. and Taylor, S. (2014) Suitable regions for date palm cultivation in Iran are predicted to increase substantially under future climate change scenarios. *The Journal of Agricultural Science* 152(4), 543–557. DOI: 10.1017/S0021859613000816.

Sheth, J.N. and Sharma, A. (2005) International e-marketing: opportunities and issues. *International Marketing Review* 22(6), 611–622. DOI: 10.1108/02651330510630249.

Statista (2021) Fruit dates market. Available at: www.statista.com/study/59958/dat es-market/ (accessed 19 December 2022).

Statista (2022) Global dates production 2010–2020. Available at: www.statista.com/s tatistics/960247/dates-production-worldwide/ (accessed 19 December 2022).

Taylor, J.W. (2005) Generating volatility forecasts from value at risk estimates. *Management Science* 51(5), 712–725. DOI: 10.1287/mnsc.1040.0355.

Varadarajan, R. (2010) Strategic marketing and marketing strategy: domain, definition, fundamental issues and foundational premises. *Journal of the Academy of Marketing Science* 38(2), 119–140. DOI: 10.1007/s11747-009-0176-7.

Varadarajan, R. (2015) Strategic marketing, marketing strategy and market strategy. *AMS Review* 5(3–4), 78–90. DOI: 10.1007/s13162-015-0073-9.

World Bank (2020) Data: World Bank Country and Lending Groups. Available at: ht tps://data.worldbank.org/income-level/high-income (accessed 15 December 2022).

Index

Note: Page numbers in **bold** refer to **figures** and page numbers in *italic* refer to *tables*.